P9-CEB-975

Environmental Process Analysis

Environmental Process Analysis

Principles and Modeling

Henry V. Mott, Professor Emeritus
Department of Civil and Environmental Engineering
South Dakota School of Mines and Technology
Rapid City, SD, USA

Published by John Wiley & Sons, Inc., Hoboken, New Jersey

Published simultaneously in Canada

Library of Congress Cataloging-in-Publication Data

Mott, Henry V., 1951–
Environmental process analysis : principles and modeling / Henry V. Mott, professor emeritus, Department of Civil and Environmental Engineering, South Dakota School of Mines and Technology, Rapid City, SD.
pages cm
Includes bibliographical references and index.
ISBN 978-1-118-11501-5 (cloth)
1. Environmental chemistry. 2. Chemical processes. I. Title.
TD193.M735 2013
577′.14–dc23

2013016208

10 9 8 7 6 5 4 3 2 1

To my deceased grandparents, Ida and Floyd Slingsby, and Ragna and Henry Mott;
to my deceased parents, Marge Marie and Henry Valentine, who raised me;
to my sisters, Jean, Judy, and Jane, with whom I shared childhood;
to my children, Harrison, Graeme, and Sarah, with whom I now share adulthood;
to my daughter-in-law, Lana, and my granddaughter, Samantha;
to Marty, my sweet bride, with whom I share a wonderful life.

Contents

Preface

This book is about mathematical and numerical modeling of processes in contexts associated with both natural and engineered environmental systems. In its assembly, I have relied on some very traditional but highly ubiquitous principles from natural and engineering science—chemical equilibria, reaction kinetics, ideal (and nonideal) reactor theory, and mass accounting. As necessary to the contexts of interest, I have incorporated principles from fluid dynamics, soil science, mass transfer, and microbial processes.

Many texts addressing introductory environmental engineering include discussions of these principles, but in opting to semiquantitatively address specific environmental contexts, never really apply them. Introductory modeling efforts seldom tread quantitatively beyond situations that are solved by single, explicit relations. This approach is fully appropriate at the entry level. Broad-based knowledge gained from an introductory course and text is essential to full appreciation of the portability of principles to myriad environmental systems. This text is not intended to replace an introductory environmental engineering textbook but to build on the contextual knowledge gained through completion of an introductory environmental engineering course.

In Chapter 2, some properties of water important to the understanding and employment of chemical equilibria are discussed. In Chapter 3, a collection of the various units describing abundance of components in gas, liquid, and solid systems is assembled. In Chapter 4, several specific conventions of the law of mass action, applicable to specific chemical "systems" are detailed. Then in Chapters 5 and 6, modeling of systems employing Henry's law and acid/base principles is examined. In Chapters 7 and 8, modeling of mixing and reactions in ideal reactors is addressed. These first eight chapters constitute the "basic" portion of this text. These topics and associated modeling work are appropriate for a third- or fourth-year undergraduate course, beyond the introductory level. I employ MathCAD as a powerful computational tool to illustrate, in the environmental contexts considered, the power of modeling in process analysis. In Chapter 9, I have extended the applications of three nonideal reactor models: completely-mixed flow reactors in series; plug-flow with dispersion; and segregated flow, beyond the level of treatment found in current texts. While containing good "food for thought" at the fourth-year undergraduate level, Chapter 9 is most appropriate for the graduate level.

Traditional water or aquatic chemistry texts introduce and discuss the chemical equilibria of acids/bases, metal complexes, solubility/dissolution, and oxidation/reduction. Mention is made of the proton balance, but this powerful tool is most often discarded or treated cursorily in favor of the seemingly much simpler charge balance. In fact, for systems that are not infinitely dilute (virtually all real systems) the charge balance most often fails at the outset. I have extended the application of the proton balance (or condition) to provide for significant advances in understandings of the acid- and base-neutralizing capacity of aqueous solutions and both solution–vapor and solution–solid systems. I have also demonstrated the relative ease with which nondilute solution principles can be incorporated into chemical equilibrium modeling.

For modeling of systems, traditional texts most often rely heavily upon simplifying assumptions, leading to graphical or approximate solutions, or upon sophisticated chemical equilibrium modeling software for quantitative description of chemical equilibria. Some recent texts have begun to chip away at the computational wall separating pencil/paper/graphical solutions from those involving sophisticated software but have not made significant headway. No other existing text known to me addresses, in transparent detail, the process of coupling mathematics with chemical equilibria and both mass and proton accounting for numerical modeling of chemical equilibrium systems.

Herein, I employ the general mathematical/numerical worksheet software MathCAD to occupy the region beyond approximate solutions and encroaching upon that of sophisticated software. A huge assembly of mathematical capability is available in a "what you see is what you get" user interface. Key to modeling of chemical equilibrium systems is ready capability to write user-defined functions, to program the solution of systems of nonlinear equations, and to create structured-code-like programs, all entirely visible in printable, portable worksheets. In fact, the vast majority of work illustrated in examples of this text has been conveniently exported into the manuscript as captures directly from worksheets. I make few, if any, simplifying assumptions beyond those associated with the first principles used in the mathematical modeling. The modeling efforts described herein, associated with the traditional water chemistry principles, are numerically as capable as those of the sophisticated software but much more flexible. These created models can be used not only to numerically model the equilibria but also to employ the equilibrium modeling to assess the consequences of perturbing the systems. Coupled with Chapters 2–6, Chapters 10–12 constitute the "advanced" portion of this text addressing chemical equilibrium modeling.

Those who will benefit from reading and studying this text are those who wish to mathematically and numerically model environmental processes and systems and who wish to fully understand the connections among the various factors leading to the results. Practitioners, depending upon their level of fundamental understandings, would benefit in a manner similar to students. No specific numerical methods skills are necessary, beyond attention to detail and an understanding that for numerical solution methods to work, they must be started in some vicinity of the final solution, assigning initial guesses to all unknowns sought. Although not absolutely necessary, it is certainly recommended that the reader obtain the MathCAD software and carefully follow through the worked examples. Such an approach promotes both understandings of the principles and mathematical modeling as well as capability for implementation of numeric solutions.

HENRY V. MOTT

Additional MathCAD files that accompany this text are available at booksupport.wiley.com by entering ISBN 9781118115015.

Additionally adopters of the text can obtain the solutions manual to the text by going to the books landing page at www.wiley.com and requesting the solutions manual.

Acknowledgments

I offer my special thanks to four former students, Zane Green, Nathan Kutil, Ulrike Lashley, and Teryl Stacey, who painstakingly reviewed the manuscript of this text, freely offering their time and abilities to make this effort as useful as possible for the students to come. I also offer my thanks to the many graduate and undergraduate students who sat in my classrooms, and with great enthusiasm engaged in the discussions and related efforts necessary to the development of the understandings manifest in the many example problems included in this text. I also offer my heartfelt thanks to my friend and colleague, Melvin Klasi, who, through my many years as a member of the Faculty of the SD School of Mines, was always willing to assist me in my understandings of mathematics and its implementation in modeling efforts.

I also must acknowledge some of my many teachers and mentors. Sam Ruzick and John Willard helped me unlock my love of chemistry, although it was to remain dormant for many of the years I studied to be and called myself a civil engineer. Hank Trangsrud taught me to ask tough questions and then to answer them. Al Wallace was, well, Al Wallace. My good friend Tom Nielsen and I learned much as we tackled the tough problems and topics with which Al charged us. Don Johnstone and Harry Gibbons were instrumental in the development of my understanding of microbes and aquatic insects as living, breathing beings. David Yonge, Erv Hinden, and Ken Hartz helped propel me onward by suggesting, at my MS thesis proposal presentation, that I extend it to a PhD dissertation, although I left Washington State to pursue my PhD. Walt Weber presented me with a challenging and relevant PhD thesis project and solid mentoring and support for its completion. Then, Walt, Don Gray, Linda Abriola, and Rane Curl helped me ensure that my work was top notch. I learned much from my common struggles alongside and interactions with my peer PhD students: Yo Chin, Lynn Katz, Domenic Grasso, Kevin Ohlmstead, Chip Kilduff, Margaret Carter, and Ed Smith. In the classrooms of Bernie Van Wie, Linda Abriola, Rich Kapuscinski, Jon Bulkley, Rane Curl, Ray Canale, Scott Fogler, and Bob Kadlec, I learned to couple mathematics with physical, chemical, and biological processes. The understandings of the portability of fundamental principles among systems quite naturally arose as an added bonus. In the classrooms of Brice Carnahan and James Wilkes, I learned that quantitative answers need not be exact, but certainly as close as reasonably possible.

I am the primary author of this text; I have no coauthors. However, I have chosen to employ the first person plural, we, in many of the discussions of the text. The knowledge and understandings employed in those discussions and companion examples arise as a consequence of the foundational work I did as assisted and guided by my many teachers and mentors. Their collective pursuit of personal and student betterment certainly contributed greatly to the expertise that I now claim as my own. In this text, when I use the term "we," it is I and my teachers and mentors to whom I refer.

Chapter 1

Introductory Remarks

1.1 PERSPECTIVE

From the outset, let us make no mistakes about the purpose and content of this textbook. The main title—*Environmental Process Analysis*—suggests that we will analyze processes. The targeted processes are those operative homogeneously in aqueous solutions, involving the gas–water interface, and involving the water–solid interface. Understandings of the behavior of environmental systems can arise from examination of both natural or engineered processes under equilibrium or near-equilibrium conditions. The effects of perturbations on systems can be determined using the initial and predicted final equilibrium conditions. In addition, understandings can arise from examination of the progress of such processes under transient or near (quasi) steady-state conditions. Then, Environmental Process Analysis is the examination of the processes operative in conjunction with perturbations of environmental systems, either natural or engineered, arising mostly from actions of our society. Certain of these perturbations beget negative consequences associated with actions that, while well-intentioned, contribute to the detriment of an environmental system. Others are intended to positively affect a compromised natural system or to implement a desired outcome within the context of an engineered system. The subtitle—*Principles and Modeling*—suggests that we will employ appropriate principles, develop models in support of our analyses, and employ these models to predict the outcomes from intended or unintended perturbations. Modeling has three distinct levels. Conceptual modeling involves identifying, understanding, and interrelating processes operative

Environmental Process Analysis: Principles and Modeling, First Edition. Henry V. Mott.

within targeted systems. Mathematical modeling involves coupling of relevant mathematical relations with processes identified by conceptual modeling efforts and assembling those mathematical relations into overall models describing behaviors of processes within systems. Lastly, numerical modeling involves work with the developed mathematical model to produce quantitative predictions of behavior.

We examine the scientific literature to understand processes and the means by which they may be mathematically described and consult resources assembled by the mathematicians to develop sets of or even single equations that might be used to describe the behavior of the system. It is not until we have collected these relations and devised means to use them to obtain quantitative answers that we have accomplished the process called modeling. A model can be as simple as a single linear relation or as complex as a set of coupled, higher-order, partial differential equations. The key is that, in either case, the conceptual, mathematical, and numerical aspects are employed. Even today, in the minds of many, numerical modeling is associated with the writing of lines and lines of structured programs that employ numerical methods in solution of sets of mathematical relations that defy closed-form analytic solution. We prefer the simpler idea that numerical modeling merely involves the production of numerical results using appropriate means to describe behaviors of processes in systems. Fortunately, with the development of the microchip, personal computers, and general computational software, the numerical part of modeling efforts has become much more conveniently accomplished. Then, in this text we illustrate and employ the modeling process to analyze effects of perturbations on both natural and engineered systems. We also illustrate the portability of key principles and concepts among the myriad contexts within which environmental engineering operates.

1.2 ORGANIZATION AND OBJECTIVES

Our prime objective with this textbook is the education of the student, interested faculty member, or practitioner in the means and methodologies to conceptually, mathematically, and numerically model processes operative in environmental systems. We begin with very basic processes and simple systems and progress to processes that are somewhat complex and to systems well beyond the simplistic. We have organized the text into 11 additional chapters beyond this introduction. Chapters 2–6 build upon each other in the general area of equilibrium aqueous chemistry. Chapters 7–9 are aligned along an alternative thread addressing reactions and reactors. Then Chapters 10–12 return to the aqueous equilibrium chemistry thread to address more advanced applications of the principles. In the following sections, we briefly describe the focus of each of the ensuing chapters.

1.2.1 Water

Although vital to environmental systems and perhaps of greatest importance relative to the future of the Earth and its inhabitants, water is somewhat ancillary to our analyses herein. We are mostly concerned about constituents within water and are mostly interested in the properties of water that contribute to the behaviors of these constituents. We have thus included a short chapter addressing the properties of water that are important in examination of the behaviors of acids and bases, cations and anions, and specifically hydronium and hydroxide in aqueous solutions. For those wishing to delve more deeply into the mechanical or other properties of water, we suggest examination of the many texts written addressing fluid properties and physical chemistry of water.

1.2.2 Concentration Units

Each scientific and engineering discipline, and subdiscipline in many cases, has its own means to specify the abundances of constituents in gases, liquids, and solids. Since environmental engineering must embrace most of the natural sciences (e.g., chemistry, physics, biology, geology, limnology, etc.) and many of the engineering disciplines (e.g., chemical, civil, geological, metallurgical, mining, etc.), we environmental engineers must be conversant with the preferred means to describe specie abundances by the many disciplines. To that end, we have included Chapter 3, in which we have assembled a database of concentration units used across these disciplines. Chapter 3 also contains a review of the means to interconvert units from one set to another using the basic chemical concepts of molecular mass and the ideal gas equation of state.

1.2.3 Chemical Equilibria and the Law of Mass Action

Over the past three plus centuries, the chemists have assembled a wonderful system with which to describe chemical processes. Tendencies for processes to proceed, rates at which they would proceed, and associated ending points (the equilibrium conditions) are all addressed within this very logical, quantitative system. In examination of perturbations of environmental systems, herein we choose to predict the final state of a system via close attention to the processes operative within. To that end, we employ chemical equilibria in combination with mass or molar accounting. Distinct styles for describing these equilibria arise from special applications of the law of mass action. Specifically, Henry's law, acid deprotonation, metal–ligand complex formation, solubility and dissolution, and oxidation/reduction half reactions all have their characteristic formulations of the law of mass action. These are reviewed in Chapter 4. For chemical equilibria, the change in standard-state Gibbs energy associated with a reaction as written is employed to define the equilibrium constant under standard conditions. The change in standard state enthalpy associated with a reaction as written is used in adjusting the magnitude of the equilibrium constant for varying temperature. We leave detailed discussions of these topics to the physical chemists and choose to employ two important results. Use of standard-state Gibbs energy changes to determine the magnitude of equilibrium constants is introduced in Chapter 10 and employed in detail in Chapter 12. Use of standard-state enthalpy changes to adjust equilibrium constants for alternative temperatures is employed in Chapter 10.

1.2.4 Henry's Law

Chapter 5 is devoted to developing understandings of the application of Henry's law to distributions of nonelectrolyte species between vapor and water. We employ Henry's law to predict abundances in the vapor from known abundances in water, and to predict abundances in water from known abundances in the vapor. We employ varying discipline-specific concentration units in these analyses. We begin our integrated modeling efforts by carrying Henry's law with us into a number of environmental contexts addressing air/water distributions in atmospheric, terrestrial, biogeochemical, and engineered systems. We showcase its portability.

1.2.5 Acids and Bases

In Chapter 6, we introduce the concept of water as an acid and a base and examine the interactions between water and the hydrogen ion (often simply called a proton) to form the hydronium ion, and begin the discussion of the hydration of cations in general, using the hydronium ion as an example. We introduce and solidify the concept that each acid has a conjugate base and that each base has a conjugate acid. Mono- and multiprotic acids are examined. Unlike many texts which focus on the carbonate system, the sulfur system, the nitrogen system, and the phosphorus system, we approach acid deprotonation from the standpoint of the general behavior of acids, employing a systematic approach to quantitate the behaviors of specific acids in defined systems. We stress that if any specie of an acid system is present in an aqueous solution, all must be present. We introduce the mole balance concept and strive toward an understanding of the idea of the *predominant specie or species* as dictated by the relation between the hydronium ion abundance within the system and the acid dissociation constant of the targeted acid system. We illustrate the connection between Henry's law and acid deprotonation equilibria. For a system that has attained the equilibrium condition, all equilibria must be simultaneously satisfied. We illustrate the prediction of aqueous speciation when the abundance of a vapor-phase specie and one critical condition of the aqueous solution are known. Similarly, from knowledge of at least two conditions relative to an acid system within an aqueous solution, we can predict the entire speciation within the aqueous solution as well as the abundance of the nonelectrolyte acid specie in vapor with which the water is in equilibrium. Employing the proton balance in the context of conjugate bases accepting protons and conjugate acids donating protons, we seek to develop beginning understandings of buffering capacity and the functional properties termed alkalinity and acidity. We make a beginning foray into the concepts of acid and base neutralizing capacity. We extend our integrated modeling efforts by carrying our understandings of acid deprotonation with us to join our understandings of Henry's law from Chapter 5 in contextual applications, again involving the atmospheric, terrestrial, biogeochemical, and engineered systems. In a manner similar to that employed in Chapter 5, we illustrate the portability of these principles and concepts.

1.2.6 Mixing

The mixing of two or more continuous streams is an important environmental process often given but cursory treatment in environmental engineering texts. While "zero volume mixing" is simple in concept, the nuances regarding when, how, and to what systems we can employ this principle often smudge the understandings of its applicability. In Chapter 7, we use continuous mixing of flows to begin our examination of the differences between transient and steady-state processes. Understandings of mixing phenomena are employed in developing beginning understandings of ideal reactors. The principles behind residence time distribution analyses are addressed and used in the definitions of completely mixed flow and plug flow reactors (CMFRs and PFRs). Impulse and step input stimuli are introduced, and exit responses for CMFRs and PFRs are examined. We introduce the process mass balance: the rate of accumulation within a control volume is the sum of the rates of input, output, and generation of a targeted substance. We employ the process mass balance to model the behavior of CMFRs receiving impulse and step input stimuli. We carry these zero-volume and transient mixing principles into environmental contexts, using them to model responses of selected natural and engineered systems to perturbations involving substances that are assumed to be nonreactive. In this manner, we illustrate the portability of these principles.

1.2.7 Reactions in Ideal Reactors

Although chemical stoichiometry is examined in preuniversity courses as well as in general chemistry courses completed by environmental engineers, the ability to employ these principles in specific environmental applications is not assured. Therefore, in Chapter 8 we begin with a review of the use of stoichiometry to determine reactant requirements and production of products using a number of common environmental engineering contexts. With these we illustrate quantitatively the conversions of one substance to another, without the complication associated with examination of the rates of transformation. We include mass–volume–porosity relations so that both the requirements for reactants and creation of products, for example, from water treatment operations can be expressed using molar, mass, and volume units. Mass–volume–porosity relations are also useful in quantitating rates of a process in natural systems considered as reactors (either ideal as examined in Chapter 8 or non-ideal as examined in Chapter 9).

We introduce two formulations of the reaction rate law: pseudo-first-order and saturation (arising from enzyme-limited microbial processes). Beyond radioactive decay, few processes rates are directly and linearly dependent only upon the abundance of the reactant. The pseudo-first-order rate law arises when certain of the reactants, aside from a target reactant, upon which the reaction rate is truly dependent, are maintained at constant abundance. If we can quantitate the abundances of these non-target reactants, we can mathematically treat the overall reaction as if it were a first-order reaction, greatly simplifying the resultant mathematics. Microbial reactions are said to be first-order in biomass abundance while, relative to a targeted substrate, they are enzyme-limited. Then, for saturation-type reactions, whose rate laws are described by Monod or Michaelis–Menton kinetics, we include the biomass abundance term in the rate law. Initially we examine systems in which the biomass abundance is considered constant in order that we can illustrate modeling of processes using closed-form analytic solutions. Then, we couple substrate utilization with microbial growth to illustrate the necessary numeric solution of such a system. We employ ideal reactor–reaction principles in multiple contexts, spanning both natural and engineered systems, thereby illustrating the portability of the principles and concepts in modeling efforts.

While not necessarily a reaction, we examine the transfer of oxygen to and from aqueous solutions, employing the concept of the mass transfer coefficient. We examine this mass transfer process in contexts appropriate for implementation of ideal reactor principles, providing a beginning understanding of the broad applicability of mass-transfer phenomena. We model transfer of oxygen across vapor–liquid interfaces of natural systems and in aeration of wastewaters. Extension of mass transfer principles to modeling of subsurface contaminant remediation systems or to modeling of gas–liquid, gas–solid, and liquid–solid contactors would be relevant and perhaps interesting to the student. These advanced systems become special cases of ideal reactors, best left to the more focused texts in which they are currently addressed. We hope the student can gain phenomenological understandings upon which competency in modeling of the more complex systems can be built later, if desired.

1.2.8 Nonideal Reactors

The ideal flow reactors mentioned in Chapter 8 comprise the extremes relative to the real reactors encountered in environmental engineering. No reactor can truly be perfectly plug flow or completely mixed flow. The engineering literature addresses three models for use in analyses of real (nonideal) reactors: CMFRs (Tanks) in series (TiS), plug-flow with dispersion (PFD), and segregated flow (SF). In Chapter 9, we examine the development and analyses of exit responses to input stimuli, useful in quantitatively describing the residence time distributions of real reactors. We employ the three nonideal reactor models to predict performance of a plug-flow like reactor and compare results with those predicted using the ideal plug-flow reactor model. The analyses and applications of the nonideal reactor models included in Chapter 9, especially for the PFD and SF models, are well beyond those included in any alternative texts known to this author.

1.2.9 Acids and Bases: Advanced Principles

In Chapter 10, we build upon the foundational principles addressed in Chapters 5 and 6. We address the hydration of cations and anions in the context of developing understandings regarding the behavior of electrolytes in nondilute solutions. Relative to these nondilute solutions, we introduce the relation between chemical activity and abundance and present a number of equations used for computing activity coefficients. We incorporate activity coefficients into our accounting system of mole balances, while preserving the unique relation among the chemical activities of reactants and products expressed by the law of mass action. Mole balances account for total abundances while chemical equilibria relate activities and the equilibrium constant. We address use of enthalpy in adjusting equilibrium constants for varying temperature and, along the way, provide an introduction to use of Gibbs energy in determination of equilibrium constants. We reserve significant application of Gibbs energy concepts for Chapter 12 in conjunction with redox half reactions that we write. We introduce the proton balance, equating evidence of protons accepted with corresponding evidence of protons donated as a consequence of proton-transfer reactions. Our treatment of the proton balance is well beyond that of any alternative text known to this author. The proton balance is a powerful tool in modeling changes in speciation as a consequence of a perturbation involving addition of an acid or base to an environmental system. The proton balance also is a critical tool in modeling acid- and base-neutralizing capacity of aqueous solutions. We present a step-wise approach to the visualization of proton-transfer reactions, leading to critical ability to define the initial conditions, upon mixing two or more solutions, prior to the occurrence of any proton transfers. We carry the proton balance along with the law of mass action and our mole balance accounting equations into a variety of environmental contexts specific to atmospheric, terrestrial, biogeochemical, and engineered systems. We complete our work in Chapter 10 by examining the behavior of water in solutions of high salt content.

1.2.10 Metal Complexation and Solubility

Many texts address coordination chemistry (metal complexation) before and separately from the solubility and dissolution of metals. Others address solubility and dissolution prior to metal complexation. We believe that the two topics are so closely related that simultaneous treatment is highly warranted. Hence, in Chapter 11 from the outset we couple formation of metal–ligand complexes and formation/dissolution of inorganic solids containing metals and ligands. We illustrate the hydrolysis of hydrated metal ions and present the correlations between cation hydrolysis and the process the chemists have termed complexation. Most importantly, in Chapter 11, we quantitatively address speciation of metals and ligands in aqueous systems, beginning with hydrolysis-dominated systems and then addressing multiple ligand systems. We illustrate the coupling of processes within mixed metal–ligand systems and provide means to quantitatively model such systems. We include metal solubility equilibria in the context of the mixed ligand systems. We illustrate the concept of solid-phase control of metal solubility and showcase multiple systems in which dual control of metal solubility, and hence control of ligand solubility is operative. We extend the concepts of acid- and base-neutralizing capacity to systems involving soluble metals and their metal–ligand solid phases. We carry these sets of principles into selected environmental systems to illustrate their portability.

1.2.11 Oxidation and Reduction

We begin our treatment of oxidation and reduction processes by writing half reactions: determining oxidation states of the element to which the reduction from the oxidized condition is ascribed, and employing the chemists' algorithm for balancing such reactions. We employ Gibbs energy to determine the equilibrium constant, in the context that much of the geochemical literature shuns equilibrium constants in the favor of tabulated values of Gibbs energy of formation. Most of the acid deprotonation and complex formation equilibrium constants have been measured or estimated and are tabulated. Similar data for redox half reactions is not so readily available. We thus waited until we really needed Gibbs energy concepts to illustrate their application. We review the addition of half reactions to produce overall oxidation–reduction reactions. The geochemical literature is rife with pE (or E_H) versus pH specie predominance diagrams. In order that these can be fully appreciated, we illustrate the process of construction: first the lines separating predominance regions and then entire diagrams. We then examine the dependence of speciation on electron availability at constant pH before investigating the determination of specie abundances in the near vicinity of predominance boundary lines. Finally, we illustrate means by which assays of the abundance of key redox species in combination with modeling of the system can provide accurate estimates of electron availability of environmental systems.

1.3 APPROACH

For this text, we did not perform exhaustive searches of the literature to uncover the detailed specific knowledge of targeted phenomena. Many fine texts have been assembled in that vein. Rather, we collected basic principles from the scientific literature, mostly chemistry-based texts, for implementation in environmental contexts. We call these first principles. Some of these principles are the detailed chemical stoichiometry and equilibria, mass (or mole) accounting, reaction rate laws, theory of ideal and nonideal reactors, thermodynamic fundamentals, and various special definitions associated with chemical systems.

We combine these fundamental principles with companion mathematical relations to quantitatively describe processes operative within environmental systems. In many cases, we have combined sets of first principles applicable to general contexts and derived usable relations. We might refer to these as second principles. These second principles relate the important parameters characteristic to the general contexts in which they would be applied. Typically, these relations have been designated as numbered equations. Intermediate results necessary to the understandings of the relations among the first principles and the general contexts in which they are applied, while important, are not intended for direct use in analysis/modeling efforts. These then are not assigned equation numbers. When we illustrate the applications of principles via an example, without fail, we begin either with first or second principles.

In this text, through the many detailed examples, we address many real processes operative in real contexts. Our process with examples is carried well beyond that traditionally employed: pose a question, with some associated reasoning select an equation for implementation, show how the numbers fit into the equation, and state the result. We wanted our examples to go much deeper, illustrating the true complexity of the mathematical/numerical methodologies necessary to obtaining quantitative results for questions posed in conjunction with complex systems. For computations, beginning with the simple linear relations associated with application of Henry's law, we have employed MathCAD as a computational tool. Then, with its "what you see is what you get" user interface, each MathCAD worksheet becomes an absolutely complete and accurate record of the mathematical/numerical processes employed. MathCAD programmers have developed a set of toolbars: arithmetic operators, graphing, vector and matrix operations, evaluation, calculus operations, Boolean operations, programming operations, Greek symbols, and symbolic operations. Then, with a click of the mouse, the user has at his or her command this entire broad and deep array of mathematical operations. A symbolic operations feature allows the user to set up integrals and derivatives and symbolically solve them. Approximately 450 intrinsic functions are available for use either by entering the function name or selecting desired functions from a drop-down list. MathCAD's help section explains each of these functions and provides examples of their use in computational efforts. Beyond these intrinsic functions, the user can define his or her own functions that employ many of the operations from the toolbars as well as employing user-defined functions and programs developed by the user. Among the intrinsic functions are several which can be employed to obtain numeric solutions of systems of (both linear and nonlinear) algebraic equations, systems of ordinary differential equations, and selected partial differential equations. The capability of solving systems of nonlinear algebraic equations is key to developing convenient models, employing chemical equilibria, mole balances, and the proton balance in examination of environmental systems. Of great utility is the fact that the aforementioned capability can be conveniently programmed using loops and logic to conveniently develop complex user-defined programs. In fact, each entire worksheet can become a program useful for analyzing the "what ifs" to predict system behavior. Huge sections of the work sheet can be "hidden," allowing the user to directly view results corresponding to manipulation of selected forcing parameters.

At this point we could go on and on about the numerical and mathematical capabilities programmed into MathCAD. Indeed, this author has moved well below the surface of MathCAD's sea of capabilities, but still has much to learn. Then, given that each MathCAD worksheet is a perfect visual record of the mathematical and numerical operations employed, we determine that for most of our examples, we would use "snippets" from our MathCAD worksheets to illustrate both the mathematics and the numerics employed in our examples. Our examples are intended to be complete logical and mathematical records of our solutions to the posed questions. It is our intent that the reader be able to follow all the mathematical and numerical operations embedded in our examples and translate them for use with mathematical/numerical modeling software alternative to MathCAD. We urge readers to adopt a favorite such software and employ that software in quantitatively understanding the processes and procedures of our examples. Perhaps 95% of the work addressed in Chapters 3 and 5–8 can be accomplished using a pencil, paper, and a calculator. In Chapter 7 we use some programming capability to conveniently generate some of our plots. In Chapter 8 we employ a root-finder in several examples and for the modeling of the rise of an air bubble emitted from an aeration diffuser, we employ the nonlinear equation solver in a looping program. In Chapter 9, we employ numerical integration techniques for large sets of data that do beg for solution using a computer. Also in Chapter 9, we write a number of short programs. Seemingly quite straightforward within the MathCAD worksheet, several of these involve the use of a root-finding process within a set of nested loops. Such a program, coded in a structured language, would require many lines of code. Then, in Chapters 10–12 we employ the nonlinear equation solver to provide numeric solutions to systems of nonlinear equations. In one example we illustrate a worksheet assembled in MS Excel that accomplishes the same solution as is performed in the immediately previous example using MathCAD. We much prefer the transparent structure of the MathCAD worksheet. This author is not well-versed in any other modern general mathematical/numerical modeling software (beyond MathCAD and Excel). Given the time demands of assembling a textbook of this nature, a decision was made to rely nearly exclusively upon the capabilities available from MathCAD for illustration of the mathematical/numerical techniques employed in *Environmental Process Analysis: Principles and Modeling*.

Chapter 2

Water

2.1 PERSPECTIVE

As the Earth's human population continues its exponential increase, the importance of water to the preservation of the standard of living we humans enjoy is becoming of utmost importance. Water is the substance without which we know life, as currently understood, could not exist. The examination of water ranges from the accounting of the vast quantities lying in the oceans and under the surface of the Earth to the minutest details of the structure of water, allowing understanding of its behavior in both natural and contrived systems. As related to environmental process analysis, water is the substance without which there could be no water chemistry. In environmental systems, it is generally water and how water might be affected by a situation or perturbation of a system that drives our desire to understand. Thus, given the importance of water to virtually all that is water chemistry, we will examine important properties of water as related to its structure.

Engineers use many of the physical properties of water in analyses of engineered systems; tables yielding values, correlated with temperature, of density, specific weight, viscosity, surface tension, vapor pressure, and bulk modulus of elasticity are found in most textbooks addressing fluid mechanics. These are mechanical properties but are often important in environmental process analysis. Consideration of the molecular structure and molecular behaviors within liquid water can yield fascinating insights as to why these mechanical properties are as they are. For example, the physical chemists (e.g., Levine, 1988; Williams et al., 1978) tell us that

Environmental Process Analysis: Principles and Modeling, First Edition. Henry V. Mott.

the ordering of the oxygen–hydrogen bonds as water freezes leads to a density of solid water (ice) that is lower than that of liquid water. Consider the alternate existence we would know if the crystallization of water behaved in a manner similar to the crystallization of many other liquids wherein the solid is more dense than the liquid.

The properties of water leading to its rather anomalous behavior relative to other liquids are those that also govern the behavior of water in interactions with solutes—constituents present in and intimately mixed within the water. The term "dissolved" seems to have functional definitions. In the past, we referred to dissolved solids as those not separable from liquid water by a particular glass microfiber filter. In another application, we "filter" sodium and other ions from seawater or brackish water using reverse osmosis. We might use a term like "solvated," suggesting that the solid somehow has a bond with water in the aqueous solution. It is the particular structure of water that leads to its ability to bond with "solvated" solids. The important properties of water stem from the unique arrangement of electron orbitals around the water molecule. Herein we could launch into a detailed investigation of the quantum chemistry surrounding the water molecule—at which point a typical engineering student's mind wanders to seemingly more relevant topics. Thus, we will restrict our discussions and associated understandings to the semiquantitative nature.

2.2 IMPORTANT PROPERTIES OF WATER

Based on Pauling's electronegativity scale (H=2.2, O=3.4), we may quite simply understand that hydrogen is quite content to contribute its lone electron to a bond with another atom while oxygen is quite intent upon acquiring two electrons to render its outer electron orbital to be like that of neon, a noble gas. Consequently, each hydrogen atom of a water molecule shares a pair of electrons with the oxygen and two remaining pairs of electrons are largely associated with the oxygen atom. A Lewis dot diagram for water is shown in Figure 2.1. When we consider the three-dimensional nature of the water molecule, the tendency for the electron pairs to orient their molecular orbitals (MOs) as far removed as possible from the other MOs ideally would lead to a tetrahedron as the base shape. Were the structure to be a regular tetrahedron, the H–O–H bond angle would be 109.5°. Attractions of the shared electron pairs to both the O and an H "thin" the MOs relative to those of the unshared pairs. Then, the unshared electron pairs exert further influence to "push" the MOs of the shared electron pairs closer together. The faces of the tetrahedron are not equilateral triangles. The electrons of the lone pairs exercise greater repulsion on each other,

FIGURE 2.1 *Lewis "dot" diagram for water.*

making the lone pair MOs "fatter" than those of the bonded pairs. Further, the lone pair MOs exert greater repulsion on each other than the bonded pair MOs and thus push the bonded pair MOs closer to each other. As a result, the bond angle from the centroid of the hydrogen atom through the centroid of the oxygen atom to the centroid of the other hydrogen atom (H–O–H bond angle) is measured to be 104.5° rather than the ideal 109.5° (Levine, 1988). In order to visualize the departure from the ideal shape, we set the tetrahedron on the table with the hydrogen atoms and one unshared molecular orbital as the base. A line through the two hydrogen atoms is north–south and the unshared molecular orbital is to the east. The remaining unshared molecular orbital then is at the apex. Then, relative to the apex of a regular tetrahedron, the true apex would be displaced upward and to the west. The north–south line connecting the two hydrogen atoms would be shorter than that of the regular tetrahedron. The west face of the tetrahedron would be an isosceles triangle with a base shorter than the other two sides. The northeast and southeast faces would be isosceles triangles with the side oriented to the east as the longest side. The base would be an isosceles triangle of shape identical to the westward oriented face.

The electronegativity of the oxygen relative to the hydrogen atoms leads to the well-known polarity of the water molecule. The bonded pair electrons exist in MOs that are associated with both the hydrogen and the oxygen. As a consequence of the greater electronegativity of oxygen, the electrons have a higher probability of residing in a portion of the MO associated with the oxygen atom than with the hydrogen atom. The consequence of this probability is the familiar partial positive (δ^+) charges assigned to the hydrogen atoms and partial negative (δ^-) charge assigned to the oxygen. The requirement for electroneutrality leads us to conclude that δ^- is twice δ^+. The positive charge is concentrated at each of the hydrogen atoms and the negative charge is concentrated along the line connecting the centroids of the two nonshared MOs. This concentration of negative charge is responsible for the capability for the bonding of a proton with a water molecule to form the hydronium ion. Were we to allow the centroids of the hydrogen and oxygen atoms to define a plane and to develop a shorthand diagram of the water molecule, we might arrive at something similar to the depiction shown in Figure 2.2.

When we examine this shorthand structure, we may easily understand that hydrogen bonding (interaction between the partial positive of the hydrogen with the partial negative of the oxygen) within liquid water can lead to the formation of a structure within the liquid. Williams et al. (1978) and Stumm and Morgan (1996) refer to "clusters" of structured water molecules within the liquid separated by regions of free, molecular water, shown pictorially in Figure 2.3. Within the clusters, water molecules have a "structure," with obviously shorter average bond distances than in crystalline ice. At the temperature of its maximum density (3.98 °C) the predominance of these clusters is at maximum. As temperature is raised, the predominance of

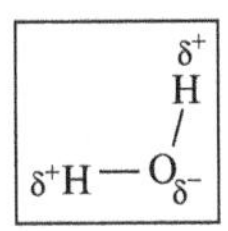

FIGURE 2.2 *Shorthand structure for the water molecule.*

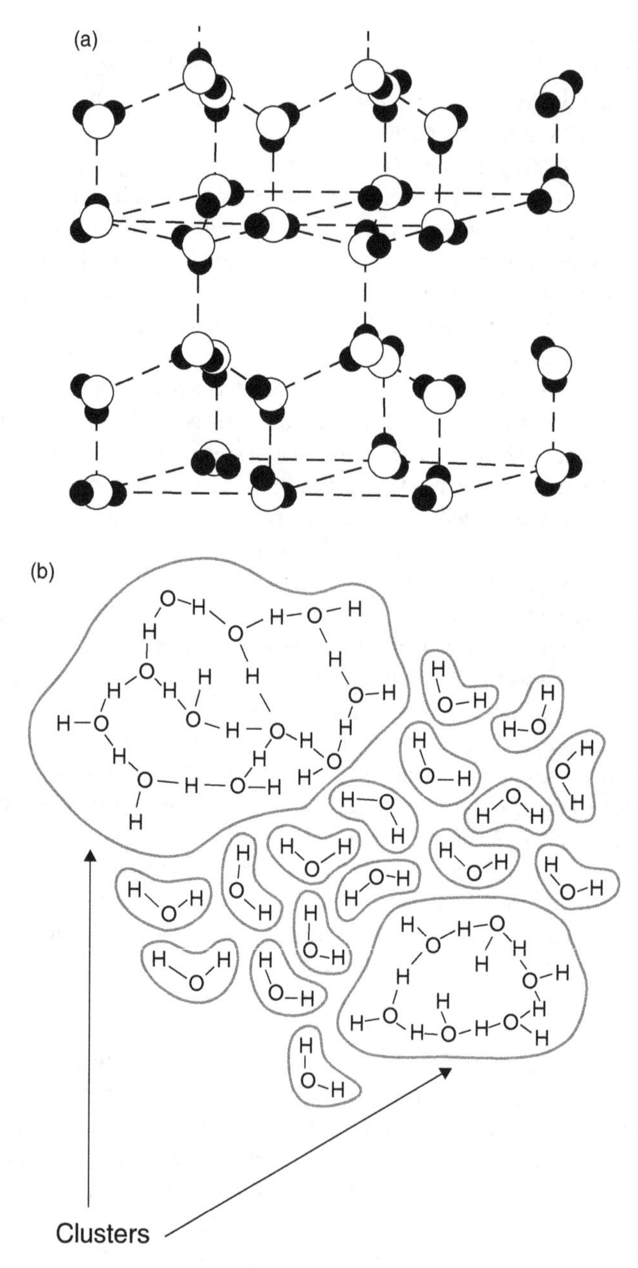

FIGURE 2.3 *(**a**) Hydrogen-bonded open tetrahedral structure of ice. (**b**) Frank–Wen flickering cluster model of liquid water. Reproduced from Stumm and Morgan (1996) with permission from John Wiley & Sons.*

clusters is decreased until at the boiling point, clustered water is at minimum. As temperature is increased from 3.98 °C, the density of water is decreased as a consequence of the longer hydrogen bonds predominant in the free water. As temperature is reduced below 3.98 °C, the ordering of the hydrogen–oxygen bonds into a structure more like that of crystalline ice renders the solution to be less dense. More detailed discussions of these "clusters" and of their "flickering" nature are presented by Williams et al. (1978) and by various texts addressing water chemistry (e.g., Brezonik and Arnold, 2011; Stumm and Morgan, 1996). The physical chemists have modeled the various properties of water using this structure in combination with the Valence Shell Electron Pair Repulsion (VSEPR) method and attained surprising agreement between model predictions and experimental observations (Levine, 1988). We will leave such endeavors to the physical and quantum chemists. Herein, we are much more interested in understanding the manifestations of these subatomic properties on the interactions of water molecules with solutes residing within the liquid water.

Of particular interest are the interactions between water and charged entities—ions—within an aqueous solution. The partial negative of the oxygen tends to orient with the positive charge of cations while the partial positive of the hydrogen tends to orient with the negative charge of anions. In each case, since the orientation of water with either the cation or anion does not satisfy the net charge, additional water molecules may be attracted. Water molecules attracted to monoatomic ions in aqueous solution would be expected to become oriented in roughly spherical shells with the nucleus of the ion situated at the centroid. This process is often referred to as hydration of ions. The result is that the effective size of a hydrated ion in aqueous solution is most often much greater than its true ionic size (Baes and Mesmer, 1976, 1981). With ordering of the water molecules about the ion, a release of energy occurs. Information relating to the "energy of hydration" for many ions is available from the scientific literature. In general, smaller ionic radii lead to greater hydrated radii, for a given base atomic structure (e.g., alkali or alkaline earth metals). The effective size of ions often can explain a great deal about the specific interactions of the ions with other dissolved substances or with solid surfaces with which aqueous solutions in which the hydrated ions reside are in intimate contact. In like manner, water molecules attracted to a solid surface of net charge would be expected to form layers of structured water associated with the surface of the solid (Bohn et al., 1979; Sposito, 1984). This "vicinal" water plays a large role in the near-surface interactions of both electrolytes and nonelectrolytes with engineered and natural solid surfaces in contact with aqueous solutions.

Perhaps the most well-known chemical property of water arises from the tendency of water molecules to take on positively charged protons, which become associated with the partial negative charges of the nonshared MOs, or to lose one of the hydrogen atoms (which then becomes a proton). The protonated water molecule is called hydronium while the deprotonated water molecule is called hydroxide. This combination of potential chemical reactions renders water to be both an acid and a base. In a later chapter, we will explore this phenomenon in greater detail, along with the basic acid/base behaviors of substances we call strong and weak acids or strong and weak bases.

Chapter 3

Concentration Units for Gases, Liquids, and Solids

3.1 SELECTED CONCENTRATION UNITS

With the exception of pure substances for which volume, density, and mass have a unique relation depending upon the nature of pure substances, in order to express the quantity (abundance) of a substance present in a solution or in a volume of soil, for example, we need to have a parameter termed concentration. Concentration is an analog of density. For a substance dissolved in a liquid, intimately mixed in a gas, or comingled with a solid or soil, the concentration and density would be identical if we held volume constant and simply removed all components other than the constituent of interest. Engineers tend to express their concentrations using mass units, scientists (here predominantly the chemists) tend to desire use of molar units, and various groups within each major area have their own pet sets of units used in their particular subdiscipline.

In Table 3.1, various units are listed and defined. These are divided into gas-phase, liquid-phase, and special categories. Further subdivisions are included for mass and molar units. Following the table, a number of examples of application/interconversion are presented.

In Table 3.2, several values of the universal gas constant (R) are presented. The first six are of course the most useful and the remainder are included in case the reader might encounter a situation in which alternative units of measure are employed.

Environmental Process Analysis: Principles and Modeling, First Edition. Henry V. Mott.

TABLE 3.1 Commonly Used Units of Concentration

Unit	Description
Gas phase units	
atm_i	Partial pressure of component *i*—the pressure exerted in a gas phase by the component of interest, component *i*. Dalton's law informs us that the total pressure of a gas is the sum of the partial pressures of each of the components. Equilibrium expressions that relate the distribution of a component between the gas phase and a liquid most often use partial pressure as the unit of abundance for the constituent of interest in the gas-phase
mol_i/mol_{tot}	Mole fraction—moles of constituent per mole of solution, the mole fractions of the components of a gas or liquid solution must sum to unity. If we subscribe to the ideal gas law (wherein a molecule of one component exerts the same pressure and occupies the same volume as a molecule of any other component, applicable in 99+% of environmental systems), in gases the mole fraction, pressure fraction, and volume fraction for a given constituent are identical.
vol_i/vol_{tot}	Volume fraction—were we to segregate all components of a whole gas into their own volumes, maintaining the pressure constant, the volume of each constituent per total volume would be the volume fraction of that component.
atm_i/atm_{tot}	Pressure fraction—the portion of the total pressure exerted by a gas attributable to component *i*. The partial pressure of a constituent of a gas divided by the total pressure. For a constituent present within an ideal gas the mole fraction, volume fraction, and pressure fraction are equal.
ppm_v	Parts per million by volume—the number of each million volume parts of the total gas volume attributable to component *i*. Given that we have an ideal gas, $ppm_v = 10^6$ times mole fraction, pressure fraction, or volume fraction, thus $ppm_{volume} = ppm_{mole} = ppm_{pressure}$. A variation is ppb_v (parts per billion by volume) such that $ppm_v = 10^3\ ppb_v$.
mol/L	Moles per liter—essentially the same as for liquids, moles of constituent per liter of gas. Gas phase concentration in moles per liter is related to partial pressure through the combination of Dalton's law and the ideal gas law.
%	Percent by volume—simply the mol (or pressure or volume) fraction times 100. Percent by volume is identical to percent by moles and percent by pressure.
$\mu g/m^3$	Micrograms per cubic meter—in many air pollution applications dealing with constituents that are present at very low concentrations, mass per volume units are sometimes employed. The easiest means to interconvert employs the ideal gas law to convert the volume of gas to moles and the molar mass of the constituent to convert mass of constituent to moles. The result is the mole fraction, which is easily converted to ppm_v or ppb_v.

(*Continued*)

TABLE 3.1 (*Continued*)

Unit	Description
Aqueous (liquid) phase units	
mg/L	Milligrams per liter—perhaps the most commonly used aqueous concentration unit, simply the mass of a solute in milligrams per liter of constituted solution. Recall that a milligram is 0.001 g. Useful variations are $g/m^3 = mg/L$ and $kg/m^3 = g/L$
μg/L	Micrograms per liter—simply the mass of a solute in micrograms per liter of constituted solution. Recall that a microgram is 0.001 mg or 10^{-6} g
ng/L	Nanograms per liter—simply the mass of a solute in nanograms per liter of constituted solution. Recall that a nanogram is 0.001 μg (10^{-6} mg or 10^{-9} g).
$mass_i/mass_{tot}$	Mass fraction, mass constituent per mass solution—The mass fraction concentrations of a solution or solid system must sum to 1.0.
ppm_m	Parts per million by mass—mass of constituent per 10^6 mass units of solution or solid, equal to mass fraction times 10^6. Herein (certainly not the case within the profession) the symbology for parts per million by mass will include the subscripted m. Note for ***dilute aqueous solutions only*** that 1 $ppm_m \approx$ 1 mg/L, as 1 L of water at 4 °C has a total mass of 10^6 mg.
ppb_m	Parts per billion (by mass) —mass of constituent per 10^9 mass units of solution or solid, equal to mass fraction times 10^9, $ppb_m = 10^3\ ppm_m$. Note for ***dilute aqueous solutions only*** that 1 $ppb_m \approx$ 1 μg/L, as 1 L of water at 4 °C has a total mass of 10^9 μg.
mol/L (M)	Moles per liter—moles of constituent per liter of constituted solution. Variations include mM, μM, and nM (millimoles per liter, micromoles per liter, and nanomoles per liter), $M = 10^{-3} mM = 10^{-6}\ \mu M = 10^{-9}$ nM.
mol/mol	Mole fraction—moles of constituent per mole of constituted solution. This unit finds most of its applicability in nonaqueous solutions and is in fact quite convenient in such applications. The mole fraction concentrations of the components of a solution must sum to unity. For conversion among mass and mole fraction concentrations for aqueous systems, the density of water (quite invariant in the range of temperature interest in environmental engineering) is taken as 1 kg/L and, hence, the molar density of water is 55.56 mol/L. This conversion factor proves immensely useful in categorizing the practical limits on the molar concentrations of components of interest in aqueous solutions.
mg/L as...	Milligrams per liter as (constituent)—most often concentrations of the various nitrogen species (NH_3/NH_4^+, NO_3^-, NO_2^-, N_{org}), and those of ortho-phosphorus ($H_3PO_4/H_2PO_4^-/HPO_4^=/PO_4^{-3}$) are expressed considering only the quantity of N or P in solution. Assay procedures do not allow for discernment of the individual species. For example, NH_3–N is nitrogen present in the NH_3/NH_4^+ system such that 14 mg/L NH_3–N would be 0.001 mol/L N; or 3.1 mg/L PO_4–P would be 0.0001 mol/L P. Other constituents typically expressed in this manner are total acetate (and other carboxylic acid systems), total cyanide (often also called weak acid dissociable), and total sulfate or sulfide.

Unit	Description
eq/L (N)	Equivalents per liter—equivalents of a substance per liter of constituted solution. An equivalent is of course a mole of replaceable protons, often simplified as a mole of charges. One must look carefully at the chemical context of the abundance before one can be absolutely sure of the conversion between equivalents and moles. A solution that is one equivalent per mole is also referred to as a one normal solution. Normality and equivalents per liter are interchangeable. A commonly used variation is meq/L (eq = 10^3 meq).
mg/L *as* $CaCO_3$	Milligrams per liter as calcium carbonate—this unit is a surrogate for expressing concentrations in meq/L or mN. Since the molar mass of $CaCO_3$ is ~100 g/mol and the calcium represents two hydrogen ions (or the carbonate has capacity to accept two hydrogen ions) the number of equivalents per mole is two, rendering the equivalent weight to be 50 g/eq (50,000 mg/meq). Then, a meq/L of a substance can be expressed as 50 mg/L *as* $CaCO_3$. In water treatment the concentrations of alkalinity, calcium, magnesium, and hardness are most often expressed using mg/L *as* $CaCO_3$. In order to convert these expressed concentrations into eq/L (N), one multiplies the value expressed in mg/L *as* $CaCO_3$ by 1 eq/L per 50,000 mg/L *as* $CaCO_3$.
Gr/gal	Grains per gallon—used in expressing water hardness along the same lines as mg/L *as* $CaCO_3$. A grain is 64.8 mg and when applied in water hardness or cation exchange a grain per gallon is 64.8 mg ($CaCO_3$) per 3.785 L of solution, or 17.1 mg/L as $CaCO_3$. One grain per gallon is then 0.3424 meq/L.
	Solid phase units
mass/mass	Mass fraction—the mass of the constituent of interest divided by the total mass of the solid, usually expressed based on the mass of the solid after drying.
% (by mass)	Percent by mass—the mass fraction times 100%. Moisture content and the organic carbon fraction of soils are most often expressed using this unit.
mg/kg	Milligrams per kilogram—the mass of constituent in milligrams per kilogram of the solid phase, most often based upon the mass of solid after drying. In expressing ultralow levels, μg/kg (=10^{-3} mg/kg) is often used.
ppm_m	Parts per million (by mass)—as is the case for liquids, the mass of constituent per million mass units of the solid. Milligrams per kilogram, micrograms per gram, and parts per million by mass are equivalent. Micrograms per kilogram and parts per billion by mass are equivalent.
	Other selected units
vol_i/vol_{tot}	Volume fraction (often called porosity)—the volume of a specific subportion of a system divided by the total volume of the system, used most often in the characterization of subsurface soils and sediments. For saturated soils, the volume fractions of the liquid and solid sum to unity. For unsaturated soils the volume fractions of gas, liquid, and solid sum to unity.

TABLE 3.2 Values of the Universal Gas Constant for Various Unit Systems[a]

8.317×10^7	$erg \cdot {}_g mol^{-1} \cdot {}^\circ K^{-1}$
1.9872	$cal \cdot {}_g mol^{-1} \cdot {}^\circ K^{-1}$
8.3144	$J \cdot {}_g mol^{-1} \cdot {}^\circ K^{-1}$
8.2057×10^5	$m^3 \cdot atm \cdot {}_g mol^{-1} \cdot {}^\circ K^{-1}$
0.082057	$L \cdot atm \cdot {}_g mol^{-1} \cdot {}^\circ K^{-1}$
82.057	$cm^3 \cdot atm \cdot {}_g mol^{-1} \cdot {}^\circ K^{-1}$
1.9869	$BTU \cdot {}_{lb} mol^{-1} \cdot {}^\circ R^{-1}$
1.314	$atm \cdot ft^3 \cdot {}_{lb} mol^{-1} \cdot {}^\circ K^{-1}$
0.7302	$atm \cdot ft^3 \cdot {}_{lb} mol^{-1} \cdot {}^\circ R^{-1}$
10.731	$psi \cdot ft^3 \cdot {}_{lb} mol^{-1} \cdot {}^\circ R^{-1}$
1545	$lb_f \cdot ft \cdot {}_{lb} mol^{-1} \cdot {}^\circ R^{-1}$
18,510	$lb_f \cdot in \cdot {}_{lb} mol^{-1} \cdot {}^\circ R^{-1}$

${}_g mol$ refers to a quantity in grams equal to the molar mass; ${}_{lb} mol$ refers to a quantity in lb_m (pounds mass) equal to the molar mass in lb_m.
[a]Adapted from Balzhiser et al. (1972).

3.2 THE IDEAL GAS LAW AND GAS PHASE CONCENTRATION UNITS

The Ideal Gas Law is presented as Equation 3.1:

$$P_{\text{tot}} V_{\text{tot}} = n_{\text{tot}} RT \tag{3.1}$$

where P, V, and n are pressure (atm), volume (L^3), and the number of moles; R is the gas constant and T is the absolute temperature. The subscript tot indicates that Equation 3.1 applies to the total gas mixture. Dalton's law of partial pressure is given by Equation 3.2:

$$\sum P_i = P_{\text{tot}} \tag{3.2}$$

P_i is the partial pressure of component i. Combination and manipulation of these two relations in Example 3.1 yields understandings of the mole fraction, pressure fraction, and volume fraction units, as well as their relationship to the parts per million by volume (ppm_v) unit.

Example 3.1 Consider an arbitrary mixture of gases. Also consider that the individual component gases of the gas mixture each occupy the total volume. We invoke the ideal gas assumption that molecules of each individual gas exert pressure and occupy volume equal to molecules of each of the other gases. We may then develop a set of relations for the molar, pressure, and volume fractions of the isolated component relative to the full gas mixture.

We imagine that we may isolate gas i from the remainder and consider it independently. We may then write:

$$P_i V_{\text{tot}} = n_i RT$$

We then divide this result by the LHS and RHS of Equation 3.1, yielding:

$$\frac{P_i V_{\text{tot}}}{P_{\text{tot}} V_{\text{tot}}} = \frac{n_i RT}{n_{\text{tot}} RT}$$

Simplifying the result leads to one of the relations we seek:

$$\frac{P_i}{P_{\text{tot}}} = \frac{n_i}{n_{\text{tot}}}$$

Hence, we may confirm a portion of the statement of Table 3.1 regarding the equality of pressure fraction and mole fraction concentrations.

Now let us stretch our imaginations and consider that we may segregate the component gases into volume compartments while holding the total pressure constant. We may then write:

$$P_{\text{tot}} V_i = n_i RT$$

Dividing by the LHS and RHS of Equation 3.1 then yields:

$$\frac{P_{\text{tot}} V_i}{P_{\text{tot}} V_{\text{tot}}} = \frac{n_i RT}{n_{\text{tot}} RT}$$

Simplifying the result leads to the second of the two relations we seek:

$$\frac{V_i}{V_{\text{tot}}} = \frac{n_i}{n_{\text{tot}}}$$

Hence, we may confirm the equality of the volume and mole fractions. The triple equality then arises, such that the volume fraction is also equivalent to the pressure fraction with the overall result:

$$\frac{V_i}{V_{\text{tot}}} = \frac{n_i}{n_{\text{tot}}} = \frac{P_i}{P_{\text{tot}}}$$

It is obviously difficult at best to physically segregate the gases in a given volume into subvolumes, but the unit ppm_v persists. When we encounter this unit, we may simply recall the result of Example 3.1 and know that $\text{ppm}_\text{v} = \text{ppm}_{\text{moles}} = \text{ppm}_{\text{pressure}}$. Most often in practice in use of this unit the subscripted v is omitted, and we must simply know that for gases the term ppm refers to ppm_v.

In Example 3.2, some important interconversions of gas-phase units are illustrated. A typical composition of stack gas emitted from a coal-fired electricity generating plant is employed.

Example 3.2 The gas exiting the stack of a coal-fired electricity generating plant, located near Gillette, WY, contains water vapor (H_2O), nitrogen (N_2), nitric oxide (NO_2), carbon dioxide (CO_2), carbon monoxide (CO), and sulfur dioxide (SO_2) at the mole percent values of 26.0, 55.2, 4.1, 12.2, 1.5, and 0.8%. The remainder of the exhaust gas consists of several other minor gas species. At the sampling point, located in the stack and 20 ft below the exit from the stack, the atmospheric pressure of the gas stream is 0.962 atm and the temperature is 87 °C.

Determine the mole fraction concentrations of each of the major components:

$$\begin{pmatrix} Ymol_{H2O} \\ Ymol_{N2} \\ Ymol_{NO2} \\ Ymol_{CO2} \\ Ymol_{CO} \\ Ymol_{SO2} \end{pmatrix} := \begin{pmatrix} 0.260 \\ 0.552 \\ 0.041 \\ 0.122 \\ 0.015 \\ 0.008 \end{pmatrix}$$

Since the above values are given as mole percents we may compute the mole fractions simply by dividing by 100. Note that the sum of the mole percents given is 99.8, meaning that the minor constituents comprise 0.2% of the gas. Here, in order to meet that requirement that the mole fraction concentrations sum to unity, we would specify an $Ymol_{minor}$. The literature most often employs the symbol Y for gas phase mole fractions.

What is the total molar concentration of the exhaust gas stream in moles per liter?

$$R := 0.082057 \ \frac{L-atm}{mol-°K} \qquad T := 87 + 273.15 = 360.15 \ °K \qquad P_{Tot} := 0.962 \ atm$$

$$C_{Tot} := \frac{P_{Tot}}{R \cdot T} = 0.033 \ \frac{mol}{L}$$

C_{Tot} is N_{Tot}/V_{Tot}

What are the concentrations of each of the major gases expressed in units of moles per liter?

We first compute the partial pressure of each gas constituent. Note that the sum of all partial pressures (including the sum of those of the minor components) must equal the total pressure:

$$\begin{pmatrix} P_{H2O} \\ P_{N2} \\ P_{NO2} \\ P_{CO2} \\ P_{CO} \\ P_{SO2} \end{pmatrix} := P_{Tot} \cdot \begin{pmatrix} Ymol_{H2O} \\ Ymol_{N2} \\ Ymol_{NO2} \\ Ymol_{CO2} \\ Ymol_{CO} \\ Ymol_{SO2} \end{pmatrix} = \begin{pmatrix} 0.25 \\ 0.531 \\ 0.039 \\ 0.117 \\ 0.014 \\ 7.696 \times 10^{-3} \end{pmatrix} \ atm$$

Since this is an ideal gas, we may compute the molar concentration of each constituent as if it were the only constituent present. The total of the individual molar concentrations must equal the total concentration:

$$\begin{pmatrix} Cmol_{H2O} \\ Cmol_{N2} \\ Cmol_{NO2} \\ Cmol_{CO2} \\ Cmol_{CO} \\ Cmol_{SO2} \end{pmatrix} := \frac{1}{R \cdot T} \cdot \begin{pmatrix} P_{H2O} \\ P_{N2} \\ P_{NO2} \\ P_{CO2} \\ P_{CO} \\ P_{SO2} \end{pmatrix} = \begin{pmatrix} 8.463 \times 10^{-3} \\ 0.018 \\ 1.335 \times 10^{-3} \\ 3.971 \times 10^{-3} \\ 4.883 \times 10^{-4} \\ 2.604 \times 10^{-4} \end{pmatrix} \quad \frac{mol}{L}$$

One minor component, anthracene, was present at 5.2 ppb_v. What are the mole fraction, molar (in mol/L and nM/m^3), and mass (in $\mu g/m^3$) concentrations of this suspected carcinogen.

Since 1 ppb_v is one part per billion by volume (or by pressure or by moles), the mole fraction concentration is simply 10^{-9} times ppb_v:

$$Ymol_{anth} := 5.2 \cdot 10^{-9} \frac{mol_{anth}}{mol_{gas}} \qquad P_{anth} := P_{Tot} \cdot Ymol_{anth} = 5.002 \times 10^{-9} \quad atm$$

$$Cmol_{anth} := \frac{P_{anth}}{R \cdot T} = 1.693 \times 10^{-10} \quad \frac{mol}{L}$$

We have 10^9 nM/mol, and 10^{-3} m^3/L:

$$Cnmol_{anth} := Cmol_{anth} \cdot \frac{10^9}{10^{-3}} = 169.27 \qquad \frac{nmol}{m^3}$$

We consult one of several sources (e.g., Dean, 1992) from which we may find the molecular structure of anthracene. We find that anthracene is a polynuclear aromatic hydrocarbon that consists of three benzene rings fused into one linear structure; anthracene has the empirical chemical formula $C_{14}H_{10}$. Recall that we have 10^6 μg/g and 10^{-3} m^3/L:

$$MW_{anth} := 14 \cdot 12 + 10 \qquad MW_{anth} = 178 \quad \frac{g}{mol}$$

$$Cmass_{anth} := Cmol_{anth} \cdot MW_{anth} = 3.013 \times 10^{-8} \quad \frac{g}{L}$$

$$C\mu mass_{anth} := Cmass_{anth} \cdot \frac{10^6}{10^{-3}} = 30.13 \quad \frac{\mu g}{m^3}$$

3.3 AQUEOUS CONCENTRATION UNITS

The major ion composition of a water of moderate hardness is used in Example 3.3 to illustrate the interconversion of some important aqueous phase units.

Example 3.3 An aqueous solution contains hardness at a level of 240 mg/L *as* $CaCO_3$, of which 80 is magnesium (Mg^{+2}) and 160 is calcium (Ca^{+2}). Hardness is the sum of all multivalent ions in solution but is normally virtually entirely due to calcium and magnesium. The anions associated with the hardness are most often bicarbonate and sulfate. In this case, the bicarbonate (HCO_3^-) is 185 mg/L *as* $CaCO_3$ and sulfate ($SO_4^=$) is 115 mg/L *as* $CaCO_3$. Note that the difference between the sum of Ca^{+2} and Mg^{+2} and the sum of bicarbonate and sulfate is most often made up by sodium and potassium, as the total charge of the cations must equal the total charge of the anions.

Express the hardness values in terms of meq/L.

Fifty grams *as* $CaCO_3$ corresponds to 1 eq, then 50 g/L *as* $CaCO_3$ = 1 eq/L (50 mg/L *as* $CaCO_3$ = 1 meq/L):

$$C_{meq.Ca} := \frac{160}{50} = 3.2 \qquad C_{meq.Mg} := \frac{80}{50} = 1.6 \qquad \frac{meq}{L}$$

What are the mass and molar concentrations of Ca^{+2} and Mg^{+2}?

A periodic table of the elements (or other suitable reference) provides the molecular weights. Both calcium and magnesium have +2 charges (each is equivalent to two hydrogen ions), so there are 2 eq/mol in each case. Several routes to the solution are available. Here, let us find the equivalent weights, use these to find the mass concentrations (using care to convert from meq/L to eq/L), and then use the mass concentrations with the molecular weights to find the molar concentrations:

$$MW_{Ca} := 40.1 \qquad MW_{Mg} := 24.3 \qquad \frac{g}{mol}$$

$$EW_{Ca} := \frac{MW_{Ca}}{2} = 20.05 \qquad EW_{Mg} := \frac{MW_{Mg}}{2} = 12.15 \qquad \frac{g}{Eq}$$

$$Cmass_{Ca} := \frac{C_{meq.Ca}}{1000} \cdot EW_{Ca} = 0.064 \qquad Cmass_{Mg} := \frac{C_{meq.Mg}}{1000} \cdot EW_{Mg} = 0.019 \qquad \frac{g}{L}$$

$$Cmol_{Ca} := \frac{Cmass_{Ca}}{MW_{Ca}} = 1.6 \times 10^{-3} \qquad Cmol_{Mg} := \frac{Cmass_{Mg}}{MW_{Mg}} = 8 \times 10^{-4} \qquad \frac{mol}{L}$$

We also could simply have converted to molar concentration by dividing the concentrations in meq/L by the number of equivalents per mole and converting from mM to M by dividing by 1000.

What are the concentrations of HCO_3^- and $SO_4^=$ in units of meq/L, mg/L, and mol/L?

Again, a periodic table is of use. Bicarbonate (HCO_3^-) has a single negative charge (equivalent to a single hydroxide) and has 1 eq/mol while sulfate ($SO_4^=$) has a divalent negative charge (equivalent to two hydroxides) and has 2 eq/mol. Conversion from mg/L *as* $CaCO_3$ is identical to that for calcium and magnesium. When we use mass concentrations in mg/L, we need to use the molecular weight in units of mg/mol:

$$C_{meq.HCO3} := \frac{185}{50} = 3.7 \qquad C_{meq.SO4} := \frac{115}{50} = 2.3 \quad \frac{meq}{L}$$

$$MW_{HCO3} := 1 + 12 + 3 \cdot 16 = 61 \qquad MW_{SO4} := 32 + 4 \cdot 16 = 96 \quad \frac{g}{mol}$$

$$EW_{HCO3} := \frac{MW_{HCO3}}{1} = 61 \qquad EW_{SO4} := \frac{MW_{SO4}}{2} = 48 \quad \frac{g}{Eq} \left(\frac{mg}{mEq}\right)$$

$$C_{mass.HCO3} := C_{meq.HCO3} \cdot EW_{HCO3} = 225.7 \quad \frac{mg}{L}$$

$$Cmol_{HCO3} := \frac{C_{mass.HCO3}}{1000 \cdot MW_{HCO3}} = 3.7 \times 10^{-3} \quad \frac{mol}{L}$$

$$C_{mass.SO4} := C_{meq.SO4} \cdot EW_{SO4} = 110.4 \quad \frac{mg}{L}$$

$$Cmol_{SO4} := \frac{C_{mass.SO4}}{1000 \cdot MW_{SO4}} = 1.15 \times 10^{-3} \quad \frac{mol}{L}$$

In liquids with density different from that of water, mass fraction units are quite handy, allowing for very convenient computations of diluted concentrations. Here we examine the dilution of a technical grade acid with water to illustrate the versatility of the mass fraction unit. Since the density of these solutions is not that of water (~1 g/cm^3, or 10^6 mg/L), 1 ppm$_m$ is not the approximate equivalent of 1 mg/L. The dilution of technical grade nitric acid to create working acid solutions is illustrated in Example 3.4.

Example 3.4 A technical grade nitric acid solution contains 97.5% nitric acid by mass (the other 2.5% is water and impurities). An aliquot of 2.00 mL of this technical grade acid is added to a 100-mL volumetric flask and the solution is diluted to exactly 100 mL with deionized, distilled water. Note that the density of 97.5% HNO_3 is 1.53 g/cm^3 (Perry and Chilton, 1973). Determine the mass fraction concentration of HNO_3 in the final diluted solution.

Let us consider that the 2.5% of the technical grade acid that is not HNO_3 is water. Let us also be careful to consider that for every gram of the technical grade acid added, only 0.975 g is HNO_3:

$$\rho_{acid} := 1.53 \qquad \rho_w := .997 \quad \frac{g}{mL} \qquad V_{acid} := 2.00 \quad mL \qquad V_{tot} := 100 \quad mL$$

$$Mass_{acid} := \rho_{acid} \cdot V_{acid} = 3.06 \quad g$$

$$Mass_{HNO3} := Mass_{acid} \cdot 0.975 = 2.983 \quad g$$

$$V_w := V_{tot} - V_{acid} + \frac{(1 - 0.975) \cdot Mass_{acid}}{\rho_w} = 98.077 \quad mL$$

$$Mass_w := \frac{V_w}{\rho_w} = 98.372 \quad g$$

$$Xmass_{HNO3} := \frac{Mass_{HNO3}}{Mass_w + Mass_{HNO3}} = 0.02944 \quad \frac{g_{HNO3}}{g_{solution}}$$

This computation neglects the fact that a volume change occurs upon mixing of the acid with the water, and perhaps slightly more than 98 mL of water is actually required to accomplish this dilution. Measuring the final mass of the diluted solution using an analytical balance would allow the determination of the exact mass of water added and hence the most accurate determination of the mass fraction. In fact, use of gravimetric measurements for preparation of aqueous solutions is indeed the most accurate means. The computed mass fraction concentration is equivalent to ~2.9%. The density of 2.9% nitric acid solution at 20 °C is ~1.014 g/mL (Perry and Chilton, 1973). Then, the mass of the prepared final solution would have been 101.4 g. An adjustment can be made to better the approximation of the mass fraction concentration:

$$Mass_{Tot} := 101.4 \quad g \qquad Mass_{w} := Mass_{Tot} - Mass_{HNO3} = 98.416 \quad g$$

$$Xmass_{HNO3} := \frac{Mass_{HNO3}}{Mass_{w} + Mass_{HNO3}} = 0.02942 \qquad \frac{g_{HNO3}}{g_{solution}}$$

We observe from this second result that the first approximation was indeed accurate to about three significant figures.

With this improved estimate of the mass fraction concentration, we can now quite accurately illustrate the computation of the mole fraction concentration of nitric acid in the final solution. The mole fraction is the ratio of the number of moles of nitric acid to the number of moles of total solution. Here, we will focus (for this approximation) upon the moles of nitric acid and the moles of water, ignoring the impurities:

$$MW_{HNO3} := 1 + 14 + 3 \cdot 16 = 63 \qquad MW_{w} := 2 + 16 = 18 \qquad \frac{g}{mol}$$

$$Mol_{HNO3} := \frac{Mass_{HNO3}}{MW_{HNO3}} = 0.047 \qquad Mol_{w} := \frac{Mass_{w}}{MW_{w}} = 5.468 \qquad mol$$

$$Xmol_{HNO3} := \frac{Mol_{HNO3}}{Mol_{HNO3} + Mol_{w}} = 8.587 \times 10^{-3} \qquad \frac{mol_{HNO3}}{mol_{solution}}$$

Oftentimes in computations with aqueous solutions we neglect the density change with increased or decreased temperature and simply use the density of water as 1 g/mL and for dilute solutions, we often assume that the solution is of density equal to that of water (1000 g/L or 55.56 mol/L). We can perform an approximate computation with little complication. Here, we need to recall that the initially specified volume of the solution was in mL and use this accordingly in the resultant computation:

$$\rho mol_{w} := \frac{1000}{MW_{w}} = 55.556 \qquad \frac{mol_{w}}{L_{w}}$$

$$Xmol2_{HNO3} := \frac{Mol_{HNO3}}{\frac{V_{tot}}{1000} \cdot \rho mol_{w}} = 8.524 \times 10^{-3} \qquad \frac{mol_{HNO3}}{mol_{solution}}$$

We see that our approximation is close to the more involved computation. As the aqueous solution becomes more dilute, the approximation using the standard density of water at 4 °C more closely approximates the true value.

For a number of acid/base systems, the assays to determine the concentrations depend upon chemical modification to enable measurement of the total quantity of the most highly protonated or deprotonated specie in the aqueous solution. Examples of this include the ammonia nitrogen, cyanide, acetate (and other carboxylic acid) systems. Thus, in dealing with such results we must be careful to examine the specified concentration value and associated information. As an example, nitrogen present in the ammonium/ammonia system is specified as ammonia nitrogen (NH_3–N), cyanide present in the hydrocyanic acid/cyanide system is expressed as total (or weak acid dissociable) cyanide, acetate present in the acetic acid/acetate system is expressed as total acetate. In Example 3.5, the determination of the concentrations of total cyanide and its two species is illustrated.

Example 3.5 A water sample was assayed and found to contain 10 mg/L total cyanide, which is the sum of cyanide (CN^-) in the two species HCN (hydrocyanic acid) and CN^- (cyanide ion). The pH of the sample was 8.2; thus, from the acid dissociation constant of hydrocyanic acid, it can be computed that the molar HCN concentration [HCN] is ten times that of the CN^- concentration [CN^-]. We will learn to perform these types of computations in a later chapter, but here, let us simply use the information.

Determine the molar concentration of total cyanide in the aqueous solution in mol/L:

$$C_{CN.T} := 10\ \frac{mg_{CN.Tot}}{L} \qquad MW_{CN} := (12 + 14)\cdot 1000 = 2.6 \times 10^{4}\ \frac{mg}{mol}$$

$$Cmol_{CN.T} := \frac{C_{CN.T}}{MW_{CN}} = 3.846 \times 10^{-4}\ \frac{mol}{L}$$

Convert the HCN concentration to mg_{HCN}/L:

$$Cmol_{HCN} := \frac{10}{11}\cdot Cmol_{CN.T} = 3.497 \times 10^{-4}\ \frac{mol}{L}$$

$$MW_{HCN} := (1 + 12 + 14)\cdot 1000 = 2.7 \times 10^{4}\ \frac{mg}{mol}$$

$$C_{HCN} := Cmol_{HCN}\cdot MW_{HCN} = 9.441\ \frac{mg_{HCN}}{L}$$

This result is of course numerically equal to ppm_m in a dilute aqueous solution.

Convert the HCN concentration to mole fraction.

Note that at 4 °C, water has a density of 1000 g/L. The formula weight of water is 18 g/mol; thus, water has a molar density of 1000/18 or 55.56 mol/L. C_{HCN} must be converted to molar units. The solution is sufficiently dilute that we may use the molar density of water as the molar density of the solution:

$\rho mol_w := 55.56 \frac{mol}{L}$ $\quad Xmol_{HCN} := \frac{C_{HCN}}{MW_{HCN} \cdot \rho mol_w}$

$Xmol_{HCN} = 6.293 \times 10^{-6} \quad \frac{mol_{HCN}}{mol_{soln}}$

Determine the mass fraction concentration of HCN and the concentration of HCN in ppm_m.

We can determine mass fraction by dividing the aqueous concentration by the mass density of water. The solution is sufficiently dilute that we may use the density of water at 4 °C (1 g/cm^3) as the density of the solution:

$\rho mass_w := 1000000 \frac{mg}{L}$ $\quad Xmass_{HCN} := \frac{C_{HCN}}{\rho mass_w}$

$Xmass_{HCN} = 9.441 \times 10^{-6} \quad \frac{g_{HCN}}{g_{soln}}$

Here we also could simply have divided the concentration in mg/L (= ppm_m in dilute aqueous solutions) by 10^6 to obtain the mass fraction (which can be translated as part per part by mass, whereas ppm_m is parts per 10^6 parts by mass).

Determine the concentration of CN^- (the cyanide ion) in the solution and express the result in mg/L, mol/L, and eq/L (N) units.

Since total cyanide is 10 mg/L and $[CN^-]$ is 1/10 of [HCN], we have:

$C_{CN} := \frac{1}{11} \cdot C_{CN.T} = 0.909 \quad \frac{mg}{L}$ $\quad Cmol_{CN} := \frac{C_{CN}}{MW_{CN}} = 3.497 \times 10^{-5} \quad \frac{mol}{L}$

The monovalent negative CN^- ion has capacity to accept one proton and therefore has 1 eq/mol. In this case, the molarity and normality concentration values are equal.

3.4 APPLICATIONS OF VOLUME FRACTION UNITS

Environmental engineers often must investigate subsurface systems consisting of soil (comprised of vapor, water, and solids). Other systems of interest include solids produced as a consequence of physical/chemical and biological processes employed for water treatment or wastewater renovation. These processes produce either chemical or biological solids and sludges, examined in Chapter 8. Understanding of the application of porosity (void fraction) is paramount to the quantitative understandings of subsurface systems. An application of the porosity concept and unit is illustrated in Example 3.6.

Example 3.6 A region of the unsaturated zone located in an area of suspected contamination by synthetic organics has a void volume fraction of 0.40. Measured moisture content, obtained by weighing a wet soil sample, drying the sample, reweighing the sample, and taking moisture as the difference, was 50 g water (ρ_{water} = 1.00 g/cm^3) per kg dry soil (5% moisture by mass). Determine the volumes of vapor, water, and solid comprising each cubic meter of the subsurface soil. Note that the term soil refers to the mixture of solids, water, and gas occupying a volume of the subsurface. The term solids refers to the portion of the soil that is comprised of solid minerals or natural organic matter, measured gravimetrically after drying.

Let us base this computation on a kg of dry solid. We note that the water and gas must reside in the soil voids. Let us specify some parameters that we would normally measure or otherwise know based on the understandings of the subsurface soil:

$$\varepsilon_{Tot} := 0.4 \quad \frac{m_{void}^3}{m_{Tot}^3} \qquad \rho_{solid} := 2.65 \quad \frac{g}{cm^3} \left(\frac{kg}{L}\right) \qquad \rho_w := 1.0 \quad \frac{g}{cm^3}$$

$$Fmass_{OC} := 0.005 \quad \frac{g_{OC}}{g_{solid}} \qquad F_w := \frac{50}{1000} = 0.05 \quad \frac{g_w}{g_{solid}}$$

$$M_{solid} := 1 \quad kg \qquad M_w := \frac{50}{1000} = 0.05 \quad kg$$

Compute the volume occupied by the kg of dry solids, using the m^3 unit for volume:

$$V_{solid} := \frac{M_{solid}}{\rho_{solid} \cdot 1000} = 3.774 \times 10^{-4} \quad m^3$$

Since the total volume must equal that of the solids plus that of the voids:

$$V_{Tot} = V_{sol} + V_{void} = V_{sol} + \varepsilon_{Tot} \cdot V_{Tot}$$

$$V_{Tot} \cdot (1 - \varepsilon_{Tot}) = V_{sol} \qquad V_{Tot} := \frac{V_{solid}}{1 - \varepsilon_{Tot}} = 6.289 \times 10^{-4} \quad m^3$$

The volume of the voids is then the total volume less the volume of the solids:

$$V_{void} := V_{Tot} - V_{solid} = 2.516 \times 10^{-4} \quad \frac{m_{void}^3}{kg_{solid}}$$

The volume of water can be obtained from the moisture content, taking care to use a consistent volume unit:

$$V_w := \frac{M_w}{\rho_w \cdot 1000} = 5 \times 10^{-5} \quad \frac{m_w^3}{kg_{solid}}$$

The volume of gas can then be obtained as the difference between the void volume and volume of water occupying the voids:

$$V_{vap} := V_{void} - V_w = 2.016 \times 10^{-4} \quad \frac{m_{vap}^3}{kg_{solid}}$$

We may now define and compute the volume fraction of gas, water, and solid:

$$\varepsilon_{vap} := \frac{V_{vap}}{V_{Tot}} = 0.32 \qquad \varepsilon_w := \frac{V_w}{V_{Tot}} = 0.08 \qquad \varepsilon_{solid} := \frac{V_{solid}}{V_{Tot}} = 0.6$$

Then each cubic meter of the whole soil in place is comprised by 0.32, 0.08, and 0.60 m^3 water, vapor, and solid, respectively.

PROBLEMS

A water sample was assayed and found to contain 10 mg/L ammonia nitrogen (NH_3– N). The pH of the solution was 7.1; thus, the NH_3–N is virtually entirely in the form of NH_4^+.

1. Convert this NH_3–N concentration to mol/L N. 7.14e–04
2. Determine the NH_4^+ concentration in mg/L NH_4^+. 1.29e+01
3. Determine the NH_4^+ concentration in ppm_m NH_4^+. 1.29e+01
4. Determine the NH_4^+ concentration in units of mole fraction NH_4^+, note that this would be a dilute aqueous solution. 1.29e–05
5. Determine the normality of NH_4^+ in the solution in terms of meq/L (milliequivalents per liter). 7.14e–01
6. Determine the mass fraction concentration of NH_4^+ in the aqueous solution. 1.29e–05
7. Determine the concentration of NH_4^+ in units of ppb_m. 1.29e+04

A water sample was assayed and found to contain 1 mg/L phosphorus as phosphate (PO_4^{-3}–P). The pH of the solution was 7.2; thus, the phosphate is equally distributed on a molar basis between the two species $H_2PO_4^-$ and $HPO4^=$.

8. Convert this PO_4^{-3}–P concentration to mol/L P. 3.23e–05
9. Determine the concentration of $H_2PO_4^-$ in mg/L. 1.565e+00
10. Determine the concentration of $HPO_4^=$ in ppm_m. 1.55e+00
11. Determine the mole fraction concentration of $H_2PO_4^-$, considering this to be a dilute aqueous solution. 2.90e–07
12. Determine the normality of $HPO_4^=$ in the solution in terms of meq/L (milliequivalents per liter). 3.23e–02
13. Determine the mass fraction concentration of $H_2PO_4^-$ in the aqueous solution. 1.565e–06
14. Determine the concentration of $H_2PO_4^-$ in units of ppb_m. 1.565e+03

The alkalinity and pH of a water sample were determined to be 300 mg/L *as $CaCO_3$* and 9.3, respectively. Application of carbonate system information leads to the result that $[HCO_3^-] = 0.005$ M and $[CO_3^=] = 0.0005$ M.

15. Express the alkalinity in units of meq/L. 6.00e+00
16. Express the alkalinity in units of moles alkalinity per liter of solution. 6.00e–03
17. Convert the concentration of HCO_3^- to units of mg/L. 3.05e+02
18. Convert the concentration of $CO_3^=$ to mg/L. 3.00e+01

19. Convert the concentration of HCO_3^- to units of normality, N (eq/L). 5.00e−03
20. Convert the concentration of $CO_3^=$ to units of normality, N (eq/L). 1.00e−03
21. Given that HCO_3^- and $CO_3^=$ together comprise the total inorganic carbon of the sample, compute the concentration of inorganic carbon in the sample in units of mg_C/L. 6.60e+01
22. Express the concentration of HCO_3^- in units of mg/L *as* $CaCO_3$. 2.50e+02
23. Express the concentration of $CO_3^=$ in units of mg/L *as* $CaCO_3$. 5.00e+01

One milliliter of sulfuric acid (H_2SO_4) is diluted to 1 L in deionized, distilled water. The acid is of 97% purity (i.e., there is 0.97 g of H_2SO_4 per gram of liquid) and the liquid has a density of 1.82 g/cm^3. The resulting solution can be considered a dilute aqueous solution.

24. Determine the resulting concentration of H_2SO_4 in units of mg/L. 1.765e+03
25. Determine the resulting concentration of H_2SO_4 in molar, M, units. 1.80e−02
26. Determine the normality, N (eq/L), of sulfuric acid in the solution. 3.60e−02
27. Determine the mass fraction concentration of sulfuric acid in the resulting solution. 1.765e−03
28. Determine the percent sulfuric acid content of the aqueous solution. 1.77e−01

A near-trophy lake trout, weighing 25 lb_m (pound mass), taken from Lake Michigan was assayed and found to contain Mirex (a fully chlorinated pesticide with MW = 540) at a concentration of 0.002 μg Mirex/g tissue.

29. Express this concentration in units of ppb_m. 2.00e+00
30. Express this concentration in units of mass fraction. 2.00e−09
31. Determine the quantity of Mirex in grams held by the tissues of the fish. Note that 1 lb_m equals 454 g. 2.27e−05
32. Determine the number of moles of Mirex held by the tissues of the fish. 4.20e−08

Rainwater sampled in Chicago, IL, contained 10 μg/L lead, Pb. The pH of the rain was 4.9; thus, the lead was virtually entirely in the form of Pb^{+2}.

33. Express this concentration in nM units. 4.83e+01
34. Express this concentration in meq/L. 9.66e−05
35. Express this concentration in mass fraction units. 1.00e−08

The normal atmosphere consists of 78.96% nitrogen (N_2), 21.00% oxygen (O_2), and 0.04% carbon dioxide (CO_2) as the three major constituents on a molar basis.

Consider that the absolute pressure, P_T, in the atmosphere is 0.95 atm and the ambient temperature is 21 °C.

36. Determine the number of moles of gas per cubic meter of the atmosphere (i.e., the total molar gas concentration in mol/m^3). 3.94e+01
37. What are the mole fraction concentrations of the three constituents? Give them in the order N_2, O_2, and CO_2 to four decimal places. 0.7896, 0.2100, 0.0004
38. Determine the partial pressure of nitrogen, P_{N_2}, in the atmosphere. 7.50e–01
39. Determine the partial pressure of carbon dioxide, P_{CO_2}, in the atmosphere. 3.80e–04
40. Determine the molar concentration of nitrogen (i.e., the concentration in mol/L). 3.11e–02
41. Determine the mass concentration of oxygen in mg/L. 2.64e+02 or 2.65e+02
42. Determine the mass concentration of carbon dioxide in μg/m^3. 6.93e+05
43. Determine the concentration of carbon dioxide in ppm$_v$. 4.00e+02

A gas sample obtained from an elevation of 1000 ft above and 1 mile downwind from the exhaust stack of an industrial boiler near Chicago, IL, contained sulfur dioxide, SO_2, at a level of 10 μg/m^3. The temperature was 8 °C (281 °K) and the atmospheric pressure was 0.94 atm.

44. Determine the concentration of SO_2 in units of mol/m^3. 1.56e–07
45. Determine the partial pressure of SO_2 in units of atm. 3.60e–09
46. Determine the mole fraction concentration of SO_2. 3.83e–09
47. Determine the concentration of SO_2 in ppb$_v$. 3.83e+00

A water sample was assayed and found to contain 21 mg/L ammonia nitrogen (NH_3– N). The pH of the solution was 8.99; thus the NH_3–N is split between NH_4^+ and NH_3^0 at a ratio of 2/3:1/3 on a molar basis (i.e., $[NH_3^0]=0.5[NH_4^+]$).

48. Convert the concentration of NH_4^+ to mol/L N. 1.00e–3
49. Convert $[NH_4^+]$ to mg/L NH_4^+. 18.0
50. Convert the concentration of NH_4^+ to ppm$_m$ NH_4^+. 18.0
51. Convert the concentration of NH_4^+ to units of mole fraction NH_4^+, note that this would be a dilute aqueous solution. 1.8e–5
52. Determine the normality of NH_4^+ in the solution in terms of meq/L (milliequivalents per liter). 1.00
53. Determine the mass fraction concentration of NH_4^+ in the aqueous solution. 1.8e–5
54. Determine the concentration of NH_4^+ in units of ppb$_m$. 18,000

A water sample was assayed and found to contain 2 mg/L phosphorus as phosphate (PO_4^{-3}–P). The pH of the solution was 8.2; thus, the phosphate is distributed on a molar basis between the two species $H_2PO_4^-$ and $HPO_4^=$ at a ratio of 1:10 (i.e., $[H_2PO_4^-]=0.1[HPO_4^=]$).

55. Convert this PO_4^{-3}–P concentration to mol_P/L. 6.452e–5
56. Determine the concentration of $H_2PO_4^-$ in mg/L. 0.569
57. Determine the concentration of $HPO_4^=$ in ppm_m. 5.865e–5
58. Determine the mole fraction concentration of $H_2PO_4^-$, considering this to be a dilute aqueous solution. 1.056e–7
59. Use the residual charge and determine the normality of $HPO_4^=$ in the solution in terms of meq/L (milliequivalents per liter). 0.117
60. Use the number of replaceable protons and determine the normality of $HPO_4^=$ in the solution in terms of meq/L (milliequivalents per liter). 0.085
61. Determine the mass fraction concentration of $H_2PO_4^-$ in the aqueous solution. 5.689e–7
62. Determine the concentration of $H_2PO_4^-$ in units of ppb_m. 569

The alkalinity and pH of a water sample were determined to be 250 mg/L *as $CaCO_3$* and 9.82, respectively. Application of carbonate system information leads to the result that $[HCO_3^-]=0.003$ M and $[CO_3^=]=0.001$ M.

63. Express the alkalinity in units of meq/L. 5.00
64. Express the alkalinity in units of moles alkalinity per liter of solution. 5.00e–3
65. Convert the concentration of HCO_3^- to units of mg/L. 183
66. Convert the concentration of $CO_3^=$ to mg/L. 60.0
67. Convert the concentration of HCO_3^- to units of normality, N (eq/L). 3.00e–3
68. Convert the concentration of $CO_3^=$ to units of normality, N (eq/L). 2.00e–3
69. Given that HCO_3^- and $CO_3^=$ together comprise the total inorganic carbon of the sample, compute the concentration of inorganic carbon in the sample in units of mg_C/L. 48.0
70. Express the concentration of HCO_3^- in units of mg/L *as $CaCO_3$*. 150
71. Express the concentration of $CO_3^=$ in units of mg/L *as $CaCO_3$*. 100

An aliquot of 2 mL of sulfuric acid (H_2SO_4) is diluted to 1 L in deionized, distilled water. The acid is of 95% purity (i.e., there is 0.95 g of H_2SO_4 per gram of liquid) and the liquid has a density of 1.79 g/cm^3. The resulting solution can be considered a dilute aqueous solution.

72. Determine the resulting concentration of H_2SO_4 in units of mg/L. 3.40e+3
73. Determine the resulting concentration of H_2SO_4 in molar, M, units. 0.035

74. Determine the normality, N (eq/L), of sulfuric acid in the solution. 0.069
75. Determine the mass fraction concentration of sulfuric acid in the resulting solution. 3.40e–3
76. Determine the percent sulfuric acid content of the aqueous solution. 0.340

A nice walleye, weighing 8 lb_m (pound mass), taken from the Cheyenne River mouth of Lake Oahe was assayed and found to contain arsenic at a concentration of 0.0002 μg arsenic/g tissue.

77. Express this concentration in units of ppb_m. 0.200
78. Express this concentration in units of mass fraction. 2.00e–10
79. Determine the quantity of arsenic in grams held by the tissues of the fish. Note that 1 lb_m equals 454 g. 7.26e–7
80. Determine the number of moles of arsenic held by the tissues of the fish. 9.70e–9

Rainwater sampled in Chicago, IL, contained 15 μg/L copper, Cu. The pH of the rain was 4.9; thus, the copper was virtually entirely in the form of Cu^{+2}.

81. Express this concentration in nM units. 236
82. Express this concentration in meq/L. 4.72e–4
83. Express this concentration in mass fraction units (g_{Cu}/g_{soln}). 1.5e–8

The gas present above the liquid in an anaerobic digester located at the Rapid City Regional Wastewater Reclamation Facility consisted of 63.45% methane (CH_4), 35.00% carbon dioxide (CO_2), 0.55% hydrogen sulfide (H_2S), and 1.00% volatile organic acids (acetic, propionic, ...) as the four major constituents on a molar basis. Consider that the absolute pressure, P_T, in the gas phase is 0.90 atm and that the tem perature is 36 °C.

84. Determine the number of moles of gas per cubic meter of the gas phase (i.e., the total molar gas concentration in mol/m^3). 35.5
85. What are the mole fraction concentrations of the first three constituents? 0.634, 0.350, 0.055
86. Determine the partial pressure of methane, P_{CH_4}, in the atmosphere. 0.571
87. Determine the partial pressure of hydrogen sulfide, P_{H_2S}, in the atmosphere. 4.95e–3
88. Determine the molar concentration of methane (i.e., the concentration in mol/L). 0.023
89. Determine the mass concentration of carbon dioxide in mg/L. 546
90. Determine the mass concentration of hydrogen sulfide in μg/m^3. 6.63e+6
91. Determine the concentration of hydrogen sulfide in ppm_v. 5.50e+3

A gas sample obtained from an elevation of 1000 ft above and 1 mile downwind from the exhaust stack of an industrial boiler near Chicago, IL, contained sulfur dioxide, SO_2, at a level of 10 ppt_v (parts per trillion by volume). The temperature was 6 °C (279 °K) and the atmospheric pressure was 0.91 atm.

92. Determine the partial pressure of SO_2 in units of atm. 9.10e−12
93. Determine the concentration of SO_2 in units of mol/m^3. 3.97e−10
94. Determine the mole fraction concentration of SO_2. 1.00e−11
95. Determine the concentration of SO_2 in μg/m^3. 0.025

Chapter 4

The Law of Mass Action and Chemical Equilibria

4.1 PERSPECTIVE

Modern chemistry has evolved primarily over the past two plus centuries to become the science it is today. Chemists concern themselves with matter—its properties, structure, and composition. They interest themselves in the changes that occur with matter and the energy that is either released or absorbed as a consequence of these changes. Chemists aim to make use of all knowledge of matter and to extend the body of knowledge concerning matter. Chemists are largely responsible for the development of the logical investigative process we call the "scientific method." Engineers, specifically environmental engineers, toward whom this text is focused, are quite concerned with and obliged to become adept at using the knowledge of matter. We must consider chemists our allies and chemistry an important tool among those with which we must develop expertise in use. Much of the chemical knowledge pertains to the conditions that arise in chemical systems that have arrived at or very near to the condition the chemists call chemical equilibrium. As opposed to chemical kinetics, which is concerned with reactions that proceed either forward or backward, chemical equilibrium addresses the condition at the end of the reaction process, when forward and reverse reactions occur at the same rate. The chemists have developed a system that allows for quantitative understandings of the equilibrium condition. Reactions are statements quantitatively describing the combining of sets of reactants to produce sets of products. The law of mass action is applied to chemical reactions to quantitatively relate the abundances of products to the abundances

Environmental Process Analysis: Principles and Modeling, First Edition. Henry V. Mott.

of reactants, under equilibrium conditions. When applied to a specific reaction, the law of mass action produces an algebraic expression, or equation, which relates the abundances of products present, the abundances of reactions present, and an equilibrium constant. Many systems involve multiple chemical reactions and, under equilibrium conditions, all statements of the law of mass action must simultaneously hold true for all reaction operative within a system. This property of chemical systems is extremely useful in environmental process analysis.

4.2 THE LAW OF MASS ACTION

This text assumes the student has completed two or more courses in general chemistry and has knowledge of the periodic table of the elements, understands the states of matter (solid, liquid, gas), and has a mental picture of the basic atomic structure (protons and neutrons comprising the nucleus, electrons occupying various orbitals surrounding the nucleus, and most particularly the potential for the exchange and sharing of electrons between and among atoms). Here we could simply refer the student to the text from his or her first-year chemistry course for a statement of the law of mass action, but for completeness, we will present the relation herein. Many of the reactions of interest have but one reactant and one product. Others may have numerous reactants as well as numerous products. The chemists typically use two reactants and two products in presenting the general reaction upon which a general statement of the law of mass action can be based:

$$aA + bB \rightleftarrows cC + dD \tag{4.1}$$

where A and B are reactants, C and D are the products, and *a*, *b*, *c*, and *d*, are the stoichiometric coefficients. The general statement of the law of mass action then follows:

$$K_{eq} = \frac{\{C\}^c \{D\}^d}{\{A\}^a \{B\}^b} \tag{4.2a}$$

where K_{eq} is the equilibrium constant, and $\{i\}$ represents the molar activity of product or reactant *i*. The concept of chemical activity and its relation to chemical concentration is not generally addressed in first-year chemistry courses. Particularly in aqueous solutions (one major focus of the applications of chemical principles) the presence of dissolved salts, which ionize to form cations and anions, renders the solution nondilute. As aqueous solutions become less and less dilute, the magnitude of the chemical activity of a reactant or product (hereafter termed a chemical specie) departs from the value of the concentration. As solutions become increasingly nondilute, for ions (species carrying a positive or negative charge) the activity decreases relative to the concentration, while for nonelectrolytes (species which carry a net charge of zero) the activity most often increases relative to the concentration. In a later chapter, the relation between activity and concentration is explored in some depth and means to quantitatively address the nonideality inherent in nondilute solutions is introduced and employed. For the discussions at hand herein and in the next several chapters, we will consider that our aqueous solutions may be approximated as infinitely-dilute and that concentration and activity in these aqueous solutions are synonymous. We will most often use the general term abundance. Equation 4.2a then may be restated in terms of molar concentrations:

$$K_{eq} = \frac{[C]^c[D]^d}{[A]^a[B]^b} \tag{4.2b}$$

where $[i]$ is the familiar symbology denoting the molar concentration of specie i. We may generalize Equation 4.2a to any number of reactants and products using some mathematical symbology:

$$K_{eq} = \frac{\prod_{n_P}[i]^{\nu_i}}{\prod_{n_R}[i]^{\nu_i}} \tag{4.2c}$$

where n_P and n_R are the numbers of products and reactants, respectively, and ν_i is the stoichiometric coefficient of product or reactant i.

In modeling environmental processes, and thus in this text, we concern ourselves with but a few basic types of reactions. These relate:

1. abundances of species in water and vapor that are in intimate contact;
2. abundances of conjugate acids and bases relative to the abundance of hydrogen ions in aqueous solutions;
3. abundances of aqueous metal–ligand complexes relative to the abundance of the free metal ion and various ligands;
4. abundances of the various aqueous ions resulting from the dissolution of a salt in water when the solid salt remains in intimate contact with the aqueous solution;
5. abundances of oxidized and reduced species relative to the abundance of hydrogen ions, other reactants and products, and the availability of electrons within aqueous solutions.

4.3 GAS/WATER DISTRIBUTIONS

The law of mass action statement that addresses the distribution of species in vapor and water is most often referred to as Henry's law. Two statements are generally used. The first simply considers a molecular specie that is present in both the vapor and aqueous phases. The second considers the combination of a water molecule with the vapor-phase specie to form an aqueous molecular pair. General reactions and statements of the law of mass action are then written:

$$i_{(g)} \rightleftarrows i_{(aq)} \tag{4.3a}$$

$$i_{(g)} + H_2O \rightleftarrows i \cdot H_2O_{(aq)} \tag{4.3b}$$

where subscripts (g) and (aq) refer to the vapor and aqueous phases. The corresponding statements of the law of mass action are as follows:

$$K_{H.i} = \frac{[i_{(aq)}]}{P_i} \tag{4.4a}$$

$$K_{H.i} = \frac{\left[i \cdot H_2O_{(aq)}\right]}{\{H_2O\}P_i} \tag{4.4b}$$

$K_{H.i}$ is the Henry's constant, simply the equilibrium constant given a special name, and P_i is the partial pressure of molecular specie i in the vapor phase, which the chemists have chosen to represent the concentration (activity) of gas-phase species in chemical equilibria. A special form of the general reaction depicted in Equation 4.3b includes water as a reactant and specifies a molecular pair ($H_2O \cdot i$) as a product. We will examine this special case in Chapter 5 when we address the dissolution of carbon dioxide and sulfur dioxide in water. Equations 4.3a and 4.3b are written as dissolution reactions. In some of the scientific literature authors have chosen to represent the air/water distribution equilibrium as a volatilization reaction, simply reversing the direction of the reaction. The associated statement of the law of mass action then has a RHS that is the inversion of the RHS of Equation 4.4a and the equilibrium constant, most often represented for volatilization reactions as H_i, is simply $1/K_{H.i}$.

4.4 ACID/BASE SYSTEMS

The deprotonation of acids (often also called dissociation) is another special case. Normally each deprotonation involves a single hydrogen ion (often called a proton). We will examine the behavior of protons in aqueous solution, which associate with a water molecule to become hydronium ions, in Chapter 6. For now, let us simply use the nomenclature we have learned previously and use the term hydrogen ion with the corresponding symbol H^+. The general acid deprotonation reaction is stated as follows:

$$H_mB^n \rightleftarrows H^+ + H_{m-1}B^{(n-1)} \tag{4.5}$$

where H_mB^n is the conjugate acid, of residual charge n, of its conjugate base $H_{m-1}B^{(n-1)}$, of residual charge $n-1$. The residual charge on the conjugate acid and, hence, its conjugate base is dependent upon the extent of the protonation of the acid and of the chemical nature of the fully deprotonated base. We will of course explore these aspects of acids and their conjugate bases further in Chapter 6. We recall that an acid is defined as any chemical specie that can donate a proton via a chemical reaction and that a base is any chemical specie that can accept a proton via a chemical reaction. The statement of the law of mass action for the general acid/base deprotonation reaction is as follows:

$$K_A = \frac{[H^+][H_{m-1}B^{(n-1)}]}{[H_mB^n]} \quad (4.6a)$$

where K_A is the acid dissociation constant, again merely the equilibrium constant given a special name. The coefficients m and n refer, respectively, to the number of protons with capacity to be donated and the overall charge on the conjugate acid. The value of n may be negative, 0, or positive, depending upon the specific acid system in question and the degree to which the conjugate acid is protonated. When we consider specific acid–base systems, Equation 4.6a will make more sense. For example, were we to consider the acetic acid–acetate system (CH_3CO_2H, $CH_3CO_2^-$), for which the fully protonated conjugate acid has a net zero charge and there is but one exchangeable proton, the general Equation 4.6a reduces to a much simpler form:

$$K_A = \frac{[H^+][B^-]}{[HB]} \quad (4.6b)$$

where HB is CH_3CO_2H and B^- is $CH_3CO_2^-$. Since the magnitudes of the K_A values for acid deprotonation reactions are generally quite small, ranging to values as low as 10^{-14}, the "p" concept is most often used when these equilibrium constants are presented in the literature. Then pK_A is $-\log_{10}(K_A)$.

4.5 METAL COMPLEXATION SYSTEMS

The formation of metal–ligand complexes is another important special case examined herein. Many metals form cations in aqueous solutions. These metal ions form chemical species by combining with ligands that are also present in the water. Typical ligands include hydroxide (OH^-), carbonate ($CO_3^=$), and phosphate (PO_4^{-3}). Further, the overall complex can include one or more protons, leading to the following general reaction statement:

$$aM^{+n} + bH^+ + cL^{-m} \rightleftarrows M_aH_bL_c^{(a \cdot n+b-c \cdot m)} \quad (4.7a)$$

where M^{+n} is the metal ion with residual charge $+n$, L^{-m} is the ligand of residual charge $-m$ (m can be 0) and $M_aH_bL_c^{(a \cdot n+b-c \cdot m)}$ is the complex with residual charge $a \cdot n+b-c \cdot m$. The law of mass action statement then follows:

$$\beta_{M_aH_bL_c} = \frac{[M_aH_bL_c^{a \cdot n+b-c \cdot m}]}{[M^{+n}]^a[H^+]^b[L^{-m}]^c} \quad (4.8a)$$

Here $\beta_{M_aH_bL_c}$, called a cumulative formation constant, is merely another special case of the equilibrium constant.

To illustrate a specific example of a complex formation reaction, let us consider the formation of the complex specie calcium bicarbonate ($CaHCO_3^+$) such that the reaction is as follows:

$$Ca^{+2} + H^+ + CO_3^= \rightleftarrows CaHCO_3^+ \tag{4.7b}$$

In this specific case, a, b, and c are all unity, $n=+2$, $m=-2$, and $a\cdot n+b-c\cdot m=+1$. The statement of the law of mass action follows easily:

$$\beta_{CaHCO_3} = \frac{[CaHCO_3^+]}{[Ca^{+2}][H^+][CO_3^=]} \tag{4.8b}$$

As we will investigate further in a later chapter, the symbol β represents a cumulative formation (often called a stability) constant, merely a special form of an equilibrium constant. As with acids and bases, a huge database of formation constant values has been assembled. Our task herein is to develop an understanding of the system in which these data are used and to learn to apply this system in developing quantitative understandings of environmental processes and operations within systems.

4.6 WATER/SOLID SYSTEMS (SOLUBILITY/DISSOLUTION)

The dissolution of inorganic solids (minerals and salts) or, conversely, the precipitation of solids is an additional important special case in which the law of mass action is applied. In its most simple form, involving a cation (most often a metal), an anion (the ligand), and the resultant solid (a complex), from the perspective of the combination of the ions to form the solid, the general reaction may be written:

$$aM^{+n} + cL^{-m} \rightleftarrows M_aL_{c(s)} \tag{4.9a}$$

The overall charge on the solid is 0 and thus the relationship between a and c is such that $a/c = |m/n|$. Equation 4.9a is written as a precipitation (formation) reaction and hence the precipitate can be viewed similarly to the complex formed as depicted in Equation 4.8a. Correspondingly, herein we use the symbol β_s as the equilibrium constant. In fact, much of the solubility/dissolution equilibrium data are included in tables with cumulative formation constants. The law of mass action statement coinciding with Equation 4.9a then can be written:

$$\beta_{s.M_aL_c} = \frac{\{M_aL_{c(s)}\}}{[M^{+n}]^a[L^{-m}]^c} \tag{4.10a}$$

Equation 4.10a is written for the formation of a simple salt, containing one metal and one ligand. Many minerals are combinations of multiple metals and multiple ligands.

For these special cases we merely incorporate the metals and ligands into the formula for the solid in the numerator and include the metals and ligands with their stoichiometric coefficients in the mathematical product of the denominator.

Equation 4.9a is also often written as a dissolution reaction such that the solid is the reactant and the ions are the products:

$$M_aL_{c(s)} \rightleftarrows aM^{+n} + bL^{-m} \tag{4.9b}$$

The corresponding statement of the law of mass action then is the inverted form of Equation 4.10a:

$$K_{sp} = \frac{[M^{+n}]^a[L^{-m}]^c}{\{M_aL_{c(s)}\}} \tag{4.10b}$$

K_{sp} (= $1/\beta_s$) is called the solubility product constant and equilibrium data are also often presented as values of K_{sp}, and often pK_{sp}. With solubility/dissolution equilibria we have a special case for which we will look ahead just a little. Since the solid, in intimate contact with the water, is not dissolved in the water we have a heterogeneous system. The solid is of uniform composition (being crystalline or microcrystalline) and exerts a thermodynamic driving force for dissolution (the reverse reaction for 4.9a and the forward reaction for 4.9b) independent of the quantity of the solid present. The chemists needed a system with which to quantitatively describe the concentration of the solid in the law of mass action. A 'reference' condition was defined—the pure solid. Generally, the chemists have determined that the activity of a component of a pure substance within that pure substance is defined to be unity. We will explore this concept of reference conditions more in later chapters. As we will discuss later, the law of mass action relates the activities of the reactants and products. Thus, since the activity of the solid in intimate contact with the aqueous solution is unity, most often, in expressions of the solubility/dissolution equilibria the activity (at this juncture of our assimilation of the concepts, the concentration) is simply omitted. The resultant expressions of the law of mass action for solubility/dissolution equilibria are then simplified:

$$\beta_s = \frac{1}{[M^{+n}]^a[L^{-m}]^c} \tag{4.10c}$$

$$K_{sp} = [M^{+n}]^a[L^{-m}]^c \tag{4.10d}$$

Since the right sides of Equations 4.10b and 4.10d are simply inverted from Equations 4.10a and 4.10c, we note that β_s is simply the reciprocal of K_{sp}. Reactions 4.9a and 4.9b as stated represent the simplest solubility/dissolution case. Multiple metals, multiple ligands, and protons can comprise solid minerals and the resultant statements of the law of mass action become quite involved. Regardless of the complexity of the solid, however, its activity is unity and most often the solid does not appear in the published statement of the equilibrium relation (i.e., the law of mass action).

4.7 OXIDATION/REDUCTION HALF REACTIONS

Oxidation/reduction (RedOx) half reactions are most often written with the oxidized specie on the LHS and the reduced specie on the RHS, hence, as reduction reactions. In their simplest form, RedOx reactions involve an oxidized specie, a reduced specie, and electrons. Most RedOx half reactions also involve protons and water:

$$a \cdot \mathrm{Ox} + n \cdot \mathrm{e}^- + c \cdot \mathrm{H}^+ \rightleftarrows d \cdot \mathrm{Red} + e \cdot \mathrm{H_2O} \tag{4.11}$$

Red is the specie containing the element after acceptance of the electrons, Ox is the specie containing the element prior to donation of electrons. The magnitudes of the stoichiometric coefficients a, n and $c - e$ are highly dependent upon the oxidation states of the element that accepts the electrons to produce the reduced specie, as well as the manner in which the element is combined in the oxidized and reduced species. RedOx reactions can describe reductions that occur in homogeneous aqueous systems, across vapor/liquid boundaries, across solution solid boundaries, and across solid/solid boundaries. Once written, the (very theoretical) equilibria can be quite conveniently represented using the law of mass action:

$$K_{\mathrm{eq}} = 10^{n \cdot \mathrm{p}E^\circ} = \frac{[\mathrm{Red}]^d \{\mathrm{H_2O}\}^e}{[\mathrm{Ox}]^a [\mathrm{e}^-]^n [\mathrm{H}^+]^c} \tag{4.12}$$

Equation 4.12 has two special cases with which we must deal. In defining the parameter p$E°$, the chemists have decided that, in departure from the normal 'p' notation, p$E°$ is $(1/n)\log_{10}(K_{eq})$ rather than $-(1/n)\log_{10}(K_{eq})$. $[e^-]$ or more appropriately $\{e^-\}$ is the electron availability of the system. Lastly, as mentioned earlier, $\{H_2O\}$ is the activity of water, and equal to unity in most aqueous systems. Then, except in the cases of extremely saline or brine solutions, $\{H_2O\}$ is appropriately omitted from the statement of the law of mass action.

We are educated that electrons cannot exist on their own (other than perhaps for extremely short durations during transfers) in chemical systems, and most particularly at equilibrium. Thus, to analyze systems involving electron transfers the chemists have devised the concept of electron availability. In galvanic systems, this availability is called a reduction potential, given the symbol E_H, and quantitated in volts. In environmental systems, the term oxidation/reduction potential, given the symbol ORP, is used and the unit of measure is the millivolt. In chemical systems, the term pE is used, with the same connotation of p as for pH or pK, p$E = -\log_{10}\{e^-\}$. The literal definition of pE would be the negative $\log_{10}$ of the electron availability in moles per liter. Many texts take advantage of the p definition and carry Equation 4.12 forward by invoking the $\log_{10}$ of both the LHS and the RHS. In this text, we will rely heavily upon the form of RedOx equilibria as stated via Equation 4.12 and will find great convenience in such use.

Chapter 5

Air/Water Distribution: Henry's Law

5.1 PERSPECTIVE

In the chemists' vernacular, Henry's law is used in the context of the relation between the abundance of a component in a vapor and the abundance of that component in a liquid. Henry's law is most often applied in examination of the air/water distribution in systems in which the component of interest comprises but a small fraction of the liquid. In fact, for aqueous systems, if we can quantitatively characterize the interactions of the component of interest within the aqueous solution, Henry's law is applicable even at high relative abundances of the component of interest. The vapor is considered to be ideal, a good assumption at temperatures and pressures near those of environmental systems. Herein, as we will deal primarily with aqueous solutions of prime interest in environmental systems and air that comprises the majority of vapor systems, we will apply Henry's law in quantitatively characterizing distributions of selected components between air and water. As discussed in Chapter 4, Henry's law is merely a special form of a statement of the law of mass action for a particular case: the equilibrium condition resulting from the distribution of a component between air and water.

When students begin the examination of the application of Henry's law, the question often arises:

When can (or must) we employ Henry's law?

Environmental Process Analysis: Principles and Modeling, First Edition. Henry V. Mott.

In reply to the student, the answer given is generally of the form:

> ...whenever (or wherever) a quantitative characterization of a component distribution between air and water is needed or desired.

Specific examples of environmental systems in which Henry's law would be highly applicable include:

- Raindrops forming in the atmosphere and falling to Earth. Natural raindrops contain dissolved nitrogen, oxygen, and carbon dioxide. Anthropogenically affected raindrops can contain a whole host of other components (e.g., carbon monoxide, oxides of nitrogen, oxides of sulfur, and volatile organic contaminants).
- The pore spaces of unsaturated porous media (e.g., natural soils, landfills, mining leach and spoils piles) wherein a vapor–liquid interface exists and contact times are long. Components of interest in these systems include (*but certainly would not be limited to*) carbon dioxide, methane, short-chain carboxylic acids, ammonia, hydrogen sulfide, and hydrogen cyanide.
- Confined spaces of engineered systems (e.g., closed and open conduits and manholes of wastewater collection systems, anaerobic digesters at wastewater treatment facilities) wherein long contact times between vapor and water of constant composition are experienced. Perhaps the component of greatest interest in wastewater collection systems is hydrogen sulfide, a highly toxic and corrosive byproduct of anaerobic biological activity in the absence of oxygen.
- Aqueous and marine sediments within which biological activity creates bubbles that reside in the pores of the sediment for long time periods. Bubbles grow in size while forces attaching them to the sediment solids are greater than those resulting from buoyancy. Upon gaining sufficient size, such that buoyant forces dominate, they are released and migrate through sediment pores to the water column above the sediment/water interface. Along with having great effect upon the character of the pore water, these bubbles can become significant sources for migration of contaminants from sediments to the waters above. In natural systems, carbon dioxide and hydrogen sulfide are abundant and in anthropogenically contaminated sediments toxic substances can be of great interest.
- The monolayers of air and water residing on either side of a vapor–liquid interface anywhere, in any system, and over short or long periods of time. This condition is used to specify interfacial conditions in interfacial mass transfer process. We will delay consideration of this concept until Chapter 8 wherein we will examine rates of transfer of various components (most notably oxygen) between gas and water.

To learn to apply Henry's law in analyses of environmental processes, we will examine its application in a variety of contexts. Our intent is that we develop an understanding of the universality of and a comfort with quantitative application of Henry's law.

5.2 HENRY'S LAW CONSTANTS

A schematic of the equilibrium distribution described by Henry's law is depicted in Figure 5.1. The air–water interface is shown to be horizontal with the water below and the air above. In macroscopic systems, this is entirely the case. However, especially in soils, the interface may have any orientation possible. For bubbles in sediments, the macroscopic interface is spherical in shape, but at the molecular level, the interface would be visualized as a plane tangent to the surface of the sphere at any point on the surface of the sphere.

The majority of the database of equilibrium constants used in water chemistry considers the forward reaction to be the dissolution of the component into water from the vapor phase. Therefore, in this chapter, we will consider the forward reaction to be as depicted by Equations 4.3 and 4.4:

$$i_{(g)} \rightleftarrows i_{(aq)} \tag{4.3a}$$

$$i_{(g)} + H_2O \rightleftarrows i \cdot H_2O_{(aq)} \tag{4.3b}$$

$$K_{H.i} = \frac{[i_{(aq)}]}{P_i} \tag{4.4a}$$

$$K_{H.i} = \frac{[i \cdot H_2O_{(aq)}]}{\{H_2O\}P_i} \tag{4.4b}$$

$K_{H.i}$ is the Henry's constant, $[i]$ is the molar concentration of component i in the aqueous solution, P_i is the partial pressure of component i in the vapor, and $i \cdot H_2O_{(aq)}$ denotes a dissolved molecular pair (e.g., $CO_2{\cdot}H_2O$ or $SO_2{\cdot}H_2O$). In many contexts (such as the examination of volatile organic contaminants, VOCs), these distributions are considered from the viewpoint of the volatilization of components from water to air. Use of Henry's

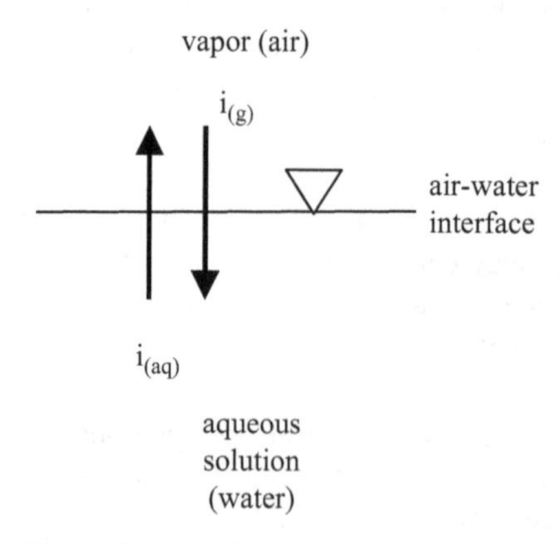

FIGURE 5.1 *Distribution of arbitrary component i between air and water.*

law from this standpoint involves the writing of Equation 4.3 in reverse, inversion of the RHS of Equation 4.4, and use of an alternate symbol for the Henry's law constant:

$$i_{(aq)} \rightleftarrows i_{(g)}; \quad H_i = \frac{P_i}{[i]}$$

This alternative Henry's constant is merely the inverse of that used in Equation 4.4 with the associated set of units also inverted.

In order that we may examine the system for the application of Henry's law, a small database of Henry's law constants has been assembled into Table 5.1. This sampling of values has been obtained from several sources and is in no way intended to constitute an exhaustive data set. Once the system for application of Henry's law has been assimilated, the student will have the ability to use Henry's constant data from any source whatsoever.

An astute eye will quickly discern that the chemical species included in Table 5.1 are all molecular species. No electrolytes (ions) are included. This exclusion is for good reason. Recall the discussions of Chapter 2 regarding the polarity of water and the tendency for ions to attract water molecules in aqueous solutions. The "shells" of water that become associated with ions in solution then absolutely prevent the distribution of the ion into vapor from the water. Since ion–vapor distributions are impossible, no Henry's law constants are available.

Most often the Henry's law distribution equilibrium involves but two species: that in the gas phase and that in the water. Perusal of Table 5.1, however, turns up two apparently special cases, carbon dioxide and sulfur dioxide. In these cases, since one of the hydrating water molecules in turn reacts with the distributed gas, it is included in the overall reaction and, hence, in the statement of the law of mass action. Specifically, for the carbon dioxide and sulfur dioxide systems, variations of Equations 4.3 and 4.4 result and are expressed as Equations 5.1 and 5.2, respectively:

$$CO_{2(g)} + H_2O_{(l)} \rightleftarrows CO_2 \cdot H_2O_{(aq)} \tag{5.1}$$

$$K_{H.CO_2} = \frac{[CO_2 \cdot H_2O_{(aq)}]}{P_{CO_2} \cdot [H_2O_{(aq)}]} \tag{5.2}$$

Then, specifically for the carbon dioxide system, it is known that the $CO_2 \cdot H_2O$ molecular pair reacts further (a rearrangement of the molecular structure occurs) to form the "true" carbonic acid specie, H_2CO_3, with the associated statements of the reaction and law of mass action:

$$CO_2 \cdot H_2O_{(aq)} \rightleftarrows H_2CO_{3(aq)} \tag{5.3}$$

$$K_m = \frac{[H_2CO_{3(aq)}]}{[CO_2 \cdot H_2O_{(aq)}]} \tag{5.4}$$

TABLE 5.1 Values of Selected Henry's Law Constants at 25 °C for 26 Air/Water Distribution Reactions.

Reactant(s)	Product	$K_H\left(\frac{mol}{L-atm}\right)$	Distributed component
1. $CO_{2(g)} + H_2O_{(l)}$ [a]	$\leftrightarrows H_2CO_3{}^*_{(aq)}$	3.39×10^{-2} ($10^{-1.470}$)	Carbon dioxide
2. $SO_{2(g)} + H_2O_{(1)}$ [a]	$\leftrightarrows H_2SO_3{}^*_{(aq)}$	1.23×10^{0} ($10^{0.08991}$)	Sulfur dioxide
3. $NH_{3(g)}$ [a]	$\leftrightarrows NH_{3(aq)}$	5.70×10^{1} ($10^{1.756}$)	Ammonia
4. $H_2S_{(g)}$ [a]	$\leftrightarrows H_2S_{(aq)}$	1.05×10^{-1} ($10^{-0.9788}$)	Hydrogen sulfide
5. $CH_3COOH_{(g)}$ [a]	$\leftrightarrows CH_3COOH_{(aq)}$	7.66×10^{2} ($10^{2.884}$)	Acetic acid
6. $CH_2O_{(g)}$ [a]	$\leftrightarrows CH_2O_{(aq)}$	6.30×10^{3} ($10^{3.800}$)	Formaldehyde
7. $N_{2(g)}$ [a]	$\leftrightarrows N_{2(aq)}$	6.61×10^{-4} ($10^{-3.180}$)	Nitrogen
8. $O_{2(g)}$ [a]	$\leftrightarrows O_{2(aq)}$	1.26×10^{-3} ($10^{-2.889}$)	Oxygen
9. $CO_{(g)}$ [a]	$\leftrightarrows CO_{(aq)}$	9.55×10^{-4} ($10^{-3.020}$)	Carbon monoxide
10. $CH_{4(g)}$ [a]	$\leftrightarrows CH_{4(aq)}$	1.29×10^{-3} ($10^{-2.889}$)	Methane
11. $NO_{2(g)}$ [a]	$\leftrightarrows NO_{2(aq)}$	1.00×10^{-2} ($10^{-2.000}$)	Nitric oxide
12. $NO_{(g)}$ [a]	$\leftrightarrows NO_{(aq)}$	1.90×10^{-3} ($10^{-2.721}$)	Nitrogen oxide
13. $N_2O_{(g)}$ [a]	$\leftrightarrows N_2O_{(aq)}$	2.57×10^{-2} ($10^{-1.590}$)	Nitrous oxide
14. $H_2O_{2(g)}$ [a]	$\leftrightarrows H_2O_{2(aq)}$	1.00×10^{5} ($10^{5.000}$)	Hydrogen peroxide
15. $O_{3(g)}$ [a]	$\leftrightarrows O_{3(aq)}$	9.40×10^{-3} ($10^{-2.027}$)	Ozone
16. $HCN_{(g)}$ [b]	$\leftrightarrows HCN_{(aq)}$	1.29×10^{1} ($10^{1.110}$)	Hydrocyanic acid
17. $C_6H_{6(g)}$ [c]	$\leftrightarrows C_6H_{6(qa)}$	1.79×10^{-1} ($10^{-0.7474}$)	Benzene
18. $C_6H_5CH_{3(g)}$ [c]	$\leftrightarrows C_6H_5CH_{3(aq)}$	1.57×10^{-1} ($10^{-0.8041}$)	Toluene
19. $C_6H_4(CH_3)_{2(g)}$ [c]	$\leftrightarrows C_6H_4(CH_3)_{2(aq)}$	1.42×10^{-1} ($10^{-0.8476}$)	Xylene
20. $C_2HCl_{3(g)}$	$\leftrightarrows C_2HCl_{3(aq)}$	1.10×10^{-1} ($10^{-0.9590}$)	Trichloroethene (TCE)
21. $C_6H_5OH_{(g)}$ [c]	$\leftrightarrows C_6H_5OH_{(aq)}$	2.20×10^{3} ($10^{-3.343}$)	Phenol
22. $CH_3OH_{(g)}$ [c]	$\leftrightarrows CH_3OH_{(aq)}$	2.23×10^{1} ($10^{-1.349}$)	Ethanol
23. $CH_3OC(CH_3)_{3(g)}$ [c]	$\leftrightarrows CH_3OC(CH_3)_{3(aq)}$	1.70×10^{0} ($10^{-0.2314}$)	Methyl-*t*-butyl ether[d]
24. $Mirex_{(g)}$ [c]	$\leftrightarrows Mirex_{(aq)}$	2.79×10^{-3} ($10^{-2.555}$)	Mirex[e]
25. $DDT_{(g)}$ [c]	$\leftrightarrows DDT_{(aq)}$	1.95×10^{0} ($10^{-0.2899}$)	DDT[f]
26. $Atrazine_{(g)}$ [c]	$\leftrightarrows Atrazine_{(aq)}$	3.86×10^{9} ($10^{-9.587}$)	Atrazine[g]

[a] From Table 5.2 of Stumm and Morgan (1996).
[b] Computed from Gibbs free energy of formation data from Table 6.3 of Dean (1992).
[c] From EPA (1990).
[d] Often referred to as MTBE.
[e] 1*a*,2,2,3,3*a*,4,5,5,5*a*,5*b*,6-dodecachlorooctahydro-1,3,4-metheno-1*H*-cyclobuta-[*cd*]pentalene.
[f] 1,1,1-Trichloro-2,2-bis(4-chlorophenyl)ethane.
[g] 2-Chloro-4-ethylamine-6-isopropylamino-*S*-triazine.

Snoeyink and Jenkins (1980) give the magnitude of K_m for Equation 5.4 to be $10^{-2.8}$. We may then conclude that under equilibrium conditions the abundance of the molecular pair specie ($CO_2 \cdot H_2O_{(aq)}$, hydrated carbon dioxide) is 630 times that of the "true" carbonic acid specie ($H_2CO_{3(aq)}$). Hence, in many references, we find that the combination of the molecular pair and the true carbonic acid specie ($[CO_2 \cdot H_2O_{(aq)}] + [H_2CO_{3(aq)}]$) is often called either dissolved carbon dioxide or simply carbonic acid and given the symbol H_2CO_3*.

Full compliance with the information given in Table 5.1 for the carbon dioxide system requires one more important piece of analysis, which is illustrated in Example 5.1, wherein the relation between the Henry's constant shown in Table 5.1 and the Henry's constant for the reaction depicted in Equation 5.1 is developed algebraically.

Example 5.1 Determine the magnitude of the overall Henry's constant, $K_{H.H_2CO_3*}$, from K_m and $K_{H.CO_2}$.

The designation of the "combined" specie accounting for both dissolved carbon dioxide and the true carbonic acid specie is H_2CO_3*, characterized by the relation:

$$[H_2CO_3*] = [CO_2 \cdot H_2O] + [H_2CO_3]$$

We then write the mass action law (equilibrium statement) for the dissolution into water with the combined specie as the product, and substitute the definition of $[H_2CO_3*]$:

$$K_{H.H_2CO_3*} = \frac{[H_2CO_3*]}{P_{CO_2}[H_2O]}; \quad K_{H.H_2CO_3*} = \frac{[CO_2 \cdot H_2O] + [H_2CO_3]}{P_{CO_2}[H_2O]}$$

From a rearrangement of Equation 5.4, we obtain:

$$[H_2CO_3] = K_m \cdot [CO_2 \cdot H_2O]$$

We may then substitute this result into the overall Henry's law relation:

$$K_{H.H_2CO_3*} = \frac{[CO_2 \cdot H_2O] + K_m[CO_2 \cdot H_2O]}{P_{CO_2}[H_2O]} = (1 + K_m)\frac{[CO_2 \cdot H_2O]}{P_{CO_2}[H_2O]}$$

We note the definition from Equation 5.2:

$$K_{H.CO_2} = \frac{[CO_2 \cdot H_2O]}{P_{CO_2}[H_2O]}$$

and substitute this result into the relation for $K_{H.H_2CO_3*}$, obtaining:

$$K_{H.H_2CO_3*} = (1 + K_m)K_{H.CO_2}$$

Since K_m is $10^{-2.8}$, we will introduce little error in employing the "combined" specie $H_2CO_3^*$.

An analysis virtually identical to that of Example 5.1 can be accomplished for the $SO_{2(g)}$/ $SO_2 \cdot H_2O/H_2SO_3$ system, leading to the magnitude of $K_{H_2SO_3*}$.

Reactions 1 and 2 in Table 5.1 present us with another opportunity to apply a reference condition. Particularly in dilute aqueous solutions, the portion of the solution that is water acts nearly as if it were a pure component. Other than in hypersaline seas, in brackish groundwaters, and in contrived solutions that are highly concentrated in one or more salts, natural waters are relatively dilute. In Chapter 10, we explore the quantitative treatment of nondilute solutions. For now, we would prefer to consider the activity of water in aqueous solutions to be unity. In support of this proposed idealization, in Example 5.2 we will examine a water that is high in total dissolved solids and would be above the EPA (U.S. Environmental Protection Agency) secondary (highly recommended but not mandatory) MCL (maximum contaminant level) for total dissolved solids in drinking water.

Example 5.2 Consider a 0.02 M solution of sodium chloride and compute the mole fraction concentration of ions in the aqueous solution.

We first check the salt concentration of the solution so we may consult a handbook, to see if the added salt might have materially affected the density of the solution. We may approximate the percent salt concentration using the standard density of water:

$$\rho_W := 1000 \ \frac{g}{L} \qquad C\%_{TDS} := \frac{0.02 \cdot (23 + 35.5)}{\rho_W} \cdot 100 = 0.117 \quad \%$$

We check our handbook (this information is available from a number of sources) for the density of sodium chloride solution as a function of the salt content and find that the first non-unity value is for 1% and that the density of the solution varies from that of pure water by less than 1%. Our salt content is about a tenth of a percent. We are quite comfortable using the standard density of water as the density of the solution and we proceed to our estimate of the mole fractions of salt and water:

$$C_{Na} = C_{Cl} = 0.02 \ \frac{mol}{L} \qquad C_{Tot.ions} := 0.04 \ \frac{mol}{L}$$

$$\rho mol_{soln} := \frac{1000}{18} = 55.556 \ \frac{mol}{L} \qquad Xmol_W := \frac{\rho mol_{soln} - C_{Tot.ions}}{\rho mol_{soln}} = 0.9993$$

$$Xmol_{salt} := \frac{C_{Tot.ions}}{\rho mol_{soln}} = 7.2 \times 10^{-4} \qquad Xmol_{salt}^{-1} = 1.389 \times 10^{3}$$

On a molar basis then of every 10,000 molecules or ions, seven would be a combination of sodium and chloride. The inversion of the salt mole fraction yields the ratio of water molecules to ions and we observe this value to be nearly 1400:1. These results confirm that in dilute aqueous solutions we may assume that the water is pure with negligible error.

We can comfortably employ the activity of water as unity in most aqueous systems. Then in application of the equilibria represented by reactions 1 and 2 of Table 5.1, we may ignore the $[H_2O]$ term in the final statements of the law of mass action. Then the equilibria for these reactions, addressing carbon dioxide and sulfur dioxide, may be expressed via Equation 4.4. For the carbon dioxide system, K_H is $K_{H.H_2CO_3*}$, $[i]$ is $[H_2CO_3*]$, and P_i is P_{CO_2}. For the sulfur dioxide system, K_H is $K_{H.H_2SO_3*}$, $[i]$ is $[H_2SO_3*]$, and P_i is P_{SO_2}.

5.3 APPLICATIONS OF HENRY'S LAW

As previously mentioned, Henry's law may be applied in any situation in which the distribution of a component (a specie) between air and water is under or very near to equilibrium conditions. The database provided by Table 5.1 is in no way exhaustive. Each and every molecular specie (often referred to as nonelectrolytes) in theory can be distributed between air and water. We will use the database in Table 5.1 as a convenient body of values with which the application of Henry's law may be illustrated.

Let us begin our applications of Henry's law with a characterization of the dissolved gases (oxygen, nitrogen, and carbon dioxide) within a raindrop.

Example 5.3 We know the oxygen and nitrogen contents of the normal atmosphere to be ~21% and ~79%, respectively. The concentration of carbon dioxide ($CO_{2(g)}$) in the atmosphere (an ever-changing value) at the time this text was under authorship was about 390 ppm_v. Consider the local ambient atmosphere at Rapid City, SD, elevation ~ 3400 ft above mean sea level. Let us also assume that the raindrops form in the atmosphere 1000 ft above the earth's surface.

Determine the normal molar concentration of the combined carbonic acid and dissolved carbon dioxide specie ($[H_2CO_3*] = [H_2CO_3] + [CO_2 \cdot H_2O]$) in an aqueous solution in equilibrium with the local atmosphere. The adjustment of equilibrium constants for nonstandard temperatures is considered in a later chapter of this book. We know the result is not fully accurate, but let us use the Henry's constant value from Table 5.1 at 25 °C in our computations.

We realize that we know something about the abundances of O_2, N_2, and CO_2 in the air and desire to use that information to determine the abundances of these components in the raindrop. Our collective set of statements of Henry's law is then:

$$\left[O_{2(aq)}\right] = K_{H.O_2} \cdot P_{O_2};$$

$$\left[N_{2(aq)}\right] = K_{H.N_2} \cdot P_{N_2};$$

$$[H_2CO_3*] = K_{H.H_2CO_3*} \cdot P_{CO_2}$$

Once we determine the partial pressures of O_2, N_2, and CO_2 in the atmosphere, the remainder of the computation is quite straightforward. We know that:

$$P_{O_2} = Y_{O_2} \cdot P_{Tot}$$

$$P_{N_2} = Y_{N_2} \cdot P_{Tot}$$

$$P_{CO_2} = Y_{CO_2} \cdot P_{Tot}$$

(*Typically we will use Y as a symbol denoting a gas-phase mole fraction*)

We also know that

$$Y_{O_2} = \%_{O_2} / 100$$

$$Y_{N_2} = \%_{N_2} / 100$$

$$Y_{CO_2} = \text{ppm}_{v.CO_2} * 10^{-6}$$

We now have only P_{Tot} to define prior to employing Henry's law.

We may seek a value for the normal atmospheric pressure at an elevation of 4400 ft from a handbook or we may employ what we have learned from fluid mechanics. Let us use fluid mechanics here and consult our fluids text. For a linear temperature lapse rate, application of the relation describing the variation of pressure with depth in conjunction with the ideal gas law, with some algebra and integral calculus, the following relation may be developed (Finnemore and Franzini, 2002):

$$\frac{P_{z2}}{P_{z1}} = \left(\frac{a + b \cdot z_2}{a + b \cdot z_1} \right)^{\frac{-g}{R \cdot b}}$$

where P_{z_1} and P_{z_2} are the total atmospheric pressures at the reference and targeted elevations, z_1 is the reference elevation (sea level), z_2 is the elevation of the system in question, g is the acceleration of gravity, R is the mass-based gas-specific gas constant (merely the product of the universal gas constant and the molecular weight of the gas with some rearrangement of units), and a (518.7 °R) and b (−0.00356 °F/ft) are the lapse rate coefficients for the relation $T = a + b \cdot z$.

From close examination of the pressure relation, we realize that even when we use the cumbersome set of English units (ft, lb_f, °R) for the RHS we may use our preferred unit of pressure (atm) on the LHS as both the LHS and RHS of the relation are dimensionless. We may rearrange the pressure equation and solve for the pressure at elevation 4400 ft above mean sea level. We will use a matrix-style organization for defining parameters and performing computations. While appearing cumbersome here, comfort with this style will provide for far superior organization of worksheets as our applications become more complex:

$$\begin{pmatrix} g \\ R_{m.air} \\ a \\ b \end{pmatrix} := \begin{pmatrix} 32.17 \\ 1716 \\ 518.7 \\ -0.00356 \end{pmatrix} \begin{pmatrix} \frac{ft}{s^2} \\ \frac{ft^2}{s^2 \cdot {}^\circ R} \\ {}^\circ R \\ \frac{{}^\circ R}{ft} \end{pmatrix} \quad \begin{pmatrix} z_1 \\ z_2 \\ P_{z1} \end{pmatrix} := \begin{pmatrix} 0 \\ 4400 \\ 1.0 \end{pmatrix} \begin{pmatrix} ft \\ ft \\ atm \end{pmatrix}$$

$$P_{Tot} := P_{z1} \cdot \left(\frac{a + b \cdot z_2}{a + b \cdot z_1} \right)^{\frac{-g}{R_{m.air} \cdot b}} = 0.851 \quad atm$$

Now we may define mole fraction concentrations and employ these with the total pressure to obtain the set of partial pressures using MathCAD's very convenient capability to multiply a matrix by a scalar:

$$\begin{pmatrix} Y_{O2} \\ Y_{N2} \\ Y_{CO2} \end{pmatrix} := \begin{pmatrix} \frac{21}{100} \\ \frac{79}{100} \\ 390 \cdot 10^{-6} \end{pmatrix} = \begin{pmatrix} 0.21 \\ 0.79 \\ 3.9 \times 10^{-4} \end{pmatrix} \frac{atm_i}{atm_{Tot}}$$

$$\begin{pmatrix} P_{O2} \\ P_{N2} \\ P_{CO2} \end{pmatrix} := \begin{pmatrix} Y_{O2} \\ Y_{N2} \\ Y_{CO2} \end{pmatrix} \cdot P_{Tot} = \begin{pmatrix} 0.179 \\ 0.672 \\ 3.318 \times 10^{-4} \end{pmatrix} \quad atm$$

Then, from the partial pressures and Henry's law, we may compute the abundances of oxygen, nitrogen, and carbon dioxide in the targeted raindrops:

$$\begin{pmatrix} K_{H.O2} \\ K_{H.N2} \\ K_{H.CO2} \end{pmatrix} := \begin{pmatrix} 0.00126 \\ 0.000661 \\ 0.0339 \end{pmatrix} \frac{mol}{L \cdot atm}$$

$$\begin{pmatrix} C_{O2} \\ C_{N2} \\ C_{H2CO3} \end{pmatrix} := \begin{pmatrix} K_{H.O2} \cdot P_{O2} \\ K_{H.N2} \cdot P_{N2} \\ K_{H.CO2} \cdot P_{CO2} \end{pmatrix} = \begin{pmatrix} 2.251 \times 10^{-4} \\ 4.443 \times 10^{-4} \\ 1.125 \times 10^{-5} \end{pmatrix} \frac{mol}{L}$$

We have now employed Henry's law to characterize the aqueous solution contained in a raindrop when we were knowledgeable of the composition of air that is presumed to be in near-equilibrium with raindrops in the atmosphere. We will learn in Chapter 10 that we may further characterize the raindrop relative to the speciation of the carbon dioxide system using acid/base equilibria and an accounting of hydrogen ions.

We can use Henry's law to determine the abundances of gas-phase species when we know the corresponding abundances in an aqueous phase. Vinegar is a substance known to almost all for its pungent odor. The active ingredient in vinegar is, of course, acetic acid. Acetic acid is an important intermediate product in the production of methane in both natural and engineered systems. Winemakers also strive to eliminate acid-forming bacteria from their musts in the fermentation process; otherwise, the wine turns to vinegar as a consequence of the production of acetic acid rather than ethanol.

Example 5.4 Distilled vinegar is an aqueous solution containing 5.00% acetic acid by mass fraction. The pH of distilled vinegar is typically in the range of 3.0, thus the acetic acid in vinegar is virtually 100% in the fully protonated form, CH_3COOH (*we will explore this type of determination in greater detail in Chapter 6*). When we open a bottle of vinegar and take a whiff, our olfactory sense is bombarded with the very pungent odor emitted by the vinegar. Let us consider a bottle of vinegar that has been opened, is ~3/4 full, and has been equilibrated with the atmosphere such that the total pressure of the gas phase above the vinegar is 0.950 atm. We know that the density of acetic acid is only slightly greater than that of water, so we may, with little error, use the density of water as that for a 5% acetic acid solution.

We realize we must obtain the gas-phase abundance from what we know of the aqueous phase. We realize that in order to use the Henry's constant for acetic acid (let us use HAc as an abbreviation) from Table 5.1 we need to determine the molar concentration of acetic acid in the vinegar. Once this is determined, the computation will be quite straightforward.

Let us first determine the mass concentration of HAc in the water:

$$Xmass_{HAc} := 0.05 \frac{g_{HAc}}{g_{soln}} \qquad \rho_{soln} := 1000 \frac{g_{soln}}{L}$$

$$Cmass_{HAc} := Xmass_{HAc} \cdot \rho_{soln} = 50 \frac{g_{HAc}}{L}$$

From the mass concentration, we may easily determine the molar concentration:

$$MW_{HAc} := 12 + 3 + 12 + 2 \cdot 16 + 1 = 60 \frac{g}{mol} \qquad C_{HAc} := \frac{Cmass_{HAc}}{MW_{HAc}} = 0.833 \frac{mol}{L}$$

We may now employ Henry's law to obtain the partial pressure of acetic acid in the vapor:

$K_{H.HAc} := 766 \ \frac{mol}{L \cdot atm}$ $\qquad P_{HAc} := \frac{C_{HAc}}{K_{H.HAc}} = 1.088 \times 10^{-3} \ atm$

The mole fraction concentration, the volume fraction, and the pressure fraction are identical so we may obtain the mole fraction as the ratio of the partial pressure of acetic acid to the total pressure:

$P_T := 0.95 \ atm$ $\qquad Y_{HAc} := \frac{P_{HAc}}{P_T} = 1.145 \times 10^{-3} \ \frac{mol_{HAc}}{mol_{gas}}$

Roughly one molecule per thousand of the gas phase is acetic acid, equating to 1145 molecules of acetic acid per million molecules of gas, 0.1145% or 1145 ppm_v.

Hydrogen sulfide is a dangerous gas that is produced by biological processes in the absence of oxygen when sulfate is available. Significant health risks arise from inhalation when abundance levels are above about 50 ppm_v, and significant danger of death arises when abundance is 300 ppm_v or greater. Workers that repair and maintain wastewater collection systems must be keenly aware of the dangers associated with this gas. In addition, large abundances of hydrogen sulfide can cause severe corrosion of concrete. Manholes and concrete pipe in wastewater collection systems can be severely affected necessitating costly replacement projects. Let us apply Henry's law to a common circumstance that is often associated with wastewater collection systems—that of a long force main that moves wastewater from a low-lying area to a portion of the collection system that drains to the central treatment facility via gravity sewers.

Example 5.5 In sewer force mains, wastewater is pumped through a closed conduit rather than flowing via gravity in an open-channel pipe. The pH of a particular wastewater after traversing a 1.5 mile force main was 6.1 and the total sulfide sulfur concentration (sum of the sulfide sulfur in H_2S and HS^-) was 33 ppm_m. At pH 6.1, the sulfide species would be in a molar ratio such that $[H_2S] = 10[HS^-]$. *We will investigate why this is so in Chapter 6, but for now, let us just use the information.* This wastewater discharges into a manhole at the upper end of a gravity collection system where a vapor phase with total pressure of 0.91 atm is present. Consider that the manhole is poorly vented such that the distribution of soluble gases between the flowing wastewater and the gas phase within the manhole can be considered to be at near-equilibrium. Would the conditions arising in this manhole present an imminent danger to a worker entering without first properly purging the gases from within?

In order to provide guidance relative to the posed question, we of course must estimate the abundance of hydrogen sulfide gas ($H_2S_{(g)}$) in ppm_v that would be in this vapor if it were in equilibrium with the aqueous solution. Again, Henry's law is of great utility and we would begin by computing the molar concentration of hydrogen sulfide in the aqueous phase ($H_2S_{(aq)}$) so that we may use Henry's law and obtain the partial pressure, and hence the abundance, of H_2S in the gas phase.

We begin by determining the total molar concentration of sulfide species in the wastewater:

$$MW_S := 32000 \frac{mg}{mol} \quad C_{Tot.S} := \frac{33}{MW_S} = 1.031 \times 10^{-3} \quad \frac{mol}{L}$$

Since $[H_2S_{(aq)}] = 10[HS^-]$ and we know that $[S^=]$ will be quite insignificant (*again, our studies in Chapter 6 will fully examine such questions*), we may determine $[H_2S_{(aq)}]$. $[HS^-]$ is 1/10 $[H_2S_{(aq)}]$ and thus $[H_2S_{(aq)}]$ is 10/11 of the total sulfide sulfur:

$$C_{H2S} := \frac{10}{11} \cdot C_{Tot.S} = 9.375 \times 10^{-4} \quad \frac{mol}{L}$$

We may now apply Henry's Law to obtain the partial pressure of hydrogen sulfide. Bisulfide, with its residual negative charge and associated hydrated "shell" of water, cannot distribute into the air:

$$K_{H.H2S} := 0.105 \frac{mol}{L \cdot atm} \quad P_{H2S} := \frac{C_{H2S}}{K_{H.H2S}} = 8.929 \times 10^{-3} \quad atm$$

With P_{H_2S} known, we may then consider the H_2S abundance in the gas phase using the unit of measure used to characterize epidemiological data (ppm_v), first by computing the mole fraction of H_2S in the vapor:

$$P_T := .91 \ atm \quad Y_{H2S} := \frac{P_{H2S}}{P_T} = 9.812 \times 10^{-3} \quad \frac{atm_{H2S}}{atm_{Tot}} = \frac{mol_{H2S}}{mol_{Tot}} = \frac{vol_{H2S}}{vol_{Tot}}$$

Since parts per million by volume is also parts per million by moles and pressure, we may simply convert the pressure fraction to mole fraction and to volume fraction, and finally to parts per million:

$$ppm_{V.H2S} := Y_{H2S} \cdot 10^6 = 9.812 \times 10^3 \quad ppm_v$$

The posed of example 5.5 situation would indeed be very dangerous for an entering worker; the hydrogen sulfide abundance is computed to be many times the threshold danger level. For many years, standard practice (OSHA confined space entry procedures) has dictated that workers who perform maintenance or repairs associated with wastewater collection systems carry portable air blowers with them and purge potentially dangerous gases from within enclosed spaces prior to entry.

The applications of Henry's law presented in this chapter are limited to computations of the distribution of the nonelectrolyte of various systems (e.g., carbon dioxide, hydrogen sulfide, and acetic acid) between air and water. Overall, computations must consider all the species of the particular systems residing in the aqueous solution to effectively capture the total character of the distribution of the various species comprising the system. In Chapter 6, along with applications of acid/base equilibria, we will examine the role of air/water equilibria in the overall distribution of the species of selected acid–base systems in air/water systems.

Later, in chapter 8, we will examine several special cases associated with the transfer of nonelectrolyte species from air to water or from water to air. These transfer processes result from a condition in which the bulk aqueous solution is not at equilibrium with the bulk gas phase. The net transfer then proceeds across the vapor/liquid interface. The direction of transfer is dependent upon the difference in the effective concentration level of the transferred component between the two phases. In order to define the effective concentration, we must employ Henry's law to characterize the gas-phase and aqueous-phase concentrations of the transferred specie in the monolayers of gas and water present on the gas side and aqueous side of the interface. We consider that at the molecular level, given the very small distances involved with movement to, across, and away from an interface, a second is a very long period of time. Consider the "nonslip" condition at interfaces associated with Newtonian fluids and realize that both water and air are considered to be Newtonian in nature. Let us use an air bubble rising through a stagnant solution of low-dissolved oxygen concentration as an example system. A microscopic depiction of the two monolayers is shown schematically in Figure 5.2. We realize that the monolayers of

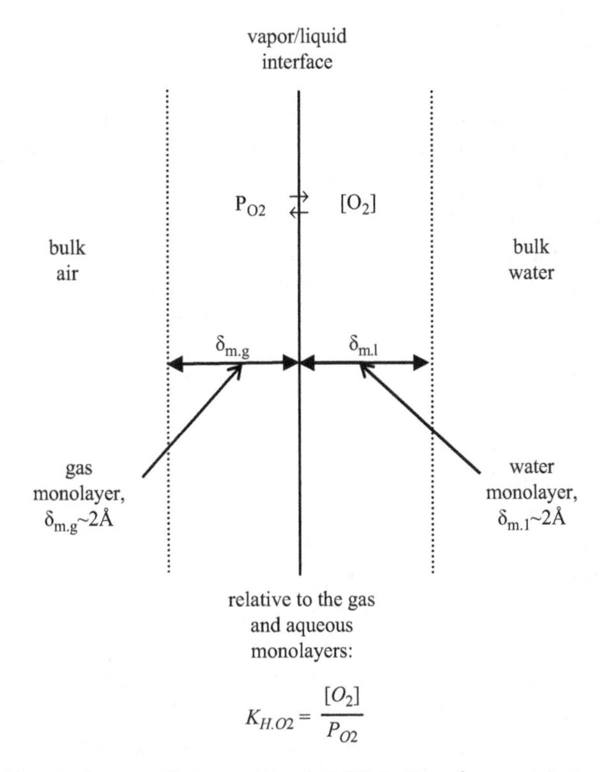

FIGURE 5.2 *Henry's law equilibrium arising for distribution of oxygen between monolayers bounding the vapor liquid interface of an air bubble.*

water and gas on either side of the vapor/liquid interface are in contact for a sufficient period to provide for an equilibrium condition to arise. We may therefore employ Henry's law to relate the abundances of oxygen in the monolayers of water and gas residing adjacent to the interface. Then, given knowledge of the concentrations of oxygen in the two bulk solutions, we can relate the concentrations in the monolayers adjacent to the interface with those of each bulk solution. If the interfacial concentration is less than that of the bulk solution, the net transfer will be toward (and then through) the interface. If the difference is positive in one phase, it must be negative in the other. We introduce this concept here and will employ it quantitatively in Chapter 8 when we examine mass transfer processes in ideal reactors.

PROBLEMS

1. The concentration of carbon dioxide ($CO_{2(g)}$) in the atmosphere is 356 ppm_v and the local ambient atmospheric pressure is 0.904 atm. Determine the molar concentration of the combined carbonic acid and dissolved carbon dioxide specie (H_2CO_3*, $[H_2CO_3^*] = [H_2CO_3] + [CO_2 \cdot H_2O]$) in an aqueous solution in equilibrium with the local atmosphere.
2. The vapor space above an anaerobic digester of a wastewater treatment facility contains 65.2% methane (CH_4) on a molar basis and the total pressure of this gas is 1.01 atm. Determine the molar concentration of methane in the aqueous solution contained in the digester considering that the aqueous solution would be in equilibrium with the vapor.
3. Hydrocyanic acid ($HCN_{(g)}$) was detected at a concentration of 10.2 ppb_v in vapor space above a tank at a local gold mine that contained aqueous spent heap leach solution. Determine the molar concentration of hydrocyanic acid in this solution, considering the vapor space to be in equilibrium with the aqueous solution. The local atmospheric pressure was measured at 0.894 atm at the time the vapor was sampled.
4. Hydrogen sulfide gas ($H_2S_{(g)}$) was detected at a level of 1.22 ppm_v in the vapor space contained within a sewage lift station after a power outage that caused the pumps to fail. According to the local weather service station, the atmospheric pressure at the time the vapor was sampled was 0.937 atm. Determine the molar concentration of hydrogen sulfide in the wastewater contained within the lift station, assuming that the aqueous solution is at equilibrium with the vapor.
5. The normal atmosphere contains nitrogen (N_2) at a mole fraction concentration of 0.788. If the local atmospheric pressure is 0.932 atm, determine the molar concentration of nitrogen in aqueous solution (as might be found in a drop of rain) in equilibrium with the normal atmosphere at the given total pressure.

6. The dissolved oxygen (O_2) concentration of water held within a tank in the laboratory of a manufacturer of aeration devices was 37.1 mg/L when the total pressure of the vapor above the tank was 0.981 atm. This resulted from a clean-water test of a particular aeration device that employed high-purity oxygen for aeration. Given that the aqueous solution and gas bubbled through the aqueous solution are in equilibrium, determine the purity of the oxygen (in percent on a molar basis) used for the test.
7. An aqueous solution contains ammonia nitrogen (NH_3–N) at a concentration of 786 mg(N)/L. The pH of the solution is 8.29, thus the ratio of the molar concentration of the NH_3 specie to that of the NH_4^+ specie is 1:10 ($[NH_3] = 0.1[NH_4^+]$). The total pressure in the vapor phase is 0.913 atm. Determine the partial pressure of ammonia, P_{NH_3}, in vapor that would be in equilibrium with this aqueous solution.
8. Groundwater is often supersaturated with carbon dioxide, much like a bottle of soda, and when exposed to atmospheric levels of CO_2, effervescence can occur. A particular water sample taken from a rather deep well had a pH of 6.3, thus half the inorganic carbon was in the form of bicarbonate, HCO_3^-, and the other half in the combined $H_2CO_3^*$ ($[H_2CO_3^*] = [CO_2 \cdot H_2O] + [H_2CO_3]$) specie. The total inorganic carbon of the sample was 0.00800 mol/L. What would be the partial pressure, P_{CO_2}, (in atm) of carbon dioxide in vapor that is in equilibrium with this groundwater?
9. The pH of a particular wastewater after traversing a 1.5 mile force main (wastewater is pumped through a closed conduit rather than flowing via gravity in an open channel) was 6.1 and the total sulfide concentration (sum of the sulfide in H_2S and HS^-) was 100 ppm_m. At pH 6.1, the sulfide species are in a molar ratio such that $[H_2S] = 10[HS^-]$. This wastewater discharges into another lift station where a vapor phase of total pressure 0.91 atm is present. Determine the concentration of hydrogen sulfide gas ($H_2S_{(g)}$) in ppm_v that would be in this vapor if it were in equilibrium with the aqueous solution.
10. Distilled vinegar contains 5.00% acetic acid by mass fraction in an aqueous solution. The pH of distilled vinegar is typically 2.4, thus the acetic acid in vinegar is virtually 100% in the undissociated form, CH_3COOH. When we open a bottle of vinegar and take a whiff, our olfactory sense is bombarded with the very pungent odor emitted by the vinegar. Determine the mole fraction concentration of acetic acid in the vapor phase above the liquid vinegar in the bottle that is responsible for this odor level. Assume the vinegar and the vapor phase above it are in equilibrium at a total pressure of 0.950 atm. You may assume that vinegar has a density equal to that of water.
11. The mole fraction concentration of carbon dioxide ($CO_{2(g)}$) in the atmosphere is 3.90×10^{-4} and the local ambient atmospheric pressure is 0.944 atm. Determine the molar concentration of the combined carbonic acid and dissolved carbon dioxide specie ($H_2CO_3^*$, $[H_2CO_3^*] = [H_2CO_3] + [CO_2 \cdot H_2O]$) in an aqueous solution in equilibrium with the local atmosphere.

12. The vapor space above an anaerobic digester of a wastewater treatment facility contains methane (CH_4) at a partial pressure of 0.631 atm and the total pressure of this gas is 1.01 atm. Determine the molar concentration of methane in the aqueous solution contained in the digester considering that the aqueous solution would be in equilibrium with the vapor.
13. Hydrocyanic acid ($HCN_{(g)}$) was detected at a concentration of 0.0254 ppm_v in vapor space above a tank at a local gold mine that contained aqueous spent heap leach solution. Determine the molar concentration of hydrocyanic acid in this solution, considering the vapor space to be in equilibrium with the aqueous solution. The local atmospheric pressure was measured at 0.869 atm at the time the vapor was sampled.
14. Hydrogen sulfide gas ($H_2S_{(g)}$) was detected at a level of 99.1 ppb_v in the vapor space contained within a sewage lift station after a power outage that caused the pumps to fail. According to the local weather service station, the atmospheric pressure at the time the vapor was sampled was 0.984 atm. Determine the molar concentration of hydrogen sulfide in the wastewater contained within the lift station, assuming that the aqueous solution is at equilibrium with the vapor.
15. The normal atmosphere contains 788,000 ppm_v nitrogen (N_2). If the local atmospheric pressure is 0.947 atm, determine the molar concentration of nitrogen in aqueous solution in equilibrium with the normal atmosphere at the given total pressure.
16. The dissolved oxygen (O_2) concentration of water held within a tank in the laboratory of a manufacturer of aeration devices was 35.9 mg/L when the total pressure of the vapor above the tank was 0.927 atm. This resulted from a clean-water test that employed high-purity oxygen for aeration. Given that the aqueous solution and gas bubbled through the aqueous solution are in equilibrium, determine the purity of the oxygen (in percent on a molar basis) used for the test.
17. An aqueous solution contains ammonia nitrogen (NH_3–N) at a concentration of 842 mg(N)/L. The pH of the solution is 8.98, thus the ratio of the molar concentration of the NH_3 specie to that of the NH_4^+ specie is 1:2 ($[NH_3] = 0.5[NH_4^+]$). The total pressure in the vapor phase is 0.946 atm. Determine the partial pressure of ammonia, P_{NH_3}, in vapor that would be in equilibrium with this aqueous solution.
18. Groundwater is often supersaturated with carbon dioxide, much like a bottle of soda, and when exposed to atmospheric levels of CO_2 effervescence can occur. A particular water sample taken from a rather deep well had a pH of 6.65, thus two-thirds of the inorganic carbon was in the form of bicarbonate, HCO_3^-, and the other third in the combined $H_2CO_3^*$ ($[H_2CO_3^*] = [CO_2{\cdot}H_2O] + [H_2CO_3]$) specie. The total inorganic carbon of the sample was 0.00900 mol/L. What would be the partial pressure, P_{CO_2}, (in atm) of carbon dioxide in vapor that is equilibrated with this groundwater?

19. The pH of a particular wastewater after traversing a 1.5 mile force main (wastewater is pumped through a closed conduit rather than flowing via gravity in an open channel) was 6.41 and the total sulfide concentration (sum of the sulfur in H_2S and HS^-) was 150 ppm_m. At pH 6.4, the sulfide species are in a molar ratio such that $[H_2S] = 5[HS^-]$. This wastewater discharges into another lift station where a vapor phase of pressure of 0.89 atm is present. Determine the concentration of hydrogen sulfide gas ($H_2S_{(g)}$) in ppm_v that would be in this vapor if it were in equilibrium with the aqueous solution.
20. A typical ammonia cleaning solution contains 2% by mass ammonia in an aqueous solution. The pH of such a solution will be around 11.7, thus the NH_3–N will be virtually entirely in the form of NH_3. When we open this bottle of solution, and have a smell, our curiosity is rewarded by both an odor that is unpleasant and, perhaps, a sharp pain in the nose. Determine the mole fraction concentration of ammonia ($NH_{3(g)}$) in the vapor (total pressure = 0.95 atm) contained within the bottle of ammonia solution, considering that the vapor would be in equilibrium with the aqueous solution.
21. The atmospheric carbon dioxide ($CO_{2(g)}$) content is 2.90 ppm_v and the local ambient atmospheric pressure is 0.958 atm. Determine the molar concentration of the combined carbonic acid and dissolved carbon dioxide specie (H_2CO_3*, $[H_2CO_3^*] = [H_2CO_3] + [CO_2 \cdot H_2O]$) in an aqueous solution in equilibrium with the local atmosphere.
22. The vapor space above an anaerobic digester of a wastewater treatment facility contains methane (CH_4) at a mole fraction concentration of 0.631 atm and the total pressure of this gas is 1.05 atm. Determine the molar concentration of methane in the aqueous solution contained in the digester considering that the aqueous solution would be in equilibrium with the vapor.
23. Hydrocyanic acid ($HCN_{(g)}$) was detected at a mole fraction concentration of 2.27×10^{-8} in vapor space above a tank at a local gold mine that contained aqueous spent heap leach solution. The local atmospheric pressure was measured at 0.926 atm at the time the vapor was sampled. Determine the molar concentration of hydrocyanic acid in this solution, considering the vapor space to be in equilibrium with the aqueous solution.
24. Hydrogen sulfide gas ($H_2S_{(g)}$) was detected at a level of 87.1 ppm_v in the vapor space contained within a sewage lift station after a power outage that caused the pumps to fail. According to the local weather service station, the atmospheric pressure at the time the vapor was sampled was 0.933 atm. Determine the molar concentration of hydrogen sulfide in the wastewater contained within the lift station, assuming that the aqueous solution is at equilibrium with the vapor.
25. The normal atmosphere contains 20.9% oxygen (O_2) on a molar basis. If the local atmospheric pressure is 0.947 atm, determine the molar concentration of

oxygen in aqueous solution in equilibrium with the normal atmosphere at the given total pressure. Express your answer in mg_{O_2}/L.

26. The dissolved oxygen (O_2) concentration of water held within a closed tank in the laboratory of a manufacturer of aeration devices was $0.09\, mg_{O_2}/L$ when the total pressure of the vapor above the tank was 0.967 atm. This resulted from a clean-water test that employed high-purity nitrogen gas bubbled through the water to deoxygenate the water. Given that the aqueous solution and gas bubbled through the aqueous solution are in equilibrium, determine the oxygen content in percent (molar basis) of the nitrogen gas used for the test. Then, if the only two components present in the gas were oxygen and nitrogen, determine the nitrogen concentration of the aqueous solution in unit of mg_{N_2}/L.
27. An aqueous solution contains total cyanide ([HCN] + [CN^-]) at a concentration of 255 mg(CN)/L. The pH of the solution is 9.51, thus the ratio of the molar concentration of the HCN specie to that of the CN^- specie is 1:2 ([HCN] = 0.5[CN^-]). The total pressure in the vapor phase is 0.926 atm. Determine the concentration in ppm_v of hydrogen cyanide, HCN, in vapor that would be in equilibrium with this aqueous solution.
28. Groundwater is often supersaturated with carbon dioxide, much like a bottle of soda, and when exposed to atmospheric levels of CO_2 effervescence can occur. A particular water sample taken from a rather deep well had a pH of 6.04, thus one-third of the inorganic carbon was in the form of bicarbonate, HCO_3^-, and the other two-thirds in the combined H_2CO_3* ([H_2CO_3*] = [$CO_2 \cdot H_2O$] + [H_2CO_3]) specie. The total inorganic carbon of the sample was 0.00600 mol/L. What would be the ratio of the equivalent partial pressure, P_{CO_2}, of carbon dioxide of this water sample to the normal atmospheric partial pressure of carbon dioxide, given that atmospheric carbon dioxide is 290 ppm_v and the normal atmospheric pressure is 0.88 atm.
29. The pH of a particular wastewater after traversing a 1.5 mile force main (wastewater is pumped through a closed conduit rather than flowing via gravity in an open channel) was 6.80 and the total sulfide concentration (sum of the sulfur in H_2S and HS^-) was 12.0 ppm_m. At pH 6.8, the sulfide species are in a molar ratio such that [H_2S] = 2[HS^-]. This wastewater discharges into terminus manhole of a gravity collection system where a vapor phase with total pressure of 0.87 atm is present. Determine the concentration of hydrogen sulfide gas ($H_2S_{(g)}$) in ppm_v that would be in this vapor if it were in equilibrium with the aqueous solution.
30. An aqueous solution contains 10.0% by mass sulfurous acid H_2SO_3* ([H_2SO_3*] = [H_2SO_3] + [$SO_2 \cdot H_2O$]). The pH of such a solution will be around 4.0, thus the sulfurous acid will be virtually entirely in the H_2SO_3* form (as opposed to HSO_3^- or SO_3^{-2}). The density of pure sulfurous acid is just slightly greater than that of water so we may approximate the density of this solution as that of water, invoking only small error. If we should (unadvisedly) open

this bottle of solution, and have a smell, our curiosity is rewarded by an odor that is unpleasant and perhaps also by a sharp pain in the nose. The ambient atmospheric pressure is 0.94 atm and the temperature is 22 °C. Determine the partial pressure and molar concentration of sulfur dioxide ($SO_{2(g)}$) in the vapor contained within the bottle of acid solution, considering that the vapor would be in equilibrium with the aqueous solution.

Chapter 6

Acid/Base Component Distributions

6.1 PERSPECTIVE

Somewhere along the educational path many of us came to a (not quite correct) belief that an acid has a pH lower than 7 and a base has a pH greater than 7. More correctly, chemists educate us that acids are constituents (*herein called chemical species*) that can donate protons and that bases are constituents that can accept protons. Protons are exactly analogous to hydrogen ions, once the single electron in the single orbital of the hydrogen atom has been donated. The ^1H atom has no neutron, and in the elemental state shares its lone electron with another ^1H atom to form the diatomic gas H_2. Hydrogen readily donates its electron in chemical reactions and once oxidized what remains after the donation of the electron is the proton. We will see in stages throughout this book that the proton, or hydrogen ion, is truly much more.

Herein, we will consider acid–base concepts as applicable in aqueous solutions. When applied qualitatively in the context of an aqueous solution, the adjectives acidic and basic might mean that the pH of the solution is below or above seven, respectively. Again, qualitatively speaking, we can have mildly, moderately, strongly, or severely acidic or basic solutions, depending upon the level of the pH. In some cases, mildly acidic solutions, depending upon the character of the dissolved substances, can have large values of acidity, indicative of large abundances of constituents that can donate protons, while moderately or even strongly acidic solutions can have low values of acidity, indicative of small abundances of proton-donating substances. In this chapter, we will delve into the details behind these happenstances. We will learn about the system the science of chemistry has

Environmental Process Analysis: Principles and Modeling, First Edition. Henry V. Mott.

provided us to quantitatively understand acid–base behavior. We will then learn how to apply these acid–base principles in the analysis of some fairly simple environmental systems. We will also combine employment of acid/base principles with employment of Henry's law, allowing us to consider heterogeneous air–water systems in an integrated context.

We should agree on a few definitions to get us started in our examination of acids and bases:

Acid: A chemical specie that can under the correct circumstances donate a proton (a hydrogen ion) via chemical reaction.

Base: A chemical specie that can under the correct circumstances accept a proton (a hydrogen ion) via chemical reaction.

Acidity: A quantitative measure of the ability of an aqueous solution to donate protons, measured by titration from an initial pH to a designated end point using a strong base.

Alkalinity: A quantitative measure of the ability of an aqueous solution to accept protons, measured by titration from an initial pH to a designated end point using a strong acid.

Buffering capacity: A mostly qualitative term used to describe the capacity of a solution to either accept or donate protons. Solutions of large buffering capacity contain species that either can accept or donate protons while the pH of the solution changes little.

Conjugate acid: The chemical specie formed from the acceptance of a proton by its conjugate base.

Conjugate base: The chemical specie formed from the donation of a proton by its conjugate acid.

Acid system: The collective set of aqueous species containing a specific base, of the fully deprotonated base and all conjugate acids of that base.

6.2 PROTON ABUNDANCE IN AQUEOUS SOLUTIONS: pH AND THE ION PRODUCT OF WATER

Chemistry has developed a methodology to measure the abundance of protons in aqueous solutions and a system for reporting the values. Generally, a pH probe and meter is used. The specific electrochemistry applied renders the pH probe to be specific to hydrogen ions. In direct (not necessarily linear) relation to the abundance of hydrogen ions in solution, an electrochemical potential between the solution and the probe arises. The potential is measured (typically in millivolts) and transmitted to the meter for display. Most pH meters have capacity to display output in mV as well as in pH units. Then, through understandings of the acid–base character of specific conjugate acid–conjugate base pairs, standard solutions of known proton abundance are created for use in calibration of pH meters. These standard solutions are used to calibrate the measured potential against a set of known standards. Most often these standard solutions, called buffers, have nominal pH values of 4.0, 7.0, and 10.0. We immerse the probe into these solutions, one at a time and set the readout of the meter to correspond with the value of the buffer being measured. Once it is calibrated, the pH meter can be employed to measure the hydrogen ion abundance in otherwise unknown solutions. The measured potential of the unknown solution is compared (via programming resident in the meter) with the calibration potentials and the pH of the solution is displayed by the meter.

Once measured, the result is displayed as the pH. The lower case p is used as a shorthand notation for the $-\log_{10}$. Thus, the pH is in fact the negative of the base 10 logarithm of the hydrogen ion abundance, in units of mol/L. Therefore, we must be careful in our use of pH as it is inversely related to the actual abundance of protons in an aqueous solution. For example, pH 7.0 refers to $[H^+]=10^{-7}$ M while pH 6.0 refers to $[H^+]=10^{-6}$ M. A **decrease** of one unit in the pH value amounts to an order of magnitude **increase** in the proton abundance. A low value of pH denotes a large proton abundance while a high value of pH denotes a small proton abundance. We often see the "p" concept applied to concentrations or reaction equilibrium constants such that $pC_i=-\log_{10}[i]$ and $pK_{eq}=-\log_{10}(K_{eq})$.

Given the electrochemical nature of the measurement technique, the measured abundance is in fact the chemical activity of the protons in the solution. We have encountered this chemical activity term earlier and determined that detailed consideration must wait until chapter 10 of this text when we are more fully armed with understandings allowing us to fully appreciate its significance. Thus, for now we will revert to the more familiar definition that $pH=-\log_{10}[H^+]$ (and the reverse that $[H^+]=10^{-pH}$), leaving examination and application of the true definition that $pH=-\log_{10}\{H^+\}$ ($-\log_{10}$ of the molar *activity* of hydrogen ions) until we embrace the advanced topics of Chapter 10.

We could delve into the specific electrochemistry of the pH meter, developing full understandings regarding the bases of the measurement process, but will not. These understandings are published in numerous sources. Although vital to the overall understanding of and refinement of methods for the measurement of pH, assimilating these understandings is beyond the scope of this textbook.

Let us examine the hydrogen ion (or proton) in light of the discussion of Chapter 2 relative to the attraction of water to ions in aqueous solution. Once its electron is donated from the hydrogen atom, what remains is the proton situated in the nucleus of the ion with a residual unit positive charge. Then, in aqueous solution, water molecules arrange themselves via short-range bonds between the partial negative of the oxygen and the positive charge of the proton. The first four water molecules tend to orient themselves in a tetrahedral structure. The bonds between the coordinating water molecules and the proton are quite strong. We will leave the computation of the bond energy to the physical chemists. Given that these first four water molecules in no way satisfy the positive charge, additional water molecules are free to orient with the resultant proton–water structure, forming another layer, still strongly bonded to the positive charge, but certainly less so than the inner four. Additional layers of water, each with successively weaker bonds, are added to the overall positively charged ionic structure. In discussions of the relation between the ionic content of water and the activity coefficient (relating molar activity to molar concentration), Dean (1992) lists the ion size parameter (the hydrated radius) of the hydrogen ion as

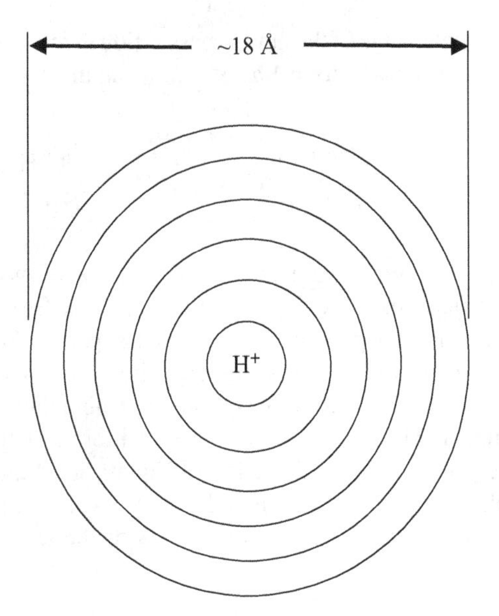

FIGURE 6.1 *Visual representation of layering of water molecules within a hydrated hydrogen ion.*

nine angstroms (Å). We can then envision (this works best if we close our eyes and imagine that we can see both the proton and the water molecules) a layered sphere, each layer being a spherical shell, of ordered water molecules surrounding the proton. Within each shell the oxygen atom of each water molecule would tend to be oriented toward the center relative to the hydrogen atoms. Figure 6.1 is a two-dimensional, quite simplistic representation of the general configuration. Given that the water molecules, as a consequence of their innate energy manifest as translational, rotational, and vibrational motion are always on the move, the structure would be dynamic with continual exchange of water molecules. Further, rather than orient in "layers" the water molecules might tend to orient in a much less ordered manner. Regardless of the details, we leave Figure 6.1 with a "picture" of the proton in aqueous solution as a rather large (given the scale of the ionic radius of the hydrogen ion) conglomeration of water surrounding the proton.

In order to simplify the treatment of the proton in aqueous solutions, the idea has been proposed that the proton "bonds" with a single water molecule (likely at the center of the hydrated sphere) to form the hydronium (H_3O^+) ion. This concept is useful in examining water as the conjugate base of the hydronium ion and as the conjugate acid of the hydroxide ion. We can write these two reactions, both in the form of Equation 4.5:

$$H_3O^+ \rightleftarrows H^+ + H_2O \tag{6.1}$$

$$H_2O \rightleftarrows H^+ + OH^- \tag{6.2}$$

We can then write the statements of the law of mass action in the form of Equation 4.6a, denoting the equilibrium constant for 6.1 as $K_{A1.H_2O}$ and that for 6.2 as $K_{A2.H_2O}$:

$$K_{A1.H_2O} = \frac{[H^+][H_2O]}{[H_3O^+]}; \quad K_{A2.H_2O} = \frac{[H^+][OH^-]}{[H_2O]}$$

Fundamentally then, the hydronium ion is a diprotic acid, with two protons to donate. Extreme difficulties arise in discerning protons from hydronium ions. The chemists inform us that protons cannot exist separate from the acids with which they are combined, except for extremely short time periods during transfers. Further difficulties arise with defining the huge abundance of water relative to those of hydronium and hydroxide. Therefore, an alternate approach necessarily was developed. We can reverse the direction of Equation 6.1 and add the result to Equation 6.2, leading to the most accepted quantitative definition of the acid–base behavior of water. The associated statements of the law of mass action are included:

$$H^+ + H_2O \rightleftarrows H_3O^+ \quad \frac{1}{K_{A1.H_2O}} = \frac{[H_3O^+]}{[H^+][H_2O]}$$

$$\underline{H_2O \rightleftarrows H^+ \pm OH^-} \quad K_{A2.H_2O} = \frac{[H^+][OH^-]}{[H_2O]}$$

$$2H_2O \rightleftarrows H_3O^+ + OH^- \quad K_w = \frac{K_{A2.H_2O}}{K_{A1.H_2O}} = \frac{[H_3O^+][OH^-]}{[H_2O]^2}$$

Chemists have reasoned that the reaction with water on the LHS and RHS can be subtracted from the result, leaving a single water on the LHS and the proton and the hydroxide on the RHS. The standard condition defined based on pure liquid water requires that the activity of water be unity. Since in most aqueous solutions of environmental interest, the solution is comprised of 99.9+% water, the activity (and hence, in this early context the concentration) of water in an aqueous solution is taken as unity. The result reduces to what we know to be the statement of the ion product of water:

$$H_2O \rightleftarrows H^+ + OH^- \tag{6.3}$$

$$K_w = [H^+][OH^-] \tag{6.4}$$

The value of K_w is known to be $\approx 1.0 \times 10^{-14}$ at 25 °C, and, like other equilibrium constants, varies with temperature.

6.3 ACID DISSOCIATION CONSTANTS

In Chapter 4, we presented the reactions and associated law of mass action statement for the general reaction describing the donation of a hydrogen ion by a conjugate acid forming the conjugate base:

$$H_mB^n \rightleftarrows H^+ + H_{m-1}B^{(n-1)} \tag{4.5}$$

$$K_A = \frac{[H^+][H_{m-1}B^{(n-1)}]}{[H_mB^n]} \tag{4.6a}$$

An often used companion reaction considers the combination of water with a conjugate base to form the conjugate acid and the hydroxide ion. If we reverse the direction of the acid deprotonation reaction and add the reaction describing the ion product of water, we arrive at the reaction that is often called a base association reaction:

$$H^+ + H_{m-1}B^{n-1} \rightleftarrows H_mB^n \quad \frac{1}{K_A} = \frac{[H_mB^n]}{[H^+][H_{m-1}B^{(n-1)}]}$$

$$\underline{H_2O \rightleftarrows H^{\pm} + OH^-} \quad K_w = \frac{[H^+][OH^-]}{[H_2O]}$$

$$H^+ + H_{m-1}B^{n-1} + H_2O \rightleftarrows H_mB^n + H^+ + OH^- \quad K_B = \frac{K_w}{K_A} = \frac{[H_mB^n][H^+][OH^-]}{[H_2O][H^+][H_{m-1}B^{(n-1)}]}$$

When we cancel chemical species appearing as both reactants and products in both the reaction and the equilibrium statements and invoke unity for the concentration (activity) of water, reaction 6.5 and its associated equilibrium statement results:

$$H_{m-1}B^{n-1} + H_2O \rightleftarrows H_mB^n + OH^- \quad K_B = \frac{[H_mB^n][OH^-]}{[H_{m-1}B^{n-1}]} \tag{6.5}$$

The relation between K_A and K_B is then that $K_B = K_w/K_A$ (and $pK_B = pK_w - pK_A$).

Equation 6.5 is developed and presented herein in the interests of completeness. Herein, we will rely strictly upon the concept that conjugate acids donate protons to form conjugate bases. The student is left to pursue the base association concept as he or she might desire.

In order that we may illustrate the systematic approach to the application of the chemistry of acid deprotonation reactions, a representative database of K_A values has been assembled and included herein. In our development of the understandings of and applications of acid–base equilibria, we will rely heavily upon the acid dissociation reactions included in Table 6.1. The student is reminded that a much larger database of reactions and associated K_A values is available and that Table 6.1 is in no way exhaustive.

Acids are categorized in terms of the number of protons that can possibly be donated under reasonable conditions. Monoprotic acids (e.g., acetic acid) can donate one proton, diprotic acids (e.g., carbonic acid) can donate two, triprotic acids (e.g., phosphoric acid) can donate three, and tetraprotic acids (e.g., ethyl-ene-diamine-tetraacetic acid) can donate four protons. Table 6.1 contains acids of all four categories. Acids capable of donating more than four protons exist, but herein we will confine our discussions to those capable of donating four or fewer protons.

6.4 MOLE ACCOUNTING RELATIONS

Before we may fully appreciate and use acid/base equilibria we need to have a means to account for the total abundance of the conjugate acids and conjugate bases of a given acid–base system. The basic premise of chemical equilibrium requires that under equilibrium conditions the forward and reverse reactions both occur at the same rate. From this fundamental requirement we conclude that when an acid–base system is present in an aqueous solution, all potential species must be present in that solution. The abundances of certain species may be insignificant relative to the total of the acid system species. Nevertheless, their presence is necessary in order for the equilibrium condition to arise. In a liter of water, we count 55.56 mol at 6.023×10^{23} molecules per mole or 3.35×10^{25} molecules of water. Were the abundance of a particular specie to be truly zero, none of those molecules would be of the target specie. We have developed special analytical instruments that can quantitate analytes to levels as low as tenths of a ppb_m (μg/L). Consider the element cadmium (MW = 112.4). At an abundance of 0.1 μg/L, there would be ~5×10^{14} atoms of cadmium present as various cadmium species, in one liter of solution. In laboratory analyses, for abundances of cadmium below this level, we would list the result as "nondetect" or "below detection limits" and be tempted to call the abundance zero. For radionuclides, when we can measure alpha, beta, or gamma particle emissions, we can extend detection limits another several orders of magnitude, but we certainly cannot analytically reach abundance levels anywhere near the true value of zero. Under certain conditions, equilibria of metal sulfides predict abundances of the metals in the presence of the metal sulfide solid to be lower than one atom in thousands or more liters aqueous solution. Even though this level of abundance is for all practical purposes zero, the system used for chemical equilibrium requires that we consider the metal to be present in the solution to maintain the equilibrium condition. Then, although we may neglect species of mathematically insignificant abundance in mole accounting equations, we must never lose sight of the fact that they are present.

Monoprotic acids have a single conjugate acid and a single conjugate base. Diprotic acids have a fully protonated conjugate acid, a fully deprotonated conjugate base, and

TABLE 6.1 Selected Acid/Base Systems and pK_A Values at 25 °C

Conjugate acid	Formula	pK_A	Conjugate Base	Formula
Hydrochloric acid	HCl	~ –3[a]	Chloride	Cl^-
Nitric acid	HNO_3	~ 0[a]	Nitrate	NO_3^-
Acetic acid	CH_3COOH	4.70[a]	Acetate	CH_3COO^-
Propanoic (propionic) acid	CH_3CH_2COOH	4.87[b]	Propanoate (propionate)	$CH_3CH_2COO^-$
Hydrocyanic acid	HCN	9.30[a]	Cyanide	CN^-
Hypochlorous acid	$HOCl$	7.5[a]	Hypochlorite	OCl^-
Ammonium	NH_4^+	9.30[a]	Ammonia	NH_3
Orthosilicic acid	H_4SiO_4	9.50[a]	Trihydrogen silicate	$H_3SiO_4^-$
Sulfuric acid	H_2SO_4	~ –3[a]	Bisulfate	HSO_4^-
Bisulfate	HSO_4^-	2.0[a]	Sulfate	SO_4^{-2}
Sulfurous acid[c]	H_2SO_3	1.89[b]	Bisulfite	HSO_3^-
Bisulfite	HSO_3^-	7.20[b]	Sulfite	SO_3^{-2}
Hydrogen sulfide	H_2S	7.10[a]	Bisulfide	HS^-
Bisulfide	HS^-	~ 14[a]	Sulfide	S^{-2}
Carbonic acid[d]	H_2CO_3*	6.35[e]	Bicarbonate	HCO_3^-
Bicarbonate	HCO_3^-	10.33[e]	Carbonate	CO_3^{-2}
Phosphoric acid	H_3PO_4	2.10[a]	Dihydrogen phosphate	H_2PO_4
Dihydrogen phosphate	$H_2PO_4^-$	7.20[a]	Hydrogen phosphate	HPO_4^{-2}
Hydrogen phosphate	$HPO_4^{=}$	12.3[a]	Phosphate	PO_4^{-3}
Citric acid	$(HO_2CCH_2)C(OH)(CO_2H)\text{-}CH_2CO_2H$	3.128	Dihydrogen citrate	$(O_2CCH_2)C(OH)(CO_2H)\text{-}CH_2CO_2H^-$
Dihydrogen citrate	$(O_2CCH_2)C(OH)(CO_2H)\text{-}CH_2CO_2H^-$	4.761	Hydrogen citrate	$(O_2CCH_2)C(OH)(CO_2H)\text{-}CH_2CO_2^{-2}$
Hydrogen citrate	$(O_2CCH_2)C(OH)(CO_2H)\text{-}CH_2CO_2^{-2}$	6.396	Citrate	$(O_2CCH_2)C(OH)(CO_2)\text{-}CH_2CO_2^{-3}$
Biselenate	$HSeO_4^-$	1.66[b]	Selenate ion	SeO_4^{-2}

(Continued)

TABLE 6.1 (*Continued*)

Conjugate acid	Formula	pK_A	Conjugate Base	Formula
Selenous acid	H_2SeO_3	2.62[b]	Biselenite	$HSeO_3^-$
Biselenite	$HSeO_3^-$	8.30[b]	Selenite ion	SeO_3^{-2}
Arsenic acid	H_3AsO_4	2.22[b]	Dihydrogen arsenate	$H_2AsO_4^-$
Dihydrogen arsenate	$H_2AsO_4^-$	6.98[b]	Hydrogen arsenate	$HAsO_4^{-2}$
Hydrogen arsenate	$HAsO_4^{-2}$	11.5[f]	Arsenate ion	AsO_4^{-3}
Arsenous acid	H_3AsO_3	8.9[f]	Dihydrogen arsenite	$H_2AsO_3^-$
Dihydrogen arsenite	$H_2AsO_3^-$	11.2[f]	Hydrogen arsenite	$HAsO_3^{-2}$
Hydrogen arsenite	$HAsO_3^{-2}$	13.4[f]	Arsenite ion	AsO_3^{-3}
Benzoic acid	$C_6H_5CO_2H$	4.20[b]	Benzoate	$C_6H_5CO_2^-$
Hydrogen aniline	$C_6H_5NH_3^+$	4.60[b]	Aniline	$C_6H_5NH_2$
Citric acid	$(CO_2HCH_2)_2CO_2HCOH$	3.128[b]	Dihydrogen citrate	$(CO_2HCH_2)_2CO_2COH^-$
Dihydrogen citrate	$(CO_2HCH_2)_2CO_2COH^-$	4.761[b]	Hydrogen citrate	$CO_2CH_2CO_2HCH_2CO_2COH^{-2}$
Hydrogen citrate	$CO_2CH_2CO_2HCH_2CO_2COH^{2-}$	6.396[b]	Citrate	$(CO_2CH_2)_2CO_2COH^{-3}$
4-Aminobenzoic acid	$H_3NC_6H_4CO_2H^+$	2.41[b]		$H_2NC_6H_4CO_2H$
	$H_2NC_6H_4CO_2H$	4.85[b]		$H_2NC_6H_4CO_2^-$
3-Aminopentanoic acid	$CH_3CH_2CHNH_4CH_3CO_2H^+$	4.02[b]		$CH_3CH_2CHNH_4CH_3CO_2$
	$CH_3CH_2CHNH_4CH_3C\ O_2$	10.399[b]		$CH_3CH_2CHNH_3CH_3CO_2^-$
4-Aminopentanoic acid	$CH_3CHNH_4(CH_2)_2CO_2H^+$	3.97[b]		$CH_3CHNH_4(CH_2)_2CO_2$
	$CH_3CHNH_4(CH_2)_2CO_2$	10.46[b]		$CH_3CHNH_3(CH_2)_2CO_2^-$
5-Aminopentanoic acid	$CH_2NH_4(CH_2)_3CO_2H^+$	4.20[b]		$CH_2NH_4(CH_2)_3CO_2$
	$CH_2NH_4(CH_2)_3CO_2$	9.758[b]		$CH_2NH_3(CH_2)_3CO_2^-$

1,3-Diamino-2-propanol	$H_3NCH_2CHOHCH_2NH_3^{+2}$	7.93[b]	$H_3NCH_2CHOHCH_2NH_2^{+}$
	$H_3NCH_2CHOHCH_2NH_2^{+}$	9.69[b]	$H_2NCH_2CHOHCH_2NH_2$
2,3-Diaminopropanoic acid	$HO_2CCHNH_3CH_2NH_3^{+2}$	1.33[b]	$O_2CCHNH_3CH_2NH_3^{+}$
	$O_2CCHNH_3CH_2NH_3^{+}$	6.674[b]	$O_2CCHNH_3CH_2NH_2$
	$O_2CCHNH_3CH_2NH_2$	9.623[b]	$O_2CCHNH_2CH_2NH_2^{-}$
Ethylene-diamine-tetraacetic acid	$H_4[(O_2CCH_2)_2NCH_2)]_2$	1.99[b]	$H_3[(O_2CCH_2)_2NCH_2)]_2^{-}$
	$H_3[(O_2CCH_2)_2NCH_2)]_2^{-}$	2.67[b]	$H_2[(O_2CCH_2)_2NCH_2)]_2^{-2}$
	$H_2[(O_2CCH_2)_2NCH_2)]_2^{-2}$	6.16[b]	$H[(O_2CCH_2)_2NCH_2)]_2^{-3}$
	$H[(O_2CCH_2)_2NCH_2)]_2^{-3}$	10.26[b]	$[(O_2CCH_2)_2NCH_2)]_2^{-4}$

[a] From Snoeyink and Jenkins (1980).
[b] From Dean (1992).
[c] $[H_2SO_3*]=[H_2SO_3]+[SO_2{\cdot}H_2O]$.
[d] $[H_2CO_3*]=[H_2CO_3]+[CO_2{\cdot}H_2O]$.
[e] From Morel and Hering (1993).
[f] From Brookins (1988).

a monoprotonated (or mono deprotonated) specie that is both a conjugate acid and a conjugate base. Triprotic acids have a fully protonated conjugate acid, a fully deprotonated conjugate base, and two partially protonated species that are both conjugate acids and conjugate bases. Tetraprotic acids have a fully protonated conjugate acid, a fully deprotonated conjugate base, and three partially protonated species that are both conjugate acids and conjugate bases. The statement for the mole balance is merely an accounting of the total abundance of the fully deprotonated base, whether free or as combined with one or more protons. The total molar abundance is simply a sum of the molar concentrations of the distinct conjugate acid and conjugate base species. These statements may be written in general form for mono- through tetraprotic acids.

$$C_{\text{Tot.B}} = [HB^{n}] + [B^{n-1}] \qquad \text{For monoprotic acids}$$

$$C_{\text{Tot.B}} = [H_2B^{n}] + [HB^{n-1}] + [B^{n-2}] \qquad \text{For diprotic acids}$$

$$C_{\text{Tot.B}} = [H_3B^{n}] + [H_2B^{n-1}] + [HB^{n-2}] + [B^{n-3}] \qquad \text{For triprotic acids}$$

$$C_{\text{Tot.B}} = [H_4B^{n}] + [H_3B^{n-1}] + [H_2B^{n-2}] + [HB^{n-3}] + [B^{n-4}] \qquad \text{For tetraprotic acids}$$

The residual charge on the fully protonated acid, n, depends upon the constitution of the acid. For most common inorganic acids, the value of n is 0. For the ammonia–nitrogen system and for organic acids that contain a single amine group, the value of n is 1. For organic acids that contain more than a single amine group, n is generally equal to the number of amine groups contained in the structure of the acid that can accept protons to become positively charged.

6.5 COMBINATION OF MOLE BALANCE AND ACID/BASE EQUILIBRIA

The accounting relations are combined with the deprotonation equilibria in the employment of acid–base principles. These concepts are perhaps best illustrated through examples.

6.5.1 Monoprotic Acids

The application of the acid–base equilibrium with mole balance for a monoprotic acid, acetic acid, is illustrated in Example 6.1.

Example 6.1 Explore the distribution of total acetate between acetic acid and the acetate ion in an aqueous solution containing 78 mg/L total acetate as a function of the pH of the aqueous solution.

We begin by defining the total molar concentration of acetate species (let us abbreviate CH_3COO^- as Ac^-) and assigning the value of the equilibrium constant to a recognizable symbol. Given its mathematical/numerical power and its convenient means for expressing mathematical relations, we will use MathCAD as the primary means for performing computations. The software does not provide flexibility of symbology to match that used in chemistry, so we will develop and conform to an alternate system of symbology:

$$MW_{Ac} := (12 + 14)\cdot 1000 = 2.6 \times 10^{4} \ \frac{mg}{mol}$$

$$C_{Tot.Ac} := \frac{78}{MW_{Ac}} = 3 \times 10^{-3} \ M \qquad K_{A.Ac} := 10^{-4.7}$$

We may then write a mole balance including the two acetate species that **must** be present in the aqueous solution. The equilibrium relation may be arranged in two distinct ways to solve for one specie in terms of the other, utilizing the hydrogen ion concentration and the equilibrium constant:

$$C_{Tot.Ac} = HAc + Ac \qquad HAc = \frac{H}{K_{A.Ac}}\cdot Ac \qquad Ac = \frac{K_{A.Ac}}{H}\cdot HAc$$

We then may use the relation for HAc in terms of Ac or for Ac in terms of HAc to write two renditions of the mole balance. One employs Ac as the master-dependent variable and the other employs HAc as the master-dependent variable:

$$C_{Tot.Ac} = HAc + \frac{K_{A.Ac}}{H}\cdot HAc \qquad C_{Tot.Ac} = \frac{H}{K_{A.Ac}}\cdot Ac + Ac$$

We may then rearrange each of these to explicitly solve for acetic acid and acetate, and create functions in a MathCAD worksheet:

$$H(pH) := 10^{-pH} \qquad HAc(pH) := \frac{C_{Tot.Ac}}{1 + \frac{K_{A.Ac}}{H(pH)}} \qquad Ac(pH) := \frac{C_{Tot.Ac}}{\frac{H(pH)}{K_{A.Ac}} + 1}$$

We may then assign a range of pH values, say 3–10, and create an *X–Y* plot in our MathCAD worksheet. We have created and populated a function to carefully place the vertical line shown at $pH = pK_{A.Ac}$. We must also order the dependent variables with the independent variables. The arguments used for the *X–Y* plot are shown in Figure E6.1.1 to illustrate this process.

We note that at pH values below $pK_{A.Ac}$ the fully protonated specie is more abundant (predominant) than the deprotonated specie, and that at pH values higher than the $pK_{A.Ac}$ the deprotonated specie is predominant. When we delve more deeply into the equilibrium relation, we can rearrange it to yield a ratio of the abundances of acetate and acetic acid:

$$\frac{K_{A.Ac}}{[H^+]} = \frac{[Ac^-]}{[HAc]}$$

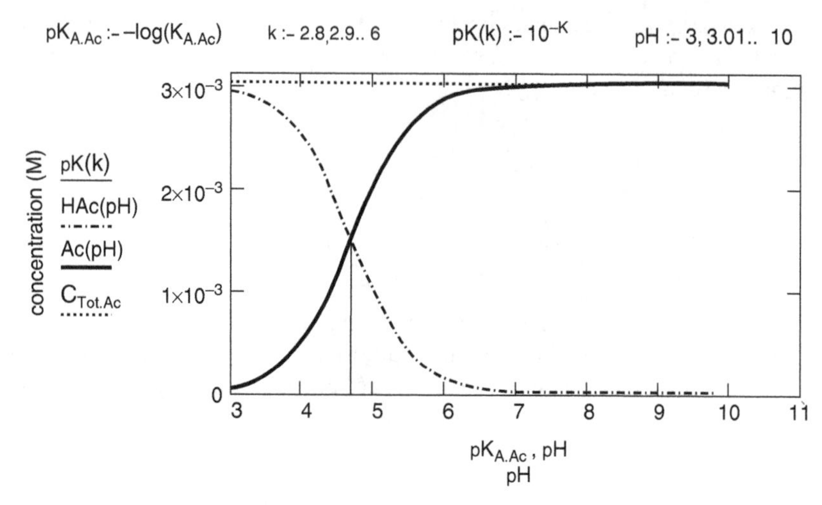

FIGURE E6.1.1 *Plot of specie predominance for the acetic acid system.*

We quickly note that when $[H^+]$ and $K_{A.Ac}$ are equal ($pH = pK_{A.Ac}$), so are $[Ac^-]$ and [HAc]. As the abundance of protons in the solution increases (pH decreases), acetic acid becomes the predominant specie. Conversely as the pH is increased (the abundance of protons in the solution decreases), the acetate ion becomes more and more predominant.

The behavior observed in Example 6.1 is not specific to the acetic acid–acetate system. We will observe this trend for every single acid deprotonation reaction. The only difference among the various acid systems will be the pH value of the equivalence of the abundances of the conjugate acid and its conjugate base. The intersection for the concentrations of each conjugate acid–conjugate base pair must be situated at $pH = pK_A$. Were we to repeat Example 6.1 for any of the monoprotic acids whose pK_A values are included in Table 6.1, we would obtain the same general result. The predominant specie concentrations at pH values well departed from the pK_A would approach (in truth asymptotically) the total concentration ($C_{Tot.B}$) of the species containing the deprotonated conjugate base. The intersection point at which the abundances of the conjugate acid and conjugate base are equal would be located at the pH value equal to the pK_A. The reader is left to verify this by completion of selected end-of-chapter problems.

6.5.2 Diprotic Acids

Diprotic acids have two protons that can be donated. Here, we may still apply the accounting equation for the total concentration of the species containing the fully deprotonated conjugate base, but now we have three species with which to deal and two equilibria with which we may work. Let us address selenous acid and investigate the similarities and differences between a diprotic acid and a monoprotic acid.

Example 6.2 Explore the distribution of selenite ($SeO_3^=$) among the three species (H_2SeO_3, $HSeO_3^-$, $SeO_3^=$) comprising the selenous acid system, again as a function of the pH of the aqueous solution. Consider an aqueous solution containing 25.38 mg/L of total selenite.

Let us first determine the total molar concentration of acetate species, collect the values of the two equilibrium constants, write the mole balance relation, and state the applicable equilibrium relations. Here the net charge on the fully protonated specie (H_2SeO_3) is zero so the deprotonated species are $HSeO_3^-$ and $SeO_3^=$:

$$MW_{SeO3} := (78.9 + 3 \cdot 16) \cdot 1000 = 1.269 \times 10^5 \ \frac{mg}{mol}$$

$$C_{Tot.SeO3} := \frac{25.38}{MW_{SeO3}} = 2 \times 10^{-4} \ M \qquad pK_{1.SeO3} := 2.62 \qquad pK_{2.SeO3} := 8.30$$

$$K_{1.SeO3} := 10^{-pK_{1.SeO3}} \qquad K_{2.SeO3} := 10^{-pK_{2.SeO3}}$$

$$C_{Tot.SeO3} = H2SeO3 + HSeO3 + SeO3$$

$$K_{1.SeO3} = \frac{H \cdot HSeO3}{H2SeO3} \qquad K_{2.SeO3} = \frac{H \cdot SeO3}{HSeO3}$$

Now, as in Example 6.1, we will perform some algebraic manipulations that will enable us to write three statements of the mole balance—one employing [H_2SeO_3], one employing [$HSeO_3^-$], and one employing [$SeO_3^=$] as the master dependent variable, along with the hydrogen ion concentration. We use some shorthand notation in MathCAD to ease the burden of variable name entry. Use of the left and right brackets in a MathCAD worksheet invokes a specific action, so we will use the chemical formulas to denote abundances in units of mol/L. Further, since we have but one opportunity to employ a formatting subscript in a variable name, we will not subscript the stoichiometric coefficients such as is done in the chemical literature.

First let us write [$HSeO_3^-$] and [$SeO_3^=$] fully in terms of [H_2SeO_3], the acid dissociation constants and [H^+]:

$$HSeO3 = \frac{K_{1.SeO3}}{H} \cdot H2SeO3$$

$$SeO3 = \frac{K_{2.SeO3}}{H} \cdot HSeO3 = \frac{K_{2.SeO3}}{H} \cdot \left(\frac{K_{1.SeO3}}{H} \cdot H2SeO3\right)$$

Then, let us write [$HSeO_3^-$] and [H_2SeO_3] in terms of [$SeO_3^=$], the acid dissociation constants and [H^+]:

$$HSeO3 = \frac{H}{K_{2.SeO3}} \cdot SeO3$$

$$H2SeO3 = \frac{H}{K_{1.SeO3}} \cdot HSeO3 = \frac{H}{K_{1.SeO3}} \cdot \left(\frac{H}{K_{2.SeO3}} \cdot SeO3\right)$$

Finally, let us write $[H_2SeO_3]$ and $[SeO_3^{=}]$ in terms of $[HSeO_3^-]$, the acid dissociation constants and $[H^+]$:

$$SeO3 = \frac{K_{2.SeO3}}{H} \cdot HSeO3 \qquad H2SeO3 = \frac{H}{K_{1.SeO3}} \cdot HSeO3$$

We can now write three statements of the mole balance each using one of the sets of algebraic substitutions from the aforementioned equations:

$$C_{Tot.SeO3} = H2SeO3 + \frac{K_{1.SeO3}}{H} \cdot H2SeO3 + \frac{K_{2.SeO3}}{H} \cdot \frac{K_{1.SeO3}}{H} \cdot H2SeO3$$

$$C_{Tot.SeO3} = \frac{H}{K_{1.SeO3}} \cdot HSeO3 + HSeO3 + \frac{K_{2.SeO3}}{H} \cdot HSeO3$$

$$C_{Tot.SeO3} = \frac{H}{K_{1.SeO3}} \cdot \frac{H}{K_{2.SeO3}} \cdot SeO3 + \frac{H}{K_{2.SeO3}} \cdot SeO3 + SeO3$$

These are then, in turn, easily rearranged to yield $[H_2SeO_3]$, $[HSeO_3^-]$, and $[SeO_3^{=}]$ as explicit functions of $[H^+]$, $K_{1.SeO_3}, K_{2.SeO_3}$, and $C_{Tot.SeO_3}$. We would like to use pH as the master-independent variable and then require a function relating $[H^+]$ to pH:

$$H(pH) := 10^{-pH} \qquad H2SeO3(pH) := \frac{C_{Tot.SeO3}}{1 + \frac{K_{1.SeO3}}{H(pH)} + \frac{K_{2.SeO3} \cdot K_{1.SeO3}}{H(pH)^2}}$$

$$HSeO3(pH) := \frac{C_{Tot.SeO3}}{\left(\frac{H(pH)}{K_{1.SeO3}} + 1 + \frac{K_{2.SeO3}}{H(pH)}\right)}$$

$$SeO3(pH) := \frac{C_{Tot.SeO3}}{\left(\frac{H(pH)^2}{K_{1.SeO3} \cdot K_{2.SeO3}} + \frac{H(pH)}{K_{2.SeO3}} + 1\right)}$$

We now may employ these mole balance relations using the pH as a master independent variable to write MathCAD functions from which we may plot the specie concentrations as a function of pH in Figure E6.2.1. Although beyond the range likely to occur in natural systems, we will use a pH range of 1–11 so we may fully illustrate the behavior of selenous acid. Details and *x*–*y* axis arguments are again shown for illustrative purposes.

We easily observe that, as was the case for the monoprotic acid of Example 6.1, the abundances of each conjugate acid–conjugate base pair are equal at $pH = pK_A$. We may rewrite the two equilibrium expressions to relate the abundance of each conjugate base to its conjugate acid:

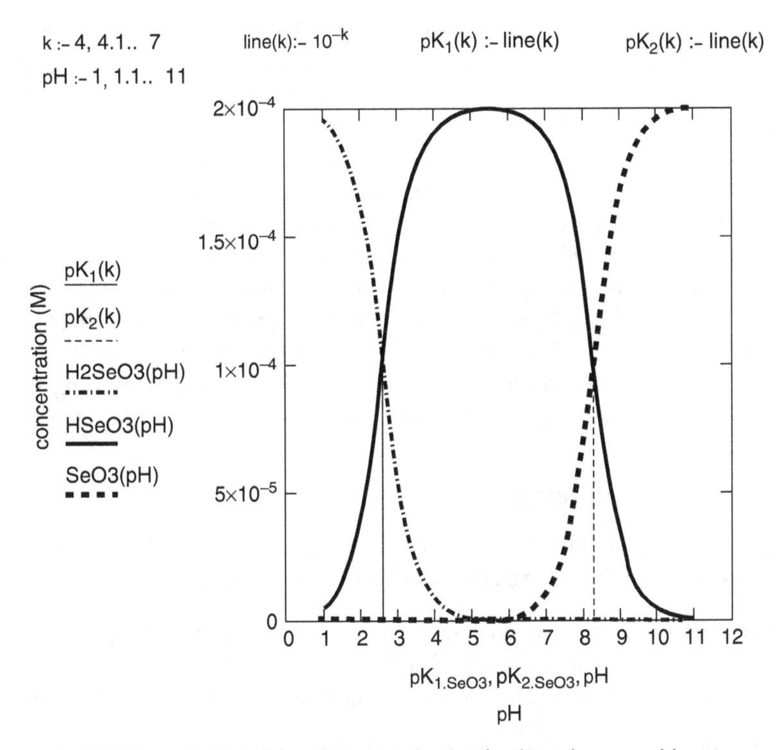

FIGURE E6.2.1 *Plot of specie predominance for the selenous acid system.*

$$\frac{K_{1.\mathrm{SeO}_3}}{[\mathrm{H}^+]} = \frac{[\mathrm{HSeO}_3^-]}{[\mathrm{H}_2\mathrm{SeO}_3]}; \quad \frac{K_{2.\mathrm{SeO}_3}}{[\mathrm{H}^+]} = \frac{[\mathrm{SeO}_3^=]}{[\mathrm{HSeO}_3^-]}$$

We observe, as expected from the $K_{1.\mathrm{SeO}_3}$ relation, that the fully protonated specie is predominant at pH below $\mathrm{p}K_{\mathrm{A}.1}$ due to the abundance of hydrogen ions in excess of $K_{\mathrm{A}.1}$ (i.e., $\mathrm{pH} < \mathrm{p}K_{1.\mathrm{SeO}_3}$), and thus $\frac{[\mathrm{HSeO}_3^-]}{[\mathrm{H}_2\mathrm{SeO}_3]} < 1$. For $\mathrm{p}K_{1.\mathrm{SeO}_3} < \mathrm{pH} < \mathrm{p}K_{2.\mathrm{SeO}_3}$, $\frac{[\mathrm{HSeO}_3^-]}{[\mathrm{H}_2\mathrm{SeO}_3]} > 1$ and $\frac{[\mathrm{SeO}_3^=]}{[\mathrm{HSeO}_3^-]} < 1$, and thus the mono-protonated specie is predominant. For $\mathrm{pH} > \mathrm{p}K_{2.\mathrm{SeO}_3}$, $\frac{[\mathrm{SeO}_3^=]}{[\mathrm{HSeO}_3^-]} > 1$, and the fully deprotonated specie is predominant.

Were we to conduct an identical analysis for any of the diprotic acids whose equilibrium constants are listed in Table 6.1 (or any diprotic acid for that matter), we would arrive at the same general result. Differences might accrue only as a consequence that the total concentration of the acid–base system might be greater or lesser than that considered in Example 6.2, and the values of $K_{1.A}$ and $K_{2.A}$ might be different, rendering the conjugate acid–conjugate base abundances to be equal at pH values different from those found in Example 6.2. The equalities of the abundances of each of the conjugate acid–conjugate base pairs indeed would occur at pH values equal to $pK_{1.A}$ and $pK_{2.A}$. Were we to convert the ordinate scale of the selenous acid specie versus pH plot to a $\log_{10}$ scale, we would find that the abundances of the nonpredominant species become increasingly smaller, but do not reach zero. In any aqueous system that contains selenous acid (we also could refer to this system as the selenite system), all three species containing selenite (H_2SeO_3, $HSeO_3^-$, $SeO_3^=$) must be present in the aqueous solution. Understanding the general behavior of acids and bases will in time allow the quick screening to identify the species of significant abundance while knowing only the system pH and the values of applicable acid dissociation constants (pK_A values).

6.5.3 Triprotic and Tetraprotic Acids

For completeness, we should address triprotic and tetraprotic acids in a manner similar to that with which we have addressed mono- and diprotic acids.

Example 6.3 Consider an aqueous solution that contains 3.1 mg/L phosphate-phosphorus (PO_4–P) and 28.8 mg/L total ethylene-diamine-tetraacetate (EDTA). Explore the behavior of these acid systems within the aqueous solution as a function of the pH.

We will approach this exercise in exactly the same manner as was employed in Examples 6.1 and 6.2. We will compute the molar concentration of total phosphate-phosphorus and EDTA, obtain values for all relevant acid dissociation constants, write the mole balance relations for phosphate and EDTA species, and employ the specific equilibrium relations for the phosphate (three equilibrium relations) and EDTA (four equilibrium relations) systems. However, given the large number of relations and the numerous algebraic manipulations necessary, only the final graphical results are presented. It is assumed that understandings of the processes employed in Examples 6.1 and 6.2 have been adequately studied by the student so that he or she may fill in the intermediate steps while examining the results of this example.

We first address the phosphate-phosphorus system, producing a plot of specie abundance versus pH as Figure E6.3.1.

We may observe behavior analogous to that for mono- and diprotic acids. Here we merely have three equivalence points and four regions in which the various conjugate acids and bases are predominant. The fully protonated phosphoric acid (H_3PO_4) has the highest abundance at $pH < pK_{1.PO_4}$. Dihydrogen phosphate ($H_2PO_4^-$, the conjugate base of phosphoric acid and the conjugate acid of hydrogen phosphate) has the greatest abundance when $pK_{1.PO_4} < pH < pK_{2.PO_4}$. Hydrogen phosphate ($HPO_4^=$, the conjugate base of $H_2PO_4^-$ and the conjugate acid of PO_4^{-3}) has the greatest abundance when $pK_{2.PO_4} < pH < pK_{3.PO_4}$. The phosphate ion ($PO_4^{-3}$) has the greatest abundance when $pH > pK_{3.PO_4}$. Certainly all triprotic acids would behave in a similar manner, with the locations (and widths relative to pH) of the regions of predominance and the equivalence points conforming with values of the dissociation constants.

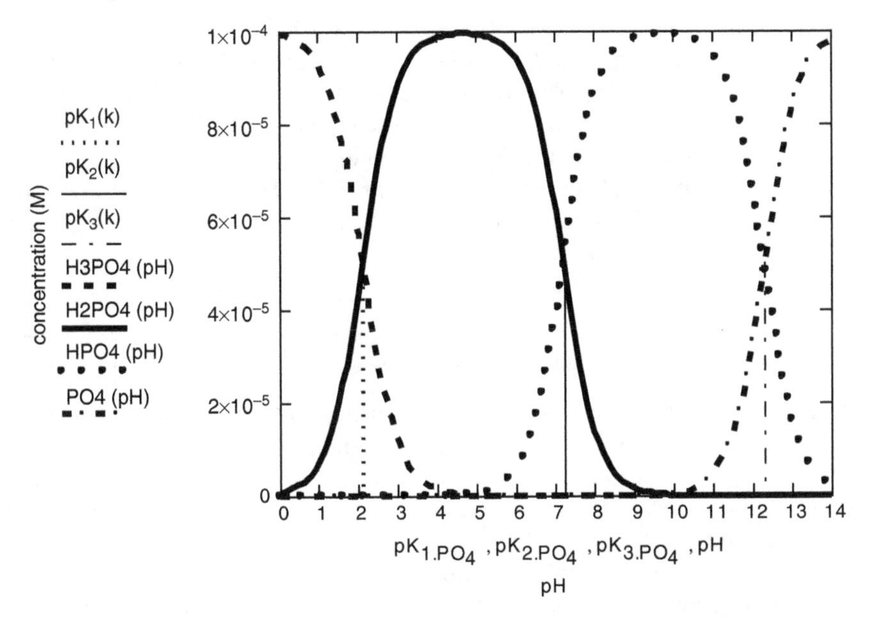

FIGURE E6.3.1 *Plot of specie predominance for the phosphoric acid system.*

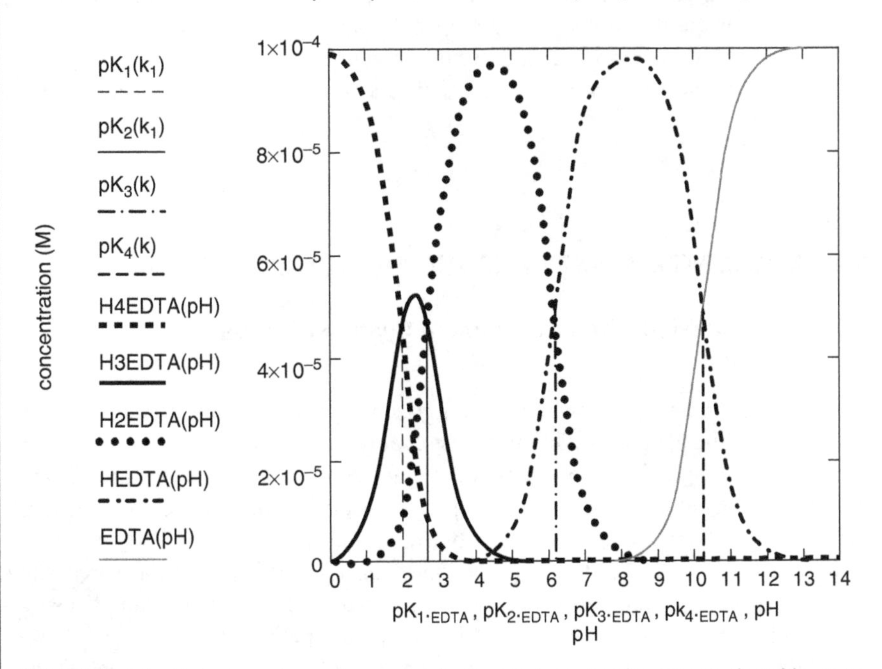

FIGURE E6.3.2 *Plot of specie predominance for the ethylene-diamine-tetraacetic acid system.*

For ethylene-diamine-tetraacetic acid in Figure E6.3.2 we have created a plot of specie abundance versus pH. We find great similarity with that for the acid systems investigated in Examples 6.1 and 6.2 as well as with the phosphate system investigated earlier.

We may easily observe from Example 6.3 that the distributions of species as functions of pH for tri- and tetraprotic acids are merely extensions of those for mono- and diprotic acids. Each conjugate acid–conjugate base pair has an equivalence at $pH = pK_A$ of the deprotonation equilibrium. The organic acid EDTA is not so "well behaved" as the three inorganic acids. The magnitudes of $K_{1.EDTA}$ and $K_{2.EDTA}$ are sufficiently close that the pH range of predominance of $HEDTA^{-3}$ is much narrower than those of the other EDTA species. We may observe this behavior upon investigation of many multiprotic organic acids.

6.5.4 Abundance (Ionization) Fractions

The algebraic manipulations we have accomplished with Examples 6.1–6.3 lead to the relations defining what others have termed ionization fractions or ratios of the abundances of the individual species to the total abundance of the acid system species. These ratios, most appropriately termed abundance fractions, are assembled for mono- through tetraprotic acids in Table 6.2.

Once programmed as general relations into a MathCAD or Excel worksheet, these become powerful tools for quick quantitative determinations of the abundance of any specie of an acid system when the total system abundance and pH are known, or quickly determining the total abundance of an acid system if one specie and pH are known. Further, with knowledge of the abundance of any two species of an acid system the pH can be found and the system can then be fully characterized.

6.6 ALKALINITY, ACIDITY, AND THE CARBONATE SYSTEM

6.6.1 The Alkalinity Test: Carbonate System Abundance and Speciation

Bases present in an aqueous solution comprise the capacity of that solution to resist depression of the pH upon addition of proton donating substances (acids). Acids present in an aqueous solution comprise the capacity of that solution to resist elevation of the pH upon addition of proton accepting substances (bases). Here we must discern between alkalinity, acidity, and buffering capacity.

Alkalinity (also called acid neutralizing capacity, [ANC]) is a measured value based on a standard laboratory test involving titration of an aqueous sample using a strong acid from its initial pH to a standard end point pH value.

Acidity (also called base neutralizing capacity, [BNC]) is a measured value based on a standard laboratory test involving titration of an aqueous solution using a strong base from its initial pH to a standard end point pH value.

TABLE 6.2 Abundance (Ionization) Fractions for Mono- Through Tetraprotic Acids

Monoprotic acids[a,b]

$$\alpha_0 = \frac{[HB^n]}{C_{T,B}} = \left(1 + \frac{K_A}{[H^+]}\right)^{-1}; \quad \alpha_1 = \frac{[B^{n-1}]}{C_{T,B}} = \left(\frac{[H^+]}{K_A} + 1\right)^{-1}$$

Diprotic acids[a,b]

$$\alpha_0 = \frac{[H_2B^n]}{C_{T,B}} = \left(1 + \frac{K_{A1}}{[H^+]} + \frac{K_{A1}K_{A2}}{[H^+]^2}\right)^{-1}$$

$$\alpha_1 = \frac{[HB^{n-1}]}{C_{T,B}} = \left(\frac{[H^+]}{K_{A1}} + 1 + \frac{K_{A2}}{[H^+]}\right)^{-1}; \quad \alpha_2 = \frac{[B^{n-2}]}{C_{T,B}} = \left(\frac{[H^+]^2}{K_{A1}K_{A2}} + \frac{[H^+]}{K_{A2}} + 1\right)^{-1}$$

Triprotic acids[a,b]

$$\alpha_0 = \frac{[H_3B^n]}{C_{T,B}} = \left(1 + \frac{K_{A1}}{[H^+]} + \frac{K_{A1}K_{A2}}{[H^+]^2} + \frac{K_{A1}K_{A2}K_{A3}}{[H^+]^3}\right)^{-1}$$

$$\alpha_1 = \frac{[H_2B^{n-1}]}{C_{T,B}} = \left(\frac{[H^+]}{K_{A1}} + 1 + \frac{K_{A2}}{[H^+]} + \frac{K_{A2}K_{A3}}{[H^+]^2}\right)^{-1}$$

$$\alpha_2 = \frac{[HB^{n-2}]}{C_{T,B}} = \left(\frac{[H^+]^2}{K_{A1}K_{A2}} + \frac{[H^+]}{K_{A2}} + 1 + \frac{K_{A3}}{[H^+]}\right)^{-1}$$

$$\alpha_3 = \frac{[B^{n-3}]}{C_{T,B}} = \left(\frac{[H^+]^3}{K_{A1}K_{A2}K_{A3}} + \frac{[H^+]^2}{K_{A2}K_{A3}} + \frac{[H^+]}{K_{A3}} + 1\right)^{-1}$$

Tetraprotic acids[a,b]

$$\alpha_0 = \frac{[H_4B^n]}{C_{T,B}} = \left(1 + \frac{K_{A1}}{[H^+]} + \frac{K_{A1}K_{A2}}{[H^+]^2} + \frac{K_{A1}K_{A2}K_{A3}}{[H^+]^3} + \frac{K_{A1}K_{A2}K_{A3}K_{A4}}{[H^+]^4}\right)^{-1}$$

$$\alpha_1 = \frac{[H_3B^{n-1}]}{C_{T,B}} = \left(\frac{[H^+]}{K_{A1}} + 1 + \frac{K_{A2}}{[H^+]} + \frac{K_{A2}K_{A3}}{[H^+]^2} + \frac{K_{A2}K_{A3}K_{A4}}{[H^+]^3}\right)^{-1}$$

$$\alpha_2 = \frac{[H_2B^{n-2}]}{C_{T,B}} = \left(\frac{[H^+]^2}{K_{A1}K_{A2}} + \frac{[H^+]}{K_{A2}} + 1 + \frac{K_{A3}}{[H^+]} + \frac{K_{A3}K_{A4}}{[H^+]^2}\right)^{-1}$$

$$\alpha_3 = \frac{[HB^{n-3}]}{C_{T,B}} = \left(\frac{[H^+]^3}{K_{A1}K_{A2}K_{A3}} + \frac{[H^+]^2}{K_{A2}K_{A3}} + \frac{[H^+]}{K_{A3}} + 1 + \frac{K_{A4}}{[H^+]}\right)^{-1}$$

$$\alpha_4 = \frac{[B^{n-4}]}{C_{T,B}} = \left(\frac{[H^+]^4}{K_{A1}K_{A2}K_{A3}K_{A4}} + \frac{[H^+]^3}{K_{A2}K_{A3}K_{A4}} + \frac{[H^+]^2}{K_{A3}K_{A4}} + \frac{[H^+]}{K_{A4}} + 1\right)^{-1}$$

[a] α_0 refers to the fully protonated acid; $\alpha_1 - \alpha_4$ refer to species resulting from donation of 1–4 protons, respectively, from the fully protonated acid.

[b] n is the residual charge on the fully protonated acid, and is generally equal to the number of $R\text{–}NH_3^+$ groups present within the fully protonated acid.

Buffering capacity is a more general or qualitative term that describes the capacity of a solution in general to resist changes in pH from additions of acids or bases. Some authors use the term buffer intensity as the specific buffering capacity of an aqueous solution at a specific value of pH.

Let us first focus upon alkalinity [ANC] as a standard laboratory test. This test was devised in most part to characterize the carbonate system species present in natural waters. The acid buffering capacity of the vast majority of natural waters is due solely to the presence of carbonate system species in the solution.

Briefly, the alkalinity laboratory test involves the measurement of the initial pH of a water sample, and titration of a known volume of that water sample to an end point pH in the range of ~4.3 using a strong acid. Historically, methyl orange has been used as an indicator in the total alkalinity titration, turning from orange to colorless at pH ~4.3. Phenolphthalein has been used in titrations of samples with initial pH above 8.3, as solutions containing phenolphthalein turn from pink to colorless at pH ~8.3. The quantity and normality of the acid added to the solution to the pH 8.3 (phenolphthalein) end point and pH 4.3 (methyl orange) end point are recorded. Most often the results are converted from $eq_{acid}/L_{solution}$ (or meq/L) into mg/L *as* $CaCO_3$ and reported. With the development of digital titrators and pH meters with fast responses, many professional laboratories now perform potentiometric alkalinity titrations, using the colorimetric indicators only for guidance that pH endpoints are near. For more details concerning the performance of the alkalinity titration, consult Standard Methods or any one of a number of water chemistry texts. As an approximation, the acid neutralized between an initial pH above 8.3 and the end point at pH 8.3 is considered to be due to the conversion of carbonate into bicarbonate and is often used as an approximation of the carbonate content of the sample. The total quantity of acid neutralized between the initial pH and the pH 4.3 end point is due to the combination of carbonate and bicarbonate in the water.

Let us examine the titration of a water sample whose ANC is due primarily to the constituents of the carbonate system.

Example 6.4 Consider an aqueous solution with total inorganic carbon (also called total carbon dioxide or total carbonate) of 0.006 M and of initial pH of 7.65. This water sample is of the approximate composition as would be drawn by the City of Rapid City, SD, from the Madison Aquifer and distributed to its customers. We will examine the chemical processes involved as the solution is titrated using a strong acid to the pH 4.3 end point.

Let us construct the abundance versus pH diagram for the carbonate system, and include plots of $[H^+]$ and $[OH^-]$. Then on this figure let us identify the initial and final concentrations of the relevant species and draw change vectors.

We must first identify and define necessary constants (Table 6.1) and other parameters:

$$\begin{pmatrix} K_{1.CO3} \\ K_{2.CO3} \\ K_w \\ C_{Tot.CO3} \end{pmatrix} := \begin{pmatrix} 10^{-6.35} \\ 10^{-10.33} \\ 10^{-14} \\ 0.006 \end{pmatrix} \quad pH_i := 7.65 \quad pH_{init}(i) := 10^{-i} \quad H(pH) := 10^{-pH}$$

$$pH_f := 4.3 \quad pH_{final}(i) := 10^{-i} \quad OH(pH) := \frac{K_w}{H(pH)}$$

First, we will write some abundance fraction relations to compute $[H_2CO_3^*]$, $[HCO_3^-]$, and $[CO_3^=]$ as functions of pH as the master-independent variable. We will use the set from Table 6.2 for a diprotic acid and employ the acid dissociation constants for the carbonate system:

$$\alpha_{0.CO3}(pH) := \left(1 + \frac{K_{1.CO3}}{H(pH)} + \frac{K_{1.CO3} \cdot K_{2.CO3}}{H(pH)^2}\right)^{-1}$$

$$\alpha_{1.CO3}(pH) := \left(\frac{H(pH)}{K_{1.CO3}} + 1 + \frac{K_{2.CO3}}{H(pH)}\right)^{-1}$$

$$\alpha_{2.CO3}(pH) := \left(\frac{H(pH)^2}{K_{1.CO3} \cdot K_{2.CO3}} + \frac{H(pH)}{K_{2.CO3}} + 1\right)^{-1}$$

We plot the abundance fractions versus pH in Figure E6.4.1.

As expected, a plot of the abundance fractions versus pH follows the typical pattern shown in Example 6.2 for a diprotic acid.

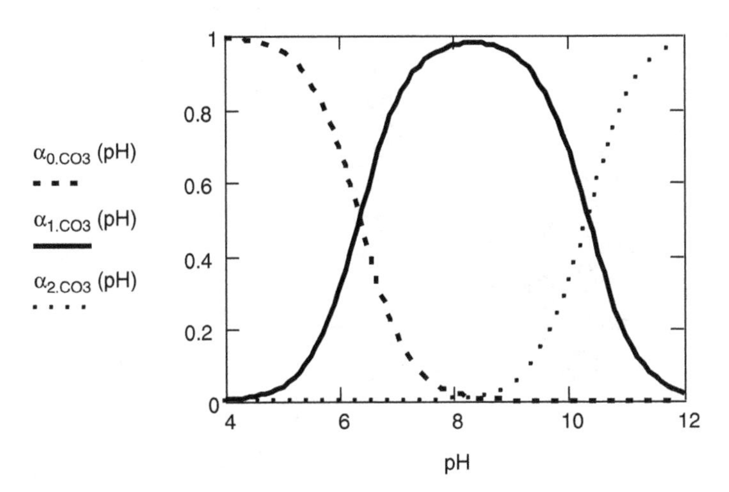

FIGURE E6.4.1 *Plot of abundance fractions for the carbonate system.*

We may then use these abundance functions with the stated value of total carbonate to write functions for the three carbonate species:

$$H2CO3(pH) := \alpha_{0.CO3}(pH) \cdot C_{Tot.CO3} \qquad HCO3(pH) := \alpha_{1.CO3}(pH) \cdot C_{Tot.CO3}$$

$$CO3(pH) := \alpha_{2.CO3}(pH) \cdot C_{Tot.CO3}$$

In Figure E6.4.2, we plot these functions as well as those for $[H^+]$ and $[OH^-]$, identify the initial and final pH values for the alkalinity titration, visually determine the changes in the abundances of the relevant species, and add the change vectors onto the plot.

We employ a logarithmic ordinate (abundance, or y axis) scale to match the logarithmic scale of the abscissa (pH, or x axis). We note that $[HCO_3^-]$ is reduced to about $0.01\times[H_2CO_3^*]$, that $[CO_3^=]$ is reduced about six orders of magnitude to $10^{-8}\times[H_2CO_3^*]$, that $[H^+]$ is increased about three orders of magnitude, to the final value of the alkalinity titration, $\sim 10^{-4.3}$ M, and that $[OH^-]$ is reduced about three orders of magnitude to the value of $\sim 10^{-9.7}$ M. Initially present HCO_3^- accepts one H^+ in its conversion to $H_2CO_3^*$. Initially present $CO_3^=$ first accepts one H^+ to become HCO_3^- and then one more H^+ to become $H_2CO_3^*$. A small quantity of H_2O accepts one H^+ to become H_3O^+ (for which, in computation, we use the shorthand notation H^+) and initially present OH^- accepts one H^+ to become H_2O. The specific reactions that occur are

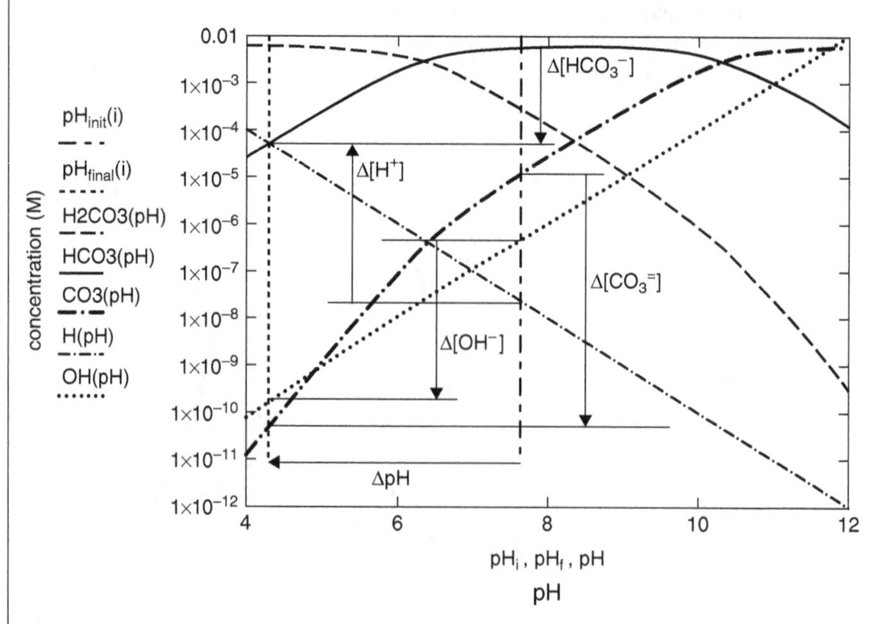

FIGURE E6.4.2 *Plot of specie predominance for the carbonic acid system and changes in speciation associated with the standard alkalinity titration for* $C_{Tot.CO_3} = 0.006$ M.

$$HCO_3^- + H^+ \rightarrow H_2CO_3^*$$

$$CO_3^= + 2H^+ \rightarrow H_2CO_3^*$$

$$OH^- + H^+ \rightarrow H_2O$$

$$H_2O + H^+ \rightarrow H_3O^+$$

We can examine the process in terms of the changes in the concentrations of the relevant species and write an equation accounting for the acceptance of the protons added. Since water is both the product of the acceptance of protons by OH^- and a reactant in the acceptance of a proton to form hydronium, we immediately see a difficulty in attempting to catalog the change in the abundance of water. However, we can easily measure the initial and final abundances of hydronium and, from the ion product of water, determine the initial and final abundances of hydroxide. We therefore opt to include the change in $[OH^-]$ along with the change in $[H_3O^+]$ in the accounting. We recall that change is the final less the initial and recognize that since hydronium is a product, we must use the negative of the change in proton abundance to represent the change in the abundance of water associated with acceptance of protons:

$$\Delta[ANC] = \Delta[Alk] = \Delta[HCO_3^-] + 2\Delta[CO_3^=] + \Delta[OH^-] - \Delta[H_3O^+]$$

We can employ the convention that $\Delta[i] = [i]_f - [i]_i$ and rewrite the proton accounting equation:

$$\begin{aligned}([ANC_f - ANC_i] = ([Alk]_f - [Alk]_i) = ([HCO_3^-]_f - [HCO_3^-]_i) \\ + 2([CO_3^=]_f - [CO_3^=]_i) + ([OH^-]_f - [OH^-]_i) \\ - ([H_3O^+]_f - [H_3O^+]_i)\end{aligned}$$

If we consider that at the end point pH, the [ANC] of the solution is fully exhausted, the acid neutralizing capacity $[ANC_i]$ of the initial solution is equal in magnitude and opposite in sign to the change in the capacity of the solution to accept protons (Δ[Alk], M) relative to the end point of the alkalinity titration.

We may use the MathCAD functions to compute the beginning and ending concentrations of the relevant species:

$$\begin{pmatrix} HCO3(pH_f) \\ CO3(pH_f) \\ OH(pH_f) \\ H(pH_f) \end{pmatrix} = \begin{pmatrix} 5.3 \times 10^{-5} \\ 4.946 \times 10^{-11} \\ 1.995 \times 10^{-10} \\ 5.012 \times 10^{-5} \end{pmatrix} \quad \begin{pmatrix} HCO3(pH_i) \\ CO3(pH_i) \\ OH(pH_i) \\ H(pH_i) \end{pmatrix} = \begin{pmatrix} 5.702 \times 10^{-3} \\ 1.191 \times 10^{-5} \\ 4.467 \times 10^{-7} \\ 2.239 \times 10^{-8} \end{pmatrix} M$$

We examine these results and readily observe that $[CO_3^=]_f$<<<< $[HCO_3^-]_f$ and that $[HCO_3^-]_f$<<$[HCO_3^-]_i$. Further, $[OH^-]_f$<<<<$[HCO_3^-]_i$ and $[H_3O^+]_i$<<<< $[HCO_3^-]_i$. We observe that these magnitudes are all quite insignificant certainly relative to that of $[HCO_3^-]_i$. If we seek a convenient value to employ for $[HCO_3^-]_f$, $[CO_3^=]_f$, $[OH^-]_f$, and $[H_3O^+]_i$ that value is zero and these terms then disappear from the relation. We also reason that at the endpoint pH of the titration, the entire [ANC] of the solution has been expended (relative to the endpoint pH of 4.3) and thus $[Alk]_f$=0:

$$-[ANC_i] = -[Alk]_i = -[HCO_3^-]_i - 2[CO_3^=]_i - [OH^-]_i - [H_3O^+]_f$$

Further observation yields the result that for this particular computation that

$$[OH^-]_i <<< [HCO_3^-]_i \text{ and } [H_3O^+]_f << [HCO_3^-]_i$$

Assigning values of zero to these additional terms of insignificant magnitude, multiplying the LHS and RHS of the relation by −1, and dispensing with the subscripts denoting initial values yields the form of the alkalinity equation with which we are most familiar. We note that this is at best an approximation:

$$[ANC] = [Alk] \approx [HCO_3^-] + 2\,[CO_3^=]$$

We may perform one more check of the analysis simply by programming our MathCAD worksheet to compute the changes in concentration:

$$\begin{pmatrix} \Delta C_{HCO3} \\ \Delta C_{CO3} \\ \Delta C_{OH} \\ \Delta C_H \end{pmatrix} := \begin{pmatrix} HCO3(pH_f) - HCO3(pH_i) \\ CO3(pH_f) - CO3(pH_i) \\ OH(pH_f) - OH(pH_i) \\ H(pH_f) - H(pH_i) \end{pmatrix} = \begin{pmatrix} -5.649 \times 10^{-3} \\ -1.191 \times 10^{-5} \\ -4.465 \times 10^{-7} \\ 5.01 \times 10^{-5} \end{pmatrix} \quad M$$

And from these results we may compute a very close approximation of the [ANC]:

$$ANC_1 := -(\Delta C_{HCO3} + 2 \cdot \Delta C_{CO3} + \Delta C_{OH} - \Delta C_H) = 5.724 \times 10^{-3} \quad M$$

If we employ the approximation ($[Alk] \approx [HCO_3^-]+2[CO_3^=]$), we can compute an additional approximation of the [ANC]. This is the method presented in Standard Methods and used quite universally:

$$ANC_2 := HCO3(pH_i) + 2 \cdot CO3(pH_i) = 5.726 \times 10^{-3} \quad M$$

We see that the error of the second approximation relative to the first is small. This will be the case when for waters of fairly high alkalinity for which [ANC] is due mainly to the carbonate system. As the inorganic carbon content of the water and hence the [ANC] due to the conversion of carbonate species is decreased, the importance of the changes in $[OH^-]$ and $[H_3O^+]$ becomes greater and greater. For waters of alkalinity measured in single digits of mg/L as $CaCO_3$ (numerous in the Adirondack Mountains of the Eastern US and in the Canadian province of Ontario), the significance of the acceptance of H^+ ions by hydroxide and water is quite high in the overall characterization of the inorganic carbon system.

We may reason with great confidence that the water samples, if obtained from surface waters, for which we measure the alkalinity are initially under equilibrium or near-equilibrium conditions. Samples that are obtained from confined systems (e.g., deep ground waters, sediment pore waters, anaerobic digesters of wastewater plants, and industrial processes) likely would not be equilibrated with the atmosphere at normal temperature and pressure. Such samples need to be preserved in a state as near as possible to that of the specific system when performing alkalinity titrations.

Then, for carbonate-dominated waters, given that the equilibrium condition prevails at the onset of the titration, we may begin with the initial quite accurate relation shown as Equation 6.6:

$$\Delta[\text{ANC}] = \Delta[\text{Alk}] = \Delta[\text{HCO}_3^-] + 2\Delta[\text{CO}_3^=] + \Delta[\text{OH}^-] - \Delta[\text{H}_3\text{O}^+] \qquad (6.6)$$

We may make use of the simplifying assumptions of Example 6.4 and utilize the acid dissociation reaction between bicarbonate and carbonate to transform Equation 6.6 into a very useful relation. The second dissociation equilibrium is solved for either carbonate in terms of bicarbonate or bicarbonate in terms of carbonate and either result is substituted into Equation 6.6 to obtain two useful forms of the alkalinity relation:

$$[\text{CO}_3^-] = \frac{K_{2.\text{CO}_3}}{[\text{H}^+]}[\text{HCO}_3^-]; \quad [\text{HCO}_3^-] = \frac{[\text{H}^+]}{K_{2.\text{CO}_3}}[\text{CO}_3^-]$$

$$[\text{Alk}] = [\text{ANC}] = [\text{HCO}_3^-]\left(1 + 2\frac{K_{2.\text{CO}_3}}{[\text{H}^+]}\right)$$

$$[\text{Alk}] = [\text{ANC}] = \left[\text{CO}_3^=\right]\left(\frac{[\text{H}^+]}{K_{2.\text{CO}_3}} + 2\right)$$

$$[\text{HCO}_3^-] = [\text{Alk}]\left(1 + 2\frac{K_{2.\text{CO}_3}}{[\text{H}^+]}\right)^{-1} \qquad (6.7\text{a})$$

$$[\text{CO}_3^=] = [\text{Alk}]\left(\frac{[\text{H}^+]}{K_{2.\text{CO}_3}} + 2\right)^{-1} \qquad (6.7\text{b})$$

These relations, of course, prove very useful in characterizations of the carbonate system speciation of natural waters containing significant abundances of carbonate species.

Generalizations regarding alkalinity have been published in the literature and widely employed. Two constraints govern the employment of these generalizations:

1. The measured alkalinity of the water is high, perhaps ≥50 mg/L *as* $CaCO_3$.
2. Beyond that contributed by water, the ANC of the water is due primarily to species of the carbonate system.

Given these two conditions are true, we may suggest for pH below about nine that $[HCO_3^-] \approx [Alk]$ and for $7.90 < pH < 8.70$ that $C_{Tot.CO_3} \approx [HCO_3^-] \approx [Alk]$. These approximations are good for "back of the napkin" considerations only and should be used with great caution.

6.6.2 Acidity

The acidity of a solution arises from the capacity of chemical species in the solution to donate hydrogen ions in neutralization of added strong base, usually measured in terms of hydroxide. In natural waters buffered by the carbonate system, these constituents would be carbonic acid, bicarbonate, hydronium, and water. In fact, if we know the initial pH of the aqueous solution and can measure the quantity of strong base added to a water sample, we can use an acidity relation, much like that developed in Example 6.4, to characterize the carbonate system species. Here our choice of end point for the acidity titration greatly affects our resultant relation.

If we seek to know the mineral acidity, due to conjugate acids of pK_A lower than $pK_{1.CO_3}$, we might choose an end point at pH 4.3. Little carbonic acid will donate protons to neutralize bases at pH below 4.3. Strong acids such as sulfuric, hydrochloric, and nitric will donate virtually all available protons to aqueous solutions at pH values of 4.3 and lower. However, acids such as acetic and propanoic will still have significant abundances of fully protonated conjugate acid at pH 4.3.

If we seek to know the carbon dioxide acidity, we might choose an end point of pH 8.3. If we observe the concentration versus pH diagram for the carbonate system in Example 6.4, we see that at pH 8.3 virtually all carbonate abundance is in the form of bicarbonate. If we employ an analysis similar to that for selenous acid of Example 6.2, we obtain the similar relations for the carbonate system:

$$[H_2CO_3^*]/[HCO_3^-] = [H^+]/K_{1.CO_3}$$

$$[CO_3^=]/[HCO_3^-] = K_{2.CO_3}/[H^+]$$

Then, at pH = 8.3 ($[H^+] = 10^{-8.3}$) both sets of these ratios are ≈0.01. Titration from an initial pH below 8.3 to pH 8.3 then converts the vast majority of initial carbonic acid to bicarbonate without creating a significant abundance of carbonate. Carbon dioxide acidity is then approximated as the base required to titrate the sample from its initial pH to the end point pH of 8.3.

Finally, if we are interested in all [BNC], we might choose a final end point pH of ~12.3. Titration to this pH value ensures that virtually all carbonic acid and bicarbonate are converted to the carbonate form. The initial pH of the solution would influence our choice of intermediate end points. Further examination of the nuances of base neutralization capacity is beyond the scope of this chapter. In Chapter 10, we examine [ANC] and [BNC] in much greater detail.

6.7 APPLICATIONS OF ACID/BASE PRINCIPLES IN SELECTED ENVIRONMENTAL CONTEXTS

6.7.1 Monoprotic Acids

One context in which acid–base principles are important is associated with anaerobic process. In such processes (digestion of wastewater biosolids, and degradation of organics in sediments beneath the sediment/water interface, to name two that are well known), the parent organic materials are microbially broken down in to simpler and simpler structures. A major group of microbial intermediates includes short-chain carboxylic acids such as acetic, propionic (also called propanoic), iso-butyric, and butyric acids. Each of these comprises a unique monoprotic acid system. Assays of aqueous samples to determine these systems most often involve adjustment of pH either upward to convert all species to the conjugate base or downward to convert all species to the conjugate acid. Thus, concentrations of carboxylic acids in aqueous systems are generally expressed in terms of the total abundance of the conjugate base, as present in both possible species. Let us examine the clarified liquid (called supernatant) from an anaerobic digester that might be employed for treatment of swine waste from a confined animal feeding operation (CAFO).

Example 6.5 Consider that the laboratory results from assay of an aqueous sample arising from the digester operation at a CAFO had the following results:

$$\text{Total acetate} = 1000\,\text{mg/L}; \text{Total propionate} = 450\,\text{mg/L}; \text{pH} = 5.40$$

Consider that the sampling was done to investigate a process upset as the digester was not operating efficiently. Determine the distribution of acetate and propionate species between the conjugate acids and bases for each system.

Let us do some data gathering first. From Table 6.1 we obtain the values of the pK_As for the two acid systems: 4.70 and 4.87 for acetic and propanoic acid, respectively.

From these values, we might reason that the abundances of the conjugate bases of each system would predominate over the abundances of the conjugate acids. We can compute the associated relevant values:

$$pK_{A.Ac} := 4.70 \qquad K_{A.Ac} := 10^{-pK_{A.Ac}} = 1.995 \times 10^{-5}$$

$$pK_{A.Pr} := 4.87 \qquad K_{A.Pr} := 10^{-pK_{A.Pr}} = 1.349 \times 10^{-5}$$

$$pH := 5.40 \qquad H := 10^{-pH} = 3.981 \times 10^{-6} \qquad \frac{mol}{L}$$

We know that the total acetate and propionate abundances must be the sums of the abundances of acetic acid and the acetate ion and of propanoic acid and the propanoate ion, respectively. We put this into the form of a mole balance. We also know that the concentrations of each conjugate acid/base pair are related through the equilibrium statement (the law of mass action) and we write these. We will use abbreviations for the conjugate bases (Ac^-=CH_3COO^-, Pr^-=$CH_3CH_2COO^-$) and simply add the proton to denote the conjugate acid. Since MathCAD is touchy about superscripts and brackets, we will use Ac to denote [Ac^-], HAc for [HAc], and so forth. We can solve the respective equilibria for the conjugate acids in terms of the conjugate bases and rewrite the mole balances:

$$C_{Tot.Ac} = HAc + Ac \qquad HAc = \frac{H \cdot Ac}{K_{A.Ac}} \qquad C_{Tot.Ac} = \frac{H \cdot Ac}{K_{A.Ac}} + Ac$$

$$C_{Tot.Pr} = HPr + Pr \qquad HPr = \frac{H \cdot Pr}{K_{A.Pr}} \qquad C_{Tot.Pr} = \frac{H \cdot Pr}{K_{A.Pr}} + Pr$$

We then have one equation with one unknown for each acid system. We can rearrange these to explicitly solve for the conjugate base in each case (we could certainly have worked with the conjugate acid) and then use the equilibrium relations to obtain the values for the concentrations of the conjugate acids. However, first we need to obtain numeric values for the total concentrations of each acid/base system using the formula weight for each of the conjugate bases:

$$FW_{Ac} := (12 + 3 + 12 + 2 \cdot 16) \cdot 1000 = 5.9 \times 10^{4} \qquad \frac{mg_{Ac}}{mol}$$

$$FW_{Pr} := (12 + 3 + 12 + 2 + 12 + 2 \cdot 16) \cdot 1000 = 7.3 \times 10^{4} \qquad \frac{mg_{Pr}}{mol}$$

$$C_{mass.Ac} := 1000 \frac{mg}{L} \qquad C_{Tot.Ac} := \frac{C_{mass.Ac}}{FW_{Ac}} = 0.0169 \qquad \frac{mol_{Tot.Ac}}{L}$$

$$C_{mass.Pr} := 450 \frac{mg}{L} \qquad C_{Tot.Pr} := \frac{C_{mass.Pr}}{FW_{Pr}} = 0.006164 \qquad \frac{mol_{Tot.Pr}}{L}$$

We may now solve for the concentrations of the conjugate bases:

$$Ac := C_{Tot.Ac} \cdot \left(\frac{H}{K_{A.Ac}} + 1 \right)^{-1} = 0.0141 \quad \frac{mol_{Ac}}{L}$$

$$Pr := C_{Tot.Pr} \cdot \left(\frac{H}{K_{A.Pr}} + 1 \right)^{-1} = 0.00476 \quad \frac{mol_{Pr}}{L}$$

Then we may solve for the conjugate acids:

$$HAc := \frac{H \cdot Ac}{K_{A.Ac}} = 0.002819 \quad \frac{mol_{HAc}}{L} \qquad HPr := \frac{H \cdot Pr}{K_{A.Pr}} = 0.001405 \quad \frac{mol_{HPr}}{L}$$

We might recognize at this point that the monoprotic acid relations from Table 6.2 could also have been very appropriately employed. For the acetic acid system $B = Ac^-$ and $n = 0$. For the propanoic acid system $B = Pr^-$ and $n = 0$. Then very simply:

$$\alpha_{1.Ac} := \left(\frac{H}{K_{A.Ac}} + 1 \right)^{-1} = 0.834 \qquad \alpha_{0.Ac} := \left(1 + \frac{K_{A.Ac}}{H} \right)^{-1} = 0.166$$

$$Ac := \alpha_{1.Ac} \cdot C_{Tot.Ac} = 0.0141 \qquad HAc := \alpha_{0.Ac} \cdot C_{Tot.Ac} = 0.002819 \quad \frac{mol}{L}$$

$$\alpha_{1.Pr} := \left(\frac{H}{K_{A.Pr}} + 1 \right)^{-1} = 0.772 \qquad \alpha_{0.Pr} := \left(1 + \frac{K_{A.Pr}}{H} \right)^{-1} = 0.228$$

$$Pr := \alpha_{1.Pr} \cdot C_{Tot.Pr} = 0.00476 \qquad HPr := \alpha_{0.Pr} \cdot C_{Tot.Pr} = 0.001405 \quad \frac{mol}{L}$$

We may observe that the relations used to compute the abundances of acetic acid and propanoic acid species could readily have been extended to become functions with capacity to output the concentrations of the respective acid/base species at the specified pH values or even over specified ranges of pH for plots as have been employed in Examples 6.1–6.4.

The ammonia-nitrogen system (ammonium, NH_4^+, and ammonia, NH_3) constitutes another acid/base system of great interest and importance in environmental systems. The discharges from many municipal waste water treatment plants (often called Publically Owned Treatment Works, POTWs) are regulated to control discharges of ammonia-nitrogen. Ammonia-nitrogen is an oxygen-demanding constituent and unionized ammonia (NH_3) is toxic to aquatic life if released into receiving waters.

Example 6.6 Consider that pH of 7.0 and water temperature of 85 °F would be the design worst-case receiving-water scenario for the discharge from a given POTW.

At what total concentration could ammonia-nitrogen be discharged under this extreme case? The design engineers have determined that a factor of safety of 2.0 should be applied to the discharge limit. Even when present at very low concentrations, unionized ammonia (NH_3^0) is quite toxic to aquatic life forms. A prolonged exposure to unionized ammonia at a concentration level of $\sim 0.002\,\text{mg}_{NH_3}/\text{L}$ at pH and water temperature of ~7.0 and ~85 °F, respectively, will result in significant death of sensitive fish species.

Let us gather the information necessary about the ammonia-nitrogen system, available from Tables 6.1 and 6.2. From Table 6.1, $\text{p}K_{A,NH_3} = 9.30$; From Table 6.2 where $B = NH_3$ and $n = 1$ we obtain the abundance fractions and modify them to be relations specific to the ammonia-nitrogen system:

$$\alpha_{0.NH_3} = \left(1 + \frac{K_{A.NH_3}}{[H^+]}\right)^{-1} ; \quad \alpha_{1.NH_3} = \left(\frac{[H^+]}{K_{A.NH_3}} + 1\right)^{-1}$$

Then we use the abundance fractions to obtain relations for the two ammonia-nitrogen species:

$$[NH_4^+] = \alpha_{0.NH_3} \cdot C_{Tot.NH_3} \text{ and } [NH_3] = \alpha_{1.NH_3} \cdot C_{Tot.NH_3}$$

Here, of course we are interested in the total ammonia-nitrogen and note that the relation for $\alpha_{1.NH_3}$ would indeed allow us to compute total ammonia-nitrogen as

$$C_{Tot.NH_3} = [NH_3] / \alpha_{1.NH_3}$$

We need then to determine the target value of $[NH_3]$, compute $\alpha_{1.NH_3}$, and then compute $C_{Tot.NH_3}$:

$$pH := 7 \quad H := 10^{-pH} = 1 \times 10^{-7}$$

$$pK_{A.NH3} := 9.3 \quad K_{A.NH3} := 10^{-pK_{A.NH3}} = 5.012 \times 10^{-10}$$

$$C_{mass.NH3} := 0.002 \frac{mg_{NH3}}{L} \quad FW_{NH3} := (14 + 3) \cdot 1000 = 1.7 \times 10^4 \frac{mg_{NH3}}{mol}$$

$$NH3 := \frac{C_{mass.NH3}}{FW_{NH3}} = 1.176 \times 10^{-7} \frac{mol_{NH3}}{L}$$

$$C_{Tot.NH3} := NH3 \cdot \left(\frac{H}{K_{A.NH3}} + 1\right) = 2.359 \times 10^{-5} \frac{mol_{NH3N}}{L}$$

Here $C_{Tot.NH_3}$ from the MathCAD worksheet refers to $[NH_3\text{–}N]$, or total ammonia-nitrogen expressed in mol/L. Most often, regulatory agencies, operations staff, and engineers prefer that ammonia-nitrogen be expressed in ppm_m (for dilute solutions, $ppm_m \approx mg/L$), thus, we will convert our answer to these units:

$$C_{mass.NH3N} := C_{Tot.NH3} \cdot FW_{NH3} = 0.401 \quad \frac{mg_{NH3N}}{L} \quad \text{or} \quad ppm_m$$

An alternate approach would involve using $[NH_3]$ to compute $[NH_4^+]$ and summing the two to obtain $C_{Tot.NH_3N}$:

$$NH3 = 1.176 \times 10^{-7} \quad \frac{mol_{NH3}}{L} \qquad NH4 := \frac{H}{K_{A.NH3}} \cdot NH3 = 2.347 \times 10^{-5} \quad \frac{mol_{NH4}}{L}$$

$$C_{Tot.NH3} := NH4 + NH3 = 2.359 \times 10^{-5} \quad \frac{mol_{NH3N}}{L}$$

Then, in light of the safety factor of 2.0, the system would be designed such that discharges of ammonia-nitrogen would not exceed half the computed $C_{Tot.NH_3}$, or ~1.2×10^{-5} mol/L (0.167 ppm_m).

We have now examined two specific types of computations with monoprotic acids: use of $C_{Tot.B}$ with pH to find $[HB^n]$ and $[B^{n-1}]$ and use of either $[HB^n]$ or $[B^{n-1}]$ with pH to find $C_{Tot.B}$. The examples examined can be generalized for use with any monoprotic acid. A third category of contexts requires combination of acid/base equilibria with Henry's law. In many systems, we have aqueous and vapor phases that are in contact for long time periods allowing both air/water and acid/base distributions to attain near equilibrium conditions. We can examine these systems from two viewpoints:

1. We know the abundance of the conjugate acid, the conjugate base or the total aqueous concentration of species containing the base, and wish to quantitate the abundance of the associated vapor phase specie, or
2. We know the abundance of the vapor phase specie and wish to quantitate the abundances of the aqueous species.

In order to illustrate the first application of the principles, let us examine a heap leach pad at a gold mine. In capture of gold from the ore, cyanide is used as a complexing agent to bind with gold in the heap leach pad and then once the gold–cyanide complex has been formed, the solution is removed from the leach pad and treated chemically to recover the gold, ideally regenerating the cyanide solution.

Example 6.7 Consider a cyanide-based gold recovery system employing a 0.05% solution prepared from sodium cyanide (NaCN) with pH in the range of 9.0–10.0. We might be interested in the dangers associated with inhalation of hydrocyanic acid (HCN) by workers in the vicinity of holding tanks containing this solution or in the vicinity of the leach pad where this solution is employed.

Although this would not truly be an equilibrium system, we would have interest in the equilibrium conditions as a worst-case scenario from which upper limits of risk to workers and the environment might be inferred. We will eventually need to use the ambient atmospheric pressure so let us consider this leach pad is at elevation 5000 ft above mean sea level.

A schematic representation of the air–water system and the associated cyanide equilibria is shown in Figure E6.7.1 to aid in visualization of the equilibrium system. From Table 6.1 we find that $pK_{A.CN} = 9.20$ and from Table 5.1 we find that $K_{H.HCN} = 12.9$ mol/L-atm. We of course need to be careful as electrolyte solutions of concentrations exceeding about 1% often have density values significantly greater than that of water. A search of our handbook yields the knowledge that the density of 0.05% NaCN solution is not significantly different from that of water. A 0.05% solution can be converted to the equivalent mass fraction concentration by considering that percent is parts per hundred (see Table 3.1).

We can compute the mass fraction ($C_{MF.NaCN}$) and mass concentrations ($C_{mass.NaCN}$) of sodium cyanide and hence the total cyanide concentration:

$$\rho_{soln} := 1000\ \frac{g}{L} \qquad C_{\%.NaCN} := 0.05 \qquad C_{MF.NaCN} := \frac{C_{\%.NaCN}}{100} = 5 \times 10^{-4}\ \frac{g_{NaCN}}{g_{soln}}$$

$$C_{mass.NaCN} := C_{MF.NaCN} \cdot \rho_{soln} = 0.5\ \frac{g_{NaCN}}{L}$$

We note that one mole of NaCN contains one mole of CN and thus, $[NaCN] = C_{Tot.CN}$:

$$FW_{NaCN} := 23 + 12 + 14 = 49\ \frac{g}{mol} \qquad C_{Tot.CN} := \frac{C_{mass.NaCN}}{FW_{NaCN}} = 0.0102\ \frac{mol_{CN}}{L}$$

We may now determine the concentrations of the aqueous cyanide species, but realize that to find the gas phase abundance of HCN we need only know

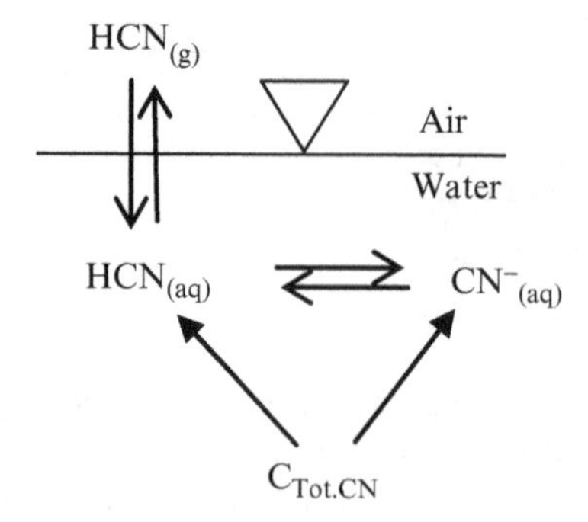

FIGURE E6.7.1 *Schematic representation of the air/water and acid/base distributions of cyanide species.*

[HCN]. The monoprotic acid α_0 relation from Table 6.2 will prove quite handy, where B=CN and n=0. Let us first investigate the upper pH value, 10.0, and carry the computation of the gas phase abundance to the unit of ppm_v (parts per million by volume), recalling that pressure, volume, and mole fractions of ideal gases are identical (Table 3.1) and that $ppm_{v.}$ = mole fraction × 10^6:

$$pH := 10 \quad H := 10^{-pH} = 1 \times 10^{-10} \quad K_{A.CN} := 10^{-pK_{A.CN}} = 6.31 \times 10^{-10}$$

$$\alpha_{0.CN} := \left(1 + \frac{K_{A.CN}}{H}\right)^{-1} = 0.137 \quad HCN := \alpha_{0.CN} \cdot C_{Tot.CN} = 0.001396 \quad \frac{mol_{HCN}}{L}$$

We can use the pressure versus elevation relation employed in Example 5.3 to determine the normal atmospheric pressure at the elevation of the site, P_{Tot}, to be 0.832 atm. We will then employ Henry's law to obtain P_{HCN} and in combination with P_{Tot}, we may find the pressure (and hence the mole or volume) fraction of $HCN_{(g)}$:

$$P_{HCN} := \frac{HCN}{K_{H.HCN}} = 1.082 \times 10^{-4} \text{ atm} \qquad P_{Tot} = 0.832 \text{ atm}$$

$$Y_{HCN} := \frac{P_{HCN}}{P_{Tot}} = 1.301 \times 10^{-4} \quad \frac{mol_{HCN}}{mol_{gas}} \qquad C_{v.HCN} := Y_{HCN} \cdot 10^6 = 130 \quad ppm_v$$

We implement the computational process from above at pH 9 and find the second desired result:

$$pH = 9 \quad HCN = 6.256 \times 10^{-3} \quad P_{HCN} = 4.85 \times 10^{-4} \quad C_{v.HCN} = 583 \quad ppm_v$$

From this analysis, we may reason that confined spaces within which the vapor phase is in contact with the specified cyanide solution could have vapor phase hydrogen cyanide concentrations in the range of hundreds of parts per million by volume and that exposure of workers to these confined spaces could be hazardous.

In order to examine the second type of analysis, we will consider a subsurface soil context. Suppose we are examining a suspected spill of anhydrous ammonia, which is normally stored in large quantities and used as fertilizer. We would like to obtain a quick and reasonably accurate estimate of the content of ammonia–nitrogen in soil water residing within the pores of the potentially contaminated soil. We bring in a "sniffer" type apparatus with which we can withdraw gas from the subsurface and pass it through a solution of sulfuric acid to measure the quantity of ammonia as proportional to the change in the acidity of the sulfuric acid solution. With the apparatus we can measure the quantity of the gas treated and from knowledge of the beginning and ending acidity of the sulfuric acid solution, we can determine with some certainty the ammonia content of the gas. We must assume that ammonia is the sole proton accepting constituent of the gas.

Example 6.8 Consider that the concentration of ammonia in vapor extracted from a subsurface soil suspected to be contaminated with anhydrous ammonia was measured to be 50 ppm_V. The ambient pressure at the time of the sampling was 0.90 atm. The pH of soil water is expected to be in the range of 8.5–9.5. Determine the ammonia–nitrogen (NH_3–N) content of the soil water. Then, consider that the moisture content of the soil is 10% by mass and determine the quantity of ammonia–nitrogen associated with a cubic meter of the subsurface soil. The moisture content is most often expressed as a percentage of the dry mass of the soil. The void fraction (total porosity) of the soil is believed to be 0.40 m_{void}^3/m^3.

We visualize the soil system and realize that the pores of the soil contain both gas and water. The water is distributed in the pores such that the vapor–liquid interfacial surface area would be minimal. As a wetting fluid water would tend to reside at or near grain-grain contact points in the soil where contact with the mineral surface area would be maximized. Water also would tend to reside with natural solid organic matter associated with the soil. Each tiny reservoir of water would be in intimate contact with the continuous vapor phase. Given long contact time, we are confident that an assumption of equilibrium conditions will not result in significant error. A schematic representation of the system is included in Figure E6.8.1 as an aid to visualization of the system.

We obtain relevant equilibrium data from Tables 5.1 and 6.1 for the ammonia–nitrogen system: $K_{H.NH_3} = 57\ \text{mol / L-atm}; pK_{A.NH_3} = 9.30$. In order to compute $[NH_3]$ we must first compute P_{NH_3} :

$$P_{Tot} := 0.90\ atm \quad C_{V.NH3} := 50\ ppm_V \quad Y_{NH3} := \frac{C_{V.NH3}}{10^6} = 5 \times 10^{-5}\ \frac{mol_{NH3}}{mol_{gas}}$$

$$P_{NH3} := Y_{NH3} \cdot P_{Tot} = 4.5 \times 10^{-5}\ atm$$

Once the partial pressure of ammonia in the gas is known, we may compute the aqueous ammonia concentration:

$$K_{H.NH3} := 57.0\ \frac{mol}{L \cdot atm} \quad NH3 := K_{H.NH3} \cdot P_{NH3} = 0.00257\ \frac{mol_{NH3}}{L}$$

We have a couple pathways for computing $[NH_3–N]$. Let us compute $[NH_4^+]$ and add the two species to obtain $[NH_3–N]$, considering the lower pH value first:

$$pH := 8.5 \quad H := 10^{-pH} = 3.162 \times 10^{-9} \quad NH4 := \frac{H}{K_{A.NH3}} \cdot NH3 = 0.016\ \frac{mol_{NH4}}{L}$$

$$C_{Tot.NH3} := NH4 + NH3 = 0.019\ \frac{mol_{NH3N}}{L}$$

The other pathway involves using the α_1 relation for a monoprotic acid from Table 6.2, with $B = NH_3$ and $n = 1$:

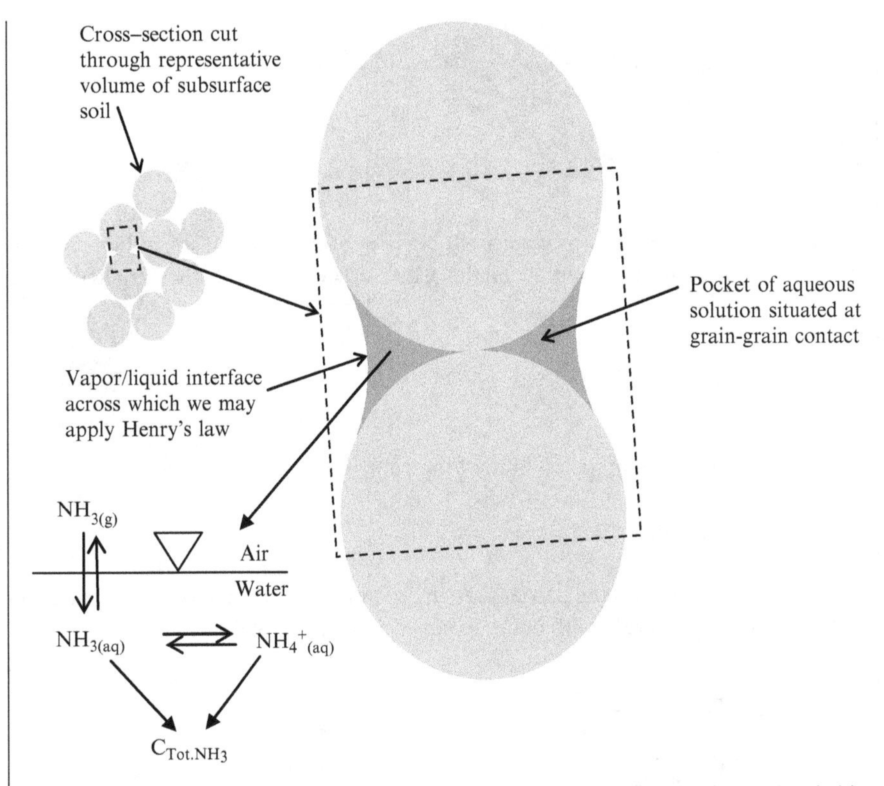

FIGURE E6.8.1 *Schematic representation of the vapor–liquid interface and associated air/water and acid/base distribution of ammonia species in a contaminated subsurface environment.*

$$\alpha_{1.NH3} := \left(\frac{H}{K_{A.NH3}} + 1\right)^{-1} = 0.137 \quad C_{Tot.NH3N} := \frac{NH3}{\alpha_{1.NH3}} = 0.019 \quad \frac{mol_{NH3N}}{L}$$

At the higher pH we have:

$$pH := 9.5 \quad H := 10^{-pH} = 3.162 \times 10^{-10} \quad \alpha_{1.NH3} := \left(\frac{H}{K_{A.NH3}} + 1\right)^{-1} = 0.613$$

$$C_{Tot.NH3N} := \frac{NH3}{\alpha_{1.NH3}} = 0.00418 \quad \frac{mol_{NH3N}}{L}$$

Now that we know about the range of concentrations expected, we can pursue the estimate of the total quantity of ammonia–nitrogen associated with the stated volume of the soil. Let us use the density of the soil solid as 2.65 g/cm^3. We have an opportunity to employ some volume fraction units examined in Chapter 3. Let us first compute the mass of the solids in the targeted cubic meter:

$V_{Tot} := 1 \quad m^3 \quad \varepsilon := 0.4 \quad \frac{m_{void}^3}{m_{Tot}^3} \quad V_{sol} := (1-\varepsilon)\cdot V_{Tot} = 0.6 \quad m^3$

$\rho_{sol} := 2.65\cdot\frac{10^6}{10^3} = 2650 \quad \frac{kg}{m^3} \quad M_{sol} := V_{sol}\cdot\rho_{sol} = 1590 \quad kg$

We know that the mass of water will be 10% of the mass of the solids and can use this to compute the volume of the water associated with the targeted cubic meter of soil:

$M_W := 0.1\cdot M_{sol} = 159 \quad kg \quad \rho_W := 1000 \quad \frac{kg}{m^3} \quad V_W := \frac{M_W}{\rho_W} = 0.159 \quad m^3$

We can then compute the volume of the cubic meter that is vapor:

$V_{vap} := \varepsilon\cdot V_{Tot} - V_W = 0.241 \quad m^3$

We then must convert the partial pressure of ammonia into a molar concentration with units commensurate with cubic meters as the unit of volume:

$R := 8.2057\cdot 10^{-5} \quad \frac{m^3\cdot atm}{mol\cdot {}^\circ K} \quad T := 25 + 273 = 298 \quad {}^\circ K$

$C_{vap.NH3} := \frac{P_{NH3}}{R\cdot T} = 1.84\times 10^{-3} \quad \frac{mol_{NH3}}{m_{vap}^3}$

We also must convert the total ammonia–nitrogen abundance of the aqueous phase for use with volume in cubic meters:

$C_{Tot.NH3N} := C_{Tot.NH3N}\cdot 1000 = 4.183 \quad \frac{mol}{m^3}$

We certainly could have used a gas constant with units of liters and computed the volumes of gas and water in liters.

Now we may perform an accounting of the ammonia–nitrogen as the sum of the quantities present in the aqueous and vapor phases. We simply need the summation of the products of abundance and volume for the gas and aqueous phases:

$Mol_{NH3N} := (V_{vap}\cdot C_{vap.NH3} + V_W\cdot C_{Tot.NH3N}) = 0.666 \quad \frac{mol_{NH3N}}{m_{soil}^3}$

Again, we would wish to communicate this result to engineers of other disciplines, managers, and technicians so we would perform one more conversion to render the result into mass units:

$MW_N := 0.014\ \frac{kg}{mol}\quad Mass_{NH3N} := Mol_{NH3N} \cdot MW_N = 9.318 \times 10^{-3}\ \frac{kg_{NH3N}}{m_{soil}^3}$

With the sniffer methodology we can investigate the extent of the plume resulting from the ammonia spill without the necessity for expensive drilling, sampling and laboratory analyses. We can perform this analysis using the result from each "sniffer" sampling of the subsurface gas phase. Each sample would be representative of a certain region. With knowledge of the areal extent of each of the regions, we can compute the quantity of ammonia–nitrogen within each region and simply sum the contents of the respective regions to obtain the total.

6.7.2 Multiprotic Acids

Multiprotic acids of course consist of two or more conjugate acid/base pairs. Each acid/base pair behaves similarly to a monoprotic acid. We may then employ each successive deprotonation equilibrium much as we would that of a truly monoprotic acid. The additional complication arises from the simultaneous existence of three or more distinct species of the acid system in any given aqueous solution. The mole accounting equation and the two or more equilibria are combined to yield a fully explicit system if the hydrogen ion abundance is known. One of the three or more species of the acid system will carry a net zero charge and in reality be distributed between air and water under equilibrium conditions. Since understandings of the carbon dioxide system (carbonic acid) are of great importance in environmental systems analysis, we will use this diprotic acid system in illustrations of the applications of the important principles.

Example 6.9 Consider raw ground water withdrawn from the alluvial aquiver of the Sheyenne River of eastern North Dakota by the City of Lisbon. The water is to be softened via lime-sodium hydroxide treatment prior to distribution to the community. From a laboratory analysis of the water we find that pH=7.65 and measured alkalinity=350 mg/L as $CaCO_3$. We desire to characterize the carbonate system speciation to enable computation of the doses of softening chemicals necessary for treatment of the water.

From Table 6.1 we find that $pK_{1.CO_3} = 6.35$ and $pK_{2.CO_3} = 10.33$. As the alkalinity value is rather high, we know that the approximate Equations 6.7a and 6.7b may be applied here. We begin by converting the stated ANC (alkalinity) to molar units. See Table 3.1 for conversions:

$pK_{1.CO3} := 6.35\quad pK_{2.CO3} := 10.33\quad C_{Alk} := 350\ \frac{mg_{asCaCO3}}{L}$

$EW_{CaCO3} := 50000\ \frac{mg}{Eq}\quad Alk := \frac{C_{Alk}}{EW_{CaCO3}} = 7 \times 10^{-3}\ \frac{Eq}{L} = \frac{mol}{L}$

(one Eq. of protons = one mole of protons)

We may now apply Equations 6.7a and 6.7b to obtain bicarbonate and carbonate concentrations:

$$H = 2.239 \times 10^{-8} \quad K_{1.CO3} = 4.467 \times 10^{-7} \quad K_{2.CO3} = 4.677 \times 10^{-11}$$

$$HCO3 := \frac{Alk}{1 + 2 \cdot \frac{K_{2.CO3}}{H}} = 6.971 \times 10^{-3} \quad CO3 := \frac{Alk}{\frac{H}{K_{2.CO3}} + 2} = 1.456 \times 10^{-5} \ \frac{mol}{L}$$

We have several choices for computing $[H_2CO_3{}^*]$.

1. We may use the $K_{1.CO_3}$ equilibrium directly, by far the most straightforward:

$$H2CO3 := \frac{H}{K_{1.CO3}} \cdot HCO3 = 3.494 \times 10^{-4} \ \frac{mol}{L}$$

2. We may use the diprotic acid α_1 from Table 6.2 ($B=CO_3$, $n-1$) to obtain $C_{Tot.CO_3}$ and then use α_0 to obtain $[H_2CO_3{}^*]$:

$$\alpha_{1.CO3} := \left(\frac{H}{K_{1.CO3}} + 1 + \frac{K_{2.CO3}}{H}\right)^{-1} = 0.95$$

$$C_{Tot.CO3} := \frac{HCO3}{\alpha_{1.CO3}} = 7.335 \times 10^{-3} \ \frac{mol}{L}$$

$$\alpha_{0.CO3} := \left(1 + \frac{K_{1.CO3}}{H} + \frac{K_{1.CO3} \cdot K_{2.CO3}}{H^2}\right)^{-1} = 0.048$$

$$H2CO3 := \alpha_{0.CO3} \cdot C_{Tot.CO3} = 3.494 \times 10^{-4} \ \frac{mol}{L}$$

3. We may use α_1 from Table 6.2 to obtain $C_{Tot.CO_3}$ and obtain $[H_2CO_3{}^*]$ as the difference between $C_{Tot.CO_3}$ and the sum of $[HCO_3{}^-]$ and $[CO_3{}^=]$:

$$H2CO3 := C_{Tot.CO3} - HCO3 - CO3 = 3.494 \times 10^{-4} \ \frac{mol}{L}$$

We see that $H_2CO_3{}^*$ is ~5% of the total carbonate for the conditions specified and indeed will consume significant lime in the overall softening reaction.

In Example 6.9, we assumed that the ANC of the water is dominated by the carbonate system. This is not always the case, especially for low alkalinity waters originating from predominantly granitic drainages such as found in the Adirondack Mountains, Ontario, and portions of the western Rocky Mountains of the United States. The question might be asked "when can we safely use the approximation (Equations 6.7a and 6.7b) and when must we use the more accurate analysis posed by Equation 6.6?" Let us examine this question in Example 6.10.

Example 6.10 Perform an analysis of the relative error associated with use of Equation 6.7 as opposed to Equation 6.6 for water for which the carbonate system comprises the major ANC beyond that of water itself. Examine alkalinity values from single digit to hundreds of mg/L as $CaCO_3$.

For this investigation, let us use three initial pH values: 6.0, 6.5, and 7.0. Let us use measured [ANC] (alkalinity) in mg/L as $CaCO_3$ as a master variable and compare the resultant $C_{Tot.CO_3}$ obtained from use of Equations 6.7a and 6.7b as the approximation against $C_{Tot.CO_3}$ obtained from use of Equation 6.6 as the "true value." We will present the result as percent Relative Error (%RE) versus measured alkalinity.

In Example 6.4, we assumed that $[H^+]_{final}$, $[OH^-]_{final}$, $[HCO_3^-]_{final}$, and $[CO_3^=]_{final}$ at the endpoint of the titration and that both $[H^+]_{init}$ and $[OH^-]_{init}$ at the beginning of the titration were all insignificant. Here, let us back away from that assumption and consider that $[H^+]_{final}$ and $[OH^-]_{init}$ are both significant. Equation 6.6 for this application reduces to a set of initial and final concentrations of relevant species:

$$[Alk] \approx [HCO_3^-]_{init} + 2[CO_3^=]_{init} + [OH^-]_{init} - [H^+]_{final}$$

We find this relation published in a number of sources, but most fail to specify that $[H^+]$ must reflect the end point pH value of the titration rather than the initial pH of the aqueous solution.

We may write a set of functions in MathCAD, employing the three pH values and the value of measured alkalinity ([Alk]) as a master-independent variable. The subscripts *A* and *T* refer to approximated and true values, respectively. The subscripts 1, 2, and 3 refer to pH 6.0, 6.5, and 7.0, respectively.

We first need to locate and define some equilibrium constants and other parameters:

$$EW_{CaCO3} := 50000 \frac{mg}{Eq}$$

$$\begin{pmatrix} K_w \\ K_{1.CO3} \\ K_{2.CO3} \\ H_f \end{pmatrix} := \begin{pmatrix} 10^{-14} \\ 10^{-6.35} \\ 10^{-10.33} \\ 10^{-4.3} \end{pmatrix} \quad \begin{pmatrix} H_{i.1} \\ H_{i.2} \\ H_{i.3} \end{pmatrix} := \begin{pmatrix} 10^{-6.0} \\ 10^{-6.5} \\ 10^{-7.0} \end{pmatrix} \begin{pmatrix} OH_{i.1} \\ OH_{i.2} \\ OH_{i.3} \end{pmatrix} := K_w \cdot \begin{pmatrix} \frac{1}{H_{i.1}} \\ \frac{1}{H_{i.2}} \\ \frac{1}{H_{i.3}} \end{pmatrix} \frac{mol}{L}$$

Then we may write functions, using Equations 6.6 and 6.7 and employing the definitions of the abundance fractions from Table 6.2. The relations written for pH 6.0 only are shown. Those for pH 6.5 and 7.0 are exactly the same except for the subscripts 2 and 3 and 6.5 and 7.0:

$$HCO3_{A.1}(Alk) := \frac{\frac{Alk}{EW_{CaCO3}}}{1 + 2 \cdot \frac{K_{2.CO3}}{H_{i.1}}} \qquad \alpha_{1.A.1}(Alk) := \left(\frac{H_{i.1}}{K_{1.CO3}} + 1 + \frac{K_{2.CO3}}{H_{i.1}}\right)^{-1}$$

$$C_{Tot.CO3.A.1}(Alk) := \frac{HCO3_{A.1}(Alk)}{\alpha_{1.A.1}(Alk)}$$

$$HCO3_{T.1}(Alk) := \frac{\frac{Alk}{EW_{CaCO3}} - OH_{i.1} - H_f}{1 + 2 \cdot \frac{K_{2.CO3}}{H_{i.1}}} \qquad \alpha_{1.T.1}(Alk) := \left(\frac{H_{i.1}}{K_{1.CO3}} + 1 + \frac{K_{2.CO3}}{H_{i.1}}\right)^{-1}$$

$$C_{Tot.CO3.T.1}(Alk) := \frac{HCO3_{T.1}(Alk)}{\alpha_{1.T.1}(Alk)}$$

$$RE_{6.0}(Alk) := \frac{C_{Tot.CO3.A.1}(Alk) - C_{Tot.CO3.T.1}(Alk)}{C_{Tot.CO3.T.1}(Alk)} \cdot 100$$

We plot the three functions against measured [ANC] (alkalinity) in Figure E6.10.1.

$Alk := 5, 5.5 .. 250 \quad \frac{mg_{asCaCo3}}{L}$

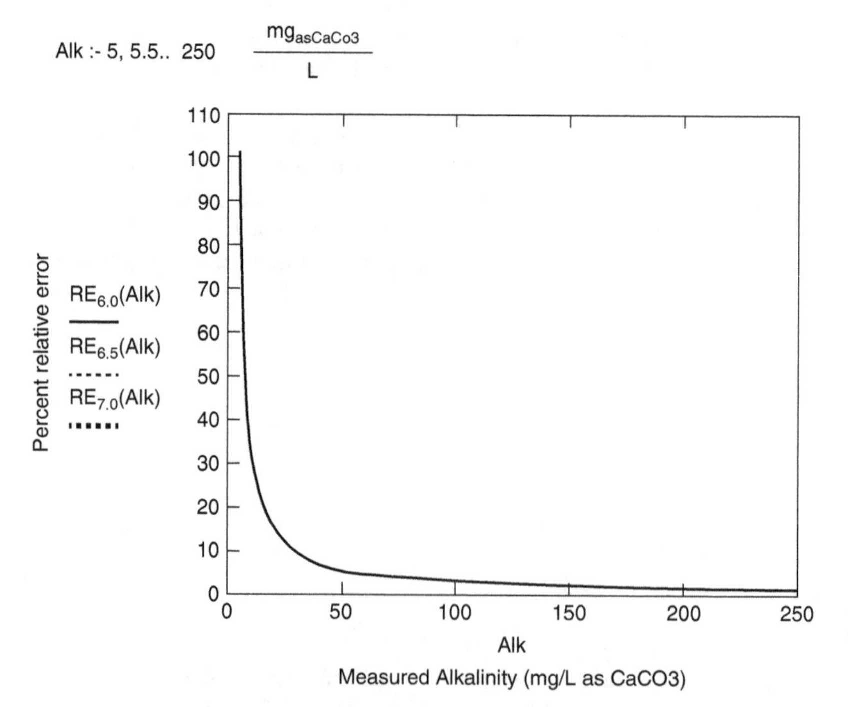

FIGURE E6.10.1 *Plot of the relative error of the computation of total inorganic carbon from alkalinity measurement associated with use of the approximate equation describing alkalinity as a function of the measured alkalinity.*

From the plot of % relative error versus measured alkalinity we can easily discern the result. As might be expected, since $[H^+]_{init}$ is quite insignificant, the initial pH of the solution has essentially no impact upon the result. There are indeed three traces on the set of axes. Alkalinity of the water, however, has a pronounced effect on the relative error introduced by the approximation (Equations 6.7a and 6.7b). Even at alkalinity values as high as 100 mg/L *as* $CaCO_3$, the error associated with the common approximation is nearly 3%. At a measured alkalinity of 5 mg/L, the error associated with use of the approximation is just over 100%. Figure E6.10.1 can be used as a rough guide regarding the necessity for implementation of the more accurate approximation when the desired accuracy of the result is known.

From Example 6.10 we are left with a somewhat uneasy feeling about computations employing measured alkalinity and using the approximate relations 6.7a and 6.7b, especially for measured alkalinity values less than 100 mg/L as $CaCO_3$. We observed the ease with which the more exact relation was applied to the measured alkalinity results. Perhaps we will resolve that, especially for low measured alkalinity values, henceforth we will employ the more accurate relation.

Sulfate–sulfur ($SO_4^=$–S) contained in wastewater and its conversion to sulfide–sulfur ($S^=$–S) via anaerobic bacteria present in wastewater collection and storage structures comprises an area of great concern. Welfare of workers servicing wastewater collection systems and the longevity of concrete pipes and structures within the collection system are primary issues. One area of concern involves outlying housing developments which are served by "lift" stations used to pump collected wastewaters over a hill to gravity-based collection systems through which flows are routed to publically owned treatment works (POTWs). Many of these housing developments are residential only and as such only small flows of wastewater occur during the nighttime hours. The pumps and pipes, of course, must be of sufficient size to efficiently carry the maximum daytime flows. Consequently, the residence time of wastewater in the pipes as well as in the wet wells can be long during times of nonpeak flow. These long residence times contribute to the depletion of oxygen by the biological processes occurring within. Microbes turn to other electron acceptors—notably sulfate-sulfur. Understandings of sulfate to sulfide conversion are useful in design of systems to vent manholes at the terminus points of force mains or to strip hydrogen sulfide from wastewater prior to discharging the flow from force mains to gravity systems. Here, let us examine the potential for the buildup of hydrogen sulfide gas in a force main terminus manhole as a consequence of the entry of sulfide-laden wastewater.

Example 6.11 Consider a lift station and force main system that is used to transport collected wastewater from a development over a hill to a gravity collection system operated and maintained by a municipality. Consider that the sulfate-sulfur content of the generated wastewater is 112 $mg_{(as\ SO_4\text{-}S)}$ /L and that under the extreme low flow conditions 10% of the sulfate-sulfur is expected to be converted to sulfide-sulfur as a consequence of residence in the collection manhole and in the force main. Typical domestic wastewater might have a pH of ~7.5. Consider that the system is located at an elevation such that the normal ambient pressure is 0.9 atm. A sketch of this system is included in Figure E6.11.1 for understanding of the context. The terminus manhole is for practical purposes isolated from the atmosphere and hydrogen sulfide gas can build up in this confined space. Here we consider that the wastewater discharged from the force main will enter the manhole and that the aqueous and vapor phases will be in contact for a time period sufficient for near equilibration of the distribution of hydrogen sulfide between the aqueous and vapor phases.

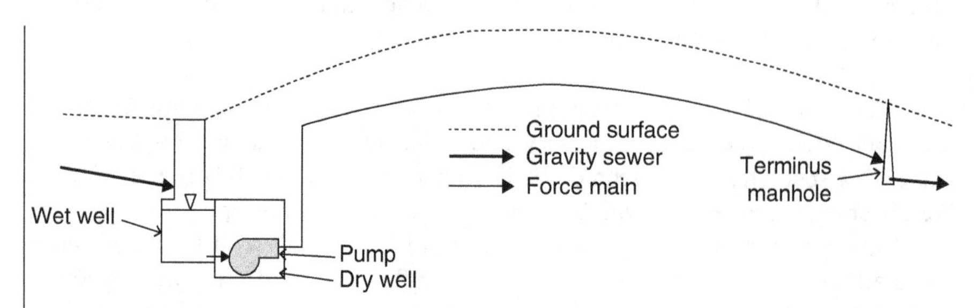

FIGURE E6.11.1 Sketch of a pump station and force main system.

We first determine the total sulfide content of the force main discharge. Ten percent of the sulfate-sulfur is converted to sulfide-sulfur and the stoichiometry of the conversion is one to one on a molar basis (Appendix, Table A.5):

$$C_{Tot.SO4S} := \frac{112}{32000} = 3.5 \times 10^{-3} \qquad C_{Tot.S} := 0.1 \cdot C_{Tot.SO4S} = 3.5 \times 10^{-4} \; \frac{mol}{L}$$

In order to eventually determine the abundance of $H_2S_{(g)}$, we will need to compute $[H_2S_{(aq)}]$, and then use the diprotic α_0 from Table 6.2 for $B = S^=$ and $n = 0$. We will also consult Table 6.1 for $pK_{1.S}$ and $pK_{2.S}$ equal to 7.10 and 14, respectively:

$$\begin{pmatrix} H \\ K_{1.S} \\ K_{2.S} \end{pmatrix} := \begin{pmatrix} 10^{-7.5} \\ 10^{-7.1} \\ 10^{-14} \end{pmatrix} \qquad \alpha_{0.S} := \left(1 + \frac{K_{1.S}}{H} + \frac{K_{1.S} \cdot K_{2.S}}{H^2}\right)^{-1} = 0.285$$

$$H2S := \alpha_{0.S} \cdot C_{Tot.S} = 9.97 \times 10^{-5} \; \frac{mol}{L}$$

From Table 5.1, we obtain the Henry's constant and can then compute the partial pressure and, hence, the abundance, in ppm_v, of hydrogen sulfide in the vapor:

$K_{H.H2S} := 0.105 \ \frac{mol}{L \cdot atm}$ $\quad P_{H2S} := \frac{H2S}{K_{H.H2S}} = 9.492 \times 10^{-4}$ atm

$P_{Tot} := 0.9$ $\quad C_{V.H2S} := \frac{P_{H2S}}{P_{Tot}} \cdot 10^6 = 1055$ ppm_V

Although our predicted abundance of H_2S in the vapor is the maximum possible under the stated conditions, we would indeed be concerned for workers entering this space. Exposure to this level of hydrogen sulfide could result in instantaneous loss of consciousness and acute danger of a quick death. We would also be concerned about corrosion of the concrete manhole.

To mitigate the conversion of sulfate-sulfur to sulfide-sulfur, little leeway is available for down-sizing the volume of the wet well or pipe diameter to decrease the residence time in the force main system. We must plan for at minimum a three foot per second velocity in the pipe during active pumping to ensure suspension of solids, but must be mindful that energy losses due to friction in the pipes are proportional to the square of the fluid velocity. The wet well must be sized so that when cycled on, the pump will remain on for several minutes while draining the wastewater from the wet well. Then, we must seek means to provide positive venting for the terminus manhole. We might also be concerned about hydrogen sulfide buildup in manholes farther down the gravity line.

Sediments deposited via the natural cycling of temperate zone dimictic lakes are microbially active. Dimictic lakes are well mixed twice each year—in the spring just after the ice recedes and in the fall just before the formation of ice. The upper several centimeters of these sediments is aerobic in most oligotrophic or mesotrophic lakes. In eutrophic lakes, depending upon the time of year, this upper layer may be either aerobic or anaerobic. The sediments beneath the aerobic layer, regardless of the trophic state of the lake, are anaerobic. In eutrophic lakes, the sediments are rich in organics. Sediments of mesotrophic lakes generally have low to moderate organic matter content. Oligotrophic lakes generally have organic-poor sediments. The sediments of wetlands (marshes, swamps) would generally be expected to have high organic matter content. Within these sediments the microbes continually mineralize the organic matter, using sulfate and carbon dioxide as electron acceptors in lieu of oxygen, producing methane and carbon dioxide as major end products. Other important intermediate and end products include sulfide-sulfur, short-chain carboxylic acids, ammonia-nitrogen, and phosphate-phosphorus. Due to its low solubility, the methane produced is preferentially distributed into a gas phase, forming bubbles. Other common neutral species (carbon dioxide, ammonia, hydrogen sulfide, various carboxylic acids) distribute into these bubbles. As microbial processes continue, bubbles grow and eventually buoyant forces exceed adhesive forces and the bubbles release, traveling through the sediment/water interface and eventually rising through the water column to the atmosphere. The rate at which bubbles form is related to the trophic state of the water body and the parent materials from which the sediments are formed. From knowledge of the composition of these bubbles, we can infer a great deal about the character of the pore water residing within the sediments (or vice versa). Let us examine the speciation of the carbonate system within typical anaerobic sediments.

Example 6.12 Consider the anaerobic sediments lying several centimeters below the sediment/water interface of a moderately eutrophic lake (i.e., Sheridan Lake, located in the Black Hills of SD). Consider the location is in the vicinity of the inlet and the water depth above the sediment is ten feet (3.05 m). The typical gas bubbles produced by anaerobic activity of sediments might contain 60–70% methane and 10–30% carbon dioxide, with the remainder (0–30%) comprised of the minor constituents mentioned previously. Herein, let us use 20% as the CO_2 content of the bubbles. The drainage into Sheridan Lake arises in terrain characterized by significant carbonate mineralogy so the water and sediments would be well buffered. Let us consider that the pH would be about neutral, 7.0. The bubbles formed remain in contact with the pore water for a sufficient time that this system can, with little error, be represented as a near-equilibrium system. A sketch of this system is included as Figure E6.12.1 to help visualize the equilibria in effect. For the stated conditions, characterize the speciation of the carbonate system within the sediment pore water.

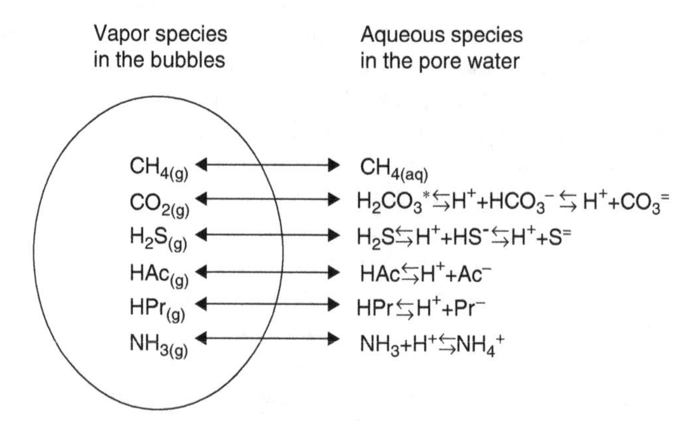

FIGURE E6.12.1 *Schematic representation of the sediment/water interface associated with a gas bubble residing in sediments, with additional representation of the gas/water and acid/base distributions of species that would be present.*

We will need the Henry's constant for CO_2 from Table 5.1 and the pK_A values from Table 6.1 for the two carbonic acid deprotonation equilibria. We will also need the specific weight of water in order, from fluid statics principles, to compute the pressure in the bubbles as they reside in the sediments. Sheridan Lake is at elevation 5000 ft above mean sea level so the normal ambient pressure is 0.832 atm (see Example 5.3). We will also need the conversion from pressure in kPa (kN/m^2) to pressure in atm. We will first compute the total pressure in the bubble and then the partial pressure of carbon dioxide:

$P_{amb} := 0.832 \ atm \quad \gamma_{H2O} := 9.79 \ \frac{kN}{m^3} \quad D := 3.05 \ m$

$1 \cdot atm = 101.3 \cdot \frac{kN}{m^2} \quad P_{Tot} := P_{amb} + \frac{\gamma_{H2O} \cdot D}{101.3} = 1.127 \ atm$

$Y_{CO2} := 0.2 \ \frac{mol_{CO2}}{mol_{gas}} \quad P_{CO2} := P_{Tot} \cdot Y_{CO2} = 0.225 \ atm$

We can then compute the dissolved carbon dioxide concentration:

$K_{H.CO2} := 0.0339 \ \frac{mol}{L \cdot atm} \quad H2CO3 := K_{H.CO2} \cdot P_{CO2} = 0.00764 \ \frac{mol}{L}$

Most conveniently then we can use the $K_{1.CO_3}$ and $K_{2.CO_3}$ equilibria to complete the carbonate system characterization:

$H := 10^{-7} \quad K_{1.CO3} := 10^{-6.35} \quad K_{2.CO3} := 10^{-10.33}$

$HCO3 := \frac{K_{1.CO3}}{H} \cdot H2CO3 = 0.034 \quad CO3 := \frac{K_{2.CO3}}{H} HCO3 = 1.596 \times 10^{-5} \ \frac{mol}{L}$

$C_{Tot.CO3} := H2CO3 + HCO3 + CO3 = 0.042 \ \frac{mol}{L}$

We could employ this process to characterize each of the acid/base systems present in the pore water and gas bubbles formed therein. The pH of the pore waster of the sediments is not necessarily the same as that of the water above the sediments. Production of carbon dioxide provides a source of protons that would tend to depress the pH. We will, in much greater detail, examine computations of this nature in Chapter 10. The remaining characterizations of the acid systems of the pore water are left as end-of-chapter exercises.

PROBLEMS

1. Supernatant (aqueous solution separated from solids in an anaerobic digester) returned from the digester to the first stage biological treatment process at the Rapid City Regional Wastewater facility had pH of 8.17 and contained ammonia nitrogen (NH_3–N) at a concentration of 950 ppm_m. Determine the concentration of unionized ammonia ($NH_{3(aq)}$) in mol/L of this aqueous solution.

2. Weak acid dissociable cyanide (WAD, the total concentration of CN^- in cyanide species) was measured in solution draining from the heap leach pad at the former Brohm Mine (now an EPA superfund site) to be 120 mg(as CN^-)/L. The

pH of the solution was 8.61. Determine the concentration in mol/L of hydrocyanic acid (HCN) in this solution.

3. Aqueous effluent from a process that will be used to anaerobically convert steer manure to methane gas, to be used for generation of energy for production of ethanol at a facility planned near Pierre, SD, contains total acetate (CH_3COO^- in acetic acid species) at a concentration of 104 mg(as CH_3COO^-)/L. The pH of this solution is expected to be 6.05. Compute the expected concentration in mol/L of the acetate ion (CH_3COO^-) in this solution.

4. The pH of wastewater leaving a force main (a closed conduit with flow induced via pumping) was 7.51 and the measured total sulfide (S^{-2} in all sulfide-containing species) concentration was 19.5 mg (as S^{-2})/L. Determine the molar concentration of hydrogen sulfide ($H_2S_{(aq)}$) in the wastewater.

5. A groundwater sample obtained from deep in an aquifer had a measured pH of 6.52 and contained total inorganic carbon (CO_3^{-2} in all carbonate containing species) at a concentration of 0.00352 M. Determine the concentration of carbonic acid (H_2CO_3*) in mol/L in this water sample.

6. Treated water leaving the reactor/clarifier of a metal-precipitation process that employs lime ($Ca(OH)_2$) for pH adjustment has a pH of 9.7 and a total carbonate (CO_3^{-2} in all carbonate containing species) concentration of 0.000547 M. Determine the concentration of bicarbonate (HCO_3^-) in units of mg(as HCO_3^-)/L.

7. A water sample taken from beneath the ice at Sylvan Lake, in Custer State Park of SD, contained dissolved reactive phosphorus (often referred to as ortho-phosphorus, which is comprised of the sum of the four phosphoric acid species) at a concentration of 1.26 mg(as P)/L. The pH of the water sample was 8.22. Determine the concentration in mol/L of dihydrogen phosphate ($H_2PO_4^-$) in this water.

8. A water sample obtained from the Dakota Maid pit at the Gilt Edge Superfund site near Deadwood, SD, had a pH of 4.62 and contained dissolved reactive phosphorus (often referred to as ortho-phosphorus, which is comprised of the sum of the four phosphoric acid species) at a concentration of 5.76 mg(as P)/L. Determine the concentration of dihydrogen phosphate ($H_2PO_4^-$) in this water. Express your answer in mg(as P)/L.

9. The measured alkalinity of a water sample was 187 mg/L as $CaCO_3$ and the pH was 6.72. Determine the concentration of H_2CO_3*, the sum of the fully protonated dissolved carbon dioxide species ($[H_2CO_3^*]=[CO_2 \cdot H_2O]+[H_2CO_3]$), and total inorganic carbon in this water sample and present your answers in units of mol/L.

10. The measured alkalinity of a water sample was 196 mg/L as $CaCO_3$ and the pH was 8.42. Determine the concentrations of H_2CO_3* in ppm$_m$ as CO_2 and of total inorganic carbon ($[H_2CO_3^*]+[HCO_3^-]+[CO_3^=]$) in mol/L.

11. A water sample from Rapid Creek was taken recently and tested by a local laboratory. Pertinent results included: alkalinity 325 mg/L as $CaCO_3$ and pH 8.05. Determine the bicarbonate, carbonate, and total inorganic carbon (also express this as total carbonate) concentrations of this water. Express your answers in both mol/L and mg/L. Remember that total carbonate (in mol/L) is the sum of carbonic acid, bicarbonate, and carbonate.

12. A water sample from Strawberry Creek near Deadwood, SD, was taken and tested by a local laboratory. Pertinent results included: alkalinity 125 mg/L as $CaCO_3$ and pH 6.05 (this is affected by acid drainage from the abandoned Gilt Edge National Priorities Listed (Superfund) site). Determine the carbonic acid, bicarbonate, carbonate and total inorganic carbon (also express this as total carbonate) concentrations of this water. Express your answers in both mol/L and mg/L. Remember that total carbonate (in mol/L) is the sum of carbonic acid, bicarbonate and carbonate.

13. Supernatant (aqueous solution separated from solids in an anaerobic digester) returned from the digester to the first stage biological treatment process at the Rapid City Regional Wastewater facility had pH of 8.17 and contained ammonia nitrogen (NH_3-N) at a concentration of 950 ppm_m. Determine the concentration of ammonium (NH_4^+) in mol/L of this aqueous solution.

14. Weak acid dissociable cyanide (WAD, the total concentration of CN^- in cyanide species) was measured in solution draining from the heap leach pad at the former Brohm Mine (now an EPA superfund site) to be 120 mg(as CN^-)/L. The pH of the solution was 8.61. Determine the concentration in mol/L of the cyanide ion (CN^-) in this solution.

15. Aqueous effluent from a process that is be used to anaerobically convert steer manure to methane gas, to be used for generation of energy for production of ethanol contains total acetate (CH_3COO^- in acetic acid species) at a concentration of 104 mg(as CH_3COO^-)/L. The pH of this solution is expected to be 6.05. Compute the expected concentration in mol/L of acetic acid (CH_3COOH) in this solution.

16. The pH of wastewater leaving a force main (a closed conduit with flow induced via pumping) was 7.51 and the measured total sulfide (S^{-2} in all sulfide-containing species) concentration was 19.5 mg(as S^{-2})/L. Determine the molar concentration of HS^- (bisulfide) in the wastewater.

17. A groundwater sample obtained from deep in an aquifer had a measured pH of 6.52 and contained total inorganic carbon (CO_3^{-2} in all carbonate containing species) at a concentration of 0.00352 M. Determine the concentration (in mg/L) of bicarbonate (HCO_3^-) in this water sample.

18. Treated water leaving the reactor/clarifier of a metal-precipitation process that employs lime ($Ca(OH)_2$) for pH adjustment has a pH of 9.7 and a total carbonate

(CO_3^{-2} in all carbonate containing species) concentration of 0.000547 M. Determine the concentration of carbonate (CO_3^{-2}) in units of mg(as CO_3^{-2})/L.

19. A water sample taken from beneath the ice at Sylvan Lake, in Custer State Park of SD, contained dissolved reactive phosphorus (often referred to as ortho-phosphorus, which is comprised of the sum of the four phosphoric acid species) at a concentration of 1.26 mg(as P)/L. The pH of the water sample was 8.22. Determine the concentration in mol/L of hydrogen phosphate (HPO_4^{-2}) in this water.

20. A water sample obtained from the heap leach pad at the Gilt Edge Superfund site near Deadwood, SD, had a pH of 9.37 and contained dissolved reactive phosphorus (often referred to as ortho-phosphorus, which is comprised of the sum of the four phosphoric acid species) at a concentration of 3.22 mg(as P)/L. Determine the concentration of hydrogen phosphate (HPO_4^{-2}) in this water. Express your answer in mg(as P)/L.

21. The measured alkalinity of a water sample was 156 mg/L *as* $CaCO_3$ and the pH was 6.93. Determine the concentration of dissolved carbon dioxide ($[H_2CO_3^*] = [CO_2 \cdot H_2O] + [H_2CO_3]$) of the water sample and present your answer in units of ppm_m CO_2.

22. The measured alkalinity of a water sample was 137 mg/L as $CaCO_3$ and the pH was 7.98. Determine the concentration of $H_2CO_3^*$, the sum of the fully protonated dissolved carbon dioxide species ($[H_2CO_3^*] = [CO_2 \cdot H_2O] + [H_2CO_3]$), in this water sample and present your answer in units of mol/L.

23. Gas bubbles formed in organic rich sediments lying 33.9 ft beneath the water surface (total pressure in these bubbles will be 2 atm absolute) of the Chesapeake Bay contain 64% methane (CH_4), 33% carbon dioxide (CO_2) and 3% other gases, on a molar basis. A sample of the sediments was obtained and a portion was centrifuged to separate aqueous solution from the sediment solids and tested to determine the content of various constituents. The pH was measured to be 7.35. Total sulfide-sulfur and total acetate were determined to be 96 ppm_m and 118 ppm_m, respectively. Use this information as necessary to answer the following questions.

    Determine the carbonate system speciation of the sediment pore water and express your answers in molar units.
    Identify and determine the concentration of the relevant sulfur specie in the gas bubbles and express its concentration in ppm_v.

24. A water sample drawn from the Madison Aquifer east of the Black Hills was tested and found to have a measured alkalinity (based on a titration of a water sample using sulfuric acid to pH 4.3) of 240 mg/L as $CaCO_3$ and an initial pH of 7.65.

    Determine the speciation of inorganic carbon in the water sample.
    The sample was obtained from a depth of 800 ft, where the pressure in the confined aquifer is 23 atm absolute. Were gas bubbles to exist in the water-bearing formation surrounding the well, what would be their carbon dioxide content in units of ppm_v?

25. A sample of leachate emanating from the collection system of a sanitary landfill was measured to have a pH of 5.8 and to have a total acetate content of 1180 mg/L. The gas phase held within the pores of the soil/solid waste mixture held within the landfill, through which water percolates to form leachate, has a carbon dioxide content of 35% at total pressure of 1.05 atm.

    Determine the concentrations of the relevant acetate species in the leachate.
    Determine the concentrations of carbonate species in the leachate.

26. A water sample was obtained from groundwater originating in an unconfined aquifer at a depth of 60 ft below the ground surface, and 35 ft below the water table, such that the hydrostatic pressure in the water was 1.05 atm gauge. The ambient atmospheric pressure is normally 0.95 atm absolute. The pH was tested and found to be 7.35 and the alkalinity was measured to be 200 mg/L as $CaCO_3$. The temperature of water at the depth of the sample was 25 °C.

    Determine the inorganic carbon speciation and total inorganic carbon concentration of this water and report your answer in mol/L.
    Were this water in equilibrium with a gas phase in the subsurface environment, determine the carbon dioxide content of that gas phase in units of mol(as CO_2)/L of gas.
    Express the carbon dioxide concentration of part b in parts per million by volume (ppm_v).

27. The gas within a cloud, in which precipitation is forming, contains sulfur dioxide at a concentration of 75 ppb_v. Otherwise, the composition of the atmospheric gas is normal—20.9% oxygen, 78.9% nitrogen, 387 ppm_v carbon dioxide, and other minor constituents. The aqueous solution within the minute droplets of precipitation that are forming within the cloud has a pH of 4.5. Consider that the temperature and pressure of the atmosphere within the cloud are 0.85 atm and 25 °C, respectively.

    Identify and determine the concentrations of the relevant nitrogen species in the droplets. Express your answers in molar units.
    Identify and determine the concentrations in molar units, of the relevant sulfur dioxide (often also referred to as sulfite or sulfurous acid) species in the droplets.
    Very briefly explain what eventually might happen to the sulfur dioxide in these droplets and the potential environmental problems that might result.

28. A force main system transports waste water from a low-lying subdivision to the gravity collection system of a major metropolitan area. Concern exists that a long residence time of the waste water in the force main (pressure conduit that conveys the waste water) will promote the conversion of sulfate in the waste water to sulfide. Modeling of the process suggests that the sulfide content of the wastewater as it exits the force main will be 4 ppm_m. The composition of the vapor space in the manhole into which the force main exits is of grave concern. Consider that the pH of the wastewater is 7.5, the temperature is 25 °C and the gas pressure inside the manhole is 0.95 atm.

Under these conditions, considering the vapor space in the manhole to be in equilibrium with the exiting wastewater, determine the level of hydrogen sulfide in the vapor space and express your answer in units of ppm_v.
Were you contemplating entering this manhole what precautions would you take prior to doing so? Why?

29. A sample was obtained from the leachate stream emanating from a rather long-standing sanitary landfill. The leachate is produced from the percolation of precipitation, incident upon the landfill, through the degraded and degrading solid waste contained within the landfill. The measured pH and total inorganic carbon content of the leachate were 8.25 and 544 mg_C/L, respectively. The temperature of the leachate was 25 °C and the pressure of the atmosphere was 0.90 atm. The mixture of degrading solid waste and cover soil within the landfill has a porosity (volume fraction of voids) of 0.45 and the voids hold moisture such that the voids are filled 1/3 on a volume basis with aqueous solution (of composition equal to that of the leachate) on a volumetric basis. The landfill covers 10 ha (1 ha is 100 m × 100 m) and has average depth of 15 m.

    Identify and determine the concentrations of the relevant carbonate (also often called carbon dioxide, or inorganic carbon) species in the leachate. Express your result in molar units.
    Identify and determine the concentrations, in moles per liter of gas, of relevant carbon dioxide species in the gas contained within the voids of the landfill.
    Compute (you may assume composition is spatially invariant) the total quantity of inorganic carbon contained in the combination of vapor and aqueous solution contained in the landfill.

30. The primary ingredient of vinegar is of course acetic acid and commercial distilled vinegar contains 50 g/L total acetate with pH of 3.70. When we remove the cap from a bottle of vinegar, we are met with its rather pungent odor. Consider the vinegar and vapor in a bottle of distilled vinegar as it resides on the shelf of the supermarket, in Rapid City, SD, where the normal atmospheric pressure is 0.90 atm and the temperature is 25 °C.

    Compute the molar concentrations of relevant acetate species in the distilled vinegar.
    Compute the partial pressure and molar concentration of relevant acetate species in the vapor residing above the vinegar in the bottle.

31. Water pumped from the Madison Aquifer and supplied to the city of Rapid City's water distribution system has a measured alkalinity of 275 mg/L as $CaCO_3$ and a pH of 7.75.

    Were this aqueous solution at equilibrium with regard to acid–base reactions, determine the molar concentrations of carbon dioxide (also called carbonate or inorganic carbon) species present in the water.
    Given that the atmosphere contains carbon dioxide at a level of 387 ppm_v, that the normal atmospheric pressure in Rapid City is 0.9 atm, and temperature of

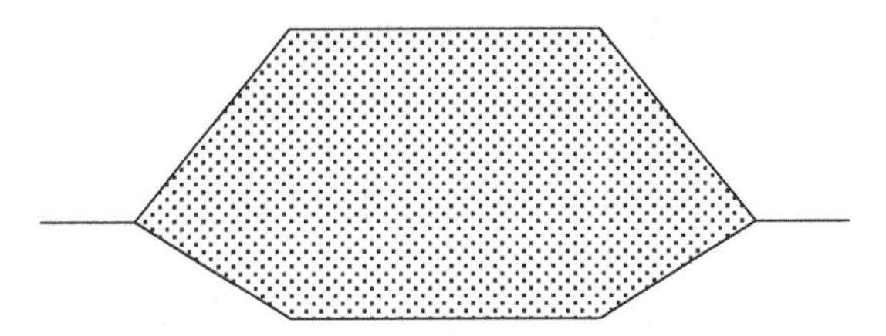

FIGURE P6.32 *A cross section through a completed solid waste landfill.*

25 °C, compute the concentration of dissolved carbon dioxide in water equilibrated with this atmospheric condition.

If the water described in part a were brought to the surface and opened to the atmosphere, what might happen with respect to carbon dioxide species contained in the water? Use the results of your computations here to support your answer.

32. A cross section through a sanitary landfill is depicted in Figure P6.32. The material inside the landfill consists of natural soil (used for daily cover of the disposed wastes), degrading solid wastes, and voids containing aqueous solution and vapor. The porosity of the mixed solid material within the landfill is 0.45. The landfill itself covers 80 ac (43,560 ft^2/ac) and is of average depth of 50 ft (Note that 1 ft^3 = 28.31 L). The moisture held in the voids of the landfill contents occupies 25% of the voids and has a pH of 8.6. The vapor held in the pores contains (among other constituents) 55% (on a molar basis) methane ($CH_{4(g)}$), 40% (on a molar basis) carbon dioxide (CO_2), and 12.0 ppm_v ammonia ($NH_{3(g)}$). The total pressure of the vapor in the voids is 1.05 atm, and the temperature inside the landfill is 25 °C.

    Determine the relevant ammonia nitrogen (NH_3–N) species in the aqueous solution of the voids and their molar concentrations.

    Determine the relevant methane species in the aqueous solution of the voids and their molar concentrations.

    Determine the quantity (in moles) of methane contained in the landfill.

33. An anaerobic digester, schematically shown in Figure P6.33, is used to convert biological solids produced at a wastewater plant into methane gas for energy production. Several other byproduct gasses are also produced. The energy from methane is used to heat the digester to the required 36 °C with excess used to heat buildings during the cold months of the year and to run electrical equipment.

    The particular digester in question is 50 ft in inside diameter with a liquid depth of 15 ft. The digester has a gas dome above the concrete tank that is a semisphere shell of radius 25 ft. At the particular condition of interest, the dome (which can move up or down on tracks, depending on the quantity of gas in the vapor space) is fully extended such that two feet of digester wall are

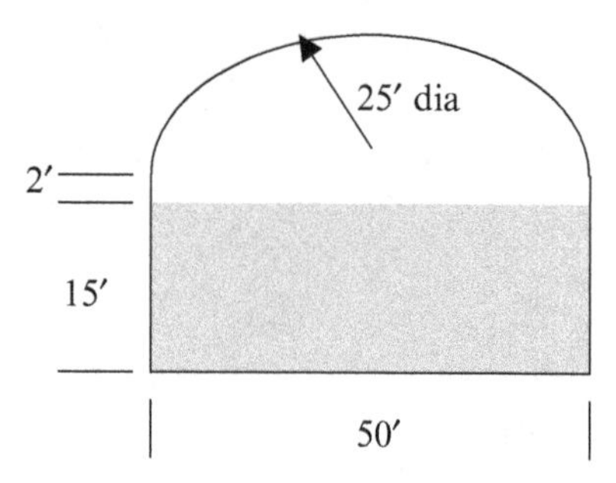

FIGURE P6.33 *Schematic diagram showing the liquid and vapor within and anaerobic digester.*

exposed to the vapor space below the dome. The digester is of course gas-tight. A configuration sketch is given at right.
Samples of the liquid and gas from the digester are obtained and analyzed as follows:

Vapor:	Liquid:
$H_2S_{(g)}$=5.7 ppm_v	NH_3–N=776 ppm_m
$CH_3CO_2H_{(g)}$=8.6 ppm_v	PO_4–P=25.6 ppm_m
$CO_{2(g)}$=34.1% (molar basis)	pH=6.42
$CH_{4(g)}$=65.2% (molar basis)	T=36 °C
P_T=1.06 atm, inside digester	

As a junior engineer at the engineering firm with which you work, you are asked by your project engineer to perform an accounting of sulfide, acetic acid, inorganic carbon (carbonate), methane, ammonia and phosphoric acid (PO_4–P) species contained in this particular digester. Upon questioning your boss about exactly what that was, you were informed that that meant a computation of the total quantity of each of these constituents contained in the vapor plus the liquid of the digester.

Therefore, compute the concentrations of all applicable (you may neglect those of insignificant concentration—but you must give reasoning when neglecting a particular specie) vapor and aqueous species in molar units and use these concentrations and the liquid and vapor volumes to determine the total quantity of each group of species. Note that values of Henry's constants and acid/base dissociation constants are given in your handouts at 25 °C. You need not attempt to convert these for the different temperature.
Approach: An accounting of vapor and liquid species can be completed by summing the products of total molar concentration and volume for each phase (gas and water):

34. A catastrophic leak of anhydrous ammonia (pure liquid NH_3) has occurred at a fertilizer distribution facility. The ammonia leaked into the soil and contaminated

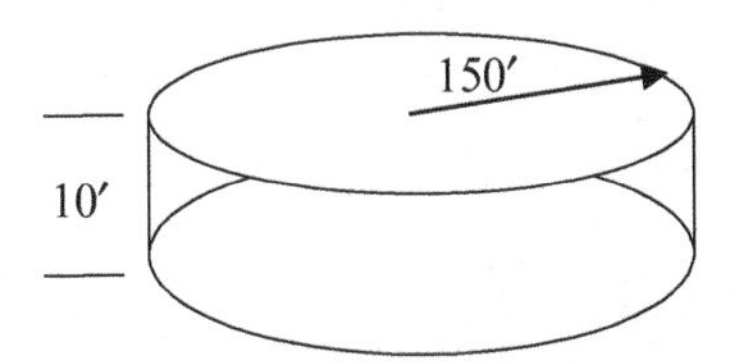

FIGURE P6.34 Idealized cylindrical shape of a hypothetical contaminated zone.

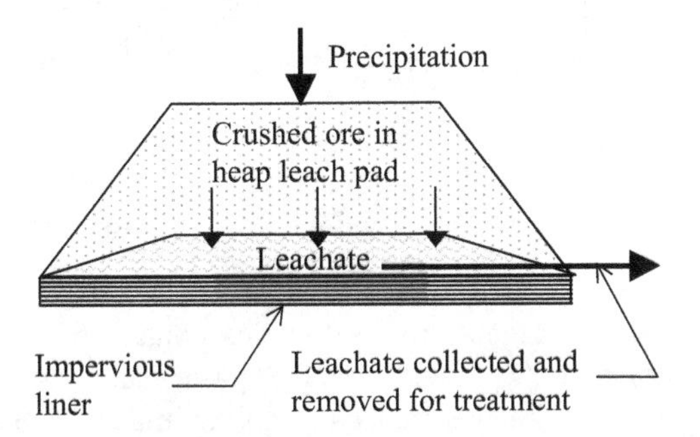

FIGURE P6.35 Schematic sketch showing a cross section of a heap leach pad.

a region of the subsurface that may be idealized as a cylinder of radius 150 ft and depth of 10 ft, depicted in Figure P6.34. The subsurface soil is 60% solid, 8% water and 32% vapor on a volume basis. The pH of the water in the soil is 9.65. The ambient pressure in the vapor within the soil pores is 0.9 atm and the temperature is 25 °C.

A vapor sample from the subsurface was obtained, assayed and found to contain 100 ppm_v $NH_{3(g)}$. How much ammonia nitrogen (in moles) is contained in the contaminated zone?

35. A schematic sketch of an abandoned heap leach pad once used for extraction of gold from low-grade ores is shown in Figure P6.35. The total volume of the heap leach pad is 700,000 m^3. Of the total volume of the pad, 35,000 m^3 is fully saturated with leachate.

The porosity (fraction of the volume that is voids) of the crushed ore is 0.40. Above the leachate pool, the voids are 12.5% moisture, thus the volume is 60% solid, 35% vapor and 5% aqueous solution. Within the leachate pool the volume is 60% solids and 40% aqueous solution.

The leachate is of pH 8.90 and total cyanide concentration of 78 ppm_m. The vapor and aqueous solution contained in the heap leach pad are in equilibrium. The temperature of the system is 20 °C and the local atmospheric pressure is 0.90 atm.

Determine the concentrations (in mol/L) of the cyanide-containing species in the leachate.

Determine the concentration (in mol/L) of cyanide-containing species in the vapor.

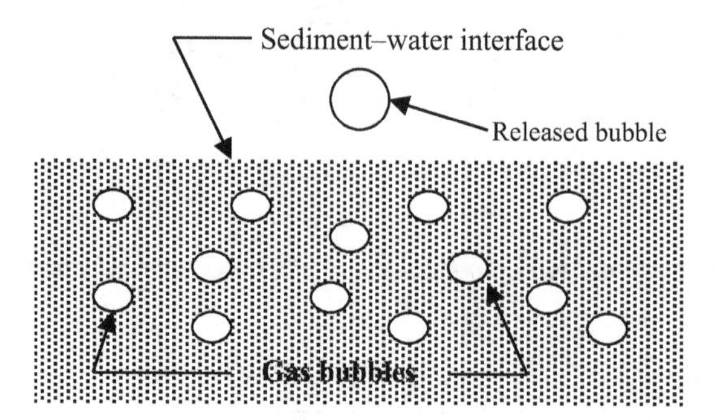

FIGURE P6.36 *Gas bubbles below the sediment/water interface.*

Express the concentration determined in part b in units of ppm_v.

36. Gas bubbles (see Figure P6.36) that form as a consequence of anaerobic biological activity in organic-rich sediments contain 64% methane ($CH_{4(g)}$), 34% carbon dioxide ($CO_{2(g)}$), and 2% other gases including 50 ppm_v hydrogen sulfide ($H_2S_{(g)}$). The sediments are located approximately 20 ft below the surface of the water such that the absolute pressure in the pores of the sediments is 1.5 atm (this will of course be the pressure inside the bubbles).

    Consider that the water and bubbles contained within the pores of the sediments are at equilibrium at a temperature of 5 °C and pH 7.5.

    Compute the significant sulfide species concentrations in the pore water.

    The bubbles of course grow as the biological process continues and when they reach a size of 5 mm in diameter, the frictional forces holding them in the sediments are overcome and they rise into the water above, eventually reaching the atmosphere. Determine the quantity in moles of methane that escapes from the sediments associated with the release of one bubble.

Chapter 7

Mass Balance, Ideal Reactors, and Mixing

7.1 PERSPECTIVE

In the previous chapters, we have invested significant effort in developing understandings of distribution equilibria in and between vapor and aqueous solutions. We should now expend some effort in understanding how environmental systems either come to the equilibrium state or how systems behave when not under equilibrium conditions. In such endeavors, we employ the principle of conservation of mass or mass accounting, of great importance in earlier chapters, in a slightly different light. We will begin with an examination of the application of mass accounting, or often called a mass balance, to control volumes. We define control volumes most often in conformance with physical boundaries. In many cases, the physical boundaries coincide with those of a defined reactor. In many other cases, we must define our control volume as some representative portion of an overall system. We will investigate the concept of ideal reactors: plug flow, completely mixed, and batch. These analyses will employ the concept of reactor dynamics: use of tracers and various input stimuli to discern the hydrodynamic character of reactors. We will then employ mass balance with the completely mixed flow reactor to develop a set of useful tools with which we may understand both steady-state and transient mixing in selected environmental systems.

Environmental Process Analysis: Principles and Modeling, First Edition. Henry V. Mott.

7.2 THE MASS BALANCE

In much the same manner as we might consider our personal financial accounts, we may consider the input, output, transformation, and accumulation of mass in the context of reactors. Normally, our first operation is to define the reactor. The reactor may be as small as a control volume at the intersection of two pipes, defined by the walls of the pipes and imagined planes that cut the pipes at right angles slightly upstream and downstream of the intersection. The reactor may be as large as a lake or the ground water lying directly beneath a cultivated agricultural field. Once we have defined the reactor, we may consider movement of mass across the reactor boundaries, transformations of components into alternative forms within the reactor, and accumulation (either positive or negative) of one or more components within the control volume. Most often, we attempt to define the geometry of the reactor such that reactor boundaries coincide with actual physical boundaries and such that the inflow and outflow of mass occur normal to areas defined in terms of the physical system.

We may relate accumulation, input, output, and transformation via the statement of mass balance applied to a component. The general form of the statement is simply a combination of the statements as shown in Figure 7.1. We may be interested in several components that are reactants which combine with other reactants to form products or components that are products resulting from the transformation of reactants. For any given reactor and comprehensive system, we may have many mass balances, typically one for each component of interest. We may have numerous inputs, outputs, and transformations of the targeted component included on the right-hand side (RHS) of our mass balance, but the left-hand side (LHS) consists of a single accumulation term. Further, depending upon the algebraic sum of the right-hand-side terms, the accumulation term can be either positive or negative—we let the right-hand side determine the sign. Transformations that create target components as a product are additive and often considered to be positive generation. Conversely, reactions that transform a target component into other products are subtractive and are most often considered to be negative

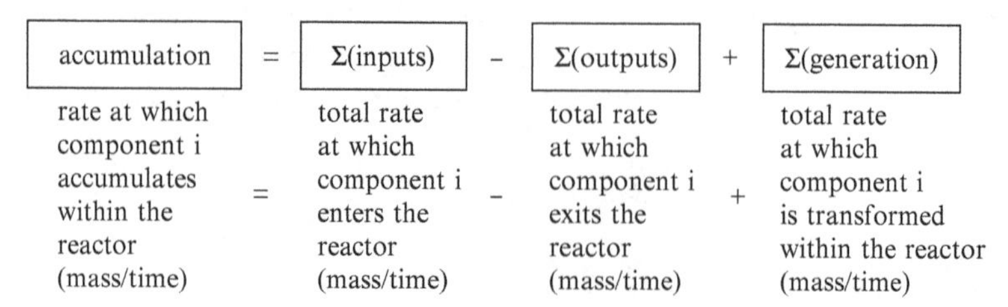

FIGURE 7.1 Word statement of the mass balance upon a targeted substance (component i) for an arbitrary reactor.

generation. Prior to quantitative consideration of transformations, we must examine the mixing conditions from which we may characterize the various types of ideal reactors.

7.3 RESIDENCE TIME DISTRIBUTION (RTD) ANALYSES

7.3.1 RTD Experimental Apparatus

A typical apparatus for performing experimental RTD analysis is shown in Figure 7.2. Any such apparatus would include a reactor with influent and effluent, a means to adjust and measure the influent flow to the reactor, an injection port in the influent line, a sampling port on the effluent line, a means to introduce the tracer as a known input, and a means to produce an experimentally derived trace of reactor effluent concentration versus time. The apparatus shown in Figure 7.2 includes a dedicated visible spectrophotometer for use with a dye tracer. Use of other tracers would require alternative means for detecting and quantitating the abundance of the tracer in the reactor effluent.

7.3.2 Tracers

A succinct review of tracers is in order before delving further into tracer analyses. In order to characterize the signature RTDs of various ideal reactors, we first need to examine tracers and their use. A tracer is any substance that may be used to characterize the flow and mixing regime (the hydrodynamics) within a reactor. Ideal tracers have the following attributes:

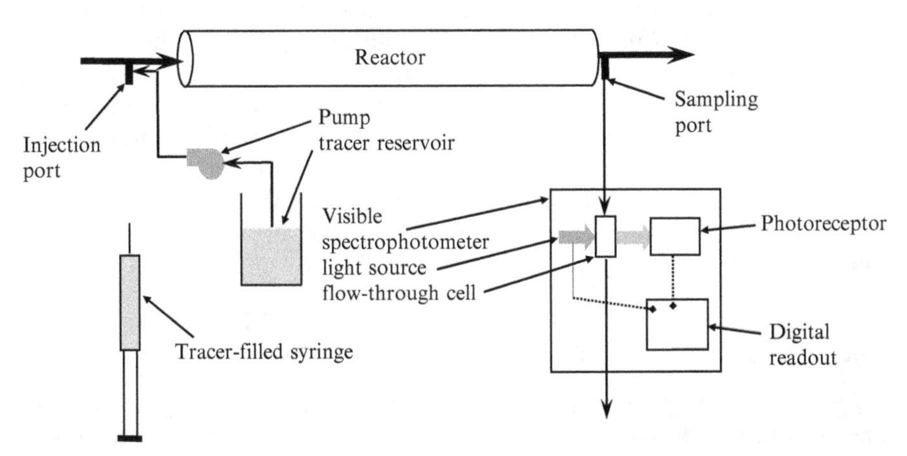

FIGURE 7.2 Schematic diagram of an apparatus for conduct of residence time distribution analyses using either impulse or step input of tracer.

1. Tracers become intimately mixed within transporting fluids. Tracers completely dissolve into liquids or become completely dispersed within gases that enter into and exit from reactors for which RTD analyses are performed.
2. Tracers can be identified and quantitated within the transporting medium. We not only can identify the presence of a tracer but we also can determine the abundance of the tracer within the transporting fluid.
3. Tracers are conservative. Reactions that might occur between tracers and the contents of a reactor are insignificant. Tracers are not significantly transformed from one form to another when present in reactors.
4. Tracers do not interact with the interior of the reactors. Often reactors consist of porous media wherein flow through the reactor occurs within the pore spaces bounded by the particle surfaces. Tracers do not interact with these surfaces nor do tracers interact with the interior surfaces of the reactor itself if the reactor in fact has physical confining boundaries.
5. Tracers most often are not normally present in the systems in which they are used to characterize.

As a consequence of these attributes of tracers, the generation term included in the mass balance statement of Figure 7.1 is reduced to 0. Then, in the absence of transformation, we may bank on the idea that the rate at which a tracer enters a reactor less the rate at which the tracer leaves the reactor must result in the rate at which the tracer is accumulated in the reactor.

Substances used as tracers in environmental systems include dyes (e.g., methylene blue and rhodamines), alkali earth metal ions (e.g., lithium, sodium, and potassium), anions (e.g., halides), and various radioisotopes (e.g., tritiated water and ^{129}iodine). There is no perfect tracer. However, depending upon the character of a reactor and its contents, use of real tracers often introduces little error. In simple flow-through reactors through which aqueous solution is to flow without the presence of porous media, basic tracers are easily employed. In such cases, dyes work well as they are visual and can be easily detected and quantitated using visible or fluorescence spectrophotometry. As reactors become more and more complicated, the requirements of the tracer become more and more stringent. In many environmental systems, tritiated water is often the choice. Although radiation counting requires sophisticated equipment, radioactive tracers can be quantitated to very low abundance levels and certainly tritiated water behaves just as natural water would.

7.3.3 Tracer Input Stimuli

In general, when we perform a tracer analysis, we introduce the tracer into the fluid entering a reactor (the influent) and record the abundance of the tracer in samples of the outflow from the reactor (the effluent). Most often we consider the input stimulus to be ideal, but may in some cases determine it necessary to treat the impulse as real, necessitating characterization of tracer abundance versus time at the reactor influent. We generally introduce the tracer by one of two ideal methods: (1) an impulse or (2) a step.

7.3.3.1 Impulse Input Stimulus

At the bench scale, a typical means for introducing a tracer as an impulse might involve use of a syringe. The syringe is loaded with an appropriate quantity of the tracer and discharged directly into the flowing stream by rapidly depressing the plunger. In contrast to the bench scale syringe, for implementation of an impulse input for a suspended growth biological process at a wastewater plant, we might use a five-gallon bucket filled with a solution containing the tracer and simply dump the contents of the bucket into the flow at a point above the actual influent to the reactor. The tracer then "tags" a fluid element, which then enters the reactor. A perfect impulse is mathematically modeled as a Dirac delta function. This function has a magnitude of 0 for all values of time other than the exact time of the impulse. The area under the concentration versus time plot is unity. Then, as the impulse is modeled to have infinitesimal duration, the magnitude is visualized as infinite. The impulse input is represented as the leftmost arrow in the plot shown in Figure 7.3. The "ideal" plot represents what we might obtain if we were to flawlessly inject the contents of the syringe and flawlessly take samples of the influent to the reactor over a time period beginning prior to the input and ending sometime after the input. In reality, the duration of the impulse cannot be infinitesimal, but must be finite. As a consequence of the hydrodynamics of the zone of mixing and finite time period over which the impulse input is completed, the real result of the impulse input is shown on the right in the plot of Figure 7.3. The "real" case reflects the result of an input impulse that would have been well done and a near-perfect sampling and analysis program. In consideration of the use of a syringe, we note that the more rapidly the plunger is depressed, the narrower the impulse will be and the greater the magnitude of the peak will be. In our analyses, later in this chapter, characterizing ideal reactors,

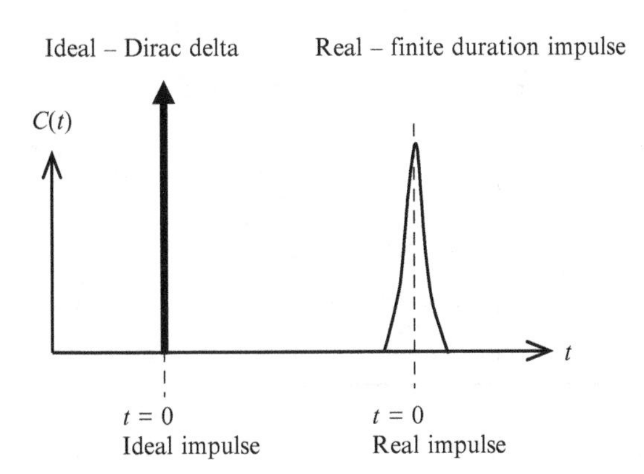

FIGURE 7.3 Ideal and real impulse inputs for tracer analyses.

we will consider that we can accomplish the ideal impulse input. In analyses of real reactors, we must account for the nonideality of the input and if the true result is desired we must characterize the time-dependent nature of the impulse input.

7.3.3.2 Step Input Stimulus

A typical means for introducing a tracer as a step input involves use of a pump and reservoir containing a concentrated solution of the tracer. This concentration is sufficiently high that dilution into the influent stream results in an easily measureable concentration. The positive "step" in theory instantaneously alters the tracer concentration from one level (perhaps zero) to another (nonzero). The step is invoked simply by turning the pump on to begin introduction of the tracer. The pumping rate is held constant and, thus, each subsequent element of the transporting fluid is "tagged" identically to the previous element. When the pump at some later time is turned off, theoretically instantaneously, the tracer concentration returns to 0, invoking a full negative step. The step counterparts to the ideal and real impulse inputs are shown pictorially in Figure 7.4. The "ideal" plot again represents what we would observe for a perfectly implemented positive step and subsequent perfectly implemented negative step through a perfectly accomplished sampling/analysis program. As perfection is impossible to attain, the "real" plot depicts what we might observe from a combination of well-executed positive and negative steps through a well-conducted sampling/analysis program. Well-executed positive steps will result in steep slopes and arcs of small radius at the beginnings and endings of the nearly vertical portions of the resultant concentration versus time plots. The use of a multi-speed pump or of multiple reservoirs containing solutions of varying tracer concentration and quick-acting valves to invoke the steps allows for implementation of partial positive and negative steps, sometimes quite useful in reactor characterizations. For our discussions of ideal reactors later in this chapter, we will consider that we are able to implement the ideal positive and negative steps. For characterizations of real reactors, consideration should be made for characterization of the time-variant nature of positive and negative step input stimuli.

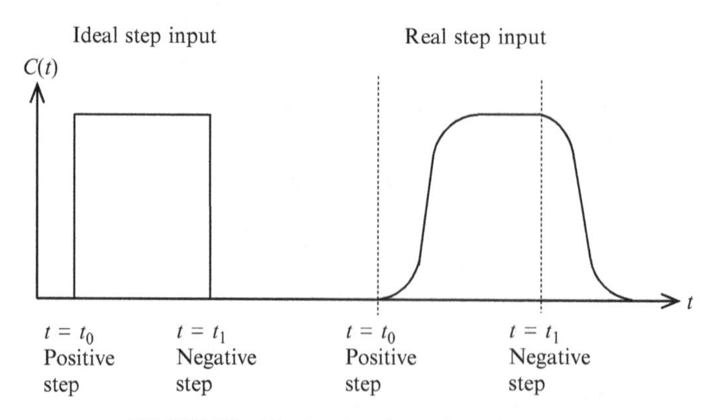

FIGURE 7.4 *Ideal and real step input functions.*

7.4 EXIT RESPONSES FOR IDEAL REACTORS

The exit response curve from a tracer test allows us to characterize the reactor. This response curve is often called the exit age distribution. Exit age distributions may be expressed in real concentrations and real times, in normalized concentrations and real times, and in concentrations and times that are both normalized. Often we normalize concentration and time in the same context that we identify dimensionless numbers for dimensional similitude—normalization renders the results applicable independent of the scales of time and concentration. A normalized concentration is determined in one of two manners, depending upon the input stimulus. To normalize concentration for an impulse, we compute the quotients of each of the real concentrations and the mass of tracer inputted. To normalize concentration for a step input, we compute the quotients of each real concentration and the value of the influent concentration for the most recent positive step. Time is normalized as the quotient of the real time and the hydraulic residence time (HRT). Normally, HRT is determined as the quotient of the reactor volume and influent flow, but for real reactors is it determined as the statistical mean residence time. Analyses of exit age distributions for real reactors are examined in Chapter 9. Herein we will concentrate on exit age distributions for two reactors: an ideal plug-flow reactor (PFR) and an ideal completely mixed flow reactor (CMFR) (also called a Continuously-Stirred Tank Reactor, CSTR).

7.4.1 The Ideal Plug-Flow Reactor (PFR)

Consider that in the laboratory we have fabricated an apparatus of configuration similar to that shown in Figure 7.2. We have included a long, slender cylinder, such as that shown in Figure 7.2, as the reactor. We have arranged this reactor to accept and discharge tap water at a known (adjustable) flow rate. The influent enters through one end and the effluent exits through the other. We have integrated the capability to input either an impulse or a step input of tracer just ahead of influent entry into the reactor. We will sample the effluent from the effluent pipe at a point just outside the downstream boundary of the reactor. This effluent sampling apparatus might be a visible spectrophotometer fitted with a flow-through cell through which we would continuously pass a small percentage of the reactor effluent. The digital readout from the spectrophotometer is routed to a computer for continuous recording of the output signal.

Consider that our reactor has a cylindrical configuration and is a perfect PFR. Then imagine that we can position ourselves inside that reactor and are able to see the individual elements of fluid entering the reactor. We would see each element instantaneously spreading across the entire cross-sectional area of the reactor upon entry. Each element is sufficiently large to cover the cross section of the reactor but sufficiently small that once spread the resultant disk would be very, very thin. We would then observe each element to traverse the length of the cylinder without mixing of its contents with the fluid in the element ahead of or behind the element under observation. The shape of each of these elements would remain constant as a thin disk covering the entire cross section. We would observe these disks to form one after another, each succeeding disk identical to the one before it, and to travel at a constant velocity through the reactor. Were we to envision fluid elements of successively smaller volume, we would observe the thickness of the disks to decrease in direct proportion to the decrease in the volume of the element under consideration, always covering the entire cross-sectional area. As the volume of the elements would approach 0, we would observe the cross-sectional area to remain constant as the cross-sectional area of the cylinder, and we would observe the thickness of the disks to become infinitesimal. Thus, an infinite number of disks would reside in the cylinder at any given time, all traversing the length of the cylinder at the same constant velocity. Were we to increase or decrease the rate of flow of the influent, we would observe the velocity of the disks to increase or decrease in direct proportion to the increase or decrease of the volumetric flow rate.

Now, let us consider that we are able to implement a perfect impulse input from the syringe and also that we may discharge the syringe in such a manner that we "mark" a single infinitesimally small element of fluid with bright red rhodamine dye. From our vantage point inside the PFR, stopwatch in hand, we observe the marked fluid element entering the reactor, spreading out across the area of the reactor immediately upon entering. We then would observe the marked disk making its way, along with the other unmarked disks, down the length of the reactor. We start the timer at the instant the marked fluid element enters and the disk forms within reactor. We measure the time required for the marked disk to travel through the reactor and stop the timer at the instant the marked disk disappears into the effluent pipe. We have measured the residence time of the fluid element in the reactor (HRT or τ). We have, of course, measured the length of the reactor (L_R) and can then relate the time of travel, the distance traveled, and the velocity of the marked fluid element (v_{FE}). We normally refer to the residence time as the HRT and often give it the symbol τ (many texts use the alternate symbol θ, but herein θ is normalized time):

$$\text{HRT (or } \tau) = \frac{L_R}{v_{FE}} \tag{7.1a}$$

Equation 7.1a is very specific to a reactor that is of constant cross-sectional area. However, if we multiply the numerator and denominator of Equation 7.1a by the cross-sectional area, A_C, we obtain the more general relation used to define HRT:

$$\text{HRT} = \frac{A_C \cdot L_R}{A_C \cdot v_{FE}} = \frac{V_R}{Q} \tag{7.1b}$$

where V_R is the volume of fluid held by the reactor and Q is the volumetric flow rate. Equation 7.1b can now be applied in cases where the reactor is not of constant cross-sectional area. In fact, Equation 7.1b can be applied to ***any reactor***, to relate the flow, volume, and nominal (or perhaps average) HRT.

Then, as we consider the journey of the marked disk through the reactor, we need to bear in mind one assumption that accompanies the definition of a PFR. Within that marked fluid element, we consider that the fluid is completely and vigorously mixed, such as that within an Erlenmeyer flask containing a magnetic stir bar when the apparatus is set to spin the stir bar at a high-angular velocity. We know that if we dripped a drop of dye into the stirred flask, we would see a virtually instantaneous dispersion of the dye throughout the content of the flask. We, of course, need to impose the shape of the disk upon our imaginings of the process we might observe within the Erlenmeyer flask.

Now, let us further imagine that we can instantaneously and perfectly measure the concentration of tracer in the effluent stream. Upon entry of the marked fluid element into the reactor influent, the effluent dye concentration is measured to be 0. The measured value continues to be 0 as the marked element traverses the length of the reactor. Then, immediately upon its exit, the recording device indicates a spike in the concentration lasting an infinitesimally short time period. Immediately following the spike, the concentration returns to 0 to remain at that level until such time as another input of tracer might traverse the reactor and exit. A pictorial representation of this impulse stimulus and associated ideal PFR response is shown in Figure 7.5. Here the Dirac delta function is represented for the perfect impulse and the perfect exit age response for a PFR. Were the input to be a real impulse, such as that shown in Figure 7.3, the theoretical exit response for an ideal PFR would be identical to the input impulse.

Let us now examine the step input using the pump, as shown in Figure 7.2. Prior to the initiation of the step, we would observe a situation identical to that previously described: fluid elements passing through the reactor as infinitesimally thick disks. The tracer pump is switched on, the tracer is instantaneously mixed with an element of fluid, and identically so with each successive element of fluid. We observe the same behavior as with the impulse stimulus. However, rather than a single disk of red color, we observe that every fluid element disk is now colored identically to the first. We watch the "plug" of red move through the reactor. Then, when the leading edge of the red dye plug is about halfway through the reactor, the tracer pump is cycled off. Coincident with this event, we observe the last of the marked disks of fluid enter the reactor and see that behind the trailing edge of the "plug" the fluid elements are all clear of red dye. We watch the leading and trailing edges of the "plug" traverse the

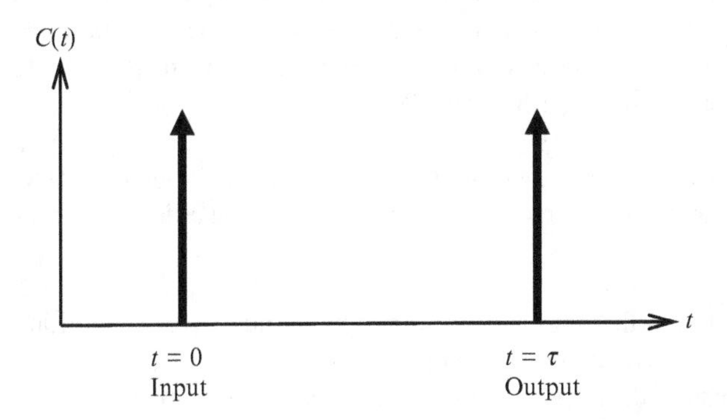

FIGURE 7.5 *Impulse input stimulus and exit response for an ideal plug flow reactor.*

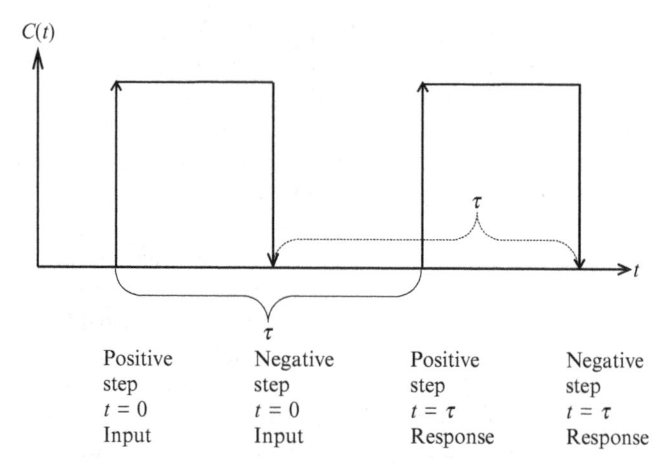

FIGURE 7.6 Positive and negative step input stimuli and exit responses for an ideal plug flow reactor.

reactor and note that both reside in the reactor a time equal to the reactor residence time. A pictorial representation of the step input stimulus and the associated ideal PFR response is shown in Figure 7.6.

7.4.2 The Ideal Completely Mixed Flow Reactor (CMFR)

Herein called a completely mixed flow reactor, the CMFR in many texts goes by a different name: the continuously stirred tank reactor (CSTR). In describing this reactor, we might return to the Erlenmeyer flask containing water, with the magnetic stir bar spinning at high angular velocity, and simply add a peristaltic pump with two identical pump heads. One head has its inlet immersed in a reservoir of tap water and its discharge directed into the flask. The other head has its inlet immersed in the contents of the flask and its outlet directed to a drain. Otherwise, the system is identical to that of Figure 7.2. We note that the influent is discharged at arbitrary location within the flask and that the intake for the effluent line is also located at an arbitrary location within the flask. Noting that the contents of the flask are vigorously (and ideally, completely) mixed, we realize that as long as the effluent intake and influent discharge lines are not directly connected, their locations matter not. This leads to an important assumption regarding the CMFR:

> The abundances of constituents residing in fluid anywhere within a CMFR are equal to the abundances of those constituents in fluid entering the effluent line emanating from a CMFR.

In consideration of the impulse and step input stimuli for introduction of tracer to an ideal CMFR, we would again take up our visualized residence inside the reactor.

Let us consider the impulse input first. Upon implementing the impulse, we would see an element of *marked* fluid entering the reactor. In much the same manner as for the PFR that marked element would spread through the reactor, not just near the

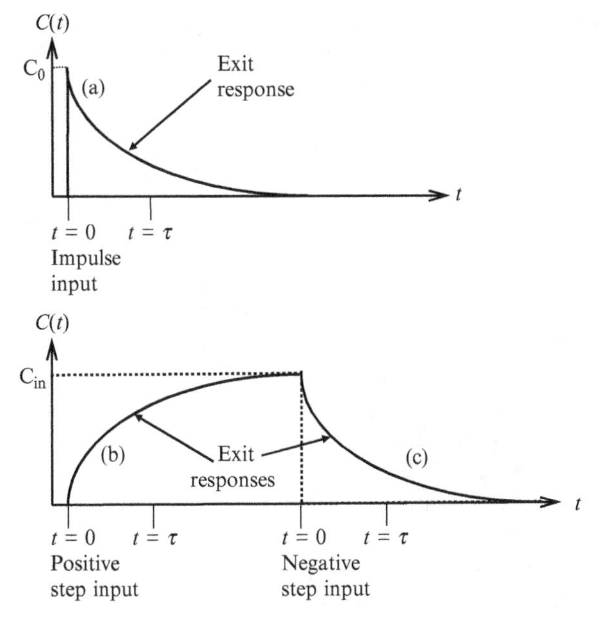

FIGURE 7.7 *CMFR exit responses for impulse (**a**) and positive (**b**) and negative (**c**) step input stimuli.*

influent, but throughout the entire reactor, instantaneously. Simultaneously, a quantity of fluid, yet unmarked, would exit the reactor through the effluent line. The next fluid element, *unmarked*, would enter and also be mixed throughout the reactor. A volume of the reactor contents equal to that of the second element would exit the reactor, carrying with it a portion of the dye that was mixed throughout the reactor by instantaneous dispersion of the marked element. Another *unmarked* element would enter the reactor while an equal volume of fluid would exit. The dye in this second element of exiting fluid would be of slightly lower concentration than that of the first exiting volume, owing to the dilution of the reactor contents by the entry of the second unmarked fluid element. And so it would go, with each new element of unmarked fluid entering the reactor, an equal volume would exit, with each exiting volume slightly diluted from that exiting immediately previously. Given sufficient time, the concentration of dye in the reactor and, hence, in the effluent from the reactor would eventually return to its initial value of 0. A pictorial representation of the impulse input stimulus and associated exit response for a CMFR is shown in Figure 7.7**a**.

We will now consider the step input, implemented exactly as we have described for the PFR. From our vantage point within the reactor, we would observe the first element of *marked* fluid entering the reactor to be dispersed throughout the entire volume of the reactor. Simultaneously, we would observe an equal volume of yet *unmarked* fluid exiting via the effluent line. With the introduction of the second *marked* element, we would observe an equal volume of fluid exiting. This fluid would be of concentration equal to the diluted value resulting from mixing of the first element throughout the entire reactor volume. Associated with the second entering *marked* fluid element, we would observe an equal volume of the reactor contents exiting. This element would be of concentration slightly higher than that of the first exiting volume owing to the dilution, throughout the reactor, of additional dye introduced with the second *marked* fluid element. And so it would go, each element of marked fluid would bring additional dye and each volume of exiting fluid would contain dye at a slightly higher concentration, until such time as the contents of the reactor contained dye at a concentration equal to that of the *marked* fluid elements entering the reactor. A representation of the exit response generated from a CMFR for a positive step input is shown in Figure 7.7**b**.

Then, sometime after the concentration of dye in the fluid within the reactor reaches the level of the *marked* influent elements, we will implement the negative step. Again, from our vantage within the reactor, we would observe *unmarked* fluid elements entering the reactor, diluting the contents of the reactor and causing each subsequent volume of effluent to be of concentration lower than that of the previously exiting volume, until such time as the concentration would return to 0. The concentration versus time trace associated with the observations is shown pictorially in Figure 7.7**c**. Beyond the initial spike, the exit concentration trace of Figure 7.7**c** would be identical in every way to that of Figure 7.7**a**. In the subsequent section addressing mixing, we will employ the mass balance to develop specific relations that can be used to correlate time-variant output responses to input stimuli for CMFRs.

7.4.3 The Ideal (Completely Mixed) Batch Reactor (CMBR)

As a visualization of the CMBR, we can return to the Erlenmeyer flask with stir bar on the magnetic stir plate. We will not add the influent and effluent lines. From a tracer standpoint, the tracer analysis is trivial. Upon adding an impulse input, the dye would be immediately dispersed and simply would reside in the reactor at the initially diluted concentration. A step input would result in immediate dilution of the influent throughout the reactor, with a steadily increasing concentration, and of course a corresponding steadily increasing volume as the batch reactor has no effluent flow. What we would have is a fed-batch reactor. Similarly to the CMFR, the contents of the CMBR are everywhere the same and it matters not where we might sample the contents to discern the abundance of constituents of interest within the reactor.

7.5 MODELING OF MIXING IN IDEAL CMFRs

7.5.1 Zero-Volume Applications

Zero-volume mixing computations might be applied to various environmental systems, including (but certainly not limited to)

- The confluence of two pipes in an industrial facility
- The confluence of two natural water courses
- The confluence of a natural watercourse with a discharge from a wastewater treatment plant or with a planned or an unplanned runoff event

The confluence of infiltrating precipitation with ground water
The combination of various gaseous inputs into a room full of equipment or people

For zero-volume mixing, let us consider that the volume of a reactor is small relative to the combination of the influent flow rates. We would have no fewer than two or perhaps several flow streams that would be mixed to form a single stream. These streams need not be physically mixed prior to entry into the CMFR in order to apply the following modeling approach. Let us begin our investigation with two streams and consider a system such as that shown in Figure 7.2. We desire to compute the concentration of tracer in the mixed influent stream based on the concentration of tracer in the reservoir, the flow rate of tap water, and the volumetric pumping rate. The statement of mass balance is our beginning point. We may shorten the statement to four keywords (all referring to mass rates) and an equal sign:

$$\text{Accumulation} = \text{in} - \text{out} + \text{generation}$$

With a reactor volume that is small relative to the combination of influent flow rates, the system attains the steady-state condition rather rapidly. From Figure 7.7b, we observe that, after sufficient elapsed time, the exit concentration reaches the level of the influent. No further input of tracer at the level of the step will result in increases in the exit concentration. The system at this stage has reached its steady-state condition. The system will remain in this condition until the concentration of the tracer in the influent flow is changed either by raising or lowering the flow rate (Q) of tap water, by raising or lowering the volumetric rate at which tracer is pumped, or by raising or lowering the concentration (C) of tracer in the tracer reservoir. Since the exit concentration is steady over time and if the volume of the reactor remains constant (as it will for the vast majority of applications of mixing principles in environmental process analysis), the mass of tracer (M_{Tr}) in the reactor is constant over time. Since M_{Tr} is constant, mathematically, the accumulation term becomes 0:

$$\frac{dM_{\text{Tr}}}{dt} = 0$$

Also, since we are employing a conservative tracer, the rate of transformation of the tracer to become an alternative constituent is 0. Considering multiple influent flows, the mass balance equation reduces to

$$0 = \sum \text{in} - \text{out}$$

We apply the mass balance equation to both the total fluid flow and to the tracer dissolved in the fluid, and put symbology to the equations resulting in a pair of mathematical relations:

$$0 = Q_1 + Q_2 - Q_{\text{Tot}} \text{ and } 0 = Q_1 \cdot C_1 + Q_2 \cdot C_2 - Q_{\text{Tot}} \cdot C_{\text{out}}$$

The subscripts 1 and 2 refer to the tracer and tap water flow, respectively. We rearrange the relations to isolate Q_{Tot} and $C_{\text{out}} \cdot C_{\text{out}}$ is the concentration of tracer in the line downstream from the control volume comprising the tee joining the two lines and bounded by plane surfaces oriented normal to each of the combined flow streams:

$$Q_{\text{Tot}} = Q_1 + Q_2 \text{ and } C_{\text{out}} = \frac{Q_1 \cdot C_1 + Q_2 \cdot C_2}{Q_1 + Q_2}$$

We realize that we may generalize the aforementioned result to include multiple (n) influent streams that are combined to form a single-output stream:

$$Q_{\text{Tot}} = \sum_{i=1}^{n} Q_i \tag{7.2}$$

$$C_{\text{out}} = \sum_{i=1}^{n} \frac{Q_i \cdot C_i}{Q_{\text{Tot}}} = \frac{\sum_{i=1}^{n} Q_i \cdot C_i}{\sum_{i=1}^{n} Q_i} \tag{7.3}$$

Let us apply Equations 7.2 and 7.3 in the context of the tracer test.

Example 7.1 Consider that the tracer reservoir of Figure 7.2 contains solution of tracer at a concentration level of 1 g/L, that the volume of the laboratory reactor is 10L, that the desired HRT is 20 min, and that the desired tracer concentration in the influent flow is 5 mg/L. Determine the flow of tap water and the flow at which tracer solution must be pumped.

We first enter our important known information assigned into MathCAD variables:

$C_{Tr} := 1000 \quad C_{in} := 5 \ \frac{mg}{L} \quad \tau := 20 \ min \quad V_R := 10 \ L$

We compute the desired total flow rate:

$Q_{Tot} := \frac{V_R}{\tau} = 0.5 \ \frac{L}{min}$

We then implement Equations 7.2 and 7.3, but let us begin with and populate the general mass balance statement:

$0 = C_{Tr} \cdot Q_{Tr} + C_w \cdot Q_w - C_{in} \cdot Q_{Tot}$

We note that since we are using tap water we would presume that its tracer concentration is 0, thus the second term disappears. We may rearrange the relations as necessary to solve for the flow rates of tracer solution and water:

$Q_{Tr} := \frac{C_{in} \cdot Q_{Tot}}{C_{Tr}} = 2.5 \times 10^{-3} \ \frac{L}{min} \quad Q_w := Q_{Tot} - Q_{Tr} = 0.4975 \ \frac{L}{min}$

Example 7.2 For the tracer pump of Example 7.1, let us consider that the pump speed is constant and we have but a single size of tubing available for the pump head. We measure the output from the pump using a beaker, which we weigh empty, fill for 1 min, and weigh again to determine the mass of water pumped. The average result, based on five tries, is 10.34 g_w/min. We desire that the flow rate of water from the tap be 0.35 L/min and that the influent tracer concentration be 10 mg/L. We have an atomic absorption spectrometer available for the overall test that we plan to conduct and decide to use potassium as a tracer. We know that the tap water contains 0.65 mg/L potassium. We wish to determine the concentration of potassium needed in the tracer reservoir and will also eventually need to know the HRT of the reactor.

We search the tables in the Appendix of our fluids text and find that at the normal temperature of the laboratory (~22 °C), water has a density of 0.997 g/mL (997 g/L). We assign that item of information and others from the problem statement into appropriate MathCAD variables:

$$Q_w := 0.35\ \frac{L}{min} \qquad C_w := 0.65\quad C_{in} := 10\ \frac{mg_K}{L} \qquad \rho_w := 997\ \frac{g_w}{L}$$

We may compute the volumetric flow rate of the tracer and the total volumetric rate using mass units and use the density of water to obtain the volumes:

$$M_{F.Tr} := 10.34\ \frac{g_w}{min} \qquad Q_{Tr} := \frac{M_{F.Tr}}{\rho_w} = 0.0104\ \frac{L}{min}$$

$$Q_{Tot} := Q_w + Q_{Tr} = 0.3604\ \frac{L}{min}$$

The statements of the overall mass balance can then be rearranged to solve for the concentration of potassium necessary in the tracer reservoir to match the desired conditions:

$$C_{Tr} := \frac{Q_{Tot} \cdot C_{in} - Q_w \cdot C_w}{Q_{Tr}} = 325.54\ \frac{mg_K}{L}$$

The HRT of the reactor may then be computed:

$$V_R := 10\ L \qquad \tau := \frac{V_R}{Q_{Tot}} = 27.749\ min$$

The zero-volume mixing concept can be readily applied to the confluence of two surface water streams. In order to invoke insignificant error, we must assume and verify that the reactor (often called the mixing zone) with its arbitrary, imagined boundaries is small relative to the influent and effluent flows. We can examine the HRT to obtain a feel for the level of truth in this assumption. Let us examine a typical mixing application—that of a discharge from a wastewater treatment plant joining a flowing watercourse. The discharge from the wastewater renovation facility operated by the City of Rapid City, SD, into Rapid Creek provides a quality example. Rapid

Creek begins in the higher elevations of the Black Hills, tumbles down out of the hills (hence the name), and as it flows through Rapid City is transformed from a turbulent mountain stream to a meandering prairie stream. Two artificial reservoirs in the Black Hills store water and allow the flow of the creek to be well controlled over most periods of the year. Much of the water is spoken for by consumptive water rights and used either for Rapid City's municipal water supply or for crop irrigation adjacent to the creek. The critical flow condition relative to the wastewater discharge occurs in the late summer of dry years when the flow in the creek above the wastewater discharge becomes as low as 5 ft^3/s (0.15 m^3/s), with an average in-stream velocity of 0.5 ft/s (0.15 m/s). As a consequence of drainage from pasture lands and irrigation return flows above the wastewater discharge, the water quality of the creek is slightly impaired even above its confluence with the wastewater plant discharge. Under the critical conditions, the creek may contain perhaps 0.15 mg/L NH_3-N and have a measured biochemical oxygen demand of 3 mg/L. Further, due to the presence of organic-rich sediments in pools, biological activity within the creek renders the dissolved oxygen (DO) to be perhaps as low as 90% of the saturation value. Under these critical conditions, the temperature might be as high as 75 °F (24 °C).

Example 7.3 Given the foregoing discussion of the character of Rapid Creek, examine the applicability of the zero-volume mixing assumption. Then based on the critical character of the discharge, compute the concentrations of the critical components associated with oxygen demand of the mixed stream on the downstream segments of the creek. From the wastewater renovation facility, the daily flow is 8.5 MGD (million gallons per day, 0.372 m^3/s), the effluent five-day biochemical oxygen demand (BOD_5) is typically 10 mg/L, the effluent NH_3-N is typically 1.0 mg/L, the temperature is 22 °C, and the DO saturation level of the discharge is 80%. At an elevation of 3400 ft above mean sea level, the normal atmospheric pressure at Rapid City is 0.88 atm (see Example 5.3).

Our first operation in performing these analyses is the construction of a detailed sketch of the system, shown in Figure E7.3.1. We are sure to indicate the flows and the reactor. We will then assign known parameters to appropriate symbology:

$$Q_{RCr} := 0.15 \quad Q_{ww} := 0.372 \quad \frac{m^3}{s}$$

$$NH3N_{RCr} := 0.15 \quad NH3N_{ww} := 1.0 \quad \frac{mg_{NH3N}}{L}$$

$$BOD_{RCr} := 3 \quad BOD_{ww} := 10 \quad \frac{mg_{BOD5}}{L} \quad u_{RCr} := 0.15 \quad \frac{m}{s}$$

In order to determine the DO levels in the creek above the confluence and in the discharged wastewater, we will need either a value of the Henry's constant for air/water distribution of oxygen at the temperatures of the two streams or a table of DO values at various temperatures. The US Geological Survey maintains an oxygen solubility table (USGS, 2013) from which saturation values may be obtained based upon temperature, ambient pressure, and salinity. For convenience, Table A.3 has

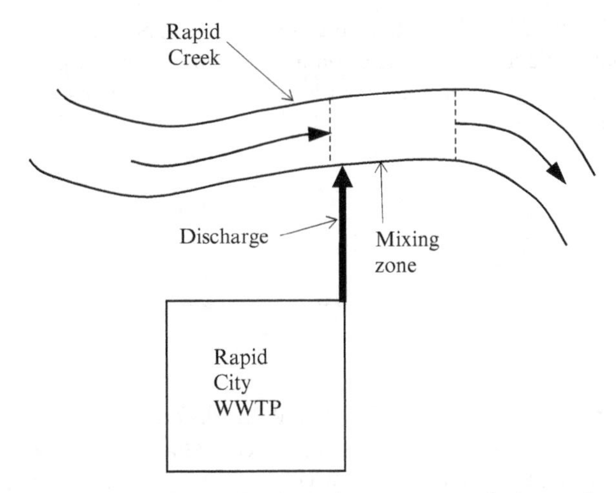

FIGURE E7.3.1 *Sketch of Rapid Creek and wastewater discharge for mixing zone computations.*

been created from that website for inclusion in this text. From this Table A.3, we find the saturation values for the two temperatures of interest and adjust for the barometric pressure (see Example 5.3) by taking the ratio of the ambient and standard pressures as a multiplier for the DO concentration from the table, which we have obtained for the stated temperature and standard pressure of 1 atm:

$SatFrac_{RC} := 0.9 \qquad SatFrac_{ww} := 0.8$

$DO_{std.22} := 8.74 \qquad DO_{std.24} := 8.42 \quad \frac{mg}{L} \qquad P_T := 0.88 \qquad P_{STD} := 1 \quad atm$

$DO_{RCr} := DO_{std.24} \cdot \frac{P_T}{P_{STD}} \cdot SatFrac_{RC} = 6.669 \quad \frac{mg}{L}$

$DO_{ww} := DO_{std.22} \cdot \frac{P_T}{P_{STD}} \cdot SatFrac_{ww} = 6.153$

The mixing zone for the wastewater discharge might comprise 100 m of the stream. The residence time of the resultant reactor would be computed as the quotient of the length of the zone and the flow velocity:

$L_{mz} := 100 \quad m \qquad \tau_{mz} := \frac{L_{mz}}{u_{RCr}} = 667 \quad sec$

A 10 min residence time is not long. We can estimate the volume of the reactor as the product of flow and residence time:

$V_{mz} := (Q_{ww} + Q_{RCr}) \cdot \tau_{mz} = 348 \qquad m^3$

We can be quite confident that little reaction occurs over this short residence time and our neglect of the generation and accumulation terms is justifiable. We may then compute the concentrations of the three constituents in the flow stream leaving the mixing zone:

$$\begin{pmatrix} NH3N_{mz} \\ BOD_{mz} \\ DO_{mz} \end{pmatrix} := \frac{\left[Q_{ww} \cdot \begin{pmatrix} NH3N_{ww} \\ BOD_{ww} \\ DO_{ww} \end{pmatrix} + Q_{RCr} \cdot \begin{pmatrix} NH3N_{RCr} \\ BOD_{RCr} \\ DO_{RCr} \end{pmatrix} \right]}{Q_{Tot}} = \begin{pmatrix} 0.756 \\ 7.989 \\ 6.301 \end{pmatrix} \quad \frac{mg}{L}$$

We might be interested in the DO deficit, the difference between the actual DO concentration level and the saturation value. The saturation level would be the DO of an aqueous solution in equilibrium with the atmosphere at the temperature of the mixed stream. We would need to determine the temperature of the mixed solution. Some transfer of heat into or from the solution would undoubtedly occur, but let us suggest that in the short residence time of the mixing zone this input or output of heat might be negligible. We can then estimate the temperature of the mixture just as we have estimated the mixed concentrations of the three targeted constituents. We write an energy balance, which is similar to the mass balance, using the mass flow rates, the thermal heat capacities and the temperatures of the streams:

$$\dot{M}_{ww} C_{p.ww} T_{ww} + \dot{M}_{RC} C_{p.RC} T_{RC} = \dot{M}_{mix} C_{p.mix} T_{mix}$$

where $\dot{M}$ is the product $Q \cdot \rho$. We might assume that the heat capacity of the wastewater stream is the same as that for the water of Rapid Creek, and also that the densities of the aqueous solutions of both streams are also of the same magnitude. We can then divide out the heat capacities and divide each term by the density of water, yielding a relation written in flow and temperature, much like that written for concentration:

$$T_{RCr} := 24 \quad T_{ww} := 22 \quad T_{mix} := \frac{Q_{ww} \cdot T_{ww} + Q_{RCr} \cdot T_{RCr}}{Q_{ww} + Q_{RCr}} = 22.6 \quad {}^{\circ}C$$

For the standard condition value of $DO_{sat,}$ we would interpolate from the DO table:

$$DO_{sat.22.6} := \frac{(22.6 - 22)}{23 - 22} \cdot (8.58 - 8.74) + 8.74 = 8.64 \quad \frac{mg}{L}$$

$$Def_{DO} := DO_{sat.22.6} - DO_{mz} = 2.34 \quad \frac{mg}{L} \qquad \%DO_{sat} := \frac{DO_{mz}}{DO_{sat.22.6}} \cdot 100 = 72.9$$

This result provides the initial or boundary condition, depending upon how we would wish to view the system, to model the DO condition downstream of the wastewater treatment discharge. Such modeling combined with beneficial use designations allows environmental regulatory agencies defensible means to set limits for various parameters in wastewater discharge effluents.

7.5.2 Time-Dependent Mixing

For time-dependent mixing, we retain the accumulation term of Figure 7.1 but for mixing-only computations consider that the component or components of interest are not transformed, hence the generation term of Figure 7.1 is not included. We employ the mass balance:

$$\text{Accumulation} = \Sigma\,\text{in} - \text{out}$$

Let us first examine a single influent stream and then generalize to several. We will assume that the reactor has constant volume and that the volumetric influent flow and hence the effluent flow are both constant. We will also assume that the reactor is subjected to a step input (either negative or positive) such that for a defined time period the concentration of the component of interest in the influent flow stream remains constant. The mass balance equation is then written for arbitrary component i (which can be any component of interest):

$$\frac{dM_i(t)}{dt} = \frac{d[V_\text{R} \cdot C_i(t)]}{dt} = Q \cdot C_{i.\text{in}} - Q \cdot C_{i.\text{out}} \tag{7.4}$$

The mass of the component in the reactor at time t, $M_i(t)$, is defined as the product $V_\text{R} \cdot C_i(t)$. The accumulation of mass is merely the time rate of change of the mass. Two simplifications arise immediately. When we apply the product rule to the LHS derivative, we note that $\frac{dV_\text{R}}{dt} = 0$ and that associated with the perfect mixing assumption $C_i(t) = C_{i.\text{out}}$. We make these simplifications, drop the i subscript, note that C is $C(t)$, and restate a much simpler form of Equation 7.4:

$$V_\text{R} \cdot \frac{dC}{dt} = Q \cdot C_\text{in} - Q \cdot C \tag{7.4a}$$

This resultant first-order ordinary differential equation is easily separated:

$$\frac{dC}{C_\text{in} - C} = \frac{Q}{V_\text{R}} \cdot dt$$

We note that at the time of the step, t_0, the concentration of our target component in the reactor is C_0. We may then integrate each side of the equation between the limits of t_0 and t with the corresponding concentration limits C_0 and C. We have inverted the argument of the natural logarithmic term to eliminate the negative sign arising from the integration:

$$\ln\left(\frac{C_{\text{in}} - C_0}{C_{\text{in}} - C}\right) = \frac{Q}{V_{\text{R}}} \cdot (t - t_0) = \frac{t - t_0}{\tau} \tag{7.5}$$

For many applications, we would prefer the exponentiated form of Equation 7.5:

$$\frac{C_{\text{in}} - C_0}{C_{\text{in}} - C} = \exp\left[\frac{Q}{V_{\text{R}}} \cdot (t - t_0)\right] = \exp\left[\frac{t - t_0}{\tau}\right] \tag{7.6}$$

Here we might further revise Equations 7.5 and 7.6 to special cases: positive or negative steps, zero initial concentration of the component of interest in the reactor, or impulse inputs. Other texts certainly have done so. Then we must memorize each of the cases. Herein, since each of these cases can be easily implemented by employing known information and employing the general relations, we will always begin with the general form (either Equation 7.5 or 7.6) in performing analyses. We need only sketch a diagram of the system, define the initial state of the reactor, and understand the type of input implemented.

Example 7.4 Consider a $0.1\,\text{m}^3$ (100 L, ~0.3 m in depth and ~$0.6\,\text{m}^2$) laboratory reactor that is outfitted with interior baffles and a powerful mixer to ensure a virtually completely mixed hydrodynamic regime within the reactor. The remainder of the system would conform with that of Figure 7.2. Consider that the tracer pump and tracer reservoir concentration are such that the rate of flow of tracer solution is insignificant relative to the flow of clear water from the tap. The influent flow rate then is unchanged with either a positive or negative input step. Let us keep the numbers simple and suggest that the influent flow rate is 1 L/min and that with a positive step the influent tracer concentration rises instantaneously to 10 mg/L. Let us ask the question: With a positive step, how long will it take for the concentration of the tracer in the effluent to reach a level of 5 mg/L? We might also ask the questions: What would the effluent tracer concentration be after elapsed time equal to the reactor residence time, what would the steady-state effluent tracer concentration be, and how long will it take for the reactor system to reach steady state?

Let us first assign some parameter values, define the initial condition of the reactor, rearrange Equation 7.6, write a MathCAD function, and plot the expected concentration versus time.

$$Q := 1 \ \frac{L}{min} \quad V_R := 100 \ L \quad \tau := \frac{V_R}{Q} = 100 \ min$$

$$t_0 := 0 \ min \quad C_0 := 0 \quad C_{in} := 10 \quad C_{end} := 5 \ \frac{mg}{L} \quad t := 0, 1 .. 100$$

$$C(t) := C_{in} - \frac{C_{in} - C_0}{\exp\left(\frac{t - t_0}{\tau}\right)}$$

We observe that the reactor, and hence the effluent concentration, will reach 5 mg/L in about 70 min, as can be observed from Figure E7.4.1. To find the time, we would employ Equation 7.5 rearranged to explicitly solve for t_{end}:

$$t_{end} := t_0 + \tau \cdot \ln\left(\frac{C_{in} - C_0}{C_{in} - C_{end}}\right) = 69.3 \ min$$

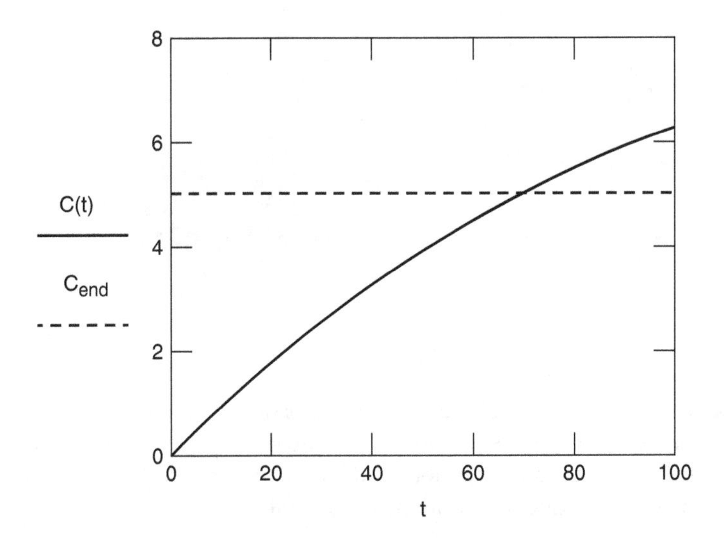

FIGURE E7.4.1 *Plot of time-variant reactor and reactor effluent concentration.*

At the reactor residence time of 100 min, the effluent tracer concentration can be found from the plotted function:

$$C(\tau) = 6.321 \ \frac{mg}{L}$$

We certainly could have inserted the value of τ for t in the rearranged relation and solved for the single value:

$$C_\tau := C_{in} - \frac{C_{in} - C_0}{\exp\left(\frac{\tau - t_0}{\tau}\right)} = 6.321$$

For the steady-state concentration, we must revert to the mass balance, eliminating the accumulation term, with the resulting relationship that the rate of mass flow in must equal that out, such that $Q \cdot C_{in} = Q \cdot C_{out}$, and hence $C_{in} = C_{out}$. Thus, in this case $C_{ss} = C_{in}$, presuming that the tracer pump remains on for a sufficient time period:

$$C_{ss} := 10 \ \frac{mg}{L}$$

In consideration of the time necessary to reach steady state, we quickly observe that if $C_{out} \rightarrow C_{in}$, the denominator of the argument of the ln() term of the equation used for t_{end} approaches 0, rendering the solution undefined. Then, let us compute the time to reach 99, 99.9, and 99.99% of the steady-state concentration. We have employed MathCAD's very convenient matrix capacity and "vectorized" the RHS of the critical relation to allow computation and output of results using matrices:

$$\begin{pmatrix} C_{99} \\ C_{99.9} \\ C_{99.99} \end{pmatrix} := C_{ss} \cdot \begin{pmatrix} 0.99 \\ 0.999 \\ 0.9999 \end{pmatrix} = \begin{pmatrix} 9.9 \\ 9.99 \\ 9.999 \end{pmatrix} \frac{mg}{L}$$

$$\begin{pmatrix} t_{99} \\ t_{99.9} \\ t_{99.99} \end{pmatrix} := \overrightarrow{\left[t_0 + \tau \cdot \ln\left[\frac{C_{in} - C_0}{C_{in} - \begin{pmatrix} C_{99} \\ C_{99.9} \\ C_{99.99} \end{pmatrix}} \right] \right]} = \begin{pmatrix} 461 \\ 691 \\ 921 \end{pmatrix} \text{min}$$

We would need to exercise a good bit of patience in our wait for steady-state conditions. Our mathematical model predicts that we would asymptotically approach the steady-state condition. Perhaps attainment of 99.99% or even 99.9% of the steady-state exit concentration is sufficient to call it a steady state.

We should investigate one additional application of the laboratory apparatus—that of the positive step followed by a negative step—before moving into environmental systems.

Example 7.5 Consider the laboratory system of Example 7.4. Suppose the reactor is flushed for sufficient time to completely expel our target component. Consider that the influent flow rate is established at $t = 0$ and the tracer pump is actuated at $t = 10$ min. The mixed concentration, influent to the reactor, instantaneously rises to a level of 10 mg/L as a consequence of the mixing of the tracer and tap water flows. Then at $t = 90$ min, the tracer pump is turned off, returning the influent concentration to 0. Develop the mathematical model with which the time-dependent behavior of the reactor effluent concentration can be predicted and graphically displayed. Then graphically display the results.

Let us first write a function that allows the presentation of the influent concentration (i.e., the step) as a continuous function of time. A useful and powerful capability of MathCAD is illustrated, embedding logical decision-making capability into the $C_{in}(t)$ function:

$$\begin{pmatrix} t_{PS} \\ C_{mix} \\ t_{NS} \end{pmatrix} := \begin{pmatrix} 10 \\ 10 \\ 90 \end{pmatrix} \begin{pmatrix} \text{min} \\ \frac{\text{mg}}{\text{L}} \\ \text{min} \end{pmatrix} \qquad C_{in}(t) := \begin{cases} 0 & \text{if } t \le t_{PS} \\ C_{mix} & \text{if } t_{PS} < t \le t_{NS} \\ 0 & \text{otherwise} \end{cases}$$

We will observe this function in the subsequent final plot:

Now we employ a rearranged form of Equation 7.6. We write an additional function for the effluent concentration based on the reactor condition at time $t = 10$ min. We need a defined influent concentration associated with the positive step:

$$C_{0.PS} := 0 \qquad C_{out.PS}(t) := \begin{cases} \left(C_{in}(t) - \dfrac{C_{in}(t) - C_{0.PS}}{\exp\left(\dfrac{t - t_{PS}}{\tau}\right)} \right) & \text{if } t \le t_{NS} \\ \text{NaN} & \text{otherwise} \end{cases} \quad \frac{\text{mg}}{\text{L}}$$

This function yields quantitative values for times up to t_{NS} and produces no values thereafter. We write a second function applicable to the time period subsequent to the negative step. No quantitative values are generated for times earlier than t_{NS}. We define the initial condition of the reactor based on the concentration in the reactor at t_{NS}:

$$C_{0.NS} := C_{out.PS}(t_{NS}) = 5.507 \qquad C_{out.NS}(t) := \begin{cases} \text{NaN} & \text{if } t \le t_{NS} \\ \left(C_{in}(t) - \dfrac{C_{in}(t) - C_{0.NS}}{\exp\left(\dfrac{t - t_{NS}}{\tau}\right)} \right) & \text{otherwise} \end{cases}$$

We plot all three functions on a single set of axes in Figure E7.5.1 to obtain the pictorial end result.

Were we to investigate further positive steps and negative steps that might be either full or partial steps relative to these first two, we would observe that each new time period following a step begins with the reactor concentration at the end of the previous step and follows as would be predicted by Equation 7.6. The MathCAD capabilities that produced the $C_{out.PS}(t)$ and $C_{out.NS}(t)$ functions are simply extended to produce four additional MathCAD functions for the desired plot. The function that produces the $C_{in}(t)$, shown in Figure E7.5.2, is an extension of

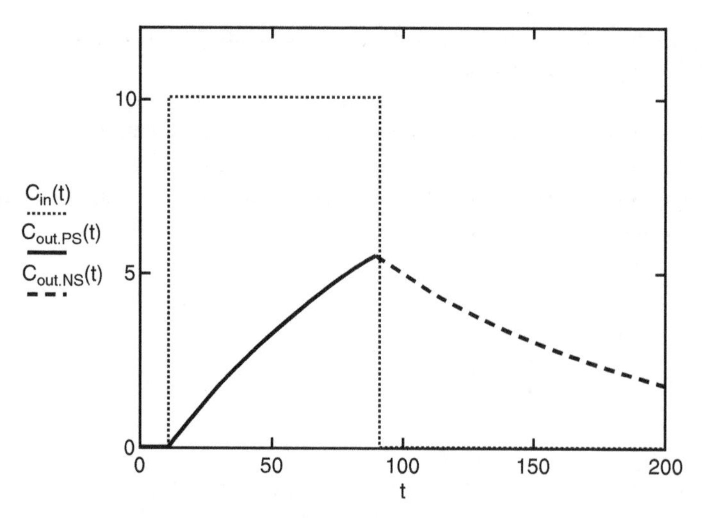

FIGURE E7.5.1 *Plot of input positive and negative step and associated exit responses for a CMFR.*

$$\begin{pmatrix} t_{PS1} \\ C_{mix1} \\ t_{NS1} \\ t_{PS2} \\ C_{mix2} \\ t_{NS2} \\ t_{PS3} \\ C_{mix3} \\ t_{NS3} \end{pmatrix} := \begin{pmatrix} 10 \\ 10 \\ 90 \\ 140 \\ 8 \\ 210 \\ 290 \\ 12 \\ 420 \end{pmatrix} \begin{pmatrix} \text{min} \\ \dfrac{\text{mg}}{\text{L}} \\ \text{min} \end{pmatrix} \qquad C_{in}(t) := \begin{vmatrix} 0 \text{ if } t \le t_{PS1} \\ C_{mix1} \text{ if } t_{PS1} < t \le t_{NS1} \\ 0 \text{ if } t_{NS1} < t \le t_{PS2} \\ C_{mix2} \text{ if } t_{PS2} < t \le t_{NS2} \\ 0 \text{ if } t_{NS2} < t \le t_{PS3} \\ C_{mix3} \text{ if } t_{PS3} < t \le t_{NS3} \\ 0 \text{ otherwise} \end{vmatrix}$$

FIGURE E7.5.2 *Capture of the short logical program defining the input concentration level.*

the function used earlier. The functions that define the concentration of tracer in the outflow, $C_{out.1}(t)$ through $C_{out.6}(t)$, are written using the logical capability of MathCAD functions and are shown in Figure E7.5.3. The first five of these six distinct functions are used in turn to compute the distinct initial concentration applicable for the next step. These six distinct functions were combined into a single $C_{out}(t)$ function to simplify the ordinate axis labeling of figure E7.5.4. This overall function merely includes the six time-conditional statements from the individual functions into a single function with the individual initial concentration values computed prior to the overall function. A single function is written for the influent concentration based on the mixed influent concentrations and the timings of the positive and negative steps.

$$C_{0.1} := 0 \qquad C_{out.1}(t) := \begin{cases} \left(C_{in}(t) - \dfrac{C_{in}(t) - C_{0.1}}{\exp\left(\dfrac{t - t_{PS1}}{\tau}\right)} \right) & \text{if } t \le t_{NS1} \\ \text{NaN} & \text{otherwise} \end{cases}$$

$$C_{0.2} := C_{out.1}(t_{NS1}) \qquad C_{out.2}(t) := \begin{cases} \text{NaN} & \text{if } t \le t_{NS1} \\ \left(C_{in}(t) - \dfrac{C_{in}(t) - C_{0.2}}{\exp\left(\dfrac{t - t_{NS1}}{\tau}\right)} \right) & \text{if } t_{NS1} < t \le t_{PS2} \\ \text{NaN} & \text{otherwise} \end{cases}$$

$$C_{0.3} := C_{out.2}(t_{PS2}) \qquad C_{out.3}(t) := \begin{cases} \text{NaN} & \text{if } t_{PS2} < t \\ \left(C_{in}(t) - \dfrac{C_{in}(t) - C_{0.3}}{\exp\left(\dfrac{t - t_{PS2}}{\tau}\right)} \right) & \text{if } t_{PS2} < t \le t_{NS2} \\ \text{NaN} & \text{otherwise} \end{cases}$$

$$C_{0.4} := C_{out.3}(t_{NS2}) \qquad C_{out.4}(t) := \begin{cases} \text{NaN} & \text{if } t \le t_{NS2} \\ \left(C_{in}(t) - \dfrac{C_{in}(t) - C_{0.4}}{\exp\left(\dfrac{t - t_{NS2}}{\tau}\right)} \right) & \text{if } t_{NS2} < t \le t_{PS3} \\ \text{NaN} & \text{otherwise} \end{cases}$$

$$C_{0.5} := C_{out.4}(t_{PS3}) \qquad C_{out.5}(t) := \begin{cases} \text{NaN} & \text{if } t_{PS3} < t \\ \left(C_{in}(t) - \dfrac{C_{in}(t) - C_{0.5}}{\exp\left(\dfrac{t - t_{PS3}}{\tau}\right)} \right) & \text{if } t_{PS3} < t \le t_{NS3} \\ \text{NaN} & \text{otherwise} \end{cases}$$

$$C_{0.6} := C_{out.5}(t_{NS3}) \qquad C_{out.6}(t) := \begin{cases} \text{NaN} & \text{if } t \le t_{NS3} \\ \left(C_{in}(t) - \dfrac{C_{in}(t) - C_{0.6}}{\exp\left(\dfrac{t - t_{NS3}}{\tau}\right)} \right) & \text{otherwise} \end{cases}$$

FIGURE E7.5.3 *The logical programs created to produce graphical output for the plot of the system behavior illustrated in Figure E7.5.3.*

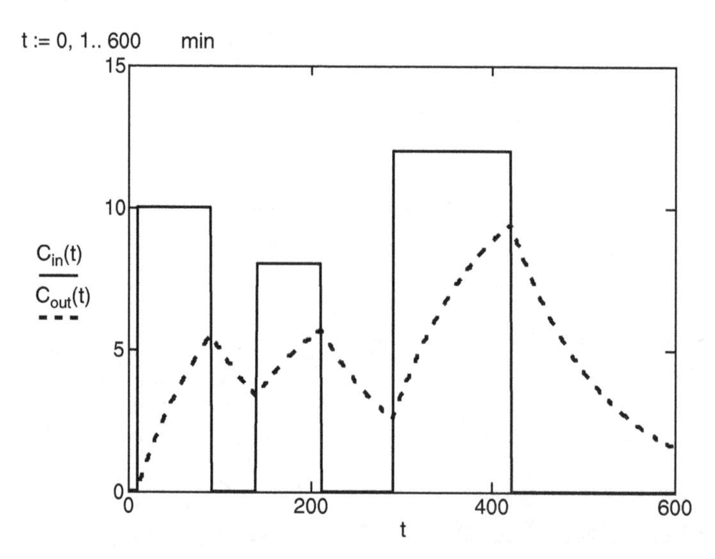

FIGURE E7.5.4 *A plot of a series of positive and negative step inputs and the resultant exit responses for a CMFR.*

Then in Figure E7.5.4 we plot the resultant functions ($C_{in}(t)$ and $C_{out}(t)$ to graphically illustrate the predicted response behavior of the reactor to a series of nonidentical positive/negative step inputs.

The remaining case yet to be considered is that of the impulse input. This case is easily addressed by noting the analogy between the impulse input and the negative step. Note that in Example 7.5 the initial tracer concentration in the reactor at the time of the negative step was computed as the output concentration arising from the influent concentration and reactor parameters for the previous positive step. For the impulse input stimulus, this initial concentration is computed as the quotient of the mass of tracer inputted and the volume of the reactor:

$$C_{0,\text{impulse}} = \frac{M_{\text{tracer}}}{V_{\text{R}}} \tag{7.7}$$

Then, the influent concentration is set to 0 for employment with either Equation 7.5 or 7.6.

7.6 APPLICATIONS OF CMFR MIXING PRINCIPLES IN ENVIRONMENTAL SYSTEMS

One context in which mixing is of great importance is in the analysis of the inputs of potentially toxic metals from inadvertent industrial discharges into collection systems of publically owned treatment works (POTWs). We can idealize the discharge stream as analogous to our combination of tracer pump and reservoir of the laboratory system of Figure 7.2. The remainder of the collection system is analogous to the tap water feed. We consider the industrial discharge to be nonreactive in the collection system and in the wastewater reactors of the POTW. We realize that a great deal of mixing and dilution occurs as the discharged industrial waste is transported by the wastewater to the POTW, but can idealize the overall process as a single stream containing the hazardous substance mixed with the entire wastewater flow immediately prior to the influent of the POTW.

Let us consider an industrial facility that employs cadmium (Cd) in one or more of its processes. The facility has an industrial waste pretreatment system designed to ensure that hazardous substances such as Cd are discharged to the wastewater collection system only at very low concentrations in streams of minimal flow. These, of course, can fail and we should understand the consequences of such failure.

Example 7.6 Cd is very toxic to biological life forms. If Cd is present in influent to a wastewater plant at sufficiently high concentrations, the biological treatment processes can be severely negatively impacted.

Consider that a valve in the process operated by the industry has malfunctioned, causing a 10 gal/min (6.31×10^{-4} m^3/s) liquid stream with a Cd concentration of 10 g/L to be routed to the sanitary sewer discharge from the facility. The total discharge from the facility is 250 gal/min (1.577×10^{-2} m^3/s), and the total flow accepted by the POTW is 5 million gallons per day (MGD, 0.2191 m^3/s), exclusive of the rogue stream containing the Cd. If the malfunction goes undetected, what will be the Cd concentrations in the industrial facility's discharge and in the influent to the POTW? Consider that all flow streams other than that from the manufacturing process have Cd levels that are too low to be detected by laboratory analysis (i.e., for this analysis they may be assigned concentration values of 0). Is the biological process at the local treatment plant in danger?

We may employ at least two approaches. We can use zero-volume mixing to compute the concentration of Cd in the industrial facility's discharge and then again to compute the concentration of Cd in the overall influent to the POTW. We also may use zero-volume mixing with assumptions detailed earlier to compute the concentration of Cd in the influent to the POTW as the mixing of two streams. Let us first convert all the relevant flow rates to units of m^3/s and then specify the relevant concentration values:

$$Q_{ID} = 6.31 \times 10^{-4} \quad Q_{mfp} = 1.577 \times 10^{-2} \quad Q_{POTW} = 2.191 \times 10^{-1} \quad \frac{m^3}{s}$$

$$C_{Cd.ID} := 10000 \quad C_{Cd.mfp} := 0 \quad C_{Cd.ww} := 0 \quad \frac{g}{m^3}$$

Now we will compute the concentration of Cd in the influent stream of the POTW using the two-stage dilution, noting that the normal flow from the manufacturing facility is already included in the total normal flow to the POTW:

$$C_{Cd.mfp} := \frac{Q_{ID} \cdot C_{Cd.ID} + Q_{mfp} \cdot C_{Cd.mfp}}{Q_{ID} + Q_{mfp}} = 385 \quad \frac{g_{Cd}}{m^3}$$

$$C_{Cd.POTW} := \frac{(Q_{ID} + Q_{mfp}) \cdot C_{Cd.mfp} + Q_{POTW} \cdot C_{Cd.ww}}{Q_{ID} + Q_{POTW}} = 28.7 \quad \frac{g_{Cd}}{m^3}$$

A computation assuming the mixing of the Cd-laden stream with that of the collection system immediately upstream of the POTW yields the same results:

$$C_{Cd.POTW} := \frac{Q_{ID} \cdot C_{Cd.ID} + Q_{POTW} \cdot C_{Cd.ww}}{Q_{ID} + Q_{POTW}} = 28.7 \quad \frac{g_{Cd}}{m^3}$$

Were this inadvertent discharge to persist over a long period of time, the Cd concentration in the biological process of the POTW could in theory reach the computed value. Hopefully, the operations staff of the manufacturing plant would detect the inadvertent discharge and correct it as quickly as possible. We would perhaps like to know how quickly the manufacturer's personnel would need to act to prevent the level of Cd in the biological process of the POTW from rising sufficiently high to be acutely toxic to the desirable bacteria of the process. Consider that this level might perhaps be 1 ppm_m (Coello Oviedo et al., 2002). We cannot address the danger to the biological process unless we know the duration of the rogue discharge and are able to compute the corresponding Cd concentration in the reactor.

Example 7.7 Consider that the biological reactor has a volume of 230,000 ft^3 (6513 m^3) and that in order to avert harm to the microbes of the activated sludge process, the concentration of Cd in the biological reactor should not rise above 1 ppm_m. Estimate the necessary response time to detect and discontinue the inadvertent discharge of Example 7.6.

This situation is exactly analogous to the laboratory reactor and pump delivering the tracer of Example 7.5. The biological process is represented by the laboratory reactor and the inadvertent discharge containing Cd is represented by the tracer pump and reservoir. Were we to plot the concentration of Cd in the biological reactor over time, along with the concentration of Cd in the influent to the POTW, our resultant plot would be very similar to that of Example 7.5. Since we desire the necessary response time and the time required for the specified washout, Equation 7.5 is the most convenient.

We specify the initial condition and limiting conditions, the influent concentration as a consequence of the positive step, and the influent flow during the positive step, and compute the HRT:

$$C_0 := 0 \quad C_{crit} := 1 \quad C_{in} = 28.717 \; \frac{g}{m^3} \quad Q_{Tot} := Q_{ID} + Q_{POTW} = 0.2197 \; \frac{m^3}{s}$$

$$V_R = 6.513 \times 10^3 \; m^3 \quad \tau := \frac{V_R}{Q_{Tot}} = 2.964 \times 10^4 \; s$$

We then rearrange Equation 7.5 to solve for the critical time, in minutes, and solve:

$$t_0 := 0 \quad t_{crit} := \left(t_0 + \tau \cdot \ln\left(\frac{C_{in} - C_0}{C_{in} - C_{crit}}\right)\right) \cdot \frac{1}{60} = 18 \; min$$

The necessary corrective action would need to occur in short order.

Example 7.8 Consider that the elapsed time expended to detect the discharge, mobilize the necessary personnel, and discontinue the inadvertent discharge would be 30 min. What would be the maximum Cd concentration reached in the biological reactor and what would be the necessary washout time for the Cd concentration in the reactor to fall back to a level of 10 ppb_m?

Here we need to first compute the Cd concentration corresponding with the 30-minute cessation of the Cd input and use this value for the initial concentration for the washout stage, subsequent to the negative step:

$$C_0 := 0 \quad t_0 := 0 \quad C_{in} = 28.717 \; \frac{g}{m^3} \quad Q_{Tot} := Q_{ID} + Q_{POTW} = 0.2197 \; \frac{m^3}{s}$$

$$V_R = 6.513 \times 10^3 \; m^3 \quad \tau := \frac{V_R}{Q_{Tot}} = 2.964 \times 10^4 \; s$$

$$t_{resp} := 30 \cdot 60 = 1.8 \times 10^3 \; s \quad C_{resp} := C_{in} - \frac{C_{in} - C_0}{\exp\left(\frac{t_{resp} - t_0}{\tau}\right)} = 1.692 \; \frac{g_{Cd}}{m^3}$$

Now we may address the washout stage:

$$C_0 := C_{resp} = 1.692 \quad C_{in} := 0 \quad C_{WO} := 0.01 \; \frac{g_{Cd}}{m^3}$$

$$t_0 := 0 \quad t_{WO} := \left(t_0 + \tau \cdot \ln\left(\frac{C_{in} - C_0}{C_{in} - C_{WO}}\right)\right) \cdot \frac{1}{60} = 2535 \; min$$

The washout stage would require a little more than 42 h.

Examples 7.6–7.8, of course, neglect the fact that Cd would be taken up by biomass in the reactor. The buildup in concentration might go somewhat as the model has predicted, but since the Cd would be taken up by the biomass, the washout period likely would be longer. The biomass would remove the Cd from the aqueous solution and hold it. Biomass are grown in the reactor and recycled back into the reactor from the settled solids removed by clarification downstream of the reactor, thus maintaining the Cd in the reactor longer than predicted by the simple mixing model. The computed concentrations are thus the maximum possible for the stated circumstances. Although they are "worst-case" approximations, analyses of this nature are valuable in analysis of flow and mixing in environmental systems.

PROBLEMS

1. The background chloride, Cl^-, level in Rapid Creek that flows through Rapid City is 5 mg/L. The City of Rapid City, in order to reduce suspended particulate matter in ambient air, has begun using a deicer consisting of magnesium and calcium chloride on its streets. On a particular snow day in the winter of 2001/2002, the flow of water at the upstream extremity of Rapid City was 15.45 ft^3/s. The total flow of runoff from the city into the creek was 3.55 ft^3/s on that day during which the chemical was applied to the streets and the concentration of chloride in this snow-melt runoff was 226 mg/L. Determine the concentration in mg/L of chloride in Rapid Creek water as it leaves Rapid City. Chloride is, of course, a conservative substance and is not transformed in the environment. Assume that the creek and runoff are well mixed.

2. An automobile mechanic is working on a poorly running engine in her shop. It is the dead of winter and the exhaust is connected to the outside via a rubber tube. The engine is idling and producing exhaust gases at a rate of 10 ft^3/min (at 0.9 atm pressure and 22 °C, the T and P_T of the garage). The carbon monoxide content of the exhaust is 2% by volume. The telephone rings at exactly the time that the exhaust tubing becomes detached from the automobile tailpipe. The mechanic hurries to the adjoining room and answers the phone—it is her husband and he engages her in a lengthy conversation. The garage is 20′ × 20′ × 10′ and is equipped with a ventilation system that exhausts 30 ft^3/min of air from the garage with input air drawn through a louver. What steady-state concentration of CO **in ppm$_v$** would be reached in the garage if the engine is left idling for a long time under the given conditions? CO degrades slowly enough in this situation to be considered conservative. Assume that the air in the garage is well mixed.

3. The water production rate from one of Rapid City's Madison Aquifer wells is 500 gal/min (note that the density of water is 8.34 lb_m/gal). The target concentration for fluoride in Rapid City's drinking water is 0.75 ppm_m. A fluoride solution (1.00% Fl^- by mass, ρ = 1.05 g/cm^3 or 8.76 lb_m/gal) is to be fed to attain the target fluoride concentration. At what volumetric rate (*in gal/h*) should this fluoride solution be fed in order to attain the 0.75 ppm_m target fluoride concentration?

4. The suspended growth biological treatment basin at the Spearfish, SD, wastewater renovation facility is completely mixed and has a volume of 6.75 million gallons. The influent (as well as effluent) flow rate is 5.23 MGD. An accident on one of the streets has caused a toxic substance to pool above a cracked manhole cover in the sewer collection system and to enter the wastewater stream through the manhole cover, producing a concentration of the toxic substance in the wastewater influent to the biological reactor at the plant of 3.3 mg/L. Given that none of the substance is initially in the reactor, how long will it take (in days), once the contaminated wastewater stream begins to enter the reactor, for the concentration to reach a level of 250 μg/L, the level at which this substance will interfere with the reaction normally occurring in the reactor?

5. The liquid volume of Pactola Reservoir, located west of Rapid City, in the Black Hills of SD, is 2.44×10^9 ft^3. The reservoir is completely filled and the springtime flow of Rapid Creek during an extremely wet year is 500 ft^3/s into and out of the reservoir. Assume that a tributary of Rapid Creek becomes affected by acid rock drainage such that the level of copper in Rapid Creek just above the Pactola Reservoir is instantaneously elevated on May 1 to 2.0 mg/L. What will be the copper concentration in the reservoir **in units of mg/L** on June 1, 31 days later, assuming the reservoir is well mixed?

6. A truck carrying concentrated cyanide solution is involved in a mishap near a bridge across Rapid Creek just above Canyon Lake (liquid volume $= 4.80 \times 10^6$ ft^3), located in western Rapid City. The tanker truck develops a leak such that the cyanide solution trickles into Rapid Creek and is mixed with the flowing water, which then flows into Canyon Lake, that for this situation can be considered to be well mixed. The flow in the creek is 120 ft^3/s into and out of the lake and the cyanide concentration in the inflow is 25.0 mg/L. How quickly (in min) must the leak from the tanker be stopped in order that the cyanide concentration in Canyon Lake does not reach 1 mg/L.

7. An automobile mechanic is working on a poorly running engine in her shop. It is the dead of winter and the exhaust is connected to the outside via a rubber tube. The engine is idling and producing exhaust gases at a rate of 10 ft^3/min (at 0.9 atm pressure and 22 °C, the T and P_T of the garage). The carbon monoxide content of the exhaust is 2% by volume. The telephone rings at exactly the time that the exhaust tubing becomes detached from the automobile tailpipe. The mechanic hurries to the adjoining room and answers the phone—it is her husband and he engages her in a lengthy conversation. The garage is $20' \times 20' \times 10'$ and is equipped with a ventilation system that exhausts 30 ft^3/min of air from the garage with input air drawn through a louver. If the phone conversation lasts for an hour, what will be the concentration in **ppm$_v$** of CO in the garage when she returns? Assume that the air in the shop is well mixed.

8. Two gas streams, mixed together, are to be used to calibrate the response of a gas chromatograph for assay of a particular gas constituent. One gas stream is ultrapure helium flowing at a rate of 65.4 mL/min while the other is ultrapure helium gas

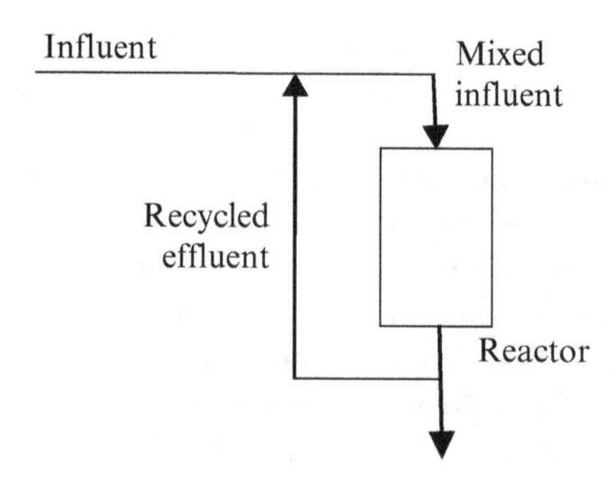

FIGURE P7.9 *Overall schematic of a recycle bioreactor depicting the influent mixing point.*

containing 10 ppm_v of the constituent and flowing at a rate of 4.72 mL/min. Both gas streams are at a temperature of 25 °C and pressure of 0.963 atm. Determine the concentration of the constituent in the mixed gas stream in units of $\boldsymbol{ppb_v}$.

9. The influent stream to an anaerobic biological process reactor, shown in Figure P7.9, flows at a rate of 500 L/min and contains 8070 mg/L of chemical oxygen demand (COD). The effluent stream from this reactor contains 455 mg/L of COD. The effluent stream is mixed with the influent stream just prior to its entry into the reactor at a ratio of 12:1 (i.e., 12 parts effluent to 1 part influent) by volume. Determine the actual concentration of the mixed influent to the reactor in units of mg/L COD.

10. A solution of alum (48.8% aluminum sulfate, $Al_2(SO_4)_3$, $\rho = 1.61\ g/cm^3$) is fed to a process used by the Mount Rushmore KOA campground to precipitate phosphorus from effluent produced by two stabilization ponds prior to discharge of the low-phosphorus effluent to an engineered wetland. The precipitation process operates at a flow rate of 50 gal/min and the dose of alum is to be 18 mg/L. At what volumetric flow rate (in mL/min) should the alum solution be fed?

11. A truck carrying concentrated cyanide solution is involved in a mishap near a bridge across Rapid Creek just above Canyon Lake (liquid volume = $4.80 \times 10^6\ ft^3$), located in western Rapid City. The tanker truck develops a leak such that the cyanide solution trickles into Rapid Creek and is mixed with the flowing water that then flows into Canyon Lake, which for this situation can be considered to be well mixed. The flow in the creek is 120 ft^3/s into and out of the lake and the cyanide concentration in the inflow is 25.0 mg/L. How quickly (in min) must the leak from the tanker be stopped in order that the cyanide concentration in Canyon Lake does not reach 1 mg/L.

12. An automobile mechanic is working on a poorly running engine in her shop. It is the dead of winter and the exhaust is connected to the outside via a rubber tube. The engine is idling and producing exhaust gases at a rate of 10 ft^3/min (at 0.9 atm pressure and 22 °C). The carbon monoxide content of the exhaust is 2% by volume. The telephone rings at exactly the time that the exhaust tubing

becomes detached from the automobile tailpipe. The mechanic hurries to the adjoining room and answers the phone—it is her husband and he engages her in a lengthy conversation. The garage is 20′ × 20′ × 10′ and is equipped with a ventilation system that exhausts 30 ft^3/min of air from the garage with input air drawn through a louver. The phone conversation lasts for an hour, when she hangs up the phone, she happens to look at the carbon monoxide detector that can be viewed through a small window from the shop office. It reads 1550 ppm_v. As she flips the switch to open the garage door to vent the shop, she ponders the condition. She knows that the automobile exhaust is ~2% CO and that the rate of gas production is about 10 ft^3/min and wonders then at what rate (in ft^3/min) the exhaust fan was drawing gas from the shop during this incident. Assume that the air in the shop is well mixed.

13. Two gas streams, mixed together, are to be used to calibrate the response of a gas chromatograph for assay of a particular gas constituent. One gas stream is ultrapure helium flowing at a rate of 115 mL/min while the other is ultrapure helium gas containing 10 ppm_v of the constituent. At what rate, in mL/min, must the second gas flow rate be set to attain a mixed gas concentration of 900 ***ppb****v*? Both gas streams are at a temperature of 25 °C and pressure of 0.963 atm.

14. The influent stream to an anaerobic biological process reactor, shown in Figure p7.9, flows at a rate of 650 L/min and contains 9040 mg/L of COD. The effluent stream from this reactor contains 762 mg/L of COD. The effluent stream is mixed with the influent stream just prior to its entry into the reactor. The desired mixed influent concentration is 1000 mg/L COD. Determine the recycle ratio (recycle flow divided by the influent flow) that will lead to this condition.

15. Chlorine is fed to water destined for potable water distribution systems in order to provide disinfection in the water plant and to provide a residual concentration of disinfectant in the pipes of the distribution system. A typical chlorine dose might be 2 mg(Cl)/L of water treated. Small systems often use 12% solutions of calcium hypochlorite ($Ca(OCl)_2$, $\rho = 1.11$ g/cm^3) rather than liquefied chlorine due to the increased safety and capital cost savings for feeding a liquid rather than chlorine gas. For a system that produces 100 gal/min of finished water, what would be the flow rate in *mL/h* of 12% $Ca(OCl)_2$ solution to attain this dose?

16. An automobile mechanic is working on a poorly running engine in her shop. It is the dead of winter and the exhaust is connected to the outside via a rubber tube. The engine is idling and producing exhaust gases at a rate of 10 ft^3/min (at 0.9 atm pressure and 22 °C). The carbon monoxide content of the exhaust is 10% by volume. The telephone rings at exactly the time that the exhaust tubing becomes detached from the automobile tailpipe. The mechanic hurries to the adjoining room and answers the phone—it is her husband and he engages her in a lengthy conversation. The garage is 20′ × 20′ × 10′ and is equipped with a ventilation system that exhausts 30 ft^3/min of air from the garage with input air drawn through a louver. A CO level of 1% (by volume) is associated with

highly acute detrimental health effects. How long (in min) after the hose is detached will it take for the CO concentration to reach this dangerous level? Assume that the air in the garage is well mixed.

17. A truck carrying concentrated cyanide solution is involved in a mishap near a bridge across Rapid Creek just above Canyon Lake (liquid volume $=4.80\times 10^6\,ft^3$), located in western Rapid City. The tanker truck develops a leak such that the cyanide solution trickles into Rapid Creek, is mixed with the flowing water that then flows into Canyon Lake, which for this situation can be considered to be well mixed. The flow in the creek is $120\,ft^3/s$ into and out of the lake and the cyanide concentration in the inflow is 25.0 mg/L. The emergency response crew requires 1 h and 25 min to stop the leak from the tanker. What is the resulting cyanide concentration in Canyon Lake?

18. The Black Hills State YellowJackets are hosting the SD Mines HardRockers in the final local matchup of the season with a berth at the national NAIA mens' basketball tournament on the line. The BHSU safety officer is concerned about carbon dioxide buildup in the gym during the game and has asked you for some computations. The seating capacity of the gymnasium in the Young Center is listed at 3800, but 4500 fans, coaches, security staff, and players are expected for the game, and due to the excitement of the evening, the average metabolic rate will be 125% of the normal resting rate (i.e., 625 mL/min per person). Some typical data on carbon dioxide emissions from humans are contained in Figure P7.18. The rate at which gas is inhaled

Characteristics of air and human exhalations

http://users.rcn.com/jkimball.ma.ultranet/BiologyPages/P/Pulmonary.html

Compo-nent	**Atmospheric air (%)**	**Expired gas(%)**
N_2	78.62	74.9
O_2	20.85	15.3
CO_2	0.04	3.6
H_2O	0.5	6.2
	100.0%	100.0%

At rest, we breathe 15–18 times a minute exchanging about 500 mL/min of air.

FIGURE P7.18 *Typical characteristics of human respiration.*

is exactly that at which it is exhaled. The exhalation rate is best considered as an influent flow while the inhalation rate is best considered as an effluent flow. The gym is 18,000 ft^2 and has a ceiling height of 30 ft. The ventilation system exhausts air from the gym at a rate of 500 ft^3/min, *concurrently drawing air in at the same rate*. Consider that normal air (drawn from the outside) contains 400 ppm_v carbon dioxide. The attendees are expected to be in the gym 2 h and 15 min total, arriving at 6:45 PM on a Wednesday evening, and exiting at 9:00 PM. Determine the concentration (in ppm_v) of carbon dioxide in the gym at game's end. Upon completion of the game the gym clears quickly and remains empty overnight. What will be the concentration (in ppm_v) of carbon dioxide in the gym when the BHSU students arrive for their 8 AM PE class the next morning?

19. Deerfield reservoir, located near Hill City, SD, contains 15,000 ac ft (1 ac-ft = 43,560 ft^3) of water. During April of one particular year, Castle Creek and other inflows to the lake are flowing at a combined rate of 100 ft^3/s. For this exercise, we may consider the reservoir to be well mixed. On April fool's day, an old tailings pond in the Castle Creek drainage bursts and a flow of 1 ft^3/s carrying zinc at a concentration of 1 g_{Zn}/L is initiated and flows into Castle Creek. The rogue stream flows under these conditions for exactly 10 days before the SD DENR discovers the problem and has it corrected. Determine the concentration of zinc reached in the reservoir. The recommended maximum contaminant level (MCL) for zinc in drinking water is 20 μg/L (or somewhere thereabouts). How long (in days) after the rogue stream flow is corrected will be required for the zinc concentration in the reservoir to drop to the MCL?

20. The wastewater treatment facility operated by the City of Rapid City relies upon a complete-mix activated-sludge basin for biological treatment to reduce the biodegradable pollution content of treated wastewater discharged to Rapid Creek. The basin has a volume of 2.5×10^6 gal. Wastewater flows into the basin at a rate of 7000 gal/min. One day a tanker truck carrying a 20% solution of sodium chromate (Na_2CrO_4, $20\ g_{Na_2CrO_4}/100\ g_{20\%solution}$, $\rho_{20\%} = 1.194$ g/mL) overturns on a city street. The tank is ruptured, leaking the contents and creating a pool of liquid on the street and causing it to flow into the sewer collection system via "pick" holes in the top of a manhole cover. The flow rate of sodium chromate solution into the manhole as a consequence of the leak is 1.5 gal/min. An emergency HAZMAT crew arrives at the spill site, contains the spill, and seals the manhole 45 min after the spill occurred. A sketch of this overall system likely will help immensely in your ensuing computations. Compute the concentration of **chromium** in the wastewater flowing into the plant during this spill event? What maximum concentration of **chromium** will be realized in the biological reactor? Ensure that your solution includes a sketched plot of the chromium concentration in both the influent to and effluent from the basin versus time. Once the influent Cr slug has fully passed into the reactor and unaffected wastewater begins entering such that the maximum

concentration level is reached, how much time will be necessary for the **chromium** concentration to be decreased to 50 ppb_m? Include a carefully sketched plot of effluent concentration versus time.

21. A door manufacturing facility located in a small city uses a batch process to first galvanize (apply a zinc coat to the steel door), pickle (contact with a solution containing phosphoric acid), and then rinse the doors prior to drying and painting. Each day, the facility produces 50,000 gal of rinse water that is contaminated with zinc at a level of 200 ppm_m. Currently the facility captures this rinse water for each day's production, holds it in a basin, and at the end of the production day removes zinc from this rinse water, prior to discharge, using a batch precipitation process. The facility manager is tired of paying for treatment chemicals and has applied for a permit to simply meter the rinse water over a 24-h period each day to the sanitary sewer discharge from the plant. The flow of sewage from the plant averages 75 gal/min over the course of a 24-h day. The flow of wastewater to the wastewater treatment plant averages 5 MGD. Compute the average concentration of zinc that would result in the discharge from the door manufacturing plant. Compute the average concentration that would result in the influent to the city's wastewater renovation facility.

22. A valve in a facility producing chrome-plated motorcycle gas tanks malfunctions and a waste stream containing hexavalent chromium (Cr^{+6}) is accidentally routed to the plant's sanitary sewer system. With the chromium-laden stream joining the plants wastewater stream and the plant stream joining the total flow entering the biological treatment basin at the city's wastewater treatment plant, the concentration of chromium in the plant influent would be 7.38 mg(Cr)/L during the time it is affected by the accidental discharge. The total flow to the wastewater treatment facility is 5.0 MGD and the completely mixed biological treatment basin has a volume of 1.25 million gallons. In order that the concentration of chromium in the biological treatment reactor not exceed a level of 0.50 ppm_m, what is the maximum amount of time the valve may be in the malfunctioning condition (i.e., how long would an emergency response team have to fix the problem?)? Given that the concentration of chromium in the biological treatment basin would reach the level of 0.50 mg/L, compute the required time for the chromium concentration to be reduced to 1 ppb_m.

23. A leak flowing at 10 gal/min develops in the dam containing a tailings pond at a mine site in western Montana. The water leaking from the pond contains Cd at a concentration of 1.0 g/L. This small stream flows only a short distance before it joins a babbling brook flowing at a rate of 2 ft^3/s. The background Cd level in the brook is 0.1 mg/L. Only a little farther downstream, this babbling brook joins a blue-ribbon trout stream with a flow of 150 ft^3/s and a background Cd level of 0.5 μg/L. Determine the resultant concentration of Cd in the trout stream below the confluence with the babbling brook?

24. The wastewater treatment facility operated by the City of Rapid City relies upon a complete-mix activated-sludge basin for biological treatment to reduce the biodegradable pollution content of treated wastewater discharged to Rapid Creek. The basin has a volume of 2.5×10^6 gal. The wastewater flows into the basin at a rate of 7000 gal/min. One day a tanker truck carrying a toxic, nonbiodegradable chemical overturns on a city street. The tank is ruptured, leaking the contents onto the street and into the sewer collection system via "pick" holes in the top of a manhole cover. The leakage into the manhole occurs at a rate that results in a concentration of the toxic substance in the overall wastewater flow stream arriving at the wastewater plant to be 10 mg/L. An emergency HAZMAT crew arrives at the spill site, contains the spill, and seals the manhole 45 min after the spill occurred. To what level will the concentration of toxic substance in the activated sludge reactor rise as a consequence of this accident? How long will be required (after the initial entry of contaminated wastewater into the reactor) for the concentration of the toxic substance in the activated sludge reactor to fall to 0.10 mg/L?

25. A door manufacturing facility located in a small city uses a batch process to first galvanize (apply a zinc coat to the steel door), pickle (contact with a solution containing phosphoric acid), and then rinse the doors prior to drying and painting. Each day, the facility produces 50,000 gal of rinse water that is contaminated with zinc at a level of 200 ppm_m. Currently the facility removes zinc from this rinse water, prior to discharge, using a batch precipitation process. The facility manager is tired of paying for treatment chemicals and has applied for a permit to simply meter the rinse water over a 24-h period each day to the sanitary sewer discharge from the plant. The flow of sewage from the plant averages 75 gal/min over the course of a 24-h day. The flow of wastewater to the waste water treatment plant averages 5 MGD. You are assigned the task of evaluating this revised permit request. In your deliberations you decide that you need to: Compute the quantity of zinc (in kg/day) that would be discharged to the city's wastewater system; compute the average concentration of zinc that would result in the discharge from the door manufacturing plant and the average concentration that would result in the influent to the city's wastewater renovation facility.

26. Due to a mechanical failure, a truck carrying a solution 50% by mass ammonium nitrate (NH_4NO_3, $\rho_{soln} = 1.22\,g/cm^3$) overturns on the SD Highway 44 bridge immediately upstream from Canyon Lake, located in west Rapid City, SD. As a consequence of the accident, a slow leak develops in the tank allowing liquid ammonium nitrate solution to flow into Rapid Creek as the water passes under the bridge. The flow rate is 20 gal/min. Canyon Lake holds 8,640,000 ft^3 of water and at the time of the incident the flow of Rapid Creek is 100 ft^3/s. The emergency response crew arrives on site within one half hour of the accident and manages to stem the flow of ammonium nitrate within 1 h after arrival at the site. If Canyon Lake were considered to be a completely mixed flow reactor, assuming no loss of ammonium nitrate from the lake through reaction or association with sediments, what maximum value would the concentration of

ammonium nitrate reach in association with this accident scenario? Under an alternate scenario, the response team requires one and one-half hours to arrive at the site (It is a summer Sunday afternoon and they were all playing golf or water skiing at Pactola reservoir.) and they require an hour to stem the flow of liquid from the tank. In this case, the maximum predicted ammonium nitrate concentration would be 26.89 ppm_m, again, assuming that there is no loss of ammonium nitrate through reaction or association with the sediments. How much time (*relative the time of occurrence of the accident*) will be required for the ammonium nitrate concentration of the lake to be reduced to 1.0 ppm_m?

Chapter 8

Reactions in Ideal Reactors

8.1 PERSPECTIVE

In previous chapters, we have dealt with reactions at equilibrium, specifically under the condition at which the rate of the forward reaction is equal in magnitude to the rate of the reverse reaction. Herein, we will embrace rates and extents of reactions as processes that involve the conversion of one or more components (reactants) into other components (products). These reactions can be either homogeneous (occurring completely within a single phase) or heterogeneous (involving two or more phases). We will first consider the overall result of particular reactions, using the chemical reaction itself to quantitatively understand the overall conversions of reactants into products. Then we will consider these overall conversions from the standpoint of the forward velocity, or rate, of the reaction. We will consider two specific forms of rate laws, used to describe the rate of reaction in terms of the abundances of reactants and products and of the magnitudes of rate constants. These cover the vast majority of reactions of interest in environmental systems. Quantitative rates of reactions are important to determinations of the necessary sizes of reactors in which these reactions would be accomplished, to define optimum conditions for carrying out these reactions, and to quantitatively understand processes that might occur in either natural or engineered systems. Since reactions need to occur in reactors, we will investigate reactions in the three specific types of "ideal" reactors—completely mixed batch reactor (CMBR), completely mixed flow reactor (CMFR), and plug flow reactor (PFR). We will also investigate some special cases in which the actual reactors have attributes of two types of ideal reactor.

Environmental Process Analysis: Principles and Modeling, First Edition. Henry V. Mott.

8.2 CHEMICAL STOICHIOMETRY AND MASS/VOLUME RELATIONS

Table 8.1 contains a listing of some chemical reactions important in environmental process analysis. These arise in both physical/chemical and in biological treatment of waters and wastewaters. Table 8.1 is by no means an exhaustive collection of reactions. Nonetheless, we will confine our illustrations of the application of stoichiometry to systems in which one or more of these reactions are operative.

TABLE 8.1 Selected Chemical Reactions Illustrating Stoichiometric Relations

Conversion of ammonium to nitrate:

1. $NH_4^+ + 2O_2 \Rightarrow NO_3^- + 2H^+ + H_2O$

Conversion of nitrate to nitrogen gas using methanol (denitrification):

2. $6NO_3^- + 5CH_3OH \Rightarrow 3N_{2(g)\uparrow} + 5HCO_3^- + 7H_2O + OH^-$

Conversion of nitrate to nitrogen gas using acetic acid (denitrification):

3. $8NO_3^- + 5CH_3COOH \Rightarrow 4N_{(g)2\uparrow} + 10HCO_3^- + 4H_2O + 2H^+$

Destruction of cyanide using free chlorine:

4. $2NaCN + 5Cl_2 + 12NaOH \Rightarrow N_{2(g)\uparrow} + 2Na_2CO_3 + 10NaCl + 6H_2O$
5. $2CN^- + 5Cl_2 + 12OH^- \Rightarrow N_{2(g)\uparrow} + 2CO_3^= + 10Cl^- + 6H_2O$
6. $2HCN + 5Cl_2 + 12OH^- \Rightarrow N_{2(g)\uparrow} + 2CO_3^= + 10Cl^- + 6H_2O + 2H^+$

Coagulation using aluminum sulfate (alum):

7. $Al_2(SO_4)_3 + 2H_2O \Rightarrow 2Al(OH)_{3(s)\downarrow} + 3SO_4^= + 3H^+$

Precipitation of phosphorus using aluminum sulfate (alum) or ferric chloride:

8. $Al_2(SO_4)_3 + 2PO_4^{-3} \Rightarrow 2AlPO_{4(s)\downarrow} + 3SO_4^=$
9. $FeCl_3 + PO_4^{-3} \Rightarrow FePO_{4(s)} + 3Cl^-$

Precipitation of divalent metals using lime ($Ca(OH)_2$):

10. $Zn^{+2} + Ca(OH)_2 \Rightarrow Zn(OH)_{2(s)\downarrow} + Ca^{+2}$
11. $Cu^{+2} + Ca(OH)_2 \Rightarrow Cu(OH)_{2(s)\downarrow} + Ca^{+2}$
12. $Cd^{+2} + Ca(OH)_2 \Rightarrow Cd(OH)_{2(s)\downarrow} + Ca^{+2}$

Microbial conversion of glucose to carbon dioxide using various electron acceptors:

13. $C_6H_{12}O_6 + 6O_2 \Rightarrow 6CO_{2\uparrow} + 6H_2O$
14. $C_6H_{12}O_6 + 4Cr_2O_7^= + 32H^+ \Rightarrow 6CO_{2\uparrow} + 8Cr^{+3} + 22H_2O$
15. $C_6H_{12}O_6 + 3SO_4^= + 3H^+ \Rightarrow 6CO_{2\uparrow} + 3HS^- + 6H_2O$
16. $C_6H_{12}O_6 \Rightarrow 3CH_{4\uparrow} + 3CO_{2\uparrow}$

Precipitation of both carbonate and non carbonate hardness from water:

17. $H_2CO_3^* + Ca(OH)_2 \Rightarrow CaCO_{3(s)\downarrow} + 2H_2O$
18. $Ca(HCO_3)_2 + Ca(OH)_2 \Rightarrow 2CaCO_{3(s)\downarrow} + 2H_2O$
19. $CaSO_4 + Na_2CO_3 \Rightarrow CaCO_{3(s)\downarrow} + 2Na^+ + SO_4^=$
20. $Mg(HCO_3)_2 + Ca(OH)_2 \Rightarrow MgCO_3 + CaCO_{3(s)\downarrow} + 2H_2O$
21. $MgCO_3 + Ca(OH)_2 \Rightarrow Mg(OH)_{2(s)} + CaCO_{3(s)\downarrow}$
22. $Mg(HCO_3)_2 + 2Ca(OH)_2 \Rightarrow Mg(OH)_{2(s)\downarrow} + 2CaCO_{3(s)\downarrow}\ 2H_2O$
23. $MgSO_4 + Ca(OH)_2 \Rightarrow Mg(OH)_{2(s)\downarrow} + CaSO_4$
24. $CaO_{(s)} + H_2O \Rightarrow Ca(OH)_2$ (quicklime conversion to hydrated lime)

Reductive dechlorination of chlorophenol:

25. $12C_6H_4OHCl + C_6H_{12}O_6 + 6H_2O \Rightarrow 12C_6H_5OH + 12Cl^- + 6CO_{2\uparrow} + 12H^+$

8.2.1 Stoichiometry and Overall Reaction Rates

Let us consider the general reaction employed earlier in the text.

$$a\mathrm{A} + b\mathrm{B} \Rightarrow c\mathrm{C} + d\mathrm{D}$$

We have written it as an irreversible reaction, considering only the forward progress of the reaction. The systems we will consider will involve abundances of reactants dictating that the reactions will proceed forward as written. Further, we will typically consider that component A is the component or specie of interest. Most often in environmental engineering we are concerned with the conversion of a contaminant to less harmful forms. However, we must also be mindful that some contaminants form transformation products that are as hazardous as the contaminant itself or more. We therefore need to develop a means to relate the appearance of products with the disappearance of reactants. If the stoichiometric coefficient of reactant A is nonunity, we can normalize the overall reaction to a mole of A by dividing through the equation by the stoichiometric coefficient of A:

$$\mathrm{A} + \frac{b}{a}\mathrm{B} \Rightarrow \frac{c}{a}\mathrm{C} + \frac{d}{a}\mathrm{D}$$

Now we define the change in the molar concentration of A as the final minus the initial concentration. The changes in B, C, and D are also related to the initial and final concentrations:

$$\Delta[\mathrm{A}] = [\mathrm{A}]_{\mathrm{final}} - [\mathrm{A}]_{\mathrm{initial}}$$

$$\Delta[\mathrm{B}] = [\mathrm{B}]_{\mathrm{final}} - [\mathrm{B}]_{\mathrm{initial}}$$

$$\Delta[\mathrm{C}] = [\mathrm{C}]_{\mathrm{final}} - [\mathrm{C}]_{\mathrm{initial}}$$

$$\Delta[\mathrm{D}] = [\mathrm{D}]_{\mathrm{final}} - [\mathrm{D}]_{\mathrm{initial}}$$

Typically $[\mathrm{A}]_{\mathrm{initial}}$ might be known and $[\mathrm{A}]_{\mathrm{final}}$ might be specified, or perhaps we would be interested in the converse or other situations. In either case, we can then relate the changes in the number of moles of the other reactants and products to $\Delta[\mathrm{A}]$.

$$\begin{aligned} \Delta[\mathrm{B}] &= \frac{b}{a}\Delta[\mathrm{A}] \\ \Delta[\mathrm{C}] &= -\frac{c}{a}\Delta[\mathrm{A}] \\ \Delta[\mathrm{D}] &= -\frac{d}{a}\Delta[\mathrm{A}] \end{aligned} \tag{8.1}$$

The negative signs in Equation 8.1 arise from the fact that when the reaction proceeds forward as written, $\Delta[A]$ and $\Delta[B]$ are negative (reactants are transformed into products) while $\Delta[C]$ and $\Delta[D]$ must be positive (products are created from reactants). This set of stoichiometric relations can be applied to reactions with as few as one reactant and one product or with many more reactants and products than considered in the general reaction. Equation 8.1 may be extended beyond simple changes in the concentrations of the reactants and products to the overall rates of the disappearance of reactants and appearance of products. For a system for which the overall rate of flow is known, the overall rate of reaction R (Mt^{-1}) may be expressed as the product of flow and change in concentration:

$$R_{\mathrm{A}} = Q \cdot \Delta[\mathrm{A}] \tag{8.2}$$

Then for constant Q the relations of Equations 8.1 and 8.3 may easily be extended:

$$\begin{aligned} R_{\mathrm{B}} &= \frac{b}{a} R_{\mathrm{A}} \\ R_{\mathrm{C}} &= -\frac{c}{a} R_{\mathrm{A}} \\ R_{\mathrm{D}} &= -\frac{d}{a} R_{\mathrm{A}} \end{aligned} \tag{8.3}$$

We can apply the sets of relations of Equation 8.1 and 8.3 in many ways.

8.2.2 Some Useful Mass, Volume, and Density Relations

In engineering systems for water treatment and wastewater renovation, we often employ chemistries that produce solid products from dissolved products. Designs must consider the disposition of these solids and therefore we need the ability to compute quantities in both mass and volumetric units. These solids streams are often referred to as "sludge." The discerning property of sludge relative to a suspension is the solids content. We specify the solids content of suspensions most often in units of $\mathrm{ppm_m}$ (mg/L if solutions are dilute and aqueous). We specify the concentrations of sludge in units of percent solids. Sludge normally can be pumped and may have solid contents greater than 10%. "Cakes" are mixtures of solids and water such that the product behaves essentially as a solid. The solid content of a "cake" is often 20% or greater.

In examining the mass–volume relations for sludge and cakes, we need to employ the concept from soil mechanics that accounts for both mass and volume of a mixture of solids and voids. Similarly to saturated soils, the voids of sludge are filled with water. Similarly to unsaturated soils, dried cakes may contain both water and air in the voids between the solid particles. We begin by simply accounting for mass (M) and volume (V):

$$V_{\text{Tot}} = V_{\text{sol}} + V_{\text{void}}$$

$$M_{\text{Tot}} = M_{\text{sol}} + M_{\text{void}}$$

We have two cases—voids fully water-filled and voids partially water-filled—for which we desire usable relations. These relations would allow us to relate bulk density (ρ_{bulk}, M/V_{Tot}) to solid density (ρ_{sol}, M/V_{Sol}) and void fraction (often called porosity, $\varepsilon_{\text{Tot}} = V_{\text{void}}/V_{\text{Tot}}$). To allow full examination of unsaturated (partially dried) solids, we need to extend the void fraction concept to include the fractions of solids and water-filled and gas-filled voids. Water-filled porosity (moisture volume fraction) is the volume of water per total volume and vapor-filled porosity (vapor volume fraction) is the volume of vapor per total volume. We use the following basic relations to create some important overall relations defining bulk density:

$$\varepsilon_{\text{sol}} = \frac{V_{\text{sol}}}{V_{\text{Tot}}}; \quad \varepsilon_{\text{w}} = \frac{V_{\text{w}}}{V_{\text{Tot}}}; \quad \varepsilon_{\text{vap}} = \frac{V_{\text{vap}}}{V_{\text{Tot}}}$$

$$\varepsilon_{\text{Tot}} = \varepsilon_{\text{w}} + \varepsilon_{\text{vap}}$$

$$\varepsilon_{\text{w}} + \varepsilon_{\text{vap}} + \varepsilon_{\text{sol}} = 1$$

Bulk density is the quotient of total mass and total volume:

$$\rho_{\text{bulk}} = \frac{M_{\text{sol}} + M_{\text{void}}}{V_{\text{sol}} + V_{\text{void}}}$$

For dried, powdered solids with voids filled with dry vapor, the relation between bulk and solid density is quite easily obtained. The mass in the void is 0:

$$\rho_{\text{bulk.dry}} = \frac{\rho_{\text{sol}}(1-\varepsilon_{\text{Tot}})V_{\text{Tot}}}{(1-\varepsilon_{\text{Tot}})V_{\text{Tot}} + \varepsilon_{\text{Tot}}V_{\text{Tot}}} = \rho_{\text{sol}}(1-\varepsilon_{\text{Tot}}) \tag{8.4}$$

For sludge with pores filled with water, the water occupies the total volume of the voids and its mass must be considered:

$$\rho_{\text{bulk.sat}} = \frac{\left[\rho_{\text{sol}}(1-\varepsilon_{\text{Tot}}) + \rho_{\text{w}}\varepsilon_{\text{Tot}}\right]V_{\text{Tot}}}{(1-\varepsilon_{\text{Tot}})V_{\text{Tot}} + \varepsilon_{\text{Tot}}V_{\text{Tot}}} = \rho_{\text{sol}}(1-\varepsilon_{\text{Tot}}) + \rho_{\text{w}}\varepsilon_{\text{Tot}} \tag{8.5}$$

For sludge that is partially dried or bulk solids that contain some free moisture, we need to know the moisture content, measured by weighing a given quantity of the material, drying in an oven to drive off the moisture, and reweighing to obtain the mass of the dried solids (M_{Dsol}). The difference between the wet mass and the dried mass is the mass of the associated water (M_{W}). The moisture content is most often normalized to the dried mass and expressed as a percentage, but the moisture mass fraction ($F_W = M_W/M_{Dsol}$) is most useful in the development of the unsaturated bulk density:

$$\rho_{\text{bulk.Usat}} = \frac{(1+F_W)\rho_{\text{sol}}(1-\varepsilon_{\text{Tot}})V_{\text{Tot}}}{(1-\varepsilon_{\text{Tot}})V_{\text{Tot}} + \varepsilon_{\text{Tot}}V_{\text{Tot}}} = (1+F_W)\rho_{\text{sol}}(1-\varepsilon_{\text{Tot}}) \tag{8.6}$$

Often we would like to know the void fractions of the water and the vapor. For such systems, we might need to take into account the abundance of constituents in the aqueous and vapor phases associated with partially dried solids (unsaturated soils). If we know (or can estimate) the total porosity and can measure the moisture content, the volume fractions of water and vapor can be defined. We begin with the definition of the moisture mass fraction:

$$F_W = \frac{M_W}{M_{\text{sol}}} = \frac{V_{\text{Tot}}\varepsilon_W \rho_W}{V_{\text{Tot}}(1-\varepsilon_{\text{Tot}})\rho_{\text{sol}}} \tag{8.7}$$

This expression can then be rearranged to yield the moisture volume fraction:

$$\varepsilon_W = F_W(1-\varepsilon_{\text{Tot}})\frac{\rho_{\text{sol}}}{\rho_W} \tag{8.8}$$

We might be tempted to cancel some units in Equation 8.7 or to employ the concept of specific gravity (SG) for the ratio of the densities of the solid to that of water. If we did so, we would lose the unit structure of the expression. The volume fraction of water must have units of volume of water per total volume. We will leave Equations 8.7 and 8.8 as they are.

The volume fraction of the vapor may then be obtained as the difference between the void volume fraction and the moisture volume fraction: $\varepsilon_{\text{vap}} = \varepsilon_{\text{Tot}} - \varepsilon_W$.

8.2.3 Applications of Stoichiometry and Bulk Density Relations

Let us consider an application in the area of wastewater renovation. Suppose we have a biological process that utilizes a mixed culture of bacteria to simultaneously reduce carbonaceous oxygen demand and convert total Kjeldahl nitrogen (the combination of organic and ammonia nitrogen) to nitrate nitrogen. Biodegradable organic matter contributes to carbonaceous oxygen demand. The combination of organic nitrogen (present in organic matter mostly as amines) and ammonia nitrogen constitutes Kjeldahl nitrogen. Many biological processes are engineered to accomplish both of these processes in a single reactor. The resultant effluent is poor in biodegradable organic matter and rich in nitrate nitrogen. The often separate process for subsequently removing the nitrate nitrogen via conversion to nitrogen gas is called denitrification. In some cases, a choice is made to address the removal of nitrate nitrogen in a process separate from the main biological process. The organic matter reactant necessary for the conversion of nitrate to nitrogen gas is often supplied as a purchased, distinct biodegradable reactant. Reaction 2 of Table 8.1 is written for the use of methanol (CH_3OH) as such a biodegradable organic reactant.

Example 8.1 Consider that the effluent from a biological process at a wastewater plant contains virtually no degradable organic matter and 40 mg/L nitrate nitrogen. The discharge permit issued to the facility is under review for assignment of a limit to reduce total effluent nitrogen to a level of 5 mg/L (as nitrogen) or below. One alternative for meeting the effluent limit would involve implementation of a denitrification process using methanol as a carbon source to convert nitrate to nitrogen gas. In order to assess chemical costs, the quantity of methanol needed per year is to be determined. The average flow rate is 10 MGD (million gallons per day) (37,860 m^3/day).

Our first action is the construction of a schematic diagram of the system, shown in Figure E8.1.1. We are particularly interested in the portion of the system enclosed by the dashed line.

We have two input streams and three output streams. Overall, the process involves the use of methanol as a food source and the use of nitrate as an electron acceptor for the combination of respiration by the microbes and growth of new microbes. Herein, we are interested in a first approximation, which would overstate the quantities of nitrate and methanol converted, for which we would neglect the fact that some of the organic carbon (from the methanol) and some of the nitrogen would be tied up in the biomass produced by the process.

Next, employing the simplifying assumption, we consider the overall chemical reaction from Table 8.1:

$$6NO_3^- + 5CH_3OH \Rightarrow 3N_{2(g)\uparrow} + 5HCO_3^- + 7H_2O + OH^-$$

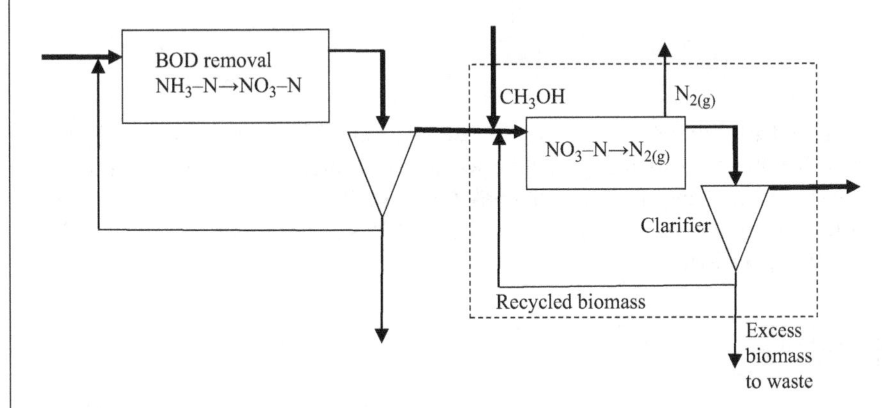

FIGURE E8.1.1 Schematic diagram of a dual-sludge system with focus on the nitrogen removal (denitrification) process.

We compute the change in the NO_3–N concentration, which must be in molar units, to apply the stoichiometry:

$$MW_N := 14 \frac{g_N}{mol_{NO3N}} \quad C_{NO3N.in} := 40 \quad C_{NO3N.out} := 5 \quad \frac{g_N}{m_{ww}^3}$$

$$NO3N_{in} := \frac{C_{NO3N.in}}{MW_N} = 2.857 \quad NO3N_{out} := \frac{C_{NO3N.out}}{MW_N} = 0.357 \quad \frac{mol}{m_{ww}^3}$$

$$\Delta NO3N := NO3N_{out} - NO3N_{in} = -2.5 \quad \frac{mol}{m_{ww}^3}$$

We may now compute the necessary change in the methanol concentration using the stoichiometry and the initial concentration. Our desire would be to expend exactly the quantity of methanol necessary to attain the necessary conversion of nitrate nitrogen to nitrogen gas. Thus, the final concentration of methanol in the effluent would desirably be 0. This change in concentration, assuming that all the methanol is consumed, would be the **dose**:

$$\Delta CH3OH := \frac{6}{5} \cdot \Delta NO3N = -3 \quad CH3OH_{in} := -\Delta CH3OH = 3 \quad \frac{mol_{CH3OH}}{m_{ww}^3}$$

Since we would purchase methanol by a unit of mass or of volume, we need to convert it back to a mass concentration:

$$MW_{CH3OH} := 32 \frac{g}{mol} \quad C_{CH3OH.in} := CH3OH_{in} \cdot MW_{CH3OH} = 96 \quad \frac{g}{m_{ww}^3}$$

We now calculate the total quantity of methanol necessary for the quantity of wastewater treated, noting that wastewater treatment (particularly that accomplished by municipal publicly owned treatment works (POTWs)) is a 24 h/day, 365 day/year operation. We might want the result in kilograms:

$$Q := 3.786 \cdot 10^4 \frac{m_{ww}^3}{d} \quad M_{CH3OH} := \frac{C_{CH3OH.in}}{1000} \cdot Q \cdot 365 = 1.327 \times 10^6 \quad \frac{kg_{CH3OH}}{yr}$$

In order to figure costs, we need to know the number of truckloads used per year and thus the corresponding volume would be of interest. We could find the mass density of methanol from a handbook (SG = 0.7913 at 20 °C). Typically, a tanker truck traveling over the highways of the United States is limited to about 5000 gal (18.9 m^3) per load:

$$\rho_{CH3OH} := 791.3 \frac{kg}{m^3} \quad V_{CH3OH} := \frac{M_{CH3OH}}{\rho_{CH3OH}} = 1.676 \times 10^3 \quad \frac{m^3}{yr}$$

$$V_{shipment} := 18.9 \quad m^3 \quad N_{shipments} := \frac{V_{CH3OH}}{V_{shipment}} = 88.7 \quad \frac{truckloads}{yr}$$

We would be accepting a shipment of methanol about every 4 days. These computations likely yield a close approximation of the minimum requirement. National Pollutant Discharge Elimination System (NPDES) discharge permits are written such that the prescribed limits cannot be exceeded. In order to ensure that limits would be met, we might use a factor of safety and plan on reducing the nitrate nitrogen well below the specified 5 mg/L. Further, methanol purchased in bulk would likely be industrial or technical grade, with purity in the range of 95–99%. The planned quantity of methanol to be expended and the mass (and hence the volume) of the actual reagent shipped would be greater than computed herein.

The removal of phosphate phosphorus (PO_4–P) from wastewaters is of great importance for POTWs with discharges into drainages that eventually flow into the Great Lakes. These huge bodies of water have experienced varying degrees of cultural eutrophication as a consequence of anthropogenic phosphorus discharges, with Lake Superior being the least affected and Lake Erie the most affected.

A particular treatment process (Phostrip©) devised for removal of PO_4–P from wastewaters employs an anaerobic chamber to stress bacteria resulting in release of their cellular phosphate. Then, luxury uptake of phosphorus occurs when the microbes are reintroduced into the aerobic reactor. Microbes need phosphorus in order to metabolize food and create cell mass. In the Phostrip© process, the bacteria are conditioned by the anaerobic contact period to take on phosphorus in amounts greater than normal. The bacteria, after exercising this luxury uptake, are separated from the wastewater and subjected to the anaerobic contact period mentioned earlier. During this contact period, the excess phosphorus taken up in the aerobic process is released by the bacteria. Separation of the bacteria from the aqueous solution produces an aqueous stream of high PO_4–P content leaving suspension of bacteria that are ready to take up phosphorus on a luxury basis in the aerobic process. Biomass (measured as volatile suspended solids (VSS)) grown in typical aerobic systems will have a phosphorus content of approximately 2.3%. By stressing the biomass as done via the Phostrip© process, the phosphorus content of the biomass can reach levels of 7–10% of the volatile solids.

The low volume, phosphorus-rich stream emanating from the anaerobic contact basin is then the target for the capture of the PO_4–P. One methodology employed is precipitation of the phosphorus using aluminum sulfate (alum). This reaction is represented as reaction 8 of Table 8.1. Let us examine the application of phosphorus precipitation using alum in the context of removal of PO_4–P from a phosphorus-rich effluent from the anaerobic contact basin of the Phostrip© process.

Example 8.2 Consider that the POTW of Example 8.1 receives wastewater containing PO_4–P at a quantity of 30 mg/L. A discharge limitation of 0.5 mg/L is believed necessary and one of the processes under consideration for accomplishing the treatment objective is Phostrip©. For evaluation purposes, on a yearly basis, we must estimate the quantity of alum (mass in kg, volume of liquid alum solution) required and the quantity of aluminum phosphate sludge (mass of solids, mass of sludge, mass of cake, and volume of cake) produced. For computations involving produced sludge, we will consider that the suspension of aluminum phosphate sludge produced by the process would have a solids content of 5%. Then, by employing solids-dewatering processes such as centrifuges, vacuum filters, or filter presses, we believe that we can recover sufficient water from the sludge stream to render the phosphorus-laden solids stream to form a "cake" of 30% solids by mass.

Our approach to this overall computation begins with the construction of a schematic depicting the overall process in Figure E8.2.1.

The boxes outlined by dashed lines represent the boundaries of selected control volumes. We have considered the aerobic bioreactor with its companion sedimentation tank and the anaerobic contact basin with its companion sedimentation tank in one control volume, and the phosphate precipitator with its companion liquid/solid separator in the second.

In order to determine the loading from the aerobic reactor system to the phosphate precipitator, we first consider the control volume containing the aerobic reactor. Phosphorus enters with the influent and a much smaller quantity exits with the effluent. The flow and concentrations are known. To maintain the steady-state

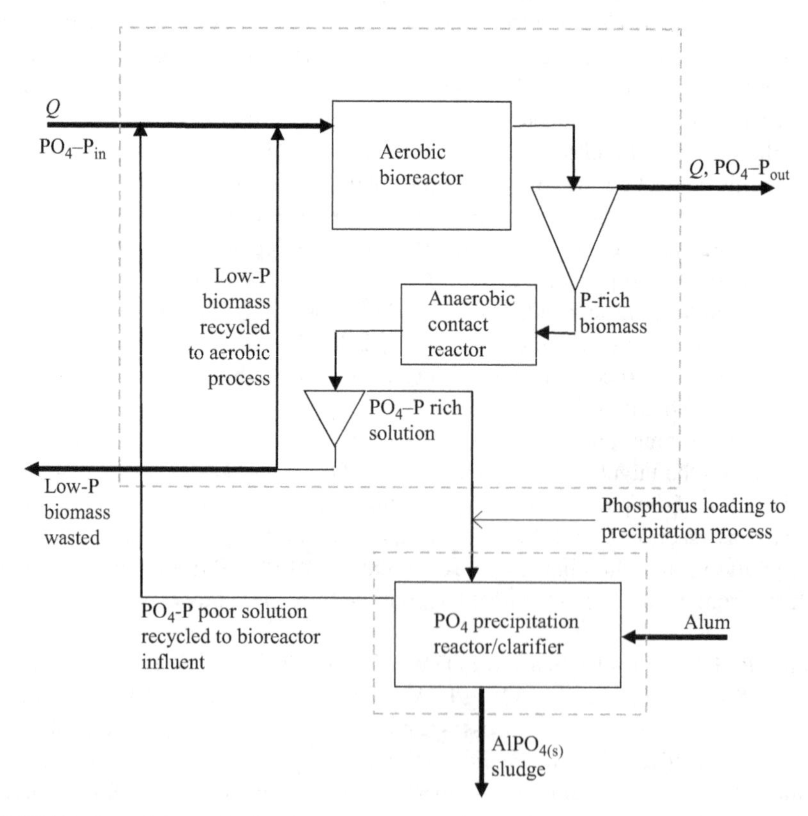

FIGURE E8.2.1 *Schematic diagram of the Phostrip® process indicating two control volumes upon which mass balances can be drawn.*

condition of the process, some biomass must be discarded (wasted). The quantity is small but perhaps significant. We have not yet considered biomass growth and are not prepared to deal with this stream. Thus, for now, we will consider that the quantity of *P* exiting with this stream is small. Our computations for quantities of alum consumed and solids produced will therefore be overstated somewhat. The phosphorus precipitation process converts nearly 100% of the applied phosphorus into aluminum phosphate, so we will neglect any *P* that might be in the solution recycled back into the bioreactor with this stream. The recycling of biomass is internal to the control volume and its *P* content need not be considered in this material balance. When we perform a balance on we consider that only a very small amount will leave the system with the aluminum phosphate sludge. We consider that losses associated with the precipitator are offset by gains from the recycling of product water from the precipitator. Further, losses of water with the small flow of wasted biomass are likely insignificant.

We may now compute the total rate at which PO_4–P is removed from the wastewater stream, as the product of flow and change in concentration (Equations 8.1 and 8.2). The phosphorus loading (mass or moles per time) to the precipitation process will be exactly that removed from the wastewater:

$$C_{PO4P.in} := 30 \quad C_{PO4P.out} := 0.5 \quad \frac{g_P}{m_{ww}^3} \quad MW_P := 31 \quad \frac{g_P}{mol}$$

$$\Delta PO4P := \frac{C_{PO4P.out} - C_{PO4P.in}}{MW_P} = -0.952 \quad \frac{mol_P}{m_{ww}^3}$$

$$Q := 37860 \quad \frac{m^3}{d} \quad R_{PO4P} := Q \cdot \Delta PO4P = -3.603 \times 10^4 \quad \frac{mol_{PO4P}}{d}$$

We may now address the precipitation process and consider reaction 8 from Table 8.1, rewriting it such that phosphate is our targeted reactant:

$$2PO_4^{-3} + Al_2(SO_4)_3 \Rightarrow 2AlPO_{4(s)} \downarrow + 3SO_4^{=}$$

One mole of alum is required for every two moles of PO_4–P. Two moles of aluminum phosphate solid are produced for every two moles of phosphate converted. Then, employing Equation 8.3 we can express the rates of alum consumption and aluminum phosphate precipitation in terms of the transformation of PO_4–P:

$$R_{alum} := R_{PO4P} \cdot \frac{1}{2} = -1.801 \times 10^4 \quad \frac{mol_{Alum}}{d}$$

$$R_{AlPO4} := -R_{PO4P} \cdot \frac{2}{2} = 3.603 \times 10^4 \quad \frac{mol_{AlPO4}}{d}$$

The negative sign preceding R_{alum} signifies that alum is consumed. The absence of the negative sign preceding R_{AlPO_4} signifies that aluminum phosphate is produced.

Using formula weights, we can readily convert these to mass units:

$$MW_{alum} := \frac{2 \cdot 27 + (32 + 4 \cdot 16) \cdot 3}{1000} = 0.342 \quad \frac{kg}{mol}$$

$$Mass_{yr.alum} := R_{alum} \cdot MW_{alum} \cdot 365 = -2.249 \times 10^{6} \quad \frac{kg_{alum}}{yr}$$

$$MW_{AlPO4} := \frac{27 + 31 + 4 \cdot 16}{1000} = 0.122 \quad \frac{kg}{mol}$$

$$Mass_{yr.AlPO4} := R_{AlPO4} \cdot MW_{AlPO4} \cdot 365 = 1.604 \times 10^{6} \quad \frac{kg_{AlPO4}}{yr}$$

We check potential suppliers of alum and find that alum may be shipped in liquid form for water and wastewater treatment operations as a 48% (as $Al_2(SO_4)_3$) solution. From our handbook, we find that the SG of 48.8% alum solution is 1.33. This formulation of aluminum sulfate might be well suited for small systems with low alum consumption rates. A decision to ship a significant quantity of water with the treatment chemical might be made to offset the capital requirement for a system to store and handle dry alum. In either case, we would require a system to accurately feed a liquid solution of alum. We find that alum is also shipped as bulk solid alum and has chemically entrained water such that the formula is $Al_2(SO_4)_3 \cdot 14H_2O$. In order that we may evaluate the costs of these two alternatives to decide which option to choose, we need to compute quantities of the chemicals shipped.

Let us first compute the mass and volume of the 48% solution of alum:

$$\rho_{48\%alum} := 1330 \frac{kg}{m^3} \qquad Xmass_{48\%alum} := 0.48 \quad \frac{kg_{Al2SO43}}{kg_{48\%alum}}$$

$$Mass_{yr.48\%alum} := Mass_{yr.alum} \cdot \frac{1}{Xmass_{48\%alum}} = -4.685 \times 10^{6} \quad \frac{kg_{48\%alum}}{yr}$$

$$Vol_{yr.48\%alum} := \frac{Mass_{yr.48\%alum}}{\rho_{48\%alum}} = -3.522 \times 10^{3} \quad \frac{m^3}{yr}$$

$$Loads_{48\%alum} := \frac{Vol_{yr.48\%alum}}{18.9} = -186.37 \quad \frac{tankerloads}{yr}$$

The negative signs are retained reminding us that these are quantities that will be consumed.

Then we will compute the mass and volume of the dry hydrated alum. We will need to know the bulk density of the hydrated alum shipped. The alum will be ground to a relatively fine powder so once on site the aqueous reagent can easily be made by quickly dissolving the fine-grained alum powder. From our handbook, we find that the bulk density of ground aluminum sulfate is ~0.80 g/cm^3 (800 kg/m^3). We will assume that the volumetric capacity of the hauling units is the same 18.9 m^3 as for the liquid product:

$$MW_{hyd.alum} := \frac{2 \cdot 27 + (32 + 4 \cdot 16) \cdot 3 + 14 \cdot (16 + 2)}{1000} = 0.594$$

$$Mass_{yr.hyd.alum} := R_{alum} \cdot MW_{hyd.alum} \cdot 365 = -3.906 \times 10^{6} \quad \frac{kg_{hyd.alum}}{yr}$$

$$\rho_{bulk.hyd.alum} := 800 \frac{kg}{m^{3}} \quad Vol_{yr.hyd.alum} := \frac{Mass_{yr.hyd.alum}}{\rho_{bulk.hyd.alum}} = -4.882 \times 10^{3}$$

$$Loads_{hyd.alum} := \frac{Vol_{yr.hyd.alum}}{18.9} = -258.308 \quad \frac{tankerloads}{yr}$$

We might have been tempted to base our decision—liquid alum or solid alum—upon "conventional wisdom," which would suggest that the mass and hence the volume of liquid alum would be greater than for the dry product. Here, we are quite surprised by the outcome of the computation. Purchase and use of the 48% alum solution is likely the more cost effective alternative, costing less to transport per unit of aluminum sulfate than the solid aluminum sulfate. The cost of dry alum storage and handling facilities would then be totally unwarranted.

We have determined the daily and yearly quantities of $AlPO_{4(s)}$ solids produced. We now need to employ some density–volume relations to compute quantities of sludge and sludge cake to be processed. First, we specify the parameters of interest, including the density of particles of pure $AlPO_{4(s)}$ and compute the mass of sludge produced daily:

$$F_{sol.sludge} := \frac{5}{100} = 0.05 \quad \frac{kg_{solids}}{kg_{sludge}} \qquad \rho_{AlPO4} := 2566 \quad \rho_{w} := 1000 \quad \frac{kg}{m^{3}}$$

$$R_{sludge} := R_{AlPO4} \cdot \frac{1}{F_{sol.sludge}} = 7.206 \times 10^{5} \quad \frac{kg_{sludge}}{d}$$

In order to determine the volume of sludge to be processed each day, we employ the density of the solids and of water to separately compute the volumes of solids and water and add them. We can then rearrange the relations as needed:

$$R_{sludge} = R_{AlPO4} + R_{w} = \frac{R_{AlPO4}}{F_{sol.sludge}}$$

$$R_{w} := \left(\frac{1}{F_{sol.sludge}} - 1\right) R_{AlPO4} = 6.845 \times 10^{5} \quad \frac{kg_{W}}{d} \qquad V_{w} := \frac{R_{w}}{\rho_{w}} = 685 \quad \frac{m_{W}^{3}}{d}$$

$$V_{AlPO4} := \frac{R_{AlPO4}}{\rho_{AlPO4}} = 14.041 \quad V_{sludge} := V_{w} + V_{AlPO4} = 698.6 \quad \frac{m_{sludge}^{3}}{d}$$

We might follow an alternate route for this computation by first computing the bulk density of the sludge:

$$\rho_{sludge} = \frac{M_{sludge}}{V_{sludge}} = \frac{\dfrac{R_{AlPO4}}{F_{sol.sludge}}}{\dfrac{R_{AlPO4}}{\rho_{AlPO4}} + \left(\dfrac{1}{F_{sol.sludge}} - 1\right) \cdot \dfrac{R_{AlPO4}}{\rho_w}}$$

$$\rho_{sludge} := \frac{\dfrac{1}{F_{sol.sludge}}}{\dfrac{1}{\rho_{AlPO4}} + \left(\dfrac{1}{F_{sol.sludge}} - 1\right) \cdot \dfrac{1}{\rho_w}} = 1.031 \times 10^3 \quad \frac{kg}{m^3}$$

Then we may compute the volume from mass and density:

$$V_{sludge} := \frac{R_{sludge}}{\rho_{sludge}} = 698.6 \quad \frac{m^3}{d}$$

Certainly, when we do not know the answer a priori, agreement between computations by a second route and our original computation allows us to walk away with that warm, fuzzy feeling. Often, with little error, we simply approximate the density of sludge as the density of water. Certainly at 5% solids the error is small, but in using this approximation we would overestimate the volumes of sludge computed.

A cake that is 30% solids would remain fully saturated with water and we may simply repeat the computations for the sludge, but with a larger solids mass fraction:

$$F_{sol.cake} := 0.3 \quad \frac{kg_{AlPO4}}{kg_{cake}} \qquad R_{cake} := \frac{R_{AlPO4}}{F_{sol.cake}} = 1.201 \times 10^5 \quad \frac{kg_{cake}}{d}$$

$$\rho_{cake} := \frac{\dfrac{1}{F_{sol.cake}}}{\dfrac{1}{\rho_{AlPO4}} + \left(\dfrac{1}{F_{sol.cake}} - 1\right) \cdot \dfrac{1}{\rho_w}} = 1.224 \times 10^3 \quad \frac{kg}{m^3}$$

$$V_{cake} := \frac{R_{cake}}{\rho_{cake}} = 98.1 \quad \frac{m^3}{d}$$

By dewatering we would reduce the mass and volume of residuals necessarily removed from the facility for further disposition by 84 and 86%, respectively.

We may easily compute the quantities necessarily processed each year, as this wastewater renovation facility would operate 365 days/year:

$$\begin{pmatrix} V_{sludge.yr} \\ M_{cake.yr} \\ V_{cake.yr} \end{pmatrix} := 365 \cdot \begin{pmatrix} V_{sludge} \\ R_{cake} \\ V_{cake} \end{pmatrix} = \begin{pmatrix} 2.55 \times 10^5 \\ 4.383 \times 10^7 \\ 3.581 \times 10^4 \end{pmatrix} \begin{pmatrix} \dfrac{m^3}{yr} \\ \dfrac{kg}{yr} \\ \dfrac{m^3}{yr} \end{pmatrix}$$

Knowledge of reagents necessary and of products produced by treatment processes is of great importance in decisions regarding process selection and in subsequent decisions regarding equipment selection.

8.3 REACTIONS IN IDEAL REACTORS

8.3.1 Reaction Rate Laws

A reaction rate law is a mathematical expression that relates the rate at which a reaction would occur with abundances of reactants and products and one or more rate coefficients. For reactions that occur in homogeneous aqueous solutions, we most often use a volume-specific rate (mass or moles reacting per unit time per unit volume of reactor, $ML^{-3}t^{-1}$). For heterogeneous systems involving solids, we might choose to express abundances relative to mass or surface area of solid reactant. For heterogeneous systems involving vapor phases, we might choose to employ the vapor–liquid surface area or the volume of gas phase present in the system. Each distinct reaction should have its own distinct reaction rate law. Reaction rates are generally dependent upon the abundances of both the reactants and products. High abundances of reactants tend to speed up reactions while high abundances of products tend to slow reactions. In the examination of reactions that occur in environmental systems, we will most often consider them to be virtually irreversible in that the reverse reaction rates are quite slow relative to forward reaction rates and we can ignore the abundances of products. We examine two forms of the reaction rate law with which we can address ~99% of all reactions of environmental interest. We apply these reaction rate laws in conjunction with CMBR, CMFR, and plug-flow reactor (PFR). We primarily consider homogeneous systems, but address others later in the chapter.

The first form of the rate law is the pseudo first order. The reaction rate is linearly dependent upon the abundance of the reactant of interest:

$$r_i = -k' \cdot C_i \tag{8.9}$$

The specific reaction rate of the arbitrary reactant i is r_i ($ML^{-3}t^{-1}$), the pseudo-first-order rate coefficient is k' (overall unit is t^{-1}), and the abundance is the concentration (ML^{-3}), which can be employed either in molar or mass units. Obviously the units of r_i and C_i must be consistent. We use the symbol k' for the overall pseudo-first-order rate coefficient since the presence of other reactants can influence the overall rate of reaction. If these other reactants are much more abundant than our component of interest or are held constant for whatever reason, we may treat the pseudo-first-order rate coefficient as a constant.

The second form of the rate law is used for descriptions of microbial reactions and begins as a first-order mathematical formulation such that the rate of growth of biomass, $r_{G.X}$, is linearly dependent upon the abundance of biomass in the system:

$$r_{G.X} = \mu \cdot X \tag{8.10}$$

where μ is a specific growth rate coefficient (volume-specific rate of biomass growth per unit of biomass abundance, $(M_{BM}/V_R \cdot t)/(M_{BM}/V_R)$), X is the abundance of biomass in the reactor (M_{BM}/V_R), and V_R is the volume of the reactor. Many texts give inverse time as the unit of μ and we are tempted to follow suit. However, such action requires that we cancel a unit of biomass growth with a unit of biomass abundance, and if we were to follow this approach, we would lose the understanding of the exact definition of the specific growth rate coefficient.

The principle often called the Michaelis–Menton equation or theory quite simply addresses the fact that even with food available in abundance, the rate at which microbes can assimilate that food is limited by the rate at which intracellular enzymes can be brought to bear in the conversion process. A very simplistic, stepwise chemical reaction may be employed to describe the general overall process:

$$\mathrm{S} + \mathrm{E} \rightarrow \mathrm{ES} \rightarrow \mathrm{E} + \mathrm{P}$$

The substrate (S, food or nutrient for the biomass) combines with an enzyme (E) to form an enzyme–substrate complex (ES), which upon further reaction produces the product (P) and the regenerated enzyme, which is recycled through the reaction. The rate of the reaction is highly dependent upon the availability of enzymes, which are often in short supply, leading to what is termed an enzyme-limited reaction rate. We may put this into the context of an apple pie. Most of us enjoy eating a slice of freshly baked apple pie, particularly à la mode. If a baker put a slice of apple pie in front of any of us each day and if there were a ready supply of ice cream available from the freezer, the pie would be readily eaten each day. However, if that baker put a whole pie or multiple pies in front of us each day, each of us would arrive at some steady consumption rate of apple pie and ice cream, our maximum specific rate of apple pie consumption. Certainly our appetites and diet selections are much more complicated than those of microbes, but the principle is quite similar. The microbial conversion of organic matter to products may involve literally thousands of specific enzyme–substrate reactions, occurring both in parallel and in series. We prefer to represent that overall reaction with a single rate law.

The Michaelis–Menton equation was developed analytically from first principles and correlated with empirical observations. In parallel, the Monod equation was developed primarily from empirical observations and correlated with first principles. Some texts choose to identify the following relation as the Michaelis–Menton equation and some choose to identify it as the Monod equation. Both are essentially the same. The equation (by either name) relates the specific growth rate to the abundance of biodegradable food (substrate, S), the maximum theoretical specific growth rate (μ_{max}), and the half reaction rate coefficient (K_S):

$$\mu = \frac{\mu_{max} \cdot \mathrm{S}}{K_S + \mathrm{S}}$$

The parameter μ_{max} is the specific biomass growth rate: the reaction rate in the presence of a large overabundance of food (i.e., the substrate S or our component i).

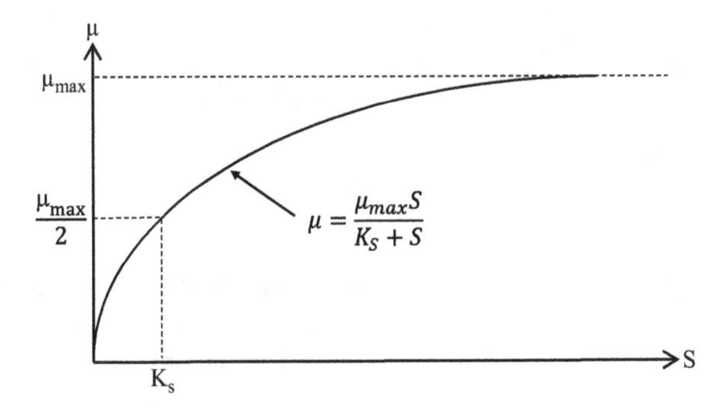

FIGURE 8.1 *The enzyme-limited microbial specific growth rate coefficient for constant biomass abundance (X).*

If we examine this relation in consideration of constant biomass abundance (constant X) and consider the value of μ as S grows large, in the limit as $S\rightarrow\infty$, $\mu\rightarrow\mu_{max}$. This relation is a hyperbola. The parameter K_S is the value of S such that $\mu = ½\cdot\mu_{max}$. This relation is shown graphically in Figure 8.1. We combine the definition of the specific growth rate coefficient with the first-order dependence of growth upon biomass abundance to fully define the true specific rate of biomass growth:

$$r_{G.X} = \frac{\mu_{max} \cdot S \cdot X}{K_S + S} \tag{8.10a}$$

In order to fully apply this concept to the conversion of substances in environmental systems, we need to consider the concept of biomass yield. Bacteria grow and divide. Newly formed bacteria then subsequently grow and divide. This process consumes food (substrate) such that the quantity of food consumed per unit of biomass grown can be defined as a yield coefficient. We can express this as the ratio of the rate of biomass production ($r_{G.X}$, mass of cells grown per time per unit volume, $M_{cells}/t/L^3$) per rate of substrate (food, our component i) consumed ($M_{food}/t/L^3$):

$$Y = -\frac{r_{G.X}}{r_S}$$

The yield coefficient (M_{cells}/M_{food}) is then the mass of biomass produced per unit (usually mass) of substrate (food) consumed, since r_X and r_S are both volume-specific rates. The negative sign arises from the fact that as biomass is produced, substrate is consumed.

We may now return to Equation 8.10 and substitute some results, arriving at a relation between the rate at which substrate is consumed, the substrate abundance, the yield coefficient, and the reaction rate coefficients:

$$r_S = -\frac{r_{G.X}}{Y} = -\frac{\mu_{max} \cdot S \cdot X}{Y \cdot \left(K_S + S\right)} \tag{8.11}$$

In parallel with the form of the pseudo-first-order reaction rate law, since for a particular reaction involving a particular type of biomass both μ_{max} and Y are constants, we may combine their quotient to obtain a rate law coefficient:

$$k' = \frac{\mu_{\max}}{Y}$$

The K_S coefficient of the Michaelis–Menton theory might be best remembered if we call it K_{half}. The substrate S is our component i and we may refer to its abundance as C_i. The resultant relation is of a form similar to that of Equation 8.9:

$$r_i = -\frac{k' \cdot X \cdot C_i}{K_{\text{half}} + C_i} \tag{8.12}$$

In the context of biomass growth and utilization of substrate, in order to be complete in our consideration of the principles, we need to address the death and decay of biomass. Biomass death is considered in the same context as growth – that the rate of biomass death, $r_{D.X}$, is directly proportional to the biomass abundance through the death/decay rate coefficient, k_D:

$$r_{\text{D.X}} = -k_{\text{D}} \cdot X$$

The net (also called observed) production of biomass is the algebraic sum of growth and death:

$$r_{\text{X.obs}} = r_{\text{G.X}} + r_{\text{D.X}} = (\mu - k_{\text{D}})X = \left(\frac{\mu_{\max} \cdot S}{K_{\text{S}} + S} - k_{\text{D}} \right) X \tag{8.13}$$

Equations 8.11 and 8.13 are often used in conjunction with the modeling of the activated sludge process for removal of organics manifested as biochemical oxygen demand (BOD) concurrent with the production of biomass.

8.3.2 Reactions in Completely Mixed Batch Reactors

A mass balance is applied to a constant volume that is considered to be perfectly mixed. The reactants are introduced into the reactor and, without further additions or withdrawals, allowed to react for some time period t. The abundance of the component of interest can be related to the rate law and time. For a batch reactor, the flow rates of the influent and effluent streams are both 0 and the mass balance reduces to the equality of accumulation and generation:

$$\frac{dM_i}{dt} = V_{\text{R}} \frac{dC_i}{dt} = V_{\text{R}} \cdot r_i \tag{8.14a}$$

The constant V_R may be divided out leaving the straightforward and familiar statement of the mass balance for the conversion of component i in a CMBR:

$$\frac{dC_i}{dt} = r_i \tag{8.14b}$$

The mathematical form for a pseudo-first-order rate law may be substituted for the specific reaction rate yielding a first-order initial value ordinary differential equation (ODE):

$$\frac{dC_i}{dt} = -k' \cdot C_i \tag{8.15}$$

We desire the relation between the abundance of our arbitrary component, the kinetic coefficient, and time. We separate Equation 8.15 and integrate from C_0 to C and t_0 to t to obtain this relation. We drop the i subscript since we know that C is C_i:

$$\int_{C_0}^{C} \frac{dC}{C} = \int_{t_0}^{t} -k'\, dt; \quad \text{and } \ln(C)\Big|_{C_0}^{C} = -k' t\Big|_{t_0}^{t}$$

The result is the logarithmic form of the relation we seek:

$$\ln\left(\frac{C}{C_0}\right) = -k' \cdot (t - t_0) \tag{8.16}$$

We exponentiate both sides of Equation 8.16 and obtain the relation which can be explicitly solved for the abundance of the target component as a function of time, the pseudo-first-order rate coefficient and the abundance at time t_0:

$$C = C_0 \cdot \exp(-k' \cdot (t - t_0)) \tag{8.17}$$

The mathematical form for a saturation-type rate law may be substituted for the specific reaction rate in Equation 8.14 yielding a second, distinct, first-order initial value ODE:

$$\frac{dC_i}{dt} = -\frac{k' \cdot X \cdot C_i}{K_{\text{half}} + C_i} \tag{8.18}$$

We separate Equation 8.18 and integrate to obtain an algebraic relation between the abundance of component i, time, the kinetic coefficients, and the abundance at time t_0. Again, we drop the i subscript:

$$-\frac{dC \cdot (K_{\text{half}} + C)}{k' \cdot X \cdot C} = dt; \rightarrow -\frac{1}{k' \cdot X}\left(K_{\text{half}} \cdot \ln(C)\Big|_{C_0}^{C} + C\Big|_{C_0}^{C}\right) = t\Big|_{t_0}^{t}$$

Employing the limits of the integration and performing a minor rearrangement leads to a relation implicit in concentration of our target substance, but still a very useful relation:

$$\frac{1}{k' \cdot X}\left(K_{\text{half}} \cdot \ln\left(\frac{C_0}{C}\right) + C_0 - C\right) = t - t_0 \tag{8.19}$$

With new capabilities of hand-held calculators, computational software such as Excel's solver and MathCAD's root() function, or given-find solve block, implicit relations such as Equation 8.18 are easily manipulated for numeric solutions.

8.3.3 Reactions in Plug-Flow Reactors

8.3.3.1 *The General PFR Mass Balance*

A schematic representation of an element of finite volume (cross-sectional area of A and thickness of Δz) arbitrarily located within a PFR is shown in Figure 8.2. Although not necessarily so, the schematic represents a reactor of constant cross-sectional area. The reactor shown may have boundaries that are physically defined such as a reactor in a chemical process, often called a tubular reactor by the chemical engineers. Alternatively, the reactor may be a cylindrical region of the subsurface with unit cross-sectional area through which ground water flows. In Chapter 7, we envisioned disk-like elements of fluid entering a PFR. Here, we are considering the volume occupied by one of those fluid elements, but keeping it fixed at an arbitrary location within the reactor. The elements of fluid passing through the reactor would, one after another, enter, occupy, and leave this volume.

We write a mass balance on an arbitrary component present in the influent to the reactor, dispensing with the subscript. Our mass balance considers the arbitrarily located volume element as the control volume. Overall, we have a full statement of the mass balance as shown in Figure 7.1:accumulation = in – out + generation

$$\text{accumulation} = \text{in out} + \text{generation}$$

$$\Delta V_R \frac{dC}{dt} = Q \cdot C\big|_z - Q \cdot C\big|_{z+\Delta z} + \Delta V_R \cdot r$$

where ΔV_R is the volume of the arbitrarily located element of the reactor volume, Δz is the thickness of the disk-shaped control volume, and $C|_z$ and $C|_{z+}\Delta_z$ are the abundances of our targeted component at the upstream and downstream boundaries of the volume element under consideration. Here we have considered that ΔV_R is constant in time and have moved it outside the differential ($d(\Delta V_R \cdot C) = \Delta V_R \cdot dC$). All other symbology is as earlier defined.

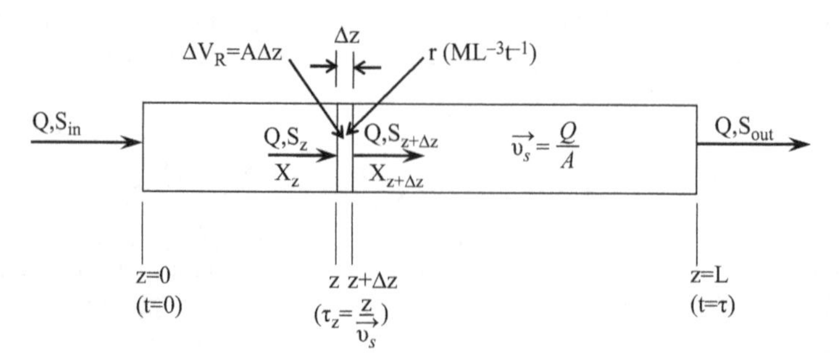

FIGURE 8.2 *Schematic representation of an arbitrarily situated finite volume element within a cylindrical plug flow reactor. Symbology is included such that the mass balance on an arbitrary component can be drawn upon the finite volume element.*

We consider that the flow, Q, and influent concentration of arbitrary component, C_{in}, remain constant in time. Then, with sufficient elapsed time, the overall process will attain a steady-state condition such that the concentration in the effluent as well as at each position within the reactor will arrive at a value steady in time. The concentration varies (between C_{in} and C_{out}) through the length of the reactor, but not with time, as long as Q and C_{in} are held constant. The rate of accumulation of mass within the control volume, and hence, the time derivative of the LHS of the mass balance relation goes to 0. We may substitute the product $A \cdot \Delta z$ for ΔV_R. Then we define the change in concentration across the control volume as the downstream concentration minus the upstream concentration:

$$\Delta C = C\big|_{z+\Delta z} - C\big|_{z}$$

The mass balance is then converted to a secondary, but not quite yet usable form:

$$0 = -Q \cdot \Delta C + A\Delta z \cdot r$$

We divide through by Δz and further rearrange this intermediate result. We state the RHS in two ways, using the quotient of area and flow or the superficial velocity, $\overrightarrow{v_s} = Q / A$:

$$\frac{\Delta C}{\Delta z} = \frac{A}{Q} \cdot r = \frac{r}{\overrightarrow{v_s}}$$

Now we take the limit of the expression as $\Delta z \to 0$, replacing Δ with d, arriving at a first-order ODE, which is of the boundary value type. This last step is exactly analogous to letting our fluid elements passing through the PFR of Chapter 7 grow sufficiently small to render their thickness infinitesimal:

$$\frac{dC}{dz} = \frac{A}{Q} \cdot r = \frac{r}{\overrightarrow{v_s}} \tag{8.20a}$$

Equation 8.20a, although in itself not yet quantitatively useful, is an important result, relating the change in concentration with position directly with the specific reaction rate for the generation/transformation of our arbitrary component. To render this relation quantitatively useful, we need only substitute each of our two rate laws for r and perform the necessary mathematics.

8.3.3.2 The Pseudo-First-Order Reaction Rate Law in PFRs

For the pseudo-first-order rate law, Equation 8.20a may be restated by substituting the rate law for the specific reaction rate, *r*:

$$\frac{dC}{dz} = -\frac{A \cdot k' \cdot C}{Q} = -\frac{k' \cdot C}{\overrightarrow{v_s}} \tag{8.20b}$$

We separate and integrate this result using the limits $C = C_{in}$ at $z = 0$ and $C = C_z$ at $z = z$:

$$\int_{C_{in}}^{C_z} \frac{dC}{C} = -\frac{A}{Q} \cdot k' \cdot \int_0^z dz = -\frac{k'}{\overrightarrow{v_s}} \int_0^z dz$$

$$\ln\left(\frac{C_z}{C_{in}}\right) = -\frac{A}{Q} \cdot k' \cdot z = -\frac{k'}{\overrightarrow{v_s}} z \tag{8.21a}$$

The logarithmic form is exponentiated to obtain a relation for C_z as a function of position:

$$C_z = C_{in} \cdot \exp\left(-\frac{A}{Q} \cdot k' \cdot z\right) = C_{in} \cdot \exp\left(-\frac{k'}{\overrightarrow{v_s}} z\right) \tag{8.22a}$$

If we consider the time an element of fluid resides in the reactor between the position $z = 0$ and $z = z$, we may substitute the quotient $t_z = \frac{A \cdot z}{Q} = \frac{z}{\overrightarrow{v_s}}$ into Equations 8.21a and 8.22b to yield relations exactly analogous to Equations 8.16 and 8.17:

$$\ln\left(\frac{C_z}{C_{in}}\right) = -k' \cdot t_z; \quad C_z = C_{in} \cdot \exp(-k' \cdot t_z)$$

where t_z is the time required for each entering fluid element to reach position z within the reactor. We may consider the PFR then in two different ways: the Lagrangian view wherein we would follow the fluid element through the PFR and realize that its reaction time t_z is related to the position within the reactor; or the Eulerian view wherein we position ourselves at a fixed point within the reactor and realize that by the time the entering fluid element reaches us, it has reacted for a time period t_z.

Most often we are interested in the concentration of the component in the effluent from the reactor and we need only substitute C_{out} for C_z and L for z:

$$\ln\left(\frac{C_{out}}{C_{in}}\right) = -\frac{A \cdot L}{Q} \cdot k' = -\frac{L}{\overrightarrow{v_s}} k' \tag{8.21b}$$

$$C_{out} = C_{in} \cdot \exp\left(-\frac{A \cdot L}{Q} \cdot k'\right) = C_{in} \cdot \exp\left(-\frac{L}{\overrightarrow{v_s}} k'\right) \tag{8.22b}$$

We recognize that the product $A \cdot L$ is V_R, and that $V_R/Q = \tau$, and also that $L / \overrightarrow{v_s} = \tau$, so we may restate Equations 8.21a and 8.22a to relate influent and effluent concentrations with the pseudo-first-order rate coefficient and the hydraulic residence time:

$$\ln\left(\frac{C_{\text{out}}}{C_{\text{in}}}\right) = -k' \cdot \tau \tag{8.21c}$$

$$C_{\text{out}} = C_{\text{in}} \cdot \exp(-k' \cdot \tau) \tag{8.22c}$$

8.3.3.3 *The Saturation Rate Law in PFRs*

For the saturation rate law, Equation 8.20a may be restated by substituting the saturation rate law for the specific reaction rate, r. Since we have examined the Lagrangian and Eulerian viewpoints with the pseudo-first-order rate law, we will consider the system only from the Eulerian viewpoint. We will also retain the biomass concentration, X, noting that this saturation rate law is most often associated with reactions that involve transformations of components of interest by biomass:

$$\frac{dC}{dz} = -\frac{A}{Q} \cdot \frac{k' \cdot X \cdot C}{K_{\text{half}} + C} \tag{8.20c}$$

We separate this first-order ODE and integrate between the limits of C_{in} and C_z and 0 and z:

$$-K_{\text{half}} \int_{C_{\text{in}}}^{C_z} \frac{dC}{C} - \int_{C_{\text{in}}}^{C_z} dC = \frac{A \cdot k' \cdot X}{Q} \int_0^z dz$$

The general relation may be rearranged (we invert the limits of the LHS integrations to discharge the negative signs) to yield a convenient form:

$$K_{\text{half}} \cdot \ln\left(\frac{C_{\text{in}}}{C_z}\right) + C_{\text{in}} - C_z = \frac{A \cdot k' \cdot X}{Q} z = k' \cdot X \cdot t_z \tag{8.23a}$$

This general relation, though implicit and necessarily solved using a numerical root-finding technique, yields a relation between C_z and z. To obtain the relation yielding C_{out}, we simply replace C_z with C_{out} and z with L. We may also again consider that $A \cdot L = V_R$ and $V_R/Q = \tau$:

$$K_{\text{half}} \cdot \ln\left(\frac{C_{\text{in}}}{C_{\text{out}}}\right) + C_{\text{in}} - C_{\text{out}} = k' \cdot X \cdot \tau \tag{8.23b}$$

8.3.4 Reactions in Completely Mixed Flow Reactors

A CMFR is shown schematically in Figure 8.3. Assumptions associated with a CMFR have been incorporated such that the effluent concentration of the arbitrary component is equal to the concentration of the arbitrary component in the reactor. The reactor volume is held constant as has been previously assumed and the boundaries of CMFRs are well defined.

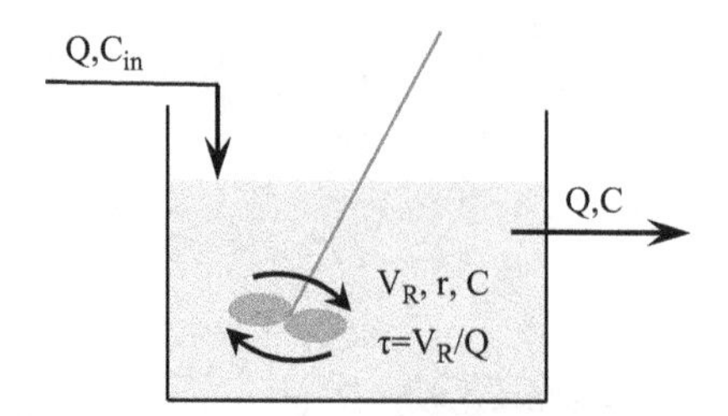

FIGURE 8.3 Schematic diagram of a CMFR with reaction of arbitrary component. A mixing propeller is shown but the mixing may be accomplished via various means.

As we did for the PFR, we perform a mass balance, but for the CMFR we must consider the entire reactor, noting that there is no spatial variation of concentration within the reactor. The overall mass balance on our arbitrary component may be written, neglecting subscripts:

$$V_R \frac{dC}{dt} = Q \cdot C_{in} - Q \cdot C + V_R \cdot r .$$

The unsteady case is interesting and important in certain applications, and we postpone its analysis for later. For now we hold C_{in} and Q constant over time and allow the reaction within the CMFR to come to a steady state such that the accumulation term becomes 0. We may further simplify the relation by dividing through by V_R and noting that $Q/V_R = 1/\tau$. We may also rearrange the result to provide a very succinct statement:

$$\frac{Q(C_{in} - C)}{V_R} = \frac{C_{in} - C}{\tau} = -r \tag{8.24}$$

Now we may substitute the two rate laws under consideration to arrive at specific relations between the effluent concentration and other parameters:

$$\frac{C_{in} - C}{\tau} = k' \cdot C \tag{8.25a}$$

$$\frac{C_{in} - C}{\tau} = \frac{k' \cdot X \cdot C}{K_{half} + C} \tag{8.26a}$$

Equation 8.25a may be rearranged to provide an explicit relation for C as a function of C_{in}, k', and τ:

$$C = \frac{C_{in}}{1 + k' \cdot \tau} \tag{8.25b}$$

Conversely, Equation 8.26a cannot be arranged to provide an explicit relation for C as a function of C_{in}, k′, X, K_{half}, and τ. However, algebraic expansion and rearrangement of Equation 8.26a yields a relation that is the classic second-order polynomial in C ($f(C) = 0 = a_2 \cdot C^2 + a_1 \cdot C + a_0$):

$$C^2+(K_{\text{half}}+\tau\cdot k'\cdot X-C_{\text{in}})\cdot C-K_{\text{half}}\cdot C_{\text{in}}=0 \qquad (8.26b)$$

where $a_2 = 1$, $a_1 = K_{\text{half}} + \tau\cdot k'\cdot X - C_{\text{in}}$, and $a_0 = -K_{\text{half}}\cdot C_{\text{in}}$. The effluent concentration is then found using the quadratic formula:

$$C=\frac{-a_1+\sqrt{a_1^2-4\cdot a_2\cdot a_0}}{2\cdot a_2}$$

Since we cannot physically have a negative value for a concentration, we seek only the positive root and the normally used ± preceding the square root is replaced by +.

8.3.5 Unsteady-State Applications of Reactions in Ideal Reactors

8.3.5.1 Unsteady-State CMFR

Let us return to our ODE describing the overall mass balance in terms of the general specific reaction rate. We may insert the pseudo-first-order rate law into the relation, divide through by V_R, and rearrange the relation to yield a succinct ODE:

$$\frac{dC}{dt}=\frac{C_{\text{in}}-C-\tau\cdot k'\cdot C}{\tau}$$

This result is separated and integrated (easily done using MathCAD's calculus palette) between the limits of C_0 and C_t and t_0 and t:

$$\frac{\ln\left(\dfrac{C_0-C_{\text{in}}+C_0\cdot\tau\cdot k'}{1+\tau\cdot k'}\right)-\ln\left(\dfrac{C_t-C_{\text{in}}+C_t\cdot\tau\cdot k'}{1+\tau\cdot k'}\right)}{1+\tau\cdot k'}=\frac{t-t_0}{\tau}$$

This intermediate result may be rearranged to yield either an explicit relation for time as a function of C_t or for C_t as a function of time:

$$\frac{t-t_0}{\tau}=\frac{\ln\left(\dfrac{C_0-C_{\text{in}}+C_0\cdot\tau\cdot k'}{C_t-C_{\text{in}}+C_t\cdot\tau\cdot k'}\right)}{1+\tau\cdot k'} \qquad (8.27a)$$

$$C_t=\frac{C_{\text{in}}+(C_0-C_{\text{in}}+C_0\cdot\tau\cdot k')\cdot\exp\left(-\dfrac{t-t_0}{\tau}(1+\tau\cdot k')\right)}{1+\tau\cdot k'} \qquad (8.27b)$$

Equations 8.27a and 8.27b are useful if we are interested in analyzing the progression of a reactor between startup and the steady-state condition or the progression from one steady-state condition to another, where influent conditions need to be changed. We can certainly observe significant similarities between Equations 8.27 and 7.5. The inclusion of reaction with mixing simply requires an enhancement of the basic structure of the mathematical result.

Use of the saturation rate law for unsteady-state analyses in CMFRs leads to a differential equation for which a closed-form analytic solution would be extremely difficult to obtain at best:

$$\frac{dC}{dt} = \frac{C_{\text{in}} - C}{\tau} - \frac{k' \cdot X \cdot C}{K_{\text{half}} + C}$$

Attempts to develop that closed-form solution are beyond the scope of the current discussion. However, numerous methods are available for obtaining approximate, numeric solutions for both single and systems of ODEs.

8.3.5.2 The Fed-Batch Reactor

For the fed-batch reactor, we simply include a flow and associated concentration as influent to the batch reactor in the mass balance. A significant complication arises in that the volume of the reactor is no longer a constant. We write the overall mass balance as an ODE, employing first the pseudo-first-order rate law.

$$\frac{d(V_{\text{R}}C)}{dt} = Q \cdot C_{\text{in}} - V_{\text{R}} \cdot k' \cdot C$$

We expand the LHS using the chain rule, use the definition that $\frac{dV_{\text{R}}}{dt} = Q$, and employ the relation that $V_{\text{R}} = V_0 + Q \cdot t$:

$$Q \cdot C + (V_0 + Q \cdot t)\frac{dC}{dt} = Q \cdot C_{\text{in}} - (V_0 + Q \cdot t) \cdot k' \cdot C$$

We rearrange the result to collect coefficients of dC/dt and C, and rearrange further such that the coefficient of $dC/dt = 1$:

$$\frac{dC}{dt} + \left(\frac{Q}{V_0 + Q \cdot t} + k'\right)C = \frac{Q \cdot C_{\text{in}}}{V_0 + Q \cdot t}$$

This result is a linear, first-order ODE. We use the integrating factor prescribed by Wylie (1966) to obtain the general solution and employ the initial condition that at $t = 0$, $C = C_0$ to evaluate the constant of integration. Some further rearranging yields the final relation for C as a function of time:

$$C = \left(\frac{C_{\text{in}} \cdot \text{e}^{k' \cdot t}}{k'} - \frac{C_{\text{in}}}{k'} + C_0 \cdot \frac{V_0}{Q}\right) \cdot \frac{Q}{V_0 + Q \cdot t} \cdot \text{e}^{-k' \cdot t} \qquad (8.28)$$

This result may then be applied for analyses of reactions in fed-batch reactors with a pseudo-first-order rate law, when k' is constant.

Consideration of the saturation rate law for a fed-batch reactor leads to a similar, but more mathematically complex, ODE than that for the pseudo-first-order rate law:

$$\frac{dC}{dt} = \frac{Q(C_{\text{in}} - C)}{(V_0 + Q \cdot t)} - \frac{k' \cdot X \cdot C}{K_{\text{half}} + C}$$

We quickly observe that the resultant ODE is nonlinear, with noninteger powers of C on the RHS. Attempts to develop a closed-form solution for this case would be beyond the scope intended herein. We realize that since we have isolated dC/dt on the LHS, we would be able to readily develop numeric approximations using Euler's or one of the Runge–Kutta methods.

8.3.5.3 *Time-Variant Analyses of the PFR*

We might ponder development of relations for the unsteady case for the PFR. These would undoubtedly result in partial differential equations (PDEs) in time and position. Development of closed-form analytic solutions or examinations of numerical approximations of these PDEs are well beyond the scope of the discussions intended herein.

8.4 APPLICATIONS OF REACTIONS IN IDEAL REACTORS

In order that we may apply reaction/reactor principles in analyses of environmental systems of interest we must first answer a few questions:

1. What is the specific system in question – are the boundaries real and physical or must we envision a representative reactor volume (RRV)?
2. What is the reactant of interest and what is the overall reaction?
3. What is the law expressing the rate of reaction in terms of rate coefficients and abundances of products and reactants?
4. What ideal reactor system can best describe the system in which the process takes place?

Often the system is well defined by real boundaries – the walls of an open biological reactor basin, the sediment–water and vapor–liquid interfaces of a water body, or the tankage associated with a fermentative process for ethanol production. In natural systems, the boundaries are not always well defined – the mixing and reaction zone beneath a solid-waste landfill or the porous medium through which contaminated groundwater flows. In these cases, we need to examine whether we need real boundaries or whether we must visualize the boundaries of the system to be investigated. For this latter case, we must apply the concept of the RRV. A RRV would comprise a unit cross-sectional area and known reactor length to which we may apply the appropriate reactor model. The most significant property of a RRV is that the RRV is identical in all ways to all other such volumes comprising the overall system represented by the RRV.

Research into the specific reactants of interest, oftentimes toxic or hazardous components or oxygen-demanding substances, is necessary to identify the specific reaction. Particularly with many pseudo-first-order reactions, the abundances of all reactants contribute to the forward rate of the reaction while abundances of products tend to reduce the forward rate of the reaction. Herein, we consider reactions with rates that are predominantly dependent upon reactants while having little or no inverse dependence upon the abundances of products. Specific reactions have been very well defined by chemists, biochemists, and microbiologists, and herein we do not attempt to compile a database of chemical and biochemical reactions. Rather, our focus will be upon the process for application of the chemistries once they are identified and understood. Most often, environmental process analysis requires specific research by the engineer to develop necessary understandings of the process or chemistry.

The science associated with chemical reaction kinetics is highly empirical in practice. The magnitudes of rate coefficients and dependences of those rate coefficients upon the abundances of reactants are derived almost wholly from experiment. Certain reactions have well-known (as a consequence of significant experimental examination) kinetics while the kinetics of others are poorly quantitatively understood. Thus, the rate laws employed in environmental process analysis often need to be defined from the scientific and engineering literature on a case-by-case basis as necessary to analyses of targeted environmental systems. Herein, we do not attempt to develop a database of kinetic information, but rather use information that we are able to identify in the examination of targeted systems. Our focus is upon the **use** of quantitative kinetic information, once known, in the analyses of environmental processes.

The choice of a reactor system is dependent upon the character of the environmental system in question. No real reactor is perfectly ideal in the context of either plug flow or complete mixing. In some cases, analyses of real reactors using ideal reactor principles would lead to significant error. In a subsequent chapter, we examine some strategies useful in analyses of nonideal reactors. Herein, however, we strive to identify the aspects of examined systems that would permit us to select one of the three ideal reactor systems (or one of the modifications discussed earlier) with which we may apply the known chemistry and quantitative understandings of reaction rates.

8.4.1 Batch Reactor Systems

In the strict sense of a batch reactor, both inputs to and outputs from the reactor after the time of initiation of a reaction are 0. If we determine that we must apply this concept as "black and white," seldom would we enable ourselves to employ the ideal batch reactor. We must back away from that strict definition and suggest that we can apply the batch reactor principle in systems whose inputs and outputs would be insignificant relative to the process of interest.

Oxidation of ferrous iron is an important water treatment process. Biological processes that occur in ground waters often deplete oxygen. Then ferric iron (Fe(III)), if present, is used as an electron acceptor and converted to ferrous iron (Fe(II)). Fe(II) is far more soluble in than Fe(III) and is considered a nuisance constituent. When present in water pumped to consumers Fe(II) is slowly oxidized back to Fe(III). Use of such water turns plumbing fixtures rust-colored and often results in plugged pipes. Iron is often removed using a specific water treatment process, involving oxidation of the ferrous iron to ferric iron, formation of a ferric hydroxide precipitate, and removal of the precipitate from the finished water by sedimentation/filtration. The oxidation of the iron is the governing rate process and is often accomplished by bubbling air through the water.

Example 8.3 Consider that groundwater is pumped to the surface, where normal ambient pressure is 0.9 atm, from an aquifer containing ferrous iron (Fe(II)) at a concentration of 5 ppm_m and with a pH of 6.4. If a sample of this groundwater were placed in a large beaker in the laboratory and had air bubbled through it when the atmospheric pressure is 0.9 atm, how long would it take to reduce the Fe(II) concentration to 5 ppb_m? Explore the time-variant behavior of the Fe(II) concentration in the water, the effect of raising the pH to 6.9 and 7.4, and the effect of using high-purity oxygen ($Y_{O_2} = 0.95$) on the time to reach the target Fe(II) concentration level.

We search the literature and find that the rate of oxidation of ferrous iron to ferric iron follows a rate law with dependence on pH and upon the partial pressure of oxygen in the gas bubbled through the water (Stumm and Morgan, 1996):

$$r_{\mathrm{Fe(II)\rightarrow Fe(III)}} = -k \cdot P_{O_2} \cdot [\mathrm{OH}]^2 \cdot [\mathrm{Fe(II)}]$$

In this relation, if we consider the partial pressure of oxygen in the gas and the hydroxide ion concentration to be constant, we have a pseudo-first-order reaction rate law where $k' = k \cdot P_{O_2} \cdot [\mathrm{OH}]^2$. We also find that the magnitude of k is ~8×10^{13} $L^2/mol^2/atm/min$. We will need to keep careful track of the units to ensure that when fully evaluated, the units of k' are min^{-1}.

We first address the system at pH 6.4 and begin by computing the magnitude of k':

$pH := 6.4 \quad H := 10^{-pH} \quad K_w := 10^{-14} \quad OH := \frac{K_w}{H} = 2.512 \times 10^{-8} \quad \frac{mol}{L}$

$P_{Tot} := 0.9 \quad atm \quad Y_{O2} := 0.209 \quad P_{O2} := P_{Tot} \cdot Y_{O2} = 0.188 \quad atm$

$k := 8 \times 10^{13} \quad \frac{L^2}{mol^2 \cdot atm \cdot min} \quad k_{pr} := k \cdot P_{O2} \cdot OH^2 = 9.495 \times 10^{-3} \quad min^{-1}$

Note that while merging units of atmospheres and moles per liter, the unit of k' is inverse time (min^{-1}) and we can choose whatever unit we wish for the concentration of Fe(II) – certainly ppm_m (mg/L) is most convenient. Let us write a MathCAD function for Fe(II) named $Fe_2(t)$ and explore the time-variant behavior of the ferrous iron concentration with a plot of $Fe_2(t)$ versus time:

$Fe_{2.0} := 5 \quad Fe_{2.target} := 0.005 \quad \frac{mg}{L} \quad Fe_2(t) := Fe_{2.0} \cdot \exp(-k_{pr} \cdot t)$

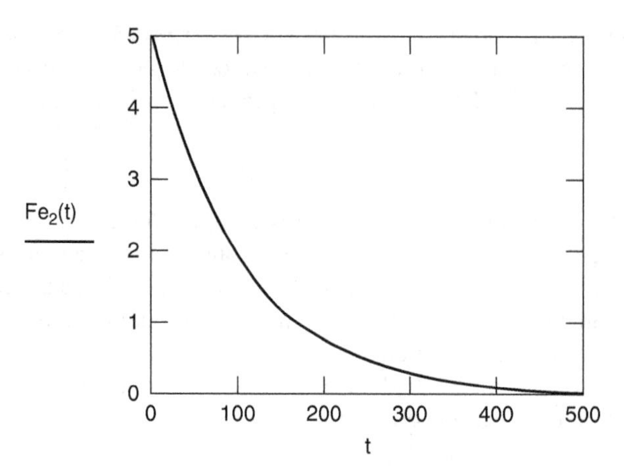

FIGURE E8.3.1 *Plot of ferrous iron versus time in a batch process using oxygen in air as an oxidant.*

We plot this function in Figure E8.3.1 and observe exponential decay of the Fe(II) concentration as expected. The time required to reach the target [Fe(II)] can be computed either from the written function or from the logarithmic form of the relation:

$$t_f := \frac{\ln\left(\frac{Fe_{2.target}}{Fe_{2.0}}\right)}{-k_{pr}} = 728 \quad \text{min}$$

We can easily assess the effects of altering the pH. MathCAD's matrix capability allows for a succinct presentation of the result:

$$pH := \begin{pmatrix} 6.9 \\ 7.4 \end{pmatrix} \quad OH := \overrightarrow{\frac{K_w}{10^{-pH}}} \quad k_{pr} := \overrightarrow{\left(k \cdot P_{O2} \cdot OH^2\right)} \quad t_f := \overrightarrow{\frac{\ln\left(\frac{Fe_{2.target}}{Fe_{2.0}}\right)}{-k_{pr}}} = \begin{pmatrix} 73 \\ 7 \end{pmatrix} \quad \text{min}$$

As expected, raising the pH has a profound effect upon the needed reaction time. We then examine the effect of employing high-purity oxygen, including the initially posed pH value and again using the matrix capability:

$$pH := \begin{pmatrix} 6.4 \\ 6.9 \\ 7.4 \end{pmatrix} \quad OH := \overrightarrow{\frac{K_w}{10^{-pH}}} \quad Y_{O2} := 0.95 \quad P_{O2} := P_{Tot} \cdot Y_{O2}$$

$$k_{pr} := \overrightarrow{\left(k \cdot P_{O2} \cdot OH^2\right)} = \begin{pmatrix} 0.043 \\ 0.432 \\ 4.316 \end{pmatrix} \quad t_f := \overrightarrow{\frac{\ln\left(\frac{Fe_{2.target}}{Fe_{2.0}}\right)}{-k_{pr}}} = \begin{pmatrix} 160 \\ 16 \\ 2 \end{pmatrix} \quad \text{min}$$

With our modeling approach we can determine that raising the pH, necessitating the employment of a reagent with its associated capital costs for chemical feed and associated material costs can drastically reduce the reaction time; thus, reducing the time necessary to accomplish the reaction, which may have financial benefits. Use of high-purity oxygen can further reduce the reaction time with associated capital and operational expenses. We can use our model to help us understand the capital and operating costs and enable a judgment regarding the most cost-effective means of implementing the process to produce water of the desired final Fe(II) content.

Let us turn our attention now to a system that employs the saturation-type reaction rate law. Phenol is a byproduct of many industrial processes and its presence in byproduct streams is often mitigated using biological processes.

Example 8.4 Consider that a wastewater containing phenol (C_6H_5OH) at a concentration level of 1.0 mmol/L is to be treated in a biological process under batch conditions. The biomass responsible for accomplishing the treatment is to be added to a 10 m^3 reactor that is originally charged with a solution containing the phenol at the specified initial abundance level. It is believed that a biomass concentration at the initiation of the reaction process of 500 mg/L will effectively render the final concentration to be 0.005 mmol/L or lower in a reasonable time period. Examine the effect of using higher or lower biomass concentrations upon the time required to reach the target final concentration level.

Biological degradation of phenol can be modeled using the saturation rate law:

$$r_{Ph} = -\frac{k' \cdot X \cdot C_{Ph}}{K_{half} + C_{Ph}}$$

Typical values of k' and K_{half} would be $\sim 20 \frac{mmol_{ph}}{g_{BM} \cdot d}$ and $\sim 0.100 \frac{mmol}{L}$.

We need to reconcile the units based on grams of biomass to our use of mg_{BM}/L:

$$C_{Ph.0} := 1 \quad C_{Ph.target} := 0.005 \frac{mmol}{L}$$

$$k := \frac{20}{1000} = 0.02 \frac{mol}{mg_{BM} \cdot d} \quad K_{half} := 0.100 \frac{mmol_{Ph}}{L} \quad X := 500 \frac{mg_{biomass}}{L}$$

In order to explore the time-variant nature of the concentration of phenol we need to consider that our governing Equation 8.19 cannot be explicitly solved for the concentration at time t. Fortunately with the use of MathCAD's root function and its rather convenient-to-use programming palette, we may write a pair of programs that will store values of time and associated values of phenol concentration in two one-dimensional matrixes (the MathCAD help resource calls them vectors)

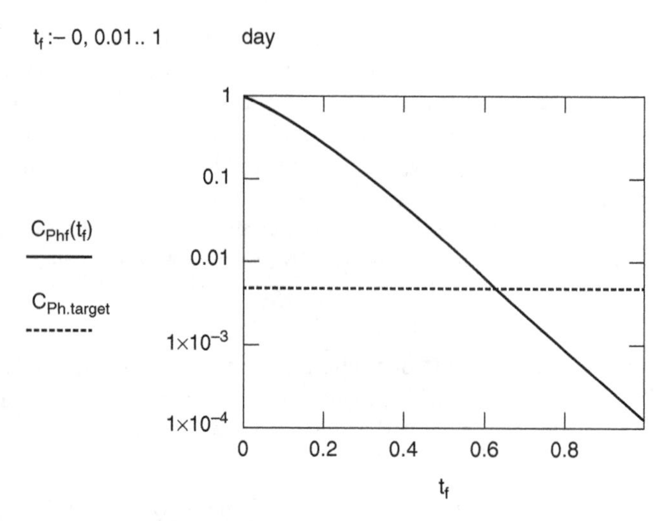

FIGURE E8.4.1 *Plot of phenol concentration versus time for a process conducted in a batch reactor.*

and hence produce a plot as Figure E8.4.1 of phenol concentration versus time for the stated set of conditions:

$C_{Ph} := 0.0001$ We need to venture an initial guess for the root() function.

$$C_{Phf}(t_f) := \text{root}\left[\frac{1}{k \cdot X} \cdot \left(\ln\left(\frac{C_{Ph.0}}{C_{Ph}}\right) + C_{Ph.0} - C_{Ph}\right) - t_f, C_{Ph}\right] \qquad \frac{\text{mmol}}{\text{L}}$$

The plot of C_{Phf} ($[C_6H_6O]$) versus time is not log-linear (note the logarithmic ordinate axis) in the region where $C_{Phf} > K_{half}$, as a consequence of the saturation-type reaction rate law. Then, once C_{Phf} is reduced below K_{half}, the influence of the phenol concentration in the denominator of the rate law is diminished and the rate law behaves as if it were pseudo-first-order. In many cases, then, if reactant concentrations are at or below the level of the half-reaction coefficient, K_{half}, we may, with minimal error, employ the saturation rate law as a pseudo-first-order rate law by neglecting the reactant concentration in the denominator. For the function employing the root(), it was necessary to adjust the magnitude of the initial guess downward from that initially employed in order that the MathCAD worksheet would converge for the full range of t_f values specified.

An alternative formulation of the solution employing MathCAD's programming palette, allowing adjustment of the initial guess for the root function, is shown in Figure E8.4.2. The values of $C_{Phf}(t)$ are stored in a 101 row vector for subsequent plotting. We plot the result in Figure E8.4.3. An alternative abscissa variable was necessarily defined in order to allow plotting of the target phenol concentration on the plot. The ~0.6 day reaction time may not be sufficiently rapid for our purposes. Adjusting the biomass abundance would allow us to examine the merits.

$$t_{Tot} := 1 \quad \text{day} \quad t_1 := 0, \frac{t_{Tot}}{100} .. t_{Tot}$$

$$\begin{pmatrix} C_{Ph.f} \\ t \end{pmatrix} := \left| \begin{array}{l} C_{Ph.f_0} \leftarrow C_{Ph.0} \\ \text{for } i \in 1..100 \\ \quad \left| \begin{array}{l} t_i \leftarrow \frac{i}{100} \cdot t_{Tot} \\ C_{Ph} \leftarrow C_{Ph.f_{i-1}} \\ C_{Ph.f_i} \leftarrow \text{root}\left[\frac{1}{k \cdot X} \cdot \left(\ln\left(\frac{C_{Ph.0}}{C_{Ph}}\right) + C_{Ph.0} - C_{Ph}\right) - t_i, C_{Ph}\right] \end{array} \right. \\ \text{return } \begin{pmatrix} C_{Ph.f} \\ t \end{pmatrix} \end{array} \right.$$

FIGURE E8.4.2 *Screen capture of a logical program to implement the Root() function to produce a concentration versus time tracesss for a CMBR.*

We need either to plot $C_{Ph.f}$ against t and $C_{Ph.target}$ against t_1, or to create a vector of 101 $C_{Ph.target}$ values to plot against t.

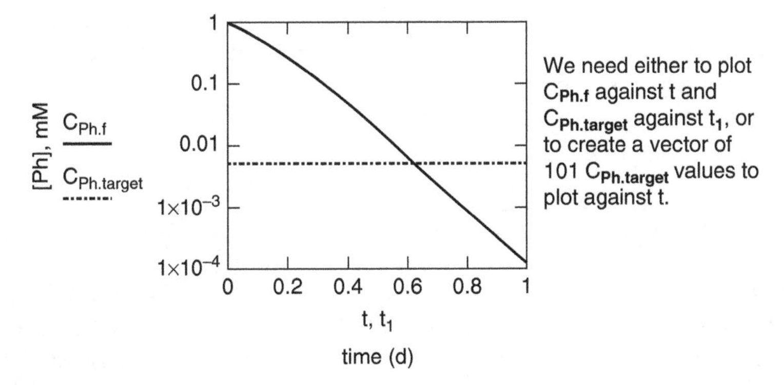

FIGURE E8.4.3 *A plot of phenol concentration identical to that of Figure E8.4.1, but generated using MathCAD's capability to employ matrix operations.*

We may investigate the effect of the biomass concentration by using the initial and target phenol concentrations with varying biomass concentration and using Equation 8.19 explicitly solved for the reaction time. A MathCAD function and the *X–Y* plot of Figure E8.4.4 work well for this purpose:

$$X := 10, 20 .. 2000 \qquad t(X) := \frac{1}{k \cdot X} \cdot \left(\ln\left(\frac{C_{Ph.0}}{C_{Ph.target}}\right) + C_{Ph.0} - C_{Ph.target}\right)$$

We might be interested in the biomass concentration associated with attainment of the target C_{Phf} in a tenth of a day. We may obtain this using at least two different paths. We may use this function in a given-find block:

$$X := 2000 \quad \text{Given} \quad t(X) = 0.1 \quad X_{0.1} := \text{Find}(X) = 3.147 \times 10^3 \quad \frac{mg_{BM}}{L}$$

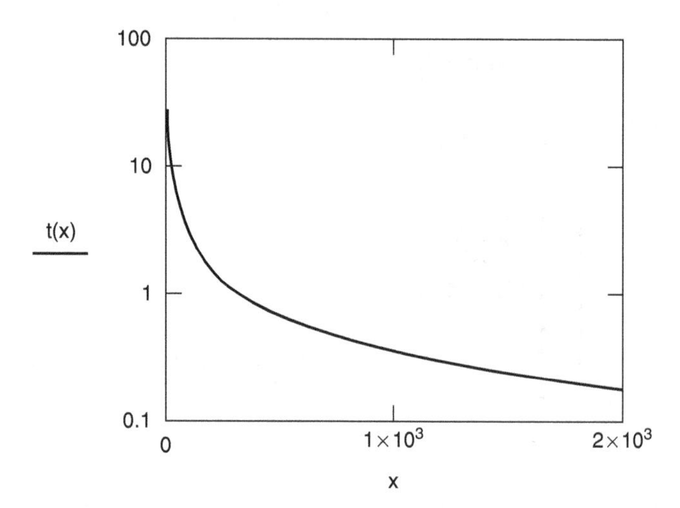

FIGURE E8.4.4 *A plot of time required to accomplish the desired phenol reduction using the biomass concentration as the independent variable.*

Alternatively, we may rearrange Equation 8.19 to solve for the biomass concentration for the stated constraints:

$t_{target} := 0.1 \quad \text{day}$

$$X_{0.1} := \frac{1}{k \cdot t_{target}} \cdot \left(\ln\left(\frac{C_{Ph.0}}{C_{Ph.target}} \right) + C_{Ph.0} - C_{Ph.target} \right) = 3.147 \times 10^3 \quad \frac{mg_{BM}}{L}$$

In these analyses of the microbial degradation of phenol in a batch reactor, we have neglected the fact that as the phenol would be converted to by-products including water and carbon dioxide, biomass would be grown. Hence, the biomass concentration would tend to increase. If this increase is small relative to the actual biomass concentration, the assumption of constant biomass concentration introduces but a small error. In order that biomass growth (and death) be quantitatively considered, a second rate law describing the growth of biomass is necessary. Even with the addition of a simple additional coupled ODE, we no longer would be able to employ any analytic solution, as none can be mathematically obtained. We will examine coupled contaminant degradation and biomass growth later in this text.

8.4.2 Plug-Flow Reactor Systems

In the strict sense of the PFR, elements of fluid enter and exit the reactor in exactly the same order. In truth, some degree of forward and backward mixing of fluid among elements occurs. This phenomenon is often referred to as dispersion. In order to create conditions that minimize dispersion, engineered reactors are generally modified in one of two ways. For closed reactors with a "straight-through" flow path, packing is used to ensure a high degree of local mixing and lateral dispersion, which minimizes longitudinal dispersion. Many varieties of manufactured packing can be obtained and each has its advantages and disadvantages. For reactors that are open basins of overall rectangular shape, baffling is employed to route the flow in pathways such that elements of fluid follow a guided, longer path through the reactor than would be the case in the absence of baffling. For natural systems, we can identify flow regimes resembling plug flow by the presence of porous media in groundwater systems or by the long, narrow (when viewed from the large-scale perspective) reactor system presented by a river or stream.

Example 8.5 A reactor of design that may, with little error, be modeled as a PFR has a bed volume of 5 m^3. The total volume of the reactor is 6.25 m^3 and the total porosity of the packing material as packed in the reactor is 0.8. The bed volume is the volume of solution contained within the pores of the packing with which the reactor is filled and is the product of total volume and total porosity. Consider that this reactor is to be employed for a pseudo-first-order reaction having an overall rate constant $k' = 0.1$/min. Consider that the design flow is to be 100 L/min and that the influent concentration of the substance targeted for removal from the flow stream is 0.1 mol/L. Let us investigate the relation between concentration of the target reactant and position within the reactor and then determine the concentration of the target substance in the effluent from the reactor.

Let us first compute the bed volume and residence time and assign some known values:

$$\varepsilon_{BED} := 0.8 \ \frac{m_{pore}^3}{m_{Tot}^3} \quad V_{Tot} := 6.25\ m^3 \quad V_R := V_{Tot} \cdot \varepsilon_{BED} \cdot 1000 = 5 \times 10^3\ L$$

$$Q := 100\ \frac{L}{min} \quad \tau_{Tot} := \frac{V_R}{Q} = 50\ min \quad k_{p.pfo} := 0.1\ min^{-1}$$

We may now invoke Equation 8.22b to compute the effluent concentration from the reactor:

$$C_{inf} = 0.1 \quad C_{eff} := C_{inf} \cdot \exp(-k_{p.pfo} \cdot \tau_{Tot}) = 6.738 \times 10^{-4}\ \frac{mol}{L}$$

Suppose that we observe this result and realize that the computed effluent concentration is lower than the design value, $C_{eff} = 0.001$ mol/L. We might wish to determine the flow rate that could be applied to the reactor to attain the target result. We would employ Equation 8.21b for this computation. We could first compute the alternative residence time and then the alternative flow rate:

$$C_{eff} := 0.001\ \frac{mol}{L} \quad \tau_{Tot} := \frac{\ln\left(\frac{C_{eff}}{C_{inf}}\right)}{-k_{p.pfo}} = 46.1\ min \quad Q := \frac{V_R}{\tau_{Tot}} = 108.6\ \frac{L}{min}$$

Here it would be of value to illustrate the concentration profile (value of concentration versus position in the reactor) across the reactor. We have stated nothing about the exact configuration of the reactor other than its strong resemblance to an ideal PFR. Regardless of the exact configuration, we may suggest that a unique flow path exists, along which each element of fluid entering the reactor follows the element ahead of it and precedes the element behind it. We would also postulate that the fluid velocity through the reactor is constant. The fraction of the total residence time at position z relative to the total length of the flow path, L_{FP}, is identical to the fraction of the total residence time associated with the travel time from the influent ($z = 0$) to position z.

We realize that the ratio $\frac{\tau_z}{\tau_{Tot}}$ is identical to $\frac{z}{L_{FP}}$. We can visualize the concentration profile as the series of effluent concentrations associated with successively increasing the length of the flow path by an incremental Δz, correspondingly increasing the reactor residence time by $\Delta\tau_z$. In this case, we will somewhat arbitrarily use $\Delta t_z = \frac{\tau_{Tot}}{50}$. We write a function using the incremental value of τ and plot the result in Figure E8.5.1:

$$\tau_z := 0, \frac{\tau_{Tot}}{50} .. \tau_{Tot} \qquad C(\tau_z) := C_{inf} \cdot \exp\left[-k_{p.pfo} \cdot \left(\frac{\tau_z}{\tau_{Tot}}\right) \cdot \tau_{Tot}\right]$$

$$C(\tau_{Tot}) = 1 \times 10^{-3} \quad \frac{mol}{L}$$

As expected, the in figure E8.5.1 profile across the reactor is "log-linear," conforming to the governing relation which is a decaying exponential.

A significant additional consideration with this context involves the sizing of the reactor itself when constrained by the influent and effluent concentrations of the target reactant and the influent flow rate. Let us suppose the influent and effluent concentrations

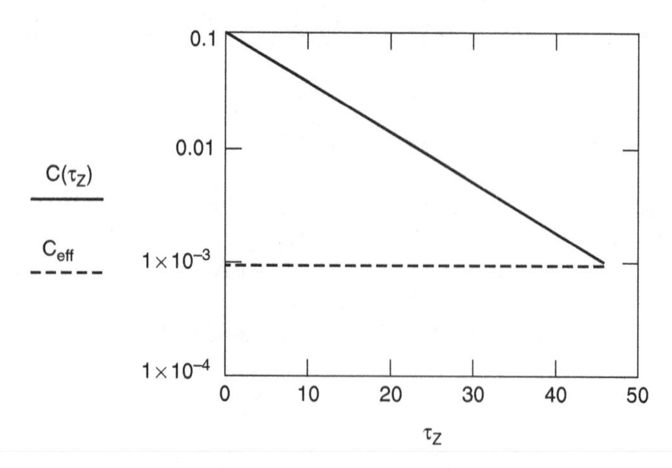

FIGURE E8.5.1 *Plot of reactant abundance versus position in an ideal PFR for arbitrary reactant with transformation governed by a pseudo-first-order rate law.*

are 0.1 and 0.0001 mol/L, respectively, and the flow is 100 L/min. We can find the necessary residence time and, hence, the volume of the reactor using Equation 8.21b.

Since many of the parameters have been specified in our MathCAD worksheet, we need only specify the new target effluent value and invoke Equation 8.21b:

$$C_{eff} := 0.0001 \quad \tau_{Tot} := \frac{\ln\left(\frac{C_{eff}}{C_{inf}}\right)}{-k_{p.pfo}} = 69.1 \text{ min} \quad V_R := Q \cdot \tau_{Tot} = 7.5 \times 10^3 \text{ L}$$

Our final consideration for this context involves examination of the pseudo-first-order rate coefficient. We have not considered a particular reaction here but merely suggested that it is pseudo-first-order, so we may have several factors potentially under our control through which we might manipulate the magnitude of the rate constant. Perhaps we can adjust the solution composition or the temperature, or employ an alternative set of supporting reactants. The necessary magnitude of the rate constant can be computed based on our final set of process constraints. Suppose the influent and effluent concentrations must again be 0.1 and 0.0001 mol/L, the flow would remain at 100 L/min, but the bed volume of the reactor would be limited to 2500 L. Let us determine the necessary value of the pseudo-first-order rate constant.

Since the MathCAD worksheet already contains many of the necessary parameter values, we need only define new constraining reactor volume, and again, Equation 8.21b serves our purpose:

$$V_R := 2500 \text{ L} \quad \tau_{Tot} := \frac{V_R}{Q} = 23.026 \text{ min} \qquad k_{p.pfo} := \frac{\ln\left(\frac{C_{eff}}{C_{inf}}\right)}{-\tau_{Tot}} = 0.3 \text{ min}^{-1}$$

Then, knowing the necessary magnitude of the rate constant, we can examine the adjustments necessary to attain the desired performance.

The high-rate activated sludge process is accomplished in a reactor that is configured to attain a near-plug-flow configuration. The activated sludge process is employed to convert both soluble and insoluble biodegradable organic matter into biomass. High-rate systems are often designed specifically to address organic carbon substrates in a manner as efficient as possible with regard to reactor sizing and consumption of oxygen. Removal of nitrogen and phosphorus, if necessary, is left to additional processes. Our next example targets such a reactor.

Example 8.6 Consider a rectangular, baffled reactor (see Figure E8.6.1) with a total volume of 10,000 m^3 (2.64 MG) that receives a flow of 0.694 m^3/s (6×10^4 m^3/day, 15.85 MGD). The contaminant of concern is an oxygen-demanding organic material (both soluble and particulate) measured as chemical oxygen demand (COD). The level of COD in the influent is 450 g/m^3. The biomass employed in the reaction is provided via a recycle line through which a concentrated stream (4500 g_{VSS}/m^3, often called return activated sludge) is returned to the influent for introduction into the bioreactor. This biomass-containing recycle stream originates from the underflow

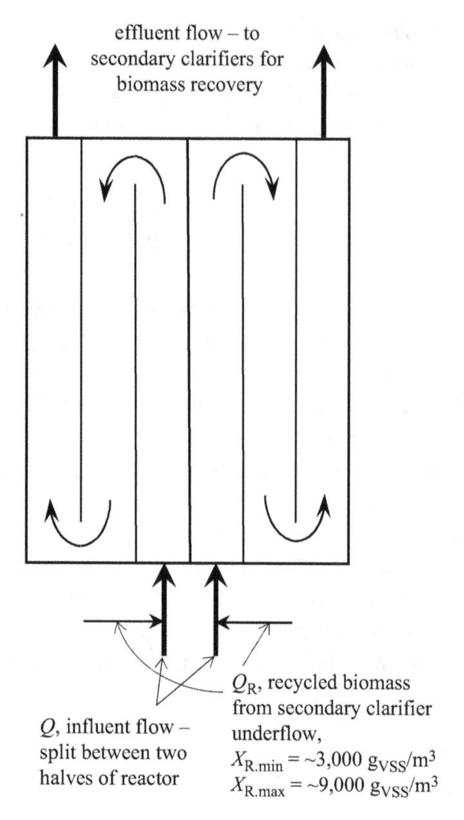

FIGURE E8.6.1 *Schematic plan view of an activated sludge reactor arranged to approximate plug flow conditions.*

of the sedimentation basin immediately following the bioreactor. The quantity of VSS is the typical measure of the abundance of biomass in biological reactors. To determine VSS, solids are filtered from a known volume of sample, dried to obtain total suspended solids (TSS), and then fired in a muffle furnace at 550–600 °C to drive off the combustible portion. The difference in mass between the dried solids and the residual, after firing, is the VSS. The overall schematic of a reactor system with cell recycle, presented in Figure E8.1.1, is modified for inclusion here as Figure E8.6.2. Rather than nitrate, the reactant of interest here is merely biodegradable organic material, manifest as COD. The clarifier serves simply to separate the biomass from the effluent stream and produce a high-concentration stream for mixing with the influent to produce a seed of biomass to increase the rate of the reaction. Examine the effect associated with the manipulation of the recycle flow rate (Q_R) upon the performance of the reactor.

The process under investigation is characterized as high-rate activated sludge. Typical magnitudes of k' and K_{half} suggested in the literature for the saturation-type reaction rate law are 10 $\frac{g_{COD}}{g_{VSS} \cdot d}$ and 40 g_{COD}/m^3, respectively (Tchobanoglous

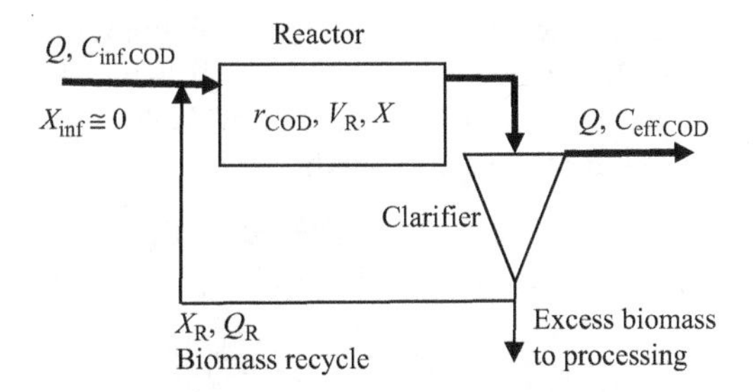

FIGURE E8.6.2 *Schematic of the reactor and clarifier system used for an activated sludge recycle reactor.*

et al., 2003). We construct a model of the bioreactor system, accounting for a variable recycle flow, using the master independent variable $R = Q_R/Q$. Given the baffling within the basin leading to a channeling of the flow, we consider the reactor to be ideal and plug-flow. We recognize that we have the flow split between two halves of the reactor and can consider half the flow introduced into half the volume, or can consider the two halves together, assuming that the total recycle flow is split exactly to each half of the reactor. Also, we recognize that as the flow passes through the reactor, biomass will grow, rendering the biomass concentration to be variable in position. For extended aeration activated sludge systems, the level of biomass in the reactor is typically much larger than the rate of biomass growth, and we can with only small error neglect this growth of biomass. For high-rate systems, this assumption perhaps leads to measureable error. However, were we to consider the growth of biomass, our model would need a second mass balance (on biomass) and the level of sophistication of the effort would need to be upped significantly. We will address biomass growth along with reduction of COD later. Here, since the presence of additional biomass is to be neglected, we realize our results will likely be conservative; however, the analyses performed here will be useful for process understanding.

We assign values to pertinent parameters:

$$k := 10 \quad \frac{g_{COD}}{g_{VSS} \cdot d} \qquad K_{half} := 40 \quad \frac{g_{COD}}{m^3} \qquad V_R := 10000 \quad m^3$$

$$C_{inf} := 450 \quad \frac{g_{COD}}{m^3} \qquad X_R := 4500 \quad \frac{g_{VSS}}{m^3} \qquad Q := 6 \cdot 10^4 \quad \frac{m^3}{d}$$

The return sludge line must be mixed with the influent line, using mixing principles from Chapter 7. Let us first try a recycle ratio ($R = Q_R/Q$) of 0.05, noting that the actual hydraulic residence time of the reactor and the biomass concentration will vary with the magnitude of R:

$$R := 0.05 \quad Q_R := R \cdot Q = 3 \times 10^3 \quad Q_{Tot} := (1 + R) \cdot Q = 6.3 \times 10^4 \quad \frac{m^3}{d}$$

$$\tau := \frac{V_R}{Q_{Tot}} = 0.159 \quad d \qquad X := \frac{Q_R \cdot X_R}{Q_{Tot}} = 214 \quad \frac{g_{VSS}}{m^3}$$

Since we are considering a PFR and we have a saturation-type rate law, we may employ Equation 8.23b. The computation of the effluent concentration cannot be explicitly accomplished, but with the use of MathCAD's given-find solve block, the effluent COD concentration is easily found:

$$C_{eff} := \frac{C_{inf}}{2} \quad \text{Given} \quad K_{half} \cdot \ln\left(\frac{C_{inf}}{C_{eff}}\right) + C_{inf} - C_{eff} = k \cdot X \cdot \tau$$

$$\text{Find}(C_{eff}) = 153 \quad \frac{g_{COD}}{m^3}$$

Of note, MathCAD's root() function might also serve for this computation. Unfortunately, for implementation of the root() function, the initial guess tendered for C_{eff} must be quite close to the final value upon which the solution converges. We note that as R becomes infinitesimally greater than 0, due to the mixing of the influent stream (of concentration C_{inf}) with the recycle stream (of concentration C_{eff}), the actual concentration of the target reactant in the influent to the reactor is reduced from C_{inf}. If we ignore this for now, we must realize that our resultant computations are conservative with respect to the degree of reduction of the reactant concentration for any given set of constraining parameters. If the value of R is low, the error will be small. As R increases, the error of course increases. We will address mitigation of this error later, when we are ready to advance our capacity for mathematical modeling.

Now we have the form of the solution. Faced with determining the recycle ratio, R, necessary to attain a desired target effluent concentration, we realize we are either into a "guess and check" situation or we must harness some additional capabilities of MathCAD. Let us accomplish the latter. We can combine the use of the given-find capability with a MathCAD function. We first convert several of the computations to become functions of R:

$$Q_R(R) := R \cdot Q \quad Q_{Tot}(R) := (1 + R) \cdot Q \quad \tau(R) := \frac{V_R}{Q_{Tot}(R)} \quad X(R) := \frac{Q_R(R) \cdot X_R}{Q_{Tot}(R)}$$

We now may invoke the solve block, assigning the find statement into a function $C_{eff}(R)$, which we may plot against R in Figure E8.6.3 to have a look at the performance of the reactor over a range of recycle ratios:

$$C_{eff1} := \frac{C_{inf}}{5} \quad \text{Given} \quad K_{half} \cdot \ln\left(\frac{C_{inf}}{C_{eff1}}\right) + C_{inf} - C_{eff1} = k \cdot X(R) \cdot \tau(R)$$

$$C_{PFR1}(R) := \text{Find}(C_{eff1}) \quad R := 0, 0.01 .. 0.15$$

We may now define $C_{eff.target}$ and make further use of the written function to identify the value of R that would yield the target effluent concentration:

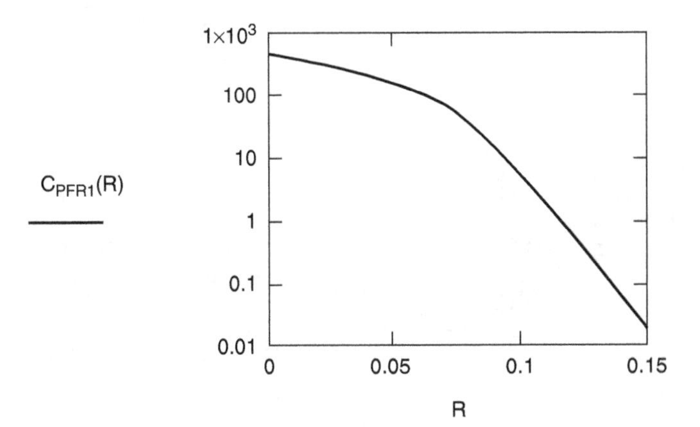

FIGURE E8.6.3 *Plot of effluent COD concentration versus recycle ratio for a plug-flow with recycle reactor.*

$C_{target} := 5 \quad \frac{g_{COD}}{m^3} \quad R := 0.05 \quad \text{Given} \quad C_{eff1}(R) = C_{target}$

$R_{target} := \text{Find}(R) = 0.101$

We double-check the effluent concentration for the determined target *R* value and of course it matches with our specification.

We may examine the effect of the dilution of the influent with the effluent by including the mass balance for mixing of the influent and recycle streams. This relation must accompany Equation 8.23b in the given-find block. With this additional complication, we are no longer able to easily write a function of *R* that can be plotted against *R*. We must invoke this solve block for specific values of *R* and collect the C_{eff} results for comparison against our conservative approximations performed initially:

$R := .11 \quad Q_R(R) := R \cdot Q \quad Q_{Tot}(R) := (1 + R) \cdot Q \quad \tau(R) := \frac{V_R}{Q_{Tot}(R)}$

$X(R) := \frac{Q_R(R) \cdot X_R}{Q_{Tot}(R)} \quad C_{eff} := \frac{C_{inf}}{5} \quad C_{in} := \frac{C_{inf} \cdot Q + C_{eff} \cdot Q_R(R)}{Q_{Tot}(R)}$

$\text{Given} \quad K_{half} \cdot \ln\left(\frac{C_{in}}{C_{eff}}\right) + C_{in} - C_{eff} = k \cdot X(R) \cdot \tau(R) \quad C_{in} = \frac{C_{inf} \cdot Q + C_{eff} \cdot Q_R(R)}{Q_{Tot}(R)}$

$C_{eff.2}(R) := \text{Find}(C_{eff}, C_{in}) \quad C_{eff.2}(R) = \begin{pmatrix} 0.542 \\ 405.459 \end{pmatrix} \quad \frac{mg_{COD}}{L}$

At this point, the most straightforward means of illustrating the comparison between this approximation and this more accurate solution is to select several values of

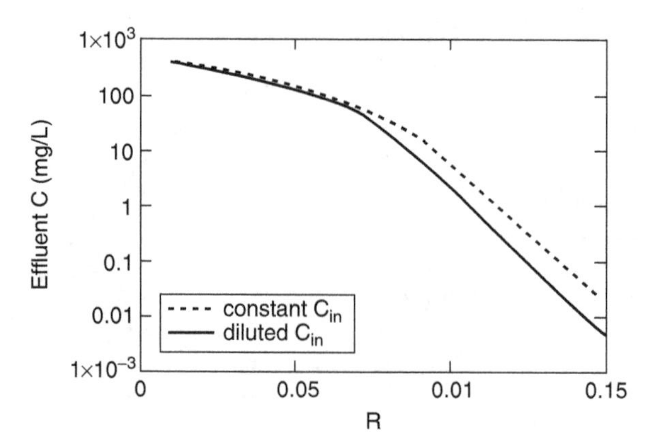

***FIGURE E8.6.4** Plot of effluent COD concentration versus recycle ratio comparing the inclusion of the recycle dilution of the influent against the constant influent assumption.*

R, define a vector (C_{eff1}) using the approximation, and define a second vector (C_{eff2}) using the solve block to obtain effluent concentrations for each of the values of R:

$$R := \begin{pmatrix} 0.01 \\ 0.03 \\ 0.05 \\ 0.07 \\ 0.09 \\ 0.11 \\ 0.13 \\ 0.15 \end{pmatrix} \quad C_{\text{eff.1}} := \begin{pmatrix} C_{\text{PFR1}}(R_0) \\ C_{\text{PFR1}}(R_1) \\ C_{\text{PFR1}}(R_2) \\ C_{\text{PFR1}}(R_3) \\ C_{\text{PFR1}}(R_4) \\ C_{\text{PFR1}}(R_5) \\ C_{\text{PFR1}}(R_6) \\ C_{\text{PFR1}}(R_7) \end{pmatrix} \quad C_{\text{eff.1}} = \begin{pmatrix} 382.933 \\ 259.877 \\ 153.013 \\ 67.393 \\ 15.81 \\ 1.777 \\ 0.177 \\ 0.02 \end{pmatrix} \quad C_{\text{eff.2}} := \begin{pmatrix} 382.272 \\ 254.509 \\ 140.383 \\ 50.446 \\ 7.219 \\ 0.542 \\ 0.043 \\ 4.028 \times 10^{-3} \end{pmatrix}$$

The plot of the more accurate result ($C_{\text{eff.2}}$), considering the dilution of the influent concentration, and the approximation ($C_{\text{eff.1}}$), considering undiluted influent concentration, are plotted in Figure E8.6.4 against the recycle ratio to illustrate the magnitude of the error. We observe that our original approximate solution, ignoring the dilution of the influent concentration by the recycle stream, is indeed in error.

8.4.3 Completely Mixed Flow Reactor Systems

Engineered CMFRs are usually quite easily identified. Typical examples include biological reactors, devoid of baffles, with large mixer-aerators situated throughout the reactor or with diffused aeration systems employing rising gas bubbles as agents for mixing. Often reactors have submersed propeller mixers that vigorously circulate solution contained in the reactor. Large earthen-lined ponds (sometimes called "aerated lagoons") are also often modeled as CMFRs. Laboratory reactors that have stirrers are dead giveaways as CMFRs. Few natural systems fall clearly into the category of CMFRs.

In an earlier chapter, we modeled lakes as CMFRs. In many cases, this might lead to significant error. In the northern and southern hemispheres, lakes of depth more than a few meters most often become stratified in summer with a layer of warm water overlying the colder waters beneath. Strongly stratified lakes might be modeled as a set of CMFRs and CMBRs. The specific geometry of the lake would dictate the number and placement of the various CMFRs and CMBRs. In temperate zones, at least twice each year for short periods, (these dimictic) lakes can be modeled as single CMFRs (or CMBRs). During these periods—cooling of the water prior to winter freeze-up and warming of the water after the thaw—the stratification is broken and the lakes become well mixed as a consequence of warming and cooling of the surface of the lake. During the day, the surface is warmed while during the night the surface is cooled. The cooled surface water becomes more dense and migrates to lower regions in the water column. These thermal gradients and associated convection lead to significant mixing. Further, during these periods of weak or the absence of stratification, wind action results in significant mixing. The period immediately prior to formation of ice cover and the period immediately after the breakup of ice cover constitute the times when temperate lakes can be considered completely mixed. Equatorial lakes, conversely, remain permanently stratified and most simply can be modeled by one CMFR (the epilimnion) and one CMBR (the hypolimnion).

In the following discussions, we will illustrate the application of the CMFR relations with both pseudo-first-order and saturation-type reaction rate laws.

Example 8.7 Let us compare the performance of a CMFR with that of a PFR for the hypothetical conditions of Example 8.5. Let us assume an influent concentration of the target reactant to be 0.1 mol/L and that the rate law and constant are equal to those of Example 8.5. For the stated influent flow of 6×10^4 m^3/day and reactor volume of 5000 L, let us find the resultant effluent concentration and compare it with that of the PFR of Example 8.5. In this situation, the CMFR would contain no packing and hence the total volume of the CMFR would equal the bed volume of the PFR. Here Equation 8.25 will prove immensely useful.

Assign values to the proper parameters:

$$k_{p.pfo} := 0.1 \ \text{min}^{-1} \quad C_{inf} := 0.1 \ \frac{\text{mol}}{\text{L}} \quad Q := 100 \ \frac{\text{L}}{\text{min}} \quad V_R := 5 \cdot 1000 \ \text{L}$$

Rearrange Equation 8.25 to yield the effluent concentration and evaluate:

$$\tau := \frac{V_R}{Q} = 50 \ \text{min} \quad C_{eff} := \frac{C_{inf}}{1 + \tau \cdot k_{p.pfo}} = 0.01667 \ \frac{\text{mol}}{\text{L}}$$

We compare this performance with that of the PFR. For the same conditions the effluent concentration was reduced in the PFR to 0.000674 mol/L, the effluent from the CMFR is nearly 25 times that from a PFR of the same active volume. This result is easily explained. Let us return to the PFR system and use the concentration profile to develop a plot of the reaction rate versus position in the reactor:

$$\tau_{PFR} := 50 \quad \text{min} \qquad C(\tau_z) := C_{inf} \cdot \exp(-k_{p.pfo} \cdot \tau_z) \qquad r_{PFR}(\tau_z) := -k_{p.pfo} \cdot C(\tau_z)$$

$$r_{PFR}(\tau_{PFR}) = -6.738 \times 10^{-5} \quad \frac{\text{mol}}{\text{L} \cdot \text{min}}$$

We now compute the volume of a CMFR necessary to match the performance of the PFR. We simply rearrange Equation 8.25 to yield the necessary hydraulic residence time:

$$C_{eff.CMFR} := C(\tau_{PFR}) = 6.738 \times 10^{-4} \qquad \frac{\text{mol}}{\text{L}}$$

$$\tau_{CMFR} := \frac{C_{inf} - C_{eff.CMFR}}{k_{p.pfo} \cdot C_{eff.CMFR}} = 1.474 \times 10^{3} \qquad \text{min}$$

As a consequence of complete mixing, the concentration of the target reactant and hence the specific reaction rate is constant throughout the CMFR. We may compute the rate using the effluent concentration:

$$r_{CMFR}(\tau_{z2}) := -k_{p.pfo} \cdot C_{eff.CMFR} \qquad r_{CMFR}(\tau_{CMFR}) = -6.738 \times 10^{-5} \qquad \frac{\text{mol}}{\text{L} \cdot \text{min}}$$

We immediately observe that the rate of reaction throughout the entire CMFR is of the same magnitude as the rate of reaction at the effluent of the PFR. From Figure E8.7.1 we gain a pictorial representation.

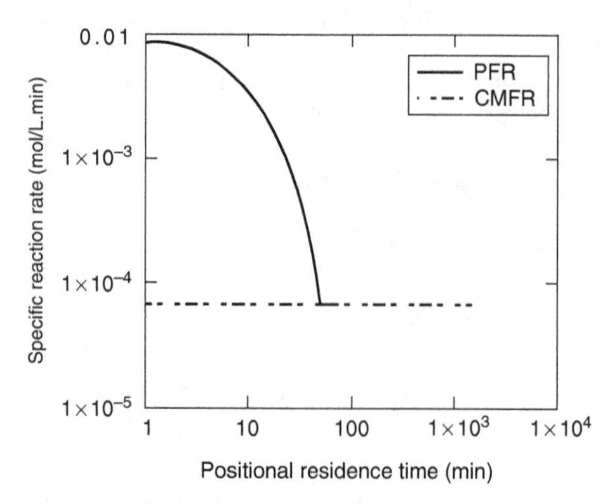

FIGURE E8.7.1 *Plot of specific reaction rate versus positional residence time, comparing an ideal CMFR with an ideal PFR.*

We observe that the reaction rate in the PFR begins at a maximum value and decreases with increasing position in the reactor. That for the CMFR is constant throughout, as the concentration of the reactant is constant throughout the reactor. Except for zeroth-order reaction rate laws, which specify at a constant rate independent of the reactant abundance, nearly all reactions taking place in PFRs will have higher rates than in CMFRs. Then, for a given treatment objective for a reaction whose rate is dependent upon the reactant concentration, the necessary volume of a PFR will nearly always be less than that of a CMFR.

We should examine one more analysis that will confirm the conclusions drawn herein. We may compute the overall rate of reaction for either the PFR or CMFR as the product of flow and change in concentration:

$$Q\cdot\left(C\left(\tau_{PFR}\right) - C_{inf}\right) = -9.933 \ \frac{mol}{min}$$

We may then integrate the specific reaction rate over the volume of each reactor. For the CMFR the computation is quite simple: the constant specific reaction rate times the reactor volume:

$$Q\cdot r_{CMFR}\left(\tau_{CMFR}\right)\cdot\tau_{CMFR} = -9.933 \ \frac{mol}{min}$$

For the PFR we must implement a definite integral, noting that $dV = Q{\cdot}d\tau$. The written function proves immensely useful for numerical integration by invoking the definite integral from MathCAD's math palette:

$$\int_0^{\tau_{PFR}} Q\cdot r_{PFR}\left(\tau_z\right) d\tau_z = -9.933 \ \frac{mol}{min}$$

Given the large volume disparity, one might then ask why a CMFR might ever be chosen over a PFR for a treatment process. Of course, the answer is that other considerations, including ease of process control and greater resistance to shock loadings, might lead designers to choose the required extra volume in order to implement the process using a CMFR.

We can also compare the performance of a CMFR with that of a PFR for a biological reaction described using the saturation rate law.

Example 8.8 Consider that the reaction of Example 8.6 is to be analyzed for employment of a CMFR with recycle. The reactor basins shown in Example 8.6 would now likely be square and the baffles would not be included. The overall schematic would be identical – the reactor would now be a CMFR rather than a PFR. We can employ the same kinetic parameters, reactor volume, flow, and influent concentration. Let us examine the effect of the recycle ratio, R, on the effluent concentration. Equation 8.26 is our beginning point and we realize that, as is the case with the PFR, computation of the effluent concentration cannot be accomplished fully explicitly. We have a choice: we can employ a given-find block as was the case for the PFR or we can employ Equation 8.26b and the associated implementation of the quadratic formula. Let us choose the latter. Along the way, we would be interested in the biomass concentration necessary to attain various target effluent concentrations.

We state the parameter values (to be complete):

$$k := 10\,\frac{g_{COD}}{g_{VSS}\cdot d} \quad K_{half} := 40\,\frac{g_{COD}}{m^3} \quad V_R := 10000\ m^3$$

$$C_{inf} := 450\,\frac{g_{COD}}{m^3} \quad X_R := 4500\,\frac{g_{VSS}}{m^3} \quad Q := 6\cdot 10^4\,\frac{m^3}{d}$$

We write a set of functions (for $Q(R)$, $\tau(R)$, and $X(R)$) that allow us to write intermediate functions for the final implementation of the quadratic formula:

$$Q_R(R) := R\cdot Q \quad Q_{Tot}(R) := (1+R)\cdot Q \quad \tau(R) := \frac{V_R}{Q_{Tot}(R)} \quad X(R) := \frac{Q_R(R)\cdot X_R}{Q_{Tot}(R)}$$

$$a_2(R) := 1 \quad a_1(R) := K_{half} + \tau(R)\cdot k\cdot X(R) - C_{inf} \quad a_0(R) := -K_{half}\cdot C_{inf}$$

We then write the final function and plot it versus R. As was the case with the first computations of Example 8.6, the resultant model is an approximation, with error arising as a consequence of neglecting the dilution of the influent with the effluent:

$$C_{CMFR1}(R) := \frac{-a_1(R) + \sqrt{(a_1(R))^2 - 4\cdot a_2(R)\cdot a_0(R)}}{2\cdot a_2(R)} \qquad \frac{mg_{COD}}{L}$$

We will include, from Example 8.6, the function written for the effluent concentration from the PFR examined therein:

$$C_{PFR} := \frac{C_{inf}}{5} \quad \text{Given} \quad K_{half}\cdot \ln\left(\frac{C_{inf}}{C_{PFR}}\right) + C_{inf} - C_{PFR} = k\cdot X(R)\cdot \tau(R)$$

$$C_{PFR}(R) := \text{Find}(C_{PFR}) \qquad C_{target} := 5\,\frac{g_{COD}}{m^3}$$

In Figure E8.8.1, we plot the two relations versus recycle ratio to compare the PFR and CMFR performance for the biological reaction.

We note that, while with the PFR we can meet the target effluent (5 g_{COD}/m^3) with a recycle ratio (R) of ~0.1, from our plot we realize that our necessary R for the CMFR is beyond 0.5. If we use the earlier model to find the magnitude of R that will result in attaining the target effluent concentration, MathCAD will inform us that the solution is undefined. We believe that we can attain an effluent concentration at or below the target value if we make the recycle rate sufficiently large. We will sharpen our computations in a later example.

R:- 0, .01.. .5

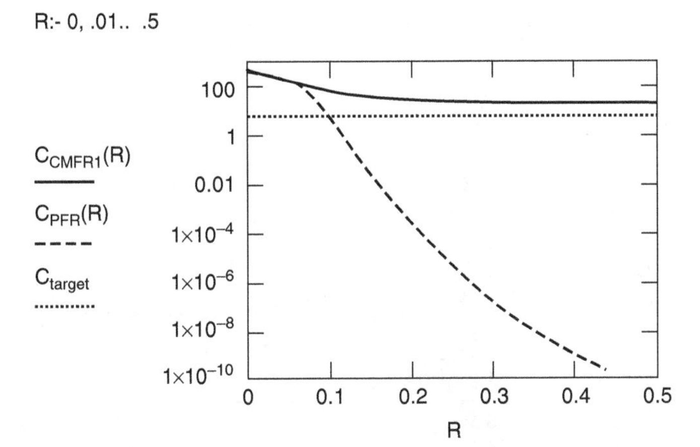

FIGURE E8.8.1 *A plot of effluent concentration versus recycle ratio for CMFR and PFR reactor, neglecting the dilution of influent concentration by the recycle flow.*

Via Example 8.8, we have verified, for the other rate law under consideration, that a PFR is again more efficient than a CMFR on the basis of meeting a desired performance level under a given constraining flow, reactor volume, influent concentration, and effluent concentration. We recall that we have neglected the error associated with the dilution of the influent by the recycle stream and realize it is important now to implement the additional modeling complexity. We will do so in Example 8.9.

Example 8.9 Consider the system of Example 8.8 and include the dilution effect on the actual influent concentration due to the mixing of the influent and recycle flows just prior to the actual influent of the reactor. Let us examine the effect of the recycle ratio on the performance of the reactor and see if we can predict attainment of the target effluent concentration with our CMFR using the previously defined constraining factors.

We can construct a given-find solve block that includes both the mixing mass balance and the CMFR performance relation. We must manually adjust the value of R to obtain the corresponding effluent concentration:

$$R := .5 \qquad Q_R(R) := R \cdot Q \quad Q_{Tot}(R) := (1 + R) \cdot Q \qquad \tau(R) := \frac{V_R}{Q_{Tot}(R)}$$

$$X(R) := \frac{Q_R(R) \cdot X_R}{Q_{Tot}(R)} \qquad C_{CMFR} := \frac{C_{inf}}{50} \quad C_{in} := \frac{Q \cdot C_{inf} + Q_R(R) \cdot C_{CMFR}}{Q_{Tot}(R)}$$

$$\text{Given} \qquad \frac{C_{in} - C_{CMFR}}{\tau(R)} = \frac{k \cdot X(R) \cdot C_{CMFR}}{K_{half} + C_{CMFR}} \qquad C_{in} = \frac{Q \cdot C_{inf} + Q_R(R) \cdot C_{CMFR}}{Q_{Tot}(R)}$$

$$C_{eff2}(R) := \text{Find}(C_{CMFR}, C_{in}) \qquad C_{eff2}(R) = \begin{pmatrix} 8.58 \times 10^{0} \\ 3.03 \times 10^{2} \end{pmatrix}$$

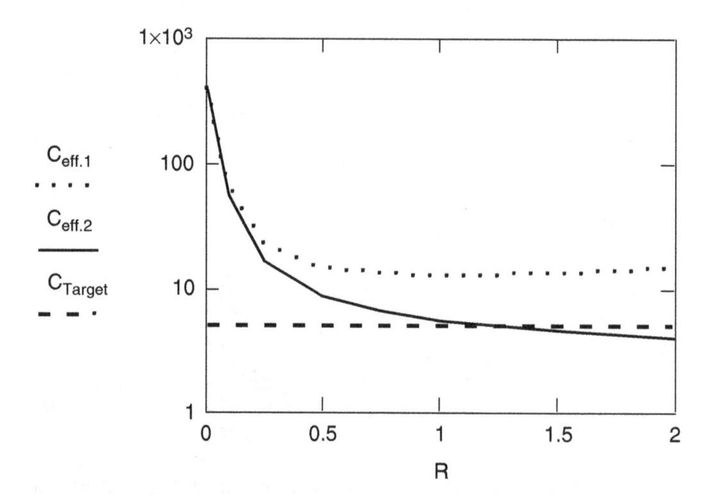

FIGURE E8.9.1 *A plot of predicted effluent concentration versus recycle ratio for two CMFR scenarios – $C_{eff.1}$ considers no dilution of influent by recycle flow while $C_{eff.2}$ is based on adjusting the influent concentration by the effluent concentration value in the recycle flow. Biomass content of the recycle flow is taken as 0.45%.*

We then construct a plot of the approximated effluent concentration ($C_{eff.1}$) and the more accurate value ($C_{eff.2}$) versus the recycle ratio. As was the case for Example 8.7, we define four vectors – R, C_{eff1}, C_{eff2}, and C_{target} – into which we assign values using appropriate assignments for R, the function written for the approximation of the CMFR effluent for C_{eff1}, the solution of the given-find solve block earlier for the selected R values for C_{eff2}, and the target concentration for C_{target}. The resultant vectors are included:

$$R := \begin{pmatrix} .01 \\ .1 \\ .25 \\ 0.5 \\ 0.75 \\ 1.0 \\ 1.5 \\ 2.0 \end{pmatrix} \quad C_{eff.1} := \begin{pmatrix} C_{CMFR1}(R_0) \\ C_{CMFR1}(R_1) \\ C_{CMFR1}(R_2) \\ C_{CMFR1}(R_3) \\ C_{CMFR1}(R_4) \\ C_{CMFR1}(R_5) \\ C_{CMFR1}(R_6) \\ C_{CMFR1}(R_7) \end{pmatrix} \quad C_{eff.2} := \begin{pmatrix} 3.83 \times 10^2 \\ 5.51 \times 10^1 \\ 1.63 \times 10^1 \\ 8.58 \times 10^0 \\ 6.4 \times 10^0 \\ 5.38 \times 10^0 \\ 4.4 \times 10^0 \\ 3.92 \times 10^0 \end{pmatrix} \quad C_{Target} := \begin{pmatrix} C_{target} \\ C_{target} \\ C_{target} \\ C_{target} \\ C_{target} \\ C_{target} \\ C_{target} \\ C_{target} \end{pmatrix}$$

From the plot produced as Figure E8.9.1, we may determine that a recycle ratio slightly greater than 1 produces the target effluent concentration.

We observe, somewhat to our dismay that the first approximation cannot be correct since the predicted value of the effluent concentration increases beyond an

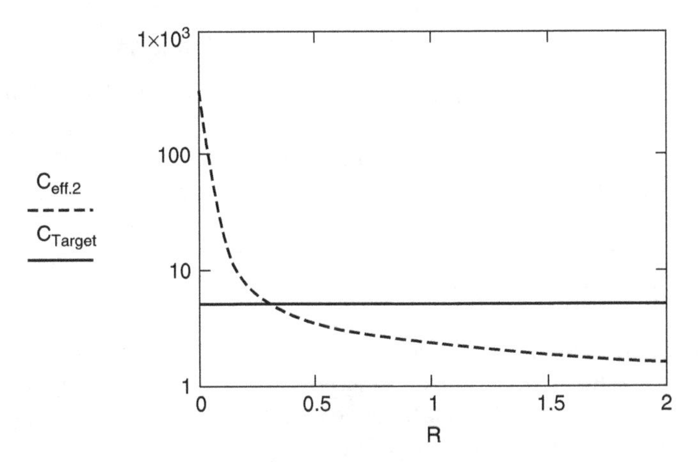

FIGURE E8.9.2 *A plot of predicted effluent concentration from a CMFR with recycle based on abundance of biomass in the recycle flow of 1%.*

R value of ~1. Thankfully, however, we do observe that the CMFR can reduce the influent concentration to the target level.

Here we have set the constraint that the concentration of biomass in the recycle stream is 4500 mg/L. Perhaps we should examine the consequences of rendering the clarifier more efficient. We would invest more capital into a better-designed clarifier and hopefully improve the performance of our CMFR. Let us suggest that we can, with a better system for collection of underflow solids from the clarifier, increase the level of biomass in the recycle stream to 1% solids (10,000 mg/L). With the model we have assembled, we can investigate this "what if" merely by reassigning the value of X_R to 10,000 mg/L, reinvoking our programmed solution and replotting the results in Figure E8.9.2.

We observe that with a recycle ratio of ~0.3 and the greater concentration of biomass, our system is predicted to produce the target effluent concentration. The benefit of increasing the recycle ratio much beyond this value is greatly diminished. This revised level of biomass in the recycle flow is perhaps about the maximum level we may obtain employing conventional clarification equipment for the recovery of biomass from the effluent streams of activated sludge basins.

We observe that the required recycle ratio, and hence the necessary biomass concentration, for the CMFR is significantly greater than that for the PFR. The improved performance due to higher reaction rates along the flow path of the PFR is most evident as the target concentration is decreased. The cost associated with the improved performance of the PFR stems from the necessity for stricter control of the operation of the process. The improved PFR performance, translating into lower capital costs for tankage might be offset by the often vastly simpler operation of a CMFR. One cannot, without detailed analyses, make a choice of reactor. The final choice must be made based on quantitative analyses of performance capability and operability.

8.4.4 Some Context-Specific Advanced Applications

Example 8.10 A wastewater treatment plant with a total flow of 3.60 m^3/s (10.98 MGD) is to be upgraded for use of ultraviolet (UV) irradiation to kill (inactivate) harmful bacteria in the effluent from the plant. Each UV disinfector of a system under consideration consists of a shallow channel in which a bank of quartz tubes containing UV lamps is submerged. The wastewater flows longitudinally through the channels, in between the quartz tubes. The configuration is schematically illustrated (certainly not to scale) in Figure E8.10.1.

The UV lamps emit radiation in the range of 254 nm to disrupt cellular DNA, preventing reproduction and thus inactivating the bacteria as they pass with the wastewater in between the quartz tubes. The degree to which bacteria are inactivated depends upon the intensity of the radiation, the spacing of the lamps, and the time of contact between the wastewater and the radiation source. Sufficient lateral dispersion occurs to bring bacteria into close proximity with the surfaces of the quartz tubes and the positioning of the tubes in the flow channel ensures a reduced degree of longitudinal dispersion. This configuration can be modeled as a PFR. The particular UV system contains quartz tubes of 2.5 cm diameter spaced uniformly vertically and horizontally at 5 cm center to center. The concrete channel in which the bank of UV lamps is to be submerged is 0.75 m in

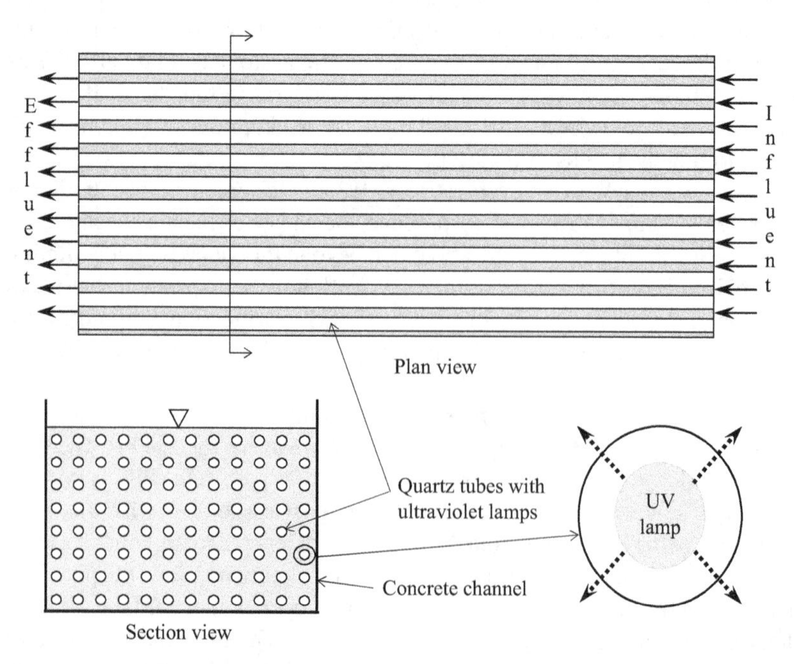

FIGURE E8.10.1 Schematic diagram of the configuration of a typical submerged UV disinfection system.

depth and 1.0 m in width. The bank of lamps is 2.0 m in length. Analysis of this configuration yields a void volume fraction of 0.80 m^3_{void}/m^3.

We would like to determine the number of these UV banks necessary to satisfactorily reduce the abundance of *Escherichia coli* bacteria in treated effluent from 10^8 CFU/L to $2 \cdot 10^3$ CFU/L. A CFU may consist of a single bacterium or a group of bacteria, forming a visible colony during the standard *E. coli* incubation test.

The rate law for inactivation of harmful bacteria (usually taken to be *E. coli*) is pseudo-first-order:

$$r_I = -k \cdot I \cdot N_{EC}$$

k is the inactivation constant (~0.012 cm²/µW/s), I is the design UV radiation intensity (variable, depending upon power input, from 0.25 to 100 µW/cm²), N_{EC} is the abundance of *E. coli* bacteria in CFU/L, and r_I is the volume-specific rate of *E. coli* inactivation (CFU/L/s). One potential design decision would reserve one-third of the available intensity of the UV system to provide capability for short-term surges in UV treatment. These are necessitated by upsets of the system employed to remove suspended solids from the treated wastewater stream. The total number of UV basin/bank combinations is to be determined.

We first determine the pseudo-first-order rate constant for the maximum intensity, reserving one-third of the power input for short-term, emergency use.

$$I_{max} := 100 \quad I := \frac{2 \cdot I_{max}}{3} = 66.7 \quad \frac{\mu W}{cm^2} \quad k := 0.012 \quad \frac{cm^2}{\mu W \cdot s}$$

$$k_{pr} := k \cdot I = 0.8 \quad s^{-1}$$

We then determine the detention time necessary to achieve the desired reduction in *E. coli* CFUs. Equation 8.21a, rearranged to solve explicitly for residence time, will serve well for this purpose:

$$N_{EC.inf} := 10^8 \quad N_{EC.eff} := 2 \cdot 10^3 \quad \frac{CFU}{L} \quad \tau := \frac{\ln\left(\frac{N_{EC.eff}}{N_{EC.inf}}\right)}{-k_{pr}} = 13.525 \quad s$$

Now we may apply the definition of the hydraulic residence time to determine the total volume of flow channel necessary:

$$Q := 3.6 \ \frac{m^3}{s} \quad V_R := Q \cdot \tau = 48.7 \quad m^3 \quad \varepsilon_{flow} := 0.8 \quad \frac{m_{flow}^3}{m^3}$$

$$V_{bank} := \varepsilon_{flow} \cdot 2.0 \cdot 1.5 \cdot 1.0 = 2.4 \quad m^3 \quad N_{units} := \frac{V_R}{V_{bank}} = 20.3$$

We of course cannot install a fraction of a unit. Since the choice of one-third reserve capacity is somewhat arbitrary, we decide to recommend 20 units and will perform a computation to determine the actual reserve fraction based on the 20 units. We simply work backward from the original computation. First we compute the required hydraulic residence time:

$$N := \text{floor}(N_{units}) = 20 \quad Q_{bank} := \frac{Q}{N} = 0.18 \; \frac{m^3}{s \cdot bank} \quad \tau := \frac{V_{bank}}{Q_{bank}} = 13.333 \; s$$

We then compute the necessary value of the pseudo-first-order rate constant and the associated UV intensity:

$$k_{pr} := \frac{\ln\left(\frac{N_{EC.eff}}{N_{EC.inf}}\right)}{-\tau} = 0.811 \; s^{-1} \qquad I := \frac{k_{pr}}{k} = 67.624 \; \frac{\mu W}{cm^2}$$

The reserve intensity (in percentage) can then easily by computed:

$$R := \frac{I_{max} - I}{I_{max}} \cdot 100 = 32.4 \; \%$$

With 20 parallel banks of UV disinfectors we may meet the design effluent requirement for allowable *E. coli*. CFUs. The design reserves 32.4% of the system intensity to provide for disinfection in the case of short-term solids removal process upsets.

Example 8.11 Consider that the insecticide lindane was applied to an agricultural plot. Shortly after the application, a "gully washer" rainstorm occurred and washed a large fraction of the applied lindane into a nearby pond. The presence of a toxic substance in the pond was first suspected by the near-total die-off of the resident fish population, appearing "belly up" on the surface of the pond. The pond covers approximately 1 ha (~2.5 ac) and has an average depth of 3 m (~10 ft). The small stream that feeds the pond has an average flow of 0.00300 m^3/s. This important piece of information would be obtained by searching the U.S. Geological Survey stream-flow records and determining the weighted average for the nearest upstream gauging station. At the urging of a local environmental group, water samples were obtained and assayed for potential toxic substances, one of which was lindane. Results of assays for all potentially toxic substances other than for lindane were below detection limits and the result for lindane was 0.5 ppm_m. A local fishing club stocks perch in the pond and is anxious to know when restocking might be safe. Means to control future runoff from the agricultural plot are under investigation. Your firm has been retained to perform the necessary time estimates.

We gain an understanding of the properties of compound lindane, an isomer of hexachlorocyclohexane (also called benzene hexachloride) with empirical chemical formula $C_6H_6Cl_6$. Lindane is mildly toxic to birds and mammals but highly toxic to fish and both terrestrial and aquatic insects. Information relevant to the commissioned study includes the following:

1. Environmental half-life of lindane ranges between 150 and 400 days. The half-life is defined as the time required for a 50% reduction of the abundance of a degradable substance in a batch reactor.
2. The 96-h LC-50 ranges from 44 to 131 ppb_m for catfish and perch. *The 96-h LC-50 is derived from a toxicity test and is the concentration level of a potentially toxic substance resulting in 50% mortality of a sample population within 96 h.*

These two items of information allow the prediction of the first-order reaction rate coefficient for the decay of lindane and definition of a potentially safe aqueous concentration level of lindane.

We must determine the reactor model to be used in the analysis. A check of the pond yields the following information:

1. The pond is fed by an ephemeral stream that flows at a trickle most of the year with the average flow rate specified earlier.
2. The inflow rate swells periodically with rainfall events when rainfall intensity greatly exceeds the infiltrative capacity of the drainage area.

The decision is made to consider that the degradation of lindane in the pond can be modeled using a first-order reaction rate law and that the pond may be treated as a batch reactor. Neglecting the inflow into and outflow from the pond will provide for a conservative recommendation.

The half-life information is used to obtain the first-order reaction rate coefficient using Equation 8.16 rearranged to yield the rate coefficient, where $k' = k$, $t_0 = 0$, and $C = C_0/2$:

$$t_{half} := \begin{pmatrix} 400 \\ 150 \end{pmatrix} \text{ days} \qquad k := \frac{-\ln\left(\frac{1}{2}\right)}{t_{half}} = \begin{pmatrix} 1.733 \times 10^{-3} \\ 4.621 \times 10^{-3} \end{pmatrix} \text{ day}^{-1}$$

The reaction rate coefficient is not precisely known and the decision is made to use the smaller of the two values (k_0 is the zeroth element of the defined k vector) to be conservative such that the local fishing club could be certain of no mortality of its stocked perch, were your firm's recommendations to be followed.

Then, the decision is made to use a final lindane concentration of one-tenth of the lowest value of the 96-h LC-50 as the recommended safe level for 100% survival of stocked perch. Then, $C_{safe} = 4.4$ ppb_m.

$$C_0 := 0.5 \qquad C_{96} := \frac{44}{1000} = 0.044 \qquad C_{safe} := 0.1 \cdot C_{96} = 4.4 \times 10^{-3} \qquad ppm_m$$

$$t_{safe} := -\left(\frac{1}{k_0}\right) \cdot \ln\left(\frac{C_{safe}}{C_0}\right) = 2.731 \times 10^3 \text{ days} \qquad \frac{t_{safe}}{365} = 7.483 \text{ years}$$

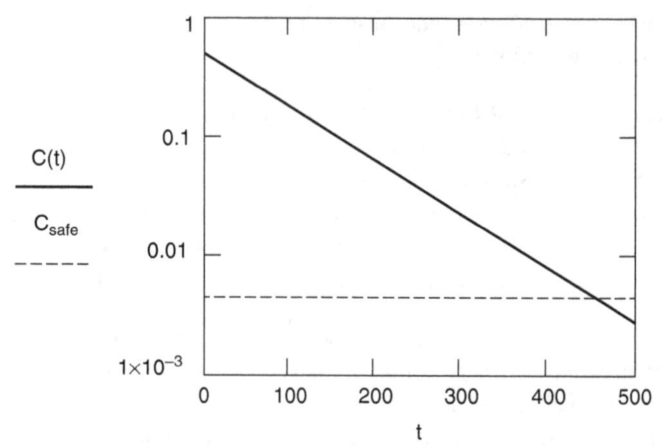

FIGURE E8.11.1 *A plot of lindane concentration versus time in a hypothetical pond affected by lindane in agricultural runoff received by the pond.*

You do not wish to inform the fishing club that perch stocking will need to wait 7.5 years. A decision is made to sharpen the proverbial pencil and see if a more realistic estimate can be attained, at the expense of a more rigorous modeling effort.

It is decided to investigate the issue by application of the non-steady-state CMFR model, considering inflow that contains no detectable lindane. Application of Equation 8.27b yields the following result, shown graphically in Figure E8.11.1:

$$V_R := 1.0 \cdot 10^4 \cdot 3 = 3 \times 10^4 \quad m^3 \qquad Q := 0.003 \cdot 86400 = 259.2 \quad \frac{m^3}{d}$$

$$\tau := \frac{V_R}{Q} = 115.741 \quad d \qquad C_{in} := 0 \qquad C_0 := .5 \quad ppm_m$$

$$C(t) := \frac{C_{in} + (C_0 - C_{in} + C_0 \cdot \tau \cdot k_0) \cdot \exp\left[\frac{(t) \cdot (\tau \cdot k_0 + 1)}{\tau}\right]}{1 + \tau \cdot k_0} \qquad t := 0, 1..500 \quad days$$

These results might be better received by the fishing club and even though they are several factors lower than the initial estimate based on the ideal batch reactor, they remain conservative as a consequence of the conservative choices for the half-life for lindane and the safe lindane abundance. A numerical computation of the necessary elapsed time is accomplished:

$$t_{safe} := 200 \quad \text{Given} \quad C(t_{safe}) = C_{safe} \quad t_{safe} := \text{Find}(t_{safe}) = 456 \quad days$$

Your firm's recommendation to the fishing club is that the pond could safely be restocked about 15 months after the introduction of lindane into the pond by the offending runoff event. A 2-year wait might be just a bit more conservative and warranted. Hopefully, the governing regulatory authority would ensure that measures are implemented to prevent future contamination events.

In Example 8.11 with a first-order rate law, we employed a rate constant that was based on the observed half-life of lindane in environmental systems. Certainly many factors can influence the magnitude of the half-life of a contaminant such as lindane in this pond, including (*but not necessarily limited to*) temperature, pH, presence of sediment solids onto which lindane might adsorb, and abundance of microbes capable of degrading lindane. Since there is a great deal of uncertainty associated with employing observed half-lives from one system to another system, we used the lower value of the derived first-order rate constant. The prediction arising from employment of the ideal CMBR is observed to be far more conservative than that arising from the unsteady-state CMFR. We note that as the influent flow rate is decreased, the prediction based on the unsteady CMFR would approach that based on the ideal CMBR.

An area of great interest in environmental process analysis is that of wastewater collection systems. These involve confined spaces through which gravity-flow sewers transport organic-rich wastewaters. Most wastewater collection systems include lift stations that employ below-ground concrete reservoirs to collect and temporarily store wastewater and pumps to move the wastewater over hills to central gravity flow systems. Of course, even though rates are fairly slow, the microbial communities present in the wastewater and in the wet wells of the lift stations go about their business of mineralizing organic carbon present in the wastewater. This mineralization process requires an electron acceptor, with oxygen enjoying the most-favored status. Unfortunately, the supply of oxygen in these confined systems is limited and the microbial community, upon depletion of the oxygen supply, turns to other substances as electron acceptors. Typically, little nitrate (next in line behind oxygen as a preferred electron acceptor) is present in wastewater. Nitrogen is present either as ammonia-nitrogen or organic nitrogen, in which forms it is fully reduced and cannot be used as an electron acceptor. Next in line is typically sulfate, usually present in raw water treated and distributed to communities, and hence present in wastewater. Then, once the oxygen is depleted, sulfate becomes the electron acceptor of choice.

To illustrate the electron acceptor, we can use glucose as a model for the organic matter in the wastewater and write the overall redox reaction for mineralization of glucose by combining two half-reactions. This analysis is a prelude to some interesting chemical principles we will address in Chapter 12. We first write (or obtain from the literature) the two half-reactions. We have chosen bisulfide as the sulfide product as it is one of the two predominant sulfide species in aqueous solutions of pH typical to wastewaters. The half-reaction for reduction of carbon dioxide to glucose is also needed:

$$SO_4^{=} + 8e^- + 9H^+ \rightleftarrows HS^- + 4H_2O$$

$$6CO_2 + 24e^- + 24H^+ \rightleftarrows C_6H_{12}O_6 + 6H_2O$$

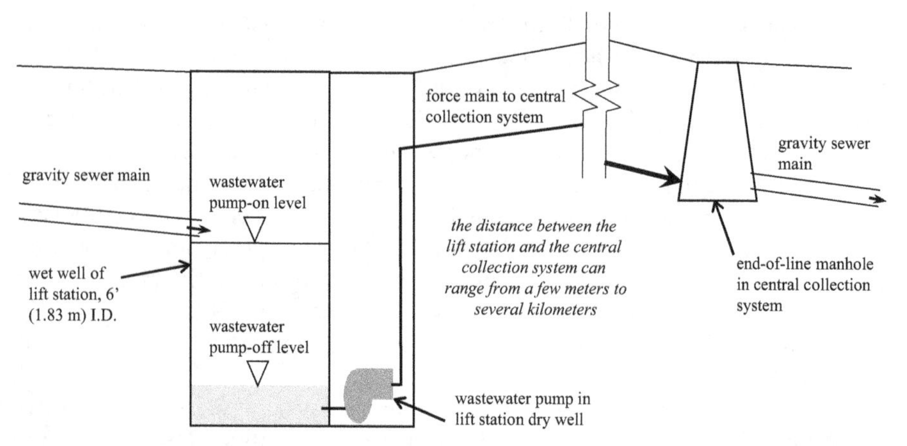

FIGURE 8.4 *Schematic diagram of a typical lift station and force main system used to transport wastewater to central collection systems from areas outside the main drainage area.*

Since the sulfur in sulfate is the acceptor of the electrons and the carbon of glucose is the donor, we reverse the second half-reaction, so we can balance the electrons on the two sides of the half-reactions before we add them:

$$3SO_4^{=} + 24e^- + 27H^+ \rightleftarrows 3HS^- + 12H_2O$$

$$C_6H_{12}O_6 + 6H_2O \rightleftarrows 6CO_2 + 24e^- + 24H^+$$

$$C_6H_{12}O_6 + 3SO_4^{=} + 3H^+ \rightleftarrows 6CO_2 + 3HS^- + 6H_2O$$

When present in wastewater, sulfide is in the form of its two conjugate acid species: hydrogen sulfide (H_2S) and bisulfide (HS^-). We should also recall that H_2S distributes between aqueous solution and vapor in relation to its Henry's constant, tabulated with values from other systems in Table 5.1. The use of sulfate as an electron acceptor by microbial populations in wastewater collection systems leads to the necessity for a set of precautions addressing worker safety, as hydrogen sulfide is highly toxic to humans. Moreover, as mentioned in Chapter 6, the presence of sulfides in collection systems leads to significant corrosion, particularly of concrete. A schematic diagram of a typical lift station and force main system is shown in Figure 8.4. We will use this system to illustrate applications of ideal reactors with both pseudo-first-order and saturation-type reaction rate laws.

In Example 8.12, we will examine the production of sulfide sulfur in a typical lift station, under conditions for which we could consider the wet well to be a batch reactor.

Example 8.12 A wastewater lift station is to serve an outlying, nearly wholly residential community. Concern exists over the potential production of hydrogen sulfide in the event of a power outage, necessitating a decision whether to provide a stand-by power generation system, at considerable capital cost. A scenario is envisioned during which a power outage occurs at a critical time during normal operation of the lift station facility. The system envisioned includes a wet well of 1.83 m (6 ft) inside diameter with a 1.22 m (4 ft) elevation differential between the pump-off and pump-on levels. The depth of the wastewater at the pump-off level is 0.305 m (1 ft). Overall, the wet well is 4.57 m (15 ft) in depth. The history of past power outages associated with both summer and winter storms reveals that on several occasions during the most recent decade, power was interrupted several times for periods as long as 24 h. A decision is made to examine sulfide production over a 1-day period.

One important scenario for investigation considers that the wet well liquid level is at the pump-on level at the instant of the power outage. Further considerations would involve an assumption that with no power, no water flow is likely and thus no wastewater would be generated for flow into the wet well. In this case, the wet well would be considered as a CMBR. As the mineralization of organic matter in wastewater is microbial, we would opt for the saturation-type reaction rate law and thus our model for the system would be Equation 8.19.

We would search the literature to ascertain magnitudes of the rate constants k' and K_{half}, and for information regarding the effective concentration of sulfate-reducing bacteria (SRB) in wastewaters. Of particular note, unless systems are included to aerate the wet well of the lift station, conditions would be favorable for the development of a population of such bacteria in the wet well, and we would seek to identify either laboratory or field studies in which the systems in concern are those where the microbial culture was acclimated by periods of low oxygen availability. Consider the following:

1. K_{half} should range between 30 and 100 μM_{SO_4-S}
2. k' should range between 0.25 and 1.0 $mmol_{SO_4-S} / g_{SRB} / h$
3. The effective concentration of SRB in the raw wastewater held within a lift station wet well should range between 2 and 5 mg_{SRB}/L.

We would also seek to establish the level of sulfate in the wastewater entering the manhole. An assay of the major ions present in the raw water source used to produce finished water distributed to the subdivision would give us our best estimates. As this is a planned facility, we are unable to obtain a sample of the wastewater and have it tested for sulfate content. The raw water source is tracked down and it is learned that sulfate abundance levels range between 80 and 100 mg_{SO_4}/L.

The first order of business is to transform all important parameters for use with a common set of units – milligrams, liters, and days. Convert the minimum and maximum values of K_{half} from μM to mg/L as SO_4:

$$MW_{SO4} := 32 + 4 \cdot 16 = 96 \quad \frac{g}{mol} \quad K_{half} := \begin{pmatrix} 30 \\ 100 \end{pmatrix} \cdot \left(MW_{SO4} \frac{10^3}{10^6} \right) = \begin{pmatrix} 2.88 \\ 9.6 \end{pmatrix} \quad \frac{mg_{SO4}}{L}$$

Convert k' from $mmol_{SO_4-S}/g_{SRB}/h$ to $mg_{SO_4-S}/mg_{SRB}/day$. Let us also assign minimum and maximum values for biomass abundance:

$$k := \begin{pmatrix} 0.25 \\ 1 \end{pmatrix} \cdot MW_{SO4} \cdot \frac{10^3}{10^3} \cdot \frac{24}{1000} = \begin{pmatrix} 0.576 \\ 2.304 \end{pmatrix} \frac{mg_{SO4}}{mg_{SRB} \cdot d} \qquad X := \begin{pmatrix} 2 \\ 5 \end{pmatrix} \frac{mg_{SRB}}{L}$$

In each of these cases, the zeroth element of each vector represents the minimum value and the first value represents the maximum value of the range.

Let us consider the scenario under which the wet well has just filled to the pump-on level with fresh wastewater when the power is interrupted. We will consider that with the power interruption, the wastewater inflow rate becomes 0. Investigation of the non-zero flow rate is left as an end-of-chapter problem. Let us determine the potential range of sulfate conversions for this batch reactor over a reaction time of 1 day. We specify the initial sulfate concentration, a guess for the final sulfate concentration, and the reaction time. Stoichiometry of the conversion is 1:1 (Table A.5), so for every unit of sulfate sulfur converted, one unit of sulfide sulfur is created. We specify the necessary parameters, including a guess for the final sulfate concentration, necessary for the root() function:

$$SO4_i := 100 \ \frac{mg}{L} \qquad SO4_f := 0.5 \cdot SO4_i \qquad t := 1 \ d$$

Equation 8.19 is then solved using MathCAD's root() function to yield the expected maximum and minimum final sulfate concentrations:

$$SO4_f := \begin{bmatrix} root\left[\frac{1}{k_1 \cdot X_1} \cdot \left(K_{half_1} \cdot \ln\left(\frac{SO4_i}{SO4_f}\right) + SO4_i - SO4_f\right) - t, SO4_f\right] \\ root\left[\frac{1}{k_0 \cdot X_0} \cdot \left(K_{half_0} \cdot \ln\left(\frac{SO4_i}{SO4_f}\right) + SO4_i - SO4_f\right) - t, SO4_f\right] \end{bmatrix} = \begin{pmatrix} 89.5 \\ 98.9 \end{pmatrix} \frac{mg_{SO4}}{L}$$

Using the 1:1 stoichiometry we may obtain the molar sulfide production as the negative of the molar sulfate transformation:

$$\Delta SO4 := \frac{SO4_f - SO4_i}{1000 \cdot MW_{SO4}} = \begin{pmatrix} -1.09 \times 10^{-4} \\ -1.166 \times 10^{-5} \end{pmatrix} \qquad \Delta S := -\Delta SO4 = \begin{pmatrix} 1.09 \times 10^{-4} \\ 1.166 \times 10^{-5} \end{pmatrix} \frac{mol}{L}$$

These values mean little until we totalize them for the wet well of the lift station. We will need the volumes of liquid and vapor:

$$d_{ww} := 1.83 \quad H_{ww} := 1.22 + .305 = 1.525 \qquad H_{Tot} := 4.57 \quad m$$

$$V_{ww} := \left(\frac{\pi}{4} \cdot d_{ww}^2 \cdot H_{ww}\right) \cdot 1000 = 4.011 \times 10^3 \ L$$

$$V_{vap} := \frac{\pi}{4} \cdot d_{ww}^2 \cdot (H_{Tot} - H_{ww}) \cdot 1000 = 8.009 \times 10^3 \ L$$

We can compute the total quantity of sulfide produced during the outage:

$$S_{Tot} := \Delta S \cdot V_{ww} = \begin{pmatrix} 0.437 \\ 0.047 \end{pmatrix} \text{mol}$$

This result does not quite finish the job. The sulfide created must be distributed between the vapor and aqueous phases and we would like to know the hazard associated with hydrogen sulfide in the vapor.

The pH of the wastewater (expected to be ~7.5) would determine the distribution between H_2S and HS^- and the Henry's Law constant would determine the distribution between $H_2S_{(aq)}$ and $H_2S_{(g)}$. From the quantity of sulfide sulfur produced, using a mass balance and equilibria from Chapters 5 and 6, we can determine the distribution to see if the vapor phase would be dangerous. We will assume that ventilation systems associated with the wet well would maintain the content of the vapor phase at or near 0 concentration of hydrogen sulfide, until the power outage. We will also consider that the wet well is vapor-tight such that all produced sulfide would remain in the wet well:

$$S_{Tot} = C_{V.H2S} \cdot V_{vap} + (C_{ww.H2S} + C_{ww.HS}) \cdot V_{ww}$$

We have not included the details for creating the final relation for P_{H_2S}. The student is encouraged to employ equilibria as appropriate and perform the algebraic work to verify the result. As the second deprotonation constant for hydrogen sulfide is ~10^{-14}, we may with little error treat hydrogen sulfide as if it were a monoprotic acid and bisulfide were the fully deprotonated conjugate base:

$$pH := 7.5 \quad H := 10^{-pH} \quad R := 0.082057 \frac{L \cdot atm}{mol \cdot {}^\circ K} \quad T := 293\ {}^\circ K$$

$$K_{H.H2S} := 0.105 \frac{mol}{L \cdot atm} \quad K_{1.S} := 10^{-7.1}$$

$$P_{H2S} := \frac{S_{Tot}}{\left[\frac{V_{vap}}{R \cdot T} + \left(K_{H.H2S} + \frac{K_{1.S} \cdot K_{H.H2S}}{H}\right) \cdot V_{ww}\right]} = \begin{pmatrix} 2.412 \times 10^{-4} \\ 2.581 \times 10^{-5} \end{pmatrix} \text{atm}$$

Then, in order to compare our results with published toxicological data for hydrogen sulfide gas, we must convert the partial pressures in atmospheres to parts per million by volume (ppm_v):

$$P_{Tot} := 1\ \text{atm} \quad ppmv_{H2S} := \frac{P_{H2S}}{P_{Tot}} \cdot 10^6 = \begin{pmatrix} 241 \\ 26 \end{pmatrix}$$

We observe that the buildup of hydrogen sulfide in the vapor space of the manhole could reach dangerous levels. For example, at 20 ppm_v exposure can cause damage to the nerves of the eyes, at 30 ppm_v it can cause injury to the blood–brain barrier, at 100 ppm_v it can render a person unconscious, and at 700 ppm_v it can cause immediate unconsciousness and quick death (Alken Murray, 2002).

Our decision then as design engineers lies between investing the capital costs associated with emergency power generation equipment and very carefully documenting this potentially dangerous condition in the operating instructions furnished to the public utility charged with the operation and maintenance of this lift station facility.

8.5 INTERFACIAL MASS TRANSFER IN IDEAL REACTORS

As a consequence of the innate, molecular-level motion associated with matter, particularly that in gaseous and liquid form, and the tendency of the molecules or ions of a given specie to repel each other, the natural tendency of the universe is in the direction of disorder. A drop of dye introduced into a beaker filled with a quiescent aqueous solution, in short order even in the absence of mechanical mixing, will become mixed throughout the beaker. If a bottle of odorous, volatile liquid is opened in a closed, nonventilated room, relatively soon after the opening, persons located in the far extremities of the room will detect the odor. At the molecular level, in the absence of mechanical mixing, the molecular activity is often called Brownian motion and the resultant mass transfer process is called diffusion. When systems are mixed at the macroscopic level, either by human design or by nature, imparting microscopic velocity gradients that affect motion beyond the molecular level, the process is most often called dispersion. As a consequence of collisions with molecules of the solvent and mutual repulsive forces between like molecules (or ions) of a solute, diffusive and dispersive processes cause the spreading of the component throughout the phase (gas, liquid, and even solid) in which the component resides. With multiple solutes present in a given solvent, the diffusive/dispersive processes tend to occur mostly independently and in parallel. Henceforth, whether we are examining diffusion, driven essentially by molecular motion, or dispersion, driven by a combination of molecular motion and mixing-induced velocity gradient, we will refer to the phenomenon as a dispersive process.

Most often in engineering, we are interested in the manifestations of dispersive processes in regard to the physical migration of target solutes toward or across either real physical or visualized surfaces. These surfaces are planar, cylindrical, and spherical in geometry. In examining dispersion of targeted solutes within large systems such as subsurface soils and aquifers, the atmosphere, and water bodies, the cylindrical and spherical geometries are often very important. Conversely, when we target dispersive processes that contribute to the transfer of target solutes across real interfaces, most often we may employ planar geometry.

For the description of the mass transfer process within a solvent we employ the term flux. Flux is the quantity (mass, moles, or volume) of a substance transported normal to or across a defined (perhaps visualized) surface per unit time per unit area of surface. We may quantitate flux in mass or molar units and also, in many cases, in volumetric units. Hence, the typical set of units for flux involves mass or moles, time, and length ($M/t/L^2$). For volumetric flux, most often used in the context of the solvent rather than the solute, we have the quantity measured in volume per time per unit area ($L^3/t/L^2$), for which we tend to cancel L^2 and L^{-2} to obtain units we know as velocity. Particularly in the case of systems involving porous media or packed reactors, the area across which the flux occurs is total area rather than void area and we call the resultant velocity "superficial" velocity. Volumetric flux is exactly analogous to specific discharge used in describing the movement of groundwater, to face velocity used to describe filter loadings in water treatment, and to overflow velocity used to describe loadings to clarifiers and sedimentation basins in both water and wastewater treatment.

8.5.1 Convective and Diffusive Flux

As a consequence of volumetric flux of a solvent containing a target solute, transport of the solute (arbitrary specie A) occurs simply due to the bulk movement of the solvent. The convective flux ($\vec{N}_A$, also called advective flux by some disciplines) can then be fully described as the product of the volumetric flux ($\vec{v}_S$, $L^3/L^2/t$) of the solvent and the concentration (C_A, ML^{-3}) of the solute in the solvent:

$$\vec{N}_A = \vec{v}_S C_A \tag{8.28}$$

In some systems, most notably in chemical reactors employing catalysts providing surfaces upon which reactions occur at high rates, and employing processes for separation of large fractions of one component from a bulk solution containing many, the diffusive process can induce volumetric flux, greatly complicating the analysis. Fortunately, in most environmental systems, we deal with solute abundances orders of magnitude lower than in these chemical engineering systems and can, with but small error, neglect the diffusion-induced convection.

Diffusive flux ($\vec{J}_A$, $M/L^2/t$) of arbitrary specie A is most often quantitatively described using Fick's law. Flux is the product of the diffusion (or dispersion) coefficient ($\mathcal{D}_{AB}$, L^2/t) and the gradient in concentration ($(\partial C_A/\partial_\eta)$, $M/L^3/L$) in the direction of the flux:

$$\vec{J}_A = -\mathcal{D}_{AB}\frac{\partial C_A}{\partial_\eta} \tag{8.29}$$

The symbol η represents the spatial coordinate normal to the plane across which flux would occur: x, y, or z in the Cartesian coordinate system, r or θ in cylindrical coordinates, and r in spherical coordinates. If we orient our principal axis parallel to the direction of diffusive flux, we may always represent diffusive flux in one dimension. Conversely, if we arbitrarily orient our coordinate system and examine flux within that coordinate system, we must consider the derivative of concentration in all principal directions. Detailed treatment of these contexts is beyond our scope. The negative sign arises as a consequence of decreasing abundance of the target specie in the direction of the flux. When we consider that abundance of a specie in an aqueous solution is directly related to chemical potential energy, we can draw the generalization that diffusion (dispersion) involves migration of a target specie downgradient, from large abundance to small abundance. If we were to plot a profile of concentration along an axis oriented parallel to the direction of diffusive flux, we would observe the profile to have a negative slope. The magnitude of the dispersion coefficient is governed by the degree of mechanical mixing, the viscosity of the solute, and the molecular size of the transported specie. In the absence of mechanical mixing, the dispersion coefficient is the diffusion coefficient. In gases, the molecular weights of the solvent and solute also affect the magnitude of the diffusion coefficient. If we examine the transport of substances of interest in various engineering disciplines, we see nearly direct analogs of Fick's law employed as Ohm's law in electron flow, Fourier's law in heat flow, and Darcy's law in groundwater flow. Excellent discussions of diffusion and the diffusion coefficient are provided by Bird et al. (1960), Treybal (1980), Cussler (1984), and Crittenden et al. (2005). Excellent treatments of the mathematics associated with diffusive processes in various systems are presented by Crank (1979). Methods for estimating diffusion coefficients for solutes in gas and water are presented by Lymann et al. (1982) and Reid et al. (1987).

Total flux ($\vec{F}_A$, $M/L^2/t$) is merely the sum of convective flux and dispersive flux:

$$\vec{F}_A = \vec{N}_A + \vec{J}_A = \vec{v}_S C_A - \mathcal{D}_{AB} \frac{\partial C_A}{\partial z} \quad (8.30)$$

In many contexts, our interest is in total flux as convective processes are significant. However, except in the context in which diffusion of a target specie at high rates of flux induces convective velocities, mass transfer in the vicinities of interfaces is dominated by dispersive processes. In this chapter, we will not address catalytic systems with high reaction rates or separations processes with high rates of transfer across interfaces. Thus, we may concentrate on diffusive and dispersive processes for which we may neglect the convective flux term, describing total flux using Fick's law.

8.5.2 Mass Transfer Coefficients

For interfacial mass transfer, we generally assume that we may deal with Newtonian fluids. The characteristic of such fluids that is of great importance herein is called the nonslip condition. We generally illustrate this condition using a fluid (water) flowing through a pipe at a fairly high rate. We make ourselves small and take up a position on the wall of the pipe so that we can view the water molecules passing by. We carefully observe that at the wall of the pipe, even though there may be exchange of water molecules between the monolayer of water at the pipe wall and the layers of water situated away from the wall, the net velocity of the water at the wall of the pipe is 0. We then inspect the system at positions apart from the wall. We find that with each successive layer, the velocity increases as we move away from the wall. Then, at some position, in molecular units, well departed from the wall, the velocity of the fluid becomes constant. This is the "bulk flow" region and often comprises more than 99% of the pipe. If we are well downstream from the initiation of the flow in the pipe, we would observe exchange of water molecules between layers residing close to the pipe wall, owing to turbulence imparted to these layers by the fluid in the bulk flow. In contrast to the bulk flow region, which would appear completely mixed, this region in close vicinity to the pipe wall is characterized by small degrees of mixing. Depending upon the transfer of turbulence from the bulk flow region, the mixing herein might be dominated by Brownian motion or by turbulence. In either case, the mixing is less intense than that in the bulk flow region.

We can apply this description to interfaces in general – between gases and liquids, between liquids and solids, and between immiscible liquids. We may go so far as to visualize mass transfer processes considering regions of complete mixing (bulk flow regions) separated by thin boundary layers characterized by small degrees of mechanical mixing. In these regions of restricted mixing, we can apply Fick's law to quantitatively characterize mass transfer from one completely mixed region to another. Let us examine this process in the context of a bubble containing oxygen and rising through water with a dissolved-oxygen deficit, relative to the composition of the bubble. Since the oxygen concentration in the water is less than would be dictated by the oxygen content of the bubble, from Le Châtelier's principle we know that the process would seek an equilibrium via the net transfer of oxygen from the bubble to the water. We will take a "snapshot" of the process at an instant in time and preserve that snapshot for our considerations. A sketch representing the bubble and associated processes at our instant in time is shown in Figure 8.5.

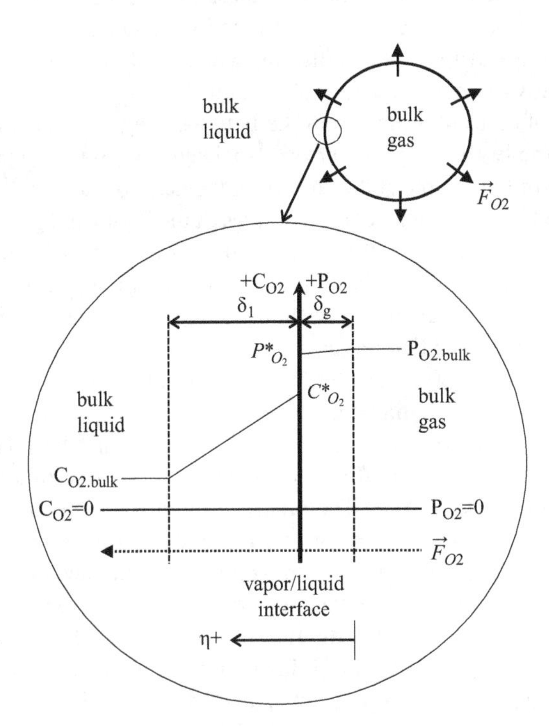

FIGURE 8.5 *Sketch representing the system and mass transfer process occurring across the gas/liquid boundary of an air bubble frozen in time during its rise through an aerated solution.*

The flux of oxygen from the bubble into the water occurs over the entire vapor–liquid interface. The bubble, perhaps a millimeter or two in diameter, is huge relative to the molecular scale (1–2 Å) so we can represent the interface and the two boundaries separating the boundary region from the bulk gas and liquid as planes. Were we able to measure the abundance of oxygen in the gas phase as partial pressure and the abundance of oxygen in the liquid phase as molar concentration at positions along our spatial coordinate, η, we would obtain the profiles plotted in Figure 8.5. The partial pressure in the bulk gas phase would be constant and would decrease across δ_g to a minimum value at the gas/liquid interface. The concentration of oxygen in the water would be maximum at the gas/liquid interface and would decline across δ_l to the constant level of the bulk liquid. We assume the bulk gas and bulk liquid to be completely mixed. δ_g and δ_l represent the thicknesses of the gas and liquid regions in the vicinity of the interface where velocities are less than those of the bulk gas and bulk liquid. These are the regions that are impacted by the requirement of a nonslip condition at the gas/liquid interface between the two Newtonian fluids, air and water. We use $P^*_{O_2}$ and $C^*_{O_2}$ to represent the partial pressure and concentration of oxygen at the interface. Given the nonslip condition, the monolayers of gas and water on either side of the interface have long contact times such that the equilibrium condition for the distribution of oxygen between gas and water is attained. This is often described as continuity of the state variable. In truth, it is continuity of energy. The combination of abundance of oxygen in the gas and the properties of the gas provide chemical energy exactly equal to the combination of abundance of oxygen in the water and properties of the water. What appears as a discontinuity in abundance is indeed a continuity of chemical energy.

The profiles of concentration across the boundary region are shown to be linear. Although not strictly the case, the error associated with this assumption is small. Upon formation of the bubble at the surface of the aeration device, the mass transfer process begins. Given the homogeneous natures of the liquid and gas, reactions occur rapidly. Within centimeters of the aerator, the mass transfer process for flux of oxygen from the bubble to the liquid is initiated and arrives at a near-steady-state condition. This is possible given that the boundary region on the gas side may be only a few molecular layers and that on the liquid side, owing to greater viscosity, is perhaps between 10 and a 100 molecular layers (1–10 nm). We will illustrate the linearity of the concentration profile in an example.

The flux of oxygen from the bubble to the liquid must be constant along its pathway, much as the flow of traffic along a controlled section of a freeway, devoid of exits and entrances would be. The interface might be likened to a state or city boundary at which the speed limit changes. The boundary region and interface comprise but a small reactor, of essentially zero volume, and thus accumulation of oxygen in these regions would be inconsequential. Then, the flux of oxygen to the gas side of the interface must exactly equal the flux of oxygen away from the liquid side of the interface. Similarly, the number of vehicles approaching the change in speed limit would exactly equal the number proceeding away from the change in speed limit.

If given the choice, we would like to reduce Fick's law to an algebraic rather than a differential relation. Given the near-linear concentration profiles we may do so. We can express the flux through the gas-side and liquid-side boundary regions, first using Fick's law directly:

$$\vec{F}_{O_2} = -\frac{\mathcal{D}_{O_2.g}}{RT}\frac{\partial P_{O_2}}{\partial_\eta} = -\mathcal{D}_{O_2.w}\frac{\partial C_{O_2}}{\partial_\eta}$$

Since the mass transfer process occurs under near-steady-state conditions and the accumulation term approaches 0, we can replace the partial derivatives with ordinary derivatives. Further, since the concentration profiles are linear, yielding constant slopes, and the slope is the derivative, we may replace the derivatives with algebraic expressions:

$$\frac{\partial P_{O_2}}{\partial_\eta} = \frac{dP_{O_2}}{d_\eta} = \frac{\Delta P_{O_2}}{\delta_g} = \frac{P^*{}_{O_2} - P_{O_2,\text{bulk}}}{\delta_g}$$

$$\frac{\partial C_{O_2}}{\partial_\eta} = \frac{dC_{O_2}}{d_\eta} = \frac{C_{O_2}}{\delta_1} = \frac{C_{O_2,\text{bulk}} - C^*{}_{O_2}}{\delta_1}$$

The relation for flux is then updated with the algebraic substitutions:

$$\vec{F}_{O_2} = -\frac{\mathcal{D}_{O_2.g}}{RT}\frac{P^*{}_{O_2} - P_{O_2,\text{bulk}}}{\delta_g} = -\mathcal{D}_{O_2.w}\frac{C_{O_2,\text{bulk}} - C^*{}_{O_2}}{\delta_1}$$

In this particular context, due to innate molecular activity of gas that is orders of magnitude greater than that of liquid and due to the viscosity of liquid that is orders of magnitude greater than that of gases, the diffusion (dispersion) coefficient of oxygen in the gas is many orders of magnitude greater than that in the water. Correspondingly, the gradient in abundance of oxygen in the gas is orders of magnitude smaller than that of oxygen in the water. As a consequence of this major inequality, most often we can neglect the change in abundance of transferred constituents across the gas-side boundary. We can define a mass transfer coefficient that is the quotient of the diffusion (dispersion) coefficient in water and the liquid-side boundary thickness. We refer to this coefficient as the liquid-side mass transfer coefficient and subscript the symbol accordingly:

$$k_{1.O_2} \approx \frac{\mathcal{D}_{O_2.w}}{\delta_1}$$

If we simply cancel units, we arrive at the base units of length per time, which is the set we most often see published. If we explore deeper, we can discern that k_1 has units of mass transferred per unit area per time per unit of concentration difference across the boundary region $(M/L^2/t)/(M/L^3)$. Then the diffusion coefficient has real physical units of mass transferred per unit area per time per unit of concentration gradient at the position of interest $(M/L^2/t)/(M/L^3/L)$.

We employ this substitution and rearrange the expression to obtain a handy algebraic formulation describing interfacial flux:

$$\vec{F}_{O_2} = k_{1.O_2}\left(C^*_{O_2} - C_{O_2,\text{bulk}}\right) \tag{8.31}$$

We can directly generalize this result to other solutes through the diffusion coefficient and degree of turbulence in the boundary region across which the transfer is to occur, both manifest in the value of the mass transfer coefficient. For processes in which the net transfer is from the liquid to the vapor, we merely reverse the assumed positive direction of the flux and our result employs the difference between the abundance in the bulk solution and that at the interface. The definition of interfacial flux as an algebraic, linear relation rather than a nonlinear differential relation will render its application much less computationally intensive. Equation 8.31 stems from the very simplest interfacial mass transfer case: that in which the resistance to mass transfer of the gas side is negligible and in which no reaction occurs. Levenspiel (1999) gives very informative discussions regarding mathematical treatment of the various special contexts in which interfacial mass transfer coefficients are applied.

Mass transfer coefficients must be experimentally determined. Results are generally specific to configurations of the mass transfer contactors and are correlated using dimensional similitude for scale-up to full-size reactors. Parameters used for correlation include diffusion coefficients in liquid and gas, liquid and gas loading rates, viscosity of gas and liquid, packing configuration, surface tension of packing materials and liquids, and liquid and gas density. Compilation of the various correlations from which mass transfer coefficients can be computed is well beyond our scope herein. Good sources from which to obtain correlations for mass transfer coefficients include chemical engineering texts addressing mass transfer and specific sources such as Weber and DiGiano (1996) and Perry and Green (2007).

For transfer from gas to water or from water to gas, we couple Equation 8.31 with Henry's law. Since the change in abundance between the bulk gas and the gas/liquid interface is negligible, we can relate the concentration of the target solute in the water at the gas/liquid interface directly to the partial pressure of the target solute in the bulk gas phase. Herein, we will develop an initial understanding of the mass transfer by employing Equation 8.31 in selected environmental contexts, all addressing the interfacial transfer of oxygen. Perhaps understandings gained from these straightforward applications will spark an interest in the examination of the many more complex contexts in which the principles associated with interfacial mass transfer can be applied.

8.5.3 Some Special Applications of Mass Transfer in Ideal Reactors

8.5.3.1 Characterization of Performance of Aeration Diffusers

Oxygen transfer in biological reactors is of great interest. Moving air from the atmosphere into distribution networks and to the diffusers that supply the air to treatment basins is the most costly aspect of aerobic suspended growth (activated

FIGURE 8.6 *Standard FlexAir T-series diffuser operating in a clean-water tank. Photo courtesy of Environmental Dynamics Inc.*

sludge) treatment systems. Manufacturers of aeration devices such as that shown in Figure 8.6 maintain their own testing facilities for use in characterizing the performance of the devices they manufacture. Most often these are near-full-scale reactors fitted with arrays of dissolved oxygen (DO) sensors and data acquisition systems that allow for real-time measurement and spatial averaging of DO concentrations during performance tests.

Testing most often involves deoxygenating the water in the tank either chemically or by contact with gas devoid of oxygen to strip out the oxygen and then aerating with atmospheric air or high-purity oxygen gas. Measurements of oxygen concentration are collected over time and analyzed to determine aerator performance characteristics. Since engineers typically use 20 °C as a standard temperature along with one atmosphere as standard pressure, most facilities also have means for controlling temperature at standard conditions. Given that atmospheric pressure within the facility is known, adjustments can easily be made for pressures that are nonstandard.

In order to appreciate the complexities of aerator performance testing, we should follow one of the air bubbles of Figure 8.6 on its journey from the orifice of the aerator from which it is created to the surface of the liquid in the reactor.

Example 8.13 Consider a reactor containing 5470 L of water with floor dimension 1.046 m^2 (3.43 ft^2) and total liquid depth of 5 m. Consider a bubble emitted from one of two diffusers, such as shown in Figure 8.6, with initial diameter of 2 mm. Compute the interfacial surface area per unit of air contained within the bubble and the liquid-side oxygen concentration corresponding with the oxygen content of the bubble as it rises through the liquid in the reactor. Estimate also the velocity of the bubble as it rises through the liquid and the corresponding residence time of the bubble in the liquid. The temperature is held constant at 20 °C and the absolute atmospheric pressure in the laboratory above the reactor basin is 0.95 atm. The centerline of the diffuser is located 0.1 m off the floor of the reactor.

From the absolute pressure within the bubble as it is emitted, we can compute the quantity of air contained in the bubble. We know that atmospheric air contains 20.9 mol-percent of oxygen. We find in our fluid mechanics text that a pressure of one atmosphere is equivalent to 101.325 kN/m^3, and that the specific weight of water at 20 °C is 9.79 kN/m^2. We need some parameter values and we will collect them at the outset:

$$\begin{pmatrix} Y_{O2} \\ \gamma_w \\ D \\ P_{amb} \end{pmatrix} := \begin{pmatrix} 0.209 \\ 9.79 \\ 4.9 \\ 0.95 \end{pmatrix} \begin{pmatrix} \frac{mol_{O2}}{mol_{air}} \\ \frac{kN}{m^3} \\ m \\ atm \end{pmatrix} \qquad \begin{pmatrix} R_{atm} \\ d_D \\ DO_{sat} \\ T \end{pmatrix} := \begin{pmatrix} 8.2057 \cdot 10^{-5} \\ 0.002 \\ 9.17 \\ 273.15 + 20 \end{pmatrix} \begin{pmatrix} \frac{m^3 \cdot atm}{mol \cdot °K} \\ m \\ \frac{mg}{L} \\ °K \end{pmatrix}$$

From the total pressure at the diffuser and the size of the bubble, we can compute the quantity of air contained in the bubble:

$$P_D := P_{amb} + \frac{D \cdot \gamma_w}{101.325} = 1.423 \quad atm \qquad V_D := \frac{4 \cdot \pi}{3} \cdot \left(\frac{d_D}{2}\right)^3 = 4.189 \times 10^{-9} \quad \frac{m^3}{bubble}$$

$$n_{air} := \frac{P_D \cdot V_D}{R_{atm} \cdot T} = 2.479 \times 10^{-7} \quad \frac{mol_{air}}{bubble}$$

We ignore depletion of the oxygen (as well as nitrogen) in the bubble and assume constant gas quantity in the bubble. In clean-water tests, the actual transfer of oxygen to the liquid occurs at 1–2% efficiency, with the highest efficiency occurring in the early stages of the test. We will invoke only a small error with this assumption. We write functions for pressure, bubble volume and diameter, and interfacial surface area per mole of gas using position as the argument:

$$P(z) := P_{amb} + \frac{z \cdot \gamma_w}{101.325} \quad atm \qquad V(z) := \frac{n_{air} \cdot R_{atm} \cdot T}{P(z)} \quad \frac{m^3}{bubble}$$

$$d(z) := 2 \cdot \left(\frac{3 \cdot V(z)}{4 \cdot \pi}\right)^{\frac{1}{3}} \quad m \qquad SA(z) := \frac{\pi \cdot d(z)^2}{n_{air}} \quad \frac{m^2}{mol_{air}}$$

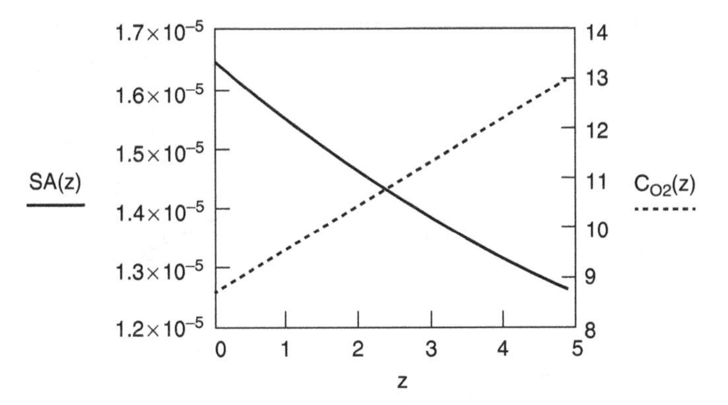

FIGURE E8.13.1 *Bubble surface area per mole of gas and saturation concentration of dissolved oxygen as a function of vertical position in a clean water tank. Transfer of O_2 and N_2 into or out of the bubble is neglected in the computation.*

We can employ the standard saturation concentration (20 °C, 1 atm) with the pressure function to create an additional function for the oxygen concentration on the water side of the gas/liquid interface:

$$C_{O2}(z) := P(z) \cdot DO_{sat} \frac{mg}{L}$$

In Figure E8.13.1 we plot surface area per mole air and interfacial concentration against position along the rise. As expected, we observe that the bubble grows and the surface area across which oxygen transfer would occur increases (~31%) as the bubble rises. Conversely, with the decrease in pressure, the interfacial oxygen concentration decreases (~33%) as the bubble rises.

We first perform an estimate of the velocity using Stokes' law. This analysis leads to velocities greater than a meter per second and associated Reynolds numbers many times the limit of unity, the maximum Reynolds number for Stokes' law applicability. We dig deeper into the principle of drag and find a plot (Figure 9.21, Munson et al., 1998) relating the drag coefficient of spheres with the Reynolds number. We need a smooth mathematical function relating the drag coefficient C_D and the Reynolds number Re. We selected five points from the plot and, using an Excel chart, plotted them in Figure E8.13.2 and fit a power law function to obtain the two necessary coefficients describing the relation.

We assume that the velocity of the bubble is very near the terminal velocity: that acceleration due to change in bubble volume, and hence buoyancy, is small relative to velocity and can be neglected. Our velocity relation will slightly overstate the magnitude, but herein, simplicity with small error trumps analysis with the

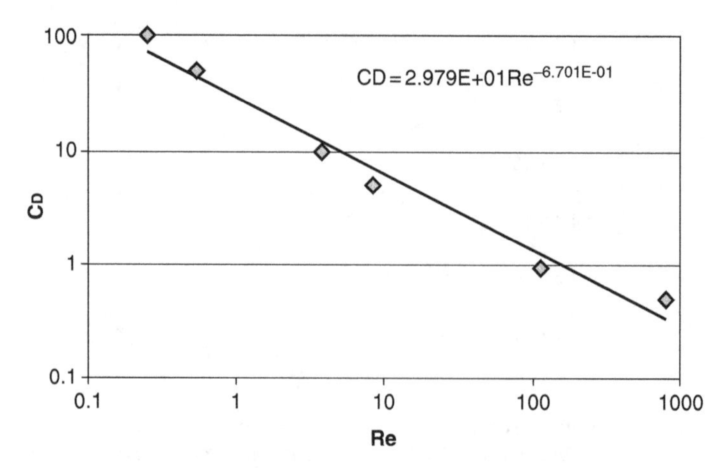

FIGURE E8.13.2 Drag coefficient versus Reynolds number for gas bubbles rising in a clean-water aeration test tank, adapted from Figure 9.21 of Munson et al. (1998).

full-blown Navier–Stokes analysis. Then the buoyant and drag forces acting on the bubble are equal. We combine three relations, force balance on the bubble, drag coefficient versus Reynolds number, and the definition of Reynolds number, into a given-find block. We then assign the find into a convenient function, retaining position as one of the arguments. U is the velocity:

$$z := 0 \qquad \begin{pmatrix} U \\ C_D \\ Re \end{pmatrix} := \begin{pmatrix} 0.253 \\ 0.438 \\ 543.157 \end{pmatrix} \qquad \text{given} \qquad \rho_f \cdot g \cdot V(z) = \frac{1}{2} \cdot \rho_f \cdot U^2 \cdot \frac{\pi}{4} \cdot d(z)^2 \cdot C_D$$

$$C_D = 29.79 \cdot Re^{-0.6701} \qquad Re = \frac{U \cdot d(z) \cdot \rho_f}{\mu} \qquad Rise(z, U, C_D, Re) := find(U, C_D, Re)$$

Our first solution is for $z = 0$ and provides the zeroth values of the U, C_D, and Re vectors we will compute to enable graphical display of the results:

$$\begin{pmatrix} U_0 \\ C_{D_0} \\ Re_0 \end{pmatrix} := Rise(z, U, C_D, Re) \rightarrow \begin{pmatrix} 0.274 \\ 0.399 \\ 624.297 \end{pmatrix}$$

We employ the function in a programmed loop to create the three resultant vectors of velocity, drag coefficient, and Reynolds number. We employ a spatial step of 0.01 m to generate the solution. Also, we employ the result from the previous position as the initial guess for each new current position, to ensure proper convergence of the given-find block within the function. The programming is illustrated in Figure E8.13.3.

$$\begin{pmatrix} z \\ U \\ C_D \\ Re \end{pmatrix} := \left| \begin{array}{l} \text{for } i \in 1..\,490 \\ \quad \left| \begin{array}{l} z_i \leftarrow \dfrac{i}{490} \cdot D \\ U_i \leftarrow U_{i-1} \\ C_{D_i} \leftarrow C_{D_{i-1}} \\ Re_i \leftarrow Re_{i-1} \\ \begin{pmatrix} U_i \\ C_{D_i} \\ Re_i \end{pmatrix} \leftarrow \text{Rise}\left(z_i, U_i, C_{D_i}, Re_i\right) \end{array} \right. \\ \text{return } \begin{pmatrix} z \\ U \\ C_D \\ Re \end{pmatrix} \end{array} \right.$$

FIGURE E8.13.3 *The programmed loop for computation of velocity with position for a bubble rising in a clean-water aeration test tank.*

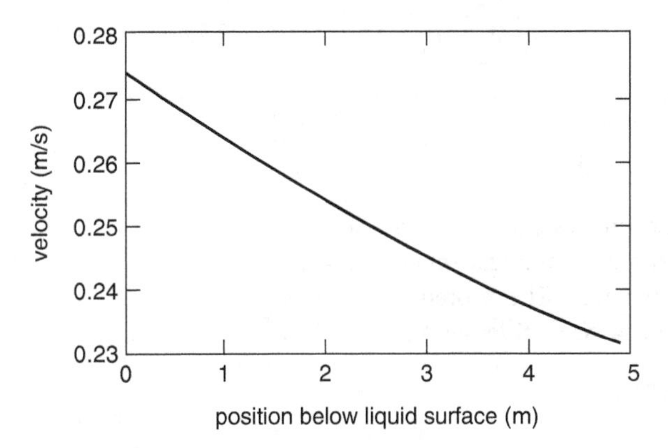

FIGURE E8.13.4 *Velocity versus position for a bubble rising in a clean-water aeration test tank.*

We have the desired vector of velocities which we plot as a function of position in Figure E8.13.4.

We observe that the bubbles do accelerate as they rise. Then, in order to determine the contact time during the rise, we compute the time required for the bubble to traverse each Δz, as the quotient of distance and average velocity. We

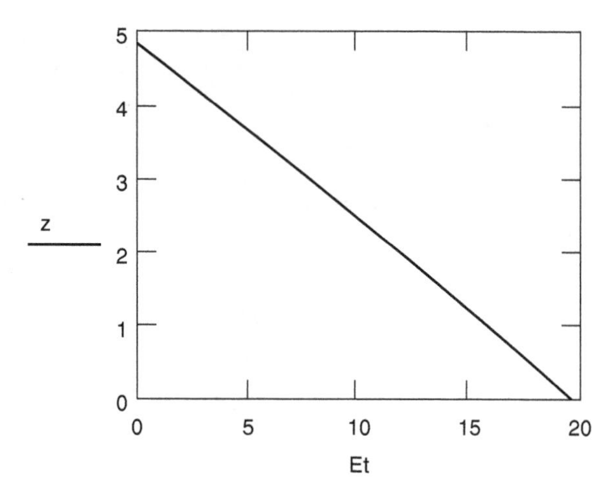

FIGURE E8.13.5 *Plot of position versus elapsed time for a bubble rising in a clean-water aeration tank.*

also compute the elapsed time (the vector Et) required to rise from the bottom of the reactor to each position along the rise of the bubble:

$$i := 0..489 \qquad t_i := \frac{\frac{D}{490}}{\frac{U_i + U_{i+1}}{2}} \; s$$

```
Et :=  | sum ← 0                                   s
       | for i ∈ 489..0
       |     | sum ← sum + t_i
       |     | Et_i ← sum
       | return Et
```

From the plot of position versus elapsed time shown in Figure E8.13.5, we observe a slightly nonlinear relation, owing to the acceleration of the bubble as it rises and expands. The aforementioned programming allows us to output the elapsed time as the bubble breaks the surface of the liquid:

$$Et_0 = 19.605 \; s$$

We predict that the total residence time of the bubble (and of course all other bubbles of initial diameter of 2 mm) would form, rise to the surface, and burst into the ambient atmosphere above the liquid in a period of about 19.6 s.

We might consider extending the analysis of Example 8.13 to include the distribution of initial bubble sizes. We could define the flux from each bubble and knowing the surface area and elapsed time, we could define the total quantity of oxygen transferred. Then, we would need to know the distribution of bubble sizes and the fraction of the total applied air that forms each distinct bubble size. This analysis assumes that we have a single bubble, unaffected by the behaviors of other bubbles and by the effects the rising bubbles exert on the hydrodynamics of the reactor. Herein, we have created an analogy to discrete particle settling, which is not correctly applied in the context of flocculant settling (primary wastewater clarifiers), flocculant suspension settling (secondary wastewater clarifiers), or hindered settling (gravity thickeners). Certainly, mixing would impact the rise of the bubble, and coalescence of colliding bubbles en route from the diffuser to the liquid surface would affect all computed parameters. Example 8.13 is indeed an approximation.

In analysis of oxygen transfer in biological processes, engineers considered the complexities associated with the changing conditions and undefined hydrodynamics of aerated reactors. Some hard decisions were made and a methodology and set of standard conditions were defined. We illustrate the analysis of the process and associated modeling in the context of an aerated batch reactor. We write the mass balance upon oxygen in the water contained within the reactor. We note that no water enters or leaves and might be tempted to discard both the input and output terms. However, since oxygen enters the water across the aggregate gas/liquid interface, we most appropriately consider the transfer as an input. Some prefer to treat the transfer as a reaction. Both views yield the exact same result. When we employ aerators in biological reactors, in which oxygen is consumed as a reactant, we realize that considering oxygen transfer as a reaction leads to conceptual complications. In fact, in such systems, we must transfer oxygen into the process as an input at exactly the same rate at which oxygen is consumed by the process. Herein, then, we will consider the oxygen transfer as an input to the CMBR without volume change:

$$V_R \frac{dC_{O_2}}{dt} = \vec{F}_{O_2} A_I$$

where A_I is the aggregate interfacial surface area provided by all bubbles residing in the reactor at any given time. The characterization of the distribution of bubble sizes and behavior within the reactor is virtually impossible and thus A_I is impossible to quantitate. Therefore, an "average" interfacial surface area per unit volume, α, is assumed to be representative of bubbles residing anywhere within the reactor. Then A_I is the product of reactor volume and interfacial surface area per unit volume ($A_I = \alpha \cdot V_R$). We substitute this relation and employ Equation 8.31 to define flux. We also drop the O_2 and bulk subscripts, noting that C is the concentration in the bulk liquid, generalizing the relation:

$$\frac{dC}{dt} = k_l \cdot \alpha \cdot (C^* - C) \tag{8.32}$$

In this relation, C^* is an aggregate value for the concentration on the liquid side of the interface. Based on the analyses of Example 8.13, we might consider using the mid-height values as the aggregate. C^* would then be the saturation concentration at the temperature of the system at the partial pressure of oxygen in the bubble under the pressure condition at the reactor mid-height. Many texts simply combine k_l and α into a single parameter $k_l\alpha$ and both discuss and mathematically treat $k_l\alpha$ as a reaction rate coefficient. The latter works just fine, but we should maintain our view that $k_l\alpha$ is truly the product of the mass transfer coefficient and the interfacial surface area per unit volume.

As we did for a pseudo-first-order reaction in a batch reactor, we may separate and integrate Equation 8.32 from t_0 to t and from C_0 to C. We will also perform some rearrangement to directly present the final result:

$$\ln\left(\frac{C^* - C_0}{C^* - C}\right) = k_1 \cdot \alpha \cdot (t - t_0) \qquad (8.33)$$

Mathematically, this is analogous to the result we found for a batch reactor and pseudo-first-order reaction rate law.

Example 8.14 Let us return to the reactor of Example 8.13. Consider that two diffusers each with 0.093 m^2 (1 ft^2) of bubble-emitting surface area are mounted on the floor of the reactor. A test was conducted using a total gas flow rate of 0.283 m_{std}^3/min (SCMM, 10 ft_{std}^3/min, SCFM) split to the two diffusers. A standard cubic meter (SCM) is equivalent to the gas contained in 1 m^3 of volume at 20 °C and 1 atm of total pressure. The laboratory temperature was held at 20 °C during the test and the ambient atmospheric pressure during the test was 0.95 atm. The liquid depth was adjusted to provide diffuser submergence of 4.57 m (15.0 ft). Spatially averaged DO concentration versus time data are shown in the following table.

Time (min)	Spatially averaged DO (mg/L)	Time (min)	Spatially averaged DO (mg/L)
0	3.95	5	9.95
0.5	5.37	5.5	10.32
1	6.35	6	10.2
1.5	7.22	6.5	10.48
2	7.9	7	10.47
2.5	8.68	7.5	10.63
3	8.81	8	10.64
3.5	9.38	8.5	10.68
4	9.49	9	10.77
4.5	9.99		

Determine the lumped overall mass transfer coefficient, $k_1\alpha$, and use this parameter to determine the value of the standard oxygen transfer efficiency (SOTE) rating of the diffuser for the conditions of the test.

When we examine Equation 8.33, we observe that a plot of $\ln\left(\frac{C^* - C_0}{C^* - C}\right)$ versus time should have a slope of $k_1\alpha$. We will transform the data using an Excel worksheet to obtain $k_1\alpha$. C_0 is merely the first usable value of DO concentration and we assign this to time 0. Parsing of the data was performed prior to final assembly in order to eliminate data that are obviously in error, particularly at the beginning of the test. We determine C^* as the saturation concentration at the mid-depth pressure, using the function of Example 8.13, to be 10.93 mg/L. We insert this value into the worksheet and, for convenience, name it. We also, for convenience, name C_0. We create a column for $\frac{C^* - C_0}{C^* - C}$ and another for $\ln\left(\frac{C^* - C_0}{C^* - C}\right)$. The Excel worksheet output is shown in Figure E8.14.1.

C.sat =	10.93	mg/L	
t (min)	C (mg/L)	(C.sat-C.0)/ (C.sat-C)	ln[(C.sat-C.0)/ (C.sat-C)]
0	3.95	1	0
0.5	5.37	1.255	0.2271
1	6.35	1.524	0.4213
1.5	7.22	1.881	0.6318
2	7.9	2.304	0.8346
2.5	8.68	3.102	1.132
3	8.81	3.292	1.191
3.5	9.38	4.503	1.505
4	9.49	4.847	1.578
4.5	9.99	7.426	2.005
5	9.95	7.122	1.963
5.5	10.32	11.44	2.437
6	10.2	9.562	2.258
6.5	10.48	15.51	2.741
7	10.47	15.17	2.719
7.5	10.63	23.27	3.147
8	10.64	24.07	3.181
8.5	10.68	27.92	3.329
9	10.77	43.63	3.776

FIGURE E8.14.1 *Output from an Excel worksheet in which concentration versus time data were processed to obtain the overall value of $k_l \cdot \alpha$.*

We then plot $\ln\left(\frac{C^* - C_0}{C^* - C}\right)$ versus t (see Figure 8.14.2) and add a linear trend line, requesting that Excel output the equation of the trend line. We have formatted the trend line label to obtain four significant figures of the output. Initially, we let Excel output the intercept to ensure that the data would support the [0, 0] origin of the relation. Satisfied the condition was met, we formatted the trend line to originate at [0, 0]. From the trend line we observe the value of the lumped parameter $k_l\alpha$ to be 0.4080 min^{-1}. The volume-specific standard oxygen transfer rate, often abbreviated as ***sotr***, is the product of the lumped $k_l\alpha$ and standard-state oxygen saturation concentration:

$$sotr := kla \cdot C_{std} = 3.741 \quad \frac{mg}{L \cdot min}$$

The *SOTE* is the quotient of the volume-specific standard oxygen transfer rate (*sotr*) and the volume-specific oxygen application rate (*oar*). We may also use the overall standard oxygen transfer rate for the reactor (*SOTR*) and the overall oxygen application rate (*OAR*). Since *sotr* and *SOTR* are related by the reactor volume as

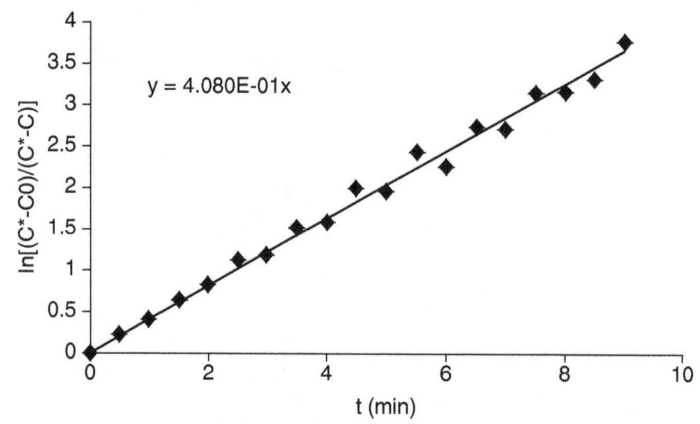

FIGURE E8.14.2 *A plot of* $\ln\left(\frac{C^* - C_0}{C^* - C}\right)$ *versus time for a clean-water test of an aeration diffuser.*

are *oar* and *OAR*, we can use either the overall or the volume-specific parameters to obtain *SOTE*. Now that we have *sotr*, let us compute *oar* and then *SOTE*:

$$Y_{O2} := 0.209 \qquad MW_{O2} := 32000 \ \frac{\text{mg}}{\text{mol}} \qquad T_{std} := 293 \ {}^{\circ}\text{K}$$

$$R_{atm} := 0.082057 \ \frac{\text{L} \cdot \text{atm}}{\text{mol} \cdot {}^{\circ}\text{K}} \qquad V_R := 5000 \ \text{L} \qquad Q_{air} := 283 \ \frac{\text{L}}{\text{min}}$$

$$Mol_{air} := \frac{1}{R_{atm} \cdot T_{std}} \cdot Q_{air} = 11.771 \ \frac{\text{mol}}{\text{min}}$$

$$oar := \frac{Mol_{air} \cdot Y_{O2} \cdot MW_{O2}}{V_R} = 15.745 \ \frac{\text{mg}}{\text{L} \cdot \text{min}}$$

$$SOTE := \frac{sotr}{oar} \cdot 100 = 23.8 \ \%$$

We note that there are two diffusers each of 1 ft^2 area and the air flow rate is 5 SCFM/ft^2, with a standard efficiency of 23.8%. This test would put one point on a set of performance curves such as that shown in Figure E8.14.3 for the Flex Air T-Series diffuser (courtesy of Environmental Dynamics, International (EDI)).

The test tank can be filled to any desired submergence (e.g., 5, 10, 15, 20, and 25 ft) for a series of tests to fully characterize the performance of a given diffuser design. Thus, the curves shown in EDI's data plot are experimental rather than predicted. Given automation of data recording, in a well-designed test facility, experimental data can be developed quite accurately and rapidly.

From the results of Example 8.13, we predict that a single bubble rising as assumed would have a mid-depth diameter of 2.1 mm. Realistically, we know that bubbles will coalesce as they are emitted from the diffuser and as they collide as a

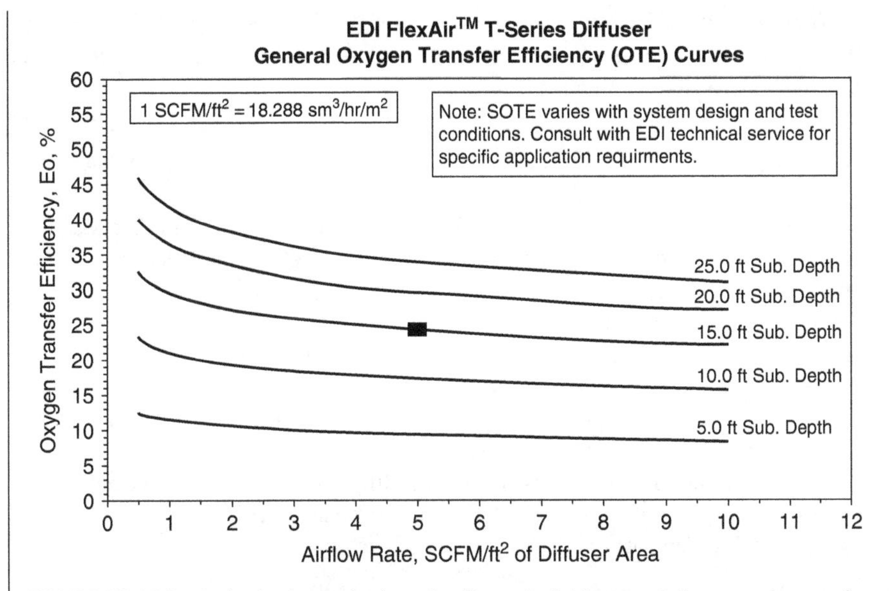

FIGURE E8.14.3 *A single data point from the Example 8.14 computation superimposed on a plot of standard oxygen transfer efficiency versus air flow per diffuser for a Flex Air, T-Series diffuser manufactured by Environmental Dynamics International. T-Series performance curve courtesy of Environmental Dynamics International.*

consequence of turbulence in the water column during their rise to the surface. We might guess that the average bubble size within the liquid would be ~3 mm, owing to coalescence of five bubbles into a single bubble during the overall residence time. We could assume that at the mid-height, half the coalescence would have occurred. In the absence of significant study using underwater photography, this value, of course, is only slightly better than a wild guess. However, we are now in a position from which we can estimate the value of the mass transfer coefficient, independently of the interfacial surface area. We may use our function from Example 8.13 to determine the specific surface area per mole of applied air, using the 3-mm bubble diameter and assuming that three bubbles have coalesced to become our "average" bubble:

$$n_{C.air} := 3 \cdot n_{air} = 7.436 \times 10^{-7} \quad \frac{mol_{air}}{bubble}$$

$$SA_C := \pi \cdot 0.003^2 = 2.827 \times 10^{-5} \quad \frac{m^2}{bubble}$$

$$sa_C := \frac{SA_C}{n_{C.air}} = 38.023 \quad \frac{m^2}{mol_{air}}$$

We will employ the computed rise time from Example 8.13 (19.6 s) to compute the quantity of air that is present in the water at any given time as the product of air flow rate and residence time:

$$t_{res} := \frac{19.6}{60} = 0.327 \quad \text{min} \qquad Inv_{air} := Mol_{air} \cdot t_{res} = 3.845 \quad mol_{air}$$

Then, from the quantity of air held in the liquid, we estimate the associated total and volume-specific surface areas. We will need to carefully consider the units and opt to use area in square decimeters to match the unit of the reactor volume, noting that a liter is a cubic decimeter:

$$SA_{inv} := Inv_{air} \cdot sa_C \cdot 100 = 1.462 \times 10^4 \quad dm^2 \qquad \alpha_{inv} := \frac{SA_{inv}}{V_R} = 2.924 \quad \frac{dm^2}{dm^3}$$

We would like to employ this result with diffusion coefficients, normally tabulated in units of cm²/s, so we will use our computed $k_l\alpha$ with α_{inv} to obtain an estimate of k_l in units of cm/s:

$$k_l := \frac{kl\alpha}{\alpha_{inv}} \cdot \frac{10}{60} = 0.0233 \quad \frac{cm}{s}$$

The molecular diffusion coefficient for oxygen in water is about 2×10^{-5} cm²/s. If we return to the simple boundary layer theory discussed earlier, we may estimate the thickness of the boundary layer surrounding the bubble as the quotient of the mass transfer and diffusion coefficients:

$$D_{O2.w} := 2 \cdot 10^{-5} \quad \frac{cm^2}{s}$$

$$\delta_l := \frac{D_{O2.w}}{k_l} = 8.6 \times 10^{-4} \quad cm \qquad \delta_{l.\mu} := 10^4 \cdot \delta_l = 8.6 \quad \mu m$$

For this boundary layer thickness, we assume that the turbulence imparted to the liquid by the rising bubbles has no effect within the boundary layer. If the turbulence within the liquid imparts turbulence within the boundary region, the dispersion coefficient would be of a greater magnitude than the diffusion coefficient and we would compute the boundary layer to be greater in thickness. In either case, we certainly can determine that the boundary layer thickness is several thousand times the nominal size of a water molecule.

Examples 8.13 and 8.14 certainly are a mix of theory and practice. We now have an understanding of some of the major complexities surrounding oxygen transfer from bubbles to liquid in the context of aeration systems. We observe how the industry has adapted to define a standard rating system for aeration diffuser performance and now have a basic understanding of the testing process, employing diffusion across a boundary layer as a rate process occurring in an ideal reactor. We may now certainly visualize the oxygen bubbles, surrounded by their associated boundary layers across which oxygen transfer occurs, as they rise from aerators to the surface of aerated reactors. We certainly can consider alternate processes using high-purity nitrogen gas to deoxygenate the water or using high-purity oxygen gas to oxygenate the water. The modeling process would be quite similar in either case.

8.5.3.2 Oxygen Transfer Across a Macroscopic Surface

The Streeter–Phelps model is published in most textbooks addressing introductory environmental engineering. The model relates the abundance of DO in water as a function of a first-order reaction rate constant, a reaeration constant, the initial DO concentration, and the initial concentration of biologically degraded substance in the water with the time of travel downstream from the confluence of a wastewater discharge with a flowing stream. Let us dig a little more deeply into the processes coupled together and develop an alternative formulation employing somewhat more descriptive parameters.

Example 8.15 Consider a rectangular concrete channel carrying return sludge from a clarifier to the reactor influent. The flow emanates from a closed pipe, flows the length of the channel, and exits into a closed pipe at the end of the channel. The channel is 1.0 m wide, 0.5 m deep, and 100 m in length, with steady, uniform flow at 0.5 m/s. The temperature at the time of the analysis is 20 °C and the normal ambient pressure at the location of the channel is 0.9 atm. Under a targeted flow scenario, the wastewater enters the channel containing 4 mg/L DO, 5 mg/L biodegradable chemical oxygen demand (bCOD), and 500 mg/L of biomass capable of degrading the dissolved organic substances manifest as the bCOD. Estimate the DO and bCOD in the wastewater that exits the channel into the downstream pipe.

We must first develop the mathematical model with which we can quantitatively solve this question. We employ Figure 8.2 with some enhancements, most specifically, to represent the rate of reaction and the rate at which oxygen would be transferred across the gas/liquid interface, to produce Figure E8.15.1.

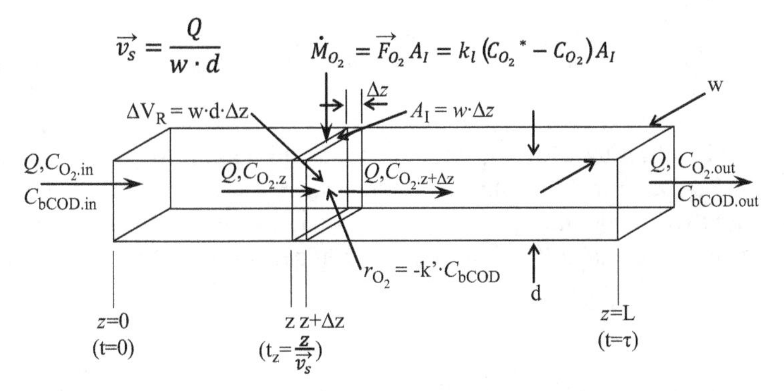

FIGURE E8.15.1 Schematic representation of a rectangular channel and arbitrarily located fluid element within the channel, upon which a mass balance is drawn.

In the same manner as we developed Equation 8.20a, we write a mass balance considering the oxygen dissolved in the water within the rectangular element bounded by z, $z+\Delta z$, the channel walls and floor, and the gas/liquid interface. We will consider that the process has been in progress sufficiently long that the process has attained a near-steady-state condition. Oxygen enters the element with the flow at z, exits the element with the flow at $z+\Delta z$, enters the element across the area of the gas/liquid interface, and is consumed by the biological process that occurs within the element:

$$0 = v_s \cdot w \cdot d\left(C_{O_2}\Big|_z - C_{O_2}\Big|_{z+\Delta z} \right) + \vec{F}_{O_2} \cdot w \cdot \Delta z + w \cdot d \cdot \Delta z \cdot r_{O_2}$$

We immediately recognize that the specific rate of oxygen consumption is directly related to the abundance of bCOD and is fully independent of the abundance of oxygen. We may use the pseudo-first-order rate law in the context of the PFR to express r_{O_2} fully in terms of the influent bCOD and position in the reactor:

$$C_{bCOD}(z) = C_{bCOD}(0) \cdot \exp\left(\frac{k' \cdot z}{v_s} \right)$$

$$r_{O_2} = -k' \cdot C_{bCOD}(\bar{z}) = -k' \cdot C_{bCOD.in} \cdot \exp\left(\frac{k' \cdot \bar{z}}{v_s} \right)$$

The position $\bar{z}$ is $z + \frac{\Delta z}{2}$, and we will see later that when $\Delta z \to 0$, $\bar{z} \to z$. We employ the definition of change in concentration, $\Delta C_{O_2} = C_{O_2}\Big|_{z+\Delta z} - C_{O_2}\Big|_z$ and the definition of interfacial flux, $\vec{F}_{O_2} = k_1 \cdot \left(C_{O_2}^* - C_{O_2} \right)$. We also slightly rearrange the resulting relation to attain the base-level mass balance on oxygen within the arbitrarily located element of fluid within the PFR:

$$v_s \cdot w \cdot d \cdot \Delta C_{O_2} = k_1 \cdot \left(C_{O_2}^* - C_{O_2} \right) \cdot w \cdot \Delta z - w \cdot d \cdot \Delta z \cdot k' \cdot C_{bCOD.in} \cdot \exp\left(\frac{k' \cdot \bar{z}}{v_s} \right)$$

The units of all terms of this relation reduce to mass of oxygen per unit time. Now we can perform some algebra and rearrange the result into a more succinct expression: we divide through by v_s, w, d, and Δz. Once done with the algebra and rearrangement, we take the limit as $\Delta z \to 0$, requiring that $\Delta z \to dz$, $\Delta C \to dC$ and $\bar{z} \to z$:

$$\frac{dC_{O_2}}{dz} = \frac{k_1}{v_s \cdot d} \cdot \left(C_{O_2}^* - C_{O_2} \right) - \frac{k'}{v_s} \cdot C_{bCOD.in} \cdot \exp\left(\frac{k' \cdot z}{v_s} \right)$$

We desire a closed-form solution of this ODE and consult our friendly local mathematician (e.g., Wylie, 1966, or one of the myriad online resources) and find that we have an inseparable, linear, first-order ODE. We can further rearrange the relation to the standard form:

$$y' + P(x)y = Q(x)$$

Where $\frac{dC_{O_2}}{dz}$ is y', C_{O_2} is y and z is x:

$$\frac{dC_{O_2}}{dz} + \frac{k_1}{v_s \cdot d} \cdot C_{O_2} = \frac{k_1}{v_s \cdot d} \cdot C_{O_2}{}^* - \frac{k'}{v_s} \cdot C_{\text{bCOD.in}} \cdot \exp\left(\frac{k' \cdot z}{v_s}\right)$$

Then $P(z) = \frac{k_1}{v_s \cdot d}$ and $Q(z) = \frac{k_1}{v_s \cdot d} \cdot C_{O_2}{}^* - \frac{k'}{v_s} \cdot C_{\text{bCOD.in}} \cdot \exp\left(\frac{k' \cdot z}{v_s}\right)$. In order to develop the solution to this type of ODE, we are instructed to create an integrating factor as $\exp(\int P(z)dz)$ and then to multiply all terms of the relation by this integrating factor. Then, we can separately integrate each side of the relation, maintaining the equality. We are instructed that the integral of the left side is merely the product of the dependent variable, in this case C_{O_2}, and the integrating factor, $\exp(\int P(z)dz)$. We must determine the integral of the right side of the relation using whatever resource we have at our disposal. We can easily perform this integration using MathCAD's calculus palette. The general solution from our MathCAD worksheet is presented as follows. In keeping with previous literature, we have used the symbol L_0 for $C_{\text{bCOD.in}}$. In addition, since MathCAD has some character-formatting inflexibility, for k' we use k_{pr} and for C^* we use C_{sat}:

$$e^{\frac{k_l \cdot z}{d \cdot v_s}} \cdot C = C_{sat} \cdot e^{\frac{k_l \cdot z}{d \cdot v_s}} - \frac{L_0 \cdot d \cdot k_{pr} \cdot e^{\frac{k_l \cdot z}{d \cdot v_s} - \frac{k_{pr} \cdot z}{v_s}}}{k_l - d \cdot k_{pr}} + K$$

We must be careful to define the constant of integration before we perform any rearrangement of the relation. We know that at the influent $z = 0$, we have $C_{O_2} = C_0$. When we insert these known parameters into the general relation, we can solve for the constant of integration, K.

$$K = C_0 - C_{sat} + \frac{L_0 \cdot d \cdot k_{pr}}{k_l - d \cdot k_{pr}}$$

When we combine the definition of K with the general solution to obtain the particular solution, we have the model that we can now employ to determine the DO concentration, as related to the suite of forcing parameters, as a function of position along the channel. We have also not shown a few algebraic simplification steps en route to the function we desire:

$$C(z) := C_{sat} - \frac{L_0 \cdot d \cdot k_{pr} \cdot e^{-\frac{k_{pr} \cdot z}{v_s}}}{k_l - d \cdot k_{pr}} + e^{-\frac{k_l \cdot z}{d \cdot v_s}} \cdot \left(C_0 - C_{sat} + \frac{L_0 \cdot d \cdot k_{pr}}{k_l - d \cdot k_{pr}} \right)$$

We define some forcing parameters:

$$\begin{pmatrix} \mu_{max} \\ K_{half} \\ Y \\ X \\ L_0 \end{pmatrix} := \begin{pmatrix} \frac{6}{86400} \\ 20 \\ 0.4 \\ 500 \\ 5 \end{pmatrix} \begin{pmatrix} \frac{mg_{vs}}{mg_{vs} \cdot s} \\ \frac{mg_{bCOD}}{dm^3} \\ \frac{mg_{vs}}{mg_{bCOD}} \\ \frac{mg_{vs}}{dm^3} \\ \frac{mg_{bCOD}}{dm^3} \end{pmatrix} \qquad \begin{pmatrix} v_s \\ d \\ w \\ k_l \\ C_{sat} \\ C_0 \end{pmatrix} := \begin{pmatrix} .5 \\ 5 \\ 10 \\ 0.0024 \\ 9.09..9 \\ 4 \end{pmatrix} \begin{pmatrix} \frac{dm}{s} \\ dm \\ dm \\ \frac{dm}{s} \\ \frac{mg_{O2}}{dm^3} \\ \frac{mg_{O2}}{dm^3} \end{pmatrix}$$

The units for this effort are milligram, decimeter, and seconds. Note that a liter is a cubic decimeter. We have decided to employ the pseudo-first-order rate law to allow for a closed-form solution to the ODE and thus must compute k' from the definition of the saturation rate law. Since our C_{bCOD} is well below the value of K_{half}, $L_0 = 5$ while $K_{half} = 20$, we can approximate the saturation rate law as a pseudo-first-order rate law with only a small error:

$$k_{pr} := \frac{\mu_{max} \cdot X}{Y \cdot K_{half}} = 4.34 \times 10^{-3} \quad \frac{1}{s}$$

We now investigate the function we created to see what happens to the DO concentration as the return sludge stream passes along the open channel. We produce Figure E8.15.2 for visual observation. We observe the classic DO sag curve. In the first 200 m, the biological oxidation of the organics in the wastewater occurs at a faster rate than the transfer of oxygen. Once the organic matter is depleted, the rate of oxygen transfer overpowers the consumption and the DO level begins to rise. Were the channel another 100 m in length, we could predict a DO level near 7 mg/L at the exit.

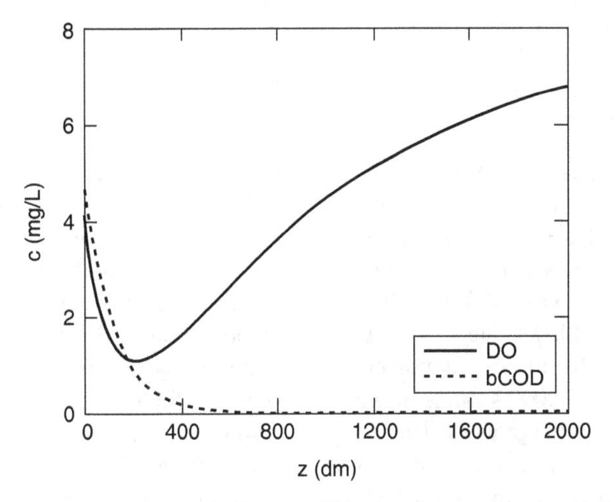

FIGURE E8.15.2 *Profiles of DO and biodegradable chemical oxygen demand along the flow path of a rectangular open channel transmitting return activated sludge from a clarifier to the influent of a recycle reactor.*

If we wanted to determine the exact position at which the DO concentration is at a minimum, we could simply set the derivative of $C(z)$ equal to 0 and solve for the position, which can be easily done with a MathCAD given-find block. Once z_{min} is defined, the minimum DO value is easily found from our $C(z)$ function:

$z := 100$ Given $\frac{d}{dz}C(z) = 0$ $z_{min} := \mathrm{Find}(z) = 213.2$ dm

$C_{min} := C(z_{min}) = 1.076$ $\frac{mg_{O2}}{L}$

Our analysis of Example 8.15 parallels that of the Streeter–Phelps equation:

$$D = \frac{k_1 L_0}{k_2 - k_1}\left(e^{-k_1 t} - e^{-k_2 t}\right) + D_0 e^{-k_2 t}$$

where D is the DO deficit ($C^*_{O_2} - C_{O_2}(t)$, M/L^3), k_1 is the deoxygenation rate coefficient (1/t), k_2 is the re-aeration rate coefficient (1/t), L_0 is the initial bCOD concentration (M/L^3), D_0 is the initial DO deficit (M/L^3), and t is the elapsed reaction time (t), usually the quotient of distance downstream (L) and stream velocity (L/t). With some algebraic manipulation, we could render our $C(z)$ function from Example 8.15 to a form exactly like that of the Streeter–Phelps model. Other than the rewards arising from the intellectual activity, we have no reason for accomplishing that task herein. If we were to use the Streeter–Phelps model directly, we would need to either guess the coefficients k_1 and k_2 or perform an analysis such as that accomplished in Example 8.15 to quantitatively understand the Streeter–Phelps' k_1 and k_2 coefficients.

PROBLEMS

CMBR Problems

1. A water treatment process is employed to remove radium from drinking water prior to its distribution to customers in the Countryside Subdivision south of Rapid City, SD. The radium removed is a combination of ^{228}Ra and ^{226}Ra in a ratio of 1:9. The removal process consists of an ion exchanger (in truth a very large home water softener) that removes radium along with hardness. The waste by-product is a brine containing sodium, potassium, calcium, magnesium, and radium. The influent radium concentration is 25 pg/L (1 pg = 10^{-12} g), the plant processes 50,000 gallons of water per day, and the radium is virtually entirely removed in the process. The removed radium must of course be safely disposed of. How much radium (in g) was removed during 2001? How much of that year's (01/01/01 through 12/31/01) removed radium 226 and 228 (respectively) will be around (presumably in the disposal location) at the beginning of the next millennium (i.e., 01/01/3001, 999 years in the future)? The literature yields the values $k_{228} = 0.1205$/year and $k_{226} = 0.0004331$/year.

2. Lindane (one of several hexachlorocyclohexane isomers, $C_6H_6Cl_6$) is an insecticide now used mostly outside the United States for the control of insects in a variety of contexts. Consider that a rainfall runoff event has occurred just after the application of lindane within the watershed of a small lake. The runoff has caused the lindane concentration in the lake to rise to a level of 1 mg/L in the water. Consider that lindane decays in the environment in accord with a first-order rate law with a decay constant of 0.014/day. What will the concentration of lindane be in this lake 1 week after the rainfall runoff incident? The small lake can be considered in this case to be a stirred batch reactor.

3. Lindane (one of several hexachlorocyclohexane isomers, $C_6H_6Cl_6$) is an insecticide now used mostly outside the United States for the control of insects in a variety of contexts. Consider that a rainfall runoff event has occurred just after the application of lindane within the watershed of a small lake. The runoff has caused the lindane concentration in the lake to rise to a level of 1 mg/L in the water. Consider that lindane decays in the environment in accord with a first-order rate law with a decay constant of 0.014/day. How long (in days) will it take for the lindane concentration in the lake to reach 100 ppb_m? The small lake can be considered in this case to be a stirred batch reactor.

4. Atrazine ($C_8H_{14}N_5Cl$), often applied to sorghum and corn after emergence to control grasses and broadleaf weeds, was applied one June day to a field adjacent to a ranch stock pond near New Underwood, SD. A rainfall runoff event occurred the day after the application, washing a significant quantity of atrazine from the field and into the stock pond. The rancher became aware of the problem when a number of calves were found to have died. In the legal wrangling that followed, an expert consultant was asked to determine the concentration level

(in mmol/L) of atrazine that might have existed in the stock pond the day of the runoff event. The concentration level in the stock pond was measured 30 days after the event and found to be 85 ppb_m. The consultant found that atrazine decays in accordance with a first-order rate law with a decay constant of 0.00693/day. The stock pond can be considered in this case to be a stirred batch reactor.

5. A groundwater is to be treated for removal of iron via the oxidation of Fe(II) to Fe(III). The groundwater has a pH of 6.25 and a Fe(II) concentration of 4.7 mg/L as pumped from the aquifer. The Fe(II) concentration is to be reduced to 0.03 mg/L or below in a batch process prior to its use in an industrial process. The oxidation of ferrous iron is known to follow a pseudo-first-order rate law such that $r_{Fe(II)\rightarrow Fe(III)} = -k[Fe(II)]P_{O_2}[OH^-]^2$, where $k = 8.0\times 10^{13}$ L^2/mol^2-atm-min, [Fe(II)] is in either mass or molar units, P_{O_2} is in atmospheres, and $[OH^-]$ is in molar units. The local atmospheric pressure is 0.95 atm. How long (in min) must this batch process be allowed to run in order that the [Fe(II)] concentration level is met? Note that pH and P_{O_2} are to be held constant at the stated levels.

6. A groundwater is to be treated for removal of iron via the oxidation of Fe(II) to Fe(III). The groundwater has a pH of 6.25 and a Fe(II) concentration of 4.7 mg/L as pumped from the aquifer. The Fe(II) concentration is to be reduced to 0.03 mg/L or below in a batch process prior to its use in an industrial process. The oxidation of ferrous iron is known to follow a pseudo-first-order rate law such that: $r_{Fe(II)\rightarrow Fe(III)} = -k[Fe(II)]P_{O_2}[OH^-]^2$, where $k = 8.0\times 10^{13}$ L^2/mol^2-atm-min, [Fe(II)] is in either mass or molar units, P_{O_2} is in atmospheres, and $[OH^-]$ is in molar units. The local atmospheric pressure is 0.95 atm. To what value must the pH be adjusted in order that this batch process can be accomplished with a reaction time of 30 min? Note that pH will be held constant in this process at the computed level and P_{O_2} is also to be held constant at the stated level.

7. A particular sewage lift station (these consist of a reservoir or wet well and a pumping system, and are used when wastewater must be lifted to a higher elevation within a collection system) serving an outlying subdivision of Rapid City, SD, has an oversized wet well such that during low-flow periods, which occur during each night, the residence time of the wastewater in the wet well is 8 h. The wastewater contains 30 mg/L sulfate ($SO_4^=$), which is 10 mg/L $SO_4^=$–S. This $SO_4^=$–S is converted biologically into $S^=$–S (i.e., sulfate ⇒ sulfide). Sulfide in water, of course, is readily converted at neutral pH to the volatile and toxic H_2S, always a danger for workers associated with wastewater collection. A consultant has been asked to advise the city about this situation. The literature pertaining to the conversion of sulfate to sulfide yielded the following:

$$r_{SO_4^= \rightarrow S^=} = -\frac{k(X)[SO_4 - S]}{K_{1/2} + [SO_4 - S]}$$

where $k = 0.10 \frac{mg_{SO_4S}}{mg_{BM} \cdot h}$, X (for a 4-ft diameter wet well) $= 10 \frac{mg_{BM}}{L}$, and $K_{1/2} = 2.0 \frac{mg_{SO_4S}}{L}$. The volume of stagnant water contained in the wet well is 3000 L. How long (in h) will it take for half the sulfate sulfur in the wet well to be converted to sulfide sulfur?

8. A particular sewage lift station (these consist of a reservoir or wet well and a pumping system, and are used when wastewater must be lifted to a higher elevation within a collection system) serving an outlying subdivision of Rapid City, SD, has an oversized wet well such that during low-flow periods, which occur during each night, the residence time of the wastewater in the wet well is 8 h. The wastewater contains 30 mg/L sulfate ($SO_4^=$), which is 10 mg/L $SO_4^=$–S (to be considered the initial concentration for a batch process). This $SO_4^=$–S is converted biologically into $S^=$–S (i.e., sulfate ⇒ sulfide). Sulfide in water, of course, is readily converted at neutral pH to the volatile and toxic H_2S, always a danger for workers associated with wastewater collection. A consultant has been asked to advise the city about this situation. The literature pertaining to the conversion of sulfate to sulfide yielded the following:

$$r_{SO_4^= \to S^=} = -\frac{k(X)[SO_4 - S]}{K_{1/2} + [SO_4 - S]}$$

where $k = 0.10 \frac{mg_{SO_4S}}{mg_{BM} \cdot h} r$, X (for a 4-ft diameter wet well) $= 10 \frac{mg_{BM}}{L}$, and $K_{1/2} = 2.0 \frac{mg_{SO_4S}}{L}$. The volume of stagnant water contained in the wet well is 3000 L. How much $S^=$–S (in g) will be produced during this stagnant period that occurs each night. Note that the molar rate of disappearance of sulfate sulfur will numerically equal the molar rate of appearance of sulfide sulfur. Note that the disappearance of sulfate sulfur and the appearance of sulfide sulfur will be of exactly the same magnitude but of opposite signs as $SO_4^=$–S disappears and $S^=$–S appears.

PFR Problems

9. A wastewater treatment plant is to use UV irradiation to kill (inactivate) harmful bacteria in the effluent from the plant. The contact basin is to be a PFR with a 30-min contact time. The discharge from the treatment facility is 5 MGD. The influent to the contact basin will have 10^8 colony-forming units (CFUs) of bacteria (one bacterium can form one colony of bacteria) per liter of solution. The first-order rate constant for inactivation of these bacteria is 0.425/min. What will be the concentration of bacteria in the effluent from the contact basin?

10. A wastewater treatment plant is to use UV irradiation to kill (inactivate) harmful bacteria in the effluent from the plant. The contact basin is to be a PFR. The discharge from the treatment facility is 5 MGD. The influent to the contact basin will have 10^8 CFUs of bacteria (one bacterium can form one colony of bacteria) per liter of solution. The first-order rate constant for inactivation of these bacteria is 0.425/min. What must the volume (in gal) of the contact basin be in order to attain a level of bacteria in the reactor effluent that is 100 CFUs/L?

11. A wastewater treatment plant is to use UV irradiation to kill (inactivate) harmful bacteria in the effluent from the plant. The contact basin is to be a PFR of volume equal to 100,000 gal. The discharge from the treatment facility is 5 MGD. The influent to the contact basin will have 10^8 CFUs of bacteria (one bacterium can form one colony of bacteria) per liter of solution. The effluent from the contact basin must be 100 CFUs/L. Note that the UV radiation level from the lamps may be adjusted in order to increase or decrease the intensity as desired. What value of the rate constant is necessary in order to attain this performance level?

12. A wastewater force main (these are used to transmit wastewater uphill in wastewater collection systems) acts like a PFR. Wastewater to be pumped through a particular force main will contain sulfate sulfur (SO_4-S) at a concentration of 10 mg($SO_4^=$–S)/L of wastewater. From the literature we find the following regarding the conversion of sulfate to sulfide in wastewater:

$$r_{SO_4^= \to S^=} = -\frac{k(X)[SO_4 - S]}{K_{1/2} + [SO_4 - S]}$$

where $k = 0.10\ \dfrac{mg_{SO_4S}}{mg_{BM} \cdot h}$ and $K_{1/2} = 2.0\ \dfrac{mg_{SO_4S}}{L}$. The effective concentration of SRB (these convert sulfate to sulfide) in the wastewater is $40\ \dfrac{mg_{BM}}{L}$. In order to maintain the production of sulfide sulfur in the force main to a level below 1 mg/L (i.e., the effluent sulfate sulfur concentration cannot be less than 9 mg/L), what is the maximum residence time in hours allowed in the force main?

13. A wastewater force main (these are used to transmit wastewater uphill in wastewater collection systems) acts like a PFR. Wastewater to be pumped through a particular force main will contain sulfate sulfur (SO_4–S) at a concentration of 10 mg(SO_4–S)/L of wastewater. From the literature we find the following regarding the conversion of sulfate to sulfide in wastewater:

$$r_{SO_4^= \to S^=} = -\frac{k(X)[SO_4 - S]}{K_{1/2} + [SO_4 - S]}$$

where $k = 0.10 \frac{mg_{SO_4S}}{mg_{BM} \cdot h}$ and $K_{1/2} = 2.0 \frac{mg_{SO_4S}}{L}$. The effective concentration of SRB (these convert sulfate to sulfide) in the wastewater is $20 \frac{mg_{BM}}{L}$. The flow rate during the critical period of the day is 50 gal/min through the force main. The force main is a 4-in. diameter pipe and is 9200 ft in length. What will the concentration (in mg/L) of sulfide sulfur be in the effluent from the force main? Note that one cubic foot is 7.48 gal. Also note that this solution will require some sort of numeric solver (MathCAD root or given-find, Excel goal seek or solver, or other) as the resultant equation cannot be solved explicitly for the effluent concentration of sulfate sulfur.

14. A wastewater force main (these are used to transmit wastewater uphill in wastewater collection systems) acts like a PFR. Wastewater to be pumped through a particular force main will contain sulfate sulfur (SO_4-S) at a concentration of 10 mg(SO_4-S)/L of wastewater. From the literature we find the following regarding the conversion of sulfate to sulfide in wastewater:

$$r_{SO_4^= \rightarrow S^=} = -\frac{k(X)[SO_4 - S]}{K_{1/2} + [SO_4 - S]}$$

where $k = 0.10 \frac{mg_{SO_4S}}{mg_{BM} \cdot h}$ and $K_{1/2} = 2.0 \frac{mg_{SO_4S}}{L}$. The flow rate during the critical period of the day is 50 gal/min through the force main. The force main is a 4-in. diameter pipe and is 9200 ft in length. Regulations set by the owner of the collection system limit the concentration of sulfide in wastewater entering its system from force mains such as this one to 0.5 mg/L sulfide sulfur ($S^=$–S). To what value (in mg/L) must the concentration of SRB (X in this rate law) be reduced in the wastewater in order for this condition to be met at the outlet of the force main?

CMFR Problems

15. A wastewater treatment plant is to use UV irradiation to kill (inactivate) harmful bacteria in the effluent from the plant. The contact basin is to be a CMFR with a 30-min contact time. The discharge from the treatment facility is 5 MGD. The influent to the contact basin will have 10^8 CFUs of bacteria (one bacterium can form one colony of bacteria) per liter of solution. The first-order rate constant for inactivation of these bacteria is 0.425/min. What will be the concentration of bacteria in the effluent from the contact basin? Given that the effluent requirement is most often about 1000 CFUs/L, is this a good design?

16. A wastewater treatment plant is to use UV irradiation to kill (inactivate) harmful bacteria in the effluent from the plant. The contact basin is to be a CMFR. The discharge from the treatment facility is 5 MGD. The influent to the contact

basin will have 10^8 CFUs of bacteria (one bacterium can form one colony of bacteria) per liter of solution. The first-order rate constant for inactivation of these bacteria is 0.425/min. What must the volume (in gal) of the contact basin be in order to attain a level of bacteria in the reactor effluent that is 100 CFUs/L? Compare this with the volume of a PFR in which the same reaction would be carried out. Which reactor system would you choose for this reaction?

17. A wastewater treatment plant is to use UV irradiation to kill (inactivate) harmful bacteria in the effluent from the plant. The contact basin is to be a CMFR of volume equal to 1,000,000 gal. The discharge from the treatment facility is 5 MGD. The influent to the contact basin will have 10^8 CFUs of bacteria (one bacterium can form one colony of bacteria) per liter of solution. The effluent from the contact basin must be 100 CFUs/L. Note that the UV radiation level from the lamps may be adjusted in order to increase or decrease the intensity as desired. What value of the rate constant is necessary in order to attain this performance level? Increasing the intensity of the UV irradiation requires a lot of energy. Is a CMFR the best choice as a reactor for this system?

18. A treatment process with the objective of biologically converting nitrate nitrogen (NO_3–N) to nitrogen gas (N_2) (i.e., the named process of denitrification) is to be completed using a CMFR. The influent to the process will contain 10 mg/L NO_3–N and the NO_3–N concentration is to be reduced to 0.10 mg/L. The process conforms to a saturation-type rate law:

$$r_N = -\frac{kX[NO_3 - N]}{K_{1/2} + [NO_3 - N]}$$

where $k = 0.250 \dfrac{mg_{NO_3N}}{mg_{BM} \cdot d}$ and $K_{1/2} = 0.2 \dfrac{mg_{NO_3N}}{L}$

If the biomass concentration in the reactor is to be 500 mg_{BM}/L and the volumetric flow rate to the process is 1.25 MGD, how large must the basin be (in gal) to accomplish this treatment objective?

19. A treatment process with the objective of biologically converting nitrate nitrogen (NO_3–N) to nitrogen gas (N_2) (i.e., the named process of denitrification) is to be completed using a CMFR. The influent to the process will contain 10 mg/L NO_3–N and the NO_3–N concentration is to be reduced to 0.10 mg/L. The process conforms to a saturation-type rate law:

$$r_N = -\frac{kX[NO_3 - N]}{K_{1/2} + [NO_3 - N]}$$

where $k = 0.250 \dfrac{mg_{NO_3N}}{mg_{BM} \cdot d}$ and $K_{1/2} = 0.2 \dfrac{mg_{NO_3N}}{L}$

The reactor to be constructed for this process has a volume of 20,000 gal and the influent flow rate will be 1.25 MGD. What level of biomass (mg_{BM}/L) must be maintained in the reactor to accomplish this treatment objective?

20. A treatment process with the objective of biologically converting nitrate nitrogen (NO_3–N) to nitrogen gas (N_2) (i.e., the named process of denitrification) is to be completed using a CMFR. The influent to the process will contain 10 mg/L NO_3–N and the NO_3–N concentration is to be reduced to a level as low as possible. The process conforms to a saturation-type rate law:

$$r_N = -\frac{kX[NO_3 - N]}{K_{1/2} + [NO_3 - N]},$$

where $k = 0.250 \dfrac{mg_{NO_3N}}{mg_{BM} \cdot d}$ and $K_{1/2} = 0.2 \dfrac{mg_{NO_3N}}{L}$

The reactor to be constructed for this process has a volume of 15,000 gal, the influent flow rate will be 1.25 MGD and the biomass concentration is limited to 3000 mg_{BM}/L based on mixing considerations. (Biomass concentrations greater than about 3000 mg/L cannot be effectively mixed using the available mixing equipment in the reactor.) To what level in mg/L will the concentration be reduced under this scenario?

21. A treatment process with the objective of biologically converting nitrate nitrogen (NO_3–N) to nitrogen gas (N_2) (i.e., the named process of denitrification) is to be completed using a CMFR. The influent to the process will contain 10 mg/L NO_3–N and the NO_3–N concentration is to be reduced to 0.10 mg/L. The process conforms to a saturation-type rate law:

$$r_N = -\frac{kX[NO_3 - N]}{K_{1/2} + [NO_3 - N]},$$

where $k = 0.250 \dfrac{mg_{NO_3N}}{mg_{BM} \cdot d}$ and $K_{1/2} = 0.2 \dfrac{mg_{NO_3N}}{L}$

The reactor biomass concentration is limited to 3000 mg_{BM}/L and the reactor volume is 50,000 gal. What maximum flow rate (in gal/day) can be applied to this reactor while the treatment objective is met?

22. A treatment process with the objective of biologically converting nitrate nitrogen (NO_3–N) to nitrogen gas (N_2) (i.e., the named process of denitrification) is to be completed using a CMFR. The influent to the process will contain 10 mg/L NO_3–N and the NO_3–N concentration is to be reduced to 1.0 mg/L. The process conforms to a saturation-type rate law:

$$r_N = -\frac{kX[NO_3 - N]}{K_{1/2} + [NO_3 - N]},$$

where $k = 0.250 \dfrac{mg_{NO_3N}}{mg_{BM} \cdot d}$ and $K_{1/2} = 0.2 \dfrac{mg_{NO_3N}}{L}$

The influent flow rate is to be 1.25 MGD and the volume of the reactor is to be 25,000 gal. The biomass concentration is to be 3000 mg_{BM}/L. Compute the number of moles of nitrogen gas (N_2) that will be produced on a daily basis.

Stoichiometry Problems

23. Wastewater flowing through a force main contains 50 mg/L sulfate. While resident in the force main, the DO content of the wastewater drops to 0 and the conversion of sulfate sulfur to sulfide sulfur occurs, converting 25% of the sulfate sulfur to sulfide sulfur. What is the concentration of sulfide sulfur in the wastewater exiting the force main?

24. A wastewater stream contains 100 mg/L chlorophenol. A special oxygen-free process is to be used to virtually entirely convert the chlorophenol to phenol. The process waste stream flows at 500 gal/min. Compute the quantity of phenol in the effluent in mg/L and the quantity (in kg/day) that will be produced each day.

25. A wastewater stream contains 30 mg/L ammonia nitrogen. The overall biological treatment process is to include the process of nitrification (conversion of ammonia to nitrate). The flow rate is 10 MGD. How much oxygen (in kg/day) will be consumed by this process?

26. A wastewater stream contains 30 mg/L nitrate nitrogen and flows at a rate of 10 MGD. What minimum rate of methanol feed, in milligrams of methanol per liter of wastewater, will be required for the conversion of all nitrate to nitrogen gas? How much methanol (in kg) must be fed each day to the process?

27. A wastewater stream contains 30 mg/L nitrate nitrogen and flows at a rate of 10 MGD. What minimum rate of acetic acid feed, in milligrams of acetic acid per liter of wastewater, will be required for the conversion of all nitrate to nitrogen gas? How much acetic acid must be fed each day to the process?

28. A process stream contains 60 mg/L total cyanide and flows at a rate of 100 gal/min. What feed rate of chlorine, in milligrams of chlorine per liter of wastewater must be maintained in order to destroy the cyanide? How much chlorine (in kg) would be required each day?

29. A water treatment process designed to remove suspended solids from raw water prior to treatment by filtration is applied to a water that contains 10 mg/L TSSs, which are virtually completely removed by the coagulation/flocculation/sedimentation process. The dose of alum ($Al_2(SO_4)_3$) is to be 20 mg/L. How much solid material will be produced per liter of water treated? If the flow rate is 1000 gal/min, what quantity of solids (in kg) will be produced each day?

30. The Phosstrip® wastewater treatment process is employed at a wastewater plant for removal of phosphorus. The plant influent contains 10 mg/L PO_4^{-3}–P

which is reduced to 0.5 mg/L via this process. The plant flow rate is 100 MGD. How much phosphorus (in kg) is removed on a daily basis?

The Phosstrip® process produces a by-product stream containing a rather high (~1 g/L) concentration of phosphate phosphorus (PO_4^{-3}–P), which is then subjected to a chemical precipitation process using aluminum sulfate to precipitate phosphorus as aluminum phosphate. How much aluminum sulfate (in kg) would be required for this precipitation operation over the course of 1 year, assuming that the precipitation process reduces the phosphorus concentration effectively to 0? How much aluminum phosphate (in kg) would be produced each year from the process?

31. A precipitation process is used at the Gilt Edge superfund site, located near Deadwood, SD, to remove metals from acid rock drainage generated at the site. Zinc, copper, and cadmium are reduced from initial concentrations of 250, 500, and 50 mg/L, respectively, to discharge concentrations in the range of 0.05 mg/L (essentially complete removal). For this process it is necessary to feed 1.25 mEq/L of lime (over and above the precipitation requirement) to adjust the pH of the solution for these precipitation reactions to occur. The process operates 24/7 at a rate of 100 gal/min. How much lime ($Ca(OH)_2$, in kg) is required on an annual basis to accomplish this treatment objective?

Advanced Problems

32. The biomass present in a wastewater force main actually colonizes the interior surface of the pipe as shown, with important geometric information in Figure p8.32. Thus, the conversion of sulfate sulfur to sulfide sulfur occurs within this biofilm, which is generally only about 1–2 mm in thickness:

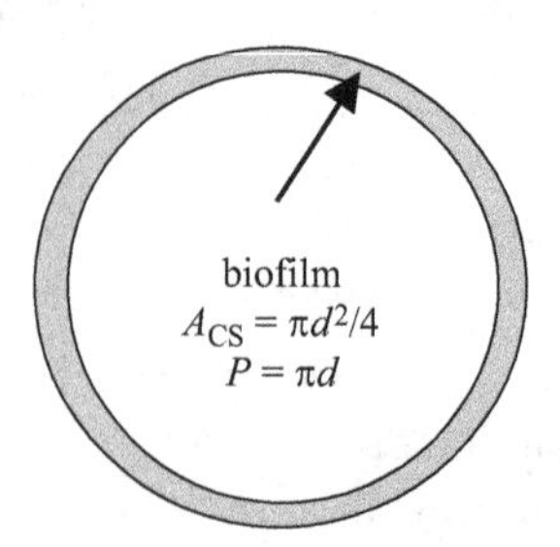

parameter	4″	6″	8″
α (dm^2/dm^3)	3.94	2.62	1.97
A_{CS} (dm^2)	0.811	1.82	3.24
P (dm)	3.19	4.79	6.38

FIGURE p8.32 Sketch of biofilm on and important geometry of the interior surface of a force main.

$$SO_4^{=} + 8e^- + 10H^+ = H_2S + 4H_2O$$

The design of a particular lift station/force main system is under consideration. A search of the literature yields the following rate law for conversion of sulfate sulfur to sulfide sulfur:

$$r_{SO_4^{=} \to S^{=}} = -\frac{k\alpha[SO_4 - S]}{K_{1/2} + [SO_4 - S]}$$

where k is a rate coefficient (1.5 mg($SO_4^{=}$–S)/dm^2(biofilm area)-h). α is the interfacial surface area of the biofilm contact with the flowing wastewater in units of dm^2/dm^3 (the quotient of the perimeter (P) and the cross-sectional area (A_{CS}) of the pipe. Values for α, A_{CS}, and P are given for 4-, 6-, and 8-in. diameter pipes in the table at the right $K_{1/2} = 2.0\ mg_{SO_4-S} / L$. Wastewater will enter the force main, 28,000 dm in length, containing sulfate sulfur at a concentration of 10 mg($SO_4^{=}$–S)/L (dm^3) of wastewater. The average flow rate over the critical period between 10 PM and 6 AM is 500 L/min.

For the stated flow and other conditions:

a. State the reactor model to be employed and concisely describe the reasoning for your selection.
b. Were a 6-in. pipe to be employed, determine the concentration of sulfide sulfur (in mg($S^{=}$–S)/L) in the wastewater exiting the force main at its termination.
c. Determine which alternative pipe size (4-in. or 8-in.) would result in a lower end-of-pipe sulfide sulfur concentration and support your assertion with appropriate computations.

33. Ammonia nitrogen applied as a fertilizer is often in the root zone far in excess of plant needs. In the ammonia form, nitrogen is mobile and can be carried with infiltrating precipitation through the unsaturated zone and into underlying groundwaters. Consider that a soaking rain has occurred on the June 10 and carried ammonia (as NH_3–N) into the shallow groundwater beneath a soybean field in eastern South Dakota, such that the concentration of ammonia nitrogen in the groundwater, after mixing of the infiltrated water with the groundwater, was 5 mg/L (g/m^3) as NH_3–N. On September 10, the concentration of ammonia nitrogen in the groundwater was 0.5 mg/L.

The law describing the rate at which nitrate is microbially converted to nitrogen gas via the reaction

$$NH_4^+ + 2O_2 = NO_3^- + H_2O + 2H^+$$

can be used as

$$r_{NH_3-N \to NO_3-N} = -\frac{k}{K_N} \cdot X \cdot C_{NH_3-N}$$

where $k = 5.00 \frac{\text{g(NH}_3 - \text{N)}}{\text{g(biomass)} - \text{day}}$, K_N is a half-reaction velocity constant $\left(= 0.75 \frac{\text{g(NH}_3 - \text{N)}}{\text{m}^3} \right)$, X is the effective biomass concentration in the groundwater $\left(\frac{\text{mg(biomass)}}{\text{L}} \right)$, and C_{NH_3-N} is the concentration of ammonia nitrogen in g/m^3 (as NH_3–N).

a. Based on the stated conditions, the overall system, and the desired information required in part (c), state the reactor model to be used herein for this computation and include specific reasoning for your choice.
b. Begin with a mass balance and develop the specific relation that is to be used computationally in part (c).
c. From the known information, determine the effective concentration of biomass in the groundwater beneath the soybean field.

34. When under ice cover, Sylvan Lake (total surface area = 18 ac @ 43,560 ft^2/ac; average depth = 12 ft), located in Custer State Park of SD, experiences depletion of DO, particularly in the deeper regions of the lake as a consequence of biological activity that occurs within its organic-rich sediments. Under ice cover, the lake becomes weakly stratified such that the temperature of water is 4 °C at the deepest portions and colder directly under the ice. Upon melting of the ice cover, the water near the surface is warmed and the weak thermal stratification is broken, causing the lake to be virtually completely mixed for a period of several days. The temperature of the water throughout the lake can be assumed to be 4 °C during this period of complete mixing.

    Sylvan Lake has a surface elevation of ~5000 ft above mean sea level, resulting in a normal ambient atmospheric pressure of 0.83 atm. Consider that at the time that the ice cover disappears and the lake becomes fully mixed, the DO concentration in the lake is 8 mg/L. Consider that the mass transfer coefficient for the transport of oxygen into water under the prevailing conditions is 1×10^{-4} dm/s.

    a. Begin with a mass balance and develop the mathematical model that can be used to relate the DO concentration in the lake under the stated conditions to elapsed time. Be sure to fully, but concisely, document the development of this mathematical model.
    b. Consider that the ice fully melts on Sylvan Lake and it becomes fully mixed at 3 PM on April 25. Consider also that the lake becomes restratified (a layer of warmer water, the epilimnion, lies above the colder water below, the hypolimnion) at 3 PM on May 1. At this time the lake is no longer completely mixed. What will be the DO concentration in the lake at 3 PM on each day between April 26 and May 1, considering that water temperature and atmospheric pressure remain constant?

35. A force main carrying wastewater discharges (critical minimum flow = 0.00100 m^3/s) into a manhole at its terminus and the flow leaves from the manhole via an 8-in. (0.2032 m) gravity sewer line, sloped at 0.004 m/m. The wastewater

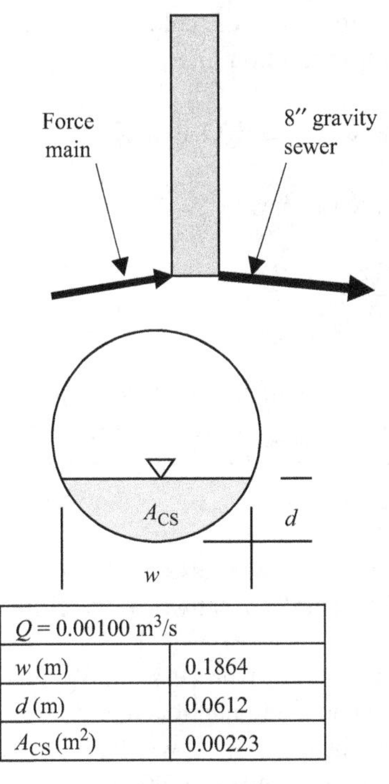

$Q = 0.00100\ m^3/s$	
w (m)	0.1864
d (m)	0.0612
A_{CS} (m^2)	0.00223

FIGURE p8.35 *Layout, cross section and geometry of a uniform flow situation in a gravity sewer.*

discharges from the force main containing virtually no DO, as biological activity in the wastewater consumes all available oxygen during transport through the force main. A profile, a cross section of the pipe, and important geometry information are given in Figure p8.35. As the wastewater collection system is somewhat isolated from the atmosphere, the gas phase above the flowing wastewater has an oxygen content of 14% by moles, pressure, or volume. The total pressure of this gas phase is 0.90 atm absolute and the system temperature is 14 °C. The next manhole in the collection system is 122 m down the pipe. Determine the oxygen content of the wastewater as it enters the downstream manhole. For this particular computation, you may consider that biological utilization of oxygen is negligible. You have determined that the mass transfer coefficient for this condition is 5.66×10^{-5} m/s.

36. The influent stream to an anaerobic biological process reactor, which is well mixed and contains no flow baffles, of volume equal to 20,000 gal flows at a rate of 500 gal/min and contains 100 mg/L of chlorophenol (C_6H_4OHCl). The concentration of chlorophenol is to be reduced to 1 mg/L in this reactor. A search of the literature surrounding the biological degradation of chlorophenol indicated that chlorophenol is converted to phenol by special bacteria that

simply replace the chlorines of the chlorophenol molecules with hydrogens by the oxidation–reduction reaction

$$C_6H_4OHCl + H^+ + 2e^- \rightarrow C_6H_5OH + Cl^-$$

and in accord with the rate law

$$r_{C_{Ph}\rightarrow Ph} = -\frac{kX[C_{Ph}]}{K_{1/2} + [C_{Ph}]},$$

where k is a rate constant equal to 0.00100 mmol(C_{Ph})/mg$_{biomass}$-h, X is the biomass concentration in mg$_{biomass}$/L, $[C_{Ph}]$ is the concentration of chlorophenol in mmol/L, and $K_{1/2}$ is the half-reaction rate constant equal to 0.14 mmol(C_{Ph})/L.

What concentration of biomass is necessary in this reactor to accomplish the treatment objective?

If the volume of the reactor were to be 10,000 gal and the biomass concentration were to be held at 2500 mg/L, what would the effluent concentration of chlorophenol be?

Consider that this reaction process simply converts the chlorophenol to phenol and that the generated phenol is not in any way biologically or chemically degraded in this process. What will be the concentration of phenol in the effluent from the reactor in units of mg(C_6H_5OH)/L?

37. A backpacker carries household bleach that will be used to disinfect water to be used for drinking while backpacking along the Appalachian Trail of the Eastern United States. The water container has a 2-gal capacity. The water is to be filtered using river sand in a foot-deep bed to remove particulate matter prior to introduction into the container and bleach is to be added to the water in the container to attain the desired concentration of chlorine for disinfection.

Disinfection occurs in accord with a pseudo-first-order rate law such that $r_N = -k[Cl_2]C_{bacteria}$, where k is the rate constant equal to 0.1 L/mg(Cl_2)-min, $[Cl_2]$ is the concentration of chlorine in mg/L, and $C_{bacteria}$ is the number of active organisms per liter of the solution undergoing disinfection. Research has shown that surface water filtered through a foot-deep bed of medium sand can typically contain 10^6 harmful organisms per liter of solution. The backpacker wishes to reduce this level to one organism per liter prior to drinking this water. The desired contact time for the chlorine to do its work is 1 h (i.e., the backpacker wishes to have the water disinfected and ready to drink in an hour).

What concentration of chlorine will be required in order to accomplish this disinfection in the desired time frame? You may assume that the chlorine concentration remains constant during the disinfection process. Disclaimer: *In truth, the chlorine will be consumed by the process, but we will keep it somewhat simple and ignore this coupled process. Also, please do not attempt producing safe drinking water using methods suggested here without thoroughly researching the complete process.*

How long should the backpacker wait to drink the water (assuming the final bacteria concentration is to be one organism per liter) if the chlorine concentration is 10 mg/L?

38. A pilot-scale reactor, consisting of a 55-gal plastic drum fitted with influent and effluent piping and a mechanical mixing apparatus, was employed to test the conversion of cyanide (CN^-) to cyanate (CNO^-, much less toxic than cyanide) by oxidation with hydrogen peroxide (H_2O_2) at elevated pH by the overall reaction

$$CN^- + H_2O_2 = CNO^- + H_2O$$

The reactor accepted an influent flow of 2.75 gal/min containing 20 ppm_m total cyanide (recall that $[CN_T] = [HCN] + [CN^-]$) at pH of 10.6. Hydrogen peroxide was fed to the reactor as a concentrated solution using a separate peristaltic pump. The effluent from the reactor contained CN_T at 10 ppm_m and H_2O_2 at 5 ppm_m, and was of pH equal to 10.5. The destruction of cyanide by hydrogen peroxide is believed to follow a rate law of the form

$$r_{CN^- \rightarrow CNO^-} = -k[H_2O_2][OH^-]CN_T$$

where $[i]$ denotes a molar concentration of component i (e.g., H_2O_2, OH^- or CN_T).

a. From these test data, determine the value of the rate constant k for the conversion of cyanide to cyanate. Be sure to correctly specify the resultant units for k. *In your solution be sure to state the reactor model used and the bases for your assertion.*
b. For this particular test, compute the concentration (in mg/L) of CNO^- in the effluent from the pilot-scale reactor.

39. A reactor configured as shown in Figure p8.39 is to be used to reduce the COD of a waste stream prior to discharge. The system, a high-rate activated sludge system, will have the capability to recycle biomass recovered from a clarifier following the process to adjust the biomass concentration in the reactor as necessary to attain the desired treatment objective. The process follows a rate law of the form

$$r_{COD} = -\frac{k(X)[COD]}{K_{1/2} + [COD]}$$

The influent will contain 500 mg_{COD}/L and the effluent can contain no more than 5 mg_{COD}/L. The reactor volume is to be 10^6 L and the influent flow rate is to be 4×10^6 L/day. The value of the rate constant k is believed to be 3 mg_{COD}/mg_{VSS}-day and the half-reaction rate constant $K_{1/2}$ is believed to be 50 mg_{COD}/L.

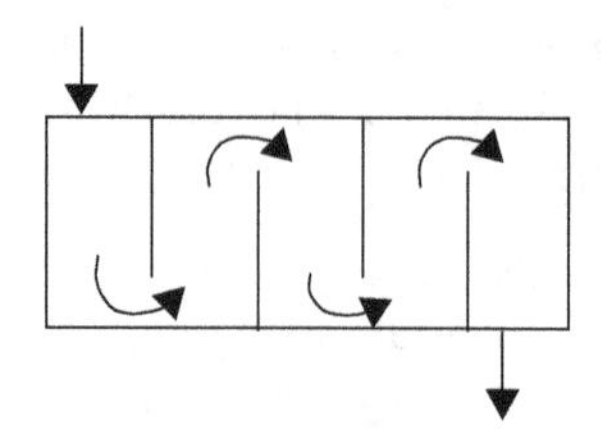

FIGURE p8.39 *Flow pattern in a reactor arranged to approximate plug flow.*

The biomass X is usually characterized in terms of volatile solids, VSS, which constitute that part of TSSs that would be lost when a TSS sample is "fired" in a muffle furnace at 550 °C.

a. Determine the concentration level of biomass (in units of mg_{VSS}/L) in the reactor necessary to attain the desired reduction in COD. Your solution is to include a statement of your selection of the appropriate reactor and the reasoning behind that selection.
b. The maximum level at which biomass may be maintained in a reactor of design under consideration herein is 2000 mg_{VSS}/L. Were the effluent constraint to remain 5 mg_{COD}/L, compute the maximum COD level (mg_{COD}/L) of the influent stream treatable, while still meeting the effluent constraint, with this reactor/reaction system.
c. Space is an issue at the location of the treatment plant and you are to consider a larger abundance of biomass in the reactor – 4000 mg_{VSS}/L.

40. A heavy, 2-day soaking rain occurred immediately following application of the herbicide atrazine to a field of emergent corn. Twenty days after the rainfall event, the groundwater beneath the field was sampled and found to contain atrazine at a level of 100 ppb_m. Atrazine is degraded through microbial action in accord with a rate law of the form

$$r_{atr} = -kC_{O_2}XC_{atr}$$

where C_{O_2} is the concentration of DO in the water in the aquifer (6.0 mg/L), X is the effective biomass concentration in the aquifer (0.05 mg_{VSS}/L), C_{atr} is the atrazine concentration in $\mu g_{atrazine}/L$, and the rate constant is defined as

$$k = 0.50 \frac{L^2}{mg_{VSS} \cdot mg_{O_2} \cdot day}$$

a. The maximum contaminant level (MCL) set for atrazine in drinking water is ~5 ppb_m. How long (give your answer in days) will it take, marking the sampling time as the initial time for the level of atrazine in the groundwater, to decay to the MCL? Specify the reactor model that would be employed for computations addressing this system and explain the reasoning behind your selection.

b. Compute the concentration level (express your answer in $mg_{atrazine}/L$) of atrazine in the groundwater immediately after the soaking rainfall event, assuming that the time of travel of infiltrating water through the unsaturated zone is negligible.

41. When conducting the measurement of the final oxygen concentration for the BOD test, it is specified that the manganous sulfate and alkali azide reagent are to be added quickly, to avoid bias of the results due to the transfer of oxygen from the atmosphere into the BOD bottle, before the DO is "fixed." Let us examine just how quickly that operation must be accomplished. The BOD bottle holds 300 mL of aqueous solution and the opening provides a vapor/liquid interface that is circular in shape and 7–8 in. in diameter. The mass transfer coefficient for this particular context would have a value of 1×10^{-3} dm/s.

   a. Write a mass balance for oxygen in the aqueous solution contained in the BOD bottle, assuming that the bottle is left open on the laboratory counter (*developing a sketch of the system is highly recommended*). Neglect the fact that the consumption of oxygen by the process within the bottle is still in progress. Changes due to biological activity would be presumed insignificant relative to changes due to the migration of oxygen into the bottle across the vapor/liquid interface from the atmosphere. Your result is to be a relation between DO concentration in the bottle, elapsed time from the opening of the bottle, DO concentration at the instant the bottle is opened, the relevant oxygen saturation value, and the mass transfer coefficient. No computations are required; your result for this part should be entirely symbolic.
   b. For the sake of argument, assume that the true oxygen concentration in a particular test is 2.0 mg_{O_2}/L. Consider that the venue of the laboratory is the EnvE lab in the C/M building and that the laboratory temperature and pressure are 23 °C and 0.89 atm, respectively. Compute the time the bottle would need to remain open in order for the DO concentration level to increase to 2.5 mg/L. *You are strongly encouraged to perform these numerical computations using dm as the length unit and, hence, dm³ (L) as the volumetric unit.*

42. Cyanide can be converted using hydrogen peroxide to the much lesser toxic substance cyanate by the following reaction:

$$CN^- + H_2O_2 \Rightarrow CNO^- + H_2O$$

The reaction rate law can be expressed as follows:

$$r_{CN^- \to CNO^-} = -k[H_2O_2][OH^-][CN_T]$$

where

$$k = 1.075 \times 10^6 \frac{L^2}{mol^2 - min}$$

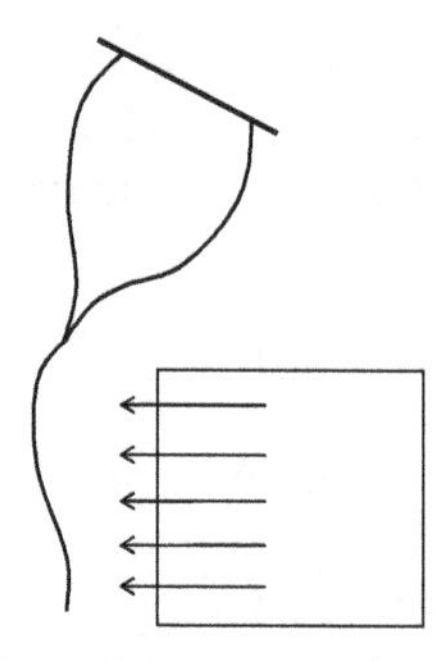

FIGURE p8.43 *Areal sketch of runoff into a stock watering pond.*

A by-product stream flows at a rate of 100 L/min from an industrial process and is expected to contain total cyanide at a concentration of 26 ppm_m. A reactor of the configuration shown in Figure p8.39 is planned. The overall footprint of the basin is to be square and the depth of the channels is intended to be twice the width of the channel.

The desired level of the effluent cyanide concentration is 2.6 ppm_m. The pH of the by-product water is 8.5 and the average peroxide concentration in the process is to be 51 ppm_m.

a. For the stated conditions and constraints, determine the required volume of the reactor.
b. How might the pH of the by-product water be adjusted (i.e., up or down) to improve the performance of the reactor system? Support your answer with specific arguments.

43. A stock watering pond located northeast of New Underwood, SD, covers an area of 16 ac (43,560 ft^2/ac) and has an average depth of 5 ft. See the sketch of Figure p8.43. The herbicide atrazine was applied to emergent sorghum in an adjacent field on June 5, 1990. A severe thunderstorm occurred on June 6, 1990, dissolving a large quantity of atrazine into runoff, which carried the atrazine into the stock pond. Sickness and death of calves in the pasture surrounding the stock pond led to sampling of the water from the stock pond on July 6, 1990, yielding a concentration of atrazine of 60 ppb_m.

The conversion of atrazine to its breakdown products is known to follow the rate law

$$r_{\text{atrazine}} = -\frac{kXC_{\text{atr}}}{K + C_{\text{atr}}}$$

where $k = 0.15$ $mg_{\text{atrazine}}/mg_{\text{biomass}}$-day, $K = 2.5$ mg_{atrazine}/L, $X = 0.5$ mg_{biomass}/L, and C_{atr} is the concentration of atrazine in appropriate units. In order to certify beef, "organic" water consumed by cattle must contain no detectable pesticides. The analytical detection limit for atrazine in water is 500 $pptr_m$ (parts per trillion by

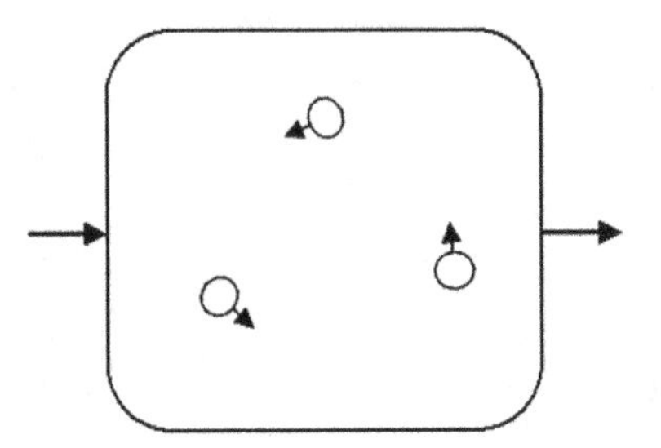

FIGURE p8.44 *Plan view sketch of aerated treatment pond with directional surface aerator/mixers.*

mass). From the known information, you are to investigate the potential for the resumption (if at all) of watering cattle from this stock pond.

a. For the "organic' designation, determine when the cattle could again drink the water from the stock pond.
b. Determine the concentration of atrazine in the water of the stock pond immediately after the rainfall runoff event on June 6.

44. A wood products manufacturing facility employs a large earthen-dike pond (sometimes these are called aerated lagoons, shown in Figure p8.44) for treatment of process waters used in the manufacturing process. The pond covers an area of 50 ac (1 ac = 43,560 ft²) and is 15 ft deep. The pond contains three aerator/mixers that both mix and aerate the contents of the pond. A layout sketch of one of the ponds is shown on the left. The arrows shown with each of the aerators indicates the direction in which fluid is propelled at ~6 ft/s by the aerator/mixer as it operates. The by-products to be treated are organic and characterized by a parameter called chemical oxygen demand (COD ≈ $BOD_{ultimate}$). The influent COD is 850 mg/L. The process is described by the rate law

$$r_{COD} = -\frac{kXC_{COD}}{K_{1/2} + C_{COD}}$$

with k = 0.5 $mg_{COD}/mg_{biomass}$-day, $K_{1/2}$ = 35 mg_{COD}/L, X = the biomass concentration in $mg_{biomass}$/L, and C_{COD} = the COD in mg_{COD}/L.
The volumetric loading of process water to be treated is 6.9×10^7 gal/day.

a. Determine the concentration of biomass that must be maintained in the pond to produce an effluent concentration of 30 mg_{COD}/L.
b. Were the biomass concentration to be 1250 $mg_{biomass}$/L, what effluent concentration could be attained?

45. Leachate leaks through the earthen barrier placed beneath the degrading solid wastes in a landfill and joins the groundwater that flows beneath the site, as depicted in Figure p8.45. The leachate contains 10 g_{BOD}/L and the total flow is 100 L/min. The groundwater upgradient of the landfill contains no BOD and flows beneath the site at a rate of 1900 L/min. Consider that the leachate and

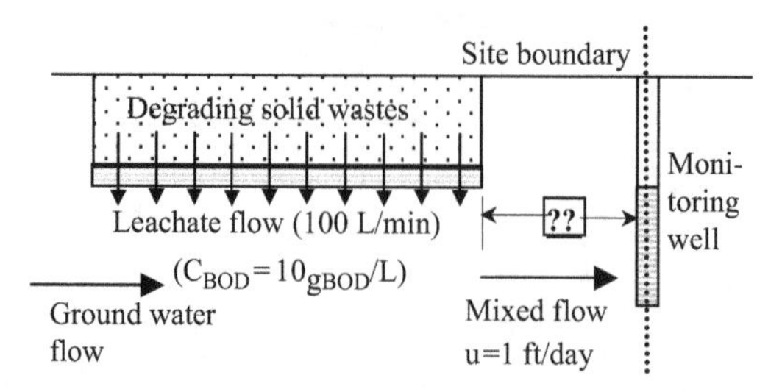

FIGURE p8.45 Sketch of landfill leachate and ground water flow system with a down-gradient monitoring well.

groundwater become completely mixed by the time the flow reaches the eastern edge of the solid waste deposition. The mixed groundwater flow direction is toward the east boundary of the landfill site. The rate law by which biodegradable organics (manifest as BOD) are biologically degraded is

$$r_{\text{BOD}} = -\frac{kXC_{\text{BOD}}}{K_{1/2} + C_{\text{BOD}}}$$

where $k = 2.0\ \text{g}_{\text{BOD}}/(\text{g}_{\text{biomass}}\cdot\text{day})$, X is the biomass concentration in the reactor in $\text{mg}_{\text{biomass}}/\text{L}$ (in this case 2.0 mg/L), and $K_{1/2} = 25\ \text{mg}_{\text{BOD}}/\text{L}$. The monitoring well shown in the sketch is used to obtain samples of the groundwater downgradient of the landfill and, given the concentration of BOD in the sampled water remains equal to or below 1 mg/L, the landfill would be in compliance with its environmental permit. The required distance between the landfill and the property boundary, necessary for determining the location of the landfill on the property or the area of land to be dedicated, is in question.

a. Specify the reactor model (batch, PFR, or CMFR) you would employ in the analysis of this problem to determine the required distance, and explain the reasons behind your choice.
b. Given the information about the nature and quantity of the leachate flowing from the landfill, the nature and quantity of the groundwater flowing beneath the site, and the kinetic information, determine the required distance between the eastern edge of the solid waste deposition and the eastern boundary of the landfill site.

46. Chlorine contact basins (see Figure p8.39 for a layout sketch) are employed with wastewater treatment facilities for inactivation of pathogenic bacteria prior to discharging the treated plant effluent to the receiving water. The typical chlorine contact basin has a hydraulic residence time of 30 min. Inactivation follows the rate law

$$r_{\text{inactivation}} = -k[\text{Cl}_2]C_{\text{bacteria}}$$

where k is 2850 L/mol-min. The governing condition for a particular system involves a chlorination basin influent containing 10^9 bacteria per liter of wastewater. The criterion upon which the process must be designed requires that the number of bacteria in the basin effluent be 2×10^3 or fewer per liter. The volumetric flow treated by the wastewater plant is 4.8×10^6 gal/day. What must be the volume of the chlorine contact basin? Determine the concentration of chlorine that must be maintained in the disinfection basin in order that the number of bacteria per liter of wastewater will be reduced to the level specified.

47. Cyanide (CN^-) is oxidized to cyanate (CNO^-) in accord with the following half-reaction:

$$CN^- + H_2O \Leftrightarrow CNO^- + 2e^- + 2H^+$$

Molecular oxygen can be the electron acceptor via the following half-reaction:

$$½O_2 + 2e^- + 2H^+ \Leftrightarrow H_2O$$

Then the two half-reactions can be combined to produce a single overall redox reaction:

$$CN^- + ½O_2 + \Leftrightarrow CNO^-$$

This overall reaction likely must be microbially driven.

The half-reaction for oxidation of cyanide to cyanate can be combined with the ion product of water to rewrite the reaction as follows:

$$CN^- + 2OH^- \Leftrightarrow CNO^- + 2e^- + H_2O$$

From this alternative statement of the half-reaction we realize that alkaline pH would favor the oxidation. Further, from the half-reaction for reduction of oxygen to water, we realize that the abundance of oxygen would favor the reaction. Then we might postulate that the rate of reaction for conversion of cyanide to cyanate could be written as a pseudo-first-order rate law:

$$r_{CN^- \to CNO^-} = -k[O_2][OH^-][CN_T]$$

The rate constant k would account for both the abundance of and kinetics associated with the microbial culture that might be present in an environmental compartment contaminated by cyanide. Consider an abandoned heap leach pad with a typical cross section shown in Figure p8.47 that has a volume of 700,000 m^3 and contains 35,000 m^3 of aqueous solution and 245,000 m^3 of vapor in the pores of the spent ore contained in the pad. The normal atmospheric pressure at the location is 0.85 atm and the vapor space in the pores has essentially the

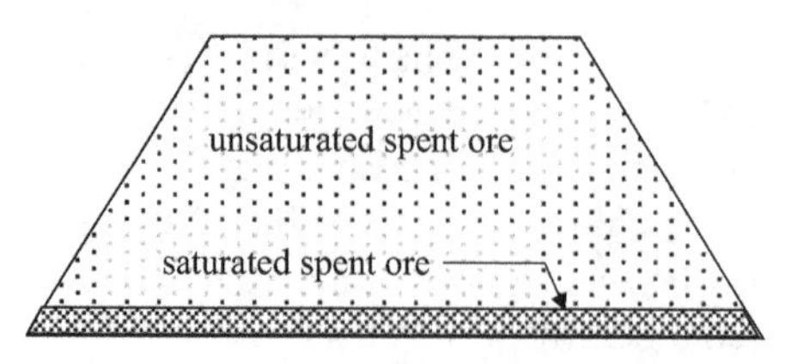

FIGURE p8.47 Schematic cross section through a heap leach pad.

same composition as atmospheric air. The pH of the aqueous solution held in the pores of the leach pad is 9.0 and the solution may be considered to be in equilibrium with the vapor held in the pores.

a. Consider that the total cyanide content of the aqueous solution in the pore spaces of the heap leach pad is 26 mg/L when the pad is abandoned. Develop a model of the system using the pseudo-first-order rate constant as a master independent variable from which you can compute the concentration of total cyanide (either in mM or mg/L) as a function of time after abandonment. (Hint: perform this computation while ignoring the HCN present initially in the vapor phase and then develop a more complete model that employs the equilibrium between the aqueous and vapor phases.) Neglect the escape of cyanide as HCN via volatilization from the heap leach pad.
b. Extend your model to enable accounting of the total quantities (in kg) of cyanide and cyanate residing in the heap leach pad as a function of time after abandonment. Consider that neither cyanide nor cyanate is lost from the heap leach pad.

48. A completely mixed flow reactor (CMFR) is to be used to treat a wastewater containing complex organic wastes characterized by BOD, the oxygen-demanding character of the waste. The rate law by which this process occurs is:

$$r_{\mathrm{BOD}} = -\frac{kXC_{\mathrm{BOD}}}{K_{1/2} + C_{\mathrm{BOD}}},$$

where $k = 2.0\ \mathrm{g_{BOD}/g_{biomass}}$-day, X is the biomass concentration in the reactor in $\mathrm{mg_{biomass}}$/L, and $K_{1/2} = 25\ \mathrm{mg_{BOD}}$/L. The reactor has a volume of one million gallons and the influent flow rate is 2780 gal/min. The influent BOD concentration is 300 mg/L.

a. Suppose the biomass concentration in the reactor were 500 mg/L, compute the effluent BOD concentration.
b. What biomass concentration would be necessary in order that the effluent BOD concentration be reduced to 20 mg/L?

49. Zero-valent iron (Fe^0) pellets with a diameter of 1 cm are used as a catalyst in an underground permeable reactive barrier (see sketch in Figure p8.49) to convert trichloroethene (TCE, C_2HCl_3) to ethene (C_2H_4) by the overall reaction

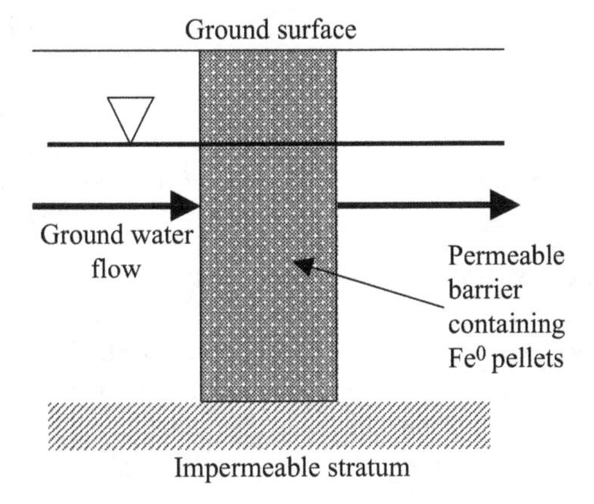

FIGURE p8.49 *Sectional sketch of a permeable reactive barrier containing zero valent iron catalyst pellets.*

$$C_2HCl_3 + 3H^+ + 6e^- \rightarrow C_2H_4 + 3Cl^-$$

It is really a lot more complicated than this, with two intermediate reactions/products (dichloroethene and chloroethene) to consider, but let us keep it simple and consider the overall conversion. Groundwater flows horizontally through the barrier, the barrier is 6 ft thick, and the flow velocity of the groundwater is 3 ft/h. The concentration of TCE in the water entering the barrier is 1 ppm_m. The reaction is believed to follow a pseudo-first-order rate law

$$r_{TCE \rightarrow ethene} = -k\alpha_{Fe}C_{TCE}$$

where k is the rate constant (0.03467 $cm^3/cm^2 \cdot min$), α_{Fe} is the surface area of iron particles per unit volume of barrier (0.7 cm^2/cm^3), and C_{TCE} is the concentration of TCE in μg/L. In this configuration, the volume fraction of the barrier occupied by the iron pellets is 0.117 (cm^3 Fe/cm^3 of the overall barrier volume).

a. Determine the concentration of TCE in the water exiting the permeable reactive barrier.
b. To what value must α_{Fe} be increased (by reducing the diameter of the iron pellets but maintaining the same total mass of iron in the barrier) in order that the effluent from this permeable reactive barrier be reduced to 5 μg/L? What diameter pellets should be employed?
c. Were the iron pellets to be reduced in diameter to 0.25 cm and the target exit concentration be reduced to 1 μg/L, what would be the necessary volume fraction of iron in the barrier?
d. Given that zero-valent iron contributes three moles of electrons per mole converted to Fe^{+3}:

$$2Fe^0 \rightarrow 2Fe^{+3} + 6e^-$$

two moles of zero-valent iron are required for each mole of TCE converted, such that

$$C_2HCl_3 + 3H^+ + 2Fe^0 \rightarrow C_2H_4 + 3Cl^- + 2Fe^+$$

A critical condition for the reactive barrier system will occur when the available zero-valent iron surface area is reduced to 90% of its value as computed in part (c). Given that the flow of TCE-containing groundwater is continuous, 24 h/day, 365 days/year, how long will this barrier function before the critical condition is reached? (*Hint: compute the number of iron pellets which will remain constant during the performance life of the barrier, and, to obtain a reasonable approximation, consider that the size and hence the specific surface area of the iron pellets will be invariant with position in the barrier.*)

50. A treated effluent flows from a closed pipe into a trapezoidal channel. The channel has a bottom width of 2.0 m and a 2:1 (horizontal to vertical) slope of the channel walls. For a particular steady, uniform flow condition, the average flow velocity is 0.33 m/s, the average depth of flow is 0.3 m, the initial DO concentration is 0.5 mg/L, the temperature is 25 °C, and the normal ambient pressure at the location of the system is 0.9 atm absolute.

    a. First, for an arbitrary steady, uniform flow condition in an arbitrarily shaped trapezoidal channel, develop the mathematical model from which the oxygen concentration can be predicted at an arbitrary position z, downstream of the entrance to the channel. Use symbology similar to that employed in Examples 8.13–8.15 and reduce the resultant model to employ the bottom width of the channel, the slope (v:h) of the channel walls, and the depth of flow as the geometric variables. Assume that the channel can be modeled as an ideal PFR.
    b. Use a mass transfer coefficient, $k_{1,20\,°C}$, of 0.25 m/h and for the stated flow conditions, produce a plot of DO concentration for a distance of 200 m downstream.

51. A manufacturer of aeration devices uses a tank at the back of the lab to test both its new and existing designs for submerged, fine-bubble aerators. The tank is 2.87 m^2 (9 ft 5 in.) with adjustable water depth.

    For a particular test, the submergence of the aerators was 4.42 m (14.5 ft) with a total depth of 4.57 m (15 ft). The test was conducted at 22 °C and the local atmospheric pressure was 0.95 atm. Air was supplied to the aeration tank at a rate of 1.31 m_{std}^3/min (46.4 ft_{std}^3/m or scfm). The aeration industry uses the unit of measure named standard cubic meter or standard cubic foot, which is equivalent in moles or mass to the gas occupying 1 m^3 or 1 ft^3, respectively, at a temperature of 20 °C (293.15K) and 1 atm absolute pressure. Eight aerators were arranged in equal areal coverage of the tank bottom. Time versus concentration data were obtained as given in the following table:

Time (min)	Dissolved oxygen (DO) (mg/L)
0	0.29
1	2.55
2	4.47
3	5.86
4	6.97
5	7.72
6	8.24
8	9.12
10	9.58
12	9.82
15	9.99
20	10.10
25	10.15
30	10.16

a. Determine the values of $k_l \cdot \alpha$ at the temperature of the test and at standard temperature for the particular aerator used in the test. Tchobanoglous et al. (2003) suggest that the conversion of the overall mass transfer coefficient for nonstandard temperatures can be accomplished using the following relation:

$$k_l \cdot \alpha_{(T)} = k_l \cdot \alpha_{(20\,°C)} \cdot \theta^{(T\text{-}20)}, \text{ wherein for diffused aeration } \theta \approx 1.024.$$

b. Determine the standard oxygen transfer rates (*sotr*) in mg/L-min and SOTR in mg/min.

c. Determine the standard oxygen transfer efficiency (SOTE) of the aeration device tested under the conditions of the test (i.e., diffuser spacing, air flow per diffuser).

d. What air application rate was used on a per diffuser basis (in scfm/diffuser)? If this rate is changed (either increased or decreased), what resultant changes would be realized in the values of the parameters determined in part (a), part (b), and part (c)? Explain.

e. A project engineer plans to use the diffusers tested via part (a), part (b), and part (c) in an activated sludge reactor. The planned basin is to be 100×30 ft by 12.5 ft in depth (12 ft diffuser submergence). The number of diffusers planned is to be 270. Is the value of $k_l \cdot \alpha$ (and hence *sotr* and SOTE) determined earlier valid for this design? Explain why or why not.

52. Consider a reach of a flowing stream below the confluence of a wastewater discharge. The region under consideration can be assumed to have constant depth and constant cross section, yielding a constant flow velocity for a given total flow rate. The sediments lying beneath the sediment/water interface are enriched in biodegradable organic matter, and hence exert an oxygen demand. That oxygen demand is satisfied by the transfer of oxygen from the water of the stream across the sediment/water interface into the sediments, where biological activity consumes the oxygen. In this case, the rate at which oxygen is transported

into the sediments can be modeled in much the same manner as the transfer across the vapor/liquid interface. We can use Fick's law and approximate the mass transfer coefficient as the quotient of the diffusion coefficient of oxygen in water and the thickness of the aerobic sediment layer directly beneath the sediment/water interface. The DO concentration beneath this aerobic layer is zero and the derivative (dC/dz) can be approximated as $C(z)/\delta_{al}$, where δ_{al} is the thickness of the aerobic layer. The mass transfer coefficient is then approximated as $D_{O_2.w} / \delta_{al}$. The rate at which oxygen would be transferred from the water into the sediments would depend upon the quantity and degradability of the organic matter in the sediments. Consider that the sediments are organic-rich such that the aerobic layer would be reduced to a thickness of 20 μm (2×10^{-4} dm).

a. Expand the model developed in Example 8.15 to include this sediment oxygen demand term, employing a rectangular channel configuration. Plot the function $C_{O_2}(z)$ to demonstrate adequacy of the developed model. Use the following input parameters to test the adequacy of your mathematical/numerical model.

$$\begin{pmatrix} \mu_{max} \\ K_{half} \\ Y \\ X \\ L_0 \end{pmatrix} := \begin{pmatrix} \frac{6}{86400} \\ 20 \\ 0.4 \\ 15 \\ 15 \end{pmatrix} \begin{pmatrix} \frac{mg_{vs}}{mg_{vs}\cdot s} \\ \frac{mg_{bCOD}}{dm^3} \\ \frac{mg_{vs}}{mg_{bCOD}} \\ \frac{mg_{vs}}{dm^3} \\ \frac{mg_{bCOD}}{dm^3} \end{pmatrix}$$

$$\begin{pmatrix} v_s \\ d \\ w_b \\ k_l \\ C_{sat} \\ C_0 \end{pmatrix} := \begin{pmatrix} .05 \\ 5 \\ 20 \\ 0.0006 \\ 9.17\cdot.9 \\ 4 \end{pmatrix} \begin{pmatrix} \frac{dm}{s} \\ dm \\ dm \\ \frac{dm}{s} \\ \frac{mg_{O2}}{dm^3} \\ \frac{mg_{O2}}{dm^3} \end{pmatrix}$$

b. Consider that the stream may be geometrically modeled as a trapezoidal channel similar to that of problem 50. Sediment oxygen demand would occur across the entire surface area of the channel bottom, biological activity would occur in the water column in a manner similar to that of Example 8.15, and oxygen transfer into the water would occur across the vapor/liquid surface area. Develop the mathematical model that can be used to describe the oxygen concentration as a function of position downstream of the confluence of the wastewater stream with the flowing natural stream. Plot the function $C_{O_2}(z)$ to demonstrate adequacy of the developed model. Use the input parameters from part (a) to test the adequacy of your mathematical/numerical model.

Chapter 9

Reactions in Nonideal Reactors

9.1 PERSPECTIVE

In the previous chapter, we have investigated ideal flow reactors, including those characterized as plug-flow (PFR) and completely mixed flow (CMFR). As discussed in Chapter 8, the CMFR, in much of the literature, is called a continuously stirred tank reactor (CSTR). These ideal conceptualizations represent the bounding extremes relative to real reactors—ranging from a single CMFR to an ideal PFR represented by an infinite number of CMFRs in series. In practice, no reactor can be completely and perfectly mixed. Similarly, no reactor can be totally devoid of forward- and back-mixing along its most probable flow path. All real reactors are somewhere in between. In many cases, the real reactor can be sufficiently close to one of the extremes that we may use the ideal model in quantitative analyses of the reactor invoking only insignificant error. In other cases, assumption of ideality can lead to significant error.

In this chapter, we will quantitatively examine the characterization of the degree of nonideality of real reactors. We will present and apply experimental methodologies for characterization in conjunction with methods for quantitative analyses and interpretation of data arising from such experimental characterizations. We will then examine three models considered in the engineering literature to be useful in quantitative analyses of processes occurring in nonideal reactors: the CMFRs (tanks) in series (TiS), segregated flow (SF), and plug-flow with dispersion (PFD) models.

Environmental Process Analysis: Principles and Modeling, First Edition. Henry V. Mott.

9.2 EXIT CONCENTRATION VERSUS TIME TRACES

In Chapter 7, we examined the impulse and step input stimuli and the responses of ideal PFRs and CMFRs to these stimuli. The ideal responses to the impulse stimulus varied from the Dirac delta function for a PFR to a decaying exponential with a long tail for a CMFR. The ideal responses to a positive step input varied from a vertical step increase for a PFR to an exponential of decreasing rate of increase for a CMFR. The ideal responses to a negative step varied from a vertical step decrease for a PFR to a decaying exponential, virtually identical to that from the impulse input, for the CMFR. Real reactors will exhibit exit responses that vary in between these extremes.

9.2.1 Impulse Stimulus

Were we to conduct a tracer analysis of a reactor, as described in Chapter 7, we would introduce a tracer as either an impulse or step input and obtain samples of the fluid exiting the reactor. In the laboratory, we would employ an arrangement, such as the one shown in Figure 7.2, employing the syringe. The set of exit concentration versus time data pairs resulting from a tracer test is often called the exit concentration curve or trace. In much of the literature, $C(t)$ is used to symbolically represent this trace and is merely the plot of observed effluent concentration versus time when the impulse is initiated at $t=t_0$.

Let us consider that we have arranged four reactors in the laboratory, each characterized by differing degrees of longitudinal dispersion. The range is from a fully, vigorously mixed reactor to reactors containing increasingly fine packing materials, which would increasingly maximize lateral dispersion and minimize longitudinal dispersion. In Figure 9.1, response curves from this residence time

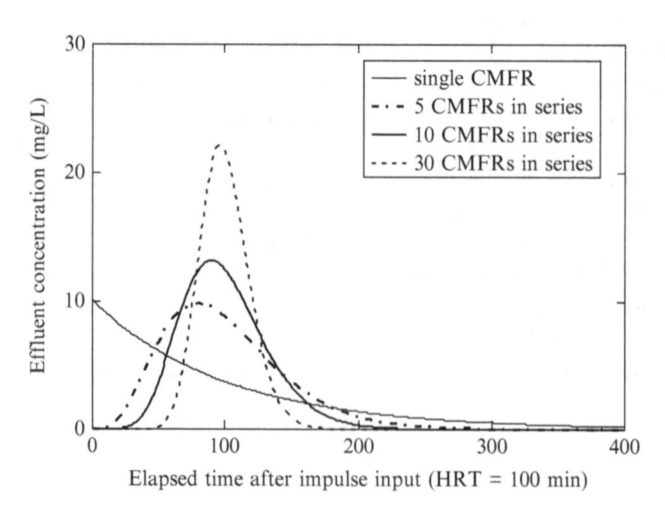

FIGURE 9.1 *Theoretical exit concentration responses of real laboratory reactors for impulse input stimuli.* $M_{tracer} = 1$ g, $V_R = 100$ L, $Q = 1$ L/min.

distribution (RTD) analysis are shown, all plotted on a single set of axes. We have arranged the reactors such that for the test the hydraulic residence times are all equal. These curves have been generated using a mathematical model we will discuss later. In Figure 9.1, the number of equivalent tanks in series represented by each reactor identifies each of the traces shown. We may easily observe, as would be expected, that the *C* curve response to the impulse stimulus from the single CMFR appears exactly like that of the ideal CMFR of Chapter 7. As the number of CMFRs in series increases, the response approaches that of the ideal PFR of Chapter 7.

9.2.2 Positive Step Stimulus

Were we to employ the tracer pump of Figure 7.2 to invoke a positive step input employing the same reactors as depicted in Figure 9.1 and under the exact same conditions, the four exit concentration curves would appear similarly to those plotted in Figure 9.2. We observe that the $C(t)$ curve for the single CMFR is exactly as we have shown in Chapter 7 for the ideal CMFR. The $C(t)$ curves for the remaining reactors increasingly approach that for an ideal PFR. Were we employ large numbers of CMFRs in series with the mathematical model used to produce these plots, we could approach very closely to the ideal response of a PFR.

9.3 RESIDENCE TIME DISTRIBUTION DENSITY

In order to quantitatively utilize the results of tracer tests such as those depicted in Figure 9.1, we must normalize the $C(t)$ curves to the mass of the tracer injected. Then we may develop subsequent functions and parameters, including the RTD density curve (E(t)), as well as the mean residence time and variance of the RTD density distribution.

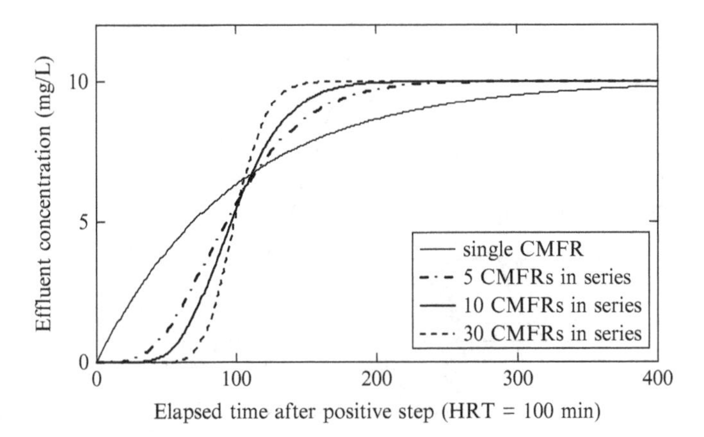

FIGURE 9.2 *Theoretical exit concentration responses of real laboratory reactors for positive step input stimuli.* $C_{in} = 10\,mg/L$, $V_R = 100\,L$, $Q = 1\,L/min$.

9.3.1 *E*(*t*) Curve and Quantitation of Tracer Mass

We may create the $E(t)$ curve from the $C(t)$ curve by normalizing each of the $C(t)$ values by the mass of tracer recovered by the RTD test. In Figure 9.3, we have plotted a typical impulse stimulus response curve, corresponding to that shown in Figure 9.1, for ten CMFRs in series. On the plot, the product of the area of the shaded element and the influent flow represents the mass of tracer, ΔM, exiting the reactor between t and $t + \Delta t$:

$$\Delta M = Q \cdot \bar{C}(t) \Delta t$$

$\bar{C}(t)$ is the average exit concentration for the time period Δt. To obtain the exact mathematical expression, we must take the limit as $\Delta t \to 0$, thus $\Delta M \to dM$ and $\Delta t \to dt$. When dt becomes infinitesimal, $\bar{C}(t) \to C(t)$. Then the mass of tracer is the integral of dM over a reasonable time period (for the RTD of Figure 9.3, $0 \le t \le 300$ min), but mathematically, we will set the upper limit at ∞:

$$M_{\text{tracer}} = \int dM = \int_{t=0}^{t=\infty} Q \cdot C(t)\, dt$$

By normalization, we consider that the quotient $\dfrac{Q \cdot C(t)\, dt}{M_{\text{tracer}}}$ is the fraction of the total quantity of tracer exiting the reactor between times t and $t + dt$. Thus, if the tracer is intimately mixed with the fluid, this quotient is also the fraction of the marked fluid

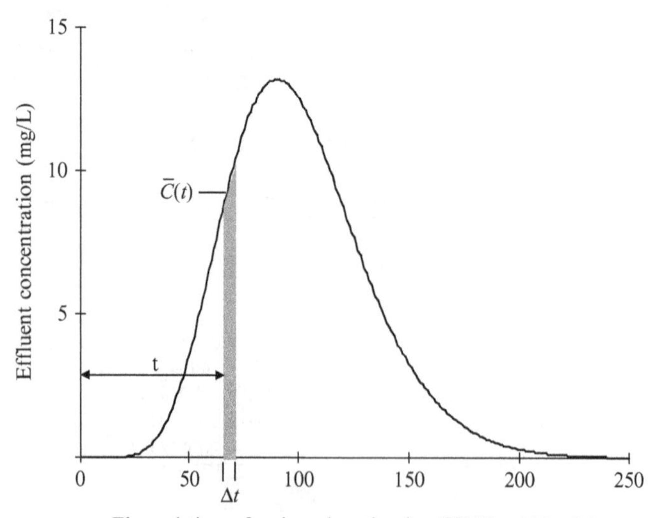

FIGURE 9.3 A typical C(t) curve from an impulse input with an element shown representing the mass of tracer exiting the reactor between t and t+Δt.

element exiting the reactor between t and $t+dt$. Then, if the reactor is operated under steady conditions, we can generalize this understanding to each and every element of fluid entering the reactor. The quotient $\frac{Q \cdot C(t)dt}{M_{\text{tracer}}}$ is then the fraction of the entering flow with residence time equal to $t+\frac{dt}{2}$, which in the limit as $dt \to 0$ is indeed t.

9.3.2 *E(t)* and *E(θ)* RTD Density Curves

Now we may define the concentration-normalized RTD density function, somewhat analogous to the probability density function:

$$E(t) = \frac{Q \cdot C(t)}{M_{\text{tracer}}} = \frac{C(t)}{\int_{t=0}^{t=\infty} C(t)dt} \tag{9.1}$$

Then the fraction of the tracer, and hence of entering flow, exiting the reactor with residence time t may be stated:

$$\frac{dM}{M_{\text{tracer}}} = E(t)dt$$

The integral over time of this concentration normalized RTD density function is the whole of the introduced tracer:

$$\int_{t=0}^{t=\infty} E(t)dt = 1$$

For some applications, we would desire to normalize the $C(t)$ curve to both concentration and time. Time would be normalized to the residence time of the reactor (HRT or τ). The normalized time is often given the symbol θ and is the quotient of elapsed time and residence time (t/τ). We must define a time-normalized concentration exit curve. We can convert the $C(t)$ shown in Figure 9.3 to obtain Equation 9.2 by defining $C(\theta)$ from $C(t)$:

$$C(\theta) = C\left(\frac{t}{\tau}\right) = C(t)$$

Then, in order to account for the full quantity of tracer introduced, we must account for the normalization of time in integrating the $C(\theta)$ curve:

$$\int_{t=0}^{t=\infty} C(t)\,dt = \int_{\theta=0}^{\theta=\infty} \tau \cdot C(\theta)d\theta$$

In reality the practical upper limit of the left integral would be finite and perhaps several times the HRT. Then the practical upper limit of the right integral would be a value between five and ten. Then we may define the RTD density function that is normalized to both mass of tracer and time:

$$E(\theta)=\frac{\tau\cdot C(\theta)}{\int_{\theta=0}^{\theta=\infty}\tau\cdot C(\theta)d\theta}=\frac{C(\theta)}{\int_{\theta=0}^{\theta=\infty}C(\theta)d\theta} \tag{9.2}$$

The integral over θ of our concentration and time normalized RTD density function is again the whole of the inputted tracer:

$$\int_{\theta=0}^{\theta=\infty}E(\theta)d\theta=1$$

Even for a single CMFR, for which the $C(t)$ curve would have the longest tail, the value of θ to which we would carry the integral is perhaps no more than five or six residence times.

Plots of $E(\theta)$ and $E(t)$ corresponding with the exit concentration trace of Figure 9.3 are shown in Figure 9.4. Note that the time axes differ by a factor of the hydraulic residence time and that the magnitudes of the corresponding normalized residence time density traces differ by the same factor. Employing dimensionless RTD curves allows us to examine the phenomena associated with longitudinal dispersion in the absence of the scale factors associated with mass of tracer and reactor size, and through the application of dimensionless time. Much mathematical analysis has been accomplished based on dimensionless RTD data.

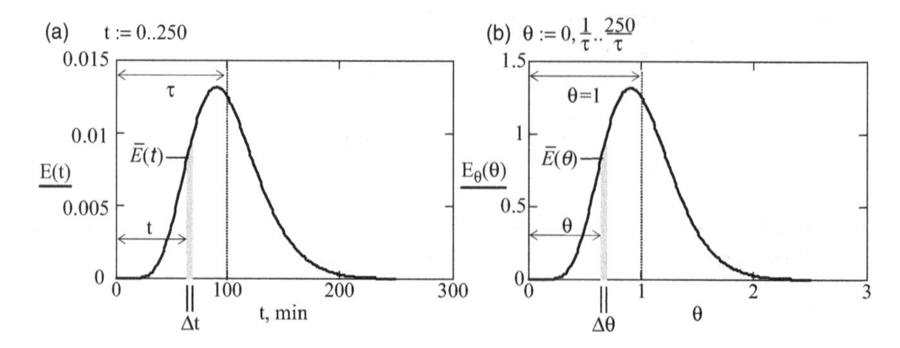

FIGURE 9.4 *Normalized residence time distribution curves for the exit concentration trace of Figure 9.3: (**a**) normalized to mass of tracer; (**b**) normalized to mass of tracer and HRT.*

Once we have a characterization using the dimensionless RTD, we may easily apply the scale factors to convert our results to reactors of real size.

In Figure 9.4 a and b respectively, single elements representing the products $\bar{E}(t)\Delta t$ and $\bar{E}(\theta)\Delta\theta$ are shown graphically. Each product represents the fraction of the tracer that exited the reactor during the time periods t and $t + \Delta t$ and θ and $\theta + \Delta\theta$, respectively. As our tracer would have been intimately mixed with its entering fluid element, we would expect each fluid element to behave exactly as that which was marked. **The extension of this principle is that the products $\bar{E}(t)\Delta t$ and $\bar{E}(\theta)\Delta\theta$ represent the fraction of each influent fluid element that would exit during the arbitrary time period *dt* or $d\theta$, respectively**. The fractions of the flow would have dimensionless residence time θ or actual residence time t. Hence the E curves of Figure 9.4 allow us not only to express the range of the residence times of fluid passing through a reactor but also to compute the fraction of the influent flow that exits at each of the identified residence times. Mathematically, when we take the limit as $\Delta t \to 0$ ($\Delta\theta \to 0$), the discrete expressions may be written as differentials: $E(t)dt$ and $E(\theta)d\theta$, representing the fraction of a fluid element with residence time t or θ.

9.4 CUMULATIVE RESIDENCE TIME DISTRIBUTIONS

The response curve from a positive step input stimulus is easily normalized to the influent concentration resulting from the step to produce a concentration normalized cumulative RTD. The cumulative RTD trace is merely the quotient of $C(t)$ and C_{in}:

$$F(t) = \frac{C(t)}{C_{in}} \tag{9.3}$$

C_{in} is the concentration of tracer resulting from the mixing of the influent stream with the continuous flow of tracer solution. Normalization to the hydraulic residence time requires adjustment for the constant ratio $\theta = \frac{t}{\tau}$. We can relate $F(\theta)$ to $F(t)$ in fashion used for $C(\theta)$:

$$F(\theta) = F\left(\frac{t}{\tau}\right) = F(t)$$

In Figure 9.5, two normalized cumulative RTD curves are shown, one normalized to the influent tracer concentration and the other normalized to both influent concentration and HRT. These are of exactly replicate shape and match that shown in Figure 9.2 for ten CMFRs in series.

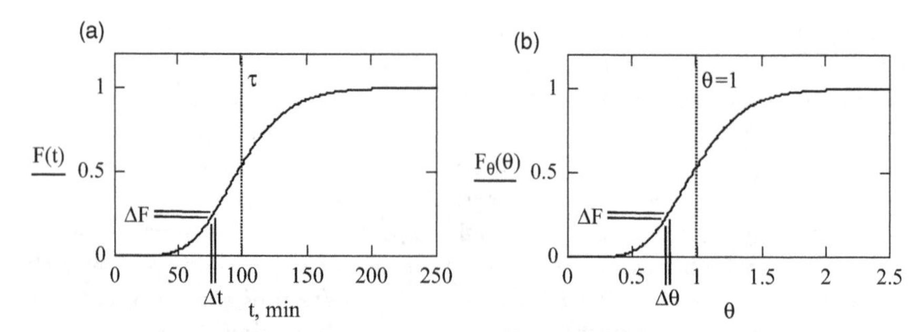

*FIGURE 9.5 Cumulative residence time distributions resulting from a positive step corresponding with the system depicted in Figure 9.4: (**a**) normalized to influent concentration; (**b**) normalized to influent concentration and HRT.*

The RTD density function and the cumulative RTD function are related in a manner very analogous to the probability density function and the cumulative probability function:

$$F(t) = \int_0^t E(t)\,dt \text{ and } F(\theta) = \int_0^\theta E(\theta)\,d\theta \tag{9.4a}$$

$$E(t) = \frac{d(F(t))}{dt} \text{ and } E(\theta) = \frac{d(F(\theta))}{d\theta} \tag{9.4b}$$

In Figure 9.5**a**, the ΔF shown is the fraction of an entering fluid element exiting the reactor during the time period Δt and exactly analogous to the product $\overline{E}(t)\Delta t$ of Figure 9.4**a**. In Figure 9.5**b**, the ΔF shown is the fraction of an entering fluid element exiting the reactor during the normalized time period $\Delta\theta$ and exactly analogous to the product $\overline{E}(\theta)\Delta\theta$ of Figure 9.4**b**.

9.5 CHARACTERIZATION OF RTD DISTRIBUTIONS

In the following sections, we will examine three models used with RTD data from real reactors to predict the extent of reactions taking place in real reactors. For two of these models, we must compute the statistical mean residence time and the variance of the RTD function. These are often called the first and second moments of the distribution. We may compute these using either the RTD density ($E(t)$ or $E(\theta)$) or cumulative RTD ($F(t)$ or $F(\theta)$) curves.

9.5.1 Mean and Variance from RTD Density

We define a distinct parameter, which is the statistical mean residence time, $\bar{t}_{SM}$. When we delve into the analysis of real reactors, we find that the statistical mean residence time and the hydraulic residence time are not necessarily equal.

We will explore this difference later. Then hydraulic residence time (HRT or τ) remains the quotient of volume and flow, while the statistical mean residence time arises from analyses of RTD data.

The statistical mean residence time is merely the weighted average of the residence times of the fractions of an arbitrary entering fluid element. Each residence time is weighted using the fraction of the entering fluid element exiting with the respective residence time. Since the E distribution consists of the fractions of entering fluid elements exiting at the various residence times, the computation is merely an integral:

$$\bar{t}_{SM} = \int_{t=0}^{t=\infty} t \cdot E(t)\,dt \tag{9.5a}$$

Most often discrete data points are available and the necessity often arises to approximate Equation 9.5a as a summation (trapezoidal rule):

$$\bar{t}_{SM} \cong \sum_{i=n}^{i=1} \bar{t}_i \cdot \bar{E}(t)_i \Delta t \tag{9.5b}$$

$\bar{t}_i$ and $\bar{E}(t)_i$ are the average residence time and average magnitude of $E(t)$, respectively, of fraction i of any arbitrary entering fluid element and n is the number of intervals into which the observed RTD density data set is subdivided. Equation 9.5a may be restated in terms of normalized time:

$$\bar{\theta}_{SM} = \int_{t=0}^{t=\infty} \theta \cdot E(\theta)\,d\theta \tag{9.6a}$$

An approximation is then available when data are discrete:

$$\bar{\theta}_{SM} \cong \sum_{i=n}^{i=1} \bar{\theta}_i \cdot \bar{E}(\theta)_i \Delta\theta \tag{9.6b}$$

The variance of the RTD density function, σ^2, is the fluid-weighted average of the squares of the departures of the residence times of each of the fractions of the entering fluid element from the statistical mean residence time:

$$\sigma_t^{\,2} = \int_{t=0}^{t=\infty} \left(t - \bar{t}_{SM}\right)^2 \cdot E(t)\,dt \tag{9.7a}$$

For discrete data, the relation often must be approximated as a summation:

$$\sigma_t^2 \cong \sum_{i=1}^{i=n} \left(\bar{t}_i - \bar{t}_{\text{SM}}\right)^2 \cdot \bar{E}(t)_i \, \Delta t \tag{9.7b}$$

These relations may also be expressed for dimensionless time:

$$\sigma_\theta^2 = \int_{\theta=0}^{\theta=\infty} \left(\theta - \bar{\theta}_{\text{SM}}\right)^2 \cdot E(\theta) \, d\theta \tag{9.8a}$$

$$\sigma_\theta^2 \cong \sum_{i=1}^{i=n} (\theta_i - \theta_{\text{SM}})^2 \cdot \bar{E}(\theta)_i \, \Delta\theta \tag{9.8b}$$

As we will understand later in this chapter, for time data that are evenly spaced, use of the RTD density function permits numerical integration using Simpson's rules, which may be more accurate than the alternative trapezoidal or rectangular rules.

9.5.2 Mean and Variance from Cumulative RTD

When the relation between RTD density and cumulative RTD is considered, Equations 9.5–9.8 may readily be converted for use with a cumulative RTD function. We need merely replace $E(t)dt$ and $E(\theta)d\theta$ with $dF(t)$ and $dF(\theta)$, respectively, in the exact mathematical relations or $\bar{E}(t)_i \, \Delta t$ and $\bar{E}(\theta)_i \, \Delta\theta$ by $\Delta F(t)_i$ and $\Delta F(\theta)_i$, respectively, in the summations. Then for the statistical mean residence time, we have the following:

$$\bar{t}_{\text{SM}} = \int_{t=0}^{t=\infty} t \cdot dF(t) \tag{9.9a}$$

$$\bar{t}_{\text{SM}} \cong \sum_{i=1}^{i=n} \bar{t}_i \cdot \Delta F(t)_i \tag{9.9b}$$

$$\bar{\theta}_{\text{SM}} = \int_{t=0}^{t=\infty} \theta \cdot dF(\theta) \tag{9.10a}$$

$$\bar{\theta}_{\text{SM}} \cong \sum_{i=1}^{i=n} \bar{\theta}_i \cdot \Delta F(\theta)_i \tag{9.10b}$$

In addition to employment of Equations 9.9 and 9.10, we may also determine $\bar{t}_{SM}$ and $\bar{\theta}_{SM}$ by determining the value of t or θ corresponding with $F(t)=0.5$ and $F(\theta)=0.5$, respectively.

For the variance of the RTD distribution, we have an additional set of relations:

$$\sigma_t^2 = \int_{t=0}^{t=\infty} \left(t-\bar{t}_{SM}\right)^2 \cdot dF(t) \tag{9.11a}$$

$$\sigma_t^2 \cong \sum_{i=1}^{i=n} \left(\bar{t}_i-\bar{t}_{SM}\right)^2 \cdot \Delta F(t)_i \tag{9.11b}$$

$$\sigma_\theta^2 = \int_{\theta=0}^{\theta=\infty} \left(\theta-\bar{\theta}_{SM}\right)^2 \cdot dF(\theta) \tag{9.12a}$$

$$\sigma_\theta^2 \cong \sum_{i=1}^{i=n} \left(\theta_i-\theta_{SM}\right)^2 \cdot \Delta F(\theta)_i \tag{9.12b}$$

σ_t^2 and σ_θ^2 are related through the statistical mean residence time:

$$\sigma_\theta^2 = \frac{\sigma_t^2}{\bar{t}_{SM}^{\,2}} \tag{9.13}$$

9.6 MODELS FOR ADDRESSING LONGITUDINAL DISPERSION IN REACTORS

Three well-known models have been employed in addressing reactions in nonideal reactors. These are the CMFRs (Tanks) in series (TiS) model, the Plug Flow with Dispersion (PFD) model, and the Segregated Flow (SF) model. The CMFRs in series and the PFD models both depend on determinations of the statistical mean residence time and the variance of the RTD function in order to quantitate the respective parameter necessary to application of the models in reaction analysis. The segregated flow model directly employs either the RTD density or cumulative RTD distribution.

9.6.1 CMFRs (Tanks) in Series (TiS) Model

For the TiS model, we simply visualize the reactor as a series of ideal CMFRs. The influent to the real reactor enters the first CMFR. The effluent from the first CMFR becomes the influent to the second. The effluent from each successive reactor

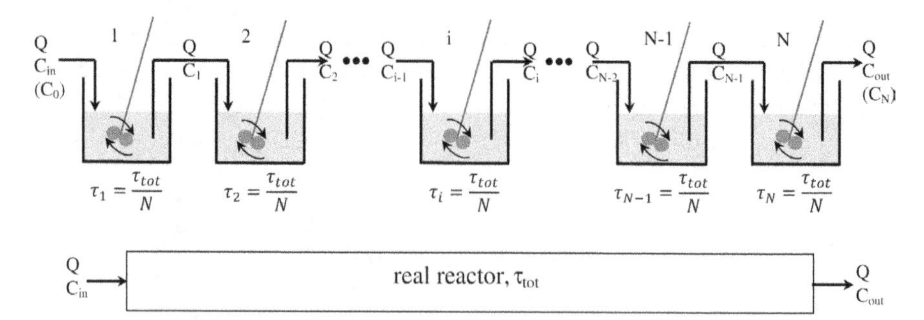

FIGURE 9.6 *Schematic representation of a non-ideal reactor as N CMFRs in series (the TiS model).*

becomes the influent to the following reactor. The effluent from the *n*th reactor is the effluent from the real reactor. A schematic is shown in Figure 9.6.

Theoretical RTD density and cumulative RTD functions for a reactor comprised of N CMFRs in series are presented by Levenspiel (1972):

$$E(\theta) = \frac{N \cdot (N \cdot \theta)^{N-1}}{(N-1)!} e^{-N \cdot \theta}$$

$$F(\theta) = 1 - e^{-N \cdot \theta} \cdot \sum_{i-1}^{i=N} \frac{(N \cdot \theta)^{i-1}}{(i-1)!}$$

N must be an integer. These relations were employed using MathCAD to generate the $C(t)$, $E(t)$, $F(t)$, $E(\theta)$, and $F(\theta)$ plots of Figure 9.1, Figure 9.2, Figure 9.3, Figure 9.4, and Figure 9.5. For the fully dimensional plots, Levenspiel's relations were modified using Equations 9.1–9.3 and appropriate intermediate results. A specific MathCAD function was developed for the exit concentration trace in response to the impulse stimulus:

$$C(t) := C_0 \cdot \frac{N \cdot \left(N \cdot \frac{t}{\tau}\right)^{N-1}}{(N-1)!} \cdot \mathrm{e}^{-N \cdot \frac{t}{\tau}}$$

Another distinct function was developed for the exit concentration trace in response to a positive step input:

$$C(t) := C_{in} \cdot \left[1 - \mathrm{e}^{-N \cdot \frac{t}{\tau}} \cdot \sum_{i=1}^{N} \left[\frac{1}{(i-1)!} \cdot \left(N \cdot \frac{t}{\tau}\right)^{i-1}\right]\right]$$

These relations use the parameter N in order to generate RTD density and cumulative RTD traces. Of great use would be the capability to obtain the value of N from an experimental $C(t)$ or $F(t)$ trace. Levenspiel (1972) presents such a means, along with its derivation. The derivation, although fascinating, is beyond the scope of the application-based focus of this chapter so we will simply present the end result:

$$N = \frac{\bar{t}_{\mathrm{SM}}^{\,2}}{\sigma_t^2} = \frac{1}{\sigma_\theta^2} \tag{9.14}$$

We realize that the value of N obtained from Equation 9.14 likely will not be an integer. Levenspiel (1972) provides little guidance for such a case. We will examine this question later, via posed examples.

9.6.2 Plug-Flow with Dispersion (PFD) Model

Dispersion in a flowing fluid arises as a consequence of turbulence caused by localized variations in momentum transfer both in the axial and lateral directions. Created as a consequence of mechanical mixing, these localized gradients in truth act in all directions. We could devote nearly an entire text in examining the phenomenon termed dispersion. Herein we must explain dispersion briefly as a macroscopic analog of molecular diffusion. At the molecular level, the translational, rotational, and vibrational motion induced by intramolecular motion results in high levels of activity at the molecular scale—the root cause of the process we call diffusion. Then, when fluid turbulence is induced within a flowing fluid, a similar process occurs at the microscopic level. The elements of fluid are battered about as a consequence of numerous collisions with other elements of fluid. Mathematically, we model dispersion in a manner quite analogous to that employed to model diffusion. For modeling of molecular diffusion, we employ Fick's first law to describe the flux of diffusing substance in terms of a diffusion coefficient and a spatial gradient in the abundance of the substance. In one dimension, for a constituent in a dilute aqueous solution, Fick's first law is quite simply stated:

$$\vec{J} = -\mathcal{D}\frac{dC}{dz}$$

$\vec{J}$ is the flux (M/L^2/t) of component across a plane normal to the direction of the flux vector (increasing z), $\mathcal{D}$ (L^2/t) is the molecular diffusion coefficient with magnitude governed by the properties of both the diffusing specie and the fluid within which the specie diffuses, and z is a spatial coordinate in a one-dimensional coordinate system.

We may employ a relation of the same form as Fick's law to describe the dispersive flux of a component in an aqueous solution. We replace the diffusion coefficient, $\mathcal{D}$, with a dispersion coefficient, D. The magnitude of D varies from that of $\mathcal{D}$, for a system characterized as having zero turbulence, to nearly ∞ in the well-mixed fluid regime of a single CMFR.

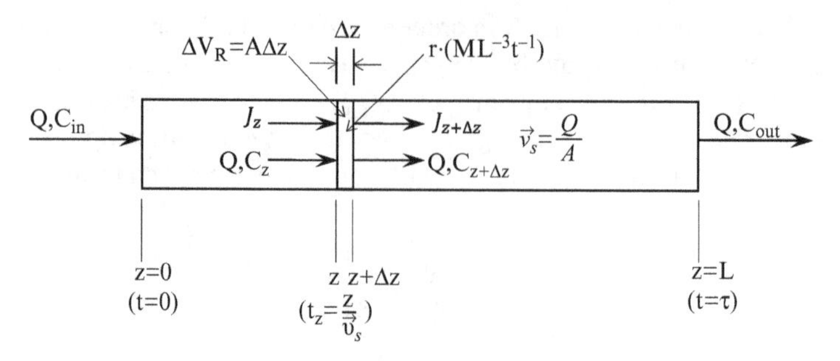

FIGURE 9.7 Schematic representation of a non-ideal reactor, resembling a PFR, with delineation of an arbitrarily located element of reactor volume upon which a mass balance may be drawn. For a conservative tracer, reaction is disregarded.

In developing the PFD model, we must arrange a reactor resembling a PFR similar to that of Figure 8.2. We must include the input to and output from the arbitrarily located disk of thickness Δz as a consequence of dispersive flux. Initially, we will consider a conservative substance and leave treatment of the reaction for later consideration. A schematic representation of this reactor is shown in Figure 9.7. Using this schematic, we may compile a mass balance on our arbitrary and conservative tracer, accumulation = Σin − Σout:

$$A \cdot \Delta z \cdot \frac{\partial C}{\partial t} = Q \cdot (C_z - C_{z+\Delta z}) + A \cdot (J_z - J_{z+\Delta z})$$

We can employ the definitions of change, write Q as the product of v_s and A, divide through by the cross-sectional area of the reactor, use the dispersion analog of Fick's law to define $\vec{J}$, and take the limit of the expression as $\Delta z \to 0$ to arrive at a second-order partial differential equation:

$$\frac{\partial C}{\partial t} = -v_s \frac{\partial C}{\partial z} - \frac{\partial J}{\partial z} = -v_s \frac{\partial C}{\partial z} + D \frac{\partial^2 C}{\partial z^2}$$

Then in order that we may employ the work accomplished by our collaborators, the mathematicians, we would, through various normalizations, seek to reduce the number of real parameters associated with the relation. We would like to use the work of Levenspiel (1972) to quantitatively characterize the dispersion for the PFD model. We normalize time and position: $t = \tau\theta$ (hence $\partial t = \tau\partial\theta$) and $z = L\xi$ (hence $\partial z = L\partial\xi$). θ is time relative to the residence time of the reactor and ξ is the relative position in the reactor, normalized to the overall length of the reactor. We define hydraulic residence time in terms of velocity and reactor length: $\tau = L/v_s$. We make these substitutions, perform some algebraic rearrangement, and arrive at the fully reduced relation for which Levenspiel (1972) has developed a means to characterize the dispersion coefficient, D:

$$\frac{\partial C}{\partial \theta} = -\frac{\partial C}{\partial \xi} + \frac{D}{v_s L}\frac{\partial^2 C}{\partial \xi^2}$$

The quotient $\dfrac{D}{v_s L}$ is called the dispersion number. It's inverse is the Peclet number. The dispersion number is dimensionless and can be applied at any scale.

For a reactor system similar to that depicted in Figure 7.2, termed a closed reactor, the relation between the variance of the RTD and the dispersion number is an implicit relation:

$$\sigma_\theta^{\,2} = \frac{\sigma_t^{\,2}}{t_{\mathrm{SM}}^{\,2}} = 2\frac{D}{v_s L} - 2\left(\frac{D}{v_s L}\right)^2 \cdot \left(1 - e^{-\frac{v_s L}{D}}\right) \tag{9.15}$$

With the convenient root() function or given-find solve block available from MathCAD, we can easily obtain a value for the dispersion number once the variance and statistical mean residence time are known from the RTD density or cumulative RTD functions. Then, once the dispersion coefficient is known for a given reactor under a given flow condition, we can (and will later in this chapter) return to the PFD reactor and include the reaction.

9.6.3 Segregated Flow (SF) Model

Figure 9.4 provides the best vantage from which to begin the explanation of the SF model. Consider that we might have a discrete set of $C(t)$ versus t data, which we may then convert to $E(t)$ versus t as depicted in Figure 9.4**a**. Then, the element of the area beneath the $E(t)$ curve shown as $\overline{E}(t)_i\,\Delta t$ is the fraction of each entering element of fluid and thus of the entering flow that has residence time $\overline{t}_i$. We can populate the remainder of the plot with similar fractions. Each fraction represents an ideal plug-flow reactor receiving a flow $Q_i = Q_{\mathrm{tot}}\overline{E}(t)_i\,\Delta t$ and of residence time $\tau_i = \overline{t}_i$. The volume of reactor i is then $V_i = Q_i \cdot \tau_i$. The distribution of the influent flow among the n reactors into which the nonideal reactor would be subdivided is shown schematically in Figure 9.8. The effluent from the n parallel PFRs is then collected and comprises the overall effluent from the nonideal reactor. To use the cumulative RTD we replace $\overline{E}(t)_i\,\Delta t$ with $\Delta F(t)_i$. A significant advantage of the segregated flow model is its direct use of the RTD distribution. Each of the subreactors is modeled as an ideal PFR and the effluents are simply mixed, based on zero-volume mixing principles described in Chapter 7.

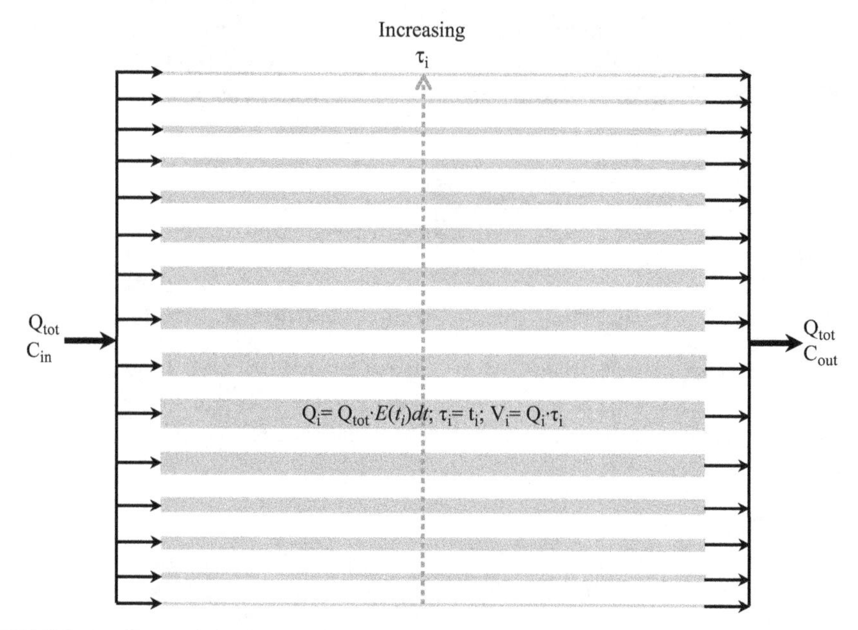

FIGURE 9.8 *Schematic representation of the segregated flow model: a real reactor resembling a PFR may be segregated into* n *sub-reactors with range of residence times in accord with the abscissa range of the RTD function and receiving fractions of the total flow in accord with RTD density or cumulative RTD function.*

9.7 MODELING REACTIONS IN CMFRs IN SERIES (TiS) REACTORS

9.7.1 Pseudo-First-Order Reaction Rate Law in TiS Reactors

For application of the pseudo-first-order rate law in reactors using the CMFRs in series model, we first rearrange Equation 8.25 to solve for the effluent concentration of the arbitrary ith reactor of the series:

$$C_i = \frac{C_{i-1}}{\left(1+\frac{\tau}{N}k'\right)} \tag{8.25}$$

We recall that k' is a pseudo-first-order rate coefficient ($1/t$) and that C is the abundance of our targeted reactant. We may write the rearranged Equation 8.25 for each of the CMFRs in the series. The result from the ith reactor provides the influent concentration for the $(i+1)$th reactor. For the first three reactors in the series, we have the following:

$$C_1 = \frac{C_0}{\left(1+\frac{\tau_{tot}}{N}k'\right)};\ C_2 = \frac{C_1}{\left(1+\frac{\tau_{tot}}{N}k'\right)} = \frac{C_0}{\left(1+\frac{\tau_{tot}}{N}k'\right)^2};$$

$$C_3 = \frac{C_2}{\left(1+\frac{\tau_{tot}}{N}k'\right)} = \frac{C_0}{\left(1+\frac{\tau_{tot}}{N}k'\right)^3}$$

C_0 is in fact C_{in} and we represent it as the zeroth effluent concentration. We may follow this progression and write the result for the *N*th reactor:

$$C_N = \frac{C_0}{\left(1+\frac{\tau_{tot}}{N}k'\right)^N} \tag{9.16}$$

9.7.2 Saturation Reaction Rate Law with the TiS Model

For the saturation rate law applied with the CMFRs in series model, we begin with Equation 8.26, again writing the relation for the ith reactor.

$$\frac{C_{i-1}-C_i}{\frac{\tau}{N}} = \frac{k\cdot X\cdot C_i}{K_{half}+C_i} \tag{8.26}$$

Recall that k is the specific substrate utilization coefficient (μ_{max}/Y, $M_{COD}/M_{VS}/t$), X is the concentration of viable biomass (M_{VS}/L^3), and K_{half} is the half-maximum reaction rate coefficient (M_{COD}/L^3). Most commonly, substrate is characterized as chemical (or biochemical) oxygen demand and biomass as volatile solids. We cannot arrange this to specifically solve for the exit concentration from any reactor and therefore cannot develop a concise relation such as Equation 9.16. Most succinctly, from the form of equation 8.26a, we employ the quadratic formula to directly solve for the exit concentration from the *i*th reactor:

$$a_2 = 1;\ a_{1,i} = K_{half} + \frac{\tau_{tot}}{N}kX - C_{i-1};\ a_{0,i} = -K_{half}C_{i-1}$$

$$C_i = \frac{-a_{1,i}+\sqrt{a_{1,i}^{\,2}-4\cdot a_2\cdot a_{0,i}}}{2\cdot a_2}$$

The C_i values then are computed by sequentially solving each of the relations beginning with $i=1$ and proceeding to $i=N$.

9.8 MODELING REACTIONS WITH THE PLUG-FLOW WITH DISPERSION MODEL

9.8.1 Pseudo-First-Order Reaction Rate Law with the PFD Model

We must return to Figure 9.7 and to model reactions in PFD reactors we must consider the reaction. We write a mass balance on an arbitrary component within an arbitrary element of volume of the reactor:

$$A\cdot\Delta z\cdot\frac{\partial C}{\partial t}=Q\cdot(C_z-C_{z+\Delta z})+A\cdot(J_z-J_{z+\Delta z})+A\cdot\Delta z\cdot r$$

We employ the definitions of change, write Q as the product of v_s and A, divide through by the cross-sectional area of the reactor, use the dispersion analog of Fick's law to define $\vec{J}$, and take the limit of the expression as $\Delta z\rightarrow 0$ to arrive at a second-order partial differential equation:

$$\frac{\partial C}{\partial t}=-v_s\frac{\partial C}{\partial z}+D\frac{\partial^2 C}{\partial z^2}+r$$

Although consideration of the unsteady state case would present interesting and challenging opportunities for intellectual pursuit, herein we will consider only the steady-state case. Most reactions in environmental systems that can be considered flow reactors are well approximated by the steady-state solution. When we set $\frac{\partial C}{\partial t}$ to 0, our mass balance will lead to a second-order ordinary, rather than a second-order partial, differential equation:

$$0=-v_s\frac{dC}{dz}+D\frac{d^2C}{dz^2}+r \qquad (9.17a)$$

In order that we may directly employ the dispersion number obtained from the RTD analyses, we will normalize the result in a manner similar to that accomplished for the conservative tracer:

$$\frac{D}{v_sL}\frac{d^2C}{d\xi^2}-\frac{dC}{d\xi}+\tau r=0 \qquad (9.17b)$$

We may substitute our pseudo-first-order rate law into Equations 9.17a and 9.17b. Although Equation 9.18a preserves the structure of the relation, in much of the previous literature, we find that the form of Equation 9.18b has been used. Of importance, the dimensionless group $\frac{D}{v_s L}$ is the dispersion number (its inverse is the Peclet number) employed in much of the literature in characterizing degrees of dispersion in both engineered and natural reactors. We will use Equation 9.18a in our analyses employing the PFD model:

$$D\frac{d^2C}{dz^2} - v_s\frac{dC}{dz} - k'C = 0 \tag{9.18a}$$

$$\frac{D}{v_s L}\frac{d^2C}{d\xi^2} - \frac{dC}{d\xi} - \tau k' C = 0 \tag{9.18b}$$

Wylie (1966) presents the methodology to develop a general, closed-form solution to Equation 9.18a:

$$C_z = c_1 e^{R_1 z} + c_2 e^{R_2 z} \tag{9.19}$$

where

$$R_1 = \frac{v_s + \sqrt{v_s^{\,2} + 4kD}}{2D} \quad \text{and} \quad R_2 = \frac{v_s - \sqrt{v_s^{\,2} + 4kD}}{2D}$$

In order to obtain the particular solution to Equation 9.18a, two boundary conditions are necessary. Hulbert (1944) proposed that the concentration at the influent must be continuous and that the reaction would cease at the effluent leading to the two conditions:

$$C_{(z=0)} = C_{\text{in}}; \quad \frac{dC}{dz}_{(z=L)} = 0$$

These were examined by Danckwerts (1953) and later by Wehner and Wilhelm (1956). Rather than concentration (state variable) continuity at the inlet, flux continuity was proposed, leading to the alternative set of boundary conditions:

$$\left[v_s C - D\frac{dC}{dz}\right]_{(z=0^+)} = \left[v_s C\right]_{(z=0^-)}; \quad \frac{dC}{dz}_{(z=L^-)} = 0$$

The resultant discontinuity of the state variable (concentration) at the influent was deemed necessary in the light that the mass rate of target reactant entering the reactor be constant across the inlet boundary. This set of boundary conditions was accepted by Levenspiel

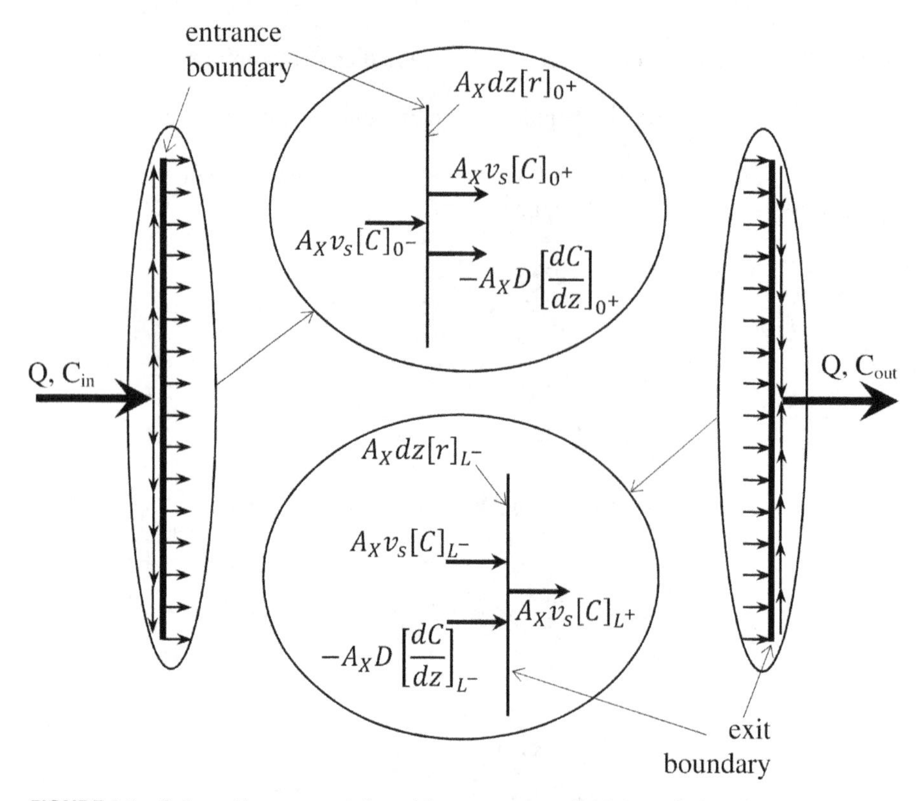

FIGURE 9.9 *Schematic representation of the entrance and exit boundaries of a reactor visualized using the plug-flow with dispersion (PFD) model. Representations of the transport and reactive processes within elements of thickness* dz *and area* A_X *on the reactor sides of the entrance and exit planes are shown.*

(1972, 1999) and later by Fogler (2005) and Froment et al. (2011). The resultant closed-form analytic solution can be found in these chemical reaction engineering textbooks. Unfortunately, the accepted boundary conditions are not correct. We cannot abide a discontinuity in the state variable at the influent and thus must conclude that the inlet boundary condition is flawed. Further, we can certainly reason that at the exit of a short reactor, the reaction is not complete and thus the outlet boundary condition is also flawed. Our subsequent analyses seek to define the profiles in target reactant concentration through the reactor. Were we to envision the overall reactor as a set of successively longer reactors, we would find ourselves successively zeroing the exit spatial derivative of reactant concentration for each implementation of the solution. Weber and DiGiano (1996) reasoned that the spatial derivative of concentration tends to zero at $z = \infty$, a correct assertion, but of little use in our analyses herein. We propose that both state variable and flux continuity must be obeyed at both the inlet and outlet boundaries of the reactor. We consider the reaction on the $z=0^+$ and $z=L^-$ planes at the inlet and outlet of the reactor, respectively. A schematic representation of the inlet and outlet boundaries and associated processes is presented in Figure 9.9.

Discontinuities in dispersion occur across the inlet and outlet boundaries and the magnitude of dispersion is considered constant throughout the reactor:

$$D\big|_{z=0^-} = D\big|_{z=L^+} = 0;\ D\big|_{0^+ \le z \le L^-} = D$$

Reaction processes begin at the reactor side of the inlet boundary and cease at the reactor side of the outlet boundary. Material balances on the target reactant across the inlet and outlet boundaries (A_X is area) are written:

$$\left[A_X \vec{v_s} C\right]_{0^-} = \left[A_X \vec{v_s} C - A_X D\frac{dC}{dz} + A_X dzr\right]_{0^+}$$

$$\left[A_X \vec{v_s} C - A_X D\frac{dC}{dz} + A_X dzr\right]_{L^-} = \left[A_X \vec{v_s} C\right]_{L^+}$$

For small reactant abundances, the total fluid flow rate is virtually constant through the entire reactor, hence also across the inlet and outlet boundaries. State variable continuity (target reactant concentration) across the boundaries must also be preserved:

$$\left[A_X \vec{v_s}\right]_{0^-} = \left[A_X \vec{v_s}\right]_{0^+} \text{ and } \left[A_X \vec{v_s}\right]_{L^-} = \left[A_X \vec{v_s}\right]_{L^+}$$

$$[C]_{0^-} = [C]_{0^+} \text{ and } [C]_{L^-} = [C]_{L^+}$$

Thus target reactant flux across the inlet and outlet boundaries is conserved:

$$\left[A_X \vec{v_s} C\right]_{0^-} = \left[A_X \vec{v_s} C\right]_{0^+} \text{ and} \left[A_X \vec{v_s} C\right]_{L^-} = \left[A_X \vec{v_s} C\right]_{L^+}$$

The gradient in concentration then is attributable solely to the reaction process, which begins and ends at the same positions as does the dispersion process:

$$0 = \left[-D\frac{dC}{dz} + dzr\right]_{0^+} \text{ and} \left[-D\frac{dC}{dz} + dzr\right]_{L^-} = 0$$

Were we to examine the conditions at the inlet and outlet boundaries of an ideal PFR, we would find the concentration gradients at both boundaries to be directly related to the respective reaction rates. We rearrange this result and employ Hulbert's inlet boundary condition to yield the boundary conditions necessary to the particular solution of Equation 9.18a.

$$C_{(z=0)} = C_{\text{in}}; \quad \frac{dC}{dz}_{(z=L)} = \frac{dz}{D} r\Big|_{z=L}$$

The concentration gradient at the effluent is indeed non-zero. Unfortunately, our result is not yet usable, since for quantitation we would necessarily know our effluent concentration to compute the reaction rate. Moreover, we are unable to assign a numerical value to dz. Of significant use, however, is the ratio of the concentration gradient at the effluent to that at the influent, for pseudo-first order kinetics, equal to the ratio of the respective concentrations

$$\frac{\left[\frac{dC}{dz}\right]_{0^+}}{\left[\frac{dC}{dz}\right]_{L^-}} = \frac{\frac{dz}{D} \cdot r\Big|_{0^+}}{\frac{dz}{D} \cdot r\Big|_{L^-}} = \frac{r|_{0^+}}{r|_{L^-}} = \frac{-k'C|_{0^+}}{-k'C|_{L^-}} = \frac{C|_0}{C|_L} \tag{9.20}$$

We employ this ratio at steady state such that $\left[\frac{dC}{dz}\right]_L$ has a constant value, K. We now have the means to develop the particular solution to Equation 9.18a. We set the derivative of concentration at $z=L$ equal to K to obtain a relation for c_1:

$$c_1 = \frac{K - R_2 C_{\text{in}} e^{R_2 L}}{R_1 e^{R_1 L} - R_2 e^{R_2 L}}$$

We employ the inlet concentration at $z=0$ to obtain a second relation for c_2:

$$c_2 = C_{\text{in}} - c_1 = \frac{C_{\text{in}} R_1 \; e^{R_1 L} - K}{R_1 e^{R_1 L} - R_2 e^{R_2 L}}$$

We still have the matter of evaluating K with which to deal. One means to approach this solution is to employ the ideal PFR solution for a given set of conditions to initialize K. The concentration gradient at the influent will be constant. Given an estimate for K, the reactor concentration profile can be solved. A new estimate of K can be obtained using the previous approximations of the exit concentration and inlet concentration gradient:

$$K_i = \left[\left[\frac{dC}{dz}\right]_{(z=0^+)}\right]_{i-1} \cdot \left[\frac{C_{\text{out}}}{C_{\text{in}}}\right]_{i-1}$$

The solution is, of course, iterative and would cease when the relative change in K reaches an acceptably small value. The iterative process involves only the positions $z=0$ and $z=L$. Once converged, we have the concentration value at the outlet and, if desired, we may produce the profile across the reactor using the alternative boundary condition $C|_{z=L}=C_{\text{out}}$. An alternative particular solution is obtained once we can specify the concentration (type 1) boundary condition at both inlet and outlet:

$$c_1 = \frac{C_{\text{out}} - C_{\text{in}}e^{R_2 L}}{e^{R_1 L} - e^{R_2 L}};\ c_2 = C_{\text{in}} - c_1 = \frac{C_{\text{in}}e^{R_1 L} - C_{\text{out}}}{e^{R_1 L} - e^{R_2 L}}$$

Most conveniently, for numerical development of the internal concentration profile through the reactor for the applicable steady-state conditions, the general solution (Equation 9.19) is left as is, and the integration constants are defined by applying the boundary conditions. We can, of course, develop the concentration profile through the reactor merely by incrementing the value of L from zero to the targeted overall flow path length.

9.8.2 Saturation Rate Law with the PFD Model

We begin with equation 9.17 and substitute the saturation rate law for the specific reaction rate. We re-arrange the result in order that the coefficient of the second order term is unity.

$$\frac{d^2C}{dz^2} - \frac{v_s}{D}\frac{dC}{dz} - \frac{k}{D}\frac{C}{K_{half} + C} = 0 \tag{9.21}$$

Upon consulting our advanced engineering mathematics text, in short order we realize that our result is a non-linear second-order ordinary differential equation. We cannot find a solution to such an ODE in the text. Should we consult our friendly local mathematician, he or she will inform us that mathematical science has not yet devised a closed-form solution for an ODE of this structure, the non-linear third term of the LHS being the major offender. We then turn to numerical approximation. Herein, we'll postulate a means to approximate the solution employing Euler's method, allowing for straightforward examination of the numeric structure.

We would invoke a definition of the concentration derivative: $\frac{dC}{dz} = p$. We would then rewrite equation 9.21 as a set of two coupled first-order ODEs.

$$\frac{dp}{dz} - \frac{v_s}{D}p - \frac{k}{D}\frac{C}{K_{half} + C} = 0;\ \frac{dC}{dz} = p$$

We would subdivide the reactor into n intervals, each of length Δz. We would define finite changes in p and C, associated with a finite change in z and would then increment the values using i as a counter. Then, when $i = 0$, $z = 0$ and when $i = n$, $z = L$. For increasingly smaller Δz, the accuracy of the approximation would be improved at the expense of an increased computational burden.

$$\Delta p_i = \left(\frac{v_s}{D} p_{i-1} - \frac{k}{D} \frac{C_{i-1}}{K_{\text{half}} + C_{i-1}} \right) \Delta z$$

$$p_i = p_{i-1} + \Delta p_i$$

We would need an initiating value of $\left(\left. \frac{dC}{dz} \right|_{z=0} \right)$ which we would obtain from the ideal PFR solution. Since the PFR model with the saturation rate law yields a governing equation that is implicit, we must obtain this beginning value of p numerically, perhaps employing a forward difference. Once p is known for a given value of i we can solve for the change in C, and thereafter increment C. $C_{i=0}$ is C_{in}.

$$\Delta C_i = p_i \Delta z;$$

$$C_i = C_{i-1} + \Delta C_i$$

Once p_i and C_i are known, we would pass these values into p_{i-1} and C_{i-1}, allowing computation of new values for p_i and C_i. This process would be repeated until $i=n$. Now we have a profile of concentration through the reactor which we would update $\left. \frac{dC}{dz} \right|_{z=0}$. We would use the relation between the ratios of the inlet and outlet gradients to the respective reaction rates expressed by Equation 9.21.

$$\left. \frac{dC}{dz} \right|_{z=0} = \frac{r|_{z=0}}{r|_{z=L}} \cdot \left. \frac{dC}{dz} \right|_{z=L}$$

$$r|_{z=0} = \frac{k}{D} \frac{C_{in}}{K_{\text{half}} + C_{in}}; \; r|_{z=L} = \frac{k}{D} \frac{C|_{z=L}}{K_{half} + C|_{z=L}}; \; \left. \frac{dC}{dz} \right|_{z=L} = \frac{C|_{z=L} - C|_{z=L-\Delta z}}{\Delta z}$$

With this new estimate of $p|_{z=0}$ we would compute new p and C profiles from which we would update the value of $p|_{z=0}$. We would iterative until the new estimate and old estimates differ by a sufficiently small relative change, and the most recent profile would be the solution we seek. We might then select successively smaller values of Δz, repeating the process until successive sets of final profiles were in substantial agreement. This methodology could be implemented with any ODE solver we might choose to employ, with its associated added complexities.

9.9 MODELING REACTIONS USING THE SEGREGATED FLOW (SF) MODEL

For implementation of the segregated-flow model, each of the individual reactors of Figure 9.8 is considered to be an ideal PFR. The influent flow is distributed among the reactors in relation to the respective fractions determined from either the RTD density or cumulative RTD function. We realize at the outset that we will not obtain any sort of closed-form solution for the overall SF reactor. Fortunately, however, we may employ either RTD function directly.

The effluent from the SF reactor is the weighted average of the effluents from each of the visualized subreactors. Each subreactor has a unique flow and residence time, leading to a unique value of the effluent concentration of targeted reactant in each effluent from each of the subreactors. We may apply the steady-state, zero-volume mixing principle from Chapter 7 to obtain the mixed concentration of targeted reactant in the total flow exiting the SF reactor:

$$C_{\text{out}} \cong \frac{\sum_{i=1}^{n} Q_i C_i}{\sum_{i=1}^{n} Q_i} \cong \sum_{i=1}^{n} \frac{Q_i}{Q_{\text{tot}}} C_i$$

The nonideal reactor is divided into n subreactors. We need know only the flow that passes through each reactor and the concentration of targeted reactant exiting each reactor to determine the weighted average.

The residence times of the reactor vary, increasing monotonically, from the minimum value (τ_0) for the zeroth sub-reactor to the maximum value (τ_n) for the nth subreactor. For an impulse stimulus, these are defined as the times corresponding with the last zero value of $E(t)$ prior to an impulse response and the first zero value of $E(t)$ subsequent to the passage of the impulse. For a positive step stimulus, we define the range of residence times as corresponding with last zero value of $F(t)$ prior to the step and the first unity (to however many SFs we choose to operate) value of $F(t)$, subsequent to the step. These are easily identified from a set of discrete $C(t)$ versus t values from any real tracer test.

We may directly apply either the cumulative RTD or the RTD density function to obtain the ratio Q_i/Q_{tot}:

$$\frac{Q_i}{Q_{\text{tot}}} = \bar{E}(t)_i \Delta\tau_i = \Delta F(\tau)_i$$

Were we to take the limit as the number of subreactors approaches infinity ($n \rightarrow \infty$), we can define a smooth function:

$$\frac{Q_i}{Q_{\text{tot}}} = E(\tau_i)dt = dF(\tau)_i$$

We can then mathematically state the relation for the effluent concentration from the SF reactor:

$$C_{\text{out}} = \int_{\tau_0}^{\tau_n} C_{pfr}(\tau)E(\tau)d\tau = \int_{\tau_0}^{\tau_n} C_{pfr}(\tau)dF(\tau) \tag{9.22}$$

The residence time of each subreactor, τ, is the time corresponding with each $E(\tau)$ or $F(\tau)$ value. The computation of $C_{\text{pfr}}(\tau)$ is accomplished for pseudo-first-order and saturation-type reactions using either Equation 8.22a or 8.23a, depending upon the applicable rate law.

If we choose to employ a higher order numerical integration method (e.g., Simpson's 1/3 rule), we may leave Equation 9.22 as written and most conveniently employ the RTD density ($E(t)$) function. In such a case, our discrete RTD data must conform to the requirements of the method or we must fit a smooth curve to the data in order to generate evenly spaced data. For application of Simpson's 1/3 rule with n $\Delta\tau$ intervals, we would use the following relation:

$$C_{\text{out}} \cong \frac{\Delta\tau}{3}\begin{pmatrix} C(\tau_0)E(\tau_0)+4C(\tau_1)E(\tau_1)+2C(\tau_2)E(\tau_2)+4C(\tau_3)E(\tau_3)+\ldots \\ +2C(\tau_{n-2})E(\tau_{n-2})+4C(\tau_{n-1})E(\tau_{n-1})+C(\tau_n)E(\tau_n) \end{pmatrix}$$

We might be tempted to use Simpson's 1/3 rule with the second form of Equation 9.22, but we would quickly remind ourselves that ΔF would not be constant, necessitating manipulations of the cumulative RTD to provide for revised residence times associated with a constant increment of $F(\tau)$.

If our discrete RTD function is of varying Δt, and we choose not to generate evenly spaced RTD data, we are obligated to employ the trapezoidal rule:

$$C_{\text{out}} \cong \sum_{i=1}^{n} C_{pfr}(\overline{\tau}_i)E(\overline{\tau}_i)\Delta\tau_{\text{i}} \cong \sum_{i=1}^{n} C_{pfr}(\overline{\tau}_i)\Delta F(\tau)_i$$

The mean residence time for sub-reactor i is $\overline{\tau}_i\left(=\frac{\tau_{i-1}+\tau_i}{2}\right)$, corresponding with the mean $E(\overline{\tau}_i)\left(=\frac{E(\tau_{i-1})+E(\tau_i)}{2}\right)$ or $\Delta F(\tau)_i$.$(=F(\tau_i)-F(\tau_{i-1}))$ for the residence time interval $\tau_i-\tau_{i-1}$. Equation 8.22a can be arranged to explicitly solve for $C_{pfr}(\overline{\tau}_i)$ while from 8.23a $C_{pfr}(\overline{\tau}_i)$ must be obtained implicitly.

9.10 APPLICATIONS OF NONIDEAL REACTOR MODELS

9.10.1 Translation of RTD Data for Use with Nonideal Models

Let us consider a reactor with hydrodynamic character that resembles that of an ideal plug-flow reactor but which may be characterized by a measurable degree of longitudinal forward- and back-mixing, i.e., longitudinal dispersion. This might well represent a reactor at a wastewater treatment facility arranged to resemble a PFR. Consider that we have accomplished a tracer test from which the concentration versus time results are listed in Table 9.1. We will employ these data to characterize the reactor for employment of the CMFRs in series, PFD and SF models. A configuration sketch is shown as Figure 9.10.

The basin consists of two parallel tanks each having two nearly full-length baffle walls. The overall dimensions of the reactor, outside to outside, are 26.42 m in width by 25.71 m in length. The exterior walls are concrete, 0.305 m thick, and the baffle walls are fiberglass composite and 0.102 m thick. We use this geometry to determine

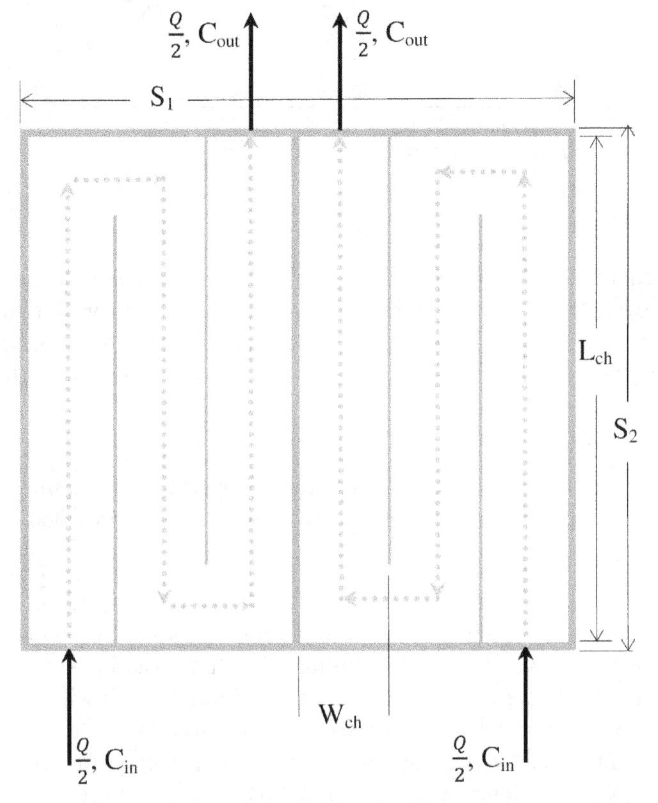

FIGURE 9.10 *Configuration sketch for a hypothetical reactor to be examined in Chapter 9 examples. (S_1 = 26.42 m, S_2 = 25.71 m, W_{ch} = 4.18 m, D_{ch} = 5 m, L_{ch} = 25.1 m, V_R = 3150 m^3 (0.833 Mgal)).*

TABLE 9.1 Hypothetical Data from RTD Analysis of a PF-Like Reactor.

t (min)	*C*(*t*) (mg/L)	
	Impulse	Step
30	0	0
45	0.102	0.005
60	0.541	0.040
75	1.404	0.158
90	2.328	0.394
105	2.832	0.723
120	2.745	1.077
135	2.237	1.391
150	1.591	1.630
165	1.013	1.792
180	0.589	1.890
195	0.317	1.945
210	0.16	1.974
225	0.076	1.988
240	0.035	1.995
255	0.015	1.998
270	0	1.999
285	0	2.000

$V_R = 3150\,m^3$ (0.833 Mgal).
$Q = 0.4375\,m^3/s$ (10 MGD).

that each channel is 4.18 m in width, and 25.1 m in total length. The openings between the baffle wall ends and the end walls of the reactors are 4.18 m in width. The reactor volume is $3150\,m^3$ (0.833 Mgal) and depth of flow is 5 m. During the time of the test, the influent flow was held at a steady rate of $0.4375\,m^3/s$ (10 MGD).

Example 9.1 Use the data from Table 9.1 to determine N, the number of CMFRs in series into which the real reactor may theoretically be subdivided.

In Figure E9.1.1, we plot the data sets to obtain a visual feel for the RTD of the reactor. C_E is the raw response from the impulse stimulus and C_F is the raw response from the positive step stimulus. We then examine the data set and observe that the discrete data are evenly spaced in time, allowing us to employ one of the Simpson's rules for numerical integration. We note further that the number of intervals for the impulse response, C_E, is even ($n = 16$) so we can employ the 1/3 rule. The values of t and $C(t)$ are used to create t, C_E, and C_F vectors in a MathCAD worksheet. The vector of necessary Simpson's rule coefficients is also generated. In this case, we have done it by hand but for larger data sets a simple program may be written and employed:

$$t := \begin{pmatrix} 30 \\ 45 \\ 60 \\ 75 \\ 90 \\ 105 \\ 120 \\ 135 \\ 150 \\ 165 \\ 180 \\ 195 \\ 210 \\ 225 \\ 240 \\ 255 \\ 270 \end{pmatrix} \text{min} \quad C_E := \begin{pmatrix} 0 \\ 0.102 \\ 0.541 \\ 1.404 \\ 2.328 \\ 2.832 \\ 2.745 \\ 2.237 \\ 1.591 \\ 1.013 \\ 0.589 \\ 0.317 \\ 0.16 \\ 0.076 \\ 0.035 \\ 0.015 \\ 0 \end{pmatrix} \frac{\text{mg}}{\text{L}} \quad C_F := \begin{pmatrix} 0 \\ 0.005 \\ 0.040 \\ 0.158 \\ 0.394 \\ 0.723 \\ 1.077 \\ 1.391 \\ 1.630 \\ 1.792 \\ 1.890 \\ 1.945 \\ 1.974 \\ 1.988 \\ 1.995 \\ 1.998 \\ 1.999 \\ 2.000 \end{pmatrix} \frac{\text{mg}}{\text{L}} \quad SR := \begin{pmatrix} 1 \\ 4 \\ 2 \\ 4 \\ 2 \\ 4 \\ 2 \\ 4 \\ 2 \\ 4 \\ 2 \\ 4 \\ 2 \\ 4 \\ 2 \\ 4 \\ 1 \end{pmatrix}$$

We address the impulse response data first. We implement the numeric integration to obtain the mass of tracer recovered per unit of flow, the denominator of Equation 9.1. We then define the $E(t)$ function as the E_t vector:

$$h := t_1 - t_0 = 15 \quad n := 16 \quad M := \frac{h}{3} \cdot \sum_{i=0}^{n} \left(C_{E_i} \cdot SR_i \right) = 239.81 \quad \frac{\text{mg} \cdot \text{min}}{\text{L}} \quad E_t := \overrightarrow{\frac{C_E}{M}}$$

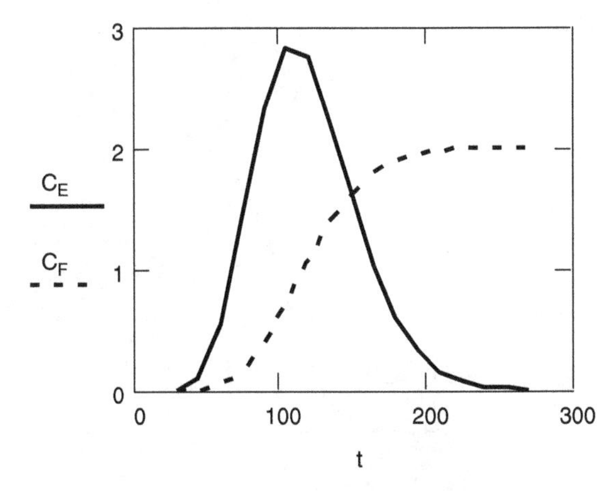

FIGURE E9.1.1 *Plot of exit response data from impulse and step inputs of Example 9.1.*

We also define the $F(t)$ function from the E_t vector as the F_t vector, most conveniently employing the trapezoidal rule:

```
Ft := | sum ← 0
      | for i ∈ 1.. n
      |   | add ← ((E_t_i + E_t_(i-1)) / 2)·(t_i − t_(i−1))
      |   | sum ← sum + add
      |   | F_i ← sum
      | return F
```

A plot in Figure E9.1.2, of F_t versus t, verifies that we have correct RTD density and cumulative RTD functions. We can now determine the statistical mean residence time, the variance of the RTD, and, hence, the value of N:

$$t_{SM} := \frac{h}{3} \cdot \sum_{i=0}^{n} \left(t_i \cdot E_{t_i} \cdot SR_i\right) = 119.9 \quad \text{min}$$

$$\sigma_s := \frac{h}{3} \cdot \sum_{i=0}^{n} \left[\left(t_i - t_{SM}\right)^2 \cdot E_{t_i} \cdot SR_i\right] = 1.184 \times 10^3 \quad \text{min}^2$$

$$N := \frac{t_{SM}^{\ 2}}{\sigma_s} = 12.15$$

Ideally, t_{SM} and τ_{PFR} would be identical. For this example, we generated the $C(t)$ plot from Levenspiel's relations and in paring the result to a discrete set of $C(t)$ versus t data, low values of $C(t)$ in the tail of the plot were truncated, resulting in a small difference between t_{SM} and τ_{PFR}.

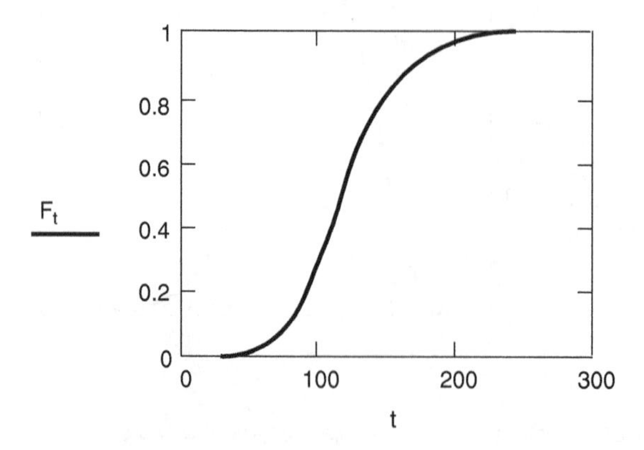

FIGURE E9.1.2 *A plot of the* F(t) *distribution as computed from* C(t).

When we perform these computations using the cumulative RTD function that we generated from E_t and employ the trapezoidal rule, we obtain a slightly different result:

$$t_{SM} := \sum_{i=1}^{n} \left[\left(\frac{t_i + t_{i-1}}{2}\right)\cdot\left(F_{t_i} - F_{t_{i-1}}\right)\right] = 119.9 \qquad \text{min}$$

$$\sigma_s := \sum_{i=1}^{n} \left[\left(\frac{t_i + t_{i-1}}{2} - t_{SM}\right)^2\cdot\left(F_{t_i} - F_{t_{i-1}}\right)\right] = 1.238\times 10^3 \qquad \text{min}^2$$

$$N := \frac{t_{SM}^{\ 2}}{\sigma_s} = 11.61$$

At the surface, the results of these computations, supposedly equivalent, are somewhat perplexing. However, when we delve into the details we find that our terminal value of $F(t)$, to five decimals, is 0.99985 while our integral of $E(t)dt$ is 1.00000. $F(t)$ was approximated using trapezoidal integration while $E(t)$ was approximated using Simpson's 1/3 rule. We could have developed better approximations of F_t by employing the trapezoidal rule to obtain the first F_t, then Simpson's 1/3 rule to obtain the second F_t, and then Simpson's 3/8 rule to obtain the third. Successive values of F_t would then be computed by employing combinations of the 1/3 rule for integrals over an even number of intervals and the 1/3 rule supplemented appropriately by the 3/8 rule for integrals over odd numbers of intervals. Better estimates of t_{SM} and σ_s then would have resulted. However, since we must employ the mean value of t for each interval and the change in F_t, integration to obtain t_{SM} and σ_s by means other than the trapezoidal rule is difficult at best. Let us resolve, when possible, to obtain RTD density data using samples evenly spaced in time so that we may directly apply Simpson's rules to integrations employing the RTD density function, $E(t)$.

Let us now work with the data from the positive step input. We could develop the $E(t)$ RTD function from the $F(t)$ data, which would involve equating each ΔF_i with each $\bar{E}_i\Delta t$. The result would be a vector of $\bar{E}_i$ values that would be no more accurate (likely less so) than the original vector of F values. We will work only with the cumulative RTD function. We create the F_t vector directly from the C_F vector and compute the statistical mean residence time, the variance of the RTD distribution, and the equivalent number of CMFRs comprising the real reactor:

$$F_t := \overrightarrow{\frac{C_F}{C_{F_n}}} \qquad t_{SM} := \sum_{i=1}^{n} \left[\left(\frac{t_i + t_{i-1}}{2}\right)\cdot\left(F_{t_i} - F_{t_{i-1}}\right)\right] = 120 \quad \text{min}$$

$$\sigma_s := \sum_{i=1}^{n} \left[\left(\frac{t_i + t_{i-1}}{2} - t_{SM}\right)^2\cdot\left(F_{t_i} - F_{t_{i-1}}\right)\right] = 1.207\times 10^3 \quad \text{min}^2$$

$$N := \frac{t_{SM}^{\ 2}}{\sigma_s} = 11.92$$

Employing the F_t vector necessitates use of the trapezoidal rule, again invoking some round-off errors and resulting in a value matching that from use of the trapezoidal rule with the generated F(t) distribution.

We observe that the RTD data of Table 9.1 lead to a characterization of the nonideal reactor as equivalent to just over 12 CMFRs in series. We will delve into the use of this characterization in a later example.

Example 9.2 Let us now use the RTD data from Table 9.1 to characterize the reactor using the PFD model. For use later on, we will extract the dispersion coefficient and thus must employ the information regarding the configuration of the reactor to define the length of the flow path.

We first examine the configuration of the reactor and determine the length of our designated flow path. A very reasonable flow path would begin at the influent wall at half depth and follow the centerline of the cross-sectional area down each channel, across from one channel to the adjacent channel through the midpoint of the opening, up the next channel, and so on:

$S_1 = 26.42 \quad S_2 = 25.71 \quad \text{m}$

$L_{ch} := S_2 - 2\cdot.305 = 25.1 \quad W_{ch} := \frac{S_1 - 3\cdot 0.305 - 4\cdot 0.102}{6} = 4.18 \quad \text{m}$

$L_{FP} := 3\cdot L_{ch} = 75.3 \quad \text{m} \quad L_{FP.2} := \frac{V_{R.SI}}{2\cdot W_{ch}\cdot H} = 75.3$

A second computation of the flow path, in full agreement with the first, is accomplished as the quotient of the reactor volume and the cross-sectional area of flow. We now can define the superficial fluid velocity (equal to the average velocity in the absence of packing):

$H = 5 \ \text{m} \quad Q = 0.4375 \ \frac{\text{m}^3}{\text{s}} \quad v_s := \frac{Q}{2\cdot W_{ch}\cdot H} = 0.01046 \ \frac{\text{m}}{\text{s}}$

We can now compute the dispersion number $N_D\left(=\frac{D}{v_s L}\right)$. We employ Equation 9.15, which is implicit and requires a numerical solution. We also use the statistical mean residence time and variance based on the RTD density distribution from Example 9.1:

$t_{SM} := 119.9 \ \text{min} \quad \sigma_s := 1.184\cdot 10^3 \ \text{min}^2$

$N_D := .02 \quad \text{Given} \quad \frac{\sigma_s}{t_{SM}^2} = 2\cdot N_D - 2\cdot N_D^2\cdot\left(1 - e^{-\frac{1}{N_D}}\right) \quad N_D := \text{Find}(N_D) = 4.303\times 10^{-2}$

$D := N_D\cdot v_s\cdot L_{FP} = 0.03389 \ \frac{\text{m}^2}{\text{s}}$

We compare our computed dispersion coefficient (D) with typical diffusivities of dissolved components in aqueous solution ($\mathcal{D}_w$ ~ 10^{-9} m/s^2) and confirm that the hydrodynamic regime of our example reactor is indeed dominated by horizontal velocity and mechanical dispersion.

We have translated the RTD data of Table 9.1 into a value of N for the CMFRs in series model and into values of N_D and D for the PFD model. We may use the RTD data directly with the segregated flow model.

9.10.2 Modeling Pseudo-First-Order Reactions

Let us continue with the reactor system of Figure 9.10 and apply the TiS, PFD, and SF models in the context of a pseudo-first-order reaction carried out in the hypothetical reactor under the flow conditions of the RTD analysis. We will consider a biological reaction for the reduction of organic matter manifest as biodegradable chemical oxygen demand (bCOD). Tchobanoglous et al. (2003) provide values of typical coefficients from which we may approximate a pseudo-first-order rate law. These are coefficients used in the Michaelis–Menton relation (Equation 8.11), which is most definitely of the saturation type. If we decide to neglect the substrate concentration term in the denominator, the overall relation reverts to pseudo-first-order form. Although some inaccuracies will accompany the modeling work accomplished, the results will be sufficient to illustrate applications of the nonideal models and to make comparisons both among them and with the ideal PFR model.

Example 9.3 Employ the ideal PFR and the nonideal TiS, PFD, and SF models to analyze a pseudo-first-order reaction carried out in the reactor system of Figure 9.10. Employ the models used with and results from Examples 9.1 and 9.2 in applying the TiS and PFD models.

We define the kinetic parameters using a growth rate coefficient that is in the low range of the stated typical values. We need to choose a biomass concentration also, and for a PFR-like reactor, we do not require a large abundance of biomass in the reactor to effect large reduction in the targeted reactant:

$$\begin{pmatrix} \mu_{max} \\ Y \\ K_{half} \end{pmatrix} := \begin{pmatrix} 3.0 \\ 0.4 \\ 20 \end{pmatrix} \begin{pmatrix} \frac{g_{VSS}}{g_{VSS} \cdot d} \\ \frac{g_{VSS}}{g_{bCOD}} \\ \frac{g_{COD}}{m^3} \end{pmatrix} \qquad X := 200 \quad \frac{g_{VSS}}{m^3} \qquad k := \frac{\mu_{max} \cdot X}{Y \cdot K_{half}} = 75 \quad day^{-1}$$

We gather flow and volume information from Examples 9.1 and 9.2, taking care to employ consistent time units:

$$Q := 0.4375 \cdot 86400 \ \frac{m^3}{d} \qquad V_R := 3150 \ m^3 \qquad \tau_{PFR} := \frac{V_R}{Q} = 0.08333 \ d$$

We compute the predicted bCOD concentration from the ideal PFR:

$$C_{in} := 300 \ \frac{g_{bCOD}}{m^3} \qquad C_{out.pfr} := C_{in} \cdot e^{-k \cdot \tau_{PFR}} = 0.579 \ \frac{g_{bCOD}}{m^3}$$

We employ the result for the CMFRs in series model that arose from application of the RTD density function and compute the predicted effluent bCOD concentration were the reactor a string of CMFRs in series. We would be tempted to simply employ N as a noninteger value with the following result:

$$N := 12.15 \qquad C_{out.TiS1} := C_{in} \cdot \left(\frac{1}{1 + \frac{k \cdot \tau_{SM}}{N}} \right)^N = 1.944 \ \frac{g_{bCOD}}{m^3}$$

Conversely, we can visualize a string of 12 CMFRs, each with residence time $\frac{t_{SM}}{N}$, followed by a thirteenth reactor of residence time $(0.15)\frac{t_{SM}}{N}$. This approach requires that we perform some programming to step through the first 12 reactors in the series employing a loop and then using the effluent from the twelfth reactor as the influent to the much smaller, thirteenth reactor. We illustrate two MathCAD functions that are quite useful for obtaining integer values from decimal values. We arrive at an alternate result, which is more in line with the theoretical visualization of the nonideal reactor as a string of CMFRs in series:

$$N_{count} := \text{trunc}(N) = 12 \qquad N_{Tot} := \text{ceil}(N) = 13 \qquad C_0 := C_{in}$$

$$C_{out.TiS} := \left| \begin{array}{l} \text{for } i \in 1..N_{count} \\ \quad C_i \leftarrow C_{i-1} \cdot \dfrac{1}{1 + \dfrac{\tau_{SM}}{N} \cdot k} \\ i \leftarrow N_{Tot} \\ C_i \leftarrow C_{i-1} \cdot \dfrac{1}{1 + (N - N_{count}) \cdot \dfrac{\tau_{SM}}{N} \cdot k} \\ C \end{array} \right. \qquad C_{out.TiS_{N_{Tot}}} = 1.921 \ \frac{g_{bCOD}}{m^3}$$

We computed (not shown) the effluent concentration with the fractional reactor as the first and sixth in the series with results identical to that shown. Of great significance is that via the programming we have accomplished, a vector containing the thirteen effluent values has been created. We will find this capacity immensely useful in later analyses.

We might visualize the TiS model using a third and a fourth configuration as a string of N_{count} CMFRs each of residence time $\frac{N}{N_{count}}\frac{t_{SM}}{N}$, or as a string of N_{Tot} CMFRs each of residence time $\frac{N}{N_{Tot}}\frac{t_{SM}}{N}$:

$$C_{out.TiS3} := C_{in} \cdot \left(\frac{1}{1 + \frac{N}{N_{count}} \cdot \frac{k \cdot \tau_{SM}}{N}} \right)^{N_{count}} = 1.966 \quad \frac{g_{bCOD}}{m^3}$$

$$C_{out.TiS6} := C_{in} \cdot \left(\frac{1}{1 + \frac{N}{N_{Tot}} \cdot \frac{k \cdot \tau_{SM}}{N}} \right)^{N_{Tot}} = 1.829 \quad \frac{g_{bCOD}}{m^3}$$

We observe the 12-reactor and 13-reactor strings to return effluent values that are, respectively, greater than and less than that from stepping through the reactors. Adjusting N to an integer value and proportionately increasing or decreasing the residence times of the individual CMFRs in the series might seem reasonable, but we are cautious that these modifications, as well as simply using N as a noninteger value, result in significantly different outcomes. We really cannot definitively judge which of the three methodologies employed with the TiS model in this example would be most correct. However, the implementation accomplished by stepping through the reactors and including a reactor whose volume is a fraction of that of the others in the series most closely follows the derivation of the CMFRs in series relation for the pseudo-first-order reaction case. With this approach, in order to quantitatively "step into the reactor" and understand the process within the reactor, we must decide the relative location of the partial reactor. Its location certainly affects the profile of the target concentration through the reactor. Given this difficulty, perhaps the most convenient and certainly the conservative approach would be to truncate N and accordingly adjust the volume of the N identical reactors. We can certainly use this approach to generate a profile of target reactant concentration through the reactor, allowing us to quantitatively "step through" the reactor with small steps rather than integer values from 1 to N. We will investigate this idea in later examples.

For the PFD model, we gather the appropriate data from Examples 9.1 and 9.2 and convert to a time unit in days:

$$L_{FP} := 75.3 \quad m \qquad \begin{pmatrix} D \\ v_s \end{pmatrix} := 86400 \cdot \begin{pmatrix} 0.03389 \\ 0.01046 \end{pmatrix} = \begin{pmatrix} 2928.1 \\ 903.7 \end{pmatrix} \quad \begin{pmatrix} \frac{m^2}{d} \\ \frac{m}{d} \end{pmatrix}$$

Rather than employing the particular solution for the PFD model as one massive relation, we will compute the R_1, R_2, C_1, and C_2 coefficients and use Equation 9.19 in its simplest form. R_1 and R_2 are functions only of the superficial velocity, dispersion coefficient, and pseudo-first-order rate coefficient so we may set their values as constants:

$$R_1 := \frac{v_s + \sqrt{v_s^2 + 4 \cdot k \cdot D}}{2 \cdot D} \qquad R_2 := \frac{v_s - \sqrt{v_s^2 + 4 \cdot k \cdot D}}{2 \cdot D} \qquad m^{-1}$$

We then set the initializing value for the concentration derivative at the reactor outlet from the PFR solution:

$$dC_{L.PFR} := \frac{-k}{v_s} \cdot \left(C_{in} \cdot e^{-k \cdot \frac{L_{FP}}{v_s}} \right) \qquad K := dC_{L.PFR} = -0.04811 \ \frac{g_bCOD}{m^3 \cdot m}$$

The remainder of the solution is most conveniently executed as a programmed loop. We use the initialized K to compute C_{out} and employ C_{out} to compute a new K from C_{in} and $\left.\frac{dC}{dz}\right|_{z=0}$, both of which are constant. We compute the relative change in K from one iteration to the next and when that relative change remains greater than a threshold value (here we chose 10^{-6}), we perform an additional iteration after updating K from K_{new}. When the relative change becomes less than the criterion for convergence, the computation exits the loop and we have our value of C_{out}. A capture of the MathCAD program is shown in Figure E9.3.1 and the output from that program is shown as the $C_{out.PFD}$ vector:

$$C_{out.PFD} = \begin{pmatrix} 1.987 \\ 1.756 \\ 1.798 \\ 1.79 \\ 1.792 \\ 1.791 \\ 1.791 \\ 1.791 \\ 1.791 \end{pmatrix} \frac{g_bCOD}{m^3}$$

We have programmed the loop to create a vector of C_{out} values, corresponding with the results from each of the iterations. Herein nine iterations are required to arrive at a suitable approximation for C_{out}. For our rather simple numeric solution here, each new K is computed using the old K, resident in the relations for C_1 and C_2.

$$
C_{out.PFD} := \left| \begin{array}{l} \text{for } i \in 0..\,20 \\ \quad \left| \begin{array}{l} C_1 \leftarrow \dfrac{K - R_2 \cdot C_{in} \cdot e^{R_2 \cdot L_{FP}}}{R_1 \cdot e^{R_1 \cdot L_{FP}} - R_2 \cdot e^{R_2 \cdot L_{FP}}} \\ C_2 \leftarrow \dfrac{C_{in} \cdot R_1 \cdot e^{R_1 \cdot L_{FP}} - K}{R_1 \cdot e^{R_1 \cdot L_{FP}} - R_2 \cdot e^{R_2 \cdot L_{FP}}} \\ C_{out_i} \leftarrow C_1 \cdot e^{R_1 \cdot L_{FP}} + C_2 \cdot e^{R_2 \cdot L_{FP}} \\ K_{new} \leftarrow \dfrac{C_{out_i}}{C_{in}} \cdot (R_1 \cdot C_1 + R_2 \cdot C_2) \\ RC \leftarrow \left| \dfrac{K - K_{new}}{K_{new}} \right| \\ K \leftarrow K_{new} \\ \text{break if } RC < 10^{-6} \end{array} \right. \\ \text{return } C_{out} \end{array} \right.
$$

FIGURE E9.3.1 Capture of MathCAD code for the iterative solution of the PFD model for a pseudo-first-order reaction rate law.

We might reduce the number of iterations were we to invoke an implicit solution for each new K value, involving simultaneous solution of the relations for K_{new}, C_1 and C_2. A given-find block would be assembled and written into a function, which can be employed within the programmed loop. Here, the savings of a few iterations of the loop at the expense of the added complexity of including the given-find block within the programmed loop are likely not warranted. A profile of concentration through the reactor can be computed using Equation 9.19 employing c_1 and c_2 as defined in section 9.8.1 based on the known influent and effluent concentrations.

For the SF model prediction, we employ the t, E_t, and SR vectors defined for Example 9.1. We find it to be most convenient to convert the pseudo-first-order rate constant for use of t in minutes. We create a vector of effluent concentrations for the seventeen parallel reactors visualized for the SF model using the relation yielding the effluent from an ideal PFR. We then employ Simpson's 1/3 rule, using h (the interval) and n (the number of intervals of the original $C(t)$ data set) from Example 9.1, to obtain the predicted effluent bCOD value:

$$
k_{SF} := \frac{k}{1440} \qquad C_t := C_{in} \cdot e^{-k_{SF} \cdot t} \qquad C_{out.SF} := \frac{h}{3} \cdot \sum_{i=0}^{n} \left(C_{t_i} \cdot E_{t_i} \cdot SR_i \right) = 1.945 \quad \frac{g_{bCOD}}{m^3}
$$

We have computed the percent error associated with the assumption of an ideal PFR relative to a nonideal reactor based on the TiS, PFD, and SF models for the system and conditions of Example 9.3, as the ratio of the difference between the PFR prediction and the real reactor prediction to the real reactor prediction. Relative errors are in the range of 70%.

The percent error associated with the prediction from the PFR model relative to the predictions from the non-ideal reactor (NIR) models was computed from the relation:

$$\%RE = 100\frac{CC_{eff.PFR} - C_{eff.NIR}}{C_{eff.NIR}}$$

The percent relative errors are in the range of 70%.

We easily observe from Example 9.3 that the TiS, PFD, and SF models all yield similar results but are certainly not in exact agreement. After all, they are three distinct methods employed to model nonideal reactors. Further, with the TiS model, we can choose among numerous specific means of application. One important consideration is that we would like to use these models to quantitatively understand processes that occur within the reactor so that we may perform enlightened design of the systems in support of the reactor.

9.10.3 Modeling Saturation-Type Reactions with the TiS and SF Models

At this juncture, we will cease our investigations into modeling nonideal reactors employing the plug-flow with dispersion model. The necessity for an iterative solution in combination with a numerical approximation for the saturation-type reaction will carry us well beyond the intended scope of this text. We will continue with the analyses of the hypothetical reactor of Figure 9.10 and employ the full saturation-type reaction for which Tchobanoglous et al. (2003) have provided typical rate law parameters. We employ the CMFRs in series and segregated flow models.

Example 9.4 Employ the saturation-type reaction rate law in the context of Example 9.3 and develop comparisons among the predicted performances of an ideal PFR, a nonideal reactor visualized as N CMFRs in series (TiS) and a nonideal reactor visualized as segregated flow (SF, a set of parallel PFRs).

For this computation, in order that we may produce results in the single-digit values for the effluent concentration, we have adjusted the biomass concentration to 600 mg_{VSS}/L. We compute the lumped reaction rate coefficient for the numerator of the saturation-type rate law:

$$\begin{pmatrix} \mu_{max} \\ Y \\ K_{half} \end{pmatrix} := \begin{pmatrix} 3.0 \\ 0.4 \\ 20 \end{pmatrix} \begin{pmatrix} \frac{g_{VSS}}{g_{VSS} \cdot d} \\ \frac{g_{VSS}}{g_{bCOD}} \\ \frac{g_{COD}}{m^3} \end{pmatrix} \quad X := 600\,\frac{g_{VSS}}{m^3} \quad k := \frac{\mu_{max} \cdot X}{Y} = 4.5 \times 10^3 \ \ day^{-1}$$

We compute the predicted effluent concentration from the ideal PFR:

$$Q := 0.4375 \cdot 86400 \quad \frac{m^3}{d} \qquad V_R := 3150 \quad m^3 \qquad \tau_{PFR} := \frac{V_R}{Q} = 0.08333 \quad d$$

$$C_{in} := 300 \qquad C_{out} := 10 \quad \frac{g_{bCOD}}{m^3}$$

$$f_{PFR}(C_{out}) := K_{half} \cdot \ln\left(\frac{C_{in}}{C_{out}}\right) + C_{in} - C_{out} - k \cdot \tau_{PFR}$$

$$C_{out.pfr} := \mathrm{root}(f_{PFR}(C_{out}), C_{out}) = 5.39 \quad \frac{g_{bCOD}}{m^3}$$

We employ the sequential solution of the CMFRs in series to obtain the predicted effluent concentration when the reactor is viewed as N CMFRs in series. We employ the statistical mean residence time rather than the computed hydraulic residence time. A logical program for performing the computations is shown in Figure E9.4.1. As expected, the performance of the nonideal reactor is predicted to be poorer than that of the ideal PFR. We have also computed the effluent concentration by truncating N and correspondingly adjusting the residence time of the 12 reactors in the series using the short logical program shown in Figure E9.4.2. As was the case for the pseudo-first-order rate law, the truncation of N leads to a more conservative prediction of the effluent concentration.

$$\tau_{SM} := \frac{119.9}{1440} = 0.08326 \quad d \qquad N := 12.15$$

$$N_{count} := \mathrm{trunc}(N) = 12 \qquad N_{Tot} := \mathrm{ceil}(N) = 13 \qquad a := 1 \qquad C_0 := C_{in}$$

```
C_TiS :=  for i ∈ 1.. N_count
              b ← K_half + (τ_SM / N)·k − C_(i−1)
              c ← −(K_half·C_(i−1))
              C_i ← (−b + √(b^2 − 4·a·c)) / (2·a)
          i ← N_count + 1
          b ← K_half + (N − N_count)·(τ_SM / N)·k − C_(i−1)
          c ← −(K_half·C_(i−1))
          C_i ← (−b + √(b^2 − 4·a·c)) / (2·a)
          C
```

$$C_{TiS_{N_{Tot}}} = 9.68 \quad \frac{g_{bCOD}}{m^3}$$

FIGURE E9.4.1 *Screen capture of a logical program for computing stepped concentrations for a saturation-type reaction for the TiS model.*

$$C_{TiS2} := \left| \begin{array}{l} \text{for } i \in 1..N_{count} \\ \quad \left| \begin{array}{l} b \leftarrow K_{half} + \frac{N}{N_{count}} \cdot \frac{\tau_{SM}}{N} \cdot k - C_{i-1} \\ c \leftarrow -(K_{half} \cdot C_{i-1}) \\ C_i \leftarrow \frac{-b + \sqrt{b^2 - 4 \cdot a \cdot c}}{2 \cdot a} \end{array} \right. \\ C \end{array} \right. \qquad C_{TiS2_{N_{count}}} = 10.03 \quad \frac{gbCOD}{m^3}$$

FIGURE E9.4.2 Screen capture of logical program for employment of a truncated N value.

We employ the SF model to obtain the predicted effluent concentration from the reactor when viewed as a set of parallel ideal PFRs. We note that some investigation into the best value for our initial guess for implementation of MathCAD's root() function was necessary to enable computation with a single programmed loop. The value of n and the vectors t, E_t, and SR have been defined in a previous example and are not repeated here. The logical program written to obtain the effluent concentration for the SF model is shown in Figure E9.4.3.

$$C_{SF} := \left| \begin{array}{l} \text{for } i \in 0..n \\ \quad \left| \begin{array}{l} C \leftarrow 10^{-10} \\ C_{SF_i} \leftarrow \text{root}\left(K_{half} \cdot \ln\left(\frac{C_{in}}{C}\right) + C_{in} - C - k_{SF} \cdot t_i, C\right) \end{array} \right. \\ \text{return } C_{SF} \end{array} \right.$$

Root() needs its guess value to be very small in this case to enable convergence for larger t values, the low value is applicable throughout all t.

$$C_{out} := \frac{h}{3} \cdot \sum_{i=0}^{n} \left(C_{SF_i} \cdot E_{t_i} \cdot SR_i\right) = 26.3 \quad \frac{gbCOD}{m^3}$$

FIGURE E9.4.3 Screen capture of short program to compute the effluent concentration based on the SF model.

The result is surprising relative to that of the CMFRs in series. A check of $C_{SF.6}$ ($t_i = \tau_{PFR}$) yields an identical result to that for the ideal PFR of residence time $\tau \approx t_{SM}$. With the saturation rate law, the rate of reaction is quite nonlinear: the rate of increase in the reaction rate diminishes with an increase of reactant concentration. Conversely, with a pseudo-first-order rate law, the rate of reaction is directly proportional to the concentration. At each position within the reactor, including at the effluent, we visualize that a set of target concentration values exists that corresponds with the extent of the reaction to that position in the reactor. As a consequence, the weighted average point-wise concentration, as well as that of the effluent, is influenced by the proportionately lower reaction rates in the reactors of larger residence time.

9.11 CONSIDERATIONS FOR ANALYSES OF SPATIALLY VARIANT PROCESSES

9.11.1 Internal Concentration Profiles in Real Reactors

Oftentimes we are interested in not only what exits from a reactor but also the progression from the influent to the effluent. One notable application for such an analysis would be the point-wise distribution of oxygen requirements in a biological process reactor. For a CMFR, the requirement is spatially invariant. On the other hand, for a PFR-like reactor, the requirement would mirror the rate at which organic matter, manifest as BOD or COD, is converted to carbon dioxide. The oxygen consumption process then depends upon the point-wise abundance of the target reactant and of biomass.

We realize that, along with the reduction of biodegradable organic matter, the biomass concentration will increase along the path of the reactor. At this juncture, we are willing to accept the error associated with assumption of constant biomass concentration to enable use of closed-form relations for prediction of reactant concentration. When we include a second relation for the variable biomass concentration, we lose the ability to use our previously developed closed-form solutions. We will address that situation later in this chapter.

Example 9.5 Develop the concentration versus position profiles for the system and process examined in Example 9.3. Employ the pseudo-first-order reaction rate law, visualizing the reactor as an ideal PFR, as N CMFRs in series (TiS), as a PFD reactor, and as a segregated flow (SF) reactor:

For the ideal PFR, we need only arrange a function that allows for variation of the point-wise residence time with position:

$$L_{FP} := 75.3 \quad m \qquad C_{pfr}(z) := C_{in} \cdot e^{-k \cdot \frac{z}{L_{FP}} \cdot \tau_{PFR}} \quad \frac{gbCOD}{m^3}$$

For the CMFRs in series model, we realize we have a discrete number of values for the concentration of our target reactant, and that the profile, conforming with the model assumptions, will have a number of steps. Short MathCAD programs shown in Figure E9.5.1 allow us to assign positions relating to the residence times of the *N* reactors in series and to relate the internal concentrations of the reactor to those positions. For this illustration, we have truncated N and correspondingly adjusted the residence times of the reactors comprising the series. Using the effluent concentrations from each of the reactors, we have written a function to assign the effluent concentration values (also the concentrations within the visualized series of reactors) to positions along the flow path within the real reactor. In reality, the profile across the reactor would be continuous. Unfortunately, the CMFRs in series model predicts that the reactant concentration decreases from the influent value to the effluent value in distinct steps. We must resist temptations to develop a "smoothed"

$$C_{CMFRs} := \left| \begin{array}{l} \text{for } i \in 1..N_{count} \\ \quad C_i \leftarrow C_{i-1} \cdot \dfrac{1}{1 + \dfrac{\tau_{SM}}{N_{count}} \cdot k} \\ C \end{array} \right.$$

$$C_{TiS}(z) := \left| \begin{array}{l} step \leftarrow \dfrac{L_{FP}}{N_{count}} \\ i \leftarrow 1 + \mathrm{trunc}\left(\dfrac{z}{step}\right) \\ i \leftarrow N_{count} \text{ if } i > N_{count} \\ C \leftarrow CCMFRs_i \\ C \leftarrow C_{in} \text{ if } z = 0 \\ \text{return } C \end{array} \right.$$

FIGURE E9.5.1 *Screen capture of logical programs to compute the concentration profile along the flow path of a reactor using the TiS model.*

concentration profile associated with the CMFRs in series model—the basic assumptions of the model would be violated.

For the PFD model, we use the predicted effluent value as the second type 1 boundary condition, define c_1 and c_2 as follows from equation 9.20, and use Equation 9.19 to predict the profile of target reactant through the reactor.

For the SF model, we must generate the values for concentration versus position in each of the visualized parallel reactors. Then, we utilize the RTD density distribution to compute the weighted average concentration at each z. We visualize that all parallel reactors have a length equal to that of the real reactor. The values for the concentrations versus position are most conveniently developed using a matrix approach. A vector of 101 values of dimensionless residence time was created and used with the vector of reactor residence times to produce a two-dimensional matrix of τ_z (17 rows, 101 columns). One hundred intervals were arranged to provide for a quality smooth plot. Then, the matrix C_{SF} (17 rows and 101 columns) was created using the τ_z matrix. In order that we assure the creation of the vector, we use MathCAD's "vectorize" function, evidenced by the vector accent above the RHS of the assignment statement for C_{SF}. Then the vector containing the weighted average concentration values, C_z (101 rows), was created using C_{SF} and the RTD density distribution. Then, in order to plot C_z versus position, a vector of z_{SF} was created by apportioning the flow path length into 100 equal segments. These operations are shown in the screen capture of Figure E9.5.2.

We plot the four series and compare the results. We note that the concentration profile for the SF reactor is a vector while those for the others are functions. We could easily convert the vector to a function, but by using multiple independent variable designations, we plot all relations on the same set of axes with an arithmetic ordinate axis in Figure E9.5.3 and a logarithmic ordinate axis in Figure E9.5.4. The stepped trace for the N CMFRs conforms with the assumption that the real reactor comprises N CMFRs in series. We immediately observe that the steps are large and will yield poor resolution of the target reactant concentration with position especially in the influent half of the reactor. We are

$$\theta := \begin{array}{|l} \text{for } k \in 0..100 \\ \quad \theta_k \leftarrow \frac{k}{100} \\ \text{return } \theta \end{array} \qquad j := 0..n \qquad k := 0..100 \qquad \tau_{z_{j,k}} := \overrightarrow{\left(t_j \cdot \theta_k\right)}$$

$$C_{SF_{j,k}} := \overrightarrow{\left(C_{in} \cdot e^{-k_{SF} \cdot \tau_{z_{j,k}}}\right)}$$

$$C_{SF.z} := \begin{array}{|l} \text{for } k \in 0..100 \\ \quad C_{z_k} \leftarrow \frac{h}{3} \cdot \sum_{j=0}^{n} \left(E_{t_j} \cdot C_{SF_{j,k}} \cdot SR_j\right) \\ \text{return } C_z \end{array} \qquad i := 0..100 \qquad z_{SF_i} := \overrightarrow{\left(\frac{i}{100} \cdot L_{FP}\right)}$$

FIGURE E9.5.2 *Screen captures of short programs to generate a matrix of concentrations along* n *parallel reactors of the SF model and collect them into a single profile along the flow path of the real reactor.*

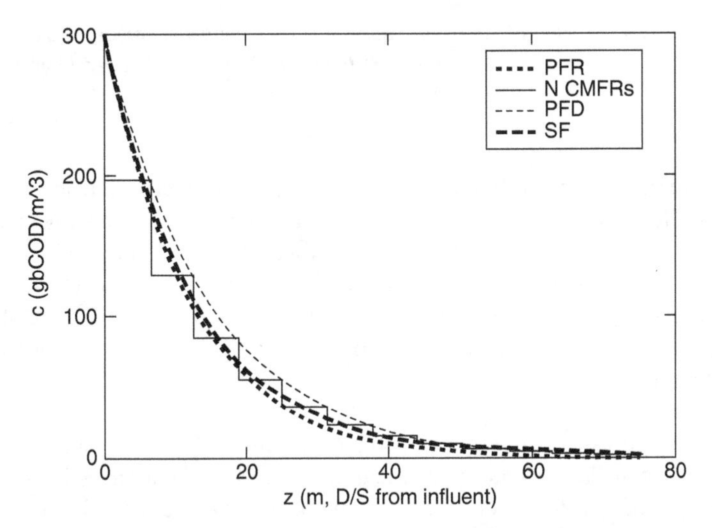

FIGURE E9.5.3 *A plot of predicted substrate concentration versus position for PFR, N-CMNRs in series, PFD and SF models.*

comforted by the passage of the SF trace essentially through that of the N CMFRs trace, indicating relative agreement between these two models. The trace for the PFD model essentially connects the discrete values predicted by the TiS model, closely approximating the result had we "smoothed" the TiS profile. Computations using the "smoothed" TiS predictions would over-predict reaction rates at all positions other than intersections of the smoothed profile with the stepped profile. And, of course, we observe that the predicted concentration profile for the PFR to be well below those of the real reactors, except for that of N CMFRs in the first 30 m of the reactor.

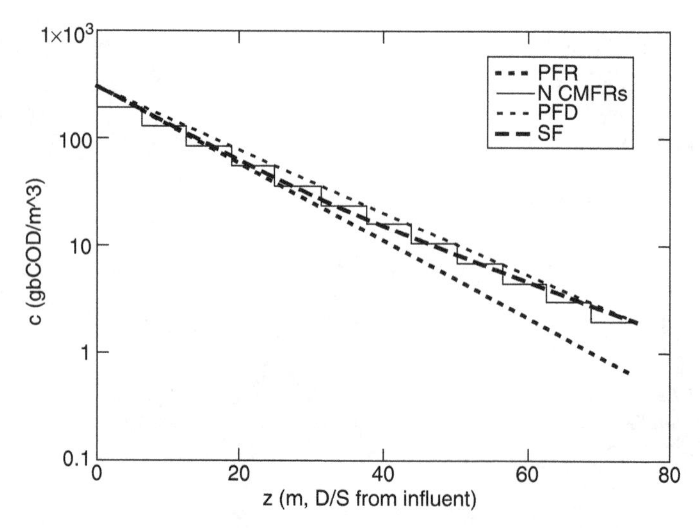

FIGURE E9.5.4 *A plot of predicted substrate concentration versus position for PFR, N-CMNRs in series, PFD and SF models, with a logarithmic ordinate scale to allow easy discernment of the differences in the exit half of the reactor.*

Conversion of the ordinate scale of the plot to a logarithmic scale in Figure E9.5.4 allows us to examine more closely the behaviors of the various models in the outlet half of the reactor. The profiles from the real reactors converge to an effluent concentration just below 2 g_{bCOD}/m^3, well above that predicted from the PFR model. We are also able to verify that the SF model predictions are in better agreement with those of N CMFRs than are those from the PFD model.

Given the necessity for iterative, numerical solution of the PFD model for reaction rate laws beyond the pseudo-first-order, we will curtail our investigations of the employment of the PFD model with the work of Example 9.5. The utility of this model does not appear to be greater than that of the TiS or SF model, and the significant efforts to develop the iterative, numeric solutions for the PFD model to employ the saturation rate law are beyond the intended scope of this text.

The reactor of Examples 9.1–9.5 really quite closely resembles a PFR. In particular in biological wastewater treatment, unless reactors are highly baffled, RTD distributions often depart profoundly from the ideal case. In the context of the application of the saturation-type rate law, let us examine a RTD density distribution that has significantly greater variance.

Example 9.6 Examine the reactor/reaction system of Example 9.4 in the case that the exit concentration trace from a RTD analysis using an impulse stimulus would be as shown in Figure E9.6.1. The actual numeric data are also presented in the following table.

t (min)	C(t) (mg/L)	t (min)	C(t) (mg/L)	t (min)	C(t) (mg/L)
15	0	120	2.086	225	0.198
30	0.097	135	1.763	240	0.122
45	0.461	150	1.383	255	0.073
60	1.079	165	1.021	270	0.043
75	1.716	180	0.718	285	0.025
90	2.137	195	0.484	300	0.014
105	2.246	210	0.314	315	0

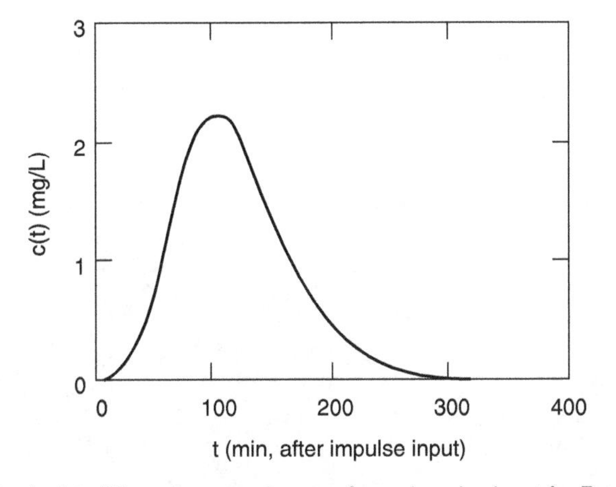

FIGURE E9.6.1 *A plot of the exit response curve for an impulse input for Example 9.6.*

Determine the predicted profiles of target reactant concentration across the reactor for the ideal PFR and nonideal TiS and SF models.

We define n and create t, $C_{E,}$ (shown graphically in Figure E9.6.1) and SR vectors, and generate the $E(t)$ versus t RTD density trace (all not shown herein). Then we compute the statistical mean residence time, the variance of the RTD distribution, and the equivalent number of CMFRs in series:

$$t_{SM} := \frac{h}{3} \cdot \sum_{i=0}^{n} \left(t_i \cdot E_{t_i} \cdot SR_i\right) = 119.8 \quad \text{min} \qquad \tau_{SM} := \frac{t_{SM}}{1440} = 0.08321 \quad \text{day}$$

$$\sigma_s := \frac{h}{3} \cdot \sum_{i=0}^{n} \left[\left(t_i - t_{SM}\right)^2 \cdot E_{t_i} \cdot SR_i\right] = 2.012 \times 10^3 \quad \text{min}^2 \qquad N := \frac{t_{SM}^{\;2}}{\sigma_s} = 7.134$$

We also define the kinetic and other parameters for immediate and later use herein. We have employed the kinetic coefficients used in Example 9.4, defined a death/decay coefficient, and defined the stoichiometry of oxygen-utilization offset by the net growth of biomass:

$$\begin{pmatrix} \mu_{max} \\ Y \\ K_{half} \end{pmatrix} := \begin{pmatrix} 3.0 \\ 0.4 \\ 20 \end{pmatrix} \begin{pmatrix} \frac{g_{VSS}}{g_{VSS} \cdot d} \\ \frac{g_{VSS}}{g_{bCOD}} \\ \frac{g_{COD}}{m^3} \end{pmatrix} \quad X := 600 \ \frac{g_{VSS}}{m^3} \quad k := \frac{\mu_{max} \cdot X}{Y} = 4.5 \times 10^3 \ day^{-1}$$

$$k_D := 0.08 \ \frac{g_{VSS}}{g_{VSS} \cdot d} \quad F_{O2.VSS} := 1.42 \ \frac{g_{O2}}{g_{VSS}}$$

The predicted effluent concentration profile for the ideal PFR model is easily written as a function of position using the root() function:

$$Q := 0.4375 \cdot 86400 \ \frac{m^3}{d} \quad V_R := 3150 \ m^3$$

$$\tau_{PFR} := \frac{V_R}{Q} = 0.08333 \ d \quad C_{in} := 300 \quad C_{out} := 10 \ \frac{g_{bCOD}}{m^3}$$

$$C_{out} := 5 \quad C_{pfr}(z) := root\left(K_{half} \cdot ln\left(\frac{C_{in}}{C_{out}} \right) + C_{in} - C_{out} - k \cdot \frac{z}{L_{FP}} \cdot \tau_{PFR}, C_{out} \right)$$

$$C_{pfr}(L_{FP}) = 5.39 \ \frac{g_{bCOD}}{m^3}$$

We compute the values of our target reactant concentration at the effluents of the visualized N_{count} reactors and created the stepped function, shown in Figure E9.6.2, which we can plot.

For the SF model, we generate the matrix of target reactant concentrations through each of the $n+1$ visualized reactors and produce the aggregate target concentration trace in much the same manner as for Example 9.5. The short codes are shown in Figure E9.6.3.

Again, relative to the TiS model we are met with a much poorer prediction of reactor performance from the SF model, owing to the nonlinearity of the rate law function and weighting of the concentration values of the visualized reactors of lesser residence times.

We plot the predicted concentration profiles from the PFR, N CMFRs, and SF models in Figure E9.6.4 for comparisons. We easily observe, in contrast to the behavior for the pseudo-first-order rate law, that prediction of reactor performance by the SF model for the saturation rate law is significantly poorer than those for the N CMFR predictions and certainly poorer than that predicted by the ideal PFR model. We attribute this to the nonlinearity of the rate law function. Reaction rates in the envisioned parallel reactors of lower residence time are proportionately lower than those in reactors of higher residence time.

$N_{count} := \text{trunc}(N) = 7 \qquad a := 1 \qquad C_0 := C_{in}$

$$C_{CMFRs} := \left| \begin{array}{l} \text{for } i \in 1..N_{count} \\ \quad \left| \begin{array}{l} b \leftarrow K_{half} + \dfrac{\tau_{SM}}{N_{count}} \cdot k - C_{i-1} \\ c \leftarrow -(K_{half} \cdot C_{i-1}) \\ C_i \leftarrow \dfrac{-b + \sqrt{b^2 - 4 \cdot a \cdot c}}{2 \cdot a} \end{array} \right. \\ C \end{array} \right.$$

$$C_{CMFRs_{N_{count}}} = 13.05 \ \frac{g_{bCOD}}{m^3}$$

$$C_{NCMFRs}(z) := \left| \begin{array}{l} step \leftarrow \dfrac{L_{FP}}{N_{count}} \\ i \leftarrow 1 + \text{trunc}\left(\dfrac{z}{step}\right) \\ i \leftarrow N_{count} \ \text{ if } \ i > N_{count} \\ C \leftarrow C_{CMFRs_i} \\ C \leftarrow C_{in} \ \text{ if } \ z = 0 \\ \text{return } C \end{array} \right.$$

FIGURE E9.6.2 *Screen captures of short logical programs for computing the concentration profile along the flow path of a real reactor using the TiS model.*

$k := 0..100 \qquad j := 0..n \qquad \theta_k := \overrightarrow{\dfrac{k}{100}} \qquad \tau_{z_{j,k}} := \overrightarrow{(t_j \cdot \theta_k)} \qquad z_{SF_k} := \overrightarrow{\dfrac{k \cdot L_{FP}}{100}}$

$$C_{SF} := \left| \begin{array}{l} \text{for } k \in 0..100 \\ \quad \text{for } j \in 0..n \\ \quad\quad \left| \begin{array}{l} C \leftarrow 10^{-15} \\ C_{SF_{j,k}} \leftarrow \text{root}\left(K_{half} \cdot \ln\left(\dfrac{C_{in}}{C}\right) + C_{in} - C - k_{SF} \cdot \tau_{z_{j,k}}, C\right) \end{array} \right. \\ \text{return } C_{SF} \end{array} \right.$$

$$C_{SF.z} := \left| \begin{array}{l} \text{for } k \in 0..100 \\ \quad C_{z_k} \leftarrow \dfrac{h}{3} \cdot \displaystyle\sum_{j=0}^{n} \left(E_{t_j} \cdot C_{SF_{j,k}} \cdot SR_j\right) \\ \text{return } C_z \end{array} \right.$$

$$C_{SF.z_{100}} = 35.7 \ \frac{g_{bCOD}}{L}$$

FIGURE E9.6.3 *Screen captures of short programs used to compute the profiles along* n+1 *parallel reactors of the SF model and collect them into an aggregate profile along the flow path of the real reactor.*

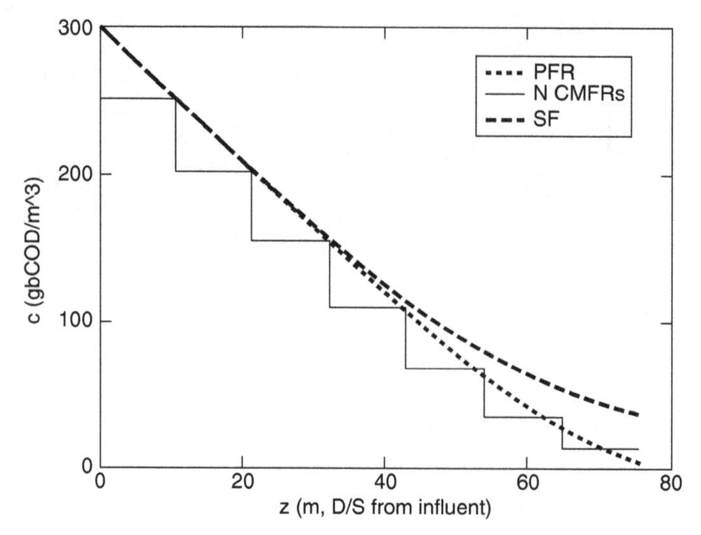

FIGURE E9.6.4 *A plot of the predicted concentration profiles across a PF-like reactor as predicted by the PFR, N-CMFRs, and SF models.*

Then, coupled with the skewed RTD density distribution, the reactors of lower residence time exert greater influence upon the effluent concentration than those of higher residence time.

We have observed and compared profiles of target reactant concentration through reactors employing the ideal PFR, N CMFRs in series, PFD and SF models. We have observed the differences in the predictions. At this point we might suggest that we are skeptical about the departure of the prediction using the SF model from that using the N CMFRs model. We certainly are not yet sure that the predictions from the ideal PFR are without merit. We will address use of these profiles to predict oxygen utilization along the flow path of PFR-like reactors in the next section.

9.11.2 Oxygen Consumption in PFR-Like Reactors

Here, we need some development regarding the prediction of the rate of oxygen consumption. In biological waste treatment, we often express the equivalent concentration of biodegradable organics as the ultimate biochemical oxygen demand (BOD_{ult}) or biodegradable chemical oxygen demand (bCOD). In theory, in the absence of inorganic oxygen demanding substances, these two parameters should be equal. Further, were we able to compute the theoretical oxygen demand (ThOD), we would find the three parameters to be essentially the same. Typically, we are not able to determine the theoretical oxygen demand of reactants in wastewaters to be treated. Most often degradable matter includes a suite of specific organic compounds, some of which have yet undefined formulae and structures. ThOD is, however, very useful in calibrating tests for chemical and biochemical oxygen demand.

Fortunately, investigators in the wastewater treatment area have examined biomass from wastewater treatment systems and determined the typical elemental content of the volatile portion of the biomass (VS). The most often published empirical formula for VS is $C_5H_7O_2N$. Biodegradable organic matter, manifest as BOD or COD is in part converted from its original state to biological cell matter. This portion of the original organic carbon, which is the energy source and carbon source for heterotrophic bacteria, when tied up in cellular mass, acts as an offset for the consumption of oxygen by the process. Thus, if we can compute the ThOD of VSS, we can determine the oxygen consumption offset associated with the net production of biomass.

For VS, we may write a half reaction for the oxidation of the organic carbon to carbon dioxide and couple that with a half reaction for the reduction of molecular oxygen to water. We will address redox half reactions in more detail in Chapter 12:

$$C_5H_7O_2N + 8H_2O \rightarrow 5CO_2 + 20e^- + 23H^+$$

$$5O_2 + 20e^- + 20H^+ \rightarrow 10H_2O$$

The sum of these reactions, after balancing the donated electrons with the accepted electrons, yields the stoichiometric relation we seek for the oxygen consumption offset:

$$C_5H_7O_2N + 5O_2 \rightarrow 5CO_2 + 3H^+ + 2H_2O$$

In this analysis, the most important aspect of the result is the ratio of five moles of oxygen offset per unit empirical VS formula. If we convert to mass, we find that the five moles of oxygen comprise 180 g and that the unit empirical formula for the VS is 113 g. The ratio (mass stoichiometry) is that 1.42 g theoretical oxygen demand (or bCOD or BOD_{ult}) is offset per gram VS produced.

We may return to Chapter 8 and use a combination of Equations 8.11 and 8.13 to produce a hybrid relation from which we may express the rate of oxygen consumption in terms of the rates at which biodegradable organic matter is utilized and VS is produced:

$$r_{O_2} = r_S + F_{O_2/VSS} r_{X.obs} \tag{9.23a}$$

r_{O_2} is the volume specific rate of oxygen consumption $\left(\frac{M_{O_2}}{V_R \cdot t}\right)$, $F_{O_2/VSS}$ is 1.42 g_{O_2}/g_{VSS}, and r_S and $r_{X.obs}$ retain the exact definitions of Chapter 8. Oxygen is a reactant and will be consumed, so we expect the rate will have a negative sign. In most cases, the net production of VS will be positive, as a product. However, in some cases, when biodegradable organic matter becomes scarce, the death/decay of biomass can occur at a rate higher than the true growth. Equation 9.23a is most useful if we employ units of biodegradable chemical oxygen demand (bCOD ≈ BOD_{ult} ≈ ThOD) as the unit of measure describing degradable organic substrate abundance.

We can populate the relation for the oxygen consumption rate using the RHSs of Equations 8.11 and 8.13 and perform some algebraic rearrangement to obtain convenient relations. We employ the pseudo-first-order approximations used in Example 9.3 to obtain one relation for the specific oxygen-utilization rate:

$$r_{O_2.\text{pfo}} = \left(F_{O_2/\text{VSS}} - \frac{1}{Y}\right)\left(\frac{\mu_{\max} XS}{K_{\text{half}}}\right) - F_{O_2/\text{VSS}} k_D X \tag{9.23b}$$

We then employ the saturation rate law of Example 9.4 to produce another relation for r_{O_2}:

$$r_{O_2.\text{sat}} = \left(F_{O_2/\text{VSS}} - \frac{1}{Y}\right)\left(\frac{\mu_{\max} XS}{K_{\text{half}} + S}\right) - F_{O_2/\text{VSS}} k_D X \tag{9.23c}$$

We can now address the point-wise distribution of oxygen consumption rates using profiles through PFR-like reactors of target reactant generated by our ideal and real reactor models.

Seemingly, we may employ Equations 9.23b and 9.23c directly with concentration profiles generated using the PFR, N CMFRs, and SF models to predict oxygen consumption along the reactor flow path. Such is truly the case for the PFR and TiS models. However, we must look deeper into the assumptions we have made with the SF model. We have subdivided the real reactor into n parallel reactors of equal length so we may correlate reactant concentrations and derived parameters with position across all n parallel reactors. In the limit as $n \rightarrow \infty$, the flow received by an arbitrary reactor i would be defined as follows:

$$Q_i = Q_{\text{Tot}} \cdot E(t_i)dt$$

The volume of the overall reactor is defined as the product of flow and statistical mean residence time:

$$V_R = Q_{\text{Tot}} \cdot \tau_{\text{SM}}$$

Then, the volume of the ith of the $n+1$ parallel reactors is similarly defined:

$$V_{R.i} = Q_i \cdot \tau_i = Q_{\text{Tot}} \cdot E(t_i)\, dt \cdot \tau_i$$

The fraction of the total volume comprising the ith reactor is then further defined:

$$\frac{V_{R_i}}{V_{R_{\text{Tot}}}} = \frac{Q_{\text{Tot}} \cdot E(t_i)dt \cdot \tau_i}{Q_{\text{Tot}} \cdot \tau_{\text{SM}}} = \frac{\tau_i}{\tau_{\text{SM}}} E(t_i)dt$$

Lastly, we require that all n reactors have flow path length of L and we choose to subdivide each of the reactors into an equal number of spatial steps. When we sum reaction rates, specific to reactor volumes, across the n reactors we must weight them by the volumes of the respective reactors. Then at any position z along the flow path of the overall reactor, we can integrate any volume-specific reaction rate across the n reactors at position z using the ratio $\frac{\tau_i}{\tau_{SM}}$ to weight the respective reactor volumes:

$$\bar{r}_z = \int_{i=0}^{n} \left(\frac{\tau_i}{\tau_{SM}} r_{z_i} E(t_i)dt \right) \tag{9.24}$$

where $\bar{r}_z$ is the weighted-average specific reaction rate at position z, and r_{z_i} is the computed specific reaction rate at position z in arbitrary reactor i. We can apply Equation 9.24 to any specific reaction rate operative along the flow path length across all of the n parallel reactors.

We are now ready to compute some oxygen consumption rates along the flow paths of our ideal and nonideal reactors.

Example 9.7 Continue with Example 9.6 and compute the profiles of oxygen consumption rates for the reactor performance predictions developed using the PFR, N CMFRs, and SF models.

We have the profiles of $C_{pfr}(z)$, $C_{NCMFRs}(z)$ and $C_{SF.z}(z)$ and the matrix of C_{SF} values from Example 9.6 that we can use with Equation 9.23b. Then, in order that we can compute the overall oxygen consumption rate by two distinct methods, we will also use these concentration predictions to compute associated predictions of biomass growth and use these in developing a check.

We compute the point-wise specific oxygen consumption rates for the ideal PFR and integrate over L_{FP} to obtain $R_{O_2.pfr}$:

$$r_{O2.pfr}(z) := \left(F_{O2.VSS} - \frac{1}{Y}\right) \cdot \frac{\mu_{max} \cdot X \cdot C_{pfr}(z)}{K_{half} + C_{pfr}(z)} - F_{O2.VSS} \cdot k_D \cdot X \qquad \frac{g_{O2}}{m^3 \cdot d}$$

$$A_X := H \cdot W_{CH} = 20.91 \quad m^2 \qquad R_{O2.PFR} := 2 \cdot \frac{A_X}{1000} \cdot \int_0^{L_{FP}} r_{O2.pfr}(z)\, dz = -5025 \quad \frac{kg_{O2}}{d}$$

We then compute the profile of VS production and integrate that along L_{FP} to obtain overall VS production were we to consider biomass growth:

$$R_{gX.PFR} := 2 \cdot \frac{A_X}{1000} \cdot \int_0^{L_{FP}} \left(\frac{\mu_{max} \cdot X \cdot C_{pfr}(z)}{K_{half} + C_{pfr}(z)} - k_D \cdot X \right) dz = 4.303 \times 10^3 \qquad \frac{kg_{VS}}{d}$$

Now we can use the difference between the effluent and influent concentrations and flow with the VS production for another prediction of the overall oxygen consumption rate:

$$R_{2.O2.pfr} := Q \cdot \frac{C_{pfr}(L_{FP}) - C_{in}}{1000} + F_{O2.VSS} \cdot R_{gX.PFR} = -5.026 \times 10^3 \qquad \frac{kg_{O2}}{d}$$

We observe reasonably good agreement between the two methods, validating them.

We now examine the stepped CMFRs data for predictions of R_{O_2}. Here, we compute the rate for each of the Ncount CMFRs in the series. The overall rate is the sum of the products of each r_{O_2} and the volume of each of the reactors. We also use the flow and difference between C_{out} and C_{in} and the overall production of VS for a second prediction:

$$r_{O2.CMFRs} := \left| \begin{array}{l} \text{for } i \in 1 .. N_{count} \\ \quad r_{O2_i} \leftarrow \left(F_{O2.VSS} - \frac{1}{Y} \right) \cdot \frac{\mu_{max} \cdot X \cdot C_{CMFRs_i}}{K_{half} + C_{CMFRs_i}} - F_{O2.VSS} \cdot k_D \cdot X \\ \text{return } r_{O2} \end{array} \right. \qquad \frac{g_{O2}}{m^3 \cdot d}$$

$$R_{O2.CMFRs} := \frac{V_R}{N_{count} \cdot 1000} \cdot \sum_{i=1}^{N_{count}} \left(r_{O2.CMFRs_i} \right) = -4.908 \times 10^3 \qquad \frac{kg_{O2}}{d}$$

$$R_{gX.CMFRs} := \frac{V_R}{N_{count} \cdot 1000} \cdot \sum_{i=1}^{N_{count}} \left(\frac{\mu_{max} \cdot X \cdot C_{CMFRs_i}}{K_{half} + C_{CMFRs_i}} - k_D \cdot X \right) = 4.194 \times 10^3 \qquad \frac{kg_{VS}}{d}$$

$$R_{2.O2.CMFRs} := Q \cdot \frac{C_{CMFRs_{N_{count}}} - C_{in}}{1000} + F_{O2.VSS} \cdot R_{gX.CMFRs} = -4.891 \times 10^3 \qquad \frac{kg_{O2}}{d}$$

We observe fairly close agreement between the two estimates. We also note that the overall consumption predicted by the N CMFRs approach yields a lower overall oxygen consumption rate than that for the PFR, consistent with the higher value of the effluent target reactant concentration. The N CMFRs in series approach using the stepped profile apparently yields accurate estimates of the oxygen consumption. However, the stepped profile is of lesser utility in point-wise predictions than would be a smooth prediction.

We turn our attention to the SF model predictions. We first generate the matrix of $r_{O_2 j,k}$ values using Equations 9.23b and 9.24:

$$j := 0..n \qquad k := 0..100$$

$$r_{O2_{j,k}} := \overrightarrow{\left[\left(F_{O2.VSS} - \frac{1}{Y}\right)\cdot\frac{\mu_{max}\cdot X\cdot C_{SF_{j,k}}}{K_{half} + C_{SF_{j,k}}} - F_{O2.VSS}\cdot k_D\cdot X\right]} \qquad \frac{g_{O2}}{m^3\cdot d}$$

We now integrate across the n parallel reactors to obtain the r_{O_2} versus z profile.

$$r_{O2.SF} := \left| \begin{array}{l} \text{for } k \in 0..100 \\ \quad r_{O2.z_k} \leftarrow \frac{h}{3}\cdot\sum_{j=0}^{n}\left(\frac{t_j}{t_{SM}}\cdot E_{t_j}\cdot r_{O2_{j,k}}\cdot SR_j\right) \quad \frac{g_{O2}}{m^3\cdot d} \\ \text{return } r_{O2.z} \end{array}\right.$$

We integrate this profile to obtain the overall oxygen consumption rate.

$$R_{O2.SF} := 2\cdot\frac{A_X}{1000}\cdot\frac{L_{FP}}{100\cdot 3}\cdot\sum_{k=0}^{100}\left(r_{O2.SF_k}\cdot SR2_k\right) = -4536 \qquad \frac{kg_{O2}}{d}$$

Then to check with our SF estimate, we first generate the matrix of $r_{GX.SF}$ and integrate that across the n parallel reactors to obtain the $r_{gXz.SF}$ versus z profile and integrate along L_{FP} to obtain $R_{gX.SF}$:

$$j := 0..n \quad k := 0..100 \qquad r_{gX.SF_{j,k}} := \overrightarrow{\left(\left(\frac{\mu_{max}\cdot X\cdot C_{SF_{j,k}}}{K_{half} + C_{SF_{j,k}}} - k_D\cdot X\right)\right)} \qquad \frac{g_{VS}}{m^3\cdot d}$$

$$r_{gXz.SF} := \left| \begin{array}{l} \text{for } k \in 0..100 \\ \quad r_{gz_k} \leftarrow \frac{h}{3}\cdot\sum_{j=0}^{n}\left(\frac{t_j}{t_{SM}}\cdot E_{t_j}\cdot r_{gX.SF_{j,k}}\cdot SR_j\right) \quad \frac{g_{VS}}{m^3\cdot d} \\ r_{gz} \end{array}\right.$$

$$R_{gX.SF} := 2\cdot\frac{A_X}{1000}\cdot\frac{L_{FP}}{100\cdot 3}\cdot\sum_{k=0}^{100}\left(r_{gXz.SF_k}\cdot SR2_k\right) = 3.85\times 10^3 \qquad \frac{kg_{VS}}{d}$$

We now combine the overall production of VS with the change in the target reactant concentration across the reactor to obtain a second estimate or $R_{O2.SF}$:

$$R_{2.O2.SF} := Q \cdot \frac{C_{SF.z_{100}} - C_{in}}{1000} + F_{O2.VSS} \cdot R_{gX.SF} = -4.522 \times 10^{3} \frac{kg_{O2}}{d}$$

We have not attained perfect agreement for the SF model, but certainly the two estimates are quite close. We also note that the overall oxygen consumption rate is lower than that predicted from the N CMFRs in series model, commensurate with the higher predicted value of the target reactant concentration by the SF model.

We are fully aware that the analyses performed in the previous two examples are simplistic and likely in error owing to the assumption that the biomass concentration along the reactor flow path is constant. Had we not embraced this assumption, the insightful analyses could not have been completed using closed-form analytic solutions for our PFR or SF reactor systems. Consideration of spatial variability due to biomass growth adds significant complexity to the effort. The CMFRs in series model, as a strictly algebraic solver, can be implemented for multiple reactions—we would simply employ a given-find solve block within the program for each reactor in the series, including relations arising from the mass balance on biomass.

The stepped N CMFRs in series approach is accurate, but of limited use owing to the stepped nature of the C versus z profile. Then by the process of elimination, we understand that our best option to accurately model processes within PFR-like reactors is through application of the segregated flow model. Examples hereafter, employing coupled substrate conversion and biomass growth will bear out that postulation.

9.12 MODELING UTILIZATION AND GROWTH IN PFR-LIKE REACTORS USING TiS AND SF

Much earlier we concluded that the PFD model would be more cumbersome than we would desire to employ with the saturation rate law owing to the necessity to iterate to a converged solution prior to "stepping into the reactor" for internal process analysis. Were we to have continued its use, we could have developed solutions alongside those employing the PFR, TiS, and SF models. At this point, in consideration of multiple reactions—the simplest case is that for bCOD uptake by heterotrophic biomass—we would have written a pair of coupled second-order ODEs each needing an iterative solution to define the exit concentrations of bCOD and VS prior to initiating any analyses of the internal process. We are content at this point to have analyzed the PFD model, its boundary conditions, and its application employing the pseudo-first-order rate law and to have left it behind.

In the previous section, we also concluded that developing a smoothed concentration profile arising from the TiS (N CMFRs in series) model would serve no good purpose beyond a false good feeling about having a point-wise rather than stepped distribution of target reactant abundance. We are also comfortable truncating noninteger values of N and adjusting the volume of the resultant identical CMFRs visualized to comprise the series. We are then left with employment of the TiS (stepped) and SF models. Of these two, only with the SF model can we produce point-wise distributions within a reactor. We will thus examine multiple reactions in nonideal reactors using these two models.

For the TiS model, we write the set of coupled relations for substrate (S, bCOD) uptake and biomass production (X, VS) as simultaneous algebraic relations and need merely to solve them. We have two choices: algebraically combine them to solve for S and then use S to solve for X; or, since we have a nonlinear equation solver available from MathCAD via a few clicks of a mouse, write a solve block into a function and use that function in a program to step through N CMFRs. Let us first write the system of equations. We will write them for the ith reactor of the N reactor series, directly employing the respective rate laws, and simplifying as appropriate:

$$\frac{S_{i-1} - S_i}{\tau_i} = \frac{\mu_{\max} \cdot S_i \cdot X_i}{Y \cdot (K_S + S_i)} \tag{9.25}$$

$$\frac{X_i - X_{i-1}}{\tau_i} = \left(\frac{\mu_{\max} \cdot S_i}{K_S + S_i} - k_D \right) X_i \tag{9.26}$$

Should we desire an explicit algebraic solution, we would most conveniently isolate X on the LHS of Equation 9.26, substitute the RHS for X in Equation 9.25, and collect the coefficients of S^2, S^1, and S^0. Final solution would be accomplished using the quadratic formula, ensuring that we take the positive root. Development of the precise relation is left as an exercise for the interested student. Hereafter, we will write the given-find block and employ it in a MathCAD function.

For the SF model, we must simultaneously solve for both S and X through each of the n parallel PFRs. A pair of coupled first-order ODEs comprises our set of relations. We employ Figure 8.2 and consider both substrate and biomass. En route, we very conveniently employ $\tau_z = \frac{z}{v_s}$ and can write the mass balances using τ_z as the independent variable:

$$\frac{dS_z}{d\tau_z} = \frac{\mu_{\max} \cdot S_z \cdot X_z}{Y \cdot (K_S + S_z)} \tag{9.27}$$

$$\frac{dX_z}{d\tau_z} = \left(\frac{\mu_{\max} \cdot S_z}{K_S + S_z} - k_D \right) X_z \tag{9.28}$$

We will not find a closed-form analytic solution anywhere in the mathematical literature for this pair of ODEs. We must turn to numerical solvers. Earlier, we illustrated the concept of a numerical solution of the set of ODEs arising from application of the saturation rate law with the PFD model. We employed Euler's method therein. Euler's method would be easily programmed but would require an extremely small independent variable step to develop accurate numerical approximations for this ODE system. We will not employ Euler's method, but will opt for the fourth-order (RK4) method. This numerical solution technique employs an efficient means to compute the change in the dependent variable for each step of the independent variable, allowing for many fewer steps. With the added complexity, the programming is more involved. The reader is directed to the many good texts (e.g., Carnahan et al., 1969) addressing numerical methods for details regarding the RK4 methodology. We will not include these details herein. Fortunately, MathCAD has two solvers employing the RK4 methodology—one employing a fixed independent variable step and one employing a variable independent variable step. The fixed-step method is our best choice for the applications to be addressed herein. Let us return to the reactor and conditions of Examples 9.6 and 9.7 and implement the TiS model to solve for the point-wise specific oxygen-utilization rate.

Example 9.8 Employ the N CMFRs in series model with the coupled relations for substrate uptake and biomass growth for the reactor and conditions of Examples 9.6 and 9.7. Consider that the specified *X*, considered constant in Examples 9.6 and 9.7, is the influent value of the biomass abundance herein.

We have the parameter assignments in Example 9.6 and we will not repeat these. So we can begin by truncating *N*, defining the volumes of the reactors in the series, and writing our given-find solve block into a function, taking care to include the influent *S* and *X* values as well as the effluent values as arguments:

$$NiN := \frac{t_{SM}^2}{\sigma_S} = 7.134 \qquad N := \mathrm{trunc}(NiN) = 7 \qquad \tau_N := \frac{\tau_{SM}}{N} = 0.012 \quad d$$

$$\begin{pmatrix} S \\ X \end{pmatrix} := \begin{pmatrix} 0.9 \cdot S_{in} \\ 1.05 \cdot X_{in} \end{pmatrix} \begin{pmatrix} \dfrac{g_{bCOD}}{m^3} \\ \dfrac{g_{VS}}{m^3} \end{pmatrix} \qquad \text{given} \qquad \frac{S_{in} - S}{\tau_N} = \frac{\mu_{max} \cdot S \cdot X}{Y \cdot (K_S + S)}$$

$$\frac{X - X_{in}}{\tau_N} = \left(\frac{\mu_{max} \cdot S}{K_S + S} - k_D\right) \cdot X \qquad TiS(S_{in}, X_{in}, S, X) := \mathrm{find}(S, X)$$

We assemble a program (shown in Figure E9.8.1) implementing the TiS() function for each of the reactors in the series. We find it most efficient to compute the volume specific and overall oxygen consumption rates right along with substrate and biomass concentrations.

Note that S_0 and X_0 are indexed using the (left bracket) matrix definition rather than the formatting (.) subscript. We output the exiting substrate and biomass concentrations and overall oxygen consumption rate. We then compute the overall

```
(S   )      | S_0 ← S_in
(X   )  :=  | X_0 ← X_in
(r_O2)      | sum ← 0
(R_O2)      | for i ∈ 1.. N
            |    | S_i ← 0.9·S_(i-1)
            |    | X_i ← 1.05·X_(i-1)
            |    | (S_i)
            |    | (X_i) ← TiS(S_(i-1), X_(i-1), S_i, X_i)
            |    | r_O2_i ← (-μmax·S_i·X_i)/(Y·(K_S + S_i)) + F_O2.VS·[((μmax·S_i)/(K_S + S_i) - k_D)·X_i]   g_O2/(m^3·d)
            |    | sum ← sum + r_O2_i·(V_R/N)
            | R_O2 ← sum/1000
            |                     (S   )
            | return              (X   )
            |                     (r_O2)
            |                     (R_O2)
```

FIGURE E9.8.1 Screen capture of the MathCAD code used to compute substrate and biomass levels as well as specific substrate utilization rates to model utilization, growth and oxygen consumption as predicted by the N-CMFRs in series model.

oxygen consumption rate from the overall changes in substrate and biomass abundances and compare the result with that computed internally:

$$S_N = 5.61\ \frac{g_{bCOD}}{m^3} \qquad X_N = 713.3\ \frac{g_{VS}}{m^3} \qquad R_{O2} = -5.056\times10^3\ \frac{kg_{O2}}{d}$$

$$R_{2.O2} := Q\cdot\left(\frac{S_N - S_0}{1000} + F_{O2.VS}\cdot\frac{X_N - X_0}{1000}\right) = -5.048\times10^3\ \frac{kg_{O2}}{d}$$

Although not in exact agreement, we are satisfied with the result. We must bear in mind that the given-find solve block is a numerical approximation. Were we to invest the effort in adjusting conversion tolerances, we might attain a closer agreement between the two computations.

Then, to complete our work with this application of the TiS model, we write functions for S, X, and r_{O_2} with position, using programs such as that written for

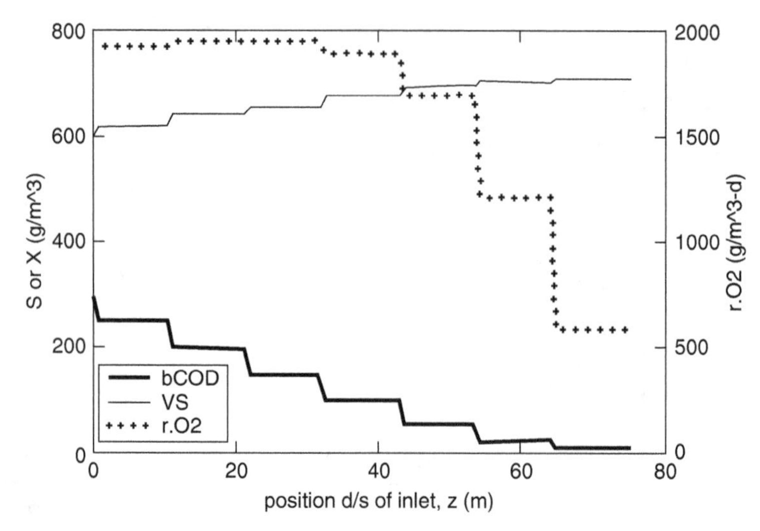

FIGURE E9.8.2 A plot of predicted substrate and biomass levels and oxygen consumption rates along the flow path of a reactor modeled using the N-CMFRS in series model.

Example 9.6, and plot them in Figure E9.8.2 for visual inspection. At best, from application of the TiS model for this reactor system and associated conditions, our resolution for arranging an aeration system to supply oxygen to this reactor would have seven regions. Beyond those results, we would need to guess.

We can now certainly employ the SF model and compare the generated results with those of Example 9.8.

Example 9.9 Employ the segregated flow model with the coupled substrate uptake and biomass production relations describing the processes within a PFR-like reactor. Use the reactor and conditions of Examples 9.6–9.8. Compare the results with those of the TiS model predictions of Example 9.8.

We first arrange a solution visualizing the reactor as an ideal PFR. We would like to have this when we compare the predictions of the SF model with those of the TiS model. All of our parameters are available from Examples 9.6–9.8, and we may dig directly into the numeric solution for the ideal PFR model.

We define the boundary conditions that for the implementation are treated as initial conditions, and then define the system of ODEs we intend to solve and implement the RK4 solution using MathCAD's rkfixed() function. The arguments define, in turn, the vector of dependent variables, the initial value of the independent variable, the ending value of the independent variable, the number of steps to be used in the numeric solution, and the vector of RHS functions for the solution. We used y as the independent variable in deference to the typical mathematical nomenclature. We instructed MathCAD to assign the rkfixed solution into the matrix Sol. The zeroth column of Sol contains the vector of progressive residence times, the first column contains the vector of substrate concentrations and the second column contains the vector of biomass concentrations.

We parse the Sol matrix into the three respective vectors. Also, for the sake of comparison, we have outputted the effluent values for substrate and biomass:

$$y := \begin{pmatrix} C_{in} \\ X_{in} \end{pmatrix} \quad D(\tau, y) := \begin{bmatrix} \dfrac{-\mu_{max} \cdot y_0 \cdot y_1}{Y \cdot (K_{half} + y_0)} \\ \left(\dfrac{\mu_{max} \cdot y_0}{K_{half} + y_0} - k_D \right) \cdot y_1 \end{bmatrix} \quad \begin{array}{l} Sol := \text{rkfixed}(y, 0, \tau_{SM}, 100, D) \\ \tau_{pfr} := Sol^{\langle 0 \rangle} \quad S := Sol^{\langle 1 \rangle} \quad X := Sol^{\langle 2 \rangle} \end{array}$$

$$S_{100} = 0.865 \ \frac{g_{bCOD}}{m^3} \qquad X_{100} = 715 \ \frac{g_{VS}}{m^3}$$

We compute the vector of specific oxygen consumption rates, create a vector of position values, and plot S, X, and r_{O_2} versus position in Figure E9.9.1. We observe a behavior similar to that predicted by the TiS model, except that the curves are smooth, and would be displaced from those predicted by the TiS model. We integrate r_{O_2} over the volume of the reactor to obtain one estimate and compute the second estimate of the overall oxygen consumption rate using the influent and effluent substrate and biomass values:

$$R_{O2.1} := \frac{2 \cdot A_X}{1000} \cdot \frac{L_{FP}}{100 \cdot 3} \cdot \sum_{k=0}^{100} \left(r_{O2_k} \cdot SR_{2_k} \right) = -5.13 \times 10^3$$

$$R_{O2.2} := \frac{Q}{1000} \cdot (S_{100} - C_{in}) + \frac{Q}{1000} \cdot (X_{100} - X_{in}) \cdot F_{O2.VS} = -5.123 \times 10^3$$

The agreement is certainly within the tolerances acceptable for these numerical approximations.

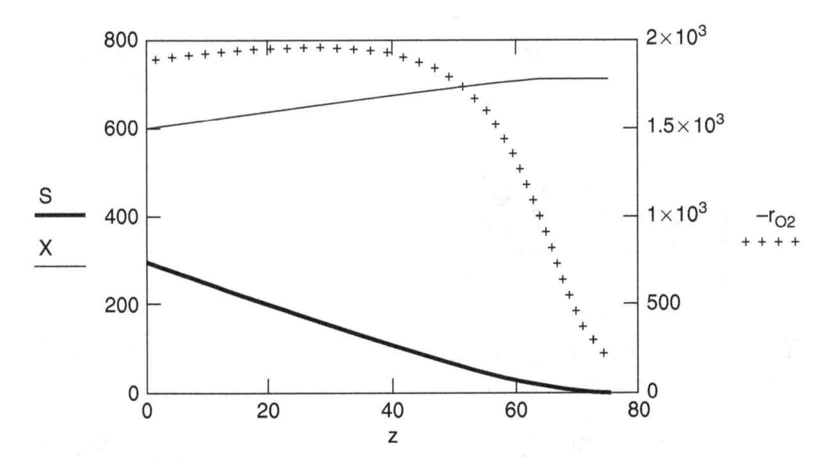

FIGURE E9.9.1 *A plot of substrate and biomass concentrations and specific oxygen consumption rates along the flow path of a PF-like reactor modeled using the ideal PFR model.*

We now turn attention to the SF model predictions. We have defined the boundary conditions and the vector of RHS functions earlier so that we can get right to the implementation of the RK4 method. Our program implements rkfixed() for each of the n reactors in turn. Then for each reactor, we collect the S and X values for the current reactor before proceeding to the next reactor of the parallel set:

```
(S_SF)  :=  | for j ∈ 0.. n
(X_SF)      |   | Sol ← rkfixed(y, 0, τ_j, 100, D)
            |   | for k ∈ 0.. 100
            |   |   | S_SF[j,k] ← (Sol<1>)_k
            |   |   | X_SF[j,k] ← (Sol<2>)_k
            | return (S_SF)
            |        (X_SF)
```

We find it most convenient to generate the matrix of r_{O_2} values once we have completed the looping operations and we include the weighting factor:

```
r_O2 :=  | for k ∈ 0.. 100
         |   for j ∈ 0.. n
         |     r_O2[j,k] ← τ_j·[(F_O2.VS − 1/Y)·(μ_max·X_SF[j,k]·S_SF[j,k])/(K_half + S_SF[j,k]) − F_O2.VS·k_D·X_SF[j,k]] / τ_SM
         | return r_O2
```

$$r_{O2_{j,k}} \leftarrow \frac{\tau_j \cdot \left[\left(F_{O2.VS} - \frac{1}{Y}\right) \cdot \frac{\mu_{max} \cdot X_{SF_{j,k}} \cdot S_{SF_{j,k}}}{K_{half} + S_{SF_{j,k}}} - F_{O2.VS} \cdot k_D \cdot X_{SF_{j,k}}\right]}{\tau_{SM}}$$

We then generate the reactor profile by integration across the n reactors at each position z:

```
r_O2.z :=  | for k ∈ 0.. 100                                        gO2/(m^3·d)
           |   r_O2.SF.z_k ← (h/3)·Σ_{j=0}^{n} (E_tj·r_O2[j,k]·SR_j)
           | return r_O2.SF.z
```

$$r_{O2.SF.z_k} \leftarrow \frac{h}{3} \cdot \sum_{j=0}^{n} \left(E_{t_j} \cdot r_{O2_{j,k}} \cdot SR_j\right) \qquad \frac{g_{O2}}{m^3 \cdot d}$$

Then we integrate along the flow path of the reactor to obtain our estimate of the overall oxygen consumption rate:

$$R_{O2} := 2 \cdot \frac{A_X}{1000} \cdot \frac{L_{FP}}{100 \cdot 3} \cdot \sum_{k=0}^{100} \left(r_{O2.z_k} \cdot SR2_k\right) = -4.677 \times 10^3 \quad \frac{kg_{O2}}{d}$$

We generate the profile of substrate and biomass concentrations along the flow path and output the effluent values. We can then check our internally generated oxygen consumption result against that employing the flow and influent and effluent substrate and biomass concentrations:

$$\begin{pmatrix} S_{SF.z} \\ X_{SF.z} \end{pmatrix} := \left| \begin{array}{l} \text{for } k \in 0..100 \\ \quad \left| \begin{array}{l} C_{SF.z_k} \leftarrow \frac{h}{3} \cdot \sum_{j=0}^{n} \left(E_{t_j} \cdot S_{SF_{j,k}} \cdot SR_j\right) \\ X_{SF.z_k} \leftarrow \frac{h}{3} \cdot \sum_{j=0}^{n} \left(E_{t_j} \cdot X_{SF_{j,k}} \cdot SR_j\right) \end{array} \right. \\ \text{return } \begin{pmatrix} C_{SF.z} \\ X_{SF.z} \end{pmatrix} \end{array} \right.$$

$$S_{SF.z_{100}} = 28.6 \; \frac{g_{bCOD}}{m^3} \qquad X_{SF.z_{100}} = 704 \; \frac{g_{VS}}{m^3}$$

$$R_{O2.2} := Q \cdot \left(\frac{S_{SF.z_{100}} - S_{SF.z_0}}{1000} + \frac{X_{SF.z_{100}} - X_{SF.z_0}}{1000} \cdot F_{O2.VS} \right) = -4.6702 \times 10^3 \; \frac{kg_{O2}}{d}$$

We are pretty satisfied with this agreement. As we did for the TiS and PFR model predictions, in Figure E9.9.2 we plot S, X, and r_{O_2} against position for comparison. We have included the predictions from the PFR model for addi-

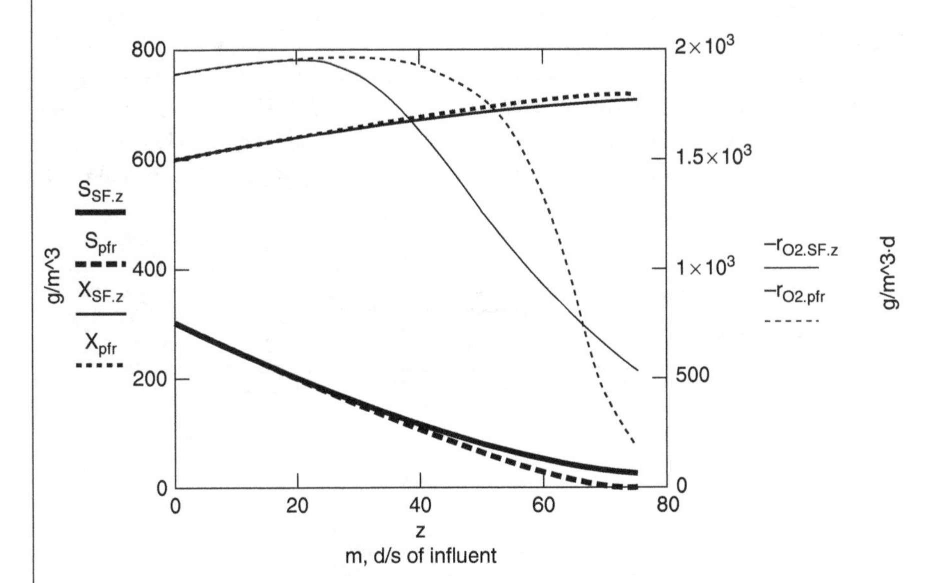

FIGURE E9.9.2 *A Plot of predicted substrate and biomass concentrations and specific oxygen consumption rates along the flow path of a PR-like reactor modeled using the ideal PFR and SF models.*

tional comparisons. We observe fairly close agreement between the SF and PFR model predictions for X but when we examine the rate of oxygen consumption we become keenly aware that the SF model predicts lower rates initially and higher rates near the effluent. Finally, we are also keenly aware that the SF predicts much poorer performance relative to the overall reduction of influent bCOD than does the PFR or the TiS model.

We need one last implementation of the segregated flow model. The comparison plot of Example 9.9 applies to the predictions from application of the PFR and SF models for what amounts to identical input conditions. We should examine a comparison between predictions from the two models to discern conditions leading to identical predicted performance. The variable with which we would choose to tinker to obtain the equality of performance prediction is the influent biomass. We could manually, by trial and error, adjust the value of X for the SF prediction of Example 9.9 until the effluent bCOD concentration was predicted to be equal to that of the PFR prediction. By such action we would still be a step away from a real PFR-like reactor. In practice, the influent biomass concentration is adjusted by recycling biomass separated from the effluent by a secondary clarifier. By recycling the biomass, the influent wastewater stream is affected. The actual influent flow to the reactor is increased and the actual influent substrate concentration is diluted in conjunction with the increase in the influent biomass. Here we have an opportunity to couple the zero-volume mixing principle with those of the PFR, TiS and SF reactor models to create an overall reactor model that better represents the real system than those we have investigated previously.

We begin by assembling a schematic of the overall system shown in Figure 9.11 (adapted from the schematic developed with Example 8.6). The clarifier underflow stream contains 99+ % of the biomass that flows from the reactor such that the X_{eff} concentration is most often as low as a few g/m^3. Then, in order that the inventory (quantity) of biomass residing within the system can be held constant, a small fraction of the underflow stream is routed to solids treatment. The remainder

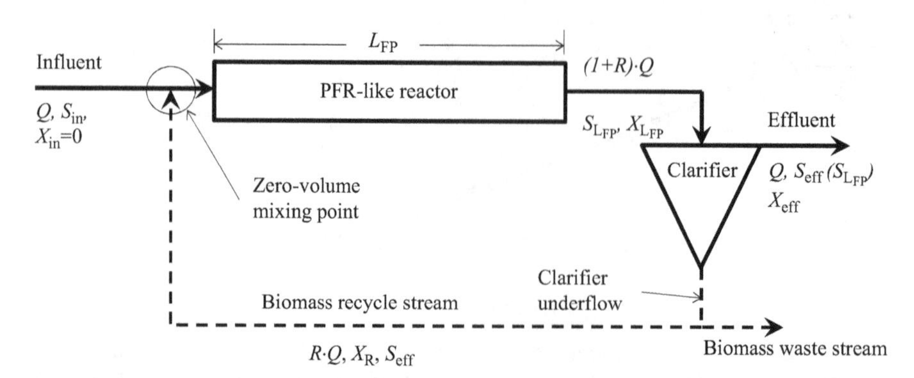

FIGURE 9.11 *Schematic of PF-like reactor system with clarifier and biomass recycle stream.*

becomes the biomass recycle stream and joins the influent to the reactor upstream of the reactor entrance. The concentration of biomass in the influent to the reactor is controlled by the recycle ratio, R, the ratio of the recycle flow to the influent flow. Then, the actual influent flow to (and effluent flow from) the reactor is $Q\cdot(1+R)$. We must account for this when modeling the process operative in the reactor. Both the influent biomass concentration and influent substrate concentration are computed using the zero-volume mixing principle addressed in Chapter 7. Given this picture of the overall system, we can delve into the process operative within the reactor.

Example 9.10 Continue with the reactor and system of Examples 9.5–9.9 and develop numerical models of the system employing the zero-volume mixing principle combined with the plug-flow, segregated flow, and CMFRs in series models to determine the conditions (e.g., the recycle ratio and associated reactor biomass levels) necessary to yield a predicted effluent concentration equivalent to that of the PFR model of Example 9.9. *A precursor analysis for a PFR with recycle is performed in Example 8.6.*

We need a few new pieces of information. Solids concentrations in the underflows of well-operated secondary clarifiers following well-operated suspended-growth reactors should easily be as high as 1% (10,000 g/m^3) The solids produced by suspended growth processes operative in PFR-like reactors should contain about 85% volatile matter. Thus, we will assume the concentration of biomass in the recycle stream, X_R, is 8500 g$_{VS}$/m^3. We realize that the addition of the recycle stream to the influent stream will increase the influent flow, decreasing the hydraulic residence time. Perhaps variations in the recycle ratio would lead to variations in the RTD. There are means to deal with this variation, but these would lead to an undue computational as well as conceptual burden at this time. We will assume that we would have the n parallel reactors arising from the original $C(t)$ versus t curve generated by the impulse stimulus and that the fraction of flow received by each of these n parallel reactors remains constant but that the overall residence time is decreased in accord with the increase in total flow.

We will address the ideal PFR first, and write a single program that will iterate from a recycle ratio of zero to the recycle ratio at which the effluent concentration equals the value predicted from the PFR model in Example 9.9. We have taken a picture of that program and inserted it herein. We have also annotated the program to explain the various operations. The program (shown in Figure E9.10.1) steps through the reactor for $R=0$ to get the computation started and then employs a while loop to increment the value of R until the target effluent concentration is met. Then, for the SF model we have updated the program (shown in Figure E9.10.2) to step through the n parallel reactors and generate matrices of S and X values along each of the reactors and return the critical value of R.

$$\begin{pmatrix} S_{PFR} \\ X_{PFR} \\ R_{PFR} \\ \tau_{PFR} \end{pmatrix} :=$$

```
R ← 0                                   We start with a no-recycle
C_0 ← C_in                              reactor
X_0 ← 0
y ← (C_0, X_0)                          We define the initial
                                        conditions
D(t,y) ← [ -μmax·y_0·y_1 / (Y·(K_half + y_0)) ;     We define the pair of
           (μmax·y_0/(K_half + y_0) - k_D)·y_1 ]    coupled ODEs
S ← rkfixed(y, 0, τ_SM, 100, D)         We implement the RK4
                                        method, fixed step, to step
                                        through the PFR
C_out ← (S^<1>)_100
while C_out > C_target                  We are pretty much assured
                                        that we'll get to increment R
    R ← R + ΔR                          Now we begin incrementing R,
                                        until the effluent is less than
                                        the target value.
    C_0 ← (R·C_out + C_in)/(1 + R)
    X_0 ← R·X_R/(1 + R)                 We mix the recycle and
                                        influent streams to get the
                                        influent values for C and X.
    y ← (C_0, X_0)
    τ_R ← τ_SM/(1 + R)                  We adjust the residence time
                                        of the reactor
    D(t,y) ← [ -μmax·y_0·y_1 / (Y·(K_half + y_0)) ;     We define the pair of
               (μmax·y_0/(K_half + y_0) - k_D)·y_1 ]    coupled ODEs
    S ← rkfixed(y, 0, τ_R, 100, D)      We again implement the RK4
                                        method, fixed step, to step
                                        through the PFR
    for k ∈ 0..100
        τPFR_k ← (S^<0>)_k              We transfer the columns of
        C_k ← (S^<1>)_k                 the S matrix into the τ, C and
                                        X vectors
        X_k ← (S^<2>)_k
    C_out ← C_100                       We ensure that the 'while'
                                        function has a comparison on
                                        which to base decisions about
                                        further iteration
return (C, X, R, τ_PFR)                 We take the C and X vectors
                                        and the recycle fraction back
                                        to the main worksheet for
                                        further computations
```

FIGURE E9.10.1 *Screen capture of a MathCAD program for implementation of the fixed-step Runge–Kutta ODE solver for approximation of the solution of coupled biomass growth and substrate utilization along the flow path of a PF-like reactor with cell recycle to control the biomass inventory of the reactor, modeled as an ideal PFR.*

```
(S, X, R, τ) := | R ← 0                                         // We start with a no-recycle reactor
                | C0 ← Cin
                | X0 ← 0
                | y ← (C0, X0)                                  // We define the initial conditions
                | D(t, y) ← [ −μmax·y0·y1 / (Y·(Khalf + y0)) ;
                |             (μmax·y0/(Khalf + y0) − kD)·y1 ]  // We define the pair of coupled ODEs
                | for j ∈ 0.. n
                |   | τj ← (t→)j / 1440                         // We implement the RK4 method, fixed step, to step through the PFR
                |   | S ← rkfixed(y, 0, τj, 100, D)
                |   | for k ∈ 0.. 100
                |   |   | CSF[j,k] ← (S<1>)k                   // We choose to work with C and X vectors rather than columns of the Solution matrix.
                |   |   | XSF[j,k] ← (S<2>)k
                | Cout ← (h/3)·Σ(j=0..n) (Etj·CSF[j,100]·SRj)  // We have our first estimate of Cout, hopefully still larger than Ctarget.
                | while Cout > Ctarget                          // Now we begin incrementing R, until the effluent is less than the target value
                |   | R ← R + ΔR
                |   | C0 ← (R·Cout + Cin)/(1 + R)               // We mix the recycle and influent streams to get the influent values for C and X.
                |   | X0 ← R·XR/(1 + R)
                |   | y ← (C0, X0)
                |   | D(t, y) ← [ −μmax·y0·y1 / (Y·(Khalf + y0)) ;
                |   |             (μmax·y0/(Khalf + y0) − kD)·y1 ]  // We define the pair of coupled ODEs
                |   | τj ← tj / (1440·(1 + R))                  // We adjust the residence times of the n reactors
                |   | for j ∈ 0.. n
                |   |   | S ← rkfixed(y, 0, τj, 100, D)          // We again implement the RK4 method, fixed step, to step through the PFR
                |   |   | for k ∈ 0.. 100
                |   |   |   | CSF[j,k] ← (S<1>)k                 // We create the jth row of the full C and X matrices for the event that Cout < Cin.
                |   |   |   | XSF[j,k] ← (S<2>)k
                |   | Cout ← (h/3)·Σ(j=0..n) (Etj·CSF[j,100]·SRj)  // We compute the estimate for Cout for each R. If Cout < Ctarget, we'll exit the while
                | return (CSF, XSF, R, τ)                       // We take the C and X matrices and the recycle fraction back to the main worksheet for further computations
```

FIGURE E9.10.2 *Screen capture of a MathCAD program for implementation of the fixed-step Runge–Kutta ODE solver for approximation of the solution of coupled biomass growth and substrate utilization along the flow path of a PF-like reactor with recycle to control the biomass inventory of the reactor, modeled using the SF model.*

For the CMFRs in series model, we write a given-find block that will solve the first reactor in the series, requiring three equations and yielding the influent bCOD to and effluent bCOD and VS from the first CMFR:

$$\begin{pmatrix} S \\ X \end{pmatrix} := \begin{pmatrix} 0.9 \cdot S_{in} \\ 1.05 \cdot X_{in} \end{pmatrix} \begin{pmatrix} \frac{g_{bCOD}}{m^3} \\ \frac{g_{VS}}{m^3} \end{pmatrix} \qquad X_{in} := \frac{R \cdot X_R}{1+R} \qquad \tau_{N.R} := \frac{\tau_N}{1+R} \qquad \text{Given}$$

$$S_{in} := \frac{C_{in} + R \cdot S}{1+R} \qquad \frac{S_{in} - S}{\tau_{N.R}} = \frac{\mu_{max} \cdot S \cdot X}{Y \cdot (K_S + S)} \qquad \frac{X - X_{in}}{\tau_{N.R}} = \left(\frac{\mu_{max} \cdot S}{K_S + S} - k_D\right) \cdot X$$

$$TiS_1(R, \tau_{N.R}, X_{in}, S_{in}, S, X) := \text{Find}(S_{in}, S, X)$$

Thereafter, since each set of input bCOD and VS values is known the solve block need only contain two equations:

$$\begin{pmatrix} S \\ X \end{pmatrix} := \begin{pmatrix} 0.9 \cdot S_{in} \\ 1.05 \cdot X_{in} \end{pmatrix} \begin{pmatrix} \frac{g_{bCOD}}{m^3} \\ \frac{g_{VS}}{m^3} \end{pmatrix} \qquad \text{Given} \qquad \frac{S_{in} - S}{\tau_N} = \frac{\mu_{max} \cdot S \cdot X}{Y \cdot (K_S + S)}$$

$$\frac{X - X_{in}}{\tau_N} = \left(\frac{\mu_{max} \cdot S}{K_S + S} - k_D\right) \cdot X \qquad TiS(\tau_N, S_{in}, X_{in}, S, X) := \text{Find}(S, X)$$

We assemble these two solve blocks in a program similar to those addressing PFR and SF recycle reactors (shown in Figure E9.10.3). The vectors of substrate and biomass, the scalar recycle ratio, and the vector of specific oxygen consumption rates are generated. Once the S, X, and r_{O_2} vectors are computed for the PFR, SF, and TiS models, using methods employed in previous examples, we can produce plots.The comparisons are most clearly illustrated if we plot substrate, biomass, and specific oxygen consumption rate in three distinct plots. First, we address substrate in Figure E9.10.4. Even though all models predict the exact same effluent substrate concentration, we observe large differences in the profiles. The stepped profile from the CMFRs in series model falls between those of the SF and PFR models.

The plot of biomass abundance versus position of Figure E9.10.5 also tells us an interesting story. The required recycle ratios, R_{PFR}, $R_{SF,}$ and R_{TiS}, are predicted to be 0.079, 0.174, and 0.102, respectively. Not surprisingly, the required biomass levels predicted from the SF model are significantly greater than predicted from the

```
( S     )
( X     )      := | R ← 0                                   We initialize R
( R_TiS )         | C_out ← C_in                            We ensure iteration
( r_O2  )         | while C_out > C_target
                  |   | R ← R + ΔR                          We increment R
                  |   | τ_N.R ← τ_N / (1 + R)               We adjust HRT
                  |   | X_0 ← R·X_R / (1 + R)               We compute influent
                  |   |                                     biomass
                  |   | S_0 ← (C_in + R·C_out) / (1 + R)
                  |   | S_1 ← 0.9·S_0
                  |   | X_1 ← 1.05·X_0
                  |   | ( S_0 )                             We solve for the
                  |   | ( S_1 ) ← TiS_1(R, τ_N.R, X_0, S_0, S_1, X_1)   influent bCOD and
                  |   | ( X_1 )                             effluent bCOD and
                  |   |                                     VS
                  |   | r_O2_1 ← -μmax·S_1·X_1 / (Y·(K_S + S_1)) + F_O2.VS·[(μmax·S_1 / (K_S + S_1) - k_D)·X_1]  g_O2/(m^3·d)
                  |   | for i ∈ 2.. N
                  |   |   | S_in ← S_(i-1)
                  |   |   | S_i ← 0.9·S_(i-1)
                  |   |   | X_i ← 1.05·X_(i-1)              We solve for each
                  |   |   | ( S_i )                         successive bCOD
                  |   |   | ( X_i ) ← TiS(τ_N.R, S_(i-1), X_(i-1), S_i, X_i)   and VS
                  |   |   | r_O2_i ← -μmax·S_i·X_i / (Y·(K_S + S_i)) + F_O2.VS·[(μmax·S_i / (K_S + S_i) - k_D)·X_i]  g_O2/(m^3·d)
                  |   | C_out ← S_N                         We ensure a
                  |                                         comparison
                  |          ( S    )                       We return the
                  | return   ( X    )                       results to the main
                  |          ( R    )                       worksheet
                  |          ( r_O2 )
```

FIGURE E9.10.3 *Screen capture of a MathCAD program for implementation of the CMFRs in series model to predict the coupled biomass growth and substrate utilization along the flow path of a PF-like reactor with recycle to control the biomass inventory of the reactor.*

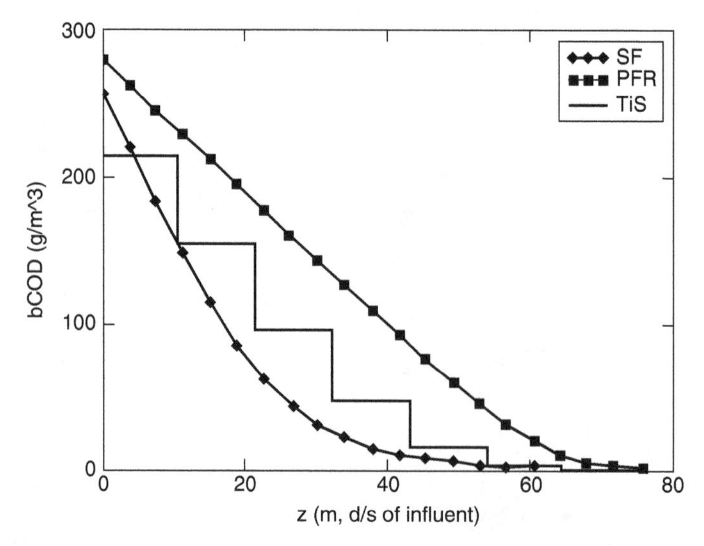

FIGURE E9.10.4 *A plot showing predicted substrate concentrations along the path of a PF-like reactor as modeled using the PFR, N-CMFRs in series and SF models.*

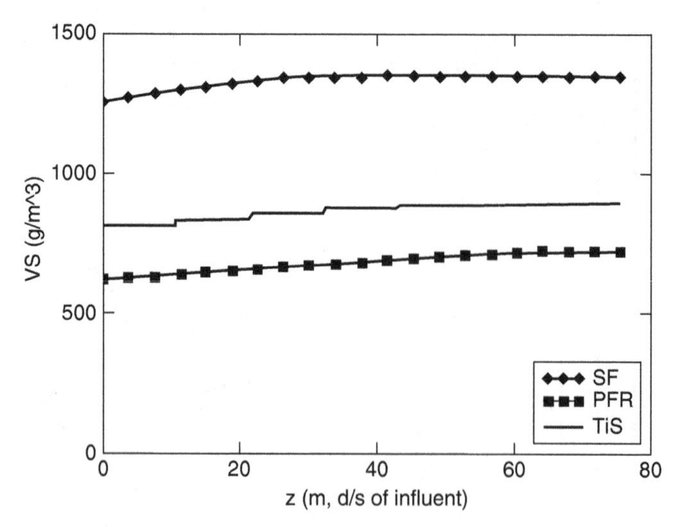

FIGURE E9.10.5 *A plot showing predicted biomass abundance along the path of a PF-like reactor as modeled using the PFR, N-CMFRs in series and SF models.*

CMFRs in series model and, of course, the required biomass level predicted by the CMFRs in series model is a good bit higher than that predicted by the PFR model.

The most telling disparity, which can be observed in Figure E9.10.6, among the predictions from the three models arises with the specific oxygen

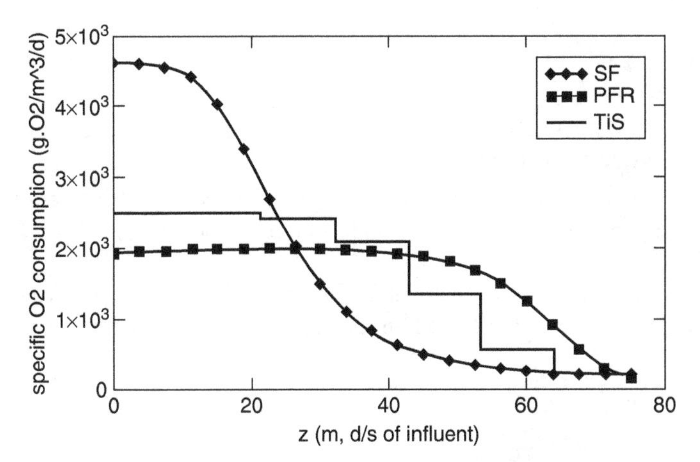

FIGURE E9.10.6 *A plot showing predicted specific oxygen consumption rates along the path of a PF-like reactor as modeled using the PFR, N-CMFRs in series and SF models.*

consumption rate. Predictions from the PFR model suggest nearly level oxygen consumption for the first two-thirds of the reactor with a gradual tapering-off in the last one-third. Specific oxygen consumption rates predicted by the TiS model suggest level consumption for the first 60% of the reactor with gradual tapering-off in the last 40%. Somewhat surprisingly, predictions from the SF model are extreme in the first fifth of the reactor, with tapering-off in the next two-fifths of the reactor and very level, low rates of consumption in the last two-fifths of the reactor. If we believe the prediction from the SF model, we will concentrate our aeration capacity in the front end of the reactor.

We will conclude our discussion into nonideal reactors with completion of this iterative analysis. Certainly this last piece of analysis has major implications for consideration in modeling biological processes in PFR-like reactors.

The principles and processes examined in Examples 9.5 through 9.10 with regard to the analysis of biological reactors employing oxygen as the terminal electron acceptor (i.e., all the various named process modifications of the activated sludge process) can be extended merely by adding the various reactions such as

Release of organic nitrogen (mostly primary amines) from organic matter during catabolism of organic matter to form the intermediates needed for production of cell mass

Conversion of ammonia-nitrogen, as an energy source, via nitrification to nitrate nitrogen

Net growth of autotrophic biomass in conjunction with utilization of ammonia nitrogen

Conversion of organic carbon to carbon dioxide or cell mass by facultative bacteria using organic carbon as carbon and energy sources and nitrate as a terminal electron acceptor

Net growth of facultative bacteria in conjunction with organic carbon utilization with nitrate as the terminal electron acceptor

Luxury uptake of phosphorus by anaerobically conditioned bacteria in systems operated for enhanced biological phosphorus removal

Certainly additional specific reactions for reduction of substrates and growth of associated biomass could be envisioned on specific case-by-case bases.

Chapter 10

Acid-Base Advanced Principles

10.1 PERSPECTIVE

In Chapter 4, we examined the definition of the equilibrium constant. In equation 4.2, the equilibrium constant was related to the chemical *activities* of reactants and products. Then in Equation 4.2a, we specified the idealized version of the relation, in which the equilibrium constant is related to the *concentrations* of reactants and products. We then carried this idealization forward in Chapters 5 and 6, examining applications of fundamental air/water and acid/base equilibria in ideal systems. Relative to the chemistry of water, an ideal system is one in which abundances of dissolved solutes are sufficiently low so as to be negligible in the chemical relations. The term used to characterize such a system is "infinitely dilute." In a system that may be considered infinitely dilute, the abundances of solute ions (often termed electrolytes) and nonelectrolytes are so small that the probability of the interaction of one solute unit with another is essentially negligible. Herein, we will be mostly concerned with the effects of ions upon the chemistry of water. All ions in infinitely dilute solutions then interact almost solely with water. We will gain a perspective regarding this condition when, later in the chapter, we consider the abundances of ions in aqueous solutions in terms of their mole fraction concentrations. We also will examine the interactions of electrolytes with water and the hydration of both cations and anions, analogous to the hydration of a proton as illustrated in Figure 6.1. In the current chapter, we will examine the relation between aqueous concentrations of dissolved ions and chemical activities of targeted chemical species. We will also introduce and illustrate

Environmental Process Analysis: Principles and Modeling, First Edition. Henry V. Mott.

a systematic approach for incorporating nonideality into the speciation of acids and bases in environmental systems.

We were careful, as are most authors, to specify that the magnitudes of equilibrium constants presented in Tables 5.1 and 6.1 are specific to the chemists' standard temperature of 25 °C. Many of the environmental systems we investigate will have temperatures either lower or higher than this standard temperature. Equilibrium constants vary with temperature. Increases in temperature tend to reduce magnitudes of equilibrium constants for exothermic reactions. Conversely, for endothermic reactions, increases in temperature tend to increase the magnitudes of equilibrium constants. We will examine understandings of the temperature dependence of equilibrium constants in sufficient depth that the reader may develop a fundamental understanding of the concept. Then we will employ the principle in the context of selected environmental systems.

In Chapter 6, we introduced the concept of the mole balance, written for all conjugate acids and bases of a given acid system. We will continue to employ this important concept in our analyses of nonideal systems. We will introduce an additional principle, a mole balance of sorts, employed for the accounting of protons involved in proton transfer (acid/base) reactions. This principle relies on the necessity that each and every proton donated in connection with a proton transfer must be accepted and we call it a proton balance (some authors use the term proton condition). The proton balance is introduced in a number of aqueous and water chemistry texts, but its usefulness is most often only cursorily examined. We will apply this powerful tool in depth in the analysis of selected environmental systems.

10.2 ACTIVITY COEFFICIENT

The relationship between the concentration of an aqueous specie and its chemical activity is quite straightforward:

$$\{i\} = \gamma_i[i] \tag{10.1}$$

where $\{i\}$ denotes the chemical activity of arbitrary specie i in molar units, γ_i is the activity coefficient specific to specie i (chemical activity per unit of concentration), and $[i]$ is the molar concentration of specie i. The relation can employ mole fraction, molal, or other units, but we will stick to molar units herein. The activity coefficient is related to the abundance of electrolyte species in the aqueous solution, the properties of specie i itself, and the specific interactions specie i might have with each of the electrolytes present in the solution, inclusive of specie i. In order to employ the concept of nonideality in analyses of environmental systems, we need only arrive at an accurate estimate of the magnitude of the activity coefficient. Levine (1988) gives excellent discussions of concept of the activity coefficient and of several means to estimate their values. Herein, we will not delve greatly into the quantitative nuances of the activity coefficient beyond a fundamental understanding allowing us to exert proper care in its application in examination of environmental systems.

Most often for consideration of electrolytes in aqueous solutions the infinitely dilute aqueous solution is taken as the reference condition. Then by definition, in infinitely dilute solutions, all activity coefficients are unity (and, hence, activity is very closely approximated by concentration). Then, as electrolyte abundance is increased, the magnitudes of ionic activity coefficients are decreased. This behavior arises as a consequence of the interactions of the individual ionic species with water and with each other. Thus, in solutions containing electrolytes, the chemical activities of aqueous ionic species are always less than their corresponding concentrations. As abundances increase, interactions become more intense, and as electrolyte abundances increase beyond the levels found in seawater, activity coefficients actually increase.

Levine (1988) suggests that nonideality of solutions containing electrolytes arises as a consequence of the hydration of ions, significantly reducing the quantity of "free" water in a given volume of solution. Natural waters are often considered by chemists to be slightly contaminated distilled water, and often the inequality of chemical activity with aqueous concentration is disregarded. We will see hereinafter that consideration of nonideality in natural fresh waters is important.

10.2.1 Computing Activity Coefficients

Three relations, adapted from Stumm and Morgan (1996), useful for estimating values of activity coefficients, are in listed in Table 10.1 as Equations 10.2 through 10.4. The relation used to compute the ionic strength of the aqueous solution is given

TABLE 10.1 Some Relations used for Estimation of Aqueous Activity Coefficients for Electrolytes (Adapted from Stumm and Morgan, 1996)

Name	Relation	Applicable Range of Ionic Strength (I)	Equation number
Extended Debye–Hückel	$\log(\gamma_i) = -\dfrac{Az_i^2\sqrt{I}}{1+B\mathring{a}_i\sqrt{I}}$	<0.10 M	(10.2)
Güntelberg	$\log(\gamma_i) = -\dfrac{Az_i^2\sqrt{I}}{1+\sqrt{I}}$	<0.10 M	(10.3)
Davies	$\log(\gamma_i) = -Az_i^2\left(\dfrac{\sqrt{I}}{1+\sqrt{I}} - 0.2I\right)$	<0.5 M	(10.4)
Ionic strength	$I = \dfrac{1}{2}\sum_i [I] z_i^2$		(10.5)

γ_i is the activity coefficient of specie i, A and B are empirical parameters, z_i is the net charge on ionic specie i, $\mathring{a}_i$ is the ion size parameter for specie i, and I is the ionic strength of the solution. $A = 1.82 \times 10^6(\varepsilon_D T)^{-3/2}$ and $B = 50.3(\varepsilon_D T)^{-3/2}$ where ε_D is the dielectric constant of water. For Equation 10.2 the ion size parameter, $\mathring{a}_i$, is roughly equivalent to the hydrated radius, in Angstroms (Å), of the hydrated ion in an aqueous solution.

TABLE 10.2 Size Parameters (Å) for Selected Ions for use with the Extended Debye–Hückel Relation (Adapted from Dean, 1992)

Ion Size Parameter (å) in Å	Ion
2.5	Rb^+, Cs^+, NH_4^+, Tl^+, Ag^+
3	K^+, Cl^-, Br^-, I^-, CN^-, NO_2^-, NO_3^-
3.5	OH^-, F^-, SCN^-, OCN^-, HS^-, ClO_3^-, ClO_4^-, BrO_3^-, IO_4^-, MnO_4^-, $HCOO^-$, H_2Cit^-
4	Na^+, Hg^{+2}, HCO_3^-, $H_2PO_4^-$, HPO_4^{-2}, PO_4^{-3}, HSO_3^-, $H_2AsO_4^-$, SO_4^{-2}, SeO_4^{-2}, CrO_4^{-2}, $H_3NCH_2COOH^+$
4.5	Pb^{+2}, CO_3^{-2}, SO_3^{-2}, CH_3COO^-, $HCit^{-2}$
5	Sr^{+2}, Ba^{+2}, Ra^{+2}, Cd^{+2}, Hg^{+2}, S^{-2}, $S_2O_4^{-2}$, Cit^{-3}
6	Li^+, Ca^{+2}, Cu^{+2}, Zn^{+2}, Sn^{+2}, Mn^{+2}, Fe^{+2}, Ni^{+2}, Co^{+2}, benzoate$^-$, phthalate^{-2}
8	Mg^{+2}, Be^{+2}
9	H^+, Al^{+3}; Fe^{+3}, Cr^{+3}. Sc^{+3}, Y^{+3}, La^{+3}, In^{+3} Ce^{+3}, Pr^{+3}, Nd^{+3}, Sm^{+3}
11	Th^{+4}, Zr^{+4}, Ce^{+4}, Sn^{+4}

also in Table 10.1 as Equation 10.5. Selected values of the ion size parameter are given in Table 10.2 (adapted from Dean, 1992).

We include use of the extended Debye–Hückel equation herein and alert the reader to the fact that each ion in a given solution will have a unique value of the activity coefficient, owing to the unique value of its ion size parameter. In analyses, where extreme accuracy is of utmost importance, use of the best available computational technique is warranted. Certainly, for solutions of high ionic strength (beyond 0.5 M) additional techniques, beyond the scope of this text, have been developed. The student is encouraged, for these applications, to seek such advanced relations. Herein, once we have illustrated the variability among ions of the same charge magnitude arising from specific ion size, we will revert to the Güntelberg approximation and the Davies equation for computation of aqueous activity coefficients. In these cases, we will generally have at most five values of the activity coefficient for any given solution: $\gamma_{\pm1}$, $\gamma_{\pm2}$, $\gamma_{\pm3}$, $\gamma_{\pm4}$, and $\gamma_{\pm5}$ (generally to be denoted as γ_1, γ_2, γ_3, γ_4 and γ_5).

In order that we may underscore the importance of employing nondilute principles for computations involving aqueous solutions, the extended Debye–Hückel equation has been employed with typical values of the ion size parameter for monovalent, divalent, trivalent, and tetravalent ions to compute the activity coefficient in aqueous solutions at temperature of 25 °C for various values of ionic strength. The results of these computations are assembled in Figure 10.1. We may easily observe from Figure 10.1 that even at ionic strength as low as 0.001 M the activities of di-, tri-, and tetravalent ions are significantly lower than corresponding molar concentrations. The error is exacerbated of course with increased ionic strength. For comparison purposes, the ionic strength of raw

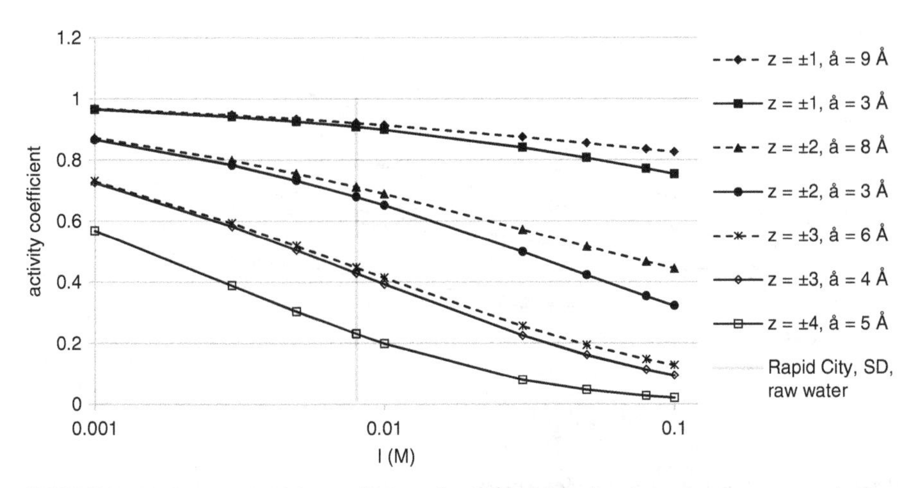

FIGURE 10.1 *Aqueous activity coefficients for electrolytes of various charge computed using the extended Debeye-Hückel equation.* $\gamma_i = 10^{-\frac{A|z_i|^2\sqrt{I}}{1+B\mathring{a}_i\sqrt{I}}}$; A *and* B *are for* T=*298.15 K. Ionic strength of Rapid City, SD, raw water (*$I \approx 0.008$*) is shown for perspective.*

water processed for municipal distribution by the City of Rapid City, SD, is shown. While relatively high in both hardness and total dissolved solids, Rapid City's water quality is well within the corresponding secondary standards for drinking water set by the EPA.

To underscore the errors associated with assuming the infinitely dilute condition for typical fresh waters, the values of the activity coefficient depicted in Figure 10.1 were used to compute the magnitude of the relative error, considering the infinitely dilute assumption would lead to the approximation and γ computed using the extended Debye–Hückel relation would represent the true value.

$$\text{Err\%} = \frac{1-\gamma_{\text{D-H}}}{\gamma_{\text{D-H}}}$$

Unity is the value assumed for the infinitely dilute condition and $\gamma_{\text{D-H}}$ is the activity coefficient computed using the Debye–Hückel equation. These results are assembled in Figure 10.2. We easily observe from Figure 10.2 that relative errors associated with neglecting ionic strength effects in specification of chemical activity range to >20% for monovalent ions, to >100% for divalent ions and to 1000% and greater for tri- and tetravalent ions. Thus, unless we are truly dealing with "only slightly" contaminated distilled water, for meaningful work we should never employ the infinitely dilute assumption, using unity values for activity coefficients.

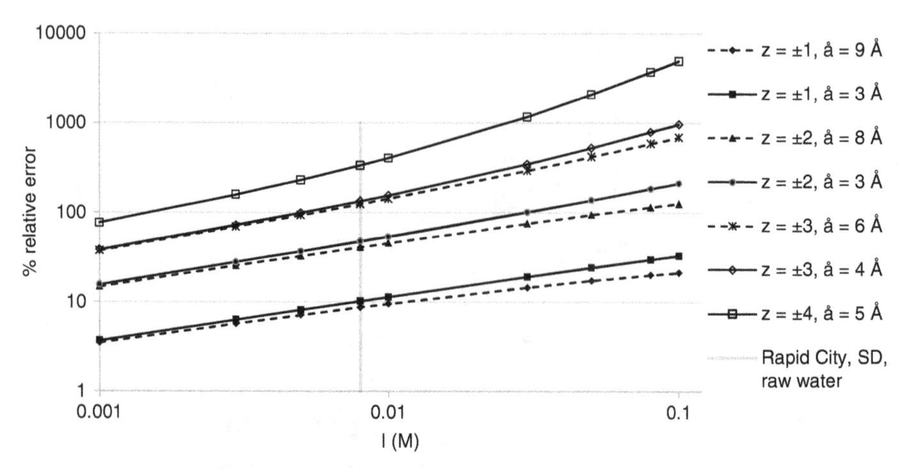

FIGURE 10.2 *Percent relative error for chemical activity of electrolytes based on the infinitely dilute assumption ($\gamma_i = 1$) relative to values computed using the Debeye-Hückel equation. Ionic strength of Rapid City, SD, raw water ($I \approx 0.008$) is shown for perspective.*

10.2.2 Activity Coefficient and Law of Mass Action

Equation 4.6 relates the equilibrium constant for an acid dissociation reaction with the abundances of protons, of the conjugate acid and of the conjugate base. These abundances are expressed as the actual molar concentrations of the respective species. In light of the foregoing, we must now correct Equation 4.6, in consideration of the principle of aqueous solution nonideality:

$$K_A = \frac{\{H^+\}\left\{H_{m-1}B^{(n-1)}\right\}}{\{H_mB^n\}} \tag{10.6a}$$

We simply replace the square brackets with curly brackets, rendering the RHS of the relation to be specie abundances expressed in units of chemical activity rather than molar concentration. Then, for example, for a general diprotic acid whose fully protonated specie is neutral, we would write:

$$K_{A1} = \frac{\{H^+\}\{HB^-\}}{\{H_2B\}} \quad \text{and} \quad K_{A2} = \frac{\{H^+\}\{B^{-2}\}}{\{HB^-\}}$$

We may use this law of mass action relation in all of our work with acids and bases. It simply supplants the initial approximate relation stated in Chapter 4 and employed in Chapter 6. In fact, we likely will find it most convenient to always express our mass action law in terms of activities. We then never require the use of an activity coefficient in a statement of chemical equilibrium. As we will describe later in this chapter, use of activity coefficients in statements of mole and proton balances most often proves to be the most straightforward approach.

However, the issue has been muddied. In the aquatic chemistry literature, we find two additional types of relations between acid dissociation constants and reactant/product abundances. These employ concentration-based and mixed acid dissociation constants. For the concentration-based approach, we simply express all abundances in units of concentration and redefine the equilibrium constant (now a function of the activity coefficients of the products and reactants).

$$K_A^C = \frac{[H^+]\left[H_{m-1}B^{(n-1)}\right]}{[H_mB^n]} \tag{10.6b}$$

For the mixed approach, we express the proton abundance in units of chemical activity and the abundances of the other products and reactants in terms of molar concentrations.

$$K_A^m = \frac{\{H^+\}\left[H_{m-1}B^{(n-1)}\right]}{[H_mB^n]} \tag{10.6c}$$

K_A^m is also now a function of the activity coefficients of the products and reactants, other than the proton.

Obviously, the true, concentration-based and mixed acidity constants are all different from each other, since concentration and activity of each of the species of the acid dissociation reaction are different from each other. We may use equation 10.1 to sort out the relation between K_A and K_A^C and between K_A and K_A^m. We may use the general diprotic acid for illustration, but must remain aware that the relations for monoprotic and triprotic acids will differ from those for the diprotic acid and that the relations will be highly sensitive to the residual charge of the fully protonated acid. We will subscribe to the Güntelberg approximation of the extended Debye–Hückel relation such that we have but two activity coefficients, γ_1 and γ_2.

Example 10.1 Develop relations between the true acid dissociation constant and both the concentration-based and mixed acidity constants for a diprotic acid whose fully protonated conjugate acid is neutral.

We begin by inserting Equation 10.1 appropriately into Equation 10.6.

$$K_{A1} = \frac{\gamma_1[H^+]\,\gamma_1[HB^-]}{[H_2B]} \quad \text{and} \quad K_{A2} = \frac{\gamma_1[H^+]\,\gamma_2[B^{-2}]}{\gamma_1[HB^-]}$$

We write Equation 10.6b for our selected general diprotic acid.

$$K_{A1}^{C} = \frac{[H^+][HB^-]}{[H_2B]} \quad \text{and} \quad K_{A2}^{C} = \frac{[H^+][B^{-2}]}{[HB^-]}$$

We then relate K_{A1} and K_{A1}^{C} by their quotient, with our new RHS being the quotient of the respective RHSs.

$$\frac{K_{A1}^{C}}{K_{A1}} = \frac{[H^+][HB^-]/[H_2B]}{\gamma_1[H^+]\gamma_1[HB^-]/[H_2B]} \quad \text{and} \quad \frac{K_{A2}^{C}}{K_{A2}} = \frac{[H^+][B^{-2}]/[HB^-]}{\gamma_1[H^+]\,\gamma_2[B^{-2}]/\gamma_1[HB^-]}$$

We quickly observe that all but the activity coefficients on each of the RHSs cancel, leading us to the final expressions, applicable ***only*** for diprotic acids whose fully protonated specie is uncharged.

$$K_{A1}^{C} = K_{A1}\frac{1}{\gamma_1^2} \quad \text{and} \quad K_{A2}^{C} = K_{A2}\frac{1}{\gamma_2}$$

We would obtain alternative definitions for K_{A1}^{C} and K_{A2}^{C} for diprotic acids whose fully protonated conjugate acid would have a +1 or +2 charge. Similarly, we would develop unique relations for K_{A1}^{C}, K_{A2}^{C}, and K_{A3}^{C} for triprotic acids and again, further unique relations for acid systems whose fully protonated conjugate acid might have +1, +2, or +3 charge. Two possibilities then exist for monoprotic acids: acid systems whose conjugate acid is neutral and whose conjugate acid has a +1 charge.

For the mixed acidity constant, our approach is similar. We write Equation 10.6c for the targeted diprotic acid.

$$K_{A1}^{m} = \frac{\{H^+\}[HB^-]}{[H_2B]} \quad \text{and} \quad K_{A2}^{m} = \frac{\{H^+\}[B^{-2}]}{[HB^-]}$$

We insert Equation 10.1 appropriately into Equation 10.6. The quotients leading to the relation between the respective K_{A}^{m} and K_A values are written.

$$\frac{K_{A1}^{m}}{K_{A1}} = \frac{\{H^+\}[HB^-]/[H_2B]}{\{H^+\}\,\gamma_1[HB^-]/[H_2B]} \quad \text{and} \quad \frac{K_{A2}^{m}}{K_{A2}} = \frac{\{H^+\}[B^{-2}]/[HB^-]}{\{H^+\}\,\gamma_2[B^{-2}]/\gamma_1[HB^-]}$$

Again, the RHSs cancel nicely and we are left with the desired relations involving only activity coefficients, applicable **only** for diprotic acids whose fully protonated specie is uncharged.

$$K_{A1}^{m} = K_{A1}\frac{1}{\gamma_1} \quad \text{and} \quad K_{A2}^{m} = K_{A2}\frac{\gamma_1}{\gamma_2}$$

As with the concentration-based acidity constants, we would arrive at alternative unique relations were our diprotic acid to have a charged fully protonated conjugate acid. Distinct unique relations would arise for the three mixed acidity constants for each potential general triprotic acid system (uncharged, +1, +2, and +3 fully protonated conjugate acid). Two distinct relations would be written for K_A^m for a monoprotic acid, depending upon the net charge of the conjugate acid.

Concentration-based acidity (or for that matter general equilibrium) constants have apparently been developed to allow direct use of concentration information, the presentation of choice for analytical laboratories and such. As we observed in Chapter 6, much of the time these items of data require interpretation using equilibrium chemical principles. The originators of the K^C concept might have believed them to be more convenient than use of the true K_{eq} values. However, use of K^C requires conversion of pH measurements from activity to concentration. Measured pH is in fact the value of the chemical activity of protons rather than the concentration. Confusion is then highly probable. Employment of mixed acidity constants appears somewhat more sensible as the chemical activity of protons is used directly. When dealing with nondilute aqueous solutions, we surely must include the activity coefficients in our computations. It is surely each worker's choice as to how to do so. The musings of Example 10.1 can be translated into general relations for defining concentration-based and mixed acidity constants in terms of the true equilibrium constant and activity coefficients.

$$K_A^C = K_A \frac{\gamma_{H_mB^n}}{\gamma_{H^+}\gamma_{H_{m-1}B^{(n-1)}}}$$

$$K_A^m = K_A \frac{\gamma_{H_mB^n}}{\gamma_{H_{m-1}B^{(n-1)}}}$$

With the work of Example 10.1 we will conclude our treatment (and utilization) of concentration-based and mixed acidity constants. This author firmly believes that the least confusing and most thoughtful approach is the implementation the true acidity (and certainly all other equilibrium) constants for equilibria and use of Equation 10.1 in mole balance relations, to express concentrations using activities and activity coefficients. This belief is sufficiently strong that this text includes no illustrations of the implementation of concentration-based or mixed acidity constants. While use of K_A^C and K_A^m is not overly challenging for acid/base work employing monoprotic and even diprotic acids, extension to the applications beyond diprotic acids and to applications of coordination (complexation) chemistry and solubility/dissolution principles renders their use very cumbersome and confusing. In later sections of this chapter, we will illustrate the use of Equation 10.1 in mole and proton balance relations.

10.3 TEMPERATURE DEPENDENCE OF EQUILIBRIUM CONSTANTS

Before we may discuss the adjustments of equilibrium constants for changes in temperature, we must address in a very basic manner, the relations among enthalpy, entropy, and Gibbs energy. Herein, our environmental systems overwhelmingly involve water, occur at environmental temperatures ($\sim 273\,^\circ K < T < \sim 315\,^\circ K$), and occur at environmental pressures ($\sim 0.5\,atm < P < \sim 1.5\,atm$). Levine (1988) for example describes in detail the many relations involving enthalpy, entropy, Gibbs energy, temperature, pressure, phase changes, and more. The scope of this text dictates that interest in these more advanced topics requires the reader to consult one of the many excellent texts addressing physical chemistry. We will address the principles directly related to adjusting magnitudes of equilibrium constants for nonstandard temperatures.

10.3.1 Standard State Gibbs Energy of Reaction

Before work addressing the physical chemistry of reactions could be standardized, physical chemists needed to determine and agree upon a reference condition. Physical chemistry and its companion engineering science, thermodynamics, deal with changes in systems associated with reactions. Having a reference condition was necessary against which to log such changes. The reference state was determined to be that of the element. Reference conditions were determined to be temperature of 298.15 °K, pressure of one atmosphere for gaseous elements, and concentration of one molal for the remainder. Bear in mind that this standard condition is imagined for many of the elements as, in their elemental state, they are insoluble in water. Then the Gibbs energy of formation of each element under the reference conditions was defined to be zero. The Gibbs energies of formation for the proton and the electron were also defined to be zero.

For our purposes herein, we begin with the relation defining the standard Gibbs energy of reaction. Equation 10.7a in our context considers that the system is under standard conditions of 298.15 °K and that all reactants and products are under equilibrium conditions at the standard abundances.

$$\Delta G_{\mathrm{rxn}}^{\circ} = \sum_{i} \nu_i \Delta G_{\mathrm{f},i}^{\circ} \tag{10.7a}$$

$\Delta G_{\mathrm{rxn}}^{\circ}$ is the standard state Gibbs energy of the reaction as written. ν_i is the stoichiometric coefficient of reactant or product i. $\Delta G_{\mathrm{f},i}^{\circ}$ is the standard Gibbs energy of formation of product or reactant i. In application of Equation 10.7a, we must understand that the creation of a product and the transformation of a reactant involve thermodynamic changes in opposite directions with regard to energy. The convention most often followed then is that, since the reactants are being converted, the changes in the Gibbs energy are the negative values of the Gibbs energy of formation. We most often see Equation 10.7a written as the difference between the products and the reactants.

$$\Delta G^\circ_{\text{rxn}} = \sum_i \nu_i \Delta G^\circ_{\text{f},i_{\text{products}}} - \sum_i \nu_i \Delta G^\circ_{\text{f},i_{\text{reactants}}} \tag{10.7b}$$

Through experiment, physical chemistry has created a database of Gibbs energy of formation values for many substances that are combinations of the elements. Measurements of the energy (heat) either consumed or released under carefully controlled and carefully monitored conditions and application of the relations among enthalpy, entropy, and Gibbs energy have allowed the assembly of this very useful database.

Enthalpy, entropy, and Gibbs energy are related, in general by Equation 10.8a.

$$\Delta G = \Delta H - T\Delta S \tag{10.8a}$$

H is enthalpy and S is entropy. Equation 10.8a is certainly applicable to the standard state and thus may be used to relate enthalpy, entropy, and Gibbs energy of formation.

$$\Delta G^\circ_{\text{f}} = \Delta H^\circ_{\text{f}} - T\Delta S^\circ_{\text{f}} \tag{10.8b}$$

Then sources such as the *CRC Handbook of Chemistry and Physics*, *Lange's Handbook of Chemistry*, and *Perry's Chemical Engineers' Handbook* contain tables of standard state Gibbs energy, enthalpy, and entropy for common substances. An ability to use this information proves valuable in understanding reaction equilibria.

Relative to equilibrium constants under standard conditions, the remaining relation from physical chemistry addresses the overall change in Gibbs energy with reaction. To be complete, we must include the partial derivative.

$$\frac{\partial G}{\partial \xi_{T,P}} = \Delta G^\circ_{\text{rxn}} + RT\ln(Q) \tag{10.9}$$

Equation 10.9 relates the change in the extent of the reaction (at arbitrary condition along the path from the initial to the final condition) to the change in standard Gibbs energy of reaction and the reaction quotient, Q. The partial derivative ($\partial G/\partial \xi_{T,P}$) is the infinitesimal change in the Gibbs energy of the system associated with an infinitesimal change in the extent of the reaction at constant temperature, T, and pressure, P. The reaction quotient, Q, is defined in a manner identical to the equilibrium constant (K_{eq}), except that the activities used are the actual (non-equilibrium) values for the reactants and products at the corresponding extent of the reaction. Equation 10.9 then applies to a reaction in process, the forward reaction is occurring at a rate different from that of the reverse reaction. We seek the relation useful for the equilibrium condition. At equilibrium, the net progress of the reaction ceases; the forward and reverse reactions occur at identical rates. Then, the partial derivative term becomes zero and since the reaction has attained the equilibrium condition, the relation for Q becomes identical to that for K_{eq}. Substitution of K_{eq} for Q and rearrangement of Equation 10.9 leads to the relation of use herein.

$$\Delta G^{\circ}_{rxn} = -RT\ln(K_{eq}) \tag{10.10a}$$

The exponentiated form is perhaps of greater use, since in our applications most often we seek the magnitude of the equilibrium constant from Gibbs energy data.

$$K_{eq} = \exp\left(-\frac{\Delta G^{\circ}_{rxn}}{RT}\right) \tag{10.10b}$$

Example 10.2 Use standard Gibbs energy data to determine the equilibrium constant for the deprotonation of phosphoric acid (H_3PO_4) to yield a proton and dihydrogen phosphate ($H_2PO_4^-$) and of dihydrogen phosphate to yield a proton and hydrogen phosphate ($HPO_4^=$).

We first must compute the standard Gibbs energy of reaction for the deprotonation reactions.

$$H_3PO_4 \Leftrightarrow H^+ + H_2PO_4^- \text{ and } H_2PO_4^- \Leftrightarrow H^+ + HPO_4^=$$

In Table A.1 of the appendix of this text, the Gibbs energy of formation values for hydrogen phosphate, dihydrogen phosphate, and phosphoric acid are given as −1089.3, −1130.4 and −1142.6 kJ/mol, respectively. The choices for hydrogen phosphate and dihydrogen phosphate are clear; we choose the aqueous specie. For phosphoric acid, we must choose the value for the undissociated, standard state condition in the aqueous state. The Gibbs energy of formation for the proton is 0. We apply Equation 10.7b with our retrieved data.

$\Delta G_{f.HPO4} := -1089.3$ $\quad \Delta G_{f.H2PO4} := -1130.4$ $\quad \Delta G_{f.H3PO4} := -1142.6$

$$\Delta G^{\circ}_{H3PO4.H2PO4} := \Delta G_{f.H2PO4} - \Delta G_{f.H3PO4} = 12.2$$

$$\Delta G^{\circ}_{H2PO4.HPO4} := \Delta G_{f.HPO4} - \Delta G_{f.H2PO4} = 41.1 \quad \frac{kJ}{mol}$$

We may now compute K_{A1} and K_{A2} (pK_{A1} and pK_{A2}) from Gibbs energy of reaction using Equation 10.10a.

$R_{kJ} := 0.0083144 \dfrac{kJ}{mol \cdot {}^\circ K}$ $\quad T := 298.15 \ {}^\circ K$

$K_{A1} := \exp\left(\dfrac{-\Delta G^\circ_{H3PO4.H2PO4}}{R_{kJ} \cdot T}\right) = 7.288 \times 10^{-3}$ $\quad pK_{A1} := -\log(K_{A1}) = 2.14$

$K_{A2} := \exp\left(\dfrac{-\Delta G^\circ_{H2PO4.HPO4}}{R_{kJ} \cdot T}\right) = 6.303 \times 10^{-8}$ $\quad pK_{A2} := -\log(K_{A2}) = 7.2$

We compare our results with those presented in Table 6.1 and determine satisfactory agreement between the two.

One justifiably might ask "Why would we employ Gibbs energy computations to obtain acidity constants that the chemists have already determined and published?" The answer to this question, of course, is that we generally would not. In this chapter, we will rely entirely upon acidity constants that are tabulated. In practice, we can find acidity constants for virtually any inorganic and most organic acids with which we might work. Conversely, when we enter the realm of geochemistry, we will find that the geochemists by and large have assembled their database of information leading to equilibrium constants as Gibbs energy of formation. Later in this text we will employ this concept heartily in our work with oxidation/reduction principles. Further, the idea of adjusting equilibrium constants for varying temperature before gaining some understanding of the application of Gibbs energy and enthalpy does not seem wise. Finally, since applications of aqueous chemistry involve use of the system that chemistry has assembled, this author believes it is wise to gain insight regarding the entire system. We now may address the correction of equilibrium constants for varying temperature.

10.3.2 Temperature Corrections for Equilibrium Constants

Levine (1988) begins with Equation 10.10a (rearranged with $\ln(K_{eq})$ on the LHS), takes the derivative with respect to T to produce the basis for temperature dependence.

$$\frac{d\left(\ln K_{eq}\right)}{dT} = \frac{\Delta G^\circ_{rxn}}{RT^2} - \frac{1}{RT}\frac{d\left(\Delta G^\circ_{rxn}\right)}{dT}$$

Then, Equation 10.7a is used to express ΔG°_{rxn} in terms of the Gibbs energies of formation.

$$\frac{d\left(\Delta G^\circ_{rxn}\right)}{dT} = \sum_i \nu_i \frac{d\left(G^\circ_{f,i}\right)}{dT}$$

From the relation $\partial\bar{G} = -\bar{S}\partial T + \bar{V}\partial P$ rearranged at constant pressure the definition $(\partial\bar{G}/\partial T) = \bar{S}$ is obtained leading to $\left(d\left(G^{\circ}_{\mathrm{f},i}\right)/dT\right) = S^{\circ}_{\mathrm{f},i}$. Then $\left(d\left(\Delta G^{\circ}_{\mathrm{rxn}}\right)/dT\right) = \sum_i v_i S^{\circ}_{\mathrm{f},i} = S^{\circ}_{\mathrm{rxn}}$. This result is inserted into the above relation to obtain the last intermediate result.

$$\frac{d\left(\ln K_{\mathrm{eq}}\right)}{dT} = \frac{\Delta G^{\circ}_{\mathrm{rxn}}}{RT^2} - \frac{S^{\circ}_{\mathrm{rxn}}}{RT} = \frac{\Delta G^{\circ}_{\mathrm{rxn}} - TS^{\circ}_{\mathrm{rxn}}}{RT^2}$$

Then, since $\bar{H} = \bar{G} - T\bar{S}$ Equation 10.11 arises.

$$\frac{d\ln(K_{\mathrm{eq}})}{dT} = \frac{\Delta H^{\circ}_{\mathrm{rxn}}}{RT^2} \tag{10.11}$$

Separation of 10.11 and integration from initial to final temperature yields a useable relation for adjusting equilibrium constants for varying temperature. However, ΔH_{f} varies with temperature and, hence, ΔH_{rxn} does also. For environmental systems, operative within a narrow range of temperatures, we may disregard this temperature variability of the enthalpy of reaction, considering it to be constant, and invoke only small (preferably negligible) error. The result is quite useful for temperature adjustments.

$$\frac{K_{\mathrm{eq},T_2}}{K_{\mathrm{eq},T_1}} = \exp\left(\frac{\Delta H^{\circ}_{\mathrm{rxn}}}{R}\left(\frac{1}{T_1} - \frac{1}{T_2}\right)\right) \tag{10.12}$$

In use of this relation, if we use the standard temperature (298.15 °K) as T_1, T_2 is our temperature of interest.

The standard state enthalpy of reaction is computed in a manner exactly analogous to the standard state Gibbs energy of reaction.

$$\Delta H^{\circ}_{\mathrm{rxn}} = \sum_i v_i \Delta H^{\circ}_{\mathrm{f},i_{\mathrm{products}}} - \sum_i v_i \Delta H^{\circ}_{\mathrm{f},i_{\mathrm{reactants}}} \tag{10.13}$$

Temperature corrections, if data are available, should be made routinely in analyses of environmental processes and systems.

Example 10.3 Investigate the error associated with neglecting to correct for temperature on the value of the second equilibrium constant of the phosphoric acid system with application at the sediment/water interface of a deep mesotrophic lake (4 °C) and within a digester employed for solids reduction at a wastewater renovation facility (36 °C).

From Table A.1 we find the following for standard enthalpy of formation:

$$\Delta H_{f.HPO4} := -1292.1 \qquad \Delta H_{f.H2PO4} := -1296.3 \qquad \frac{kJ}{mol}$$

We compute the standard enthalpy of reaction using Equation 10.13.

$$\Delta H^{\circ}_{H2PO4.HPO4} := \Delta H_{f.HPO4} - \Delta H_{f.H2PO4} = 4.2 \qquad \frac{kJ}{mol}$$

We compute the two adjusted values for the equilibrium constant using Equation 10.12.

$$T_{std} := 25 + 273.15 \quad T_4 := 4 + 273.15 \quad T_{36} := 36 + 273.15 \quad {}^{\circ}K$$

$$K_{A2.4} := K_{A2} \cdot \exp\left[\frac{\Delta H^{\circ}_{H2PO4.HPO4}}{R_{kJ}} \cdot \left(\frac{1}{T_{std}} - \frac{1}{T_4}\right)\right] = 5.477 \times 10^{-8}$$

$$K_{A2.36} := K_{A2} \cdot \exp\left[\frac{\Delta H^{\circ}_{H2PO4.HPO4}}{R_{kJ}} \cdot \left(\frac{1}{T_{std}} - \frac{1}{T_{36}}\right)\right] = 6.614 \times 10^{-8}$$

Now we compute the two relative errors and express our result as percentage.

$$err_4 := \frac{K_{A2} - K_{A2.4}}{K_{A2.4}} \cdot 100 = 13.7 \qquad err_{36} := \frac{K_{A2} - K_{A2.36}}{K_{A2.36}} \cdot 100 = -5.85 \qquad \%$$

Failure to adjust for temperature at the sediment/water interface would lead to 13.7% error in the equilibrium constant, which of course would be propagated onto ensuing computations associated with the analysis. In this case, the abundance of the product, hydrogen phosphate, would be overstated. Conversely, in the digester the error would be nearly negative 6%, resulting in the overstatement of the abundance of the reactant, dihydrogen phosphate.

We an important observation from the results of Example 10.3. The state change in enthalpy is positive, meaning that the standard state enthalpy of the products is greater than that of the reactants. Under standard conditions, the system must gain energy for conversion of reactants to products. Hence, this reaction would be considered endothermic. The values of the equilibrium constant at 4 and 36 °C are lower and higher, respectively, than the standard state value. At the lower temperature less energy is available from the surroundings to move the reaction forward. Conversely, at the higher temperature more energy is available from the surroundings to move the reaction forward. Were the standard enthalpy of reaction negative (the products would be of lower combined energy) the opposite set of situations would be true.

10.4 NONIDEAL CONJUGATE ACID/CONJUGATE BASE DISTRIBUTIONS

In Chapter 6, we performed significant investigation into the combination of acid/base equilibria with mole balances. If we choose to accept the errors associated with invoking the infinitely dilute assumption, these relations work quite well. However, if we choose to avoid the potentially significant errors associated with assuming that $\gamma \approx 1$ for all species of interest, these relations need some work.

Given that we choose to employ true, activity-based equilibrium constants and use chemical activities, we must alter the formulation of the mole balance relations to express the specie concentrations using chemical activities. We simply rearrange Equation 10.1 to yield the perfect result.

$$[i] = \frac{\{i\}}{\gamma_i} \tag{10.1a}$$

Then, given we choose to employ the Güntelberg approximation of the Extended Debeye–Hückel equation, our selection of activity coefficients is reduced (as previously mentioned) to γ_1, γ_2, γ_3, γ_4 and γ_5. We may employ these in the mole balance equations, accounting for the abundances of species associated with an acid/base system by replacing each specie concentration with the appropriate application of Equation 10.1. We will assemble the mole accounting equations for mono- and diprotic acids for various residual charges of the fully-protonated conjugate acid. The pattern will hopefully become obvious to the reader and assembly of the mole balances for multiprotic acids beyond diprotic is left as an exercise for the reader. The resulting relations are presented in Table 10.3.

In Table 10.3, we have included the activity coefficients of the nonelectrolytes (uncharged species) as γ_0. If we subscribe to Debye–Hückel theory, for a nonelectrolyte ($z=0$), the argument of the power of 10 reverts to 0 and $\gamma_0 \approx 1$. Typically, for solutions of low to moderate ionic strength, this assumption leads to small errors. Later in this chapter we will address the concept termed "salting out" for which a relation and small set of data are available for certain important nonelectrolytes. Typically, for aqueous solutions of ionic strength significantly lower than that of seawater, the activity coefficients for nonelectrolytes are not significantly different from unity. Thus, hereafter, in this chapter γ_0 will mostly be omitted from developed relations. Then also, for most computations with aqueous solutions, since $\gamma_0 \approx 1$, $\{i^0\} \approx [i^0]$.

TABLE 10.3 Mole Balance Relations for Typical Mono- and Diprotic Acid Systems for Noninfinitely Dilute Aqueous Solutions

Acid Type	Fully Protonated Charge[a]	Example Acid System	Non-Ideal Mole Balance Equation[b]
Monoprotic	$n=0$	HCN $B=CN^-$	$C_{Tot,B}=\frac{\{HB^0\}}{\gamma_0}+\frac{\{B^-\}}{\gamma_1}$
	$n=1$	NH_4^+ $B=NH_3$	$C_{Tot,B}=\frac{\{HB^+\}}{\gamma_1}+\frac{\{B^0\}}{\gamma_0}$
Diprotic	$n=0$	$H_2CO_3^*$ $B=CO_3^=$	$C_{Tot,B}=\frac{\{H_2B^0\}}{\gamma_0}+\frac{\{HB^-\}}{\gamma_1}+\frac{\{B^=\}}{\gamma_2}$
	$n=1$	$H_3NC_6H_4CO_2H^+$ $B=H_2NC_6H_4CO_2^-$	$C_{Tot,B}=\frac{\{H_2B^+\}}{\gamma_1}+\frac{\{HB^0\}}{\gamma_0}+\frac{\{B^-\}}{\gamma_1}$
	$n=2$	$H_3NCH_2CHOHCH_2NH_3^{+2}$ $B=H_2NCH_2CHOHCH_2NH_2^0$	$C_{Tot,B}=\frac{\{H_2B^{+2}\}}{\gamma_2}+\frac{\{HB^+\}}{\gamma_1}+\frac{\{B^0\}}{\gamma_0}$

[a] n is the residual charge on the fully protonated conjugate acid.
[b] These results assume that activity coefficients may be computed using the Güntelberg approximation (of the extended Debye–Hückel equation or the Davies equation. Use of the extended Debye–Hückel or other ion-specific equation necessitates computation of activity coefficients for the individual species of the specific acid system.

Equipped now with our corrected mole balance relations we may update the relations presented in Table 6.3. Now, however, we have two sets of relations for monoprotic acids, three sets of relations for diprotic acids, four potential sets of relations for triprotic acids and five potential sets of relations for tetraprotic acids. We also must carefully consider that the abundance fractions (α values) are ratios of specie concentrations to the total concentration of the entire set of species in an aqueous solution that contain the deprotonated base. Further, in application of these relations we find that our employment of the relations favors isolation of the chemical activity of each specie on the LHS of each of the relations. We have performed algebraic manipulations of the nature completed in Examples 6.1 through 6.3 to arrive at the relations specifying chemical activity of the various species of mono- and diprotic acids for the fully protonated conditions of Table 10.3. These are assembled in Table 10.4. Again, the algebraic manipulations to produce the relations for additional multiprotic acids are left as an exercise for the reader.

We quickly observe from Table 10.4 that tri- and tetraprotic acids would have a large number of associated relations. Herein, the relations of Table 10.4 for

TABLE 10.4 Relations Employing Proton Activity, Equilibrium Constants, and Total Acid Specie Concentrations for Specie Activities and Abundance Fractions in Noninfinitely Dilute Aqueous Solutions

Acid Type		Activity of Specie $i,\{i\}$	Abundance Fraction $\left(\alpha = \frac{[i]}{C_{Tot,B}}\right)$
Monoprotic	$n=0$	$\{HB\} = C_{Tot,B}\left(1+\frac{K_A}{\{H^+\}\gamma_1}\right)^{-1}$	$\alpha_0 = \left(1+\frac{K_A}{\{H^+\}\gamma_1}\right)^{-1}$
		$\{B^-\} = C_{Tot,B}\left(\frac{\{H^+\}}{K_A}+\frac{1}{\gamma_1}\right)^{-1}$	$\alpha_1 = \frac{1}{\gamma_1}\left(\frac{\{H^+\}}{K_A}+\frac{1}{\gamma_1}\right)^{-1}$
	$n=1$	$\{HB^+\} = C_{Tot,B}\left(\frac{1}{\gamma_1}+\frac{K_A}{\{H^+\}}\right)^{-1}$	$\alpha_0 = \frac{1}{\gamma_1}\left(\frac{1}{\gamma_1}+\frac{K_A}{\{H^+\}}\right)^{-1}$
		$\{B^0\} = C_{Tot,B}\left(\frac{\{H^+\}}{K_A\gamma_1}+1\right)^{-1}$	$\alpha_1 = \left(\frac{\{H^+\}}{K_A\gamma_1}+1\right)^{-1}$
Diprotic	$n=0$	$\{H_2B\} = C_{Tot,B}\left(1+\frac{K_{A1}}{\{H^+\}\gamma_1}+\frac{K_{A1}K_{A2}}{\{H^+\}^2\gamma_2}\right)^{-1}$	$\alpha_0 = \left(1+\frac{K_{A1}}{\{H^+\}\gamma_1}+\frac{K_{A1}K_{A2}}{\{H^+\}^2\gamma_2}\right)^{-1}$
		$\{HB^-\} = C_{Tot,B}\left(\frac{\{H^+\}}{K_{A1}}+\frac{1}{\gamma_1}+\frac{K_{A2}}{\{H^+\}\gamma_2}\right)^{-1}$	$\alpha_1 = \frac{1}{\gamma_1}\left(\frac{\{H^+\}}{K_{A1}}+\frac{1}{\gamma_1}+\frac{K_{A2}}{\{H^+\}\gamma_2}\right)^{-1}$
		$\{B^=\} = C_{Tot,B}\left(\frac{\{H^+\}^2}{K_{A1}K_{A2}}+\frac{\{H^+\}}{K_{A2}\gamma_1}+\frac{1}{\gamma_2}\right)^{-1}$	$\alpha_2 = \frac{1}{\gamma_2}\left(\frac{\{H^+\}^2}{K_{A1}K_{A2}}+\frac{\{H^+\}}{K_{A2}\gamma_1}+\frac{1}{\gamma_2}\right)^{-1}$

$n=1$	$\{H_2B^+\} = C_{Tot,B}\left(\frac{1}{\gamma_1} + \frac{K_{A1}}{\{H^+\}} + \frac{K_{A1}K_{A2}}{\{H^+\}^2\gamma_1}\right)^{-1}$	$\alpha_0 = \frac{1}{\gamma_1}\left(\frac{1}{\gamma_1} + \frac{K_{A1}}{\{H^+\}} + \frac{K_{A1}K_{A2}}{\{H^+\}^2\gamma_1}\right)^{-1}$
	$\{HB\} = C_{Tot,B}\left(\frac{\{H^+\}}{K_{A1}\gamma_1} + 1 + \frac{K_{A2}}{\{H^+\}\gamma_1}\right)^{-1}$	$\alpha_1 = \left(\frac{\{H^+\}}{K_{A1}\gamma_1} + 1 + \frac{K_{A2}}{\{H^+\}\gamma_1}\right)^{-1}$
	$\{B^-\} = C_{Tot,B}\left(\frac{\{H^+\}^2}{K_{A1}K_{A2}\gamma_1} + \frac{\{H^+\}}{K_{A2}} + \frac{1}{\gamma_1}\right)^{-1}$	$\alpha_2 = \frac{1}{\gamma_1}\left(\frac{\{H^+\}^2}{K_{A1}K_{A2}\gamma_1} + \frac{\{H^+\}}{K_{A2}} + \frac{1}{\gamma_1}\right)^{-1}$
$n=2$	$\{H_2B^{+2}\} = C_{Tot,B}\left(\frac{1}{\gamma_2} + \frac{K_{A1}}{\{H^+\}\gamma_1} + \frac{K_{A1}K_{A2}}{\{H^+\}^2}\right)^{-1}$	$\alpha_0 = \frac{1}{\gamma_2}\left(\frac{1}{\gamma_2} + \frac{K_{A1}}{\{H^+\}\gamma_1} + \frac{K_{A1}K_{A2}}{\{H^+\}^2}\right)^{-1}$
	$\{HB^-\} = C_{Tot,B}\left(\frac{\{H^+\}}{K_{A1}\gamma_2} + \frac{1}{\gamma_1} + \frac{K_{A2}}{\{H^+\}}\right)^{-1}$	$\alpha_1 = \frac{1}{\gamma_1}\left(\frac{\{H^+\}}{K_{A1}\gamma_2} + \frac{1}{\gamma_1} + \frac{K_{A2}}{\{H^+\}}\right)^{-1}$
	$\{B^0\} = C_{Tot,B}\left(\frac{\{H^+\}^2}{K_{A1}K_{A2}\gamma_2} + \frac{\{H^+\}}{K_{A2}\gamma_1} + 1\right)^{-1}$	$\alpha_2 = \left(\frac{\{H^+\}^2}{K_{A1}K_{A2}\gamma_2} + \frac{\{H^+\}}{K_{A2}\gamma_1} + 1\right)^{-1}$

Notes: n is the magnitude of the residual charge of the fully protonated acid specie; γ_1 and γ_2 are the aqueous activity coefficients for species of residual charge ± 1 and ± 2, respectively, as would be computed using the Güntelberg or Davies equations from Table 10.1; and α_0, α_1, and α_2 are the abundance fractions for the species resulting from the zeroth, first and second deprotonations of the acid.

mono- and diprotic acids are included as a follow-up to the relations presented in Table 6.3. As we will demonstrate, rather than tabulate the entire set of potential relations and have the reader choose the correct set for the acid system in question, we often simply use the mole balance relations in our work and employ algebraic substitutions therewith to arrive at our final models, used for the desired quantitative results.

Before moving forward, we should examine the errors associated with the assumption of infinite dilution relative to consideration of nonzero ionic strength. We will revisit Example 6.2 and employ nonunity activity coefficients to selected computations therefrom.

Example 10.4 investigates the effects of ionic strength ranging to ~0.1 M on the speciation of selenous acid. Consider the $0.0002\,\mathrm{M}C_{\mathrm{Tot.SeO_3}}$ concentration of Example 6.2.

In Example 6.2, we used relations from Table 6.3 and developed MathCAD functions relating the concentration (for $I \approx 0$, $\{i\} \approx [i]$) of each of the selenous acid species with pH, the equilibrium constants and the total aqueous selenite concentration.

$$H(pH) := 10^{-pH} \quad H2SeO3(pH) := \frac{C_{Tot.SeO3}}{1 + \frac{K_{1.SeO3}}{H(pH)} + \frac{K_{2.SeO3} \cdot K_{1.SeO3}}{H(pH)^2}}$$

$$HSeO3(pH) := \frac{C_{Tot.SeO3}}{\left(\frac{H(pH)}{K_{1.SeO3}} + 1 + \frac{K_{2.SeO3}}{H(pH)}\right)} \qquad \frac{mol}{L}$$

$$SeO3(pH) := \frac{C_{Tot.SeO3}}{\left(\frac{H(pH)^2}{K_{1.SeO3} \cdot K_{2.SeO3}} + \frac{H(pH)}{K_{2.SeO3}} + 1\right)}$$

We will use these again to compare the effects associated with the real solution relative to the infinitely dilute solution. These are easily expanded for the corresponding relations from Table 10.4 ($n=0$). Further, we will take advantage of the capability in MathCAD to develop functions using both pH and ionic strength (I) as arguments. Thus we may write the activity coefficients as functions of I and use them in the functions (of both pH and I) for activities of the selenous acid species. As we must define new functions for the activities of the species and we wish to retain the original definitions, we discern the real-case ($I \neq 0$) from the infinitely dilute case by the symbol a subscripted by the chemical formula of the targeted specie.

$$A := 0.5115 \quad I := 0.01 \quad \gamma_1(I) := 10^{\frac{-A\cdot\sqrt{I}}{1+\sqrt{I}}} \quad \gamma_2(I) := 10^{\frac{-A\cdot 4\cdot\sqrt{I}}{1+\sqrt{I}}}$$

$$a_{H2SeO3}(pH, I) := \frac{C_{Tot.SeO3}}{1 + \frac{K_{1.SeO3}}{H(pH)\cdot\gamma_1(I)} + \frac{K_{2.SeO3}\cdot K_{1.SeO3}}{H(pH)^2\cdot\gamma_2(I)}}$$

$$a_{HSeO3}(pH, I) := \frac{C_{Tot.SeO3}}{\left(\frac{H(pH)}{K_{1.SeO3}} + \frac{1}{\gamma_1(I)} + \frac{K_{2.SeO3}}{H(pH)\cdot\gamma_2(I)}\right)} \qquad \frac{mol}{L}$$

$$a_{SeO3}(pH, I) := \frac{C_{Tot.SeO3}}{\left(\frac{H(pH)^2}{K_{1.SeO3}\cdot K_{2.SeO3}} + \frac{H(pH)}{K_{2.SeO3}\cdot\gamma_1(I)} + \frac{1}{\gamma_2(I)}\right)}$$

Now we examine the pH range from 4 to 10, in which the fully protonated specie, selenous acid, is of negligible abundance. Thus we may concentrate upon the biselenite and selenite ions. We investigate two levels of ionic strength: $I=0.01$ and $I=0.10$, plot the result in figure E10.4.1 and compare the results.

We observe significant divergence between the specie activity values predicted from the infinitely dilute assumption and the real, nondilute, cases. We investigate the magnitudes of these errors using the % relative error of the ideal case prediction relative to the real case prediction.

$$err_1(pH, I) := \frac{HSeO3(pH) - a_{HSeO3}(pH, I)}{a_{HSeO3}(pH, I)}\cdot 100$$

$$err_2(pH, I) := \frac{SeO3(pH) - a_{SeO3}(pH, I)}{a_{SeO3}(pH, I)}\cdot 100$$

We apply these two relations with $I=0.001$, 0.01 and 0.1, respectively and plot the resultant % relative error versus pH of the solution in Figure E10.4.2. The result is rather surprising. We find that the relative error of the prediction for a stated ionic strength is identical for both $HSeO_3^-$ and $SeO_4^=$ for all values of pH. Were we to sharpen our algebraic pencils, perhaps we could equate the two % relative error relations. That is left to the interested reader. We will take the result at face value. Investigations of higher $C_{Tot.SeO_3}$ concentrations did not alter the equality of the result for $HSeO_3^-$ and $SeO_3^=$. We observe that even for $I=0.001$, well below that of most natural waters, the error ranges from about 5% to upwards of 15%.

The identical computations were accomplished for the carbonate system using total inorganic carbon concentration of 0.006 M (approximately that of Rapid City,

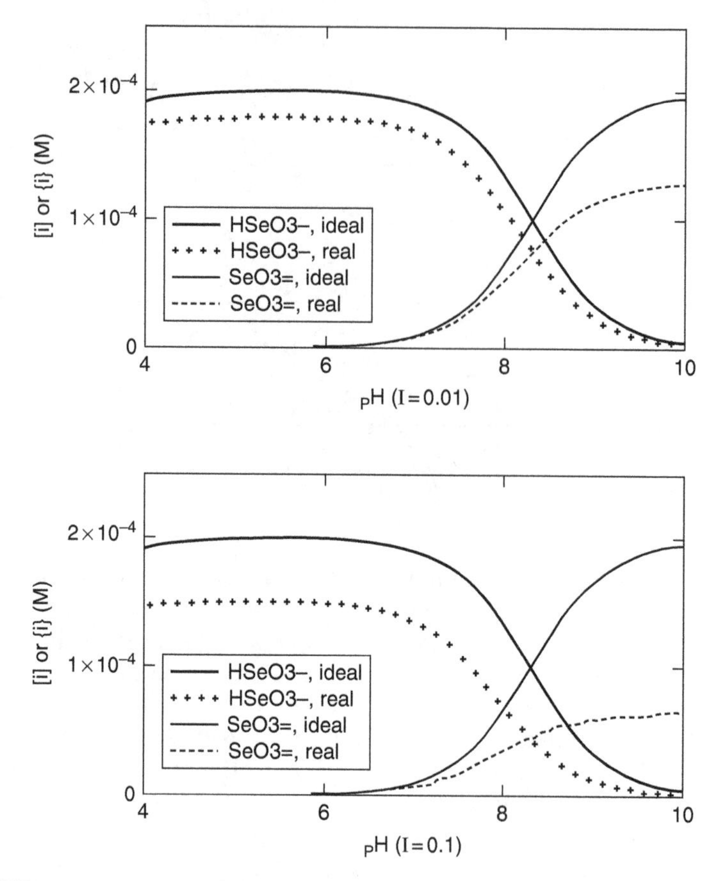

FIGURE E10.4.1 *Plots of activity versus pH for the selenous acid system at* $Se_{Tot} = 2 \times 10^{-4}$ *M for ionic strengths of 0.01 and 0.10 M.*

SD, raw water). The worksheet assembled for the selenous acid computations was used directly with the sole alterations being the values of C_{Tot}, K_1, and K_2. The relative percent error results for the carbonate system are plotted in Figure E10.4.3 Carbonic acid is significantly weaker than selenous acid and in the pH range of interest deprotonation occurs to a much lesser extent, thus the % relative error values are lower. However, the errors remain significant, particularly in the mildly alkaline pH range. Not surprisingly, the % relative error increases as ionic strength increases. Also not surprisingly, the % relative error increases as pH increases, in conjunction with greater ionization of the targeted acid species, for $I = 0.001$ M ranging to nearly 10% in the highly alkaline pH range.

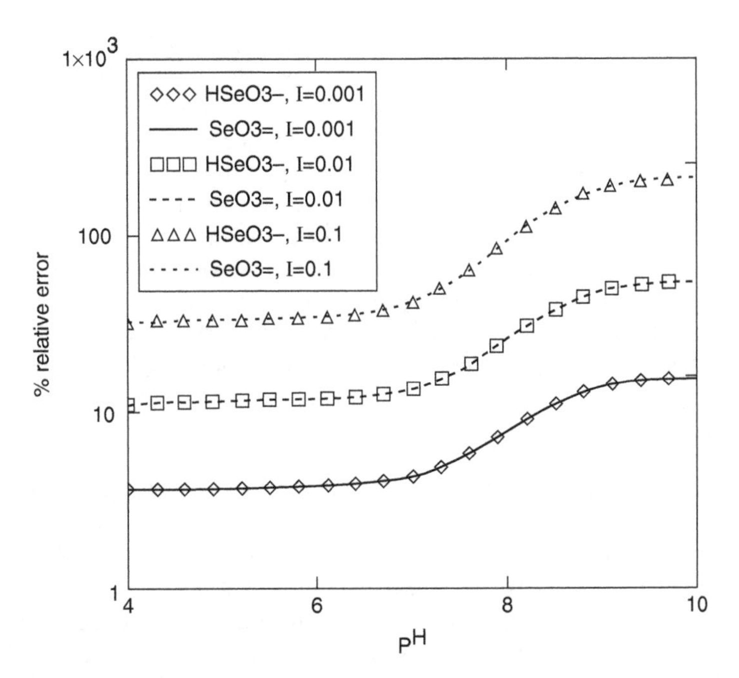

FIGURE E10.4.2 *A plot of the error of activities of hydrogen selenite and selenite predicted using the dilute-solution assumption relative to predictions employing ionic strength and the Güntelberg equation for* 2×10^{-4} *M* Se_{Tot}.

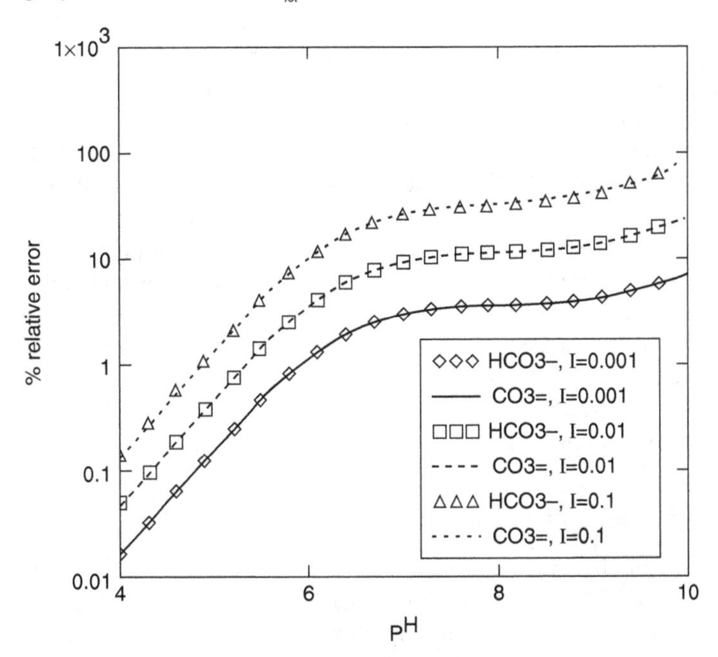

FIGURE E10.4.3 *A plot of the error of activities of bicarbonate and carbonate predicted using the dilute-solution assumption relative to predictions employing ionic strength and the Güntelberg equation for* 2×10^{-4} *M* CO_{3Tot}.

In the examination of ionic strength and associated effects on the activity/concentration relationships in aqueous solutions, the question that nearly always comes up is: "At what ionic strength value should I use nondilute activity coefficients?" The answer, of course, is typically another question: "How accurate do you wish to be?" If we are content with errors in the single digit percent range for waters of low ionic strength, then by all means use the infinitely dilute assumption. Conversely, if accuracy is of prime importance, the implementation of acid/base computations using nonunity activity coefficients is paramount. In the long-run, implementation using activity coefficients is not significantly more difficult than neglecting ionic strength and the recommendation herein is: ***To be sure, always use nonideal computations if the ionic strength of the water in question is known or can be closely approximated.***

10.5 THE PROTON BALANCE (PROTON CONDITION)

Our last and perhaps most powerful tool applicable to computations of acid speciation in aqueous systems is the proton balance. Some authors call it the proton condition. Either is fine, but herein we will use the proton balance. The proton balance is of course nothing more than an accounting of the acceptance and donation of protons during proton transfer reactions. In general, since the chemists tell us that protons cannot, under equilibrium conditions, exist in and of themselves, every proton donated by an acid via a proton transfer reaction must be accepted by a base. The accepting base is typically not the conjugate base of the proton-donating acid. This condition would make the idea of the proton transfer trivial. Rather, the proton donated by the conjugate acid of the resultant base must be accepted by the conjugate base of an alternative acid system, forming the alternative conjugate acid. Thusly, we can account for each and every proton transferred. Application of the proton balance permits us to determine the final state of a mixture of two or more solutions from the initial states of the solutions that are mixed. In applying the proton balance, we employ the evidence of protons accepted and the evidence of proton donated. Most conveniently we employ the changes in the **concentrations** of proton-donating and proton accepting species that might be present in or formed as a consequence of mixing two or more aqueous solutions. In applying the proton balance, we use our imagination and a theoretical very, very fast camera to view intermediate results along the thermodynamic path from the initial states of the individual solution to the final state of the resultant mixture.

10.5.1 The Reference Conditions and Species

As for a thermodynamic process, with the mixing of solutions we need a reference condition from which to track changes in proton-donating and proton-accepting species. Most conveniently, this reference condition is that of each of the contributing solutions prior to mixing. We use principles from Chapter 6, combined with the noninfinitely dilute activity coefficients to fully characterize the compositions of each of the mixed solutions, prior to the mixing process. We also must specify a reference specie for each acid system donating or accepting protons. The selection of the reference specie for homogenous aqueous systems is arbitrary, with the exception of water as an acid/base system. Choosing either hydronium or hydroxide as the reference specie necessitates computation of the change in the concentration of water—a tall order. Thus, the rule that we may always follow is that water is the reference specie for the hydronium–water–hydroxide system. In some cases, the selection of one specie over another may allow a simplification of the computational process as the reference abundance of the selected specie might effectively be zero. We must be careful, in this regard, that we do not oversimplify the accounting and greatly diminish the utility of the proton balance. The proton balance has its utility in any computational endeavor involving proton transfers and changes in the abundance of the hydronium ion in aqueous solutions.

10.5.2 The Proton Balance Equation

The fundamental tenet of the proton balance is that for every proton donated, one must be accepted. A proton may be donated from water to form hydroxide and conversely, a proton may be accepted by water to form hydronium. Each of these reactions may occur in either direction. More specifically, for every proton donated by an acid one proton must be accepted by a base, and conversely, for every proton accepted by a base one must be donated by an acid. This leads to the mathematical rule for the proton balance, applicable in the analysis of aqueous systems, that the evidence of protons accepted can be equated with the evidence of protons donated.

We need a systematic means through which to apply this mathematical rule. Thus, evidence of protons accepted would be manifest in increases in concentrations of species whose proton status is greater than that of the corresponding reference specie— the conjugate acids of the reference specie. Conversely, evidence of protons donated would be manifest in increases in concentrations of species whose proton status is lesser than that of the corresponding reference specie—the conjugate bases of the reference specie. We must also note that, depending upon the choice of the reference specie, for multiprotic acids, the number of protons gained or lost relative to the reference specie can be greater than 1. We therefore must account for this eventuality in our relation. The generalized mathematical statement is then as follows.

$$\sum_{N_j} \nu_j \Delta[\text{acid}]_j = \sum_{N_k} \nu_k \Delta[\text{base}]_k \tag{10.14}$$

The terms [acid] and [base] are the molar concentrations of respective conjugate acids and bases of each reference specie, ν is the absolute value of the proton difference between the conjugate acid or base specie and the reference specie for each respective acid system, N_j and N_k are the numbers of significant conjugate acid and conjugate base species considered, and $\Delta[i] = [i]_{\text{final}} - [i]_{\text{initial}}$.

Understanding the carbonate system is vital to analyses of most natural waters, as they are buffered by the carbonate system. Thus, our first foray into the application of the proton balance will address the carbonate system.

Example 10.5 Write the various proton balances applicable to the addition of a known quantity of carbon dioxide to water. Consider the solution to be nondilute, necessitating the inclusion of activity coefficients where necessary.

Here we need just a bit of imagination. Let us consider we have arranged the water in a chamber and bubbled a known quantity of pure carbon dioxide gas into the mixture and allowed the carbon dioxide to fully dissolve, such that the final volume of the solution is 1 liter and the total carbonate concentration of the solution is known.

Our first step is to identify the aqueous species that would be present in the solution under the final equilibrium conditions. We have two acid systems: carbonic acid and hydronium–water–hydroxide. Under equilibrium conditions, all species of all acid systems must be present in the aqueous solution, although in many cases several would have insignificant abundances. Let us call this the species identification step. Herein, we would have $H_2CO_3^*$, HCO_3^-, and $CO_3^=$ for the carbonate system: and $H_3O^+(H^+)$, H_2O, and OH^- for the water system.

Our next step is to select the reference specie for each acid system. For water, as mentioned earlier, we choose water. Our choice for the carbonate system is arbitrary and we may choose any of the three species. In fact, we will write proton balances using each of the carbonate species as the reference. Let us first choose carbonic acid as the reference specie for the carbonate system.

Now we may determine which species constitute evidence of protons gained and which are evidence of protons donated. Relative to carbonic acid, the proton status of bicarbonate is lesser than that of carbonic acid by one proton and the status of carbonate is lesser than that of carbonic acid by two protons. Thus increases in the concentrations of bicarbonate and carbonate would constitute evidence of protons donated by carbonic acid. Hydronium has one more proton than water and hydroxide has one fewer. An increase in the concentration of hydronium is evidence of protons accepted by water while an increase in the concentration of hydroxide is evidence of protons donated by water. We may now identify the magnitudes of the proton differences between the species of evidence and the reference species (i.e., the ν values of Equation 10.14). The proton statuses of bicarbonate, hydronium, and hydroxide all differ from that of the reference specie by one ($\nu = 1$), while that of carbonate differs from that of the reference specie by two ($\nu = 2$). The proton statuses of carbonic acid water are identical to those of the reference species, carbonic acid and water, such that the value of ν is zero and neither carbonic acid nor water will appear in the proton balance.

We can now write the base proton balance equation for the solution under consideration. In order that we are systematic, let us always place evidence of protons accepted on the RHS and evidence of protons donated on the LHS. This is of course arbitrary. Let us also revert to the use of H^+ to denote the hydronium ion (H_3O^+).

$$\Delta[H^+] = \Delta[OH^-] + \Delta[HCO_3^-] + 2\Delta[CO^{3=}]$$

This is the base proton balance for the specified solution and reference species.

Were we to choose bicarbonate as the reference specie, an increase in carbonic acid abundance would be evidence of protons accepted and an increase carbonate abundance would be evidence of protons donated. Since the proton status of bicarbonate is equal to that of the reference specie, bicarbonate, ν is zero and it is not included in the balance. The proton statuses of carbonic acid and carbonate both differ from that of bicarbonate by one, thus the respective values of ν are one.

$$\Delta[H^+] + \Delta[H_2CO_3*] = \Delta[OH^-] + \Delta[CO^{3=}]$$

This proton balance is entirely correct and when applied, will result in a solution identical to the one before, for which carbonic acid is the reference specie.

Let us be complete and write the remaining possible proton balance for this system, considering carbonate to be the reference specie. The proton status of carbonic acid is greater than that of carbonate by two and that of bicarbonate is greater than that of carbonate by one, thus the ν values are 2 and 1, respectively.

$$\Delta[H^+] + 2\Delta[H_2CO_3*] + \Delta[HCO^{3-}] = \Delta[OH^-]$$

Again, this proton balance is equally as correct as the first two and when applied will yield the identical result. In fact, these proton balances are independent of the means by which the carbonate would be introduced into the aqueous solution. These then are also applicable to the addition of a bicarbonate or carbonate salt to water or to the mixing of two or more waters in which the carbonate and water systems account for all significant proton transfers. The differences with regard to the overall applications then lie in the definition of the reference conditions.

The results of Example 10.5, although vital, are not yet ready for application. Before moving onto such application, let us examine systems that would contain multiple acid systems. The rules for assembling proton balance equations remain the same. We simply apply them to each acid system in turn. A system that is of prime importance in wastewater treatment is the anaerobic digester, in which we typically have, along with the carbonate system, carboxylic acid systems, and the ammonia system. Let us explore the proton balance that might be written for such a system.

Example 10.6 Consider that carbonate, acetate (a representative carboxylic acid) and ammonia have been added to water to produce a final solution in which the total carbonate, total acetate, and total ammonia are known. Write the proton balances that would result.

We will extend the result of Example 10.5. First, we must identify all possible additional species and then define the appropriate proton statuses relative to the selected reference species. For the acetate system, we would have acetic acid (CH_3CO_2H, often abbreviated as HAc) and acetate ($CH_3CO_2^-$, often abbreviated as Ac^-) and for the ammonia system we have NH_4^+ and NH_3. Depending on which of the species for each of these acid systems is chosen as the reference, the proton status of the other species is either plus or minus one relative to the reference. Then, for carbonic acid, acetic acid, and ammonium as reference species, we have the following.

$$\Delta[H^+] = \Delta[OH^-] + \Delta[HCO_3^-] + 2\Delta[CO^{3=}] + \Delta[Ac^-] + \Delta[NH_3]$$

For carbonic acid, acetate and ammonium as reference species the proton balance differs slightly.

$$\Delta[H^+] + \Delta[HAc] = \Delta[OH^-] + \Delta[HCO_3^-] + 2\Delta[CO^{3=}] + \Delta[NH_3]$$

We can use carbonic acid, acetate and ammonia as reference species.

$$\Delta[H^+] + \Delta[HAc] + \Delta[NH_4^+] = \Delta[OH^-] + \Delta[HCO_3^-] + 2\Delta[CO^{3=}]$$

In fact, we have a total of 12 correct proton balance statements. All possibilities are listed in the following table.

Reference Species	**Proton Balance**
H_2CO_3*, HAc, NH_4^+	$\Delta[H^+] = \Delta[OH^-] + \Delta[HCO_3^-] + 2\Delta[CO^{3=}] + \Delta[Ac^-] + \Delta[NH_3]$
H_2CO_3*, Ac^-, NH_4^+	$\Delta[H^+] + \Delta[HAc] = \Delta[OH^-] + \Delta[HCO_3^-] + 2\Delta[CO^{3=}] + \Delta[NH_3]$
H_2CO_3*, HAc, NH_3	$\Delta[H^+] + \Delta[NH_4^+] = \Delta[OH^-] + \Delta[HCO_3^-] + 2\Delta[CO^{3=}] + \Delta[Ac^-]$
H_2CO_3*, Ac^-, NH_3	$\Delta[H^+] + \Delta[HAc] + \Delta[NH_4^+] = \Delta[OH^-] + \Delta[HCO_3^-] + 2\Delta[CO^{3=}]$
HCO_3^-, HAc, NH_4^+	$\Delta[H^+] + \Delta[H_2CO_3^*] = \Delta[OH^-] + \Delta[CO^{3=}] + \Delta[Ac^-] + \Delta[NH_3]$
HCO_3^-, Ac^-, NH_4^+	$\Delta[H^+] + \Delta[H_2CO_3^*] + \Delta[HAc] = \Delta[OH^-] + \Delta[CO^{3=}] + \Delta[NH_3]$
HCO_3^-, HAc, NH_3	$\Delta[H^+] + \Delta[H_2CO_3^*] + \Delta[NH_4^+] = \Delta[OH^-] + \Delta[CO^{3=}] + \Delta[Ac^-]$
HCO_3^-, Ac^-, NH_3	$\Delta[H^+] + \Delta[H_2CO_3^*] + \Delta[HAc] + \Delta[NH_4^+] = \Delta[OH^-] + \Delta[CO^{3=}]$
$CO^{3=}$, HAc, NH_4^+	$\Delta[H^+] + 2\Delta[H_2CO_3^*] + \Delta[HCO_3^-] = \Delta[OH^-] + \Delta[Ac^-] + \Delta[NH_3]$
$CO^{3=}$, Ac^-, NH_4^+	$\Delta[H^+] + 2\Delta[H_2CO_3^*] + \Delta[HCO_3^-] + \Delta[HAc] = \Delta[OH^-] + \Delta[NH_3]$
$CO^{3=}$, HAc, NH_3	$\Delta[H^+] + 2\Delta[H_2CO_3^*] + \Delta[HCO_3^-] + \Delta[NH_4^+] = \Delta[OH^-] + \Delta[Ac^-]$
$CO^{3=}$, Ac^-, NH_3	$\Delta[H^+] + 2\Delta[H_2CO_3^*] + \Delta[HCO_3^-] + \Delta[HAc] + \Delta[NH_4^+] = \Delta[OH^-]$

When employed appropriately with reference conditions, mole balances, and equilibria, any and all of the statements of the proton balance will yield the identical, correct result in analysis of aqueous proton transfers.

We observe that the proton balance is relatively easily written, once the rules are solidly in mind. We must then combine the proton balance with the reference condition before we can use it quantitatively.

10.5.3 The Reference and Initial Conditions for the Proton Balance

With myriad ways to formulate the proton balance, we must have a systematic means by which to establish the initial and reference conditions to render the proton balance useful. As mentioned previously, the reference conditions are the initial, unmixed solutions at their previous equilibrium conditions. For Example 10.5, the reference condition would be the water to which the carbon dioxide is added and the pure carbon dioxide gas added to the water. For Example 10.6, we have not stated the means by which the carbonate, acetate, and ammonia would be combined with the water so we cannot state the reference conditions.

Let us first consider the case that the aqueous solution is initially freshly distilled water (FDW). We now can define the reference condition for the hydronium–water–hydroxide system. Then we must envision the initial condition for acids, salts, or bases that might be added to the FDW. This condition (or set of conditions) arises as the state of the system after the introduction of the various conjugate acids or bases into the aqueous solution, but immediately prior to transfer of any protons whatsoever. For the addition of a nonelectrolyte acid (e.g., carbon dioxide, acetic acid), or base (e.g., ammonia) to water, we envision that the nonelectrolyte has fully dissolved and is initially present only as its nonelectrolyte specie. For the addition of a salt to water, we envision that the salt has dissolved and split entirely into its cations and anions. Use of our fast camera allows us to visualize this theoretical condition and use it quantitatively in our subsequent analyses. Let us put this camera to work and visualize the initial conditions associated with the aqueous solutions resulting from the addition of carbonate to water as three distinct alternative forms, in continuation of Example 10.5.

Example 10.7 Consider that the addition of carbonate to the aqueous solution of Example 10.5 would be carbonic acid, sodium bicarbonate, or sodium carbonate, resulting in total aqueous carbonate concentrations of 0.01 M. Determine and specify the associated initial conditions.

Carbonic acid is a nonelectrolyte and would simply dissolve in the aqueous solution to comprise the initial condition of the solution. The initial condition would be:

$$[H_2CO_3^*]_{init} = 0.01\,M;\ \left[HCO_3^-\right]_{init} = 0;\ \text{and}\ \left[CO_3^=\right]_{init} = 0$$

Both sodium bicarbonate ($NaHCO_3$) and sodium carbonate (Na_2CO_3) would first dissolve and then ionize ($NaHCO_3 \rightarrow Na^+ + HCO_3^-$; $Na_2CO_3 \rightarrow 2Na^+ + CO_3^=$).

For addition of carbonate as sodium bicarbonate, the initial conditions would be:

$$[H_2CO_3*]_{init} = 0;\ [HCO_3^-]_{init} = 0.01\,M;\ \text{and}\ [CO_3^=]_{init} = 0$$

For addition of carbonate as sodium carbonate, the initial conditions would be:

$$[H_2CO_3*]_{init} = 0;\ [HCO_3^-]_{init} = 0;\ \text{and}\ [CO_3^=]_{init} = 0.01\,M$$

In the case that sodium (or any other alkali metal) is the cation associated with the salt, we may neglect it from the specification of the initial condition as we know that sodium (or any other alkali metal) has negligible potential to participate in proton transfer reactions.

Before we move to the quantitative applications of these principles, let us consider the various means by which acetate and ammonia may be introduced into the aqueous solution.

Example 10.8 Consider the addition of acetate as acetic acid and ammonium acetate and the addition of ammonia as ammonium acetate, and anhydrous ammonia. Determine and specify the associated initial conditions were the total acetate and total ammonia concentrations to be 0.01 M.

Acetic acid and anhydrous ammonia are nonelectrolytes and after dissolution would initially be present as the fully protonated acetic acid and fully deprotonated ammonia, respectively. Ammonium acetate would dissolve and ionize to yield the ammonium cation and acetate anion.

For addition of acetate as acetic acid, the initial conditions would be:

$$[HAc] = 0.01M;\ [Ac^-] = 0$$

For addition of ammonia as anhydrous ammonia, the initial conditions would be:

$$[NH_4^+] = 0;\ [NH_3] = 0.01M$$

For addition of ammonium acetate, the initial conditions would be:

$$[NH_4^+] = 0.01M;\ [NH_3] = 0;\ [HAc] = 0;\ [Ac^-] = 0.01M$$

With regard to the initial condition, we may treat each acid system independently of the others, assembling the matrix of initial conditions based on the exact compositions of the acids, salts or bases added to the aqueous solution. We will see next, how we can utilize these initial conditions to compute the final speciation of the aqueous solution to which the acids, salts, and bases have been added.

10.6 ANALYSES OF SOLUTIONS PREPARED BY ADDITION OF ACIDS, BASES, AND SALTS TO WATER

10.6.1 Additions to Freshly Distilled Water (FDW)

Let us keep it simple to begin. We will examine the dilution of an acid, a base, and then a salt into a known volume of aqueous solution whose final ionic strength would be sufficiently low that we may neglect errors associated with the assumption of an infinitely dilute solution. We will begin with freshly distilled water (FDW). FDW is distilled, purified using ion exchange, carbon adsorption, and filtration (0.45 μm, or finer) then stored under an inert gas to prevent dissolution of atmospheric carbon dioxide. The pH of FDW is 7.0 (at 25 °C) and the initial composition is known to be water containing 10^{-7} M H^+ and OH^-.

Example 10.9 Consider that 0.001 mole of acetic acid, then 0.001 mole of anhydrous ammonia, and then 0.001 mole of ammonium acetate are each in turn diluted into 1 L of water whose final ionic strength will be essentially 0. In each case, determine the final speciation of the solution, including the pH and activities (~ concentrations) of each significant conjugate acid and base of each acid system.

We begin by assembling the known information for the acid systems and for the character of the final solution. Acid dissociation constants are available from Table 6.1.

$$pK_{A.NH3} := 9.30 \quad pK_{A.Ac} := 4.70 \quad pK_W := 14 \quad K_W := 10^{-pK_W} = 1 \times 10^{-14}$$

$$K_{A.NH3} := 10^{-pK_{A.NH3}} = 5.012 \times 10^{-10} \qquad K_{A.Ac} := 10^{-pK_{A.Ac}} = 1.995 \times 10^{-5}$$

$$C_{Tot.Ac} := 0.001 \quad C_{Tot.NH3} := 0.001 \quad M$$

Now we will use our knowledge of each of the acid systems to determine, without invoking any simplifying assumptions, the list of potential species resulting for each of the three cases.

Case 1: acetic acid → H^+, OH^-, H_2O, HAc, Ac^-
Case 2: anhydrous ammonia → H^+, OH^-, H_2O, NH_4^+, NH_3
Case 3: ammonium acetate → H^+, OH^-, H_2O, NH_4^+, NH_3, HAc, Ac^-

We list water knowing that we certainly will not, herein, need to consider its presence either in mole balances or equilibria. However, for systems involving waters of sufficiently high dissolved solids content, the activity of water is reduced by the dissolved salts and must be considered in the overall equilibria. This topic is held for later.

We recognize that in cases 1 and 2 we have four unknowns and in case 3 we have six unknowns. Then for cases 1 and 2 we need four independent equations for each and for case 3 we need six independent equations in order to attain complete solutions for each system.

For case 1, these equations are:

the mole balance on acetate,
the deprotonation equilibrium for acetic acid,
the ion product of water, and
a proton balance.

For case 2, the equations are:

the mole balance on ammonia,
the deprotonation equilibrium for ammonium,
the ion product of water, and
a proton balance.

For case 3, the equations are:

the mole balance on ammonia,
the mole balance on acetate,
the deprotonation equilibrium for ammonium,
the deprotonation equilibrium for acetic acid,
the ion product of water, and
a proton balance.

For case 1, we find it most convenient to solve the law of mass action statement either for acetic acid or acetate and use the result in the mole balance. The decision is arbitrary and herein, we will solve for acetic acid and use that result. For case 2, we arbitrarily choose to solve for ammonia in terms of ammonium and use that result in the mole balance. For case 3, we use the work accomplished with the equilibria for cases 1 and 2 in the acetate and ammonia mole balances. Further, in each of the three cases, we use the ion product of water to solve for hydroxide in terms of hydronium. Thus, in cases 1 and 2 the systems of equations are reduced from four to two and in case 3 the system of equations is reduced from six to three. We do not include the algebra herein and the student is encouraged to work it through. We certainly can simply adapt the appropriate general relations assembled in Table 10.4. The process used herein is essentially that used in assembling the abundance fractions in Tables 6.2 and (for nondilute solutions) Table 10.4.

Now to the proton balances; let us use acetic acid and ammonia for the reference species. We will write these in a MathCAD worksheet so we may use the results directly in our algebraic manipulations leading to the final solution. MathCAD is touchy about parentheses, brackets and curly brackets. Each has a special mathematical significance. Thus, in our worksheet we will simplify the work, using the chemical formula, without subscripting stoichiometric coefficients, as the symbol for the activity of each respective specie. Note that we have invoked the infinitely dilute assumption and $[i] \approx \{i\}$. Later, when we need to express specie concentrations, we will use an upper case C with the specie formula as a subscript.

For case 1, our two-equation system is:

$$C_{Tot.Ac} = \frac{H}{K_{A.Ac}} \cdot Ac + Ac \qquad \Delta H = \Delta OH + \Delta Ac$$

For case 2, our system is:

$$C_{Tot.NH3} = NH4 + \frac{K_{A.NH3}}{H} \cdot NH4 \qquad \Delta H + \Delta NH4 = \Delta OH$$

For case 3, the set of relations is:

$$C_{Tot.NH3} = NH4 + \frac{K_{A.NH3}}{H} \cdot NH4 \qquad C_{Tot.Ac} = \frac{H}{K_{A.Ac}} \cdot Ac + Ac \qquad \Delta H + \Delta NH4 = \Delta OH + \Delta Ac$$

Note that we have retained hydroxide in the proton balance. The change in hydroxide must employ the initial value of the hydroxide concentration and we will see later that we must be careful about when we use the ion product of water to express the hydroxide activity. We will use the ion product of water at the final algebraic stage, just prior to implementing the solution.

Now we must convert all of the Δs to final minus initial. Here we use our fast camera to view each system at its critical initial state, just prior to the beginning of any of the proton transfer reactions. For cases 1 and 2, the acetic acid and anhydrous ammonia have dissolved. For case 3, the ammonium acetate has dissolved and the ammonium cations have split from the acetate anions, but neither ionized specie has undergone any acceptance or donation of protons. Then, for the acid systems the initial conditions are:

Case 1:

$$Ac_i := 0 \qquad HAc_i := 0.001 \quad \frac{mol}{L}$$

Case 2:

$$NH3_i := 0.001 \qquad NH4_i := 0 \quad \frac{mol}{L}$$

Case 3:

$$Ac_i := 0.001 \quad HAc_i := 0 \qquad NH3_i := 0 \quad NH4_i := 0.001 \quad \frac{mol}{L}$$

For water, of course, the initial conditions are the same for all three cases:

$$H_i := 10^{-7} \qquad OH_i := 10^{-7} \quad \frac{mol}{L}$$

Note that in the previous statements, we have employed two different equal signs. In MathCAD, the colon-equal (▮ := ▮) is an assignment, while the bold equal (▮ **=** ▮) is the Boolean equal. Use of the colon-equal instructs MathCAD to assign the value of the RHS into the scalar or vector variable on the LHS. The Boolean equal is used in solve blocks. Since there is no explicit instruction associated directly with the Boolean equal, we may use it in algebraic manipulations within MathCAD worksheets without the error messages arising from instructing MathCAD to perform some operation for which all necessary information has not been provided.

We are now ready to solve the three systems. Let us first examine the three fully populated proton balances. We might subscript H, Ac, and NH_4 with an f to denote the final value, but let us simply use the aforementioned symbol for activity as the final value.

$$H - H_i = \frac{K_W}{H} - OH_i + Ac - Ac_i$$

$$H - H_i + NH4 - NH4_i = \frac{K_W}{H} - OH_i$$

$$H - H_i + NH4 - NH4_i = \frac{K_W}{H} - OH_i + Ac - Ac_i$$

In all three cases, we note that H_i and OH_i appear on each side of each equation. We must resist our temptation to realize that, for this particular system, their values cancel and then omit them from further consideration. Previous authors of aquatic chemistry texts have performed this simplification at this step and carried it forward thus rendering their results usable only for computations which begin with freshly distilled, ultrapure water. Environmental systems are uncooperative in this regard. At best, natural water is affected by carbon dioxide in the atmosphere when falling as precipitation. We will therefore not make this simplification. We also note that the initial concentrations of acetate and ammonium for cases 1 and 2 are 0. Again, previous work has tended to claim that the initial concentrations of acid and base species which are conjugates of added acids or bases, can be neglected. Such action then renders the resultant technique specific to solutions to which neatly packaged acids or bases are added. In fact, it is at this point that most authors set aside the proton balance in favor of the charge balance for solution of aqueous equilibria. As a consequence of such action, a powerful and flexible tool has lain dormant and unused for many years.

We may now complete the full, uncompromised solution for each of the three cases. Please note the complete absence of any simplifying assumptions.

For case 1, we specify the initial concentrations and give MathCAD reasonable guesses for the unknowns for which we desire final values. We would expect acetic acid to donate a few protons which would be accepted by water to form hydronium and by hydroxide to form water, such that the final pH would be mildly acidic. We have also instructed MathCAD to assign the results of the given-find block into the vector containing H and Ac such that the values are resident in these scalar variables for future use within the worksheet.

Known values: initial guesses:

$Ac_i := 0 \quad H_i := 10^{-7} \quad OH_i := 10^{-7} \quad H := 10^{-5} \quad Ac := 10^{-5}$

Given $\quad C_{Tot.Ac} = \frac{H}{K_{A.Ac}} \cdot Ac + Ac \quad H - H_i = \frac{K_W}{H} - OH_i + Ac - Ac_i$

$$\begin{pmatrix} H \\ Ac \end{pmatrix} := Find(H, Ac) \rightarrow \begin{pmatrix} 1.316 \times 10^{-4} \\ 1.316 \times 10^{-4} \end{pmatrix} \qquad pH := -log(H) \rightarrow 3.881$$

Were this result our final destination on the road to analyses of environmental systems, we would have neglected the hydroxide and invoked the simplifying assumption to yield the final relation that H ≈ Ac. Since this is our beginning, we have retained the entire structure of the solution. For the implementation of the given-find block, we must ensure that the given statement is just that. Note that we may enter text into a MathCAD worksheet (i.e., Known values: and initial guesses:). If we want to convert any statement to text, we need simply hit the space bar. If we do that for the given statement, it becomes text and no longer would be recognized by MathCAD as the invocation of the solve block. Note also that each given requires a single find and that MathCAD is not yet programmed to enable nesting of a given-find block within a given-find block.

For case 2, we have a similar solution. We know that ammonia is a rather strong base and that our final solution perhaps will be moderately alkaline. For the next solution, if we initially specify a pH that is too low, MathCAD will yield a correct mathematical result that is certainly not physically correct. Were we to continue with algebraic substitutions by solving the mole balance for ammonium and substitute that result into the proton balance, our result would be a quadratic, with two mathematically correct solutions. Next, we observe the solution that we must discard as specie concentrations cannot be negative.

$NH4_i := 0 \quad H_i := 10^{-7} \quad OH_i := 10^{-7} \quad H := 10^{-8} \quad NH4 := 10^{-5}$

Given $\quad C_{Tot.NH3} = NH4 + \frac{K_{A.NH3}}{H} \cdot NH4 \quad H - H_i + NH4 - NH4_i = \frac{K_W}{H} - OH_i$

$$\begin{pmatrix} H \\ NH4 \end{pmatrix} := Find(H, NH4) \rightarrow \begin{pmatrix} -1 \times 10^{-3} \\ 1 \times 10^{-3} \end{pmatrix} \qquad pH := -log(H) \rightarrow 3 - 1.364i$$

Here we must adjust the initial guesses, and after adjusting the guess for H to a sufficiently high value of pH, the solution that we desire is accomplished.

$NH4_i := 0 \quad H_i := 10^{-7} \quad OH_i := 10^{-7} \quad H := 10^{-10} \quad NH4 := 10^{-5}$

$$\text{Given} \quad C_{Tot.NH3} = NH4 + \frac{K_{A.NH3}}{H} \cdot NH4 \qquad H - H_i + NH4 - NH4_i = \frac{K_W}{H} - OH_i$$

$$\begin{pmatrix} H \\ NH4 \end{pmatrix} := \text{Find}(H, NH4) = \begin{pmatrix} 7.597 \times 10^{-11} \\ 1.316 \times 10^{-4} \end{pmatrix} \qquad pH := -\log(H) = 10.119$$

$$OH := \frac{K_W}{H} = 1.316 \times 10^{-4}$$

Here, had we opted for simplification of the system we might have written that $\Delta[NH_4] \approx \Delta[OH^-]$, used the initial zero concentration values and written our proton balance as $[NH_4^+] \approx K_W/[H^+]$. But, again, since this is our beginning work with the proton balance, we retained the full solution.

Then for case 3 we again specify the necessary initial conditions, venture guesses for the desired unknowns and solve using the given-find block.

Known values: initial guesses:

$$\begin{pmatrix} H_i \\ OH_i \\ NH4_i \\ Ac_i \end{pmatrix} := \begin{pmatrix} 10^{-7} \\ 10^{-7} \\ 0.001 \\ 0.001 \end{pmatrix} \qquad \begin{pmatrix} H \\ NH4 \\ Ac \end{pmatrix} := \begin{pmatrix} 10^{-7} \\ 10^{-5} \\ 10^{-5} \end{pmatrix}$$

Given

$$C_{Tot.NH3} = NH4 + \frac{K_{A.NH3}}{H} \cdot NH4 \qquad C_{Tot.Ac} = \frac{H}{K_{A.Ac}} \cdot Ac + Ac$$

$$H - H_i + NH4 - NH4_i = \frac{K_W}{H} - OH_i + Ac - Ac_i$$

$$\begin{pmatrix} H \\ NH4 \\ Ac \end{pmatrix} := \text{Find}(H, NH4, Ac) = \begin{pmatrix} 1 \times 10^{-7} \\ 9.95 \times 10^{-4} \\ 9.95 \times 10^{-4} \end{pmatrix} \qquad pH := -\log(H) = 7$$

As we might have expected, the addition of ammonium acetate to water yields proton donation and proton acceptance reactions that essentially cancel each other. We see that the number of protons donated by ammonium (0.001 – 0.000995 = 0.000005 mol/L) is almost exactly the number accepted by acetate as $\Delta[NH_4^+] \approx \Delta[Ac^-]$. Of note, the final $[H^+]$ is 1×10^{-7} to 15 or more significant figures.

As we step away from Example 10.9, in which we have examined the proton balance in simplistic systems, we can take with us some additional understandings of the power of MathCAD. Its creators are mathematicians and to use their tool we must play by their rules. Early in our education, ironically by mathematicians, we were instructed in solving problems to "list all our givens" en route to the solution. Here, of course, all the "givens" appear ahead of the invocation of the given-find block by the given statement. Within the block, our purposes will be best suited by inserting only the equations that comprise the system which must be solved. We must have a number of equations exactly equal to the number of unknowns we seek. We must define all variables used and must furnish initial guesses for all variables for which we desire solutions. We find that assignment of the find() function of the given-find block into a matrix containing the names of the variables for which we seek the solution will serve us well. We will use this capability to great benefit in worksheets associated with examples to come later. Lastly, we observe that for assignment of values (and later even relations) into variables, the matrix capacity built into MathCAD allows for assembly of compact, well-organized worksheets.

10.6.2 Dissolution of a Weak Acid in Water

For the next item of work with the proton balance and associated mole balance and equilibrium relations, let us consider the addition of carbon dioxide to freshly distilled water (FDW). We might equate this process to the natural equilibration of precipitation with atmospheric carbon dioxide or the equilibration over time of distilled water held in an open jug in the laboratory with the carbon dioxide in the atmosphere of the room. Distilled water held in such open jugs is suitable for routine use, but as we will see, is not equivalent to the freshly distilled water considered in the previous example.

Example 10.10 Consider that a liter of water is drawn into a beaker from a system used to render tap water to be ultrapure and set on the counter in the laboratory located in one of Minneapolis, Minnesota's suburbs, at an elevation of 230 m (755 ft) above mean sea level. Assume that the beaker is left where substances other than the atmosphere would not come in contact or otherwise affect the nature of the resultant solution. Determine the final pH and chemical speciation of the resultant water. Make no simplifying assumptions en route to the final answer.

We know that the atmosphere contains nitrogen, oxygen, and carbon dioxide along with several minor constituents. Neither nitrogen nor oxygen will participate in proton transfer reactions, and other minor constituents either will not donate or accept protons or are present at abundances too small to consider. We then will focus upon carbon dioxide. Under the final state of the solution the air/water distribution of carbon dioxide and the proton transfer reactions associated with carbonic acid, bicarbonate, and carbonate will have attained equilibrium conditions.

We begin by determining the species that would potentially exist in the final solution:

H^+, OH^-, H_2O, $H_2CO_3^*$, HCO_3^-, $CO_3^=$

Again, the activity of water is not really an unknown in this application so we are left with five unknowns, necessitating a system of five independent equations. We write the mole balance on carbonate species, and are jolted by the fact that the total carbonate concentration is unknown, bringing the grand total to six unknowns. We require a sixth equation and turn to Henry's law to enable specification of the aqueous activity of carbonic acid from the air/water distribution coefficient. These six equations are:

a mole balance upon carbonate species,
the ion product of water,
the two equilibria for the carbonate system
proton balance
the Henry's law equilibrium for gaseous carbon dioxide and the combined carbonic acid specie

The current average concentration of carbon dioxide in the atmosphere is 390 ppm_v. We will solve the system, assuming infinitely dilute conditions for now, opting to return later in the chapter to see if our assumption results in sufficiently small error. For completeness at this juncture of our examination of the solution of acid/base problems, we will state our full original system of equations, as they appear in our MathCAD worksheet. The first five are as follows.

$$K_W = H \cdot OH \qquad C_{Tot.CO3} = H2CO3 + HCO3 + CO3$$

$$K_{1.CO3} = \frac{H \cdot HCO3}{H2CO3} \qquad K_{2.CO3} = \frac{H \cdot CO3}{HCO3} \qquad K_{H.CO2} = \frac{H2CO3}{P_{CO2}}$$

The sixth, the proton balance, requires just a little analysis. Since carbonic acid will be in equilibrium with atmospheric carbon dioxide, its dissolution by itself into the aqueous solution is not associated with proton transfer reactions. Carbonic acid is therefore a solid choice as a reference specie for the carbonate system. Water, of course, is our other reference specie. Our proton balance then, equating evidence of protons accepted with evidence of protons donated is as follows.

$$\Delta H = \Delta HCO3 + 2 \cdot \Delta CO3 + \Delta OH$$

We will use the same strategies as employed in Example 10.9 to reduce our final system of equations to two, opting to solve for the activities of bicarbonate and hydronium.

For the known parameters and initial conditions, since we began with FDW, we have the following.

$$\begin{pmatrix} K_W \\ K_{1.CO3} \\ K_{2.CO3} \\ K_{H.CO2} \end{pmatrix} := \begin{pmatrix} 10^{-14} \\ 10^{-6.35} \\ 10^{-10.33} \\ 0.0339 \end{pmatrix} \qquad \begin{pmatrix} H_i \\ OH_i \\ HCO3_i \\ CO3_i \end{pmatrix} := \begin{pmatrix} 10^{-7} \\ 10^{-7} \\ 0 \\ 0 \end{pmatrix} \frac{mol}{L}$$

Were we to specifically examine rain water, we would consider that the initial formation of each droplet via condensation onto a minute solid particle would indeed be pure water, equivalent of the FDW we consider here.

For Henry's law, we need the partial pressure of carbon dioxide for the atmospheric condition within the laboratory. We recall a relation used in Example 5.3 and use the elevation at the laboratory to compute total pressure and then use the definition of ppm_v to obtain the partial pressure of carbon dioxide. We solve the Henry's law equilibrium for H_2CO_3* in terms of P_{CO_2} and arrive at a known value for H_2CO_3*.

$$P_{Tot} = 0.973 \text{ atm} \qquad P_{CO2} := 390 \cdot 10^{-6} \cdot P_{Tot} = 3.795 \times 10^{-4} \text{ atm}$$

$$H2CO3 := K_{H.CO2} \cdot P_{CO2} = 1.286 \times 10^{-5} \frac{mol}{L}$$

We may write the mole and proton balances in terms of carbonic acid, specify the initial guesses for $\{H^+\}$ and $C_{Tot.CO_3}$ and invoke the given-find block. For completeness, the entire solution is included.

initial guesses:

$$\begin{pmatrix} H \\ C_{Tot.CO3} \end{pmatrix} := \begin{pmatrix} 10^{-6} \\ 2 \cdot 10^{-5} \end{pmatrix} \frac{mol}{L}$$

Given $\quad C_{Tot.CO3} = H2CO3 \cdot \left(1 + \frac{K_{1.CO3}}{H} + \frac{K_{1.CO3} \cdot K_{2.CO3}}{H^2}\right)$

$$H - H_i = \frac{K_{1.CO3} \cdot H2CO3}{H} - HCO3_i + 2 \cdot \left(\frac{K_{1.CO3} \cdot K_{2.CO3} \cdot H2CO3}{H^2} - CO3_i\right) + \frac{K_W}{H} - OH_i$$

$$\begin{pmatrix} H \\ C_{Tot.CO3} \end{pmatrix} := \text{Find}(H, C_{Tot.CO3}) = \begin{pmatrix} 2.399 \times 10^{-6} \\ 1.526 \times 10^{-5} \end{pmatrix} \qquad pH := -\log(H) = 5.62$$

$$\begin{pmatrix} OH \\ HCO3 \\ CO3 \end{pmatrix} := \begin{pmatrix} \frac{K_W}{H} \\ \frac{K_{1.CO3} \cdot H2CO3}{H} \\ \frac{K_{1.CO3} \cdot K_{2.CO3} \cdot H2CO3}{H^2} \end{pmatrix} = \begin{pmatrix} 4.168 \times 10^{-9} \\ 2.395 \times 10^{-6} \\ 4.669 \times 10^{-11} \end{pmatrix} \frac{mol}{L}$$

Had we chosen to neglect hydroxide and carbonate in the proton balance and carbonate in the mole balance we would have written the proton balance as $\Delta[H^+] \approx \Delta[HCO_3^-]$, which would not have been greatly in error.

From Example 10.10, we leave with the understandings that there is no such thing as freshly distilled water in environmental systems, thus use of the FDW assumption in computations involving water chemistry of environmental systems can be fraught with significant error. Moreover, since rainwater (or snowmelt) is the source of all the Earth's freshwater, we must consider the carbonate system in virtually every environmental system we might encounter.

10.6.3 Dissolution of a Basic Salt in Water

We have been waxing poetic about the importance of activity coefficients but have not yet applied them. Perhaps we are ready now to do so. We also should do one more example of a salt added to water, with the potential for significant transfers of protons. The chemical sodium carbonate (Na_2CO_3, often called soda ash) is used extensively in softening of water for removal of noncarbonate hardness. In order for calcium to be precipitated from water, it must be as the carbonate. Soda ash is added to water that is poor in natural carbonates to allow this reaction to occur. The addition occurs as a solution and the solution must be made up by operators of water treatment facilities prior to addition as a softening reagent. Let us investigate a solution of sodium carbonate. First, we will address the FDW and dilute solution idea, then we will consider that the water to which the reagent is added has the character of that addressed by Example 10.10, and then we will apply ionic strength and activity and see whether the simplifying assumption of dilute solution yields satisfactory results when compared against the more complete solution.

Example 10.11 Consider that a 0.01 M solution of sodium carbonate is to be prepared by the addition of soda ash to water. Perform an approximate solution assuming the resultant solution would be infinitely dilute. Then, refine that solution to address the reality that in water treatment systems, technical grade distilled (one pass through a commercial distiller) or even tap water would be used for solution make-up. Finally, apply ionic strength and activity coefficients to yield a solution for the nondilute case. For the second and third cases, assume that a supply of technical grade distilled water (distilled and held in an open reservoir) is available for the makeup of the sodium carbonate solution. Also assume that the facility needing the prepared solution is located in Minneapolis, MN, at an elevation equal to that of the laboratory of Example 10.10.

We will streamline somewhat the presentation here, including important new details only and, of course, present the final solution. We will perform the steps outlined in Examples 10.9 and 10.10 but simply will not burden the presentation with the details. Of note, we will choose carbonic acid as our reference specie

known values:

$$\begin{pmatrix} K_W \\ K_{1.CO3} \\ K_{2.CO3} \end{pmatrix} := \begin{pmatrix} 10^{-14} \\ 10^{-6.35} \\ 10^{-10.33} \end{pmatrix}$$

initial conditions:

$$\begin{pmatrix} H_i \\ OH_i \\ H2CO3_i \\ HCO3_i \\ CO3_i \\ C_{Tot.CO3} \end{pmatrix} := \begin{pmatrix} 10^{-7} \\ 10^{-7} \\ 0 \\ 0 \\ 0.01 \\ 0.01 \end{pmatrix} \frac{mol}{L}$$

initial guesses:

$$\begin{pmatrix} H \\ CO3 \end{pmatrix} := \begin{pmatrix} 10^{-11} \\ 0.009 \end{pmatrix} \frac{mol}{L}$$

$$\text{Given} \quad H - H_i = \frac{K_W}{H} - OH_i + \frac{H}{K_{2.CO3}} \cdot CO3 - HCO3_i + 2 \cdot (CO3 - CO3_i)$$

$$C_{Tot.CO3} = CO3 \cdot \left(\frac{H^2}{K_{1.CO3} \cdot K_{2.CO3}} + \frac{H}{K_{2.CO3}} + 1 \right)$$

$$\begin{pmatrix} H_{dil} \\ CO3 \end{pmatrix} := \text{Find}(H, CO3) \rightarrow \begin{pmatrix} 7.357 \times 10^{-12} \\ 8.641 \times 10^{-3} \end{pmatrix} \quad M \qquad pH := -\log(H_{dil}) = 11.133$$

$$\begin{pmatrix} HCO3 \\ H2CO3 \end{pmatrix} := \begin{pmatrix} \dfrac{H_{dil}}{K_{2.CO3}} \cdot CO3 \\ \dfrac{H_{dil}^2}{K_{2.CO3} \cdot K_{1.CO3}} \cdot CO3 \end{pmatrix} = \begin{pmatrix} 1.359 \times 10^{-3} \\ 2.239 \times 10^{-8} \end{pmatrix} \frac{mol}{L}$$

FIGURE E10.11.1 *Screen capture of a solve block for speciation of the carbonate system when soda ash is added to freshly distilled water considering infinitely dilute conditions.*

(since carbonic acid is equilibrated with CO_2 in the vapor, use of bicarbonate or carbonate invokes some undesirable complications). If we designate carbonic acid as our reference specie for the carbonate system, our mole, and proton balances here are identical to those written for Example 10.11, but total carbonate and the initial conditions are quite different.

The results considering addition of soda ash to FDW and implementation of the infinitely dilute assumption, are shown in Figure E10.11.1. The initial concentration for carbonate arises when we visualize that, just prior to the proton transfer reactions, a mole of dissolved sodium carbonate ionizes to produce two moles of sodium ions and one mole of carbonate ions.

Let us now employ the technical grade makeup water, retaining the infinitely dilute assumption. Since we have solved for the initial condition of the water in Example 10.10, we need only alter the set of initial conditions and adjust the value of the concentration of total carbonate species, as shown in Figure E10.11.2. We see that the results differ only slightly from the FDW case. The preparation of a rather high-concentration solution of sodium carbonate overwhelms the initial conditions.

known values:

$$\begin{pmatrix} K_W \\ K_{1.CO3} \\ K_{2.CO3} \end{pmatrix} := \begin{pmatrix} 10^{-14} \\ 10^{-6.35} \\ 10^{-10.33} \end{pmatrix}$$

initial guesses:

$$\begin{pmatrix} H \\ CO3 \end{pmatrix} := \begin{pmatrix} 10^{-11} \\ 0.009 \end{pmatrix} \frac{mol}{L}$$

initial conditions:

$$\begin{pmatrix} H_i \\ OH_i \\ H2CO3_i \\ HCO3_i \\ CO3_i \\ C_{Tot.CO3} \end{pmatrix} := \begin{pmatrix} 2.399 \cdot 10^{-6} \\ 4.168 \cdot 10^{-9} \\ 1.286 \cdot 10^{-5} \\ 2.395 \cdot 10^{-6} \\ 0.01 \\ 0.0100156 \end{pmatrix} \frac{mol}{L}$$

Given $H - H_i = \frac{K_W}{H} - OH_i + \frac{H}{K_{2.CO3}} \cdot CO3 - HCO3_i + 2 \cdot (CO3 - CO3_i)$

$$C_{Tot.CO3} = CO3 \cdot \left(\frac{H^2}{K_{1.CO3} \cdot K_{2.CO3}} + \frac{H}{K_{2.CO3}} + 1 \right)$$

$$\begin{pmatrix} H \\ CO3 \end{pmatrix} := Find(H, CO3) = \begin{pmatrix} 7.442 \times 10^{-12} \\ 8.641 \times 10^{-3} \end{pmatrix} M \qquad pH := -\log(H) = 11.128$$

$$\begin{pmatrix} HCO3 \\ H2CO3 \end{pmatrix} := \begin{pmatrix} \frac{H}{K_{2.CO3}} \cdot CO3 \\ \frac{H^2}{K_{2.CO3} \cdot K_{1.CO3}} \cdot CO3 \end{pmatrix} = \begin{pmatrix} 1.375 \times 10^{-3} \\ 2.291 \times 10^{-8} \end{pmatrix} \frac{mol}{L}$$

FIGURE E10.11.2 *Screen capture of a solve block for speciation of the carbonate system when soda ash is added to technical grade distilled water (equilibrated with the atmosphere) considering infinitely dilute conditions.*

Let us now include the effects of the nondilute solution. We need three additional equations as I, γ_1, and γ_2 are unknown. We will need to include the sodium ion, introduced into the solution with the soda ash in the computation of ionic strength. We will update the carbonate mole balance using the general mole balance relation from Table 10.3 for the diprotic acid and $n=0$, appropriate to the carbonate system.

$$C_{\text{Tot.CO}_3} = \frac{\{H_2CO_3^*\}}{\gamma_0} + \frac{\{HCO_3^-\}}{\gamma_1} + \frac{\{CO_3^=\}}{\gamma_2}$$

We will write the proton balance using specie concentrations.

$$\Delta C_H = \Delta C_{OH} + \Delta C_{HCO_3} + 2\Delta C_{CO_3}$$

We will then use Equation 10.1 in combination with the definition of the change in concentration to express the concentrations of the final unknowns using specie activities.

known values:

$$\begin{pmatrix} K_W \\ K_{1.CO3} \\ K_{2.CO3} \\ A \end{pmatrix} := \begin{pmatrix} 10^{-14} \\ 10^{-6.35} \\ 10^{-10.33} \\ 0.5115 \end{pmatrix}$$

initial conditions:

$$\begin{pmatrix} C_{H.i} \\ C_{OH.i} \\ C_{Na} \\ C_{HCO3.i} \\ C_{CO3.i} \\ C_{Tot.CO3} \end{pmatrix} := \begin{pmatrix} 2.399 \times 10^{-6} \\ 4.168 \times 10^{-9} \\ 0.02 \\ 2.395 \times 10^{-6} \\ .01 \\ 0.01001526 \end{pmatrix} \frac{mol}{L}$$

initial guesses:

$$\begin{pmatrix} H \\ CO3 \\ I \\ \gamma_1 \\ \gamma_2 \end{pmatrix} := \begin{pmatrix} 10^{-12} \\ 0.009 \\ 0.05 \\ .9 \\ .6 \end{pmatrix} \frac{mol}{L}$$

Given

$$I = \frac{1}{2}\left(\frac{H}{\gamma_1} + C_{Na} + \frac{K_W}{H \cdot \gamma_1} + \frac{H}{K_{2.CO3} \cdot \gamma_1} \cdot CO3 + 4 \cdot \frac{CO3}{\gamma_2}\right) \qquad \gamma_1 = 10^{\frac{-A \cdot \sqrt{I}}{1+\sqrt{I}}}$$

$$C_{Tot.CO3} = CO3 \cdot \left(\frac{H^2}{K_{1.CO3} \cdot K_{2.CO3}} + \frac{H}{K_{2.CO3} \cdot \gamma_1} + \frac{1}{\gamma_2}\right) \qquad \gamma_2 = 10^{\frac{-A \cdot 4 \cdot \sqrt{I}}{1+\sqrt{I}}}$$

$$\frac{H}{\gamma_1} - C_{H.i} = \frac{K_W}{H \cdot \gamma_1} - C_{OH.i} + \frac{H}{K_{2.CO3} \cdot \gamma_1} \cdot CO3 - C_{HCO3.i} + 2 \cdot \left(\frac{CO3}{\gamma_2} - C_{CO3.i}\right)$$

$$\begin{pmatrix} H \\ CO3 \\ I \\ \gamma_1 \\ \gamma_2 \end{pmatrix} := \text{Find}(H, CO3, I, \gamma_1, \gamma_2) \rightarrow \begin{pmatrix} 1.037 \times 10^{-11} \\ 4.462 \times 10^{-3} \\ 0.029 \\ 0.843 \\ 0.505 \end{pmatrix} \frac{mol}{L} \qquad pH := -\log(H) \rightarrow 10.98$$

$$\begin{pmatrix} HCO3 \\ H2CO3 \end{pmatrix} := \begin{pmatrix} \frac{H}{K_{2.CO3}} \cdot CO3 \\ \frac{H^2}{K_{2.CO3} \cdot K_{1.CO3}} \cdot CO3 \end{pmatrix} \rightarrow \begin{pmatrix} 9.896 \times 10^{-4} \\ 2.298 \times 10^{-8} \end{pmatrix} \frac{mol}{L}$$

$$\begin{pmatrix} C_H \\ C_{CO3} \\ C_{HCO3} \\ C_{H2CO3} \end{pmatrix} := \begin{pmatrix} \frac{H}{\gamma_1} \\ \frac{CO3}{\gamma_2} \\ \frac{H}{K_{2.CO3} \cdot \gamma_1} \cdot CO3 \\ H2CO3 \end{pmatrix} \rightarrow \begin{pmatrix} 1.231 \times 10^{-11} \\ 8.841 \times 10^{-3} \\ 1.174 \times 10^{-3} \\ 2.298 \times 10^{-8} \end{pmatrix} \frac{mol}{L}$$

FIGURE E10.11.3 *Screen capture of a solve block for speciation of the carbonate system when soda ash is added to technical grade distilled water (equilibrated with the atmosphere) considering non-dilute conditions.*

$$[i]=\frac{\{i\}}{\gamma_i};\quad \Delta[i]=[i]_{\text{final}}-[i]_{\text{initial}}$$

The populated result includes the activities of the unknowns we seek and the corresponding initial concentrations.

$$\frac{\{H^+\}}{\gamma_1}-C_{\text{H,init}}=\frac{\{\text{OH}^-\}}{\gamma_1}-C_{\text{OH,init}}+\frac{\{\text{HCO}_3^-\}}{\gamma_1}-C_{\text{HCO}_3\text{,init}}+2\left(\frac{\{\text{CO}_3^=\}}{\gamma_2}-C_{\text{CO}_3\text{,init}}\right)$$

The initial concentrations as shown in the proton balance are known. Some additional algebraic substitutions are performed, but not shown here. The final solution is then implemented using MathCAD's given-find, as illustrated in Figure E10.11.3. At first glance, the result employing nondilute solution principles appears to be relatively close to the infinitely dilute case, with pH value of 10.98 versus. In consideration of these results, we must recall that pH is a logarithmic value and to compute the relative error of the infinitely dilute result with the nondilute result we must compare the predicted proton **concentrations**.

$$\text{RelErr}_{\%} := \frac{H_{dil} - C_H}{C_H}\cdot 100 = -40.2$$

Relative error of 40% is hardly acceptable.

Perhaps we are comfortable with the level of the error associated with neglect of nonzero ionic strength in computations such as those of Examples 10.9–10.11, perhaps not. However, as demonstrated in Example 10.11 the inclusion of activity coefficients in these computations does not unduly complicate them. In most cases, the major ions in aqueous solutions are determined via assay and thus concentrations are known. When proton transfer reactions are of sufficient magnitude to affect ionic strength, the inclusion of the relevant species in computation of ionic strength and activity coefficients, as shown in Example 10.11, is relatively straightforward.

We will step away from Examples 10.9 to 10.11 and carry with us a deeper knowledge of and capacity to implement solutions of multiple, nonlinear equations using MathCAD's given-find block. Certainly, the solver may be used with many more than the five equations we have now employed and we wonder about the practice of performing algebraic substitutions, certainly potential sources of errors. This author has solved systems of up to 13 equations. In that particular endeavor, several hours of work with the initial guesses was necessary to attain a correct solution. With algebraic substitutions, that particular system was eventually reduced to four equations, necessitating only a couple combinations of initial guesses to attain the correct solution. For acid–base systems of the nature addressed in Examples 10.9–10.11, the system of equations can be reduced to a number that includes one mole balance for each acid system involved, the definition of ionic strength, the definition of each activity coefficient needed, and one proton balance. Algebra is a great ally in solving aqueous acid/base speciation systems.

10.6.4 A Few Words about the Charge Balance

Authors of texts addressing aquatic chemistry (e.g., Snoeyink and Jenkins, 1980; Stumm and Morgan, 1996; and more recently Brezonik and Arnold, 2011) have relied heavily upon application of a charge balance for solution of equilibrium problems such as those of Examples 10.9–10.11. The charge balance is merely a statement that the total charge of all cations in solution must numerically equal the total charge of all anions.

$$\sum_{N_i}\left([A_i]\cdot z_i\right)=\sum_{N_j}\left([C_j]\cdot z_j\right)$$

A_i is anion i, C_j is cation j, z is the residual ionic charge and N is the number of anions or cations in the system. As long as the entire dissolved content of an aqueous solution is known with great accuracy, the charge balance works well. Its application requires little knowledge beyond that of solution electroneutrality. As long as we examine only solutions that we prepare by adding acids, bases, and salts to freshly (or even technical grade) distilled water, the charge balance perhaps would serve us well. However, once we venture into real environmental systems, for which we must depend on chemical assay for characterization of the solution, the charge balance becomes of much less utility. For example, when examining a water sample for treatment options, we would obtain assays of the major ions present in the water. Our first task would be to check the condition of electroneutrality. We can define the relative error of the electroneutrality condition relative either to cations or anions. The error being computed as the difference between the total anion and cation abundances relative to either the cation or anion abundance.

$$\text{err}_{\text{rel}}=100\,\frac{\sum_{N_i}\left([A_i]\cdot z_i\right)-\sum_{N_j}\left([C_j]\cdot z_j\right)}{\sum_{N_j}\left([C_j]\cdot z_j\right)}=100\,\frac{\sum_{N_j}\left([C_j]\cdot z_j\right)-\sum_{N_i}\left([A_i]\cdot z_i\right)}{\sum_{N_i}\left([A_i]\cdot z_i\right)}$$

Most often we would feel comfortable if the relative error of the electroneutrality condition were less than say 5%. Employing a charge balance in the case that electroneutrality were violated by 5%, or even 1%, would lead to great uncertainty in the results of the computations.

This textbook is dedicated to applications of principles to modeling of environmental processes in environmental systems. In order to illustrate these applications, we must begin with well-defined laboratory type aqueous systems. We gain our understandings therefrom and find that we may then take these understandings along with us when we examine environmental systems. Then, assimilation of the capacity to employ the proton balance will serve the student (who later will become the practitioner or educator) far better than reliance solely upon the charge balance. Certainly, Examples 10.9–10.11 may be solved by replacing the proton balances in each case with charge balances. The student is encouraged to verify this. Hereafter, we will employ the proton balance in contexts that will increasingly become more representative of real environmental systems.

10.7 ANALYSIS OF MIXED AQUEOUS SOLUTIONS

We are ready now to consider the next level in the application of acid/base principles to computations of speciation in aqueous solutions—mixing of two or more solutions with each other. One fundamental tenet arises from the zero-volume mixing principles described and applied in Chapter 7. In Chapter 7, these were applied to flow systems. We find that we may directly apply them to volume systems as flow is simply volume per time. We will bring forward the fundamental relations from Chapter 7.

$$Q_{\text{Tot}} = \sum_{k=1}^{n} Q_k \tag{7.2}$$

$$C_{\text{mix},i} = \sum_{k=1}^{n} \frac{Q_k \cdot C_{k,i}}{Q_{\text{Tot}}} = \frac{\sum_{k=1}^{n} Q_k \cdot C_{k,i}}{\sum_{k=1}^{n} Q_k} \tag{7.3}$$

Herein we have subscripted C_{out} as C_{mix} and note that Q (flow) is easily replaced by V (volume). We also recall that for Equations 7.2 and 7.3 k refers to each of the particular flows or volumes, i refers to the ith constituent present in one or more of the solutions, and n is the number of flows or volumes to be mixed. In order that we may simplify nomenclature and resulting algebraic manipulations, we define f_k as the fraction of the total mixture contributed by solution k.

$$f_k = \frac{Q_k}{Q_{\text{Tot}}} \tag{10.15}$$

Equation 7.3 is particularly useful in our current context for determining mixed values for substances that are conservative, that do not change as a consequence of proton transfer (or other) reactions. These particular substances include major ions and total abundance of acid species for targeted acid systems. We will also find the application of these mixing relations to be vital to the computation of the initial conditions commensurate with our characterization of a mixed solution just prior to occurrence of proton transfer reactions. With our "snapshot" idea, we can treat each of the reacting species as if it were conservative.

10.7.1 Mixing Computations with Major Ions

Since we have determined that we will employ nondilute principles with our examinations of aqueous speciation, the employment of ionic strength will be ubiquitous in our ensuing analyses. As long as major ions are not significantly involved in proton transfer reactions, we may short-cut the computation of ionic strength for implementation in computations involving mixing of solutions containing acids and bases.

Example 10.12 Consider two solutions each containing calcium sulfate and sodium chloride at different concentration levels and develop a relationship that allows use of mixing fractions from Equation 10.15 directly with ionic strength. We write the relations for the ionic strength of the two individual and the mixed solutions.

$$I_1 = \frac{1}{2} \cdot \left(C_{1.Na} + C_{1.Cl} + 4 \cdot C_{1.Ca} + 4 \cdot C_{1.SO4}\right)$$

$$I_2 = \frac{1}{2} \cdot \left(C_{2.Na} + C_{2.Cl} + 4 \cdot C_{2.Ca} + 4 \cdot C_{2.SO4}\right)$$

$$I_{mix} = \frac{1}{2} \cdot \left(C_{Na} + C_{Cl} + 4 \cdot C_{Ca} + 4 \cdot C_{SO4}\right)$$

We then define the ion concentrations of the mixture using Equations 7.3 and 10.15.

$$\begin{pmatrix} C_{Na} \\ C_{Cl} \\ C_{Ca} \\ C_{SO4} \end{pmatrix} = \begin{pmatrix} f_1 \cdot C_{1.Na} + f_2 \cdot C_{2.Na} \\ f_1 \cdot C_{1.Cl} + f_2 \cdot C_{2.Cl} \\ f_1 \cdot C_{1.Ca} + f_2 \cdot C_{2.Ca} \\ f_1 \cdot C_{1.SO4} + f_2 \cdot C_{2.SO4} \end{pmatrix}$$

We substitute the relations for mixed concentrations into the statement for the ionic strength of the mixed solution.

$$I_{mix} = \frac{1}{2} \left[f_1 \cdot C_{1.Na} + f_2 \cdot C_{2.Na} + f_1 \cdot C_{1.Cl} + f_2 \cdot C_{2.Cl} + 4 \cdot \left(f_1 \cdot C_{1.Ca} + f_2 \cdot C_{2.Ca}\right) \cdots \atop + 4 \cdot \left(f_1 \cdot C_{1.SO4} + f_2 \cdot C_{2.SO4}\right) \right]$$

We collect terms.

$$I_{mix} = f_1 \cdot \frac{1}{2} \cdot \left(C_{1.Na} + C_{1.Cl} + 4 \cdot C_{1.Ca} + 4 \cdot C_{1.SO4}\right) \cdots$$
$$+ f_2 \cdot \frac{1}{2} \cdot \left(C_{2.Na} + C_{2.Cl} + 4 \cdot C_{2.Ca} + 4 \cdot C_{2.SO4}\right)$$

We replace the sums in parentheses by I_1 and I_2 and we have the relation we desire.

$$I_{mix} = f_1 \cdot I_1 + f_2 \cdot I_2 \quad \frac{mol}{L}$$

Then, for computations for mixing of solutions where transformations of major ions are not significant as a consequence of resultant proton transfer reactions, we can use this extension of Equation 7.3 (carefully of course) in ensuing analyses. The generalization of the result leads to Equation 10.16.

$$I_{\text{mix}} = \sum_{k=1}^{n} f_k I_k \tag{10.16}$$

Wherein k again refers to solution k and n is the number of solutions to be mixed.

10.7.2 Final Solution Composition for Mixing of Two or More Solutions

We have now set the stage for application of our acid/base equilibria, mole balance, proton balance, and nondilute solution principles in mixing of two or more aqueous solutions. In many environmental contexts, it is most often, relative to risk assessment, the contemplation of mixing two or more aqueous solutions that garners our attention. Slugs of solutions containing undesirable constituents entering wastewater collection systems, overflows of reservoirs containing acid rock drainage, leakage of leachate from landfills into underlying ground waters, combining effluent streams from various industrial processes, and flushing of storm runoff into flowing water courses are some of the various contexts to which these computational efforts can be directly applied. If we work on the basis of volumes of solution, we may easily translate the results to actual flows in environmental systems. Then, herein we will, as those before us have, initially work with solutions in beakers mixed with other solutions in beakers with resultant solutions simply contained in beakers of sufficient volume to hold all the volumes of the mixed solutions. We will begin with a simple system and then move to another that is more complex.

Example 10.13 Consider a solution of pH 8 and ionic strength of 0.008 containing total sulfide at a concentration of 0.001 M, and a solution of pH 4 and ionic strength 0.005 containing total acetate at a concentration of 0.002 M. Develop the model that may be used to predict the final pH and speciation for mixing of the two solutions in varying fractions and use that model to predict the pH and concentration of hydrogen sulfide as a function of the fraction of the solution that is contributed from the sulfide-containing solution.

In our response to the charges of Example 10.13, we will first fully characterize each of the initial solutions, then we will use our "snapshot" of the initially mixed solution to compute the initial condition of the mixed solution immediately prior to the occurrence of proton transfers, and then we will use our newly assimilated tools to determine the final composition. We will build the computational model so we may easily alter the mixing fractions to attain the desired relation. We will use relations from Table 10.4, adapted for the target acid system.

As the pH of this system certainly cannot rise above 8, that of the sulfide solution, certainly in this case, we may neglect the second deprotonation of hydrogen sulfide and treat H_2S as a monoprotic acid. Relevant equilibrium constants are available from Table 6.1. Characterization of the sulfide-containing solution (denoted by the subscript 1) yields:

$$\begin{pmatrix} K_{A.HS} \\ K_W \\ A \\ C_{Tot.S} \\ H_1 \\ I_1 \end{pmatrix} := \begin{pmatrix} 10^{-7.1} \\ 10^{-14} \\ 0.5115 \\ 0.001 \\ 10^{-8} \\ 0.008 \end{pmatrix} \qquad \gamma_{1.1} := 10^{\frac{-A\cdot\sqrt{I_1}}{1+\sqrt{I_1}}} = 0.908$$

$$H2S_1 := C_{Tot.S}\cdot\left(1 + \frac{K_{A.HS}}{H_1\cdot\gamma_{1.1}}\right)^{-1} \qquad HS_1 := C_{Tot.S}\cdot\left(\frac{H_1}{K_{A.HS}} + \frac{1}{\gamma_{1.1}}\right)^{-1}$$

$$\begin{pmatrix} C_{HS.1} \\ C_{H2S.1} \\ C_{H.1} \\ C_{OH.1} \end{pmatrix} := \begin{pmatrix} \frac{HS_1}{\gamma_{1.1}} \\ H2S_1 \\ \frac{H_1}{\gamma_{1.1}} \\ \frac{K_W}{H_1\cdot\gamma_{1.1}} \end{pmatrix} = \begin{pmatrix} 8.974\times10^{-4} \\ 1.026\times10^{-4} \\ 1.102\times10^{-8} \\ 1.102\times10^{-6} \end{pmatrix} \frac{mol}{L}$$

Characterization of the acetate-containing solution (denoted by the subscript 2) yields:

$$\begin{pmatrix} K_{A.Ac} \\ K_W \\ A \\ C_{Tot.Ac} \\ H_2 \\ I_2 \end{pmatrix} := \begin{pmatrix} 10^{-4.7} \\ 10^{-14} \\ 0.5115 \\ 0.002 \\ 10^{-4} \\ 0.005 \end{pmatrix} \qquad \gamma_{1.2} := 10^{\frac{-A\cdot\sqrt{I_2}}{1+\sqrt{I_2}}} = 0.925$$

$$HAc_2 := C_{Tot.Ac}\left(1 + \frac{K_{A.Ac}}{H_2\cdot\gamma_{1.2}}\right)^{-1} \qquad Ac_2 := C_{Tot.Ac}\left(\frac{H_2}{K_{A.Ac}} + \frac{1}{\gamma_{1.2}}\right)^{-1}$$

$$\begin{pmatrix} C_{Ac.2} \\ C_{HAc.2} \\ C_{H.2} \\ C_{OH.2} \end{pmatrix} := \begin{pmatrix} \frac{Ac_2}{\gamma_{1.2}} \\ HAc_2 \\ \frac{H_2}{\gamma_{1.2}} \\ \frac{K_W}{H_2\cdot\gamma_{1.2}} \end{pmatrix} = \begin{pmatrix} 3.548\times10^{-4} \\ 1.645\times10^{-3} \\ 1.081\times10^{-4} \\ 1.081\times10^{-10} \end{pmatrix} \frac{mol}{L}$$

Now that we have characterized the solutions that are to be mixed, we can determine the initially mixed composition. Since we will write our proton balance a couple ways to illustrate a point expressed earlier, we will simply compute the initial concentrations of all species. We begin with the conservative substances. In order that we may arrive at and check a solution, we will initially suggest that $f_1=0.25$ and $f_2=0.75$.

$$\begin{pmatrix} I \\ C_{Tot.S} \\ C_{Tot.Ac} \end{pmatrix} := \begin{pmatrix} f_1\cdot I_1 + f_2\cdot I_2 \\ f_1\cdot C_{Tot.S} \\ f_2\cdot C_{Tot.Ac} \end{pmatrix} = \begin{pmatrix} 5.75\times10^{-3} \\ 2.5\times10^{-4} \\ 1.5\times10^{-3} \end{pmatrix} \qquad \gamma_1 := 10^{\frac{-A\cdot\sqrt{I}}{1+\sqrt{I}}} = 0.92$$

We now compute the initially mixed concentrations of the acid/base species.

$$\begin{pmatrix} C_{H.i} \\ C_{OH.i} \\ C_{HS.i} \\ C_{H2S.i} \\ C_{Ac.i} \\ C_{HAc.i} \end{pmatrix} := \begin{pmatrix} f_1\cdot C_{H.1} + f_2\cdot C_{H.2} \\ f_1\cdot C_{OH.1} + f_2\cdot C_{OH.2} \\ f_1\cdot C_{HS.1} \\ f_1\cdot C_{H2S.1} \\ f_2\cdot C_{Ac.2} \\ f_2\cdot C_{HAc.2} \end{pmatrix} = \begin{pmatrix} 8.107\times10^{-5} \\ 2.755\times10^{-7} \\ 2.244\times10^{-4} \\ 2.564\times10^{-5} \\ 2.661\times10^{-4} \\ 1.234\times10^{-3} \end{pmatrix} \frac{mol}{L}$$

Note that we compute the concentration of hydroxide using Equation 7.3 rather than using the ion product of water. We must bear in mind that the composition we compute as the set of initially mixed concentrations is not under equilibrium conditions. This set of concentrations, adjusted to yield the corresponding activities, would perhaps be used in the reaction quotient discussed in our treatment of Gibbs energy. This set of concentrations is in fact the reference condition, imaginary as it is.

We are now ready to solve the system and must designate a reference specie for each acid system, enabling us to write the proton balance. Let us choose hydrogen sulfide, acetic acid, and of course water.

$$\Delta C_H = \Delta C_{OH} + \Delta C_{HS} + \Delta C_{Ac}$$

Incorporation of the proton balance with the two mole balances and using algebra to simplify the system to three equations in three unknowns, we implement a given-find block to accomplish the solution.

$$\begin{pmatrix} H \\ HS \\ Ac \end{pmatrix} := \begin{pmatrix} 10^{-5} \\ 0.1 \cdot C_{Tot.S} \\ 0.9 \cdot C_{Tot.Ac} \end{pmatrix} = \begin{pmatrix} 1 \times 10^{-5} \\ 2.5 \times 10^{-5} \\ 1.35 \times 10^{-3} \end{pmatrix}$$

$$\text{given} \qquad \frac{H}{\gamma_1} - C_{H.i} = \frac{K_W}{H \cdot \gamma_1} - C_{OH.i} + \frac{HS}{\gamma_1} - C_{HS.i} + \frac{Ac}{\gamma_1} - C_{Ac.i}$$

$$C_{Tot.S} = HS \cdot \left(\frac{H}{K_{A.HS}} + \frac{1}{\gamma_1} \right) \qquad C_{Tot.Ac} = Ac \cdot \left(\frac{H}{K_{A.Ac}} + \frac{1}{\gamma_1} \right)$$

$$\begin{pmatrix} H \\ HS \\ Ac \end{pmatrix} := \text{find}(H, HS, Ac) = \begin{pmatrix} 4.869 \times 10^{-5} \\ 4.071 \times 10^{-7} \\ 4.253 \times 10^{-4} \end{pmatrix} \qquad \begin{pmatrix} pH \\ C_{H2S} \end{pmatrix} := \begin{pmatrix} -\log(H) \\ \frac{H \cdot HS}{K_{A.HS}} \end{pmatrix} = \begin{pmatrix} 4.313 \\ 2.496 \times 10^{-4} \end{pmatrix}$$

Now to verify the correctness of alternative forms of the proton balance, we will now rewrite the proton balance using hydrogen sulfide and acetic acid as the reference species.

$$\Delta C_H + \Delta C_{H2S} + \Delta C_{HAc} = \Delta C_{OH}$$

We have our reference condition already computed. We will stick with bisulfide and acetate as the target dependent variables, otherwise we must restructure the set of equations used in the solution. The solve block is shown in Figure E10.13.1.

Separate, parallel solutions in which the sulfide specie is considered, and using, in turn, hydrogen sulfide, bisulfide, and sulfide as the sulfide system reference specie with acetate as the acetate system reference specie all return results that are identical to those using hydrogen sulfide and acetic acid as the reference species. The corresponding initial statements of these three additional proton balances are included for completeness.

$$\Delta C_H + \Delta C_{HAc} = \Delta C_{OH} + \Delta C_{HS} + 2 \cdot \Delta C_S$$

$$\Delta C_H + \Delta C_{H2S} + \Delta C_{HAc} = \Delta C_{OH} + \Delta C_S$$

$$\Delta C_H + 2 \cdot \Delta C_{H2S} + \Delta C_{HS} + \Delta C_{HAc} = \Delta C_{OH}$$

We can rest assured that the proton balance is as flexible as advertised earlier in this chapter.

$$\begin{pmatrix} H \\ HS \\ Ac \end{pmatrix} := \begin{pmatrix} 10^{-5} \\ 0.1\cdot C_{Tot.S} \\ 0.9\cdot C_{Tot.Ac} \end{pmatrix} = \begin{pmatrix} 1\times 10^{-5} \\ 2.5\times 10^{-5} \\ 1.35\times 10^{-3} \end{pmatrix}$$

Given $$\frac{H}{\gamma_1} - C_{H.i} + \frac{H\cdot HS}{K_{A.HS}} - C_{H2S.i} + \frac{H\cdot Ac}{K_{A.Ac}} - C_{HAc.i} = \frac{K_W}{H\cdot \gamma_1} - C_{OH.i}$$

$$C_{Tot.S} = HS\cdot\left(\frac{H}{K_{A.HS}} + \frac{1}{\gamma_1}\right) \qquad C_{Tot.Ac} = Ac\cdot\left(\frac{H}{K_{A.Ac}} + \frac{1}{\gamma_1}\right)$$

$$\begin{pmatrix} H \\ HS \\ Ac \end{pmatrix} := \text{Find}(H, HS, Ac) = \begin{pmatrix} 4.869\times 10^{-5} \\ 4.071\times 10^{-7} \\ 4.253\times 10^{-4} \end{pmatrix} \qquad \begin{pmatrix} pH \\ C_{H2S} \end{pmatrix} := \begin{pmatrix} -\log(H) \\ \dfrac{H\cdot HS}{K_{A.HS}} \end{pmatrix} = \begin{pmatrix} 4.313 \\ 2.496\times 10^{-4} \end{pmatrix}$$

FIGURE E10.13.1 *Screen capture of the solve block for computation of the final solution composition for mixing of acetate and sulfide solutions.*

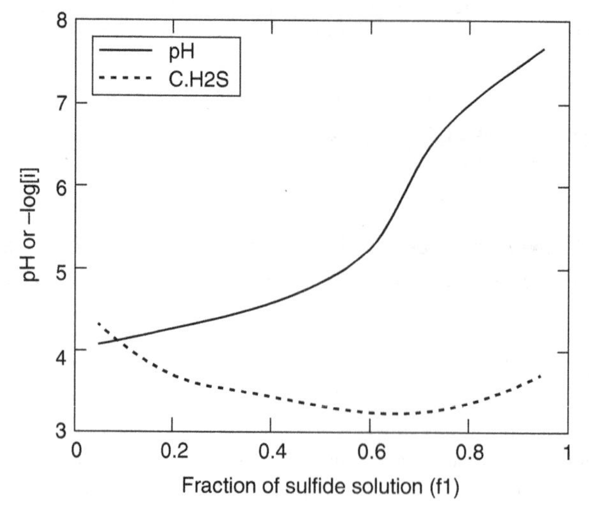

FIGURE E10.13.2 *A plot of predicted pH or $-\log_{10}[H_2S]$ for mixing of sulfide and acetate solution of different pH in varying fractions.*

We continue toward the stated objective, which is to develop a relation between the fractional make-up of the mixture and the final pH and hydrogen sulfide concentration. We simply need to use the developed model with varying f_1 values ($f_2 = 1 - f_1$), record the results, and display them graphically. We chose f_1 values ranging from 0.05 to 0.95, implemented the solve block of Figure E10.11.1 and collected the resultant pH and $-\log(C_{H_2S})$ values in corresponding vectors for display. The requested graphical output is obtained by plotting pH and $-\log[H_2S]$ against f_1 in Figure E10.13.2. We could have written the final solve block into a function such as the following.

$Sol_{Ac.HS}(f_1, f_2, I_{mix}, \gamma_1, C_{H.i}, C_{OH.i}, C_{H2S.i}, C_{HAc.i}, H, HS, Ac) = find(H, HS, Ac)$

We then would have programmed the mixing computations into a looping program to successively solve the system for incremented values of f_1. We will have some opportunities later in this text to illustrate programming of this nature.

To review, we have presented and demonstrated the application of an algorithm for the characterization of aqueous solutions resulting from proton transfers when two solutions are mixed. The algorithm involves characterization of the initial solutions, mixing computations, and application of equilibria, mole balances, and a proton balance to secure the final result.

Let us address one more example system before moving on to our next topics. Consider that a spill of anhydrous ammonia has laden the unsaturated zone above a phreatic (water table) aquifer with unionized ammonia such that the vapor phase within the unsaturated zone contains ammonia gas at a level of 50 ppm_v. The spill is located in some of the rich agricultural land of southern Minnesota at an elevation of ~900 ft above mean sea level. Precipitation, of course, would result in infiltration, percolation of water through the contaminated zone and transport of ammonia-laden water to the ground water. We would be interested to know the range of effects upon the composition of the ground water. This is in truth a transport problem, but we may perform some "bounding calculations" using chemical equilibrium principles.

Example 10.14 Develop a mathematical model addressing the mixing of ammonia-laden infiltrating water with ground water beneath the contamination zone described. Consider that we know that near-surface ground water in this general region has pH of 7.2, measured alkalinity of 250 mg/L *as* $CaCO_3$ and a high level of dissolved solids, leading to an ionic strength of 0.01 M. Let us assume that the infiltrating water picks up major ions, leading to ionic strength of 0.005 M, and total inorganic carbon of 0.002 M. We will also assume that the equilibrium condition is attained between the ammonia-laden vapor and the aqueous solution. We will also assume that dissolution of ammonia into the percolating water does not significantly reduce the quantity of ammonia held in the pores of the soil, hence this is an initial bounding calculation for the overall transport process. Relative to the carbonate abundance in percolating water, in Chapter 11 we will examine dissolution of solids, so, rather than address how the percolating water arrives at the stated total inorganic carbon abundance, we will render an assumption here. We also must consider that the water-bearing zone directly beneath the spill can be visualized as a zero-volume mixing zone. A profile sketch depicting the contaminated zone, the ground water, and the infiltrative process is included in Figure E10.14.1 as an aid for visualization of the system.

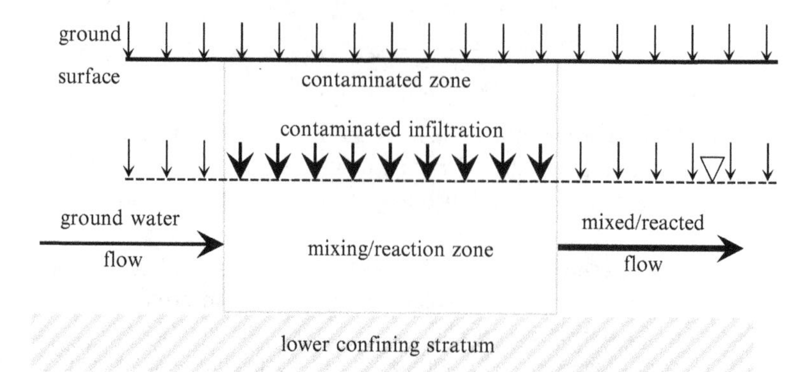

FIGURE E10.14.1 *Sketch depicting infiltration of precipitation through a contaminated zone of the unsaturated soil and mixing with ground water.*

We will apply the algorithm described in the previous example. We must first characterize the two streams that are to be mixed. We need to employ the alkalinity principle from Chapter 6, and include nondilute solution principles in characterization of the ground water. For the infiltrating water, we realize that we must somehow compute the pH in order to characterize the solution. Once the two solutions are characterized, we can perform the mixing computations to determine the initial condition and then apply mole and proton balances with equilibria to compute the final pH and speciation.

We begin by assembling (assignment statements are not shown) in vectors:

assignments of values for five equilibrium constants (two for carbonate, two for ammonia, one for water);

assignments for seven known parameters: the value of A for the Güntelberg equation, two ionic strengths, total carbonate for the infiltrating water, alkalinity for the ground water; the proton activity of the ground water and the abundance of ammonia in the gas held in the pores of the contaminated soil;

computations for four activity coefficients (γ_1 and γ_2 for both waters, results shown below);

a computation for the normal atmospheric pressure from the yielding the partial pressure of ammonia within the soil vapor and a characterization of the ammonia content of the contaminated infiltrating water (results shown below).

$$\begin{pmatrix} \gamma_{1.IW} \\ \gamma_{2.IW} \\ \gamma_{1.GW} \\ \gamma_{2.GW} \end{pmatrix} = \begin{pmatrix} 0.92 \\ 0.718 \\ 0.89 \\ 0.627 \end{pmatrix} \qquad P_{Tot} = 0.968 \quad \text{atm}$$

$$P_{NH3} := P_{Tot} \cdot ppm_{v.NH3} \cdot 10^{-6} = 4.839 \times 10^{-5} \quad \text{atm}$$

$$NH3 := K_{H.NH3} \cdot P_{NH3} = 2.7593 \times 10^{-3} \quad \frac{\text{mol}}{\text{L}}$$

We now are ready to characterize the infiltrating water. Here, we might begin with rain water, and for this effort the small initial bicarbonate and carbonate abundances. We expect bicarbonate abundance of the contaminated infiltrating water to be ~10^{-2}M and the abundance in rain water of ~10^{-6}M will not be significant. Carbonate is much less significant. The initial conditions for proton and hydroxide conditions must employ the activity coefficients.

We consider the proton balance and determine that carbonic acid and ammonia are the best choices for reference species. Carbonic acid should be insignificant in this analysis and since ammonia is at equilibrium in the gas and water, we know we should use this as our reference specie. Our full proton balance equation is then written.

$$\Delta C_{H} + \Delta C_{NH4} = \Delta C_{OH} + \Delta C_{HCO3} + 2 \cdot \Delta C_{CO3}$$

Then we may approximate the initial concentrations of ammonium, bicarbonate, and carbonate as zero and consider that initial proton and hydroxide abundances are also negligible. We have chosen to illustrate this approximation herein, commensurate with the assumption that the precipitation is FDW. We will address the error later.

$$\begin{pmatrix} pH_{IW} \\ C_{H.IW} \\ C_{OH.IW} \\ C_{NH4.IW} \\ C_{HCO3.IW} \\ C_{CO3.IW} \end{pmatrix} := \begin{pmatrix} -\log(H_{IW}) \\ \dfrac{H_{IW}}{\gamma_{1.IW}} \\ \dfrac{K_W}{H_{IW} \cdot \gamma_{1.IW}} \\ \dfrac{H_{IW}}{K_{A.NH3} \cdot \gamma_{1.IW}} \cdot NH3 \\ \dfrac{HCO3_{IW}}{\gamma_{1.IW}} \\ \dfrac{K_{2.CO3}}{H_{IW} \cdot \gamma_{2.IW}} \cdot HCO3_{IW} \end{pmatrix} = \begin{pmatrix} 9.414 \\ 4.164 \times 10^{-10} \\ 2.806 \times 10^{-5} \\ 2.292 \times 10^{-3} \\ 1.733 \times 10^{-3} \\ 2.657 \times 10^{-4} \end{pmatrix} \frac{mol}{L}$$

The infiltrating water then is characterized using the proton balance with mole balances for ammonia and carbonate, with the ammonia activity, and total inorganic carbon content known. We solve for the proton and bicarbonate activities and the total concentration of ammonia species, assigning the result into desired variables. The solve block is shown in Figure E10.14.2. We have included seven significant figures in the result for comparison with an alternative computation to be accomplished next. Once we have solved for the three master unknowns, we can compute the remainder of the characterization.

We used the results of Example 10.10 as approximations of the initial conditions for proton, hydroxide, and bicarbonate abundances. The proton balance statement from above is altered by subtracting $C_{H.init}$, from the LHS and $C_{OH.init}$ and $C_{HCO_3.init}$ from the RHS. When we compare the result with the assumption that bicarbonate is zero and that proton and hydroxide abundances cancel, we observe differences between the solutions in the seventh or eighth significant figures of the results. We can assure

$$\begin{pmatrix} H \\ C_{Tot.NH3} \\ HCO3 \end{pmatrix} := \begin{pmatrix} 10^{-9} \\ 10^{-2} \\ 0.01 \end{pmatrix} \frac{mol}{L} \qquad \text{Given} \quad C_{Tot.NH3} = NH3 \cdot \left(\frac{H}{K_{A.NH3} \cdot \gamma_{1.IW}} + 1 \right)$$

$$C_{Tot.CO3.IW} = HCO3 \cdot \left(\frac{H}{K_{1.CO3}} + \frac{1}{\gamma_{1.IW}} + \frac{K_{2.CO3}}{H \cdot \gamma_{2.IW}} \right)$$

$$\frac{H}{\gamma_{1.IW}} + \frac{H \cdot NH3}{K_{A.NH3} \cdot \gamma_{1.IW}} = \frac{K_W}{H \cdot \gamma_{1.IW}} + \frac{HCO3}{\gamma_{1.IW}} + 2 \cdot \frac{K_{2.CO3} \cdot HCO3}{H \cdot \gamma_{2.IW}}$$

$$\begin{pmatrix} H_{IW} \\ C_{Tot.NH3.IW} \\ HCO3_{IW} \end{pmatrix} := \text{Find}(H, C_{Tot.NH3}, HCO3) \rightarrow \begin{pmatrix} 3.852196 \times 10^{-10} \\ 5.051688 \times 10^{-3} \\ 1.603229 \times 10^{-3} \end{pmatrix} \frac{mol}{L}$$

FIGURE E10.14.2 *Screen capture of a solve block for characterization of a solution containing carbonate and ammonia nitrogen in equilibrium with soil gas.*

ourselves that for many computations, we can consider natural precipitation as roughly equivalent to freshly distilled water.

We turn our attention to the ground water. We recall Example 6.4 and the two relations (Equations 6.7a and 6.7b) that arose from Example 6.4. We realize these were developed using the infinitely dilute assumption and may need some touching up prior to our use here. The measured alkalinity results from the titration of a water sample from the initial pH value to the endpoint of pH ~4.3, and the result is a measured quantity of acid necessary to accomplish that task. Consequently, the *concentrations* of carbonate and bicarbonate in the initial solutions must be used in the analysis. Moreover, again, the relation between carbonate and bicarbonate must employ specie activities. Equation 6.6 is correctly written in concentration units and when we employ Equation 10.1 and neglect hydronium and hydroxide for this water of rather high alkalinity (an end of chapter problem addresses the magnitude of this error) we obtain the desired nondilute results for Equations 6.7a and 6.7b.

$$\{HCO_3^-\} = [Alk]\left(\frac{1}{\gamma_1} + 2\frac{K_{2.CO_3}}{\{H^+\}\cdot\gamma_2}\right)^{-1}; \quad \{CO_3^=\} = [Alk]\left(\frac{\{H^+\}}{K_{2.CO_3}\cdot\gamma_1} + \frac{2}{\gamma_2}\right)^{-1}$$

We have converted the measured alkalinity to molar units, necessary with Equations 6.7a and 6.7b using the definition of mg/L *as* $CaCO_3$ from Table 3.1.

$$Alk_{GW} = 5 \times 10^{-3} \quad M$$

Characterization of the ground water is then quite straightforward.

$$HCO3_{GW} := Alk_{GW} \cdot \left(\frac{1}{\gamma_{1.GW}} + 2 \cdot \frac{K_{2.CO3}}{H_{GW} \cdot \gamma_{2.GW}} \right)^{-1} = 4.483 \times 10^{-3}$$

$$CO3_{GW} := Alk_{GW} \cdot \left(\frac{H_{GW}}{K_{2.CO3} \cdot \gamma_{1.GW}} + \frac{2}{\gamma_{2.GW}} \right)^{-1} = 3.323 \times 10^{-6}$$

$$C_{Tot.CO3.GW} := HCO3_{GW} \cdot \left(\frac{H_{GW}}{K_{1.CO3}} + \frac{1}{\gamma_{1.GW}} + \frac{K_{2.CO3}}{H_{GW} \cdot \gamma_{2.GW}} \right) = 5.628 \times 10^{-3}$$

$$\begin{pmatrix} C_{H.GW} \\ C_{OH.GW} \\ C_{HCO3.GW} \\ C_{CO3.GW} \end{pmatrix} := \begin{pmatrix} \dfrac{H_{GW}}{\gamma_{1.GW}} \\ \dfrac{K_W}{H_{GW} \cdot \gamma_{1.GW}} \\ \dfrac{HCO3_{GW}}{\gamma_{1.GW}} \\ \dfrac{CO3_{GW}}{\gamma_{2.GW}} \end{pmatrix} = \begin{pmatrix} 7.023 \times 10^{-8} \\ 1.764 \times 10^{-7} \\ 4.99 \times 10^{-3} \\ 5.1 \times 10^{-6} \end{pmatrix} M$$

We certainly could have computed the activity of carbonate from that of bicarbonate using the pH of the ground water with the equilibrium from the second deprotonation.

We now perform the mixing computations to define the reference condition for the proton transfer reaction that would occur upon mixing of the two solutions. We will consider a ¼ – ¾ infiltrating water: ground water constitution of the mixture. The results are presented as a reference for those who would work through this example (highly recommended).

$$f_{IW} := 0.25 \qquad f_{GW} := 1 - f_{IW}$$

$$\begin{pmatrix} I \\ C_{Tot.CO3} \\ C_{Tot.NH3} \end{pmatrix} = \begin{pmatrix} 8.75 \times 10^{-3} \\ 4.721 \times 10^{-3} \\ 1.263 \times 10^{-3} \end{pmatrix} \frac{mol}{L} \qquad \begin{pmatrix} C_{H.i} \\ C_{OH.i} \\ C_{NH4.i} \\ C_{HCO3.i} \\ C_{CO3.i} \end{pmatrix} = \begin{pmatrix} 5.277 \times 10^{-8} \\ 7.147 \times 10^{-6} \\ 5.731 \times 10^{-4} \\ 4.176 \times 10^{-3} \\ 7.025 \times 10^{-5} \end{pmatrix} \frac{mol}{L}$$

$$\begin{pmatrix} \gamma_1 \\ \gamma_2 \end{pmatrix} = \begin{pmatrix} 0.904 \\ 0.668 \end{pmatrix}$$

We may use the proton balance from the characterization of the infiltrating water, but as is evident here, the initial abundances for the ammonia and carbonate species are not zero and certainly the initial abundances of hydronium and hydroxide will not numerically cancel. Since all the initial abundances are known, we may collect them and include the result on the LHS of the proton balance equation.

$$\Sigma C_i := C_{OH.i} + C_{HCO3.i} + 2 \cdot C_{CO3.i} - C_{H.i} - C_{NH4.i} - 3.75 \times 10^{-3}$$

We may use the carbonate mole balance exactly as stated with the infiltrating water computations. The remainder of our system of equations is exactly that employed in characterization of the infiltrating water, except now we seek the final activities of hydronium, ammonia, and bicarbonate, as shown in Figure E10.14.3.

Now that we have this computational model assembled, we may employ it to predict the system response to various forcing parameters such as the mixing fractions, the abundance of ammonia in the soil vapor, the alkalinity of the ground water and the relevant mineralogy of the infiltrating water. We have collected them in vectors, so they are easy to find and alter. Were our objective to be the investigation of this set of parameters, we might even rearrange the worksheet to collect these in a single vector. For use of the model, we can define an area containing the computations between the input and results and view only the input and output, by collapsing the area. A screen capture is shown in Figure E10.14.4. An example of an alternate solution for mixing infiltrating water with ground water on a 50:50 basis is shown. We may then easily alter forcing input and collect the results to examine the behavior for a range of mixing ratios.

This example is very appropriate for illustration of Excel for solution of problems addressing systems involving one or more implicit relations. To this point

$$\begin{pmatrix} H \\ NH3 \\ HCO3 \end{pmatrix} := \begin{pmatrix} 10^{-8} \\ 0.1 \cdot C_{Tot.NH3} \\ 0.9 \cdot C_{Tot.CO3} \end{pmatrix} \frac{mol}{L} \qquad \text{Given} \quad C_{Tot.NH3} = NH3 \cdot \left(\frac{H}{K_{A.NH3}} + \frac{1}{\gamma_1} \right)$$

$$C_{Tot.CO3} = HCO3 \cdot \left(\frac{H}{K_{1.CO3}} + \frac{1}{\gamma_1} + \frac{K_{2.CO3}}{H \cdot \gamma_2} \right)$$

$$\frac{H}{\gamma_1} + \frac{H \cdot NH3}{K_{A.NH3} \cdot \gamma_1} + \Sigma C_i = \frac{K_W}{H \cdot \gamma_1} + \frac{HCO3}{\gamma_1} + 2 \cdot \frac{K_{2.CO3} \cdot HCO3}{H \cdot \gamma_2}$$

$$\begin{pmatrix} H \\ NH3 \\ HCO3 \end{pmatrix} := \text{Find}(H, NH3, HCO3) = \begin{pmatrix} 2.04389 \times 10^{-9} \\ 2.43615 \times 10^{-4} \\ 4.12392 \times 10^{-3} \end{pmatrix} \frac{mol}{L}$$

$$pH := -\log(H) = 8.69 \quad \begin{pmatrix} C_{HCO3} \\ C_{CO3} \\ C_{NH4} \end{pmatrix} := \begin{pmatrix} \frac{HCO3}{\gamma_1} \\ \frac{K_{2.CO3}}{H \cdot \gamma_2} \cdot HCO3 \\ \frac{H \cdot NH3}{K_{A.NH3} \cdot \gamma_1} \end{pmatrix} = \begin{pmatrix} 4.561 \times 10^{-3} \\ 1.412 \times 10^{-4} \\ 1.099 \times 10^{-3} \end{pmatrix} \frac{mol}{L}$$

FIGURE E10.14.3 *Screen capture of the solve block for characterization of the final composition of mixed solutions containing ammonia nitrogen and carbonate.*

$$\begin{pmatrix} K_{1.CO3} \\ K_{2.CO3} \\ K_{A.NH3} \\ K_{H.NH3} \\ K_W \end{pmatrix} := \begin{pmatrix} 10^{-6.35} \\ 10^{-10.33} \\ 10^{-9.30} \\ 10^{1.756} \\ 10^{-14} \end{pmatrix} \quad \begin{pmatrix} A \\ I_{GW} \\ I_{IW} \\ C_{Tot.CO3.IW} \\ Alk_{GW} \\ H_{GW} \end{pmatrix} := \begin{pmatrix} 0.5115 \\ 0.01 \\ 0.005 \\ 0.002 \\ \frac{250}{50000} \\ 10^{-7.2} \end{pmatrix} \quad \begin{matrix} f_{IW} := 0.5 \\ f_{GW} := 1 - f_{IW} \\ ppm_{v.NH3} := 50 \end{matrix}$$

$$\begin{pmatrix} H \\ NH3 \\ HCO3 \end{pmatrix} := Find(H, NH3, HCO3) = \begin{pmatrix} 7.61007 \times 10^{-10} \\ 9.65231 \times 10^{-4} \\ 3.20622 \times 10^{-3} \end{pmatrix} \qquad pH := -\log(H) = 9.1186$$

$$\begin{pmatrix} C_{HCO3} \\ C_{CO3} \\ C_{NH4} \end{pmatrix} := \begin{pmatrix} \frac{HCO3}{\gamma_1} \\ \frac{K_{2.CO3}}{H \cdot \gamma_2} \cdot HCO3 \\ \frac{H \cdot NH3}{K_{A.NH3} \cdot \gamma_1} \end{pmatrix} = \begin{pmatrix} 3.522 \times 10^{-3} \\ 2.869 \times 10^{-4} \\ 1.61 \times 10^{-3} \end{pmatrix}$$

FIGURE E10.14.4 *Screen capture of collapsed worksheet illustrating convenient arrangement of the worksheet for additional computations.*

almost everything we have done in MathCAD's "what you see is what you get" platform can also be done using a MS Excel worksheet. Certainly we can program the equivalent of MathCAD functions using Excel and we also certainly can create functions of these functions and so on. Parallel to MathCAD's given-find solver, Excel has its own solver that works in much the same manner as the given-find, with one major exception—the solver can employ but one master independent variable, whose value must reside in a single worksheet cell. In invoking the Excel solver for computations such as those of Example 10.14, we must write our set of relations differently. We will revisit Example 10.14 and illustrate a parallel solution using the capacities of Excel.

Example 10.15 Repeat Example 10.14 using Excel and its solver to perform the computations.

We begin by specifying all that we know. We have brought the partial pressure and hence the abundance of ammonia in the infiltrating water from Example 10.14. Rather than simply showing output, we show the formulas programmed into each of the cells.

K.1	=10^-6.35	A	0.5115
K.2	=10^-10.33	I.GW	0.01
K.NH3	=10^-9.3	I.IW	0.005
K.H.NH3	=10^1.756	C.Tot.CO3.IW	0.002
K.W	=-10^-14	Alk.GW	=250/50000
ppmv.NH3	50	H.GW	=10^-7.2
P.NH3	0.0000483949	C.NH3.IW	=P.NH3*K.H.NH3

We arrange for the fraction of infiltrating water to be a master independent variable.

f.IW	=0.25	f.GW	=1-F.IW

We compute activity coefficients.

g.1.IW	=10^(-A*SQRT(I.IW)/(1+SQRT(I.IW)))	g.1.GW	=10^(-A*SQRT(I.GW)/(1+SQRT(I.GW)))
g.2.IW	=10^(-A*4*SQRT(I.IW)/(1+SQRT(I.IW)))	g.2.GW	=10^(-A*4*(I.GW)^(1/2)/(1+(I.GW)^(1/2)))

We now can characterize the infiltrating water and the ground water.

H.IW	3.85219548058537E-10	We need an intial guess for {H⁺}, which is replaced by the solver solution	
C.Tot.NH3.IW	=C.NH3.IW*(H.IW/(K.NH3*g.1.IW)+1)	HCO3.GW	=Alk.GW*(1/g.1.GW+2*K.2/(H.GW*g.2.GW))^-1
HCO3.IW	=C.Tot.CO3.IW/(H.IW/K.1+1/g.1.IW+K.2/(H.IW*g.2.IW))	CO3.GW	=Alk.GW*(H.GW/(K.2*g.1.GW)+2/g.2.GW)^-1
B16	=K.W/(H.IW*g.1.IW)	C.Tot.CO3.GW	=HCO3.GW*(H.GW/K.1+1/g.1.GW+K.2/(H.GW*g.2.GW))
B17	=HCO3.IW/g.1.IW	C.H.GW	=H.GW/g.1.GW
B18	=K.2*HCO3.IW/(H.IW*g.2.IW)	C.OH.GW	=K.W/(H.GW*g.1.GW)
B19	=H.IW/g.1.IW	C.HCO3.GW	=HCO3.GW/g.1.GW
B20	=H.IW*C.NH3.IW/(K.NH3*g.1.IW)	C.CO3.GW	=CO3.GW/g.2.GW
PB equation	=B16+B17+2*B18-B19-B20	solver, three equations in three unknowns, to find H.I.W	
C.H.IW	=H.IW/g.1.IW		
C.OH.IW	=K.W/(H.IW*g.1.IW)		
C.NH4.IW	=H.IW*C.NH3.IW/(K.NH3*g.1.IW)		
C.HCO3.IW	=HCO3.IW/g.1.IW		
C.CO3.IW	=K.2*HCO3.IW/(H.IW*g.2.IW)		

We began with an initial guess for $\{H^+\}$ at ~10^{-9} M and the solver failed to obtain a solution. A second initial guess of ~10^{-10} M resulted in a successful solution. The relations for C.Tot.NH_3.IW and HCO_3.IW were written with the unknowns on the left hand sides while each right hand side contains only known parameters and $\{H^+\}$. We wrote the proton balance in the form of $f(\{H^+\})=0$ and the target value for the solver is then 0. In modeling of systems, the debugging of the model is often a very significant portion of the effort. With Excel "divide and conquer" is often the best approach, therefore, we have separated each of the terms of the proton balance into separate cells. Note that we've also named every variable (names appear immediately to the left of the cells) in order to ease the burden of matching cell formulae to the companion mathematical expressions. Once the solution to the implicit system is gained, we may use the result to compute the values of the remaining variables. The characterization of the ground water is explicit.

We may now complete the mixing computations, addressing ionic strength, total abundances of acid species and activity coefficients first.

I	=F.IW*I.IW+f.GW*I.GW	C.Tot.CO3	=F.IW*C.Tot.CO3.IW+f.GW*E16
g.1	=10^(-A*SQRT(I)/(1+SQRT(I)))	C.Tot.NH3	=F.IW*C.Tot.NH3.IW+f.GW*0
g.2	=10^(-A*4*SQRT(I)/(1+SQRT(I)))		

We compute initial concentrations of species of interest and sum them for ease of use later on.

C.H.i	=F.IW*C.H.IW+f.GW*C.H.GW	C.HCO3.i	=F.IW*C.HCO3.IW+f.GW*C.HCO3.GW
C.OH.i	=F.IW*C.OH.IW+f.GW*C.OH.GW	C.CO3.i	=F.IW*C.CO3.IW+f.GW*C.CO3.GW
C.NH4.i	=F.IW*C.NH4.IW+f.GW*0	sum.C.i	=C.OH.i+C.HCO3.i+2*C.CO3.i-C.H.i-C.NH4.i

K.1	4.46684E-07	A	0.5115
K.2	4.67735E-11	I.GW	0.01
K.NH3	5.01187E-10	I.IW	0.005
K.H.NH3	57.01642723	C.Tot.CO3.IW	0.002
K.W	1E-14	Alk.GW	0.005
ppmv.NH3	50	H.GW	6.30957E-08
P.NH3	4.8395E-05	C.NH3.IW	0.002759304
f.IW	0.25	f.GW	0.75
g.1.IW	0.925166904	g.1.GW	0.898462593
g.2.IW	0.732622672	g.2.GW	0.651628394
H.IW	3.8522E-10	We need an intial guess for $\{H^+\}$	
C.Tot.NH3.IW	0.00505169	HCO3.GW	0.004483148
HCO3.IW	0.001603229	CO3.GW	3.3234E-06
	2.8059E-05	C.Tot.CO3.GW	0.005628161
	0.001732908	C.H.GW	7.02263E-08
	0.000265709	C.OH.GW	1.76401E-07
	4.16378E-10	C.HCO3.GW	0.0049898
	0.002292386	C.CO3.GW	5.10015E-06
PB equation	-1.09116E-09	solver, three equations in three unknowns	
C.H.IW	4.16378E-10		
C.OH.IW	2.8059E-05		
C.NH4.IW	0.002292386		
C.HCO3.IW	0.001732908		
C.CO3.IW	0.000265709		
I	0.00875	C.Tot.CO3	0.004721121
g.1	0.904162179	C.Tot.NH3	0.001262923
g.2	0.668321367		
C.H.i	5.27738E-08	C.HCO3.i	0.004175577
C.OH.i	7.14704E-06	C.CO3.i	7.02525E-05
C.NH4.i	0.000573097	sum.C.i	3.75008E-03
H.f	2.04E-09	initial guess for $\{H^+\}$	
NH3.f	0.000244065		
HCO3.f	0.004123669		
	2.25524E-09	pH.f	8.691
	0.001098241	C.NH4.f	0.001098241
	5.42394E-06	C.HCO3.f	0.004560763
	0.004560763	C.CO3.f	0.000141534
	0.000141534		
PB equation	-9.31745E-07	solver, three equations in three unknowns	

FIGURE E10.15.1 *Numeric output from an excel worksheet in solution of the system of Example 10.14.*

We may now complete the final solve block to determine the final speciation of the mixed solution.

H.f	2.03910064697266E-09	initial guess for {H⁺}, replaced by solver solution		
NH3.f	=C.Tot.NH3*(H.f/K.NH3+1/g.1)^-1			
HCO3.f	=C.Tot.CO3*(H.f/K.1+1/g.1+K.2/(H.f*g.2))^-1			
B40	=H.f/g.1		pH.f	=-LOG(H.f)
B41	=H.f*NH3.f/(K.NH3*g.1)		C.NH4.f	=H.f*NH3.f/(K.NH3*g.1)
B42	=K.W/(H.f*g.1)		C.HCO3.f	=HCO3.f/g.1
B43	=HCO3.f/g.1		C.CO3.f	=K.2*HCO3.f/(H.f*g.2)
B44	=K.2*HCO3.f/(H.f*g.2)			
PB equation	=B40+B41+sum.C.I-B42-B43-2*B44	solver, three equations in three unknowns, to solve for H.f		

Several tries (10^{-8}, 10^{-9}, and finally 2×10^{-9} M) at the initial guess for $\{H^+\}$ were necessary in order to obtain convergence of the solver. The output sheet is included here, with numerical values of intermediate and final results. Here we've shown the final solution, with very small residual values in the cells labeled equation. At the outset, these cells had nonzero values. Thus, the choices of initial values can be examined by observing the values in these cells. Once a guess renders the value in the cell sufficiently close to 0, the solver can confidently be invoked. The numeric output from the programmed cells shown earlier is shown in Figure E10.15.1. Then, with the numeric model in hand we may examine the effects of the master independent variables (e.g., fraction of infiltrating water and characteristics of each solution prior to mixing) on the final speciation of the mixed solution. In contrast to the MathCAD worksheet of Example 10.14, in order to complete the effort for each new set of independent variables, we must manually invoke the solver for both of the implicit computations.

10.8 ACID AND BASE NEUTRALIZING CAPACITY

10.8.1 ANC and BNC of Closed Systems

In Chapter 6, we examined acid neutralizing capacity (ANC) and base neutralizing capacity (BNC). These two parameters result directly from the constituents in water that can accept protons and donate protons, respectively. When we refer to the acid neutralizing capacity as a measurement, most often this is the measured alkalinity. Similarly, when we refer to the base neutralizing capacity as a measurement, most often this is the measured acidity. Complicating the terminology, often we find literature that refers to the alkalinity and acidity in more general terms, as general descriptions of the capacity of water to resist changes in pH due to the addition of (usually strong) acids or bases. Our analyses in Chapter 6 were quite specific to the alkalinity and acidity tests and to the resultant characterizations of the system behavior relative to the presence of carbonate system species in the water. In this section, we will generalize the concepts of [ANC] and [BNC]. We will develop expressions for the "instantaneous" [ANC] and [BNC] as well as methodologies for understanding the [ANC] and [BNC] in the contexts of finite changes in system character. We will employ $\{H^+\}$ as the master independent variable.

We may consider Equations 6.6 and 6.8 through 6.12 and express two (one for [ANC] and one for [BNC]) general relations that may be used to analyze the proton-accepting or proton-donating character of a given aqueous solution. Accordingly, we use the proton balance concept employed in earlier sections. We might be tempted to select the fully protonated acid of each acid/base system as the reference specie for [ANC]. Similarly, we might be tempted to select the fully deprotonated base of each acid/base system as the reference specie for [BNC]. Most often these choices would be advantageous. However, in the same context as for the proton balance, we can choose any set of species as the references. Then we need only account for whether it acts as an acid or a base during the examined set of proton transfer reactions. For [ANC] and [BNC], we would again be most wise to choose water as the reference specie for the hydronium–water–hydroxide system. General relations for [ANC] and [BNC] can then be easily written.

$$\Delta\,[\mathrm{ANC}] = \sum_i (\nu_{B,i} \Delta\,[B_i]) \tag{10.17a}$$

$$\Delta\,[\mathrm{BNC}] = \sum_i (\nu_{A,i} \Delta\,[A_i]) \tag{10.17b}$$

Δ[ANC] and Δ[BNC] are the capacities, in molar units of an aqueous solution to assimilate protons and hydroxides, respectively, while proceeding from an initial state to a final state. When we choose the fully protonated and fully deprotonated species as references for [ANC] and [BNC], respectively, $\nu_{B,i}$ is the number of protons accepted by specie i in its conversion to the reference conjugate acid, and $\nu_{A,i}$ is the number of number of protons donated by specie i in its conversion to the reference conjugate base. $[B_i]$ is the abundance of conjugate base i expressed as the molar concentration. $[A_i]$ is the abundance in molar units of conjugate acid i. In Chapter 6, we used the final state for the alkalinity titration as the pH ~ 4.3 endpoint while we suggested several pH values as various final states for the acidity titration. For Equations 10.16 and 10.17, the initial state is the condition prior to the addition of protons or hydroxides, while the final state is arbitrary. When we choose reference species other than the fully protonated (for [ANC]) and fully deprotonated (for [BNC]) species, $\nu_{B,i}$ and $\nu_{A,i}$ are the differences between the proton status of B_i or A_i and the reference specie and certainly can be negative as well as positive.

Equations 10.17a and 10.17b relate finite changes in [ANC], [BNC], and abundances of corresponding base or acid species. We may simply write these as infinitesimal changes, replacing Δ with ∂. Then, since we wish to employ $\{H^+\}$ as our master independent variable, we may dispense with the partial derivative, replacing ∂ with d. Since we seek the changes in [ANC] and [BNC] associated with changes in pH, we may rewrite Equations 10.17a and 10.17b as ordinary differential equations.

$$\frac{d[\mathrm{ANC}]}{d\{\mathrm{H}^+\}} = \sum_i \left(\nu_{B,i} \frac{d[B_i]}{d\{\mathrm{H}^+\}} \right) \tag{10.18}$$

$$\frac{d[\mathrm{BNC}]}{d\{\mathrm{H}^+\}} = \sum_i \left(\nu_{A,i} \frac{d\,[A_i]}{d\{\mathrm{H}^+\}} \right) \tag{10.19}$$

Authors of traditional texts addressing water chemistry (e.g., Snoeyink and Jenkins, 1980; Stumm and Morgan, 1996; Brezonik and Arnold, 2011) have defined a buffer intensity. Here we've extended their definition slightly to produce βs for [ANC] and for [BNC].

$$\beta_{\text{ANC}} = \frac{d[\text{ANC}]}{d(\text{pH})} \tag{10.20}$$

$$\beta_{\text{BNC}} = \frac{d[\text{BNC}]}{d(\text{pH})} \tag{10.21}$$

Equations 10.18 and 10.19 can be transformed into the form of 10.20 by noting the following.

$$\frac{d[\text{ANC}]}{d(\text{pH})} = \frac{d[\text{ANC}]}{d\{\text{H}^+\}}\frac{d\{\text{H}^+\}}{d(\text{pH})};\quad \frac{d[\text{BNC}]}{d(\text{pH})} = \frac{d[\text{BNC}]}{d\{\text{H}^+\}}\frac{d\{\text{H}^+\}}{d(\text{pH})} \quad \text{and}$$

$$\frac{d\{\text{H}^+\}}{d(\text{pH})} = \frac{d\{\text{H}^+\}/d\{\text{H}^+\}}{d(\text{pH})/d\{\text{H}^+\}} = -\{\text{H}^+\}\ln(10)$$

Equations 10.18 and 10.19 are then transformed to define β_{ANC} and β_{BNC}.

$$\beta_{\text{ANC}} = \sum_i \left(\nu_{B,i} \frac{d[B_i]}{d\{\text{H}^+\}} \frac{d\{\text{H}^+\}}{d(\text{pH})} \right) = -\{\text{H}^+\}\ln(10) \sum_i \nu_{B,i} \frac{d[B_i]}{d\{\text{H}^+\}} \tag{10.22}$$

$$\beta_{\text{BNC}} = \sum_i \left(\nu_{A,i} \frac{d[A_i]}{d\{\text{H}^+\}} \frac{d\{\text{H}^+\}}{d(\text{pH})} \right) = -\{\text{H}^+\}\ln(10) \sum_i \nu_{A,i} \frac{d[A_i]}{d\{\text{H}^+\}} \tag{10.23}$$

To quantitatively define each buffer intensity, we identify the conjugate bases and conjugate acids associated with a given system, write them as functions using pH as the master independent variable. We then can quantitatively evaluate β_{ANC} and β_{BNC} at any given value of pH.

Now we are in good position to define the values of the [ANC] or [BNC] of an aqueous solution. We must note that Equations 10.17a and 10.17b relate changes in [ANC] and [BNC] to changes in abundances of conjugate bases and conjugate acids, respectively. When we populate Equations 10.22 and 10.23 for a given aqueous solution, we are addressing the differential change in the [ANC] or [BNC]. Then by integrating the resultant functions from the initial pH to the final pH we compute the change in [ANC] or [BNC] between the initial and final states. If we routinely define the [ANC] and [BNC] to be zero at the end point of the targeted adjustment in pH, we must define the initial [ANC] and [BNC], relative to the end point of the process, as the negatives of the integrated results.

$$[\mathrm{ANC}]_{\mathrm{pH_i} \to \mathrm{pH_f}} = -\int_{\mathrm{pH_i}}^{\mathrm{pH_f}} \beta_{\mathrm{ANC}}\, d\mathrm{pH} \tag{10.24}$$

$$[\mathrm{BNC}]_{\mathrm{pH_i} \to \mathrm{pH_f}} = -\int_{\mathrm{pH_i}}^{\mathrm{pH_f}} \beta_{\mathrm{BNC}}\, d\mathrm{pH} \tag{10.25}$$

Example 10.16 Let us revisit Example 6.4, wherein we investigated an aqueous solution with $C_{\mathrm{Tot.CO_3}} = 0.006\,\mathrm{M}$ and pH of 7.65. Therein, we assumed infinitely dilute conditions. We employed the rigorous proton balance, for prediction of the results of a titration from the initial pH to the pH 4.3 end point. We computed the initial [ANC] to be 5.724 × 10^{-3} M. Herein, we'll compare that result with one arising from the employment of Equation 10.24. We will also investigate application of Equation 10.25 and compare that result with those we will generate using the approach employed in Example 6.4, except that we will compute the initial [BNC].

Let us first develop the set of relations that we would employ for use of Equations 10.24 and 10.25. We will make use of the powerful capabilities at our disposal via a MathCAD worksheet. We'll include important intermediate results.

$$\beta_{ANC} = -\ln(10)\cdot H\cdot\left(\frac{1}{\gamma_1}\cdot\frac{dHCO3}{dH} + \frac{2}{\gamma_2}\cdot\frac{dCO3}{dH} + \frac{1}{\gamma_1}\cdot\frac{dOH}{dH} - \frac{1}{\gamma_1}\cdot\frac{dH}{dH}\right)$$

$$\beta_{BNC} = -\ln(10)\cdot H\cdot\left(2\cdot\frac{dH2CO3}{dH} + \frac{1}{\gamma_1}\cdot\frac{dHCO3}{dH} + \frac{1}{\gamma_1}\cdot\frac{dH}{dH} - \frac{1}{\gamma_1}\cdot\frac{dOH}{dH}\right)$$

We have shown the relations with the concentration abundances converted using Equation 10.1. We have included activity coefficients here in order to be complete, but for this example we'll set them all to unity.

We develop the derivatives of each of the conjugate bases and conjugate acids and write associated MathCAD functions. We have defined the equilibrium constants and other known parameters in the worksheet but for brevity have not shown them herein. For the symbology that follows d_i(pH) is used to represent $(d\{i\}\ /\ dH)$ and written using H(pH) representing $\{H^+\}$.

$$H(pH) := 10^{-pH} \qquad d_H(pH) := \frac{1}{\gamma_1} \qquad d_{OH}(pH) := \frac{-K_W}{\gamma_1\cdot H(pH)^2}$$

$$d_{H2CO3}(pH) := \frac{C_{Tot.CO3}}{\left(1 + \frac{K_1}{H(pH)\cdot\gamma_1} + \frac{K_2\cdot K_1}{H(pH)^2\cdot\gamma_2}\right)^2}\cdot\left(\frac{K_1}{H(pH)^2\cdot\gamma_1} + \frac{2\cdot K_2\cdot K_1}{H(pH)^3\cdot\gamma_2}\right)$$

$$d_{HCO3}(pH) := \frac{-C_{Tot.CO3}}{\left(\frac{H(pH)}{K_1} + \frac{1}{\gamma_1} + \frac{K_2}{H(pH)\cdot\gamma_2}\right)^2}\cdot\left(\frac{1}{K_1} - \frac{K_2}{H(pH)^2\cdot\gamma_2}\right)$$

$$d_{CO3}(pH) := \frac{-C_{Tot.CO3}}{\left(\frac{H(pH)^2}{K_1\cdot K_2} + \frac{H(pH)}{K_2\cdot\gamma_1} + \frac{1}{\gamma_2}\right)^2}\cdot\left(\frac{2\cdot H(pH)}{K_1\cdot K_2} + \frac{1}{K_2\cdot\gamma_1}\right)$$

We then assemble the derivatives into the master expressions for β_{ANC} and β_{BNC}.

$$\beta_{ANC}(pH) := -\ln(10)\cdot H(pH)\cdot\left(\frac{d_{HCO3}(pH)}{\gamma_1} + 2\cdot\frac{d_{CO3}(pH)}{\gamma_2} + \frac{d_{OH}(pH)}{\gamma_1} - \frac{d_H(pH)}{\gamma_1}\right)$$

$$\beta_{BNC}(pH) := -\ln(10)\cdot H(pH)\cdot\left(2\cdot d_{H2CO3}(pH) + \frac{d_{HCO3}(pH)}{\gamma_1} + \frac{d_H(pH)}{\gamma_1} - \frac{d_{OH}(pH)}{\gamma_1}\right)$$

For [ANC], we set the limits of the integration as 7.65 and 4.30 and perform the integration.

$$pH_i := 7.65 \quad pH_f := 4.30 \quad ANC := -\int_{pH_i}^{pH_f} \beta_{ANC}(pH)\, dpH = 5.724\times 10^{-3}$$

We certainly are pleased that our integrated value of [ANC] matches that from Example 6.4.

Similarly for [BNC] we set the limits of the integration as 7.65 and 12.3 and perform the integration.

$$pH_i := 7.65 \quad pH_f := 12.30 \quad BNC := -\int_{pH_i}^{pH_f} \beta_{BNC}(pH)\, dpH = 0.026 \quad M$$

We now need to employ the process used in Example 6.4 to the computation of [BNC]. This is, of course, a direct application of Equation 10.17b. We've updated the relations used, including activity coefficients, to reflect newly gained understandings. The results would be identical to those for which infinitely dilute conditions of Example 6.4 would be assumed, since we have set all our γs to unity. Again, important results are shown.

$$H(pH) := 10^{-pH} \qquad OH(pH) := \frac{K_W}{H(pH)} \qquad H2CO3(pH) := \frac{C_{Tot.CO3}}{\left(1 + \frac{K_1}{H(pH)} + \frac{K_1 \cdot K_2}{H(pH)^2}\right)}$$

$$HCO3(pH) := \frac{C_{Tot.CO3}}{\left(\frac{H(pH)}{K_1} + 1 + \frac{K_2}{H(pH)}\right)} \qquad pH_i := 7.65 \qquad pH_f := 12.3$$

$$\begin{pmatrix} \Delta C_{HCO3} \\ \Delta C_{H2CO3} \\ \Delta C_{OH} \\ \Delta C_H \end{pmatrix} := \begin{pmatrix} \frac{HCO3(pH_f)}{\gamma_1} - \frac{HCO3(pH_i)}{\gamma_1} \\ H2CO3(pH_f) - H2CO3(pH_i) \\ \frac{OH(pH_f)}{\gamma_1} - \frac{OH(pH_i)}{\gamma_1} \\ \frac{H(pH_f)}{\gamma_1} - \frac{H(pH_i)}{\gamma_1} \end{pmatrix} = \begin{pmatrix} -5.639 \times 10^{-3} \\ -2.858 \times 10^{-4} \\ 0.02 \\ -2.239 \times 10^{-8} \end{pmatrix}$$

$$BNC := -(2 \cdot \Delta C_{H2CO3} + \Delta C_{HCO3} + \Delta C_H - \Delta C_{OH}) = 0.026 \qquad M$$

We are again pleased that the results from both methods are the same for [BNC].

Before we depart this example, we should investigate the pH-related behaviors of β_{ANC} and β_{BNC} as well the contributions from the various conjugate bases and conjugate acids. We have plotted β_{ANC} and β_{BNC} against pH in Figure E10.16.1. The result is quite interesting. β_{ANC} and β_{BNC} are exactly symmetric relative to each other about the abscissa axis. This is the case since we have chosen our reference species to be the fully protonated acid and fully deprotonated base of the carbonate system for the definitions of [ANC] and [BNC], respectively. This of course makes perfect sense, since the acid neutralizing capacity of a conjugate base is exactly mirrored by the base neutralizing capacity of its conjugate acid. We have also plotted the equivalent β functions for each of the five potential species contributing to either β_{ANC} or β_{BNC} in Figure E10.16.2. For both $CO_3^=$ and H_2CO_3*, we have included the stoichiometric coefficients to depict the total contribution of each specie to β_{ANC} or β_{BNC}, respectively. From these graphical depictions we may take away with us several key points about [ANC]. As might be expected, the contribution of each conjugate base to [ANC] is at maximum at pH equal to its pK_A for the protonation reaction with its conjugate acid. [ANC] is dominated by:

water, via its protonation to form hydronium, below pH 4.5;
bicarbonate, via its protonation to form carbonic acid, between pH 4.5 and 8.5;
carbonate, via its protonation to form bicarbonate, between pH 8.5 and 11;
water, via protonation of hydroxide to form water, above pH 11.

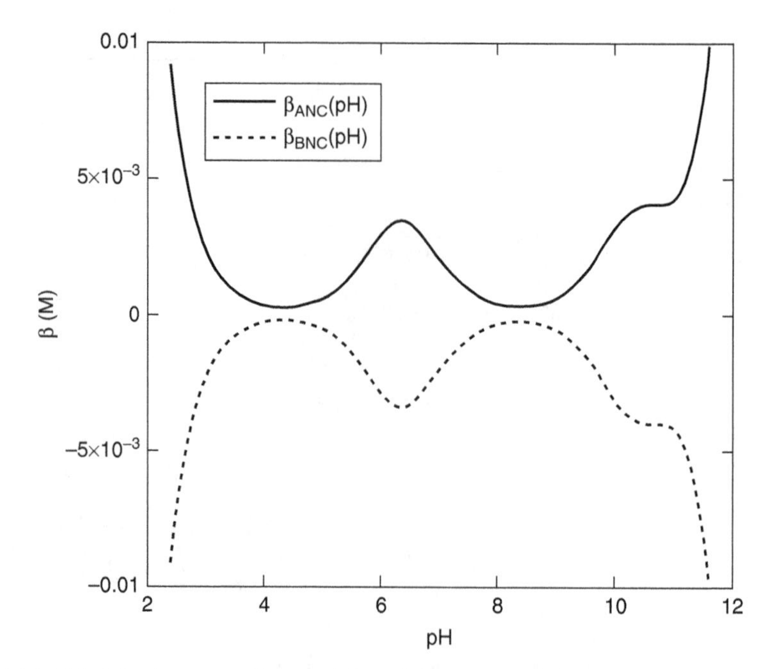

FIGURE E10.16.1 *A plot of β_{ANC} and β_{BNC} (M) versus pH for an aqueous solution of 0.006 M CO_{3Tot}.*

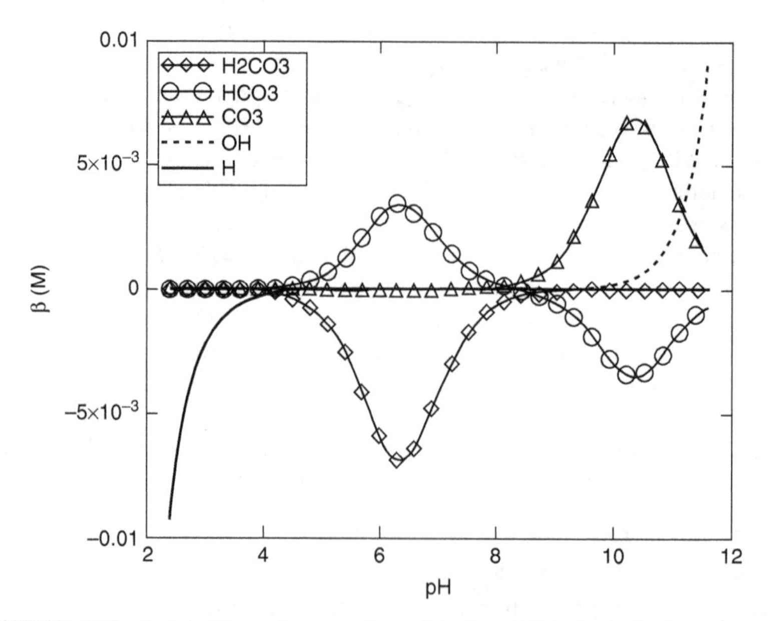

FIGURE E10.16.2 *A plot of the various specie contributions (M) to the buffer intensity against pH for the carbonate system with CO_{3Tot} = 0.006 M.*

We also may take away several key points about [BNC]. The contribution of each conjugate acid to [BNC] is at maximum at pH equal to its pK_A for the deprotonation reaction with its conjugate base. [BNC] is dominated by:

water, via deprotonation to form hydroxide, above pH 11;

bicarbonate, via deprotonation to form carbonate, between pH 8.5 and 11;

carbonic acid, via deprotonation to form bicarbonate, between pH 4.5 and 8.5;

water, via deprotonation of hydronium to form water, below pH 4.5.

Using the relations developed with Example 10.16, we could determine the [ANC] and [BNC] values associated with any given sets of initial and final pH. Given that we can characterize ionic strength and accurately compute associated activity coefficients, the methodologies employed are exact. We realize from the results of Example 10.16 that we have two choices as means to compute [ANC] and [BNC]. We may use the integral approach of Equations 10.24 and 10.25 or we may employ the two Equations, 10.17a and 10.17b, termed hereafter the "difference" approach, from which the integral approach has been derived. Either pathway will yield accurate results. For the integral approach, we must formulate the proper set of derivatives of specie concentration with pH and assemble them into the overall buffer intensity relations. We then we can either integrate them or use them as point-wise functions to gain insight into system behavior. For the difference approach, we must formulate relations for the species of interest and evaluate them at final and initial conditions to determine the differences. Certainly, also, with the capability within MathCAD worksheets, we can develop sets of continuous functions allowing their use as point-wise functions for investigations of system behavior. Example 10.16 addressed a "closed" system in which the abundances of the individual acid/base systems within the aqueous solution remain constant over the course of the process.

10.8.2 ANC and BNC of Open Systems

Herein, an open system is one in which an aqueous solution, of finite volume, is in direct contact with a gas phase, of essentially infinite volume. For aqueous solutions in contact with solids or nonaqueous liquids, we would use the term heterogeneous. In the case of the open system, the gas phase serves as an infinite source or infinite sink, relative to the distribution equilibrium of nonelectrolyte species. For environmental systems, these gas phases consist mainly of: the Earth's atmosphere whether indoor or outdoor, vapors held within the pores of the soil except in certain extreme cases, and vapors produced in either aquatic or marine sediments as a consequence of anaerobic biological activity. In this last case, we consider the vapor phase, distributed as a multitude of minute gas bubbles that appear and grow while held in sediment pores and eventually are released when buoyant forces exceed those from adhesion. Given that pore waters within sediments are rather confined, with little opportunity for exchange with the overlying water column, the gas phase produced from the biological processes can in most cases be considered an infinite source, while consideration as an infinite sink might be just a bit dangerous.

In examination of [ANC] and [BNC] associated with open systems, we must assume that the equilibrium for distribution of the nonelectrolyte between water and gas is continuously under conditions so close to equilibrium that assumption of continuous equilibrium would result in insignificant error. Thus, we must consider the consequences of the movement of nonelectrolyte into or from the aqueous solution. Further, as with our analysis of the distilled water held in the laboratory, earlier in this chapter, we consider that Henry's law will continuously govern all gas/water equilibria. Let us also employ the "difference" method for computation of [ANC] and [BNC].

Example 10.17 Consider the aqueous solution of Example 10.16, containing 0.006 M $C_{T.CO_3}$ and having initial pH of 7.65. Imagine that a liter of such solution is held in a beaker in a controlled environment such that the relative humidity of the atmosphere surrounding the beaker is continually refreshed with water-saturated air, of composition ($C_{v.CO_2} = 8782\,\text{ppm}_v$) roughly 22 times that of the normal atmosphere. This water sample may certainly have arisen from a confined aquifer with carbonate system mineralogy or from a subsurface region characterized by significant aerobic biological activity. Consider that the lab is in a suburb of Minneapolis, MN, with normal ambient pressure of 0.96 atm. Now, let us instrument the system with a pH meter and automatic digital titrator, controlled by a feedback signal from the pH meter. Acid or base would be added drop-wise into the solution and, after each addition the attainment of each step-wise equilibrium condition would be communicated to the titrator as a steady pH reading, signaling for addition of another drop. Each of the steady pH readings as well as the incremental volume of acid or base solution added would be recorded. Develop a model that can be used to describe the behavior of this system, with capacity to predict acid or base addition from projected final pH.

We will apply the difference approach and collect MathCAD functions from Example 10.15 that can be used, with minor modification. Note that the activity of carbonic acid is a constant, as a consequence of the equilibrium across the gas/water interface.

$$P_{CO2} := C_{v.CO2} \cdot 10^{-6} \cdot P_{Tot} = 8.431 \times 10^{-3} \quad \text{atm}$$

$$H2CO3 := K_{H.CO2} \cdot P_{CO2} = 2.858 \times 10^{-4} \quad \text{M}$$

$$H(pH) := 10^{-pH} \qquad C_H(pH) := \frac{H(pH)}{\gamma_1} \qquad C_{OH}(pH) := \frac{K_W}{\gamma_1 \cdot H(pH}$$

$$C_{HCO3}(pH) := \frac{K_1 \cdot H2CO3}{\gamma_1 \cdot H(pH)} \qquad C_{CO3}(pH) := \frac{K_1 \cdot K_2 \cdot H2CO}{\gamma_2 \cdot H(pH)^2}$$

$$C_{Tot.CO3}(pH) := H2CO3 + C_{HCO3}(pH) + C_{CO3}(pH)$$

We first characterize the initial conditions.

$$pH_i := 7.65 \quad \begin{pmatrix} C_{HCO3}(pH_i) \\ C_{CO3}(pH_i) \\ C_{Tot.CO3}(pH_i) \\ C_H(pH_i) \\ C_{OH}(pH_i) \end{pmatrix} = \begin{pmatrix} 5.702 \times 10^{-3} \\ 1.191 \times 10^{-5} \\ 6 \times 10^{-3} \\ 2.239 \times 10^{-8} \\ 4.467 \times 10^{-7} \end{pmatrix} \frac{mol}{L}$$

We consider the first steady pH reading of 7.60 and characterize the system.

$$pH_1 := 7.60 \quad \begin{pmatrix} C_{HCO3}(pH_1) \\ C_{CO3}(pH_1) \\ C_{Tot.CO3}(pH_1) \\ C_H(pH_1) \\ C_{OH}(pH_1) \end{pmatrix} = \begin{pmatrix} 5.082 \times 10^{-3} \\ 9.464 \times 10^{-6} \\ 5.378 \times 10^{-3} \\ 2.512 \times 10^{-8} \\ 3.981 \times 10^{-7} \end{pmatrix} \frac{mol}{L}$$

We compute the [ANC] value associated with the first acid addition. All protons added were accepted and the computed [ANC] would be the number of protons introduced with the acid, per liter of aqueous solution.

$$\begin{pmatrix} \Delta C_{HCO3.1} \\ \Delta C_{CO3.1} \\ \Delta C_{Tot.CO3.1} \\ \Delta C_{H.1} \\ \Delta C_{OH.1} \end{pmatrix} := \begin{pmatrix} C_{HCO3}(pH_1) - C_{HCO3}(pH_i) \\ C_{CO3}(pH_1) - C_{CO3}(pH_i) \\ C_{Tot.CO3}(pH_1) - C_{Tot.CO3}(pH_i) \\ C_H(pH_1) - C_H(pH_i) \\ C_{OH}(pH_1) - C_{OH}(pH_i) \end{pmatrix} = \begin{pmatrix} -6.201 \times 10^{-4} \\ -2.45 \times 10^{-6} \\ -6.226 \times 10^{-4} \\ 2.732 \times 10^{-9} \\ -4.858 \times 10^{-8} \end{pmatrix} \frac{mol}{L}$$

$$ANC_1 := -(\Delta C_{HCO3.1} + 2 \cdot \Delta C_{CO3.1} + \Delta C_{OH.1} - \Delta C_{H.1}) = 6.251 \times 10^{-4} \ \frac{mol}{L}$$

We observe that the [ANC] is dominated, as expected by conversion of bicarbonate to carbonic acid. We also observe that the lion's share of the [ANC] is accounted for by the decrease in total carbonate. For each mole reduction of total carbonate, two moles of protons exit the solution. Since the vast majority of the proton assimilation has resulted from the conversion of bicarbonate to carbonic acid, the magnitudes of the changes in bicarbonate and total carbonate each roughly equal that of the [ANC]. We observe from this first step-wise result that we may easily construct a set of functions that will yield [ANC] as a function of final pH.

$\Delta C_{HCO3}(pH) := C_{HCO3}(pH) - C_{HCO3}(pH_i)$ $\quad\Delta C_{OH}(pH) := C_{OH}(pH) - C_{OH}(pH_i)$

$\Delta C_{CO3}(pH) := C_{CO3}(pH) - C_{CO3}(pH_i)$ $\quad\Delta C_{H}(pH) := C_{H}(pH) - C_{H}(pH_i)$

$ANC_O(pH) := -(\Delta C_{HCO3}(pH) + 2\cdot\Delta C_{CO3}(pH) + \Delta C_{OH}(pH) - \Delta C_{H}(pH))$

Then, to compare, we have revisited Example 10.15 and converted the relation for final [ANC] to a point-wise function of pH (simply by redefining it). The ANC_C(pH) function outputs the value of the [ANC] relative to a change in the pH to the value in question.

$$ANC_C(pH) := -\int_{pH_i}^{pH} \beta_{ANC}(pH)\, dpH$$

We have plotted these two relations on a set of axes in Figure E10.17.1. We easily observe that, given the solution is in equilibrium with carbon dioxide at ~22 times the normal atmospheric abundance, the open system has higher [ANC] than the closed system. In the open system, protons may leave the solution with fleeing carbon dioxide, whereas in the closed system, a buildup of carbonic acid occurs and, as a consequence of Le Chatelier's principle, impedes the conversion of bicarbonate to carbonic acid. The ultimate ANCs of the two solutions are about equal, owing to the initial abundances of carbonate and bicarbonate. Were we to investigate a solution whose initial condition was that of equilibrium with the normal atmosphere, the resultant [ANC] would be significantly lower, as a consequence of much lower abundances of carbonate species at the initial condition. In fact, whole families of

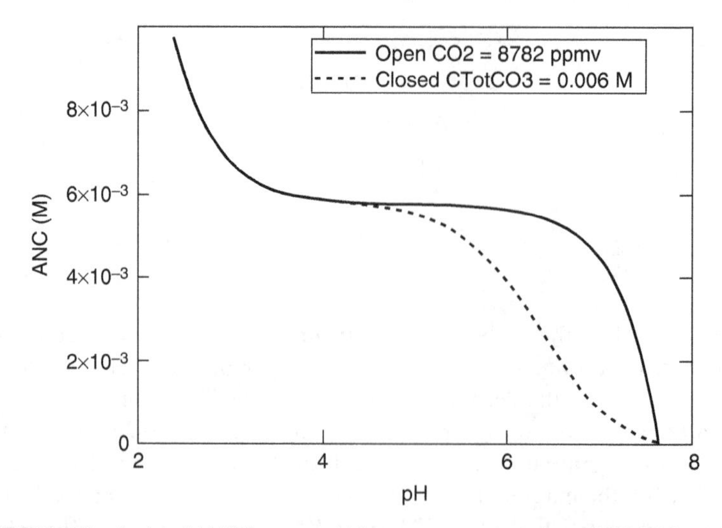

FIGURE E10.17.1 *A plot of [ANC] versus pH for an aqueous solution of initial CO_{3Tot} = 0.006 M at pH 7.65 held in an open system of known and constant P_{CO_2} and titrated with a strong base. A plot for a closed system of CO_{3Tot} = 0.006 M is shown for comparison.*

curves could be generated, each resulting from a unique value of atmospheric CO_2 abundance for the open system and $C_{Tot.CO_3}$ for the closed system.

For computation of base neutralization capacity consider that a strong base would be added dropwise into the aqueous solution. We need to examine our definition of the carbonate system reference specie. Since the solution will be in continuous equilibrium with carbon dioxide, the abundance of carbonic acid in the solution will never change, remaining at the magnitude computed above. In fact, we should use carbonic acid as the reference specie. Then, positive changes in the abundances of bicarbonate and carbonate will be indicative of protons donated and negative signs must accompany the respective changes in abundance, mirroring the situation for [ANC].

$$\Delta BNC_O = -\Delta C_{HCO3} - 2\cdot\Delta C_{CO3} + \Delta C_H - \Delta C_{OH}$$

$$BNC_O(pH) := -\left(-\Delta C_{HCO3}(pH) - 2\cdot\Delta C_{CO3}(pH) - \Delta C_{OH}(pH) + \Delta C_H(pH)\right)$$

We have plotted the BNCs for the open system and corresponding closed system, along with the total carbonate of the open system against pH in Figure E10.17.2. We observe that, with capacity to take on carbon dioxide (two moles of protons per mole of CO_2), the open system has orders of magnitude greater theoretical capacity to neutralize bases. Again, the aqueous carbonic acid abundance is governed, and held constant, by the abundance of carbon dioxide in the gas. The total carbonate increases, and certainly by pH 8 we have violated our infinitely dilute condition. In a real system, eventually we would begin precipitation of a carbonate salt, assuming we would add the base as an alkali metal hydroxide. We have not yet discussed the solubility of metal hydroxides and salts, so determination of the extent to which we could add a metal hydroxide to this system needs to wait until the next chapter.

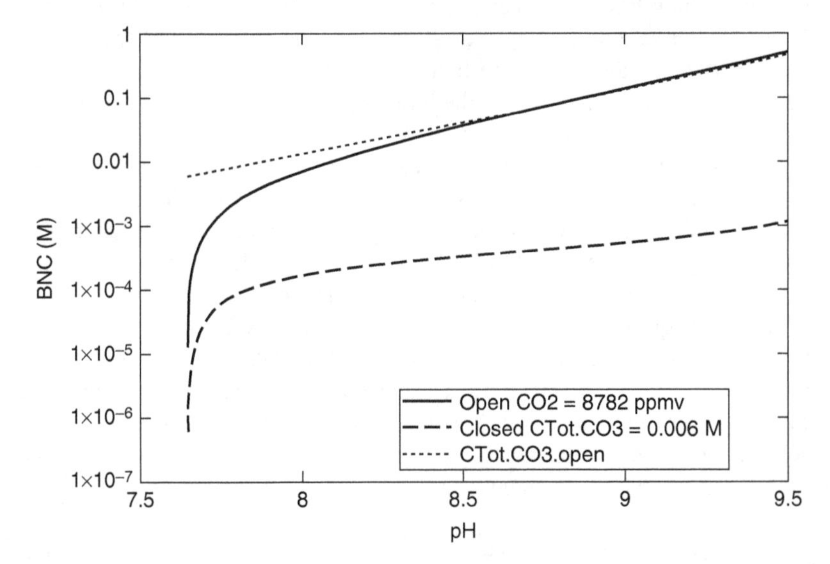

FIGURE E10.17.2 *A plot of [BNC] versus pH for an aqueous solution of initial* $CO_{3Tot} = 0.006\,M$ *at pH 7.65 held in an open system of known and constant* P_{CO_2} *and titrated with a strong base. A plot for a closed system of* $CO_{3Tot} = 0.006\,M$ *is shown for comparison.*

10.8.3 ANC and BNC of Semi-Open Systems

We certainly could also refer to semiopen systems as semiclosed systems. Nevertheless, in our discussions regarding semiopen systems we will address systems that contain both aqueous solution and a distinct gas phase, such that the equilibrium across the air/water interface must be considered alongside our acid/base equilibria. A prime example of such a system would be an anaerobic digester at a wastewater renovation facility. A semiopen system has a confining set of boundaries effectively preventing transfer of mass into or from the system in which processes are underway. Thus, a major premise, similar to that of the closed system is that the total abundance of the species of an acid system remains constant within the system. However, the distributions of relevant species across the air/water interface may then change. We then write our mole balances in terms of the total number of moles of each base and corresponding conjugate acids present within the system.

$$M_{\mathrm{Tot.B}} = M_{\mathrm{B}}^{\mathrm{vap}} + M_{\mathrm{B}}^{\mathrm{w}} \tag{10.26a}$$

M is the number of moles, B denotes base B, vap denotes the vapor phase and w denotes the aqueous phase. We would know or certainly be interested in the volumes of the vapor and water phases, and Equation 10.26a may be expanded accordingly.

$$M_{\mathrm{Tot.B}} = C_{\mathrm{B}}^{\mathrm{vap}} V_{\mathrm{vap}} + C_{\mathrm{Tot.B}}^{\mathrm{w}} V_{\mathrm{w}} \tag{10.26b}$$

V_{vap} and V_{w} are the volumes of the vapor and aqueous solution, respectively. The lone specie of base B present in the vapor would of course be the nonelectrolyte.

We can employ Equation 10.26b to characterize the equilibrium distributions of acid/base species in semiopen systems if we know the volumes and any two critical parameters of the system, including abundance of the nonelectrolyte in the vapor, pH of the solution, total abundance of the base species in the aqueous solution, or the abundance of any one of the base species in the aqueous solution. We can extend the concept of the abundance fraction to include the gas phase. For a semiopen system in which we have a monoprotic acid system present, we may write expressions for α_{g}, α_0, and α_1, retaining the idea that we would apply this with nondilute solutions. But, of potentially greater use would be the set of expressions relating the abundances of the three species (one in the vapor and two in the aqueous solution) to the total abundance of B and equilibrium constants. We accomplish this in much the same manner as that employed to obtain the relations of Table 10.4. We write the mole balance in terms of the gas phase specie and each aqueous phase specie, and then solve the resultant expression for the target specie. Herein we have included the relation for a monoprotic acid system for which $n=1$ (e.g., the ammonia system) such that the fully deprotonated base is the nonelectrolyte.

$$C_{\mathrm{B}}^{\mathrm{vap}} = \frac{M_{\mathrm{Tot.B}}}{V_{\mathrm{vap}} + V_{\mathrm{w}}\left(\dfrac{\{\mathrm{H}^+\}K_{\mathrm{H.B}}}{K_{\mathrm{A.B}}\gamma_1} + \dfrac{K_{\mathrm{H.B}}}{\gamma_0}\right)} \tag{10.27}$$

$$\{HB^+\} = \frac{M_{Tot.B}}{V_{vap}\dfrac{K_{A.B}}{\{H^+\}K_{H.B}} + V_w\left(\dfrac{1}{\gamma_1} + \dfrac{K_{A.B}}{\{H^+\}\gamma_0}\right)} \tag{10.28}$$

$$\{B^0\} = \frac{M_{Tot.B}}{\dfrac{V_{vap}}{K_{H.B}} + V_w\left(\dfrac{\{H^+\}}{K_{A.B}\gamma_1} + \dfrac{1}{\gamma_0}\right)} \tag{10.29}$$

We certainly could write a set of relations of the form of Equations 10.27–10.29 for each combination of acid (i.e., mono- or diprotic) and fully protonated charge, n, of Table 10.4. This effort is left as a potential exercise for the student. These rearranged forms of the mole balance prove immensely useful in developing computational models for target systems. Let us examine a semiopen system from the standpoint of mixing two aqueous solutions of different composition, such that a proton transfer reaction would occur. We will accomplish this by considering closed vessels in the laboratory, realizing that volumetric additions can be directly converted to flow additions and that real vessels simply would be of greater volume than our laboratory vessels.

Example 10.18 Consider one-half liter of aqueous solution, initially FDW and then contacted with vapor of total pressure equal to 0.96 atm containing CO_2 at 5% by volume. Once equilibrated, the resultant solution is held in a 2-L Erlenmeyer flask fitted with a stopper and valve through which reagent may be added with zero loss of vapor from the flask. The system is then closed to inputs or outputs other than 0.1 N sodium hydroxide, which is added (with no addition or release of other liquid or gas constituents) to raise the pH to targeted levels.

We will make no simplifying assumptions other than that we may use the Güntelberg equation with acceptable error. We will first characterize the solution equilibrated with carbon dioxide. We will then compute the quantity of sodium hydroxide solution necessary to raise the pH to a target level. Along the way we will compute the ionic strength and applicable activity coefficients. For the final mixture, we will compute the initial conditions as a function of the volume of NaOH solution added and use mole and proton balances to compute the final pH and speciation.

Since we are dealing specifically with the carbonate system, to streamline the symbology for some of the complex relations we will write K_1, K_2, and K_H are used for the acid/base equilibrium and Henry's law constants for the carbonate system. Known information and computations for the partial pressure of carbon dioxide and associated aqueous phase carbonic acid abundance are shown next.

$$\begin{pmatrix} K_1 \\ K_2 \\ K_H \end{pmatrix} := \begin{pmatrix} 10^{-6.35} \\ 10^{-10.33} \\ 10^{-1.470} \end{pmatrix} \qquad \begin{pmatrix} K_W \\ A \end{pmatrix} := \begin{pmatrix} 10^{-14} \\ 0.5115 \end{pmatrix} \qquad \begin{pmatrix} P_{Tot.0} \\ C_{V.CO2.0} \end{pmatrix} := \begin{pmatrix} 0.96 \\ 50000 \end{pmatrix} \begin{pmatrix} atm \\ ppm_v \end{pmatrix}$$

$$P_{CO2.0} := P_{Tot.0} \cdot C_{V.CO2.0} \cdot 10^{-6} = 0.048 \quad atm$$

$$H2CO3_0 := K_H \cdot P_{CO2.0} = 1.626 \times 10^{-3} \quad \frac{mol}{L}$$

$$\begin{pmatrix} C_{H.i} \\ C_{OH.i} \\ C_{HCO3.i} \\ C_{CO3.i} \end{pmatrix} := \begin{pmatrix} 10^{-7} \\ 10^{-7} \\ 0 \\ 0 \end{pmatrix} \frac{mol}{L}$$

To characterize the initially equilibrated solution, we write five equations: a proton balance, a mole balance on carbonate species, a relation for ionic strength and definitions of two activity coefficients. Our proton and mole balances are nearly identical to those of Example 10.10, except that herein, we have populated the activity coefficients with non unity values. These constitute the system of equations necessary to characterize the initial conditions of the aqueous solution.

$$\begin{pmatrix} \gamma_1 \\ \gamma_2 \\ I \\ H \\ C_{Tot.CO3} \end{pmatrix} := \begin{pmatrix} .9 \\ .7 \\ .005 \\ 10^{-5} \\ 2 \cdot 10^{-5} \end{pmatrix}$$

$$\text{Given} \quad \gamma_1 = 10^{\frac{-A \cdot \sqrt{I}}{1+\sqrt{I}}} \qquad \gamma_2 = 10^{\frac{-A \cdot 4 \cdot \sqrt{I}}{1+\sqrt{I}}}$$

$$I = 0.5 \cdot \left(\frac{H}{\gamma_1} + \frac{K_W}{H \cdot \gamma_1} + \frac{K_1 \cdot H2CO3_0}{H \cdot \gamma_1} + 4 \cdot \frac{K_1 \cdot K_2 \cdot H2CO3_0}{H^2 \cdot \gamma_2} \right)$$

$$C_{Tot.CO3} = H2CO3_0 \cdot \left(1 + \frac{K_1}{H \cdot \gamma_1} + \frac{K_1 \cdot K_2}{H^2 \cdot \gamma_2} \right) \qquad \frac{H}{\gamma_1} = \frac{K_W}{H \cdot \gamma_1} + \frac{K_1 \cdot H2CO3_0}{H \cdot \gamma_1} + \frac{K_1 \cdot K_2 \cdot H2CO3_0}{H^2 \cdot \gamma_2}$$

We have written the target relations with variables that are somewhat generic and set the results to the zeroth (initial) condition using the assignment of the find into a vector of defined variables. We have of course fully characterized the initial condition of aqueous solution for later use. Of note, the subscripts appearing later are of the formatting variety and do not indicate that we have yet established vectors for the various parameters.

$$\begin{pmatrix} \gamma_{1.0} \\ \gamma_{2.0} \\ I_0 \\ H_0 \\ C_{Tot.CO3.0} \end{pmatrix} := \text{Find}(\gamma_1, \gamma_2, I, H, C_{Tot.CO3}) = \begin{pmatrix} 0.994 \\ 0.976 \\ 2.712 \times 10^{-5} \\ 2.695 \times 10^{-5} \\ 1.654 \times 10^{-3} \end{pmatrix}$$

$$\begin{pmatrix} pH_0 \\ C_{HCO3.0} \\ C_{CO3.0} \end{pmatrix} := \begin{pmatrix} -\log(H_0) \\ \dfrac{K_1}{H_0 \cdot \gamma_{1.0}} \cdot H2CO3_0 \\ \dfrac{K_1 \cdot K_2}{H_0^{\,2} \cdot \gamma_{2.0}} \cdot H2CO3_0 \end{pmatrix} = \begin{pmatrix} 4.569 \\ 2.712 \times 10^{-5} \\ 4.793 \times 10^{-11} \end{pmatrix}$$

We also now may define the total number of moles of carbonate species in the entire semiopen system.

$$R_{atm} := 0.082057 \frac{L \cdot atm}{mol \cdot {}^{\circ}K} \quad T := 298.15$$

$$V_{w.0} := 0.5 \qquad V_{Tot} := 2 \quad V_{vap.0} := V_{Tot} - V_{w.0}$$

$$M_{Tot.CO3} := \frac{P_{CO2.0}}{R_{atm} \cdot T} \cdot V_{vap.0} + V_{w.0} \cdot C_{Tot.CO3.0} = 3.7697 \times 10^{-3} \quad mol$$

We are now ready to consider the addition of the sodium hydroxide solution.

We write a system of five equations: proton balance for [BNC], mole balance on carbonate species, ionic strength, and two activity coefficients. Our reference specie is carbonic acid, owing to its tie to the vapor via Henry's law. Then our proton balance must relate the change in the [BNC] to changes in abundances of bases within the entire semiopen system, thus Equation 10.17b is modified to employ abundances of bases. Further, since the volume of the aqueous solution will increase with the added NaOH solution, our proton balance addressing [BNC] must address the total number of protons accepted and donated. Then rather than simply summing changes in concentration we must sum changes in the number of moles of bases. The changes in the number of moles are then expressed as changes in the product of volume and concentration.

$$\Delta\,[\text{BNC}] = -\sum_i \nu_{\text{B},i} \Delta \text{Mol}_i = -\sum_i \left(\nu_{\text{B},i}\, \Delta\left(V_{\text{w}}\left[B_i\right]\right)\right)$$

We then write the proton balance for [BNC] specific to the carbonate system as a base relation in our MathCAD worksheet and then may use it as the starting point for the necessary subsequent algebraic manipulations.

$$\Delta BNC = -BNC = -V_{NaOH} \cdot N_{NaOH} = \Delta Mol_H - \Delta Mol_{OH} - \Delta Mol_{HCO3} - 2 \cdot \Delta Mol_{CO3}$$

Δ[BNC] is the negative of the total base added to the solution.

We utilize Equation 10.26b for the mole balance on carbonate species, opting (a somewhat arbitrary choice) to retain the partial pressure of carbon dioxide in the vapor as the target carbonate unknown. The volume of sodium hydroxide solution added is a target unknown and we write the final volumes of solution and vapor using the added liquid volume. The three remaining equations are for ionic strength and the two activity coefficients. We use the given-find block for further illustrations so we have written the block into a named [BNC] function with six arguments: the five unknowns and the master independent variable, $\{H^+\}$, for which we use H_1 in the worksheet.

$$\begin{pmatrix} V_{NaOH} \\ P_{CO2} \\ I \\ \gamma_1 \\ \gamma_2 \end{pmatrix} := \begin{pmatrix} 0.02 \\ 0.005 \\ 0.001 \\ 0.95 \\ 0.8 \end{pmatrix} \qquad pH_1 := 8 \qquad H_1 := 10^{-pH_1} = 1\times 10^{-8}$$

$$\text{Given} \quad \gamma_1 = 10^{\frac{-A\cdot\sqrt{I}}{1+\sqrt{I}}} \qquad \gamma_2 = 10^{\frac{-A\cdot 4\cdot\sqrt{I}}{1+\sqrt{I}}}$$

$$I = 0.5\cdot\left(\frac{V_{NaOH}\cdot N_{NaOH}}{V_{w.0}+V_{NaOH}} + \frac{H_1}{\gamma_1} + \frac{K_W}{H_1\cdot\gamma_1} + \frac{K_1\cdot K_H\cdot P_{CO2}}{H_1\cdot\gamma_1} + \frac{4\cdot K_1\cdot K_2\cdot K_H\cdot P_{CO2}}{H_1^{\,2}\cdot\gamma_2}\right)$$

$$M_{Tot.CO3} = P_{CO2}\cdot\left[\frac{V_{vap.0}-V_{NaOH}}{R_{atm}\cdot T} \cdots + \left(V_{w.0}+V_{NaOH}\right)\cdot\left(K_H + \frac{K_1\cdot K_H}{H_1\cdot\gamma_1} + \frac{K_1\cdot K_2\cdot K_H}{H_1^{\,2}\cdot\gamma_2}\right)\right]$$

$$-\left(V_{NaOH}\cdot N_{NaOH}\right) = \left[\frac{\left(V_{w.0}+V_{NaOH}\right)\cdot H_1}{\gamma_1} - \frac{V_{w.0}\cdot H_0}{\gamma_{1.0}} \cdots + -\left[\frac{\left(V_{w.0}+V_{NaOH}\right)\cdot K_W}{H_1\cdot\gamma_1} - \frac{V_{w.0}\cdot K_W}{H_0\cdot\gamma_{1.0}}\right] \cdots + -\left[\left(V_{w.0}+V_{NaOH}\right)\cdot\frac{K_1\cdot K_H\cdot P_{CO2}}{H_1\cdot\gamma_1} - V_{w.0}\cdot C_{HCO3.0}\right] \cdots + -\left[2\cdot\left[\left(V_{w.0}+V_{NaOH}\right)\cdot\frac{K_1\cdot K_2\cdot K_H\cdot P_{CO2}}{H_1^{\,2}\cdot\gamma_2}\right] - V_{w.0}\cdot C_{CO3.0}\right]\right]$$

$$BNC\left(H_1, V_{NaOH}, P_{CO2}, I, \gamma_1, \gamma_2\right) := Find\left(V_{NaOH}, P_{CO2}, I, \gamma_1, \gamma_2\right)$$

For formatting of the solve block, we have employed the capability to write the mole and proton balance equations using multiple lines. Observe that the standard format requires that each new line begin with a +. Thus each term representing a $-\Delta[i]$ must include the leading negative. We output the computations for a final pH value of 8.0.

$$\begin{pmatrix} V_{NaOH} \\ P_{CO2.1} \\ I_1 \\ \gamma_{1.1} \\ \gamma_{2.1} \end{pmatrix} := BNC(H_1, V_{NaOH}, P_{CO2}, I, \gamma_1, \gamma_2) = \begin{pmatrix} 0.035 \\ 3.896 \times 10^{-3} \\ 6.559 \times 10^{-3} \\ 0.916 \\ 0.703 \end{pmatrix}$$

$$\begin{pmatrix} C_{H.1} \\ C_{OH.1} \\ C_{HCO3.1} \\ C_{CO3.1} \end{pmatrix} := \begin{pmatrix} \frac{H_1}{\gamma_{1.1}} \\ \frac{K_W}{H_1 \cdot \gamma_{1.1}} \\ \frac{K_1}{H_1 \cdot \gamma_1} \cdot K_H \cdot P_{CO2.1} \\ \frac{K_1 \cdot K_2}{H_1^{\,2} \cdot \gamma_2} \cdot K_H \cdot P_{CO2.1} \end{pmatrix} = \begin{pmatrix} 1.092 \times 10^{-8} \\ 1.092 \times 10^{-6} \\ 6.207 \times 10^{-3} \\ 3.447 \times 10^{-5} \end{pmatrix}$$

To raise the pH of the aqueous solution to pH 8 we compute a requirement for 35 mL of the sodium hydroxide reagent. This seems like a rather arduous task to complete for a single volume calculation. Note the aforementioned, however, that we have written the find into a function we have named BNC(). BNC() has six arguments, one for each unknown we wish to find and one for each parameter we wish to vary. Our interest certainly would involve understanding of the volume of reagent necessary for varying final pH values, a titration. Our function, combined with some worksheet programming provides the necessary flexibility. We begin by defining the zeroth values, as those associated with the FDW equilibrated with the 5% carbon dioxide vapor. We use the matrix ("[" as opposed to the formatting ".") subscript to set the zeroth set of values into the zeroth elements of the six vectors. The assignment shown on the left is copied after we "clicked into" the assignment statement. The assignment shown on the right is simply copied by dragging the mouse over the assignment. Note that on the left the "[" is not evident but the "." is. Conversely, on the right the formatting "." is not shown while the periods used in the subscript themselves do appear.

$$\begin{pmatrix} V._{NaOH_0} \\ P._{CO2_0} \\ I_0 \\ \gamma._{1_0} \\ \gamma._{2_0} \\ pH_0 \end{pmatrix} := \begin{pmatrix} 0 \\ P._{CO2.0} \\ I._0 \\ \gamma._{1.0} \\ \gamma._{2.0} \\ pH._0 \end{pmatrix} \quad \begin{pmatrix} V_{NaOH_0} \\ P_{CO2_0} \\ I_0 \\ \gamma_{1_0} \\ \gamma_{2_0} \\ pH_0 \end{pmatrix} := \begin{pmatrix} 0 \\ P_{CO2.0} \\ I_0 \\ \gamma_{1.0} \\ \gamma_{2.0} \\ pH_0 \end{pmatrix}$$

We now use the solve block in a programmed loop to compute the values of V_{NaOH}, P_{CO_2}, I, γ_1 and γ_2 resulting from the various additions of the 0.1 N sodium hydroxide solution. Each incremental solution yields the volume of NaOH solution necessary to raise the pH of the solution from the current to the incremented pH level. A screen capture of the MathCAD code implementing this process is shown in Figure E10.18.1.

```
( V_NaOH )
( P_CO2  )
(   I    )  :=  | ΔpH ← 0.2
(   γ1   )      | for i ∈ 1 .. 25
(   γ2   )      |     | pH_i ← pH_(i-1) + ΔpH
(   pH   )      |     | H_i ← 10^(-pH_i)
                |     | V_NaOH_i ← V_NaOH_(i-1)
                |     | P_CO2_i ← P_CO2_(i-1)
                |     | I_i ← I_(i-1)
                |     | γ1_i ← γ1_(i-1)
                |     | γ2_i ← γ2_(i-1)
                |     | ( V_NaOH_i )
                |     | ( P_CO2_i  )
                |     | (   I_i    )  ← BNC(H_i, V_NaOH_i, P_CO2_i, I_i, γ1_i, γ2_i)
                |     | (   γ1_i   )
                |     | (   γ2_i   )
                |
                |          ( V_NaOH )
                |          ( P_CO2  )
                | return   (   I    )
                |          (   γ1   )
                |          (   γ2   )
                |          (   pH   )
```

FIGURE E10.18.1 *MathCAD program for computation of incremental sodium hydroxide dose to model titration in a semi-open system of a solution equilibrated with 5% CO_2 and titrated with sodium hydroxide to a pH end point of 9.6.*

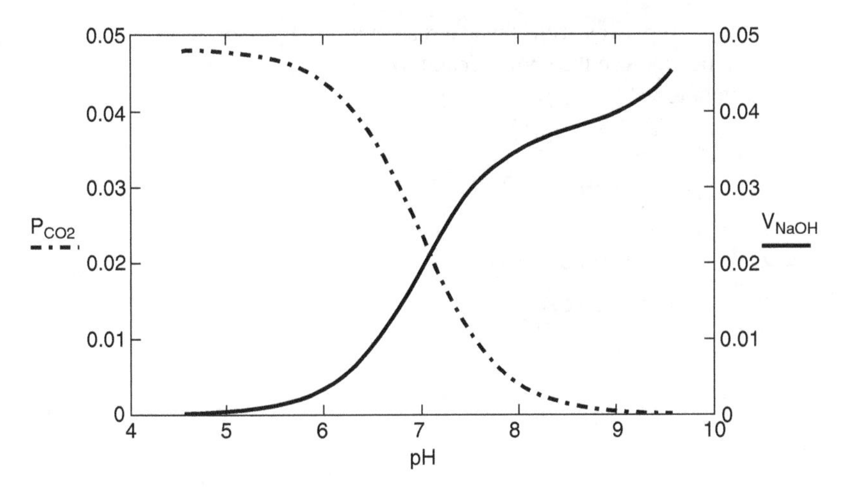

FIGURE E10.18.2 A plot of P_{CO_2} and V_{NaOH} for titration in a semi-open system, of aqueous solution initially equilibrated with 5% CO_2, to a pH end point of ~9.6.

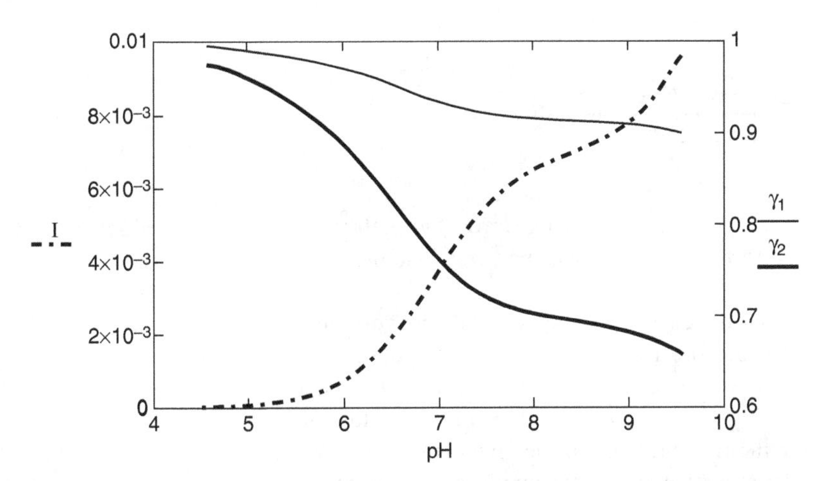

FIGURE E10.18.3 A plot of ionic strength, γ_1 and γ_2 predicted for titration in a semi-open system, of aqueous solution initially equilibrated with 5% CO_2, to a pH end point of ~9.6.

We incremented pH from the initial value to ~9.6, at which point the partial pressure of carbon dioxide in the vapor approached zero and collected the solution. The partial pressure of carbon dioxide (atm) and the volume of sodium hydroxide added (L) are plotted against pH in Figure E10.18.2. By pH 9.6 virtually all of the initial carbon dioxide has been absorbed into the aqueous solution.

We have also tracked the ionic strength and activity coefficients and plotted their predicted values against pH in Figure E10.18.3.

In order to step away from this exercise with that proverbial "warm and fuzzy" feeling, we need to use the ending result to verify that we can account for all of the initial carbonate.

$$\begin{pmatrix} H_f \\ P_{CO2.f} \\ V_{w.f} \\ \gamma_{1.f} \\ \gamma_{2.f} \end{pmatrix} := \begin{pmatrix} 10^{-pH_{25}} \\ P_{CO2_{25}} \\ V_{w.0} + V_{NaOH_{25}} \\ \gamma_{1_{25}} \\ \gamma_{2_{25}} \end{pmatrix}$$

$$M_{Tot.CO3.f} := P_{CO2.f} \cdot \left[\frac{V_{Tot} - V_{w.f}}{R_{atm} \cdot T} \ldots + (V_{w.f}) \cdot \left(K_H + \frac{K_1 \cdot K_H}{H_f \cdot \gamma_{1.f}} + \frac{K_1 \cdot K_2 \cdot K_H}{H_f^{\,2} \cdot \gamma_{2.f}} \right) \right] = 3.7697 \times 10^{-3}$$

$$\frac{M_{Tot.CO3.f} - M_{Tot.CO3}}{M_{Tot.CO3}} \cdot 100 = 1.15 \times 10^{-14} \quad \%$$

Our computed percent relative error is virtually zero. We can indeed step away from this effort with that "warm and fuzzy" feeling.

We could easily have employed the volume of NaOH solution added as a master variable rendering the pH as a dependent variable. Such an alteration would have been accomplished merely by assigning the value of V_{NaOH} and employing H_1 as a variable, requiring its inclusion in the matrix of initial guesses and in the find function of the given-find block. We then would have incremented the volume of sodium hydroxide added to solve for the set of the unknowns, including pH.

We have observed from our work earlier that the computation of either acid neutralizing capacity or base neutralizing capacity is straightforward once we characterize the initial condition of the aqueous solution with regard to the acids and bases that can donate or accept protons. Further, since each acid has a conjugate base and, of course, each base has a conjugate acid, we find that [ANC] is simply the negative of the [BNC] for any given adjustment of the proton abundance. We also find that we can compute [ANC] by addressing changes in abundance of either bases or acids and that, in similar fashion we can compute [BNC] by addressing changes in abundance of either acids or bases. In our treatment of [ANC] and [BNC] in this chapter, we have endeavored to choose either the fully protonated acid or the fully deprotonated base for each acid/base system as our reference specie. We certainly can, as was illustrated in our examinations of the proton balance in mixing of solutions, choose any of the species of a given system as the reference specie for that system. We would however be foolish to attempt to define either hydronium or hydroxide as the reference specie for the water system or to choose an electrolyte over an air/water distributed non-electrolyte in an open or semi-open system. In the examples we have illustrated, we have retained both hydronium and hydroxide in the [ANC] or [BNC] relations. In environmental systems, where the abundances of acid or base systems are 10^{-3} M or greater, most often changes in the abundances of hydronium and hydroxide may be neglected. We should exercise great care in decisions to neglect the acid/base character of water. Certainly, we have shown that its inclusion results in small additional computational complexity.

10.9 ACTIVITY VERSUS CONCENTRATION FOR NONELECTROLYTES

10.9.1 The Setschenow Equation

In analyses of the effect of nonzero ionic strength upon the activity of electrolytes, we embraced three equations, all employing the charge of the ion in a power relation. Thus, when charge was zero, the power went to zero and γ_0 reverted to unity. We have even omitted γ_0 from our relations. Certainly, none of the relations developed to compute the activity coefficients of electrolytes can be applied to nonelectrolytes. This, of course, does not mean we are without means to address nondilute effects upon the activities of nonelectrolytes. Indeed, the concept that ions in solution affect the activity of nonelectrolytes has been well studied relative to seawater. The relation most often used, generally attributed to Setschenow, relates the ratio of solubility of a nonelectrolyte in a target solution to that in an infinitely dilute solution via a "salting out" coefficient and the abundance of salt in the aqueous solution.

$$\log\left(\frac{\gamma_i^S}{\gamma_i^{\mathrm{FDW}}}\right) = \log\left(\frac{S_i^{\mathrm{FDW}}}{S_i^{\mathrm{S}}}\right) = K_{\mathrm{S}}\,[\mathrm{salt}] \tag{10.30}$$

γ_i^{S} is the activity coefficient of nonelectrolyte i in water containing dissolved salt, γ_i^{FDW} is that of i in a dilute solution (our FDW), S_i^{FDW} is the solubility of i in FDW, S_i^{S} is the solubility of i in water containing dissolved salts, K_{S} is a "salting out" coefficient and [salt] is the abundance of dissolved salts in the target water. From the structure of Equation 10.30 we observe that the activity coefficient is unity under dilute conditions. Then, as the abundance of salt in the aqueous solution increases, we observe a decrease in the solubility of the nonelectrolyte. Hence, the range of values for activity coefficients of nonelectrolytes in aqueous solution is bounded by $\gamma = 1$ as a lower bound. We must be careful to recognize that the form of Equation 10.30 is consistent with infinite dilution as the reference state for the system. A significant body of knowledge exists regarding the behavior of nonpolar synthetic organic compounds (SOCs) in water. We can employ Equation 10.30 in examination of the salting out phenomenon relative to SOCs. One detail we must consider is that in much of the literature addressing SOCs the reference state is the pure component at the temperature and pressure of the system. Nonpolar SOCs do not mingle well in aqueous solutions. The polarity of water combined with the nonpolarity of the typical SOC works to develop significant molecular-level repulsive forces. These repulsive forces between water and the SOC are many times greater than repulsive forces occurring within the pure component SOC solution. As a consequence, the solubility of the SOC in water can be very low. Then, the activity coefficient for the SOC in FDW can be very high, for some SOCs as high as 10^6. Fortunately, for work with SOCs, employing the alternative pure component reference state, Equation 10.30 works the same way as for the infinite dilution reference state. The value of γ_i^{FDW} is not unity. Then as the salt added to the water reduces the solubility of the SOC relative to the level for FDW, γ_i^S increases. Robinson and Stokes (1959) address this topic as well as numerous others relative to the behaviors of organic compounds in environmental systems. Further discussion of the interactions of SOCs in aqueous solution is beyond the intended scope of this text.

The term solubility is also often referred to as the saturation concentration. In order to determine the solubility, we must make careful measurements of the abundance of the target nonelectrolyte in aqueous solutions of varying salt content. Typically, we need to equilibrate across a solid, nonaqueous liquid, or vapor phase boundary to dissolve as much of nonelectrolyte i in the aqueous solution as is possible. The composition of the solid, nonaqueous liquid, or gas is held constant or otherwise known. The saturation concentration is nothing more than the final state under equilibrium conditions. The table of dissolved oxygen saturation values in the Appendix of this text is a perfect example of this idea. FDW was equilibrated with atmospheric air at one atmosphere of pressure and at the listed temperatures. The resultant equilibrium (saturation concentrations, or solubilities) of oxygen in water were then measured using wet chemical methods. Were we to take those same measurements using gas of higher or lower oxygen content than that of the normal atmosphere, our set of saturation concentrations would be higher or lower, corresponding with the abundance of oxygen in the vapor. Then, we could return to the standard pressure and standard atmosphere and repeat the oxygen solubility measurements using solutions of nonzero salt content. With each positive increment of the salt content, we would observe an incrementally lower oxygen saturation value for each of the temperatures. Use of gases such as oxygen, nitrogen, the noble gases, and hydrogen allows for straightforward experiments, analyses, and interpretations. Conversely, work with gases such as ammonia, acetic acid, sulfur dioxide, hydrogen sulfide, and carbon dioxide is not so straightforward. Dissolved ammonia is a strong base. Dissolved acetic acid is a moderately strong acid. Dissolution of sulfur dioxide results in formation of sulfurous acid and depending upon pH, significant abundances of bisulfite, and sulfite. Dissolved hydrogen sulfide is a weak acid and upon dissolution will deprotonate, depending upon solution pH to form bisulfide. Carbon dioxide acts in much the same manner as sulfur dioxide, except that the degree to which deprotonation of carbonic acid occurs is lesser than that for sulfurous acid. Then, direct measurement of salting out coefficients is difficult at best for nonelectrolytes that are part of an acid system. So many factors influence the solubility of the nonelectrolyte that the experiments become extremely involved, if even possible.

10.9.2 Definitions of Salt Abundance

In a search of the literature, one may turn up a number of definitions of [salt]. The value of K_S is specific to the definition used by the workers who measured it. In understanding the phenomenon of salting out, we should examine some of the means used to define [salt].

The composition of seawater varies slightly depending upon location of the particular ocean or sea, but the proportions of the major ions comprising the Total Dissolved Solids (TDS) remain fairly constant. Thus, the unit of measure most often employed to describe the abundance of dissolved solids in seawater is salinity. Stumm and Morgan (1996) give the definition of salinity as:

> "...the weight in grams of the dissolved inorganic matter in 1 kg of seawater after all Br^- and I^- have been replaced by the equivalent quantity of Cl^- and all HCO_3^- and $CO_3^=$ are converted to oxide."

The unit of salinity is then g/kg or parts per thousand (ppth$_m$). The symbol often used is ‰.

Brezonik and Arnold (2011) describe the "practical salinity scale" as:

> "...the ratio of the conductivity of a sample to that of a standard solution containing 32.4356 g KCl in 1 kg of solution (0.4452 M KCl)."

Presumably the salinity of normal seawater is accurately represented by this KCl solution.

The TDS abundance in seawater is also defined by a term called "chlorinity". Stumm and Morgan (1996) describe the most recent definition of chlorinity (also given the symbol ‰) as:

> "...the mass in grams of Ag necessary to precipitate the halogens (Cl^- and Br^-) in 328.5233 g of seawater

Wagner et al. (2006) have presented a means by which to convert conductivity measurements to both practical salinity units (PSU) and salinity in ppth$_m$. Of note, at specific conductance of 53,000 μS/cm (close to the specific conductance of "typical" seawater) the equivalent salinity in PSU and ppth$_m$ are 34.935 and 35.008, respectively. A six-parameter relation is presented that is used to convert the ratio, R, of conductivity of a target solution to that of standard seawater to PSU.

$$S = \sum_{i=1}^{6} K_i \cdot R^{\left(\frac{i-1}{2}\right)} \tag{10.31}$$

The respective K_i values are: 0.0120, −0.2174, 25.3283, 13.7714, −6.4788, and 2.5842. Use of this relation to obtain salinity values from conductivity measurements for a target water other than seawater could lead significant error. The specific conductance, and hence the predicted salinity, associated with a given TDS value varies significantly with the specific dissolved salt or combination of salts.

Unfortunately, work involving composition of saline waters has not embraced the wide use of ionic strength. We may compute the ionic strength of "typical" seawater to be ~0.7 M. As long as the relative proportions of major ions in target waters which are more or less saline than seawater mimic those of seawater, we can confidently relate ionic strength to the salinity of the water. However, with many brackish waters that are not derived from seawater, abundances of calcium, magnesium, bicarbonate, and sulfate can comprise a major portion of the TDS. Conversions to ionic strength from characterizations of the solution or measurements of specific conductance to yield values of salinity could result in significant error.

What then are we to do about aqueous activity coefficients of nonelectrolytes? If we choose to address the behavior of nonelectrolytes in the oceans of the world, we may simply work within the established framework, using salinity as our [salt]. Extension of the established framework to freshwaters, as mentioned, carries the potential for error. Perhaps we can establish some bounds upon that error.

Example 10.19 Brezonik and Arnold (2011) compiled information relative to "salting out" and report that salting out coefficients, K_S, vary in magnitude from 0.01 to 0.1, presumably using salinity as the measure of [salt] for Equation 10.30. In their Table 2.3, they report the major ion composition of the Salt River, AZ, at Lake Roosevelt including the specific conductance of 1377 μS/cm. Wagner et al. (2006) give the conductance of normal seawater as 53,087 μS/cm. Compute the range of activity coefficients that would be associated with nonelectrolytes in this fresh water of relatively high TDS content.

We convert the conductivity of the Salt River water to salinity using Equation 10.31. We define R as the ratio of the conductivity of a target fresh water to the conductivity of standard seawater.

$$K := \begin{pmatrix} 0.0120 \\ -0.2174 \\ 25.3283 \\ 13.7714 \\ -6.4788 \\ 2.5842 \end{pmatrix} \quad R := \frac{1377}{53087} = 0.026 \quad S_{SR} := \sum_{i=1}^{6} \left(K_i \cdot R^{\frac{i-1}{2}} \right) = 0.687 \quad \text{PSU}$$

We may compute the range of activity coefficients based on the stated range of salting out coefficients.

$$K_S := \begin{pmatrix} 0.01 \\ 0.1 \end{pmatrix} \quad \gamma_{SR} := 10^{K_S \cdot S_{SR}} = \begin{pmatrix} 1.016 \\ 1.171 \end{pmatrix}$$

We see that salting out of nonelectrolytes could range from insignificant to quite significant for Salt River water, depending on the magnitude of K_S.

We can take this analysis a little further by developing a plot of activity coefficient for variable salting out coefficient, K_S, values as a function of the conductance ratio, R. For activity coefficients in general nondilute solutions, we conveniently develop a function for salinity using R as an argument and use that function in a subsequent function of γ using both R and K_S as arguments.

$$S_{ND}(R) := \sum_{i=1}^{6} \left(K_i \cdot R^{\frac{i-1}{2}} \right) \qquad \gamma_{ND}(K_S, R) := 10^{K_S \cdot S_{ND}(R)}$$

We then define five distinct values of K_S and a range for R.

$$K_S := \begin{pmatrix} 0.01 \\ 0.02 \\ 0.03 \\ 0.05 \\ 0.1 \end{pmatrix} \qquad R := 0, .001 .. .05$$

We plot γ_{ND} versus the conductance ratio in Figure E10.19.1. The vertical line is the conductance ratio of Salt River water. We quickly observe that for the higher values of the salting out coefficient, activity coefficients become significantly larger than unity.

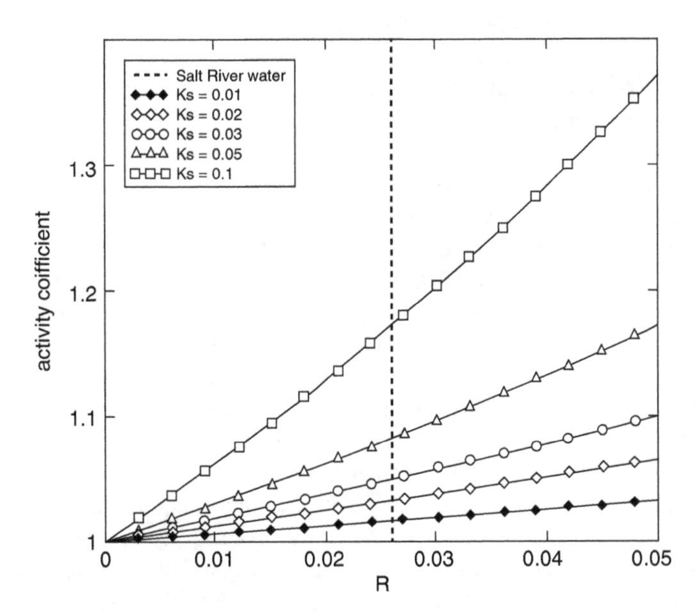

FIGURE E10.19.1 *A plot of the activity coefficient for arbitrary non-electrolyte as a function of salting out coefficient and conductance ratio (ratio of the specific conductance of a target aqueous solution to that of seawater).*

Brezonik and Arnold (2011) inform us that for nonelectrolytes whose dielectric constants are less than that of water, increases in salt concentration will increase the salting out effect and hence, increase the magnitude of the activity coefficient. We interpret this using Equation 10.1 and determine that salting out renders nonelectrolytes to become less soluble as salt content increases, hence the term "salting out" appropriately arises. Then, in contrast to the behavior of electrolytes, concentrations of nonelectrolytes in water are always less than their respective activities.

Particularly for waters whose ionic strength approaches that of Salt River water, caution should be exercised in deciding to disregard the activity coefficients of nonelectrolytes. Unfortunately, the database of quantitative data for use in numeric analyses involving "salting out" of nonelectrolytes in aqueous solution is not nearly as well-developed as that for predicting activities of electrolytes.

10.9.3 Activity of Water in Salt Solutions

In Chapters 5 and 6 as well as earlier in this chapter, we employed the assumption that the activity of water in the aqueous solutions we examined would for all practical purposes be unity. Even in systems where the solution could not be considered infinitely dilute we employed this assumption, greatly simplifying the analysis of targeted systems. We must conclude this chapter with a brief discussion addressing the activity of water in nondilute aqueous solutions. We will quantitatively examine aqueous solutions containing three electrolytes and one nonelectrolyte.

The activity of water is most generally and certainly most conveniently measured using its equilibrium distribution between aqueous solution and an ideal vapor phase, most often normal atmospheric air. We can find the vapor pressure of water tabulated against system temperature in any book addressing fluid mechanics. Certainly this property is important in understanding and designing pumping systems to prevent cavitation. Cavitation occurs under the condition that the absolute pressure in the suction line leading to a pump falls below the vapor pressure of water. Long suction lines, large static lifts, and high velocity all contribute to this condition. As long as the solution being pumped is sufficiently close to an infinitely dilute aqueous solution, the tabulated values for pure water suffice. If we pump solutions of high solute content, then, of course we should consider adjustment for the effects of the solute on the vapor pressure.

Since the vapor pressure of water, or any liquid for that matter, is a manifestation of the distribution equilibrium of the component between its liquid and gaseous state, we consider the distribution from the standpoint of chemical equilibrium. Thus when be measure the abundance of water in vapor equilibrated with an aqueous solution of known composition we have the reactant and product of the distribution reaction defined. We can examine this in light of Henry's law, written as a volatilization reaction.

$$H_2O_{(aq)} \Leftrightarrow H_2O_{(g)}$$

$$H_{H_2O} = \frac{P_{H_2O}}{a_w} \tag{10.32}$$

Where H_{H_2O} is the Henry's law constant and a_w is the activity of water expressed as moles per mole. We might be tempted to use $\{H_2O\}$, but this notation might become confused with the molar abundance of water (55.56M at 4 °C) in pure water. As mentioned previously, in employing the abundance of water in chemical equilibria, we consider pure water at the temperature and pressure of the system to be the reference state. This concept is carried forth in the examination of the vapor–liquid equilibria of synthetic organic compounds, beyond the scope of this text. Then, when we employ mole fraction to characterize the abundance, by definition the concentration, activity, and activity coefficient of water in pure water are all unity.

$$\text{For} \quad X_w = 1, \quad a_w = 1, \quad \text{and} \quad \gamma_w = 1$$

Then, rather than employ an equilibrium constant in vapor–liquid equilibrium for water, H_{H_2O} is replaced by $P^v_{H_2O}$, defined as the vapor pressure and assigned units identical to those of the partial pressure of water in the vapor phase.

$$P^v_{H_2O} = P_{H_2O} \tag{10.33}$$

Then, by simply measuring the abundance of water in vapor equilibrated with pure water (or other liquid for that matter) at various temperatures, the magnitude of the equilibrium constant, called the vapor pressure, can be characterized as a function of system temperature.

When we turn our attention to aqueous solutions of high-solute content we can rearrange Equation 10.32 to isolate the activity of water on the LHS and generalize the numerator of the RHS to become the measured partial pressure of water in vapor equilibrated with aqueous solutions containing solutes.

$$a_w^{ND} = \frac{P^{ND}_{H_2O}}{P^{v,T}_{H_2O}} \tag{10.34}$$

The ND superscript refers to a nondilute aqueous solution and we must be sure that the temperature dependence of the overall relation is included by employing the temperature-specific value of the vapor pressure of water. Since Equation 10.34 arises from a condition of chemical equilibrium, values of the activity, and hence the activity coefficient, of water obtained from partial pressure measurement are applicable in all other chemical equilibria.

We combine this result with Equation 10.1 to define the nondilute activity coefficient $\left(\gamma_w^{ND}\right)$.

$$\gamma_{\mathrm{w}}^{\mathrm{ND}} = \frac{a_{\mathrm{w}}^{\mathrm{ND}}}{X_{\mathrm{w}}^{\mathrm{ND}}} = \frac{P_{\mathrm{H_2O}}^{\mathrm{ND}}}{P_{\mathrm{H_2O}}^{\mathrm{v},T} X_{\mathrm{w}}^{\mathrm{ND}}} \tag{10.35}$$

Where $X_{\mathrm{w}}^{\mathrm{ND}}$ is the mole fraction abundance of water in the nondilute solution. In computation of this mole fraction abundance, we employ the ionization stoichiometry of solutes to determine the total molar abundance of solute species. We find it helpful to define the total solute molarity of nondilute solutions as a means to compare the activity of water among solutions containing various nonelectrolytes and salts.

$$M_{\mathrm{Tot}}^{\mathrm{sol}} = \sum_{i=1}^{k} M_i \tag{10.36}$$

M_i is the molar concentration of specie i and k is the total number of dissolved species. We must consider the stoichiometry of the ionization of salts in computing M_i. Nonelectrolytes such as sucrose form a single dissolved specie. Ionization of one to one salts such as NaCl and KCl results in two species. Ionization of $CaCl_2$ results in two species but the molarity of Cl^- is twice that of Ca^{+2}. We now can define the mole fraction abundance $\left(X_{\mathrm{w}}^{\mathrm{ND}}\right)$ of water using the total solute molarity.

$$X_{\mathrm{w}}^{\mathrm{ND}} = \frac{M_{\mathrm{w}}}{M_{\mathrm{w}} + M_{\mathrm{Tot}}^{\mathrm{sol}}} \tag{10.37}$$

M_{w} is the molar abundance of water in the nondilute solution. Equations 10.36 and 10.37 are easily extended to multisolute solutions.

We can gain perspective regarding the activity of water in the nondilute solutions considered heretofore in this chapter by examining the water activity data of Robinson and Stokes (1965). They measured water abundance in vapors equilibrated with aqueous solutions containing known abundances of sucrose, potassium chloride, sodium chloride, calcium chloride, and sulfuric acid and converted these to water activities using Equation 10.34. Since we are somewhat unsure of the exact distribution between the H_2SO_4 and HSO_4^- abundances in their sulfuric acid solutions we cannot confidently employ Equation 10.36 and therefore will not include sulfuric acid in our examination. Their measure of solute abundance was recorded in molal units. Then in order to compare directly with our discussions herein, we converted their molal abundances to molar abundances. We employed solution density versus solute concentration relations for sucrose from Asadi (2005), for KCl, $NaNO_3$, and $CaCl_2$ from Haynes (2012) and for NaCl from Rogers and Pitzer (1982) in converting from molal to molar concentration values. Since the hydrolysis reactions for sodium, potassium, and calcium each produce one complex for each cation which undergoes hydrolysis, the number of solute units is independent of the hydrolysis reactions. Lastly, the tendency of Na^+, K^+, and Ca^{+2} to form complexes with chloride are very weak. We did not consider the resultant metal-chloride complexes in computing the total molarity of the aqueous solutions. Our values of mole fraction abundance of water are then (perhaps insignificantly) understated.

Plots of water activity and activity coefficient versus total solute molarity are presented in Figure 10.3. We quickly observe, to our relief, that the vast majority of our fresh water systems fall left of the dot-dash line for the total molarity of the Salt River and we can assure ourselves that our use of unity for the activity of water in such systems is quite resonable. For the electrolytes, we observe that at total molarity levels in the range of that for seawater, the activity of water is only slightly below unity and the activity coefficient is very near unity. We observe a significant divergence of the activity coefficient values at total molarity values beyond that of seawater. We recall that, based on the Debye–Hückel ion size parameter, sodium hydrates more strongly than does potassium and that calcium hydrates more strongly than does sodium. Sucrose exhibits behavior much more severe than that of the electrolytes. Its structure is that of a glucose molecule joined to a fructose molecule, resulting in a much larger molecular presence in the aqueous solution than that of even $CaCl_2$. We have no volumetric explanation for the behavior of $NaNO_3$ relative to the other 1:1 salts. The interactions between sodium and nitrate and those between nitrate and sodium and water must be investigated to understand the behavior. Significant research efforts have been invested in developing a capability to predict the activity of water from the molal (or molar) composition of aqueous solutions. Factors such as ionic hydration and specific interactions between ions and between ions and water are duly considered.

In natural systems, (other than perhaps the Dead Sea, the Great Salt Lake at low stage, or subsurface brines we encounter in drilling for oil) we introduce little error in using both the activity and activity coefficients for water as unity. Conversely, in engineered systems, most notably those employed for absorption of carbon dioxide from industrial flue gas, we encounter solutions whose electrolyte concentrations are many times that of seawater. In such systems, we need not only address the activity of water, but we must also turn to much more involved methods for computation of the activity coefficients of electrolytes in these aqueous solutions of high salt content. Further detailed analysis of this set of phenomena is well beyond the scope intended herein and the reader is directed to works by Kojima and Tochigi (1979), Correa et al (1997), and Dutkiewicz and Jakubowska (2002) to gain additional quantitative understandings of the prediction of the activity of water in nondilute solutions from solution composition.

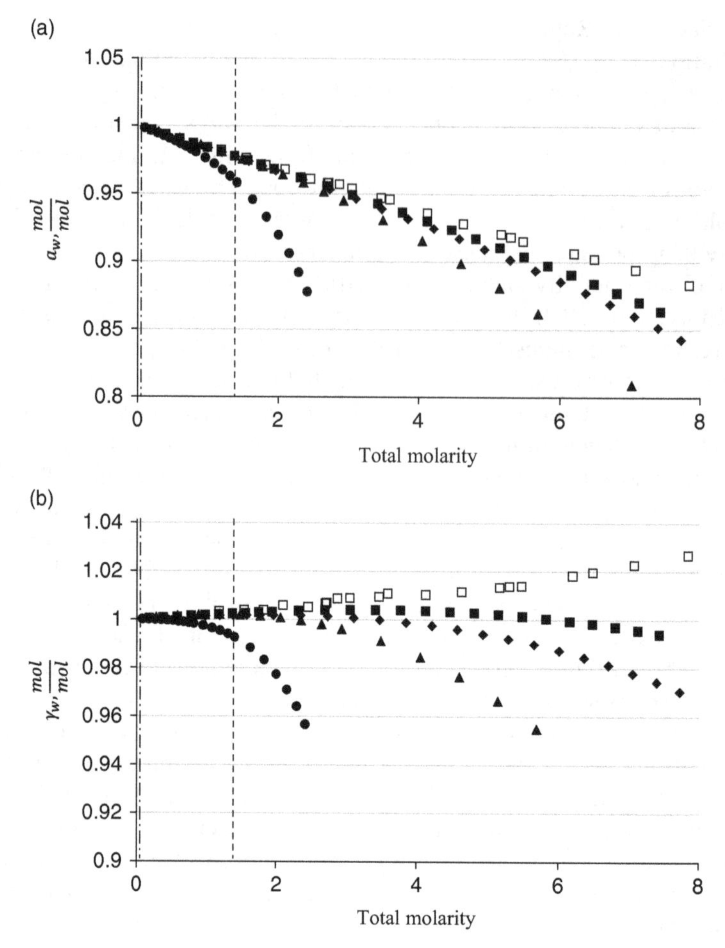

FIGURE 10.3 *Activity (**a**) and activity coefficient (**b**) of water versus total solute molarity for $NaNO_3$ (□), KCl (■), NaCl (♦), $CaCl_2$ (▲), and sucrose (●) solutions. Dot-dashed and dashed vertical lines are the approximate total molarity values for Salt River water (Brezonik and Arnold, 2011) and for seawater (Stumm and Morgan, 1996), respectively. KCl, NaCl, $CaCl_2$ and sucrose activity data are from Robinson and Stokes (1965) and $NaNO_3$ data are from Correa et al (1977). Activity coefficient is computed as the quotient of activity and mole fraction concentration.*

PROBLEMS

For all end of chapter problems, the most convenient platform from which to assemble the mathematical models for each of the problems is a MathCAD worksheet. MS Excel or other software may certainly be employed but additional algebraic manipulations or structured programming may be necessary. Certainly, also, simplifying assumptions may be invoked to render pencil/paper/calculator approximations. Certainly, even graphical approximations of the solution may be assembled.

Problems 1–8 address simplistic systems in which an acid, base or salt is diluted into freshly distilled water, with the overall solution composition sufficiently dilute that the infinitely dilute assumption is applied. These problems are intended to provide experience with the mass and proton balances and an understanding of the relations between the acidity constants and the effects of the acids and bases upon the resultant character of aqueous solutions.

1. A known quantity (0.001 mol) of acid is added to freshly distilled water, held in a glove box under a nitrogen atmosphere, and the final mixture is diluted to exactly one liter. Use mass and proton balances along with the appropriate equilibria, making no simplifying assumptions, to assemble a system model from which the final pH and speciation of the resultant aqueous solution can be computed. Assume that the final solution may be considered infinitely dilute. Consider the following acids, whose acid deprotonation constants may be found in Table 6.1.

   a. Nitric acid
   b. Acetic acid
   c. Hydrocyanic acid
   d. Sulfuric acid
   e. Selenous acid
   f. 4-Aminobenzoic acid
   g. Phosphoric acid

   In developing the solution ensure that a rigorous accounting of aqueous species is documented, that the mass balances include all species that are present, and that the proton balance is written in the form of Equation 10.14. (*Hint: construct the mathematical model such that the respective acid systems may each be investigated simply by inserting the appropriate equilibrium constants.*)

2. For the system and acids of Problem 1, based on understandings of the distributions of the respective sets of conjugate acid and base species as a function of pH, identify species whose abundances would be insignificant (say two or more orders of magnitude lower than those of predominance) and appropriately remove them from mass and proton balances. Perform the revised computations and check the results against those of problem 1 to verify that assumptions made are reasonably error free. Assume that the final solution may be considered infinitely dilute.

3. A known quantity (0.001 mol) of base is added to freshly distilled water, held in a glove box under a nitrogen atmosphere, and the final mixture is diluted to exactly 1 L. Use mass and proton balances along with the appropriate equilibria, making no simplifying assumptions, to assemble a system model from which the final pH and speciation of the resultant aqueous solution can be computed. Assume that the final solution may be considered infinitely dilute. Consider the following bases, whose acid deprotonation constants may be found in Table 6.1.

a. Aniline
b. Ammonia
c. 1,3-diamino-2-propanol

In developing the solution ensure that a rigorous accounting of aqueous species is documented, that the mass balances include all species that are present, and that the proton balance is written in the form of Equation 10.14. (*Hint: construct the mathematical model such that the respective acid systems may each be investigated simply by inserting the appropriate equilibrium constants.*)

4. For the system and bases of Problem 3, based on understandings of the distribution of the respective conjugate acid and base species as a function of pH, identify species whose abundances would be insignificant (say two or more orders of magnitude lower than those of predominance) and remove appropriately remove them from mass and proton balances. Perform the revised computations and check the results against those of Problem 3 to verify that assumptions made are reasonably error free. Assume that the final solution may be considered infinitely dilute.

5. A known quantity (0.001 mol) of an alkali metal salt is added to freshly distilled water, held in a glove box under a nitrogen atmosphere, and the final mixture is diluted to exactly 1 L. Use mass and proton balances along with the appropriate equilibria, making no simplifying assumptions, to assemble a system model from which the final pH and speciation of the resultant aqueous solution can be computed. Assume that the final solution may be considered infinitely dilute. Consider the following salts. Applicable acid deprotonation constants may be found in Table 6.1.

a. Sodium (or potassium) chloride
b. Sodium (or potassium) acetate
c. Sodium (or potassium) cyanide

In developing the solution ensure that a rigorous accounting of aqueous species is documented, that the mass balances include all species that are present, and that the proton balance is written in the form of Equation 10.14. (*Hint: construct the mathematical model such that the respective acid systems may each be investigated simply by inserting the appropriate equilibrium constants.*)

6. A known quantity (0.001 mol) of an alkali metal or ammonium salt is added to freshly distilled water, held in a glove box under a nitrogen atmosphere, and the final mixture is diluted to exactly one liter. Use mass and proton balances along with the appropriate equilibria, making no simplifying assumptions, to assemble a system model from which the final pH and speciation of the resultant aqueous solution can be computed. Assume that the final solution may be considered infinitely dilute. Consider the following salts. Applicable acid deprotonation constants may be found in Table 6.1.

a. Sodium (or potassium) bisulfate
b. Disodium (or dipotassium) sulfate

c. Sodium (or potassium) biselenite
d. Disodium (or dipotassium) selenite
e. Ammonium bisulfate
f. Diammonium sulfate

In developing the solution ensure that a rigorous accounting of aqueous species is documented, that the mass balances include all species that are present, and that the proton balance is written in the form of Equation 10.14. (*Hint: construct the mathematical model such that the respective acid systems may each be investigated simply by inserting the appropriate equilibrium constants.*)

7. A known quantity (0.001 mol) of a salt is added to freshly distilled water, held in a glove box under a nitrogen atmosphere, and the final mixture is diluted to exactly one liter. Use mass and proton balances along with the appropriate equilibria, making no simplifying assumptions, to assemble a system model from which the final pH and speciation of the resultant aqueous solution can be computed. Assume that the final solution may be considered infinitely dilute. Consider the following salts. Applicable acid deprotonation constants may be found in Table 6.1.

   a. Ammonium chloride
   b. Hydrogen aniline chloride

   In developing the solution ensure that a rigorous accounting of aqueous species is documented, that the mass balances include all species that are present, and that the proton balance is written in the form of Equation 10.14. (*Hint: construct the mathematical model such that the respective acid systems may each be investigated simply by inserting the appropriate equilibrium constants.*)

8. A known quantity (0.001 mol) of a salt is added to freshly distilled water, held in a glove box under a nitrogen atmosphere, and the final mixture is diluted to exactly one liter. Use mass and proton balances along with the appropriate equilibria, making no simplifying assumptions, to assemble a system model from which the final pH and speciation of the resultant aqueous solution can be computed. Assume that the final solution may be considered infinitely dilute. Consider the following salts. Applicable acid deprotonation constants may be found in Table 6.1.

   a. Sodium (or potassium) dihydrogen phosphate
   b. Disodium (or dipotassium) hydrogen phosphate
   c. Trisodium (or tripotassium) phosphate

   In developing the solution ensure that a rigorous accounting of aqueous species is documented, that the mass balances include all species that are present, and that the proton balance is written in the form of Equation 10.14. (*Hint: construct the mathematical model such that the respective acid systems may each be investigated simply by inserting the appropriate equilibrium constants.*)

Problems 9–23 address systems for which the infinitely dilute assumption would be inappropriate, unless the computations were truly to be "back of the napkin" approximations. The Güntelberg equation is certainly quite adequate for computation of activity coefficients. Consider that all temperatures are standard (25 °C).

9. Consider that 0.002 mol of ammonium nitrate, NH_4NO_3, is diluted to 1 L in water of background ionic strength equal to 0.03 M. Compute the final speciation for the ammonia system. The acid base character of the nitric acid system is insignificant for this computation. The final pH will be moderately acidic. Consider that the background water has pH near neutral and that initial abundances of hydronium and hydroxide in the background water are insignificant.

10. Consider that 0.002 mol of potassium dihydrogen phosphate, KH_2PO_4, is diluted to 1 L in water of background ionic strength equal to 0.02 M. Compute the final speciation of the phosphate system. The final pH will be fairly acidic. Consider that the background water has pH near neutral and that initial abundances of hydronium and hydroxide in the background water are insignificant.

11. Consider that 0.001 mol of potassium cyanide, KCN, and 0.002 mol of acetic acid, HO_2CCH_3, are diluted to 1 L in water of background ionic strength equal to 0.015 M. Compute the speciation of the cyanide and acetate systems. The final pH will be fairly acidic. Consider that the background water has pH near neutral and that initial abundances of hydronium and hydroxide in the background water are insignificant.

12. Consider that 0.001 mol of calcium hypochlorite $Ca(OCl^-)_2$, is diluted to 1 L in water of background ionic strength equal to 0.02 M. Compute the speciation of the hypochlorite system. Consider that calcium ionizes completely from the hypochlorite. Consider that the background water has pH near neutral and that initial abundances of hydronium and hydroxide in the background water are insignificant. The final pH will be fairly alkaline.

13. Consider that 0.001 mol of potassium bicarbonate, $KHCO_3$, is diluted to 1 L in water of background ionic strength equal to 0.01 M. Compute the speciation of the carbonate system. Consider that the background water has pH near neutral and that initial abundances of hydronium and hydroxide in the background water are insignificant. The final pH will be moderately alkaline.

14. Consider that 0.001 mol of ammonium acetate, $NH_4O_2CCH_3$, is dissolved in water of background ionic strength equal to 0.015 M. Consider that the background water has pH near neutral and that initial abundances of hydronium and hydroxide in the background water are insignificant. The final pH will be near neutral.

15. Consider that 0.002 mol of potassium cyanide, KCN, and 0.002 mol of phosphoric acid, H_3PO_4, are diluted to 1 L in water of background ionic strength equal to 0.02 M. Consider that the background water has pH near neutral and that initial abundances of hydronium and hydroxide in the background water are insignificant. The final pH will be fairly acidic.

16. Consider that 0.001 mol of dipotassium hydrogen phosphate, K_2HPO_4, and 0.001 mol of sodium hydroxide, NaOH, are diluted to 1 L in water of background ionic strength equal to 0.01 M. Consider that the background water has pH near neutral and that initial abundances of hydronium and hydroxide in the background water are insignificant. The final pH will be quite alkaline.

17. Consider that 0.001 mol of sodium carbonate, Na_2CO_3, and 0.001 mol of potassium dihydrogen phosphate, KH_2PO_4, are diluted to 1 L in water of background ionic strength equal to 0.015 M. Consider that the background water has pH near neutral and that initial abundances of hydronium and hydroxide in the background water are insignificant. The final pH will be moderately alkaline.

18. Consider that 0.001 mol of sodium acetate, NaO_2CCH_3, and 0.001 mol of ammonium carbonate, $(NH_4)_2CO_3$, are diluted to a volume of 1 L in water of background ionic strength equal to 0.03 M. Consider that the background water has pH near neutral and that initial abundances of hydronium and hydroxide in the background water are insignificant. The final pH will be fairly alkaline.

19. Consider that 0.001 mol of ammonium sulfate, $(NH_4)_2SO_4$, is diluted to a volume of 1 L in water of background ionic strength equal to 0.01 M. Consider that the background water has pH near neutral and that initial abundances of hydronium and hydroxide in the background water are insignificant. The final pH will be a bit acidic. Compare your results with those of Problem 6f.

20. Consider that 0.001 mol of sodium bicarbonate, $NaHCO_3$, and 0.001 mol of ammonium chloride, NH_4Cl, are diluted to a volume of 1 L in water of background ionic strength equal to 0.025 M. Consider that the background water has pH near neutral and that initial abundances of hydronium and hydroxide in the background water are insignificant. The final pH will be slightly alkaline.

21. Consider that 0.001 mol of ammonium carbonate, $(NH_4)_2CO_3$, is diluted to a volume of 1 L in water of background ionic strength equal to 0.015 M. Consider that the background water has pH near neutral and that initial abundances of hydronium and hydroxide in the background water are insignificant. The final pH will be moderately alkaline.

22. Consider that 0.001 mol of pure sulfur dioxide gas, $SO_{2(g)}$, is carefully dissolved (leaving no residual vapor phase) into one liter of water of background ionic strength equal to 0.02 M. Consider that the background water has pH near neutral and that initial abundances of hydronium and hydroxide in the background water are insignificant. The final pH will be fairly acidic.

23. Consider that 0.001 mol of pure carbon dioxide gas, $CO_{2(g)}$, is carefully dissolved (leaving no residual vapor phase) into one liter of water of background ionic strength equal to 0.02 M. Consider that the background water has pH near neutral and that initial abundances of hydronium and hydroxide in the background water are insignificant. The final pH will be somewhat acidic.

Problems 24–29 address mixing of two or more solutions, requiring application of the algorithm outlined in Section 10.7.2. For these systems the infinitely dilute assumption would be inappropriate, unless the computations were truly to be "back of the napkin" approximations. The Güntelberg equation is certainly quite adequate for computation of activity coefficients. Further, consider that the reactions themselves do not affect the overall ionic strength. Consider that all temperatures are standard (25 °C).

24. Consider that 0.3 L of water of characteristics: pH = 5.5, I = 0.02, total acetate = 0.002 M, is to be mixed with 0.7 L of water of characteristics: pH = 8.5, I = 0.01, total cyanide = 0.001 M. Find the final pH and speciation.

25. Consider that 1 L of water of pH 10.5, containing total cyanide of 0.01 M and of ionic strength equal to 0.05 M, is mixed with 9 L of water of pH 7.35, containing total inorganic carbon of 0.003 M and of ionic strength of 0.01 M. Find the final pH and associated speciation.

26. Consider that 2 L of water of pH 9 containing total phosphate-phosphorus of 0.001 M and ionic strength of 0.03 M, are mixed with 1 L of water containing total acetate of 0.003 M and of ionic strength of 0.05 M. Find the final pH and associated speciation.

27. Ground water flowing beneath an unlined sanitary landfill (see sketch in Figure p10.27) has an alkalinity of 200 mg/L *as* $CaCO_3$, ionic strength of 0.008 M, pH of 8.35, and temperature of 25 °C. Leachate passing through the bottom of the landfill contains total acetate of 5900 mg/L, is in equilibrium with the gas phase (60% methane, 30% carbon dioxide, and 10% other gases at total pressure of 1 atm) contained in the landfill at pH 5.35, has an ionic strength of 0.10 M, and is of temperature equal to 25 °C.

    a. Determine the concentrations of the relevant water and carbonate species of the ground water.
    b. Determine the concentrations of the relevant water, acetate, and carbonate species of the leachate.

    The ground water enters the mixing zone beneath the landfill at a rate of 3600 gal/min and the leachate flows from the landfill into the mixing zone at a rate of 400 gal/min. These two streams become mixed beneath the landfill and

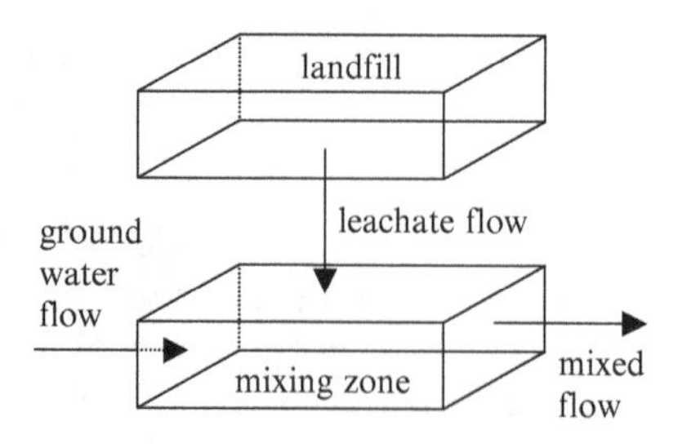

FIGURE p10.27 *Schematic diagram of landfill—ground water mixing system.*

arrive at an equilibrium condition just shortly beyond the down-gradient extremity of the landfill. You may consider the mixing zone beneath the landfill to be a closed system, which is not open to the atmosphere.

c. Determine the equilibrium pH and the equilibrium values of the activities and concentrations of carbonate and acetate species of the mixed solution.
d. Generalize the mathematical model such that the final pH and speciation can be determined based on the fraction of the final mixture contributed from leachate.

28. Consider that water collected in an acid rock drainage pit (see Figure p10.28) at the Gilt Edge Superfund site near Deadwood, SD, has a pH of 2.9 and contains sulfate sulfur at a level of 1000 mg/L (as $SO_4^{=}$–S).

Due to the presence of calcareous minerals in the exposed rock at the site, the acid rock drainage also contains inorganic carbon at a level of 0.002 mol/L. The ionic strength of the ARD solution is estimated to be 0.10 M.

Under a particular hydrologic risk scenario, which is predicted to occur in the event of an extremely wet spring, the ARD flow from the pit into the tributary is predicted to be 0.50 ft^3/s.

Under the selected scenario, the unnamed tributary of Strawberry Creek flowing adjacent to the site, and into which site drainage would flow, would have a volumetric discharge rate of 2.0 ft^3/s. The water is predicted to have a pH of 7.6, an alkalinity of 150 mg/L *as* $CaCO_3$, and a sulfate sulfur concentration of 32 mg/L (again, as $SO_4^{=}$–S). The ionic strength of the water flowing in the tributary is estimated to be 0.004 M.

a. Determine the expected pH and associated carbonate and sulfate speciation of the impacted tributary, downstream from the hypothetical discharge, under the specified scenario. You may assume the temperature is 25 °C.

FIGURE p10.28 *An acid rock drainage pit at an abandoned gold mine in sulfide-bearing rock.*

b. Generalize the mathematical model to allow investigation of varying the fraction of the total flow contributed from the ARD.

Hint: Characterize each of the solutions to be mixed, assuming acid/base reactions in each are under equilibrium conditions, perform the computations associated with mixing and finally perform the computations associated with re-equilibration of proton transfer reactions within the mixed solution.

29. A waste stream from a pickling process (metal doors are soaked in a solution of phosphoric acid to remove rust, welding scale, and other impurities before the galvanizing step) of a door manufacturing operation has a total phosphate concentration of 0.01 M at a pH of 2.6 and an ionic strength of 0.05 M. This waste stream is to be combined with effluent from a metal precipitation process containing total inorganic carbon (carbonate system) of 0.003 M, at pH of 9.3 and of ionic strength equal to 0.03 M. The pickling process generates a waste stream equal to 10,000 gal/day, which is to be combined with the metal treatment effluent, produced at a rate of 40,000 gal/day.

   a. Estimate the value of the pH (and associated phosphate and carbonate system speciation) of the final combined stream. You may assume that the carbonate content of the phosphoric acid stream and the phosphoric acid content of the metal treatment effluent are both negligible and also that the system temperatures are both 25 °C.
   b. Generalize the mathematical model to allow investigation of varying the fraction of the total flow contributed from the pickling process.
   c. Compare the nondilute result with that obtained based on the assumption that the pickling and metal precipitation solutions (and hence the mixture) may be treated as infinitely dilute. Compare the relative difference in both the hydronium ion activity and the associated concentrations.

Problems 30 address computations of acid and base neutralizing capacity. For certain of these the infinitely dilute assumption may be employed while for others, the solutions must be considered nondilute. The Güntelberg equation is certainly quite adequate for computation of activity coefficients. Further, consider that the proton transfer reactions themselves do not affect the overall ionic strength. Consider that all temperatures are standard (25 °C).

30. Consider an aqueous solution containing 0.002 M ammonium nitrate, of ionic strength 0.03 M and initial pH of 6.072.
   a. Determine the acid neutralization capacity between the initial pH and pH 4.3, the end point of the standard alkalinity titration.
   b. Determine the base neutralization capacity between the initial pH and pH 8.3, the end point of the standard acidity titration.

31. Consider an aqueous solution containing 0.002 M potassium dihydrogen phosphate, of ionic strength 0.02 M and initial pH of 4.945.

a. Determine the acid neutralization capacity between the initial pH and pH 4.3, the end point of the standard alkalinity titration.
b. Determine the base neutralization capacity between the initial pH and pH 8.3, the end point of the standard acidity titration.

32. Consider an aqueous solution containing 0.001 M ammonium acetate, of ionic strength 0.015 M and of initial pH 7.

 a. Determine the acid neutralization capacity between the initial pH and pH 4.3, the end point of the standard alkalinity titration.
 b. Determine the base neutralization capacity between the initial pH and pH 8.3, the end point of the standard acidity titration.

33. Consider an aqueous solution containing 0.001 M sodium carbonate and 0.001 M potassium dihydrogen phosphate, of ionic strength 0.015 M and of pH 8.598.

 a. Determine the acid neutralization capacity between the initial pH and pH 4.3, the end point of the standard alkalinity titration.
 b. Determine the acid neutralization capacity between the initial pH and pH 8.3, the end point of the standard phenolphthalein alkalinity titration.
 c. Determine the base neutralization capacity between the initial pH and pH 10.33, the second pK_A for the carbonate system.
 d. Determine the base neutralization capacity between the initial pH and pH 12.3, the third pK_A for the phosphate system.
 e. One might be tempted to neglect the changes in the hydronium and hydroxide concentrations in computations of [ANC] and [BNC]. Compute the error associated with neglecting these for the four end point pH values.

34. A 0.001 M solution (of 0.01 M ionic strength) of 5-aminopentanoic acid (5APA) has a pH of 7.7 and temperature of 25 °C. Compute the quantity of acid or base necessary (in eq./L of solution) to adjust the pH of the solution to 10.7, 9.7, 8.7, 6.7, 5.7, and 4.7. Information pertinent to 5APA is contained in Table 6.1. You may assume that the strong acid or strong base solution is of sufficiently high normality that added volume would be insignificant.

35. A 0.001 M solution (of 0.01 M ionic strength) of 2,3-diaminopropanoic acid (23DAPA) has a pH of 7.7 and temperature of 25 °C. Compute the quantity of acid or base necessary (in eq./L of solution) to adjust the pH of the solution to 10.7, 9.7, 8.7, 6.7, 5.7, and 4.7. Information pertinent to 23DAPA is contained in Table 6.1. You may assume that the strong acid or strong base solution is of sufficiently high normality that added volume would be insignificant.

36. Consider two lakes of equal volume each with an initial pH of 6.5. The first has a measured alkalinity of 200 mg/L *as* $CaCO_3$ and the second has a measured alkalinity of 20 mg/L as $CaCO_3$. Assume a temperature of 25 °C. Consider that the lake volume in each case is $1.7 \times 10^6\, m^3$ of water. The hypothetical lake would have area equal to 48.5 ha (120 acres) with an average depth of 3 m (10 ft).

1. Compute C_{T,CO_3} for both of these waters based on Equation 6.6 (the true value).
2. Compute C_{T,CO_3} for both of these waters based on Equation 6.6, neglecting the water system (the approximation).
3. Compute the relative percent error in the computed C_{T,CO_3} of the assumption versus the true value in each case?
4. For the low alkalinity water described earlier, compute the quantity (in kg) of acid input (as SO_2, the constituent that goes up the stack and is measured as an emission) that would cause the pH of the lake to drop to a value of 5.5, the threshold level for the survival of trout fry. Employ Equation 6.6.
5. For the higher alkalinity water described earlier, compute the quantity (in kg) of acid input (as SO_2, the constituent that goes up the stack and is measured as an emission) that would cause the pH of the lake to drop to a value of 5.5, the threshold level for the survival of trout fry. Employ Equation 6.6.

Perform the computations for parts (4) and (5) considering that carbon dioxide or other carbonate species would neither move into nor out of the lake during the addition of the acid. Once computed from alkalinity, consider that C_{T,CO_3} remains constant throughout the acid addition

37. A sample of supernatant from centrifuged liquid suspension obtained from an anaerobic digester operated at a local waste water plant was subjected to a standard alkalinity titration (potentiometrically to the endpoint pH of 4.3). The resultant alkalinity was measured to be 2100 mg/L *as* $CaCO_3$. A subsample was tested to determine the volatile fatty acid (VFA) content and found to contain total acetate $[CH_3COO^-]_{Tot}$ and total propionate $[CH_3CH_2COO^-]_{Tot}$, at levels of 885 and 584 mg/L, respectively. No other VFAs were detected above limits of quantitation. A further subsample was found to contain dissolved reactive phosphorus (also called ortho-phosphate) and ammonia nitrogen at levels of 155 mg/L as PO_4–P and 800 mg/L NH_3–N. The pH and temperature of the aqueous solution were measured to be 6.65 and 25 °C, respectively. The ionic strength of the aqueous solution is estimated to be 0.023 M.

1. Compute the carbonate system speciation of the digester supernatant based on the measured parameters.
2. The total pressure in the digester when the sample was obtained was 1 atm and the vapor is known to contain 30% carbon dioxide on a molar basis. Reconcile the result of part (1) with this information.
3. The digester has a liquid volume of 250 m^3. The optimum pH for operation of such an anaerobic digester is 7.1. Neglect the gas phase above the digester liquid and determine the quantity of strong base (in equivalents and kg NaOH) necessary to adjust the pH to the optimum level.
4. Were gas withdrawn from above the digester liquid prior to pH adjustment, the total volume of gas phase would be 50 m^3. Use the result of part (1) and the known initial composition (at pH 6.65) of the digester gas and compute

the quantities of base (in equivalents and kg NaOH) necessary to adjust the pH to the optimum level of 7.1.

Problems 38–52 address systems for which the infinitely dilute assumption would be inappropriate, unless the computations were truly to be "back of the napkin" approximations. The Güntelberg equation is certainly quite adequate for computation of activity coefficients. Consider the nonstandard temperature.

38. Consider that 0.002 mol of ammonium nitrate, NH_4NO_3, is diluted to 1 L in water of background ionic strength equal to 0.03 M at temperatures of 4 and 36 °C. Compute the final speciation for the ammonia system. The acid base character of the nitric acid system is insignificant for this computation. The final pH will be moderately acidic. Consider that the background water has pH near neutral and that initial abundances of hydronium and hydroxide in the background water are insignificant. Compare your results with those of Problem 9.

39. Consider that 0.002 mol of potassium dihydrogen phosphate, KH_2PO_4, is diluted to 1 L in water of background ionic strength equal to 0.02 M at temperature of 4 and 36 °C. Compute the final speciation of the phosphate system. The final pH will be fairly acidic. Consider that the background water has pH near neutral and that initial abundances of hydronium and hydroxide in the background water are insignificant. Compare your results with those of Problem 10.

40. Consider that 0.001 mol of potassium cyanide, KCN, and 0.002 mol of acetic acid, HO_2CCH_3, are diluted to 1 Liter in water of background ionic strength equal to 0.015 M at temperature of 4 and 36 °C. Compute the speciation of the cyanide and acetate systems. The final pH will be fairly acidic. Consider that the background water has pH near neutral and that initial abundances of hydronium and hydroxide in the background water are insignificant. Compare your results with those of Problem 11.

41. Consider that 0.001 mol of ammonium acetate, $NH_4O_2CCH_3$, is dissolved in water of background ionic strength equal to 0.015 M at temperature of 4 and 36 °C. Consider that the background water has pH near neutral and that initial abundances of hydronium and hydroxide in the background water are insignificant. The final pH will be near neutral. Compare your results with those of Problem 14.

42. Consider that 0.002 mol of potassium cyanide, KCN, and 0.002 mol of phosphoric acid, H_3PO_4, are diluted to 1 L in water of background ionic strength equal to 0.02 M at temperature of 4 and 36 °C. Consider that the background water has pH near neutral and that initial abundances of hydronium and hydroxide in the background water are insignificant. The final pH will be fairly acidic. Compare your results with those of Problem 15.

43. Consider that 0.001 mol of sodium carbonate, Na_2CO_3, and 0.001 mol of potassium dihydrogen phosphate, KH_2PO_4, are diluted to 1 L in water of

background ionic strength equal to 0.015 M at temperature of 4 and 36 °C. Consider that the background water has pH near neutral and that initial abundances of hydronium and hydroxide in the background water are insignificant. The final pH will be moderately alkaline. Compare your results with those of Problem 17.

44. Consider that 0.001 mol of sodium acetate, NaO_2CCH_3 and 0.001 mol of ammonium carbonate, $(NH_4)_2CO_3$, are diluted to a volume of 1 L in water of background ionic strength equal to 0.03 M at temperature of 4 and 36 °C. Consider that the background water has pH near neutral and that initial abundances of hydronium and hydroxide in the background water are insignificant. The final pH will be fairly alkaline. Compare your results with those of Problem 18.

45. Consider that 0.001 mol of ammonium carbonate, $(NH_4)_2CO_3$, is diluted to a volume of 1 L in water of background ionic strength equal to 0.015 M at temperature of 4 and 36 °C. Consider that the background water has pH near neutral and that initial abundances of hydronium and hydroxide in the background water are insignificant. The final pH will be moderately alkaline. Compare your results with those of Problem 21.

46. Consider that 0.001 mol of pure sulfur dioxide gas, $SO_{2(g)}$, is carefully dissolved (leaving no residual vapor phase) into 1 L of water of background ionic strength equal to 0.02 M at temperature of 4 and 36 °C. Consider that the background water has pH near neutral and that initial abundances of hydronium and hydroxide in the background water are insignificant. The final pH will be fairly acidic. Compare your results with those of Problem 22.

47. Consider that 0.001 mol of pure carbon dioxide gas, $CO_{2(g)}$, is carefully dissolved (leaving no residual vapor phase) into 1 L of water of background ionic strength equal to 0.02 M at temperature of 4 and 36 °C. Consider that the background water has pH near neutral and that initial abundances of hydronium and hydroxide in the background water are insignificant. The final pH will be somewhat acidic. Compare your results with those of Problem 23.

Chapter 11

Metal Complexation and Solubility

11.1 PERSPECTIVE

In Chapters 6 and 10, we examined applications of acid/base principles in various environmental contexts. Acids and bases become involved in proton transfers. The next logical step in our quest to become proficient at modeling environmental processes is to examine the behaviors of chemical ligands. The chemists define a ligand as a specie that can become arranged in a close-range bond with a metal. Metals are those elements that reside below and to the left of a line on the periodic table from boron through astatine. The three elements (Ge, Sb, and Po) lying just below and to the left of the line can often be referred to as metalloids, owing to their capability to act as both metals and nonmetals. Aluminum is most often classed as highly metallic, but does have the capacity to combine with oxygen in a manner consistent with nonmetals. Ligands include hydroxide, the halogens, cyanide, oxyanions of nonmetals, ammonia, amines, and numerous organic acids. Humic and fulvic substances found in environmental systems can also act as ligands. It is not our intent herein to place metals or ligands by their behaviors and properties into the various classes or to assimilate the specific behaviors of any particular metals and ligands. Rather, it is our goal to develop an understanding of the system assembled by the chemists so that we may utilize that system in our efforts to analyze and model environmental processes and systems.

A single close-range bond formed between a metal and a ligand is most often called a complex. Formation of such bonds is called complexation. If the ligand

Environmental Process Analysis: Principles and Modeling, First Edition. Henry V. Mott.

forms two or more close-range bonds with a single metal ion, the resultant chemical specie is most often called a chelate, hence defining the process called chelation. In most environmental systems, the complexation of metals with ligands occurs simultaneously with the formation or dissolution (hence existence) of insoluble metal–ligand complexes: salts or minerals. A large portion of these salts and minerals, found in environmental systems, involve hydroxide, carbonate, phosphate, silicate, and sulfide. The Earth's crust also contains a nearly innumerable variety of minerals that are comprised of solids that are of highly complex elemental composition and structure. The geochemists have assembled an entire, distinct body of scientific literature regarding these minerals. Our intent herein is that the student develops competence with both soluble metal–ligand interactions and the formation of solids containing metals and ligands. From this competence then, forays into the quantitative understandings of geochemical systems can be successfully launched. If we are comfortable employing the chemical system, we need then only know the specific chemistry of target substances in order to successfully quantitatively model their behavior in environmental processes and systems.

We will not invest a great deal of this chapter in attempts to understand the bases for the specific behavior of target metals and ligands. We will leave the correlation of atomic structure with behavior to the chemists. We will however use the manifestations of those behaviors—chemical equilibria—with great interest. In Chapter 4, we illustrated two special cases of the law of mass action: application to metal–ligand complexes and to solubility-dissolution. In this chapter, we will address applications of these special cases of the law of mass action first in simplistic laboratory-type systems to gain basic quantitative understandings and then in more involved engineered and natural environmental systems. We certainly will not lose sight of the quantitative acid/base understandings developed from Chapters 6 and 10.

11.2 HYDRATION OF METAL IONS

In Chapter 6, we envisioned the shell of water that would form around a hydronium ion. Four water molecules become arranged in a tetrahedral shape with the proton at its centroid. Then successive shells of water become oriented, at progressively longer range, with the still fully positively charged proton. In aqueous solution that is infinitely dilute, the radius of this hydrated proton sphere is about nine Å (Dean, 1992). In similar fashion, all ions (both cations and anions) become hydrated to some degree in aqueous solution. Chemists have measured the enthalpy of hydration of many ions. This quantity is the release of energy upon dissolution of a mole of the ion in sufficient water such that the resultant solution may be considered infinitely dilute. Were we to compare the published values of the enthalpy of hydration with the ion size parameters listed in Table 10.2, we would likely find direct correlations between hydrated ion size and enthalpy of hydration. Each class of elements likely would have its own correlation between hydrated radius and enthalpy of hydration, dependent upon ionic charge and position in the periodic table. The degree to which an ion becomes hydrated affects the interactions with other species in aqueous solutions and certainly the influence of the ion upon the overall behavior of the aqueous solution. Then, also the degrees to which ions hydrate affect their behaviors in myriad natural and engineered environmental systems. One notable correlation of hydrated radius with ionic behavior is the lyotropic series—the smaller the hydrated radius of the cation or anion, for a given ionic charge, the more strongly attracted the ion would be to a cation or anion exchange site of a an ion exchange resin (Weber, 1972) or, certainly, to a site of negative or positive charge on a natural surface (Bohn et al., 1979). Were we to continue along this thread, we would investigate the properties of water in the vicinity of the charged sites, judge whether the ions would shed their waters of hydration, examine the diffuse layer within the context of the triple layer model, and examine what the soil chemists call "intrinsic" equilibrium coefficients. This examination would perhaps require another book or two—a great deal has been published about this general topic. Such examination is beyond the scope of this text. Let us become comfortable with interactions between metals and ligands in homogeneous aqueous solutions and then address some of those interactions with the solids they form.

Our goal herein is not the development of quantitative understandings of the molecular-level properties and specific interactions between metals and ligands. Again, we will leave those efforts to the chemists. Our goal is to become aware that metal ions and (certainly to a lesser degree) anions (many of which are complexing ligands) are associated at the molecular level with water in aqueous solutions. Manifestations of these behaviors are included in the system of chemical equilibria developed by the chemists. We would choose herein to become proficient in the use of that system for modeling environmental processes and systems.

11.3 CUMULATIVE FORMATION CONSTANTS

11.3.1 Deprotonation of Metal/Water Complexes

While hydration of metals is certainly related to the hydrolysis of metals, the two phenomena are distinct from each other. Hydration addresses the association of water with ions while hydrolysis addresses the behavior of certain of those water molecules, once associated with ions, most typically metal cations. In addressing hydrolysis of cations, we work with a principle the chemist call the coordination number. It is effectively the number of short-range bonds that a metal cation may have with water molecules in an aqueous solution. Coordination numbers are typically two, three, four, and six, as the coordinated metal–water entities must be symmetric about the nucleus of the ion, where the positive charge arises once outer shell electrons have been shed. Excellent treatises on the intricate details of the hydrolysis of cations and of the associated thermodynamics were published by Baes and Mesmer (1976, 1981). The first is a book while the second is a journal article.

In general, a metal becomes associated with water molecules via short range bonds. From our work with acids and bases, we know that water is the conjugate acid of hydroxide. These bound water molecules then can donate protons to species present in the surrounding aqueous solution. In theory, since water is the conjugate base of hydronium, these bound water molecules could accept protons. This behavior likely would occur only under conditions of large proton abundance, atypical of natural and certainly most engineered aqueous systems. We will consider only the donation of protons by these bound water molecules. Then, the hydrated metal becomes in effect a multiprotic acid. Since coordination numbers typically range from two to six, from two to six short-range bonds can be formed between the metal and water molecules. In general, the number of protons that can be donated from these bound water molecules determines whether the metal–ligand complex would act similarly to a diprotic, triprotic, tetratprotic, pentaprotic, or hexaprotic acid.

We will use zinc as a model metal to examine the hydrolysis phenomenon as successive deprotonations of bound water molecules. Zinc has a coordination number of six, thus when present as the ionized "free metal" in aqueous solution we may write it as $Zn(H_2O)_6^{+2}$. For the overall zinc-water complex the literature suggests four successive deprotonation reactions, manifest as four published values of the formation constant (which we will address once we have illustrated the acid/base behavior).

$$Zn(H_2O)_6^{+2} \Leftrightarrow Zn(H_2O)_5OH^+ + H^+ \qquad K_{A1}$$
$$Zn(H_2O)_5OH^+ \Leftrightarrow Zn(H_2O)_4(OH)_2^0 + H^+ \qquad K_{A2}$$
$$Zn(H_2O)_4(OH)_2^0 \Leftrightarrow Zn(H_2O)_3(OH)_3^- + H^+ \qquad K_{A3}$$
$$Zn(H_2O)_3(OH)_3^- \Leftrightarrow Zn(H_2O)_2(OH)_4^{-2} + H^+ \qquad K_{A4}$$

We likely will not find evidence of a fifth or sixth deprotonation reaction in the literature. Much of the literature refers to these proton-donation reactions as hydrolysis reactions. Baes and Mesmer's (1976) text addresses, among other very relevant topics, hydrolysis reactions between metals and ligands. The reader is directed to this source for additional details beyond discussions contained herein.

11.3.2 Metal Ion Hydrolysis (Formation) Reactions

We can write the reaction for the ion product of water in reverse, in essence a formation reaction such that a proton plus a hydroxide yield water. As discussed in Chapter 4, when we reverse a reaction, the resultant equilibrium constant is the reciprocal of the initial equilibrium constant (the RHS of the law of mass action statement also becomes the reciprocal of the original). Thus, for this formation reaction, the equilibrium constant, $K_{FW} = 1/K_w = 10^{14}$, at 25 °C.

$$H^+ + OH^- \Leftrightarrow H_2O \qquad K_{FW}$$

We can then add the reaction for the formation of water to that for the first deprotonation of the free zinc ion to arrive at an expression for the formation of a zinc–hydroxide complex via the addition of a hydroxide to the free zinc ion.

$$
\begin{array}{lll}
Zn(H_2O)_6^{+2} & \Leftrightarrow Zn(H_2O)_5OH^{+} + H^{+} & \\
H^{+} + OH^{-} & \Leftrightarrow H_2O & \\
\hline
Zn(H_2O)_6^{+2} + OH^{-} & \Leftrightarrow Zn(H_2O)_5OH^{+} + H_2O & K_{F1} = K_{A1}.K_{FW}
\end{array}
$$

The equilibrium constant for the resultant formation reaction is the product of the acid deprotonation constant and the water formation constant. We may accomplish this procedure for each successive deprotonation reaction leading to the following set of complex formation reactions.

$$
\begin{array}{lll}
Zn(H_2O)_5OH^{+} + OH^{-} & \Leftrightarrow Zn(H_2O)_4(OH)_2^{0} + H_2O & K_{F2} = K_{A2} \cdot K_{FW} \\
Zn(H_2O)_4(OH)_2^{0} + OH^{-} & \Leftrightarrow Zn(H_2O)_3(OH)_3^{-} + H_2O & K_{F3} = K_{A3} \cdot K_{FW} \\
Zn(H_2O)_3(OH)_3^{-} + OH^{-} & \Leftrightarrow Zn(H_2O)_2(OH)_4^{-2} + H_2O & K_{F4} = K_{A4} \cdot K_{FW}
\end{array}
$$

We then subtract six, five, four, and three waters from each side of each of the written reactions, respectively, to yield the formation reactions most often found in the chemical literature. Algebraically this amounts to writing the reactions that six, five, four, and three water molecules on the LHS form six, five, four, and three water molecules on the RHS. In each case the equilibrium constant is unity. Then subtracting one reaction from another is the inverse of addition, so the resultant formation constant is the quotient of that for the fully stated reaction and unity.

$$
\begin{array}{llll}
Zn^{+2} + OH^{-} & \Leftrightarrow ZnOH^{+} & \text{zinc (mono) hydroxide} & K_{F1} \\
ZnOH^{+} + OH^{-} & \Leftrightarrow Zn(OH)_2^{0} & \text{zinc (di) hydroxide} & K_{F2} \\
Zn(OH)_2^{0} + OH^{-} & \Leftrightarrow Zn(OH)_3^{-} & \text{zinc (tri) hydroxide} & K_{F3} \\
Zn(OH)_3^{-} + OH^{-} & \Leftrightarrow Zn(OH)_4^{-2} & \text{zinc (tetra) hydroxide} & K_{F4}
\end{array}
$$

K_{F1} through K_{F4} for the zinc system are called the stepwise formation constants. Each represents an incremental addition of a hydroxide to the zinc ion, representing the four deprotonation reactions as complexation (complex formation) reactions.

We could perform similar analyses for each and every metal. In theory, the potential number of protons that could be donated would be equal to the coordination number. In practice, for many metals, the number of protons that can be donated is less than the coordination number. In our analysis of metal complexation and solubility/dissolution we will find it most convenient to employ reactions as formation reactions. A huge database of chemical equilibrium constants considers metal–ligand reactions as formation reactions.

11.3.3 Cumulative Hydrolysis (Formation) Reactions

Generally, since the database of chemical equilibrium constants is so arranged, we find it convenient to write the four formation reactions for the zinc system as cumulative formation reactions. To create the second cumulative reaction, we add the first and second reactions, thus the final equilibrium constant is the product of the two stepwise constants.

We then add the first three stepwise reactions to create the third cumulative reaction; the resultant equilibrium constant is the product of the three stepwise constants. The fourth cumulative reaction is the sum of all four stepwise reactions and the equilibrium constant is the product of the four stepwise constants. Herein we will use the term cumulative formation constant and the symbol β. Some of the literature refers to formation constants as stability constants, either stepwise or cumulative.

$$Zn^{+2} + OH^- \Leftrightarrow ZnOH^+ \qquad \beta_1 = K_{F1}\ (= K_{FW} \cdot K_{A1})$$
$$Zn^{+2} + 2OH^- \Leftrightarrow Zn(OH)_2^0 \qquad \beta_2 = K_{F1} \cdot K_{F2}\ (= K_{FW}^{2} \cdot K_{A1} \cdot K_{A2})$$
$$Zn^{+2} + 3OH^- \Leftrightarrow Zn(OH)_3^- \qquad \beta_3 = K_{F1} \cdot K_{F2} \cdot K_{F3}\ (= K_{FW}^{3} \cdot K_{A1} \cdot K_{A2} \cdot K_{A3})$$
$$Zn^{+2} + 4OH^- \Leftrightarrow Zn(OH)_4^{-2} \qquad \beta_5 = K_{F1} \cdot K_{F2} \cdot K_{F3} \cdot K_{F4}\ (= K_{FW}^{4} \cdot K_{A1} \cdot K_{A2} \cdot K_{A3} \cdot K_{A4})$$

We can write similar sets of deprotonation and hence formation reactions for all metals, based on the number of bound water molecules that can donate their protons, available from numerous chemical databases. Morel and Hering (1993) have assembled a table of cumulative formation constants, collecting published values for many common metal–ligand systems from various works. We have included this, with permission, as Table A.2. Herein, we will rely upon this table and tabulated values assembled by Dean (1992) as our database. Should information beyond these resources be desired, the sources consulted their compilation are provided therein.

Now that we have a means by which to correlate cumulative formation constants with stepwise formation constants and hence stepwise acid deprotonation constants, for the zinc system, let us relate the cumulative formation constants to the acid dissociation constants to gain perspective on the strengths of zinc and two other metals as acids.

Example 11.1 determine the values of the stepwise deprotonation constants for the zinc–water, manganese–water, and chromium–water systems.

From the Table A.2 we collect the values of the cumulative formation constants for each metal–ligand system. With a single possible formation reaction, $\beta_W = K_{FW}$. The values in Table A.2 are given as $\log_{10}\beta$:

Zinc

$$\begin{pmatrix} \beta_{ZnOH} \\ \beta_{ZnOH2} \\ \beta_{ZnOH3} \\ \beta_{ZnOH4} \\ \beta_{W} \end{pmatrix} := \begin{pmatrix} 10^{5.0} \\ 10^{11.1} \\ 10^{13.6} \\ 10^{14.8} \\ 10^{14} \end{pmatrix}$$

Manganese

$$\begin{pmatrix} \beta_{MnOH} \\ \beta_{MnOH2} \\ \beta_{MnOH3} \\ \beta_{MnOH4} \end{pmatrix} := \begin{pmatrix} 10^{3.4} \\ 10^{5.8} \\ 10^{7.2} \\ 10^{7.7} \end{pmatrix}$$

Chromium(III)

$$\begin{pmatrix} \beta_{CrOH} \\ \beta_{CrOH2} \\ \beta_{CrOH3} \\ \beta_{CrOH4} \end{pmatrix} := \begin{pmatrix} 10^{10} \\ 10^{18.3} \\ 10^{24} \\ 10^{28.6} \end{pmatrix}$$

Note that we have named each formation constant using a subscript that identifies the exact metal–ligand complex. Certainly for MathCAD worksheets, this system allows us to assemble a large database of formation constants assigned to the appropriate symbols. We may then include this database in any worksheet we create, and use the defined symbols in the relations we consider. In any modeling effort, significant resources are necessarily expended in "debugging" the numeric system representing the mathematics. A standardized symbol recognition system will be a significant step in conserving valuable time resources when modeling metal–ligand systems. Note that in MathCAD worksheets, we cannot use the left and right parentheses in subscripts as identifiers. We must imagine that they are present with our subscripts.

For the zinc system, we work backward from the relations given with the hydrolysis reactions earlier. A general relation can be written for the acid dissociation constant of the ith deprotonation:

$$K_{A.i} = \frac{K_{F.i}}{\beta_W} = \frac{\beta_i}{\beta_{i-1}} \cdot \frac{1}{\beta_W}$$

We employ this relation for zinc, manganese, and chromium in turn, opting for a matrix-type organization of the computations and presentation:

Zinc

$$\begin{pmatrix} K_{A.1} \\ K_{A.2} \\ K_{A.3} \\ K_{A.4} \end{pmatrix} := \frac{1}{\beta_W} \cdot \begin{pmatrix} \beta_{ZnOH} \\ \dfrac{\beta_{ZnOH2}}{\beta_{ZnOH}} \\ \dfrac{\beta_{ZnOH3}}{\beta_{ZnOH2}} \\ \dfrac{\beta_{ZnOH4}}{\beta_{ZnOH3}} \end{pmatrix} = \begin{pmatrix} 1 \times 10^{-9} \\ 1.259 \times 10^{-8} \\ 3.162 \times 10^{-12} \\ 1.585 \times 10^{-13} \end{pmatrix} \qquad \begin{pmatrix} pK_{A1} \\ pK_{A2} \\ pK_{A3} \\ pK_{A4} \end{pmatrix} := \begin{pmatrix} -\log(K_{A.1}) \\ -\log(K_{A.2}) \\ -\log(K_{A.3}) \\ -\log(K_{A.4}) \end{pmatrix} = \begin{pmatrix} 9 \\ 7.9 \\ 11.5 \\ 12.8 \end{pmatrix}$$

Manganese

$$\begin{pmatrix} K_{A.1} \\ K_{A.2} \\ K_{A.3} \\ K_{A.4} \end{pmatrix} := \frac{1}{\beta_W} \cdot \begin{pmatrix} \beta_{MnOH} \\ \dfrac{\beta_{MnOH2}}{\beta_{MnOH}} \\ \dfrac{\beta_{MnOH3}}{\beta_{MnOH2}} \\ \dfrac{\beta_{MnOH4}}{\beta_{MnOH3}} \end{pmatrix} = \begin{pmatrix} 2.512 \times 10^{-11} \\ 2.512 \times 10^{-12} \\ 2.512 \times 10^{-13} \\ 3.162 \times 10^{-14} \end{pmatrix} \qquad \begin{pmatrix} pK_{A1} \\ pK_{A2} \\ pK_{A3} \\ pK_{A4} \end{pmatrix} := \begin{pmatrix} -\log(K_{A.1}) \\ -\log(K_{A.2}) \\ -\log(K_{A.3}) \\ -\log(K_{A.4}) \end{pmatrix} = \begin{pmatrix} 10.6 \\ 11.6 \\ 12.6 \\ 13.5 \end{pmatrix}$$

Chromium(III)

$$\begin{pmatrix} K_{A.1} \\ K_{A.2} \\ K_{A.3} \\ K_{A.4} \end{pmatrix} := \frac{1}{\beta_W} \cdot \begin{pmatrix} \beta_{CrOH} \\ \dfrac{\beta_{CrOH2}}{\beta_{CrOH}} \\ \dfrac{\beta_{CrOH3}}{\beta_{CrOH2}} \\ \dfrac{\beta_{CrOH4}}{\beta_{CrOH3}} \end{pmatrix} = \begin{pmatrix} 1 \times 10^{-4} \\ 1.995 \times 10^{-6} \\ 5.012 \times 10^{-9} \\ 3.981 \times 10^{-10} \end{pmatrix} \qquad \begin{pmatrix} pK_{A1} \\ pK_{A2} \\ pK_{A3} \\ pK_{A4} \end{pmatrix} := \begin{pmatrix} -\log(K_{A.1}) \\ -\log(K_{A.2}) \\ -\log(K_{A.3}) \\ -\log(K_{A.4}) \end{pmatrix} = \begin{pmatrix} 4 \\ 5.7 \\ 8.3 \\ 9.4 \end{pmatrix}$$

We observe that the "free" zinc ion is a weak acid that manganese is weaker than zinc, and that chromium(III) would be a stronger acid than either manganese or zinc.

We must take care with analyses such as those of Example 11.1. In many cases, the database is not complete or yields results that are inconsistent with our knowledge of successive deprotonations of multiprotic acids. For example, for copper no β_2 value is included in Table A.2 and for aluminum the process of Example 11.1 returns a computed pK_{A2} that is lower than the computed pK_{A1}, which we believe should be impossible. Further, many metals form hydroxo-complexes involving more than a single metal ion (e.g., $Cu_2(OH)_2$, $Cr_3(OH)_4^-$, $Al_3(OH)_4^-$) so the acid-base behavior, particularly of these metals, is somewhat more complicated than that of the multiprotic acid/base systems examined in Chapter 10.

11.3.4 The Cumulative Formation Constant for Metal/Ligand Complexes

We might wonder why the chemists traded in the acid dissociation constants of metal ions for cumulative formation constants. Of course, the reasoning is sound. When we consider hydroxide as a ligand that binds with the metal ion, we can consider it alongside all other ligands. We may consider the reactions written by adding the water formation reaction to the metal ion deprotonation reactions as ligand exchange reactions. Then, we would consider that each hydroxide simply replaces, in succession, one of the initially bound water molecules. Then, when we subtract the requisite number of water molecules from each side of the reaction, we are left with a set of formation reactions. The deprotonation reactions for zinc were rewritten using M^{+2} and one through four hydroxides as the combining ligand, in parallel with the remainder of the known metal–ligand formation reactions. One reaction is possible for each published cumulative formation constant for each metal–ligand system. For example, we may extend this visualization to the formation of zinc chloride and zinc cyanide complexes. Chloride and cyanide become the ligands, replacing bound water molecules, to form zinc chloride or zinc cyanide complexes. We should be careful to retain this clear view of the formation process as an exchange so that we are not complacent to view the complex formation process simply as the association of the metal with a ligand. There must always be an exchange. The chemists have extended this system to all of the ligands considered in Table A.2 and certainly well beyond.

Protons are considered as if they were a second metal, participating in the formation of the complex. In fact, in Table A.2 we find that a number of acid/base systems are represented by their acid formation reactions. For example, consider the carbonate system. Two values of β are given: $\log_{10}\beta_{HL} = 10.33$ and $\log_{10}\beta_{H_2L} = 16.68$. The acid formation reactions for the carbonate system are simply the reverse of the deprotonation reactions with the second formation reaction being the sum of the two.

$$CO_3^{-2} + H^+ \Leftrightarrow HCO_3^- \qquad \beta_{HCO_3} = 1/K_{2.CO_3}$$

$$CO_3^{-2} + 2H^+ \Leftrightarrow H_2CO_3^* \qquad \beta_{H_2CO_3} = 1/K_{2.CO_3} \cdot 1/K_{1.CO_3}$$

Formation constants for other acid/base systems follow the same system.

In general, the system considers the fully deprotonated form of the ligand system as the base ligand. Do notice that in nearly all cases, the ligand system is in fact a strong or weak acid system. For example carbonate is considered as a ligand, but bicarbonate is not. Phosphate is considered a ligand but hydrogen phosphate and dihydrogen phosphate are not. Peruse Table A.2 and find that each ligand is indeed the fully deprotonated conjugate base of an acid/base system. For use with the database contained in Table A.2, we may employ the general reaction expressed by Equation 4.7, leading to the specialized statement of the law of mass action presented as Equation 4.8:

$$aM^{+n} + bH^{+} + cL^{-m} \rightleftarrows M_aH_bL_c^{(a\cdot n+b-c\cdot m)} \tag{4.7}$$

M^{+n} is the metal ion with residual charge $+n$, L^{-m} is the ligand of residual charge $-m$ (m can be 0) and $M_aH_bL_c^{(a\cdot n+b-c\cdot m)}$ is the complex with residual charge $a\cdot n+b-c\cdot m$:

$$\beta_{M_aH_bL_c} = \frac{\left[M_aH_bL_c^{\,a\cdot n+b-c\cdot m}\right]}{[M^{+n}]^a[H^+]^b[L^{-m}]^c} \tag{4.8}$$

The values of $\log_{10}\beta$ populating Table A.2 conform to the general complexation reaction written as Equation 4.7 and to the general law of mass action statement of Equation 4.8. We might seek values for metals and complexing ligands beyond those of Table A.2. In doing so, we must pay very close attention to the system (i.e., the exact law of mass action statement) in which the formation constants were developed and are presented.

11.4 FORMATION EQUILIBRIA FOR SOLIDS

We bring the general statement for the formation of metal–ligand solids from Chapter 4 to be complete:

$$aM^{+n} + bL^{-m} \rightleftarrows M_aL_{b\,(s)} \tag{4.9}$$

Equation 4.10 is the law of mass action statement in accord with the reaction of Equation 4.9. Herein we use the symbol $\beta_{S.M_aL_b}$ to indicate that the equilibrium constant applies to a solid formation reaction. We also find that to improve our ability to keep our computations organized we can append the subscript with the actual chemical formula of the specific solid, less the parentheses:

$$\beta_{S.M_aL_c} = \frac{\left[M_aL_{c(s)}\right]}{[M^{+n}]^a[L^{-m}]^c} \tag{4.10}$$

When we consider that the activity of a solid in contact with an aqueous solution is unity, Equation 4.10b arises:

$$\beta_{\mathrm{S.M}_a\mathrm{L}_c} = \frac{1}{[\mathrm{M}^{+n}]^a[\mathrm{L}^{-n}]^c} \tag{4.10b}$$

Some literature considers the dissolution of solids, hence reversing the direction of the reaction of Equation 4.9:

$$\mathrm{M}_a\mathrm{L}_{b\,(\mathrm{s})} \rightleftarrows a\mathrm{M}^{+n} + b\mathrm{L}^{-m} \tag{4.9a}$$

Then, the corresponding law of mass action statement, employing the solubility product constant, K_{sp}, follows from Equation 4.10b:

$$K_{\mathrm{sp.M_aL_c}} = [\mathrm{M}^{+n}]^a[\mathrm{L}^{-m}]^c \tag{4.10c}$$

We recall that Equations 4.10b and 4.10c are manifestations that the activity of the pure solid in heterogeneous equilibrium with the aqueous solution is unity. Many authors use the reasoning that "Well, the activity of the solid never changes and we can just ignore it." We could indeed ignore the chemical activity of the solid but in doing so we would be denying ourselves the use of a hugely powerful condition. When we take full advantage that the activities of any and all solid phases with which an aqueous solution is in equilibrium are in fact unity, we have a powerful set of equilibrium relations available for application in our analyses.

Equation 4.10 is written for a single metal and a single ligand. Many of the metal–ligand systems we would investigate fall into this category. However, when we begin investigation of minerals found in environmental systems, we quickly realize that many of these are comprised of more than one metal and multiple ligands, and that many contain protons. We can, when necessary, expand Equation 4.10 and its companion general equilibrium reaction to include multiple metals, protons, and multiple ligands.

11.5 SPECIATION OF METALS IN AQUEOUS SOLUTIONS CONTAINING LIGANDS

11.5.1 Metal Hydroxide Systems

We should begin our examination of metal–ligand interactions by considering the distribution of a target metal in an aqueous solution in which hydroxide is the sole complexing ligand. In practice, these solutions would be at best very difficult to prepare. Normally, dissolution of a metal into an aqueous solution requires its addition as a salt or as the pure metal with subsequent addition of strong acid to dissolve the metal. Nitrate is often used as the companion anion for a metal salt or nitric acid is used as the dissolving acid. If we peruse Table A.2 we would quickly observe that its creators chose not even to include formation constants for nitrate as a ligand—perhaps such equilibrium data does not even exist or the formation reactions are sufficiently weak to be considered negligible. Then for analyses of metal speciation in the absence of ligands other than hydroxide, we will consider that nitrate is present at very low levels and its complexing behavior can be ignored.

Example 11.2 Develop abundance fractions for the hydroxo- complexes of zinc and examine the relative abundances of the five resultant species as a function of the master variable pH. Consider an aqueous solution containing 10^{-5} M total zinc.

We first write a mole balance on zinc species:

$$C_{Tot.Zn} = C_{Zn} + C_{ZnOH} + C_{ZnOH2} + C_{ZnOH3} + C_{ZnOH4}$$

We write this using the specie activities, including activity coefficients for completeness but, for quantitation, set the values to unity:

$$C_{Tot.Zn} = \frac{Zn}{\gamma_2} + \frac{ZnOH}{\gamma_1} + ZnOH2 + \frac{ZnOH3}{\gamma_1} + \frac{ZnOH4}{\gamma_2}$$

We have the four complex formation equilibria available, rearranged as necessary, to rewrite our mole balance with free zinc as the master dependent variable:

$$C_{Tot.Zn} = \left[\frac{Zn}{\gamma_2} + \frac{\beta_{ZnOH} \cdot Zn \cdot OH}{\gamma_1} + \beta_{ZnOH2} \cdot Zn \cdot (OH)^2 \ldots + \frac{\beta_{ZnOH3} \cdot Zn \cdot OH^3}{\gamma_1} + \frac{\beta_{ZnOH4} \cdot Zn \cdot OH^4}{\gamma_2} \right]$$

We then employ the ion product of water, $\{OH^-\} = K_W/\{H^+\}$, to replace hydroxide as a function of $\{H^+\}$:

$$C_{Tot.Zn} = \left[\frac{Zn}{\gamma_2} + \frac{\beta_{ZnOH} \cdot Zn \cdot K_W}{H \cdot \gamma_1} + \beta_{ZnOH2} \cdot Zn \cdot \left(\frac{K_W}{H}\right)^2 \ldots + \frac{\beta_{ZnOH3} \cdot Zn \cdot K_W^3}{H^3 \cdot \gamma_1} + \frac{\beta_{ZnOH4} \cdot Zn \cdot K_W^4}{H^4 \cdot \gamma_2} \right]$$

We factor out the zinc ion and rearrange the relation, defining the activity of free zinc as a function only of the total aqueous zinc, the various cumulative formation constants, K_W, and the proton activity. We write the function: $H(pH) = 10^{-pH}$, which we use in the final function for Zn(pH). An alternative relation would have β_W (or K_{FW}) in the denominators of the last four terms of the overall divisor:

$$Zn(pH) := \frac{C_{Tot.Zn}}{\frac{1}{\gamma_2} + \frac{\beta_{ZnOH} \cdot K_W}{H(pH) \cdot \gamma_1} + \beta_{ZnOH2} \cdot \left(\frac{K_W}{H(pH)}\right)^2 + \frac{\beta_{ZnOH3} \cdot K_W^3}{H(pH)^3 \cdot \gamma_1} + \frac{\beta_{ZnOH4} \cdot K_W^4}{H(pH)^4 \cdot \gamma_2}}$$

We then define five additional functions employing that written for free zinc, to express the abundances of the five zinc species:

$$C_{Zn}(pH) := \frac{Zn(pH)}{\gamma_2}$$

$$C_{ZnOH}(pH) := \frac{\beta_{ZnOH} \cdot Zn(pH) \cdot K_W}{H(pH) \cdot \gamma_1} \qquad C_{ZnOH2}(pH) := \frac{\beta_{ZnOH2} \cdot Zn(pH) \cdot K_W{}^2}{H(pH)^2}$$

$$C_{ZnOH3}(pH) := \frac{\beta_{ZnOH3} \cdot Zn(pH) \cdot K_W{}^3}{H(pH)^3 \cdot \gamma_1} \qquad C_{ZnOH4}(pH) := \frac{\beta_{ZnOH4} \cdot Zn(pH) \cdot K_W{}^4}{H(pH)^4 \cdot \gamma_2}$$

We then employ these abundance relations to define the respective abundance fractions:

$$\alpha_{Zn}(pH) := \frac{C_{Zn}(pH)}{C_{Tot.Zn}} \qquad \alpha_{ZnOH}(pH) := \frac{C_{ZnOH}(pH)}{C_{Tot.Zn}} \qquad \alpha_{ZnOH2}(pH) := \frac{C_{ZnOH2}(pH)}{C_{Tot.Zn}}$$

$$\alpha_{ZnOH3}(pH) := \frac{C_{ZnOH3}(pH)}{C_{Tot.Zn}} \qquad \alpha_{ZnOH4}(pH) := \frac{C_{ZnOH4}(pH)}{C_{Tot.Zn}}$$

We could have developed the abundance fractions for the zinc system in the same manner as was used for the acid systems of Chapters 6 and 10, but these would find much less utility than the functions of the type we have written. A plot of abundance fractions versus pH for zinc hydrolysis species is shown in Figure E11.2.1.

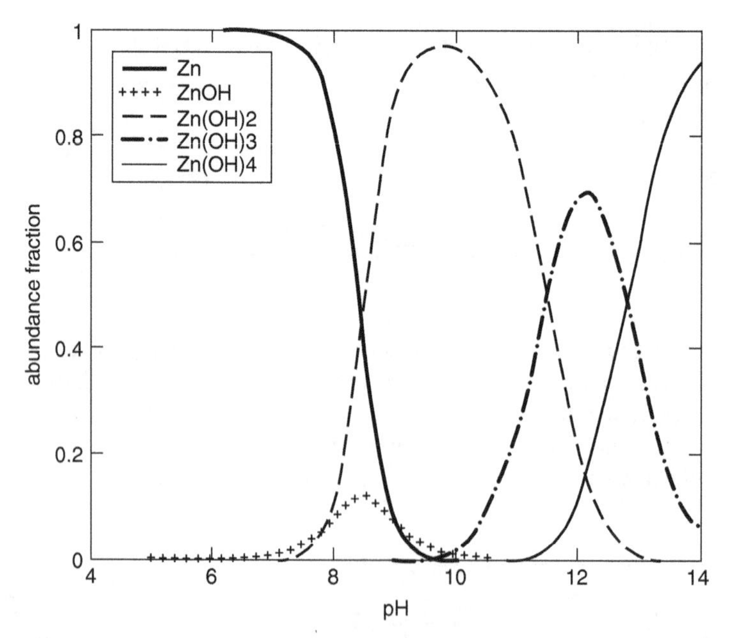

FIGURE E11.2.1 A plot of hydrolysis specie abundance fraction versus pH for the zinc metal system.

From our plot of the abundance fractions we easily observe that $ZnOH^+$ is relatively unimportant and that, in environmental systems of most interest ($5<pH<10$) the free zinc ion and $Zn(OH)_2$ are the predominant species. Relative to hydrolysis, each metal would have its signature set of abundance fractions, obtained in a manner exactly analogous to these results for zinc. Thus, this worksheet we have developed, with minor modifications could be used for similar analyses of any metal hydroxide system whose hydrolysis formation constants are available.

It is often said that the two things in life with which one must always deal are death and taxes. In similar fashion, we may conclude that in examination of metal speciation in aqueous systems one can never ignore metal hydroxide complexes. If the metal resides in water, the deprotonation equilibria will be operative and, hence, any efforts directed toward modeling behaviors of metals in aqueous solutions **must** include the metal hydroxides. The approach for consideration would be no different from that of Example 11.2.

11.5.2 Metals with Multiple Ligands

The next level of complexity relative to modeling metal–ligand complexes in aqueous solutions is examination of a system in which we have a significant complexing ligand, along with hydroxide. In general, we would write mole balances to account for the entire set of metal species and to account for the entire set of ligand species. Certainly, many of the metal species will appear in the ligand balance and vice versa.

Example 11.3 Examine an aqueous solution containing 10^{-4} M zinc and 10^{-4} M total phosphate. Produce a plot, similar to that of Example 11.2 depicting the abundances of the various zinc and phosphate species. Consider that the background ionic strength is 0.01 M and is not significantly affected by changes in metal or phosphate speciation.

We retain the applicable portions of the work completed for Example 11.2, include the formation constants for the phosphate ligand and include the effects of ionic strength:

$$\begin{pmatrix} \beta_{ZnOH} \\ \beta_{ZnOH2} \\ \beta_{ZnOH3} \\ \beta_{ZnOH4} \\ K_W \end{pmatrix} := \begin{pmatrix} 10^{5.0} \\ 10^{11.1} \\ 10^{13.6} \\ 10^{14.8} \\ 10^{-14} \end{pmatrix} \quad \begin{pmatrix} \beta_{H3PO4} \\ \beta_{H2PO4} \\ \beta_{HPO4} \end{pmatrix} := \begin{pmatrix} 10^{21.70} \\ 10^{19.55} \\ 10^{12.35} \end{pmatrix} \quad \begin{pmatrix} \beta_{ZnHPO4} \\ \beta_{ZnH2PO4} \\ \beta_{Zn3PO42} \end{pmatrix} := \begin{pmatrix} 10^{15.7} \\ 10^{21.2} \\ 10^{35.3} \end{pmatrix}$$

$$\begin{pmatrix} C_{Tot.Zn} \\ C_{Tot.PO4} \\ I \\ A \end{pmatrix} := \begin{pmatrix} 10^{-4} \\ 10^{-4} \\ 0.01 \\ 0.5115 \end{pmatrix} \quad \begin{pmatrix} \gamma_1 \\ \gamma_2 \\ \gamma_3 \end{pmatrix} := \begin{pmatrix} 10^{\frac{-A\cdot\sqrt{I}}{1+\sqrt{I}}} \\ 10^{\frac{-A\cdot 4\cdot\sqrt{I}}{1+\sqrt{I}}} \\ 10^{\frac{-A\cdot 9\cdot\sqrt{I}}{1+\sqrt{I}}} \end{pmatrix} = \begin{pmatrix} 0.898 \\ 0.652 \\ 0.382 \end{pmatrix}$$

We write mole balances for zinc and phosphate:

$$C_{Tot.Zn} = C_{Zn} + C_{ZnOH} + C_{ZnOH2} + C_{ZnOH3} + C_{ZnOH4} \cdots + C_{ZnHPO4} + C_{ZnH2PO4} + 3 \cdot C_{Zn3PO42}$$

$$C_{tot.PO4} = C_{H3PO4} + C_{H2PO4} + C_{HPO4} + C_{PO4} \cdots + C_{ZnHPO4} + C_{ZnH2PO4} + 2 \cdot C_{Zn3PO42}$$

When we write the two equations using the two master dependent variables of $\{Zn^{+2}\}$ and $\{PO_4^{-3}\}$, we very quickly realize that the two relations are coupled and we can solve neither independently. We therefore arrange a given-find solve block. We write the solve block into a function, that allows the manipulation of the proton activity as one of the arguments along with the activities of free zinc and phosphate:

$$pH := 4.8 \qquad H := 10^{-pH} \qquad \begin{pmatrix} Zn \\ PO4 \end{pmatrix} := \begin{pmatrix} 10^{-8} \\ 10^{-11} \end{pmatrix} \qquad \text{Given}$$

$$C_{Tot.Zn} = \frac{Zn}{\gamma_2} + \frac{\beta_{ZnOH} \cdot Zn \cdot K_W}{H \cdot \gamma_1} + \beta_{ZnOH2} \cdot Zn \cdot \left(\frac{K_W}{H}\right)^2 + \frac{\beta_{ZnOH3} \cdot Zn \cdot K_W^{\,3}}{H^3 \cdot \gamma_1} \cdots + \frac{\beta_{ZnOH4} \cdot Zn \cdot K_W^{\,4}}{H^4 \cdot \gamma_2} + \beta_{ZnHPO4} \cdot Zn \cdot H \cdot PO4 + \frac{\beta_{ZnH2PO4} \cdot Zn \cdot H^2 \cdot PO4}{\gamma_1} \cdots + 3 \cdot \beta_{Zn3PO42} \cdot Zn^3 \cdot PO4^2$$

$$C_{Tot.PO4} = \beta_{H3PO4} \cdot H^3 \cdot PO4 + \frac{\beta_{H2PO4} \cdot H^2 \cdot PO4}{\gamma_1} + \frac{\beta_{HPO4} \cdot H \cdot PO4}{\gamma_2} + \frac{PO4}{\gamma_3} \cdots + \beta_{ZnHPO4} \cdot Zn \cdot H \cdot PO4 + \frac{\beta_{ZnH2PO4} \cdot Zn \cdot H^2 \cdot PO4}{\gamma_1} + 2 \cdot \beta_{Zn3PO42} \cdot Zn^3 \cdot PO4^2$$

$$\text{zincsp}(H, Zn, PO4) := \text{Find}(Zn, PO4)$$

In order to obtain graphical output, we write our results into several parallel vectors. The zeroth element of each vector is computed by solving the originally stated solve block for the initial pH desired for the range of values:

$$\begin{pmatrix} Zn_0 \\ PO4_0 \end{pmatrix} := \text{zincsp}(H, Zn, PO4) \rightarrow \begin{pmatrix} 5.83 \times 10^{-5} \\ 9.294 \times 10^{-15} \end{pmatrix}$$

We write a short program incrementing pH and successively implementing the solve block to obtain vectors of pH, $\{Zn^{+2}\}$, and $\{PO_4^{-3}\}$, shown in Figure E11.3.1. Then for a pH range of 5–10, we invoke this set of worksheet programming 26 times to populate the Zn and PO_4 vectors. Once the master vectors are populated

$$\Delta pH := 0.2 \qquad \begin{pmatrix} pH \\ Zn \\ PO4 \end{pmatrix} := \left| \begin{array}{l} \text{for } i \in 1..26 \\ \quad \left| \begin{array}{l} pH_i \leftarrow pH_{i-1} + \Delta pH \\ H \leftarrow 10^{-pH_i} \\ \begin{pmatrix} Zn_i \\ PO4_i \end{pmatrix} \leftarrow \begin{pmatrix} Zn_{i-1} \\ PO4_{i-1} \end{pmatrix} \\ \begin{pmatrix} Zn_i \\ PO4_i \end{pmatrix} \leftarrow zincsp(H, Zn_i, PO4_i) \end{array} \right. \\ \text{return } \begin{pmatrix} pH \\ Zn \\ PO4 \end{pmatrix} \end{array} \right.$$

FIGURE E11.3.1 *Screen capture of a logical program to compute zinc and phosphate ion activities over a selected pH range.*

$$i := 1..26$$

$$\begin{pmatrix} c_{Zn_i} \\ c_{ZnOH2_i} \\ c_{ZnOH3_i} \\ c_{ZnHPO4_i} \\ c_{ZnH2PO4_i} \\ c_{Zn3PO42_i} \end{pmatrix} := \overrightarrow{\begin{bmatrix} \dfrac{Zn_i}{\gamma_2} \\ \beta_{ZnOH2} \cdot Zn_i \cdot \left(\dfrac{K_W}{10^{-pH_i}} \right)^2 \\ \dfrac{\beta_{ZnOH3} \cdot Zn_i \cdot K_W^{\ 3}}{\left(10^{-pH_i}\right)^3 \cdot \gamma_1} \\ \beta_{ZnHPO4} \cdot Zn_i \cdot 10^{-pH_i} \cdot PO4_i \\ \dfrac{\beta_{ZnH2PO4} \cdot Zn_i \cdot \left(10^{-pH_i}\right)^2 \cdot PO4_i}{\gamma_1} \\ \beta_{Zn3PO42} \cdot (Zn_i)^3 \cdot (PO4_i)^2 \end{bmatrix}}$$

FIGURE E11.3.2 *Screen capture of vector computations for zinc phosphate complex speciation.*

we may write relations allowing the population of vectors containing the concentrations of six species of interest, shown in Figure E11.3.2. Finally, we may use these six vectors to create six additional vectors each containing the 26 additional values of the abundance fraction for each specie, shown in Figure E11.3.3.

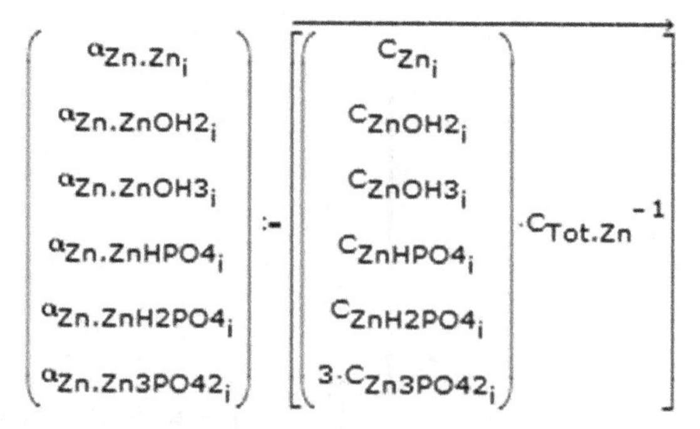

FIGURE E11.3.3 *Vector computations for zinc complex abundance fractions.*

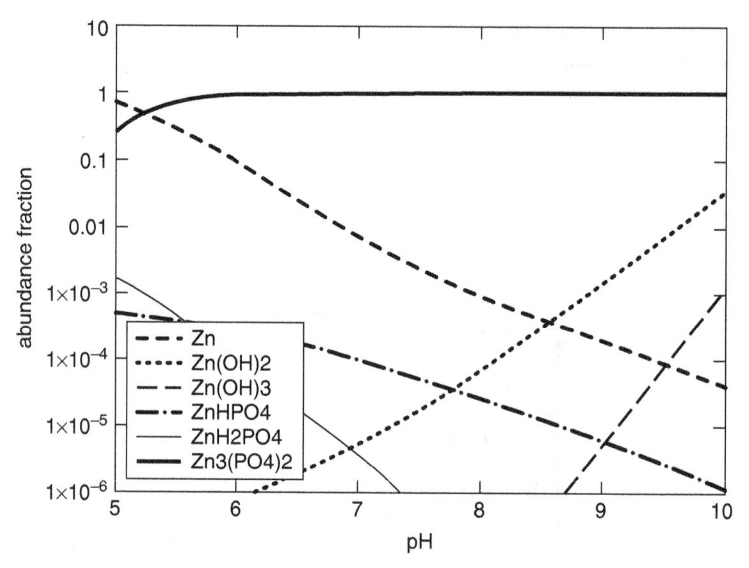

FIGURE E11.3.4 *A plot of zinc specie abundance fractions versus pH for zinc hydrolysis and zinc phosphate complex species.*

A plot of the results shown in Figure E11.3.4 allows us to gain an understanding of the speciation of zinc in aqueous solution with phosphate present. We observe that the zinc ion is predominant at low pH while the polynuclear $Zn_3(PO_4)_2$ complex is predominant in this system at virtually all pH values between 5.5 and 10, the range of interest for most environmental systems. We might think it obvious that the $Zn_3(PO_4)_2$ complex would also dominate the abundance diagram for the phosphate ligand. To be sure, however, we have computed the abundance fractions in much the same manner as those for the zinc species (not shown) and plotted them in Figure E11.3.5 to assure ourselves of our hunch. Were we to employ Excel to model

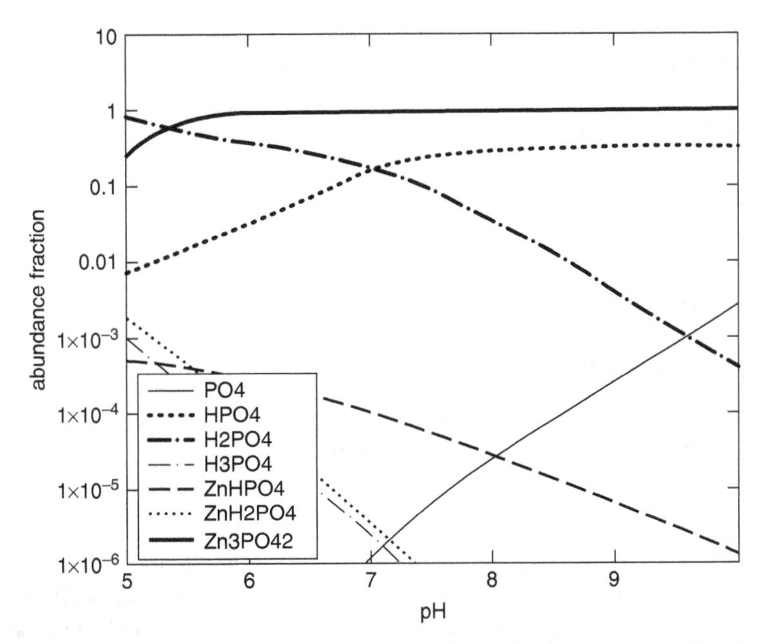

FIGURE E11.3.5 A plot of phosphate specie abundance fractions versus pH for the zinc-phosphate complex formation system.

this system we would use $\{Zn^{+2}\}$ as the master dependent variable, write the mole balance for the phosphate ligand, isolating the activity of the base ligand on the LHS. Then we would write the mole balance on zinc using the LHS of the phosphate mole balance wherever the phosphate ion would appear. We would algebraically zero this function and use the solver to find the value of zinc ion activity that satisfies the zeroed condition. We would necessarily solve the system for each pH value. We could step into the macroenvironment available from Excel and perhaps create a VBA program that would increment pH and successively solve the system and store the results. The tricky part might be the invocation of the solver from the programming environment. Certainly, the "what you see is what you get" visual interface of the MathCAD worksheet seems to this author to be a better choice.

We move away from Example 11.3 with an understanding that we can quantitatively model virtually any metal–ligand system, given we have values for the cumulative formation constants and for the acid/base equilibria of the ligand. Were additional ligands present in the system, we would of necessity write a mole balance for each additional ligand and include the appropriate terms from the ligand mole balances in the mole balance written for the metal. Regardless of the complexity or the simplicity of each metal–ligand system, we necessarily must solve the mole balance equations simultaneously. Were we of necessity to include a second or third metal, with interactions among the common ligands, we would write mole balances for the additional metals. Solution of systems with two or more metals would be straightforward using MathCAD. Unfortunately, for modeling with Excel, we would need to identify the activity of each free metal ion as a master dependent variable, and, unfortunately, Excel's solver can adjust but one variable at a time. Perhaps we would need to implement the solver with a number of target cells equal to the number of metals and manually iterate the solver among the defined target cells. MathCAD is indeed the more convenient of the two computational software packages for these computations.

11.6 METAL HYDROXIDE SOLUBILITY

11.6.1 Solubility in Dilute Solution

We continue on our journey toward competence in modeling metal ligand interactions. The next logical step is examination of the solubility of metals in aqueous systems containing the metal, water, and, of course, metal hydroxide complexes.

Example 11.4 Examine the solubility of zinc as a function of pH when a solid phase zinc hydroxide is present as a suspension in the aqueous solution. In order that the focus be squarely upon the metal–ligand system, use the infinitely dilute solution assumption, but retain the solution structure employing activity coefficients.

We need not know how much $Zn(OH)_{2(S)}$ is present, only that it is present. Its activity will be unity regardless of the abundance of the solid in the system. Were we interested in the in the rate at which the system would approach the equilibrium condition upon the addition of the solid to the solution, then we would be interested in the surface area of the solid liquid contact, across which the transfer of zinc and hydroxide would occur as the reaction progressed either forward or in reverse. Herein, we have interest only in the final equilibrium condition. Theoretically the quantity of the solid present under the equilibrium conditions can be infinitesimal or as large as several grams, and the resultant speciation would be invariant.

We assemble the database of equilibrium and other constants. We include the molar density of water and the secondary MCL for zinc simply as references. In Table A.2, we find two values for $\log_{10}\beta_S$ for the zinc hydroxide system. The question often comes up "Which value should be used?" In order to answer that question we would retrace the steps of Morel and Hering (1983) in their efforts to assemble the table. We could seek their sources, review them and determine which value was most defensible, based on the methods used by those who measured and then published the value. Perhaps this was accomplished and it was found that both values were highly defensible. Morel and Hering determined that both should be included in their table. Unless we are willing to go the extra mile to determine conclusively which value is correct we either use both and develop bounding behavior for the system, or perhaps average the two values. In averaging them, we must use the β_S value rather than the $\log_{10}\beta_S$ value. Herein, we have used both values and our efforts yield bounding sets of behaviors:

$$\begin{pmatrix} \beta_{ZnOH} \\ \beta_{ZnOH2} \\ \beta_{ZnOH3} \\ \beta_{ZnOH4} \\ K_W \end{pmatrix} := \begin{pmatrix} 10^{5.0} \\ 10^{11.1} \\ 10^{13.6} \\ 10^{14.8} \\ 10^{-14} \end{pmatrix} \quad \begin{pmatrix} \beta_{S1.ZnOH2} \\ \beta_{S2.ZnOH2} \end{pmatrix} := \begin{pmatrix} 10^{15.5} \\ 10^{16.8} \end{pmatrix} \quad \begin{pmatrix} \gamma_1 \\ \gamma_2 \end{pmatrix} := \begin{pmatrix} 1 \\ 1 \end{pmatrix}$$

$$H(pH) := 10^{-pH} \qquad \rho mol_W := \frac{1000}{18} = 55.556$$

$$MCL_{Zn} := \frac{5}{65380} = 7.648 \times 10^{-5}$$

We solve the solid-formation equilibrium for the activity of the metal. We then write two functions for the activity of free zinc based on the presence of zinc hydroxide, each using a bounding value of β_S:

$$Zn_1(pH) := \left[\beta_{S1.ZnOH2} \cdot \left(\frac{K_W}{H(pH)}\right)^2\right]^{-1} \qquad Zn_2(pH) := \left[\beta_{S2.ZnOH2} \cdot \left(\frac{K_W}{H(pH)}\right)^2\right]^{-1}$$

If zinc hydroxide solid is present, this equilibrium must be obeyed by the system components. We then write two functions for the total solubility of zinc as a function of the proton abundance:

$$C_{Tot1.Zn}(pH) := Zn_1(pH) \cdot \left(\frac{1}{\gamma_2} + \frac{\beta_{ZnOH} \cdot K_W}{H(pH) \cdot \gamma_1} + \frac{\beta_{ZnOH2} \cdot K_W^2}{H(pH)^2} + \frac{\beta_{ZnOH3} \cdot K_W^3}{H(pH)^3 \cdot \gamma_1} \cdots + \frac{\beta_{ZnOH4} \cdot K_W^4}{H(pH)^4 \cdot \gamma_2}\right)$$

$$C_{Tot2.Zn}(pH) := Zn_2(pH) \cdot \left(\frac{1}{\gamma_2} + \frac{\beta_{ZnOH} \cdot K_W}{H(pH) \cdot \gamma_1} + \frac{\beta_{ZnOH2} \cdot K_W^2}{H(pH)^2} + \frac{\beta_{ZnOH3} \cdot K_W^3}{H(pH)^3 \cdot \gamma_1} \cdots + \frac{\beta_{ZnOH4} \cdot K_W^4}{H(pH)^4 \cdot \gamma_2}\right)$$

The plot of total zinc solubility versus pH shown in Figure E11.4.1 gives us the limits of the solubility based on the two published values of the formation constant. We note that below pH6 both predictions eventually rise above the molar concentration of water. Certainly, we cannot dissolve more zinc in water than there is water. Most likely, the limit on the solubility of zinc in water would be less than ten moles per liter. Let us suggest that the limit is indeed ten moles per liter and find the pH at which this total solubility would be exceeded:

$$Zn_{limit} := 10 \qquad pH := 5$$

$$\text{given} \quad C_{Tot1.Zn}(pH) = Zn_{limit} \qquad pH_{limit.1} := \text{find}(pH) = 5.75$$

$$\text{Given} \quad C_{Tot2.Zn}(pH) = Zn_{limit} \qquad pH_{limit.2} := \text{Find}(pH) = 5.1$$

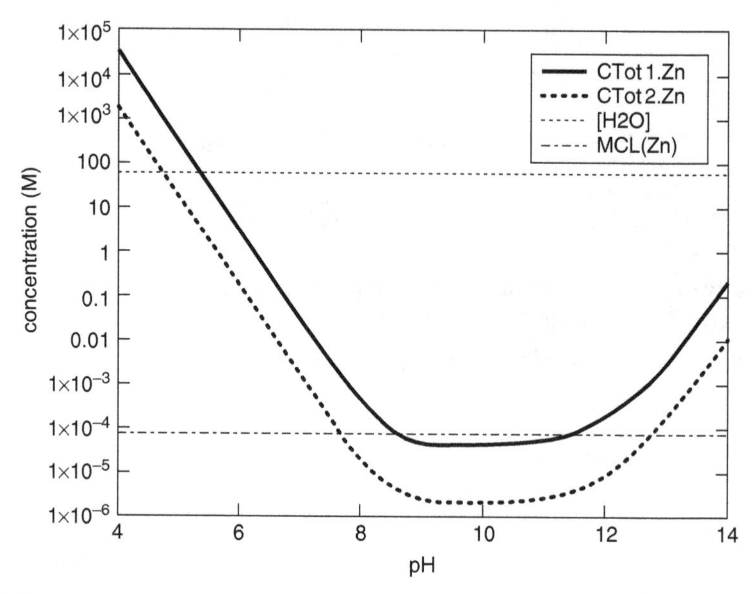

FIGURE E11.4.1 *A plot of zinc solubility based on the presence of zinc hydroxide solid in equilibrium with an aqueous solution considered to be infinitely dilute.*

Were we to hang our hats on $\log_{10}\beta_S = 15.5$ we would suggest that zinc hydroxide solid could not exist in aqueous solution below pH ~5.75. Were we to believe that $\log_{10}\beta_S = 16.8$, we would suggest that the limiting pH is ~5.1. Recall that in Example 11.2 we suggested that the total molar concentration of zinc would be 10^{-5} M. Were the $\log_{10}\beta_S$ equal to 16.8, our work of Example 11.2 would be in violation of the solubility of zinc in the presence of zinc hydroxide at pH values between about 8 and 12.

The foregoing is but an approximation of the true case. We have set activity coefficients to unity. In truth, as a consequence of the manipulation of the pH, the solution would be nondilute and activity coefficients would be less than unity, resulting in greater predicted total zinc concentration. To adjust the pH downward, we would likely use nitric acid, with nitrate becoming a significant contributor to ionic strength. Adjusting the pH upward might involve sodium or potassium hydroxide with the alkali metal ion then contributing significantly to ionic strength.

We should see which of the zinc- hydroxide complexes are most important. We can garner a pretty good idea from Example 11.2 but let us consider $\beta_{S.1}$ and develop the plot shown in Figure E11.4.2 in which the abundances of the four zinc-hydroxide complexes are shown along with the zinc ion and total zinc solubility. We extract the relations for the four complexes from the mole balance for zinc and write them into functions. We could have accomplished this with example 11.2 and summed the five relations to yield the function for total zinc solubility as a function of pH. As might be extrapolated from the result of Example 11.2, each specie, with the exception of $ZnOH^+$, takes its turn as the predominant dissolved zinc specie.

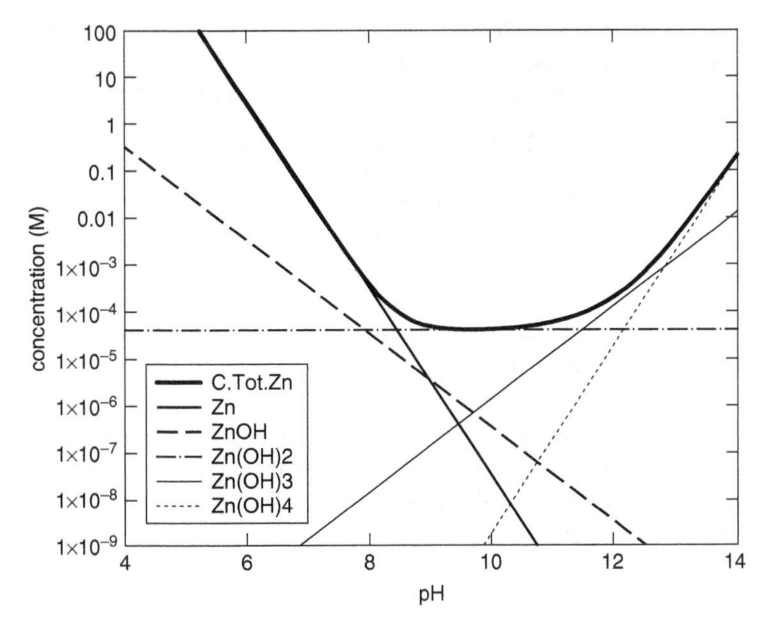

FIGURE E11.4.2 *A plot of zinc solubility and zinc hydrolysis specie abundance versus pH for the zinc hydrolysis system in equilibrium with zinc hydroxide solid.*

We can complete analyses similar to those of Example 11.4 for any metal that undergoes hydrolysis reactions in aqueous solution, given we can secure values for the respective cumulative formation constants for the hydrolysis reactions.

So, what if we wanted to know the result of equilibrating a metal hydroxide solid with distilled water? Once we found that condition, we could determine the quantities of acid and base necessary to adjust the pH downward or upward from that initial value to various target values, such as we did with Example 10.18.

Example 11.5 Determine the speciation of the zinc-hydroxide system when sufficient zinc hydroxide solid is added to freshly distilled water to ensure presence of the solid upon attainment of the equilibrium condition. We would prefer to step away with a single value, so let us use the second β_S value ($\log_{10}\beta_S = 16.8$) for the computations.

We know from the beginning here that since we must determine the final pH of the solution, we will need to employ a proton balance. We also know that the mathematical model will include at least one nonlinear equation among the several equations necessary to define the system. A given-find block (or implementation of the Excel solver) will be necessary. For the proton balance, $Zn(OH)_2$ is the most convenient reference specie. If we visualize the first step in the dissolution of

$Zn(OH)_{2(S)}$ as the formation of one $Zn(OH)_{2(aq)}$, the initial concentrations of all other zinc species are 0. Also, since we are beginning with FDW, the initial concentrations of hydronium and hydroxide will cancel. We state the corresponding proton balance:

$$2 \cdot \Delta C_{Zn} + \Delta C_{ZnOH} + \Delta C_{H} = \Delta C_{OH} + \Delta C_{ZnOH3} + 2 \cdot \Delta C_{ZnOH4}$$

We also can use any of the other zinc species as the reference. For example, if free zinc ion is used as the reference specie we would write an alternative proton balance:

$$\Delta C_{H} = \Delta C_{OH} + \Delta C_{ZnOH} + 2 \cdot \Delta C_{ZnOH} + 3 \cdot \Delta C_{ZnOH3} + 4 \cdot \Delta C_{ZnOH4}$$

We again visualize that the first step in the dissolution of zinc hydroxide would be the formation of the $Zn(OH)_{2(aq)}$ specie and, of course, the initial value for C_{ZnOH_2} would be $C_{Tot.Zn}$.

We will solve the system using each proton balance, in turn, and once again verify the flexibility inherent with the proton balance. We use complexation equilibria to reduce the system to three master equations with $\{H^+\}$, $\{Zn^{+2}\}$, and $C_{Tot.Zn}$ as the master dependent variables:

$$\begin{pmatrix} H \\ Zn \\ C_{Tot.Zn} \end{pmatrix} := \begin{pmatrix} 10^{-9} \\ 10^{-5} \\ 10^{-5} \end{pmatrix} \qquad \text{Given} \qquad Zn = \left[\beta_{S.ZnOH2} \cdot \left(\frac{K_W}{H}\right)^2\right]^{-1}$$

$$C_{Tot.Zn} = Zn \cdot \left(\frac{1}{\gamma_2} + \frac{\beta_{ZnOH} \cdot K_W}{H \cdot \gamma_1} + \frac{\beta_{ZnOH2} \cdot K_W^2}{H^2} + \frac{\beta_{ZnOH3} \cdot K_W^3}{H^3 \cdot \gamma_1} + \frac{\beta_{ZnOH4} \cdot K_W^4}{H^4 \cdot \gamma_2}\right)$$

$$2 \cdot \frac{Zn}{\gamma_2} + \frac{\beta_{ZnOH} \cdot Zn \cdot K_W}{H \cdot \gamma_1} + \frac{H}{\gamma_1} = \frac{K_W}{H \cdot \gamma_1} + \frac{\beta_{ZnOH3} \cdot Zn \cdot K_W^3}{H^3 \cdot \gamma_1} + 2 \cdot \frac{\beta_{ZnOH4} \cdot Zn \cdot K_W^4}{H^4 \cdot \gamma_2}$$

$$\begin{pmatrix} H \\ Zn \\ C_{Tot.Zn} \end{pmatrix} := \text{Find}(H, Zn, C_{Tot.Zn}) = \begin{pmatrix} 3.001 \times 10^{-9} \\ 1.428 \times 10^{-6} \\ 3.901 \times 10^{-6} \end{pmatrix} \qquad pH := -\log(H) = 8.5227$$

To employ the second proton balance, we merely replace the proton balance from the solve block above with the fleshed out second statement, and obtain the same result:

$$\frac{H}{\gamma_1} = \left[\frac{K_W}{H \cdot \gamma_1} + \frac{\beta_{ZnOH} \cdot Zn \cdot K_W}{H \cdot \gamma_1} + 2 \cdot \left(\frac{\beta_{ZnOH2} \cdot Zn \cdot K_W^2}{H^2} - C_{Tot.Zn}\right) \cdots \atop + 3 \cdot \frac{\beta_{ZnOH3} \cdot Zn \cdot K_W^3}{H^3 \cdot \gamma_1} + 4 \cdot \frac{\beta_{ZnOH4} \cdot Zn \cdot K_W^4}{H^4 \cdot \gamma_2}\right]$$

These results are corroborated with those of Example 11.4 by inserting the resultant pH value into the Zn_2(pH) and $C_{Tot2.Zn}$(pH) functions of Example 11.4:

$Zn_2(8.5227) = 1.428 \times 10^{-6}$ $\quad C_{Tot2.Zn}(8.5227) = 3.901 \times 10^{-6}$

We have examined the distribution of a target metal in systems with hydroxide as the sole ligand and with ligands beyond hydroxide. We have then examined the solubility of a target metal as a function of pH and as if it were added as the solid metal hydroxide to freshly distilled water. At this juncture it would be very appropriate to examine the acid and base neutralizing capacity of the overall system, considering that the metal hydroxide solid is ever present in the aqueous solution. We must be careful that we do not examine limits of the system for which the solid could not be present, specifically conditions of very low or very high pH.

Example 11.6 Develop the model that would allow computation of the ANC and BNC of the zinc, water, zinc hydroxide solid system of Example 11.5. Consider pH adjustment downward for ANC to pH = 7.0 and upward for BNC to pH = 10.0.

For this effort, we should begin with the statements for computing the ANC and BNC and decide how we will populate them. For ANC, we might choose the fully protonated free zinc ion as the reference:

$$ANC = -(\Delta C_{ZnOH} + 2 \cdot \Delta C_{ZnOH2} + 3 \cdot \Delta C_{ZnOH3} + 4 \cdot \Delta C_{ZnOH4} + \Delta C_{OH} - \Delta C_H - 2 \cdot \Delta C_{Tot.Zn})$$

$Zn(OH)_{2(S)}$ is at a proton level two below that of the free zinc ion. We must therefore include the change in its abundance. Since we have no measure of the quantity of solid present, the change in the abundance of the solid is taken as the negative of the change in the total dissolved zinc, hence the negative sign.

Similarly, for BNC we might choose the fully deprotonated $Zn(OH)_4^{-2}$ as the reference specie:

$$BNC = -(4 \cdot \Delta C_{Zn} + 3 \cdot \Delta C_{ZnOH} + 2 \cdot \Delta C_{ZnOH2} + \Delta C_{ZnOH3} + \Delta C_H - \Delta C_{OH} - 2 \cdot \Delta C_{Tot.Zn})$$

The solid is at a proton level two above that of the reference specie, and changes in its abundance are indicative of protons accepted. But since we must consider changes in the total dissolved zinc, we must again use the negative.

Alternatively, we can again visualize the first step of the dissolution process as dissolution of $Zn(OH)_2$ and use this specie as our reference for both ANC and BNC. Relations for both ANC and BNC may then be written:

$$ANC_2 = -(-2 \cdot \Delta C_{Zn} - \Delta C_{ZnOH} - \Delta C_H + \Delta C_{ZnOH3} + 2\Delta C.ZnOH4 + \Delta C_{OH})$$

$$BNC_2 = -(2 \cdot \Delta C_{Zn} + \Delta C_{ZnOH} + \Delta C_H - \Delta C_{OH} - \Delta C_{ZnOH3} - 2 \cdot \Delta C_{ZnOH4})$$

In each case, the solid zinc hydroxide is at the same proton level as the reference specie and is not considered in the proton balance.

We have written functions for the specie concentrations:

$$H(pH) := 10^{-pH} \qquad Zn(pH) := \left[\beta_{S.ZnOH2} \cdot \left(\frac{K_W}{H(pH)}\right)^2\right]^{-1} \qquad C_{Zn}(pH) := \frac{Zn(pH)}{\gamma_2}$$

$$C_{ZnOH}(pH) := \frac{\beta_{ZnOH} \cdot Zn(pH) \cdot K_W}{H(pH) \cdot \gamma_1} \qquad C_{ZnOH2}(pH) := \frac{\beta_{ZnOH2} \cdot Zn(pH) \cdot K_W^{\,2}}{H(pH)^2}$$

$$C_{ZnOH3}(pH) := \frac{\beta_{ZnOH3} \cdot Zn(pH) \cdot K_W^{\,3}}{H(pH)^3 \cdot \gamma_1} \qquad C_{ZnOH4}(pH) := \frac{\beta_{ZnOH4} \cdot Zn(pH) \cdot K_W}{H(pH)^4 \cdot \gamma_2}$$

$$\Delta C_H(pH) := \frac{H(pH) - H(pH_i)}{\gamma_1} \qquad \Delta C_{OH}(pH) := \frac{K_W}{H(pH) \cdot \gamma_1} - \frac{K_W}{H(pH_i) \cdot \gamma_1}$$

$$C_{Tot.Zn}(pH) := C_{Zn}(pH) + C_{ZnOH}(pH) + C_{ZnOH2}(pH) + C_{ZnOH3}(pH) + C_{ZnOH4}(pH)$$

We define concentration changes as the differences in concentrations at the target pH and the initial pH (= 8.5227 from Example 11.5):

$$\Delta C_{Zn}(pH) := C_{Zn}(pH) - C_{Zn}(pH_i)$$

$$\Delta C_{ZnOH}(pH) := C_{ZnOH}(pH) - C_{ZnOH}(pH_i)$$

$$\Delta C_{ZnOH2}(pH) := C_{ZnOH2}(pH) - C_{ZnOH2}(pH_i)$$

$$\Delta C_{ZnOH3}(pH) := C_{ZnOH3}(pH) - C_{ZnOH3}(pH_i)$$

$$\Delta C_{ZnOH4}(pH) := C_{ZnOH4}(pH) - C_{ZnOH4}(pH_i)$$

$$\Delta C_{Tot.Zn}(pH) := C_{Tot.Zn}(pH) - C_{Tot.Zn}(pH_i)$$

We populate the four aforementioned relations with the resultant functions, creating two distinct functions for ANC and two for BNC. Then for the target pH_{ANC} and pH_{BNC} we have our results:

$$ANC(pH) := -\begin{pmatrix}\Delta C_{ZnOH}(pH) + 2 \cdot \Delta C_{ZnOH2}(pH) + 3 \cdot \Delta C_{ZnOH3}(pH) \ldots \\ + 4 \cdot \Delta C_{ZnOH4}(pH) + \Delta C_{OH}(pH) - \Delta C_H(pH) - 2 \cdot \Delta C_{Tot.Zn}(pH)\end{pmatrix}$$

$$ANC_2(pH) := -\begin{pmatrix}-2 \cdot \Delta C_{Zn}(pH) - \Delta C_{ZnOH}(pH) - \Delta C_H(pH) + \Delta C_{ZnOH3}(pH) \ldots \\ + 2 \cdot \Delta C_{ZnOH4}(pH) + \Delta C_{OH}(pH)\end{pmatrix}$$

$$pH_{ANC} := 7.0 \qquad ANC(pH_{ANC}) = 3.1856 \times 10^{-3} \qquad ANC_2(pH_{ANC}) = 3.1856 \times 10^{-3}$$

$$BNC(pH) := -\left(\begin{array}{l}4\cdot\Delta C_{Zn}(pH) + 3\cdot\Delta C_{ZnOH}(pH) + 2\cdot\Delta C_{ZnOH2}(pH) + \Delta C_{ZnOH3}(pH) \ldots \\ + \Delta C_{H}(pH) - \Delta C_{OH}(pH) - 2\cdot\Delta C_{Tot.Zn}(pH)\end{array}\right)$$

$$BNC_2(pH) := -\left(\begin{array}{l}2\cdot\Delta C_{Zn}(pH) + \Delta C_{ZnOH}(pH) + \Delta C_{H}(pH) - \Delta C_{OH}(pH) \ldots \\ + -\Delta C_{ZnOH3}(pH) - 2\cdot\Delta C_{ZnOH4}(pH)\end{array}\right)$$

$pH_{BNC} := 10$ $\quad BNC(pH_{BNC}) = 1.0004 \times 10^{-4}$ $\quad BNC_2(pH_{BNC}) = 1.0004 \times 10^{-4}$

We observe again that we may write proton balances using whichever of the species we wish. Certainly, we could have written several more relations for both ANC and BNC and when correctly implemented all would have yielded output identical to that obtained herein.

The largest share of the acid neutralizing capacity arises as a result of the dissolution of zinc hydroxide. The base neutralizing capacity is dominated by the deprotonation of the waters associated with the free zinc ion and zinc monohydroxide complex.

Behaving similarly to the gas phase in our investigation of open and semiopen systems, the solid phase constitutes either a source or sink with capacity to neutralize either hydronium or hydroxide. Dissolution of the metal hydroxide allows for neutralization of hydronium. Reprecipitation of metal hydroxide as pH is increased beyond the minimum solubility level allows for the assimilation of hydroxide added to the system.

11.6.2 Solubility in the Presence of Ligands other than Hydroxide

The formation constants for a number of ligands are included in Table A.2. Certainly, there exist far more potential complexing ligands than those of Table A.2. We cannot hope to assimilate the behavior of all important ligands. We can, however, develop a quantitative strategy to employ such behavior. In previous examples, we employed the formation constants for metal-hydroxide complexes and for the formation of metal hydroxide solids. We also introduced a second ligand. The next step is the combination of additional complexing ligands with solubility-dissolution equilibria. For modeling of solubility-dissolution, we employ the solid formation equilibria along with the hydrolysis (hydroxide ligand) formation equilibria. Our system consists of a mass balance on the metal, equilibria, and, if the final pH of the solution is unknown, a proton balance. If we use pH as a master independent variable, we need not employ the proton balance. Let us continue with the thread of Examples 11.3–11.6 and examine the solubility of zinc in an aqueous system in which phosphate is present.

Example 11.7 Use pH as a master independent variable to determine the solubility of zinc metal in an aqueous solution of total phosphate abundance of 10^{-4} M. Consider that major ions in the solution that play no significant role in either the speciation of zinc or the capacity to neutralize hydronium or hydroxide are present such that the ionic strength is 0.01 M. Consider that the ionic strength will not vary significantly with adjustment of pH or with changes in the solubility of zinc.

We have our database of assigned parameters available from previous examples. To model this aqueous system we write functions for hydronium activity, zinc ion activity, and phosphate ion activity with pH as the argument and then use these in the mole balance relation for zinc to compute the total solubility and abundances of the various zinc species as a function of pH. The functions for hydronium and zinc are easily written:

$$H(pH) := 10^{-pH} \qquad \beta_{S.ZnOH2} = 6.31 \times 10^{16} \qquad Zn(pH) := \left[\beta_{S.ZnOH2} \cdot \left(\frac{K_W}{H(pH)}\right)^2\right]^{-1}$$

When we consider the mole balance for phosphorus and the substitution of the various equilibria for the individual phosphate species, we are met with $\{Zn_3(PO_4)_2\}$ and realize that substitution of the equilibrium relation will include $\{PO_4^{-3}\}^2$:

$$C_{Tot.PO4} = \beta_{H3PO4} \cdot H^3 \cdot PO4 + \frac{\beta_{H2PO4} \cdot H^2 \cdot PO4}{\gamma_1} + \frac{\beta_{HPO4} \cdot H \cdot PO4}{\gamma_2} + \frac{PO4}{\gamma_3} \ldots$$
$$+ \beta_{ZnHPO4} \cdot Zn \cdot H \cdot PO4 + \frac{\beta_{ZnH2PO4} \cdot Zn \cdot H^2 \cdot PO4}{\gamma_1} + 2 \cdot \beta_{Zn3PO42} \cdot Zn^3 \cdot PO4^2$$

We also realize that we can not simply factor out the $\{PO_4^{-3}\}$ and write the relation for $\{PO_4^{-3}\}$ in similar fashion to our old friend the abundance fraction. What we have is a second order polynomial with $\{PO_4^{-3}\}$ as the independent variable. We can use the quadratic formula, thus we algebraically zero the function and collect terms that are multipliers of $\{PO_4^{-3}\}^2$, $\{PO_4^{-3}\}^1$, and $\{PO_4^{-3}\}^0$:

$$f(PO4) = \beta_{H3PO4} \cdot H^3 \cdot PO4 + \frac{\beta_{H2PO4} \cdot H^2 \cdot PO4}{\gamma_1} + \frac{\beta_{HPO4} \cdot H \cdot PO4}{\gamma_2} + \frac{PO4}{\gamma_3} \ldots = 0$$
$$+ \beta_{ZnHPO4} \cdot Zn \cdot H \cdot PO4 + \frac{\beta_{ZnH2PO4} \cdot Zn \cdot H^2 \cdot PO4}{\gamma_1} \ldots$$
$$+ 2 \cdot \beta_{Zn3PO42} \cdot Zn^3 \cdot PO4^2 - C_{Tot.PO4}$$

In the context of "divide and conquer," we choose to write the a_2, a_1 and a_0 relations as functions, to be employed in the overall quadratic formula, which we also write as a function:

$$a(pH) := 2 \cdot \beta_{Zn3PO42} \cdot Zn(pH)^3 \qquad c := -C_{Tot.PO4}$$

$$b(pH) := \beta_{H3PO4} \cdot H(pH)^3 + \frac{\beta_{H2PO4} \cdot H(pH)^2}{\gamma_1} + \frac{\beta_{HPO4} \cdot H(pH)}{\gamma_2} + \frac{1}{\gamma_3} \ldots$$
$$+ \beta_{ZnHPO4} \cdot Zn(pH) \cdot H(pH) + \frac{\beta_{ZnH2PO4} \cdot Zn(pH) \cdot H(pH)^2}{\gamma_1}$$

pH :- 4, 4.1.. 12

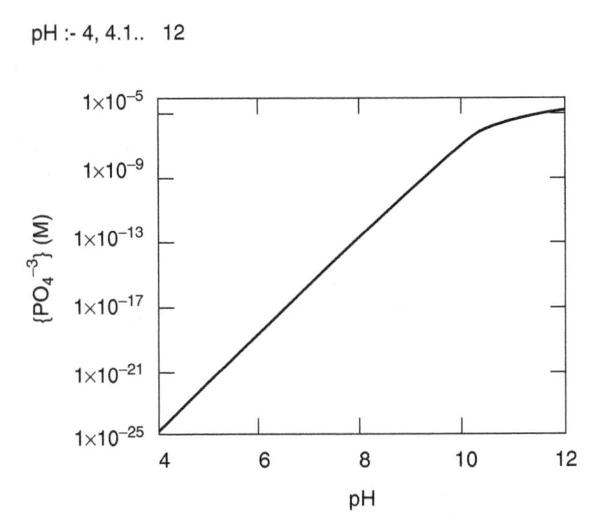

***FIGURE E11.7.1** Free phosphate ion activity versus pH for an aqueous solution of ionic strength 0.01 M containing zinc and phosphate and in equilibrium with zinc hydroxide solid phase.*

Since we are interested in the positive root (aqueous concentrations cannot be negative, and hopefully the argument of the square root will allow for a real result) we will drop the minus from the plus or minus preceding the square root term:

$$PO4(pH) := \frac{-b(pH) + \sqrt{b(pH)^2 - 4 \cdot a(pH) \cdot c}}{2 \cdot a(pH)}$$

Before moving on, we should investigate the relation between $\{PO_4^{-3}\}$ and pH. We certainly hope the function will be defined throughout the range of pH values we choose to investigate. Perhaps with certain systems, the relation would become undefined at some pH value. Such a condition will lead to an apparent discontinuity of the overall result and very anomalous apparent behavior. Our PO_4(pH) relation is well behaved in this case, as may be observed from Figure E11.7.1.

We may now write the function of pH from which we may compute the total solubility of zinc, employing the several functions previously written (definitions leading from left to right and then down are obligatory with MathCAD):

$$C_{Tot.Zn}(pH) := Zn(pH) \cdot \left[\frac{1}{\gamma_2} + \frac{\beta_{ZnOH} \cdot K_W}{H(pH) \cdot \gamma_1} + \beta_{ZnOH2} \cdot \left(\frac{K_W}{H(pH)} \right)^2 \cdots + \frac{\beta_{ZnOH3} \cdot K_W^3}{H(pH)^3 \cdot \gamma_1} + \frac{\beta_{ZnOH4} \cdot K_W^4}{H(pH)^4 \cdot \gamma_2} \cdots + \beta_{ZnHPO4} \cdot H(pH) \cdot PO4(pH) \cdots + \frac{\beta_{ZnH2PO4} \cdot H(pH)^2 \cdot PO4(pH)}{\gamma_1} \right] \cdots$$
$$+ 3 \cdot \beta_{Zn3PO42} \cdot Zn(pH)^3 \cdot PO4(pH)^2$$

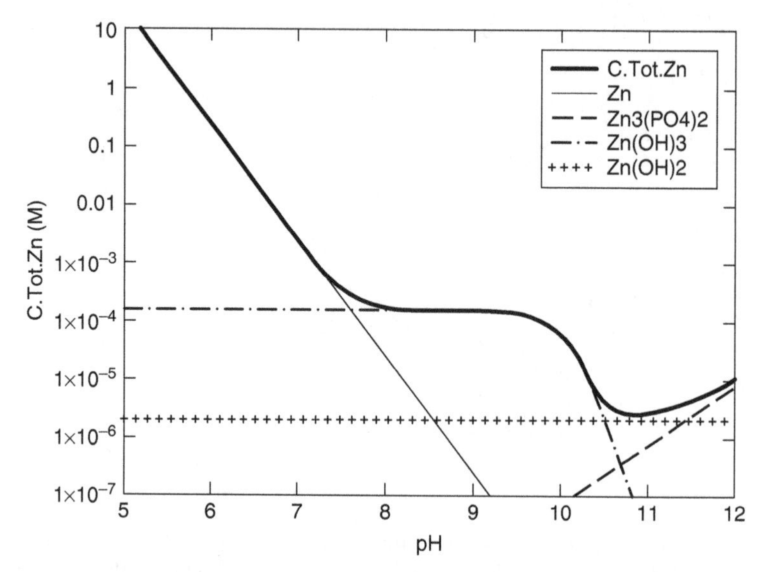

FIGURE E11.7.2 *A plot of total soluble zinc in an aqueous solution of 0.01 M ionic strength containing 10^{-4} M total phosphate-phosphorus and in equilibrium with zinc hydroxide solid.*

In Figure E11.7.2, we have plotted total zinc along with the four zinc species that are the most predominant in the four pH regions of the solubility versus pH plot. Free zinc dominates the solubility at neutral to acidic pH, the polynuclear $Zn_3(PO_4)_2$ dominates the solubility in the moderately alkaline region, $Zn(OH)_2^0$ dominates the solubility in the alkaline region and $Zn(OH)_3^-$ (and at still higher pH $Zn(OH)_4^{-2}$) dominates the solubility as pH is raised into the strongly alkaline range. We also must note that as pH drops below about 7, we begin to observe conditions that would violate our assumption of constant ionic strength. Further, we observe that zinc hydroxide solid could not exist in this aqueous system at pH much below about 5.8, owing to predicted total concentration well above 1 M. Lastly, at pH 6 or so the ionic strength would be greater than the maximum value for which the Davies equation is said to be valid. For pH values extending to about 6, we could employ ionic strength (and hence activity coefficients) as an unknown. Such an analysis would require that we implicitly solve the resultant system of equations. Examination of the system below pH 6 would require extraordinary measures in the definition of the activity coefficients based on the abundances of the electrolytes present.

Example 11.7 is specific to the zinc–phosphate system, but from these modeling efforts we realize that we can employ any number of ligands. We simply need to write mole balances and solve them for the activities of each of the base ligands. The activity of the metal ion is defined fully by the solubility-dissolution equilibrium and for most systems the model is entirely explicit. We must carefully watch for a potential caveat, however. The $Zn_3(PO_4)_2$ complex yields a second order term when we populate the mole balance with the equilibria. Certain ligands, notably cyanide (CN^-) will form complexes containing three and four ligand ions. When corresponding mole balances are populated, third and fourth order terms result. We then find ourselves of necessity solving the ligand and metal mole balances simultaneously, necessitating employment of a given-find block or Excel's solver. Significant effort is associated with employing the implicit solution and we might wish to investigate the importance of the higher order terms before we commit to the more rigorous implicit solution. We might assume that the higher order terms are insignificant and then solve the system. Once we have our approximate solution, we can investigate the validity of our assumptions. Functions are easily written and manipulated. If we find our neglect of higher order terms to be prudent, we may go forth with our functions and model the system. Conversely, if we find our assumptions to be invalid (for any range of pH) we must choose whether to accept the errors or to pursue the implicit solution. Even though we may write given-find blocks into functions, these are not so easily manipulated, requiring programming in the worksheet well beyond the writing of the functions, as we have observed in some previous examples.

11.7 SOLUBILITY OF METAL CARBONATES

In general, carbonate as a ligand forms 1:1 solids ($ML_{(s)}$) with divalent metal ions. Inspection of Table A.2 yields $\log_{10}$ values of formation constants for carbonate solids formed with Ca^{+2}, Mg^{+2}, Sr^{+2}, Ba^{+2}, Mn^{+2}, Fe^{+2}, Co^{+2}, Ni^{+2}, Cu^{+2}, Zn^{+2}, Pb^{+2}, Hg^{+2}, and Cd^{+2}. Many of these have been named by the geochemists, but, for clarity, when appropriate we will refer to them by their chemical names. Additionally, we find $\log_{10}$ values of formation constants for $Ag_2CO_{3(s)}$, $Cu_2(OH)_2CO_{3(s)}$, and $Cu_3(OH)_2(CO_3)_{2(s)}$. Two minerals that are extremely important in environmental systems but whose formation constants are not included in Table A.2 are ordered and disordered dolomite, both of the same general stoichiometry.

$$Ca^{+2} + Mg^{+2} + 2CO_3^{-2} \Leftrightarrow CaMg(CO_3)_{2(s)}$$

Stumm and Morgan (1996), in their appendix (Table 1C), give values of $\log_{10}K_{sp}$ (solubility product) as −17.09 and −16.54, respectively, for the reactions written as dissolution of calcium, magnesium, and carbonate from dolomite. For the system employed herein, we could convert these to cumulative formation constants, noting that $\beta_s = K_{sp}^{-1}$, as the formation reaction is the reverse of the solubility reaction. Thus, for ordered dolomite $\beta_s = 17.09$ and for disordered dolomite $\beta_s = 16.54$.

Examination of the formation reaction and associated formation constant for 1:1 metal carbonates yields a typical reaction of the form of Equation 4.7:

$$M^{+2} + CO_3^{-2} \Leftrightarrow MCO_{3(s)}$$

The reaction equilibrium is then a typical form of Equation 4.8:

$$\beta_S = (\{M^{+2}\}.\{CO_3^{-2}\})^{-1}$$

Then, we may solve the formation equilibrium for the activity of the free metal ion:

$$\{M^{+2}\} = (\beta_S.\{CO_3^{-2}\})^{-1}$$

The activity of the free metal ion is inversely proportional to the activity of the carbonate ion. The formation constant is the proportionality constant. The specifics of the metal-carbonate interactions seem rather straightforward, which they indeed are. The complications arise in consideration of the activity of the carbonate ion.

11.7.1 Calcium Carbonate Solubility

Calcium and carbonate are rather ubiquitous in environmental systems. Certain geographic regions are rich in limestone (calcite, dolomite) while others are not. We would be wise to develop capability to model the solubility of calcium and carbonate. Let us first examine the calcium-carbonate system from the standpoint of the equilibration of rainfall with calcite. Once we have a model for this system, the extension to include other calcium carbonate minerals into the model would not be unreasonably difficult.

Example 11.8 Develop the mathematical model that may be used to determine the speciation of aqueous solution, originating as precipitation at the location of the laboratory of Example 10.10 and which then becomes equilibrated with calcite in a closed system.

We define some formation constants and use the results of Example 10.10 to define the initial concentrations of carbonate system species:

$$\begin{pmatrix} K_W \\ \beta_{HCO3} \\ \beta_{H2CO3} \\ K_{H.CO2} \end{pmatrix} := \begin{pmatrix} 10^{-14} \\ 10^{10.33} \\ 10^{16.68} \\ 0.0339 \end{pmatrix} \quad \begin{pmatrix} \beta_{S.CaCO3} \\ \beta_{CaCO3} \\ \beta_{CaHCO3} \\ \beta_{CaOH} \end{pmatrix} := \begin{pmatrix} 10^{8.35} \\ 10^{3.2} \\ 10^{11.59} \\ 10^{1.15} \end{pmatrix}$$

$$\begin{pmatrix} c_{H.i} \\ c_{OH.i} \\ c_{H2CO3.i} \\ c_{HCO3.i} \\ c_{CO3.i} \\ c_{Tot.CO3.i} \end{pmatrix} := \begin{pmatrix} 2.399 \times 10^{-6} \\ 4.168 \times 10^{-9} \\ 1.286 \times 10^{-5} \\ 2.395 \times 10^{-6} \\ 4.669 \times 10^{-11} \\ 1.526 \times 10^{-5} \end{pmatrix}$$

All calcium species have initial concentrations equal to 0. We have five master unknowns: the proton activity, the calcium activity, the carbonate activity, the total calcium abundance, and the total carbonate abundance. We then require five independent equations. The first three are the mole balances on calcium and carbonate and the solubility equilibrium for calcite:

$$Ca = \left(\beta_{S.CaCO3} \cdot CO3\right)^{-1}$$

$$c_{Tot.Ca} = c_{Ca} + c_{CaOH} + c_{CaCO3} + c_{CaHCO3}$$

$$c_{Tot.CO3} = c_{H2CO3} + c_{HCO3} + c_{CO3} + c_{CaHCO3} + c_{CaCO3}$$

For the fourth equation we note that for every mole of calcium that dissolves, one mole of carbonate must also dissolve, leading to the equality of the changes in total concentration of the calcium and carbonate systems:

$$\Delta c_{Tot.Ca} = \Delta c_{Tot.CO3}$$

Then for the fifth equation we write a proton balance to account for proton transfers as a consequence of the dissolution of calcite. Most conveniently we choose carbonate as the reference specie for the carbonate system and $CaCO_3^0$ as the reference specie for the calcium system. Dissolution of $CaCO_{3(s)}$ to yield $CaCO_3^0$ involves no proton transfers, but subsequent ionization of $CaCO_3^0$ to yield Ca^{+2} and $CO_3^=$ involves hydration of the calcium ion and hence capacity for Ca hydrolysis. The free calcium ion is at a proton level two greater than that of $CaCO_3^0$ and calcite:

$$\Delta c_H + 2 \cdot \Delta c_{Ca} + \Delta c_{CaOH} + \Delta c_{CaHCO3} + 2 \cdot \Delta c_{H2CO3} + \Delta c_{HCO3} = \Delta c_{OH}$$

We write the system of equations using activities of hydronium, free calcium, and carbonate along with total carbonate and total calcium and employ them in the solve block shown in Figure E11.8.1. We see that since water itself has little buffering capacity, the pH is rather high but little calcium carbonate actually dissolves. Our assumption of infinitely dilute conditions remains fairly applicable.

$$\begin{pmatrix} H \\ Ca \\ CO3 \\ C_{Tot.CO3} \\ C_{Tot.Ca} \end{pmatrix} := \begin{pmatrix} 10^{-10} \\ 10^{-5} \\ 10^{-4} \\ 2 \cdot 10^{-4} \\ 1.9 \cdot 10^{-4} \end{pmatrix} \qquad \Sigma C_i := C_{H.i} + 2 \cdot C_{H2CO3.i} + C_{HCO3.i} - C_{OH.i}$$

Given $\qquad Ca = (\beta_{S.CaCO3} \cdot CO3)^{-1}$

$$C_{Tot.Ca} = \frac{Ca}{\gamma_2} + \frac{\beta_{CaOH} \cdot Ca \cdot K_W}{H \cdot \gamma_1} + \beta_{CaCO3} \cdot Ca \cdot CO3 + \frac{\beta_{CaHCO3} \cdot Ca \cdot H \cdot CO3}{\gamma_1}$$

$$C_{Tot.CO3} = \beta_{H2CO3} \cdot H^2 \cdot CO3 + \frac{\beta_{HCO3} \cdot H \cdot CO3}{\gamma_1} + \frac{CO3}{\gamma_2} + \beta_{CaCO3} \cdot Ca \cdot CO3 \ldots + \frac{\beta_{CaHCO3} \cdot Ca \cdot H \cdot CO3}{\gamma_1}$$

$$\frac{Ca}{\gamma_2} + \frac{\beta_{CaOH} \cdot Ca \cdot K_W}{H \cdot \gamma_1} = \beta_{H2CO3} \cdot H^2 \cdot CO3 + \frac{\beta_{HCO3} \cdot H \cdot CO3}{\gamma_1} + \frac{CO3}{\gamma_2} - C_{Tot.CO3.i}$$

$$\frac{H}{\gamma_1} + 2 \cdot \frac{Ca}{\gamma_2} + \frac{\beta_{CaOH} \cdot Ca \cdot K_W}{H \cdot \gamma_1} + \frac{\beta_{CaHCO3} \cdot Ca \cdot H \cdot CO3}{\gamma_1} \ldots = \frac{K_W}{H \cdot \gamma_1} + 2 \cdot \left(\beta_{H2CO3} \cdot H^2 \cdot CO3\right) + \frac{\beta_{HCO3} \cdot H \cdot CO3}{\gamma_1} - \Sigma C_i$$

$$\begin{pmatrix} H \\ Ca \\ CO3 \\ C_{Tot.CO3} \\ C_{Tot.Ca} \end{pmatrix} := \mathrm{Find}(H, Ca, CO3, C_{Tot.CO3}, C_{Tot.Ca}) = \begin{pmatrix} 4.982 \times 10^{-11} \\ 8.86 \times 10^{-5} \\ 5.041 \times 10^{-5} \\ 1.113 \times 10^{-4} \\ 9.602 \times 10^{-5} \end{pmatrix}$$

$$pH := -\log(H) = 10.303$$

FIGURE E11.8.1 *Screen capture of the solve block used for speciation of calcium carbonate equilibrated with natural precipitation.*

In Example 11.8, we illustrated how mole balance, solid formation equilibria, and proton balance can be combined to determine speciation in aqueous solution when a carbonate solid is equilibrated with water. The process would be much the same for any of the carbonate solids whose formation constants are given in Table A.2.

We must move to the next level, however, with the solubility of calcite. In many cases, the equilibration of aqueous solutions with calcite (and of course other minerals) occurs in subsurface environments, separated from the atmosphere by hundreds of feet of soil and perhaps a water-bearing zone or two. Then, as a consequence of biological activity, any gas phase present (or that we might envision) would become enriched in carbon dioxide. Certainly, in the Midwestern United States, ground waters typically contain higher levels of calcium and carbonate, while their pH values are much nearer to neutral. Let us expand upon Example 11.8 and involve the carbon-dioxide-rich gas phase in the computation of the solution speciation.

Example 11.9 Examine partial pressure of carbon dioxide as a master independent variable relative to the solubility of calcite in aqueous solution. Assume that the initial condition of the water is that of the rainwater of Examples 10.10 and 11.8. Also, since with increased partial pressure of carbon dioxide, we might expect greater solubility of calcite, consider a nondilute solution. Certainly, in environmental systems the presence of sodium, potassium, and other salts of sulfate and chloride will tend to increase the ionic strength. For this effort, let us ignore these potential contributions of major ions and consider that the ionic strength is due solely to the dissolution of calcite and carbon dioxide.

We retain nearly the entire structure of the model of Example 11.8 adding couple significant enhancements. We include independent equations for ionic strength and activity coefficients. Then rather than write the mole and proton balances using the activity of carbonate, we can write the activity of carbonate in terms of the partial pressure of carbon dioxide, which is known once we set the value for this master independent variable. Our eight unknowns are now: $\{H^+\}$, $\{Ca^{+2}\}$, ΔCO_2, $C_{Tot.CO_3}$, $C_{Tot.Ca}$, I, γ_1, and γ_2. ΔCO_2 is the increase in the total dissolved carbonate as a consequence of the dissolution of carbon dioxide. We must somehow discern the dissolved carbonate that arises from dissolution of calcite from that arising from dissolution of carbon dioxide. Then, the relation between the changes in calcium and carbonate abundance may be revised:

$$\Delta C_{Tot.Ca} + \Delta_{CO2} = \Delta C_{Tot.CO3}$$

When we define the reference specie for acid neutralizing capacity as the fully deprotonated conjugate base and choose to employ changes in abundances of conjugate acid species in definition of ANC, we must rewrite Equation 10.17a using the conjugate acids:

$$-\Delta[\text{ANC}] = \sum_i \left(\nu_{\text{A},i}\Delta[\text{A}_i]\right)$$

Then, since ΔCO_2 is essentially the addition of an acid similar to a laboratory titration, we equate ΔCO_2 with the negative of ΔANC. When fleshed out with acid species, the relation for ANC becomes our proton balance:

$$\begin{pmatrix} \Delta C_H + 2 \cdot \Delta C_{Ca} + \Delta C_{CaOH} + \Delta C_{CaHCO3} \cdots \\ + 2 \cdot \Delta C_{H2CO3} + \Delta C_{HCO3} \end{pmatrix} = \Delta C_{OH} + 2 \cdot \Delta_{CO2}$$

$$\begin{pmatrix} H \\ Ca \\ CO3 \\ C_{Tot.CO3} \\ C_{Tot.Ca} \\ \Delta_{CO2} \end{pmatrix} := \begin{pmatrix} 10^{-7} \\ 10^{-3} \\ 10^{-4} \\ 2\cdot10^{-2} \\ 1.9\cdot10^{-2} \\ 0.001 \end{pmatrix} \qquad \begin{pmatrix} I \\ \gamma_1 \\ \gamma_2 \\ A \end{pmatrix} := \begin{pmatrix} .01 \\ .9 \\ .6 \\ 0.5115 \end{pmatrix}$$

Given $$Ca = \left(\beta_{S.CaCO3}\cdot\frac{K_{H.CO2}\cdot P_{CO2}}{\beta_{H2CO3}\cdot H^2}\right)^{-1}$$

$$C_{Tot.Ca} = \frac{Ca}{\gamma_2} + \frac{\beta_{CaOH}\cdot Ca\cdot K_W}{H\cdot\gamma_1} + \frac{K_{H.CO2}\cdot P_{CO2}}{\beta_{H2CO3}\cdot H^2}\cdot\left(\beta_{CaCO3}\cdot Ca\cdot\frac{\beta_{CaHCO3}\cdot Ca\cdot H}{\gamma_1}\right)$$

$$C_{Tot.CO3} = \frac{K_{H.CO2}\cdot P_{CO2}}{\beta_{H2CO3}\cdot H^2}\cdot\left[\beta_{H2CO3}\cdot H^2 + \frac{\beta_{HCO3}\cdot H}{\gamma_1} + \frac{1}{\gamma_2} \ldots + Ca\cdot\left(\beta_{CaCO3} + \frac{\beta_{CaHCO3}\cdot H}{\gamma_1}\right)\right]$$

$$Ca\cdot\left(\frac{1}{\gamma_2} + \frac{\beta_{CaOH}\cdot K_W}{H\cdot\gamma_1}\right) + \Delta_{CO2} = \left[\frac{K_{H.CO2}\cdot P_{CO2}}{\beta_{H2CO3}\cdot H^2}\cdot\left(\beta_{H2CO3}\cdot H^2 + \frac{\beta_{HCO3}\cdot H}{\gamma_1} + \frac{1}{\gamma_2}\right) \ldots + -C_{Tot.CO3.i}\right]$$

$$\left[\frac{H}{\gamma_1} + 2\cdot\frac{Ca}{\gamma_2} + \frac{\beta_{CaOH}\cdot Ca\cdot K_W}{H\cdot\gamma_1} - \Sigma C_i \ldots + \frac{K_{H.CO2}\cdot P_{CO2}}{\beta_{H2CO3}\cdot H^2}\cdot\left[\frac{\beta_{CaHCO3}\cdot Ca\cdot H}{\gamma_1} + 2\cdot\left(\beta_{H2CO3}\cdot H^2\right) + \frac{\beta_{HCO3}\cdot H}{\gamma_1}\right]\right] = \frac{K_W}{H\cdot\gamma_1} \ldots + 2\cdot\Delta_{CO2}$$

$$I = \frac{1}{2}\cdot\left[\frac{H}{\gamma_1} + Ca\cdot\left(\frac{4}{\gamma_2} + \frac{\beta_{CaOH}\cdot K_W}{H\cdot\gamma_1}\right) + \frac{K_{H.CO2}\cdot P_{CO2}}{\beta_{H2CO3}\cdot H^2}\cdot\left(\frac{\beta_{CaHCO3}\cdot Ca\cdot H}{\gamma_1} \ldots + \frac{\beta_{HCO3}\cdot H}{\gamma_1} + \frac{4}{\gamma_2}\right)\right]$$

$$\gamma_1 = 10^{\frac{-A\cdot\sqrt{I}}{1+\sqrt{I}}} \qquad \gamma_2 = 10^{\frac{-A\cdot 4\cdot\sqrt{I}}{1+\sqrt{I}}}$$

$$\text{calcite}\left(P_{CO2}, H, Ca, \Delta_{CO2}, C_{Tot.CO3}, C_{Tot.Ca}, I, \gamma_1, \gamma_2\right) := \text{Find}\left(H, Ca, \Delta_{CO2}, C_{Tot.CO3}, C_{Tot.Ca}, I, \gamma_1, \gamma_2\right)$$

FIGURE E11.9.1 *Screen capture of the solve block for computation of calcium carbonate solubility as a function of varying partial pressure of carbon dioxide.*

For every mole of carbon dioxide that dissolves, two moles of protons are carried into the aqueous solution. Certainly then, we are comfortable that ΔCO_2 appears opposite evidence of protons accepted.

Carbonate activity is written in terms of partial pressure of carbon dioxide and these two relations are fleshed out via algebraic substitutions. The relations for ionic

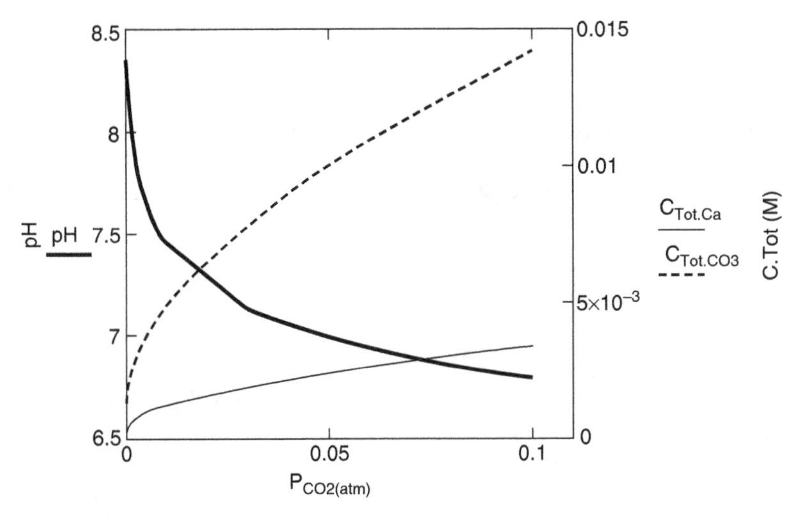

FIGURE E11.9.2 *A plot of predicted pH, total calcium and total carbonate versus carbon dioxide partial pressure for ground water in equilibrium with calcite.*

strength and activity coefficients are inserted and the given-find solve block of Figure E11.9.1 is invoked. In order to collect the various values of P_{CO_2} and resultant pH, $C_{Tot.CO_3}$, and $C_{Tot.Ca}$, the given-find is written into a function. We then use this function to solve for the eight unknowns at selected values of carbon dioxide partial pressure and collect these is eight resultant vectors. For illustration of the behavior of the system, in Figure E11.9.2 we have plotted pH, $C_{Tot.Ca}$, and $C_{Tot.CO_3}$ against P_{CO_2}. We observe that the presence of carbon dioxide in subsurface environments during equilibration of ground waters with calcite would have a very profound influence upon the character of the ground water. We have not included the contribution to ionic strength of salts that percolating water might pick up along the percolation path. Each case would be specific to the media through which the water would percolate and ionic strength would be greater.

So then, what of water that might initially be equilibrated with calcite and then subjected to either natural or anthropogenic inputs of trace metals such as zinc or cadmium. We know that these metals can form either hydroxide or carbonate solids. We would hope that in most systems the trace metal would be of much lesser abundance than calcium and carbonate. As long as the solution would experience no significant further inputs (or losses) of calcium or carbonate, we should be able, with but small error, to model the calcium-carbonate system independent from the remaining solution constituents aside from ionic strength. Let us develop such a model for the mineral calcite. The concepts and process can be applied to any dominant calcium-carbonate mineral.

Example 11.10 Develop a model from which the solubility of calcite can be predicted. Include the effects of the initial abundances of calcium and carbonate as well as those of nonzero ionic strength. Use pH as a master independent variable and develop a function that would be portable for use with computations of metal solubility in general aqueous systems that are in equilibrium with calcite.

Here we can use much of our work from Example 11.9, but since pH will be a master independent variable, we need not employ the proton balance. Our four relations are then the formation equilibria, mole balances on calcium, and carbonate and the equality of changes in total calcium and total carbonate:

$$Ca = (\beta_{S.CaCO3} \cdot CO3)^{-1}$$

$$C_{Tot.Ca} = C_{Ca} + C_{CaOH} + C_{CaCO3} + C_{CaHCO3}$$

$$C_{Tot.CO3} = C_{H2CO3} + C_{HCO3} + C_{CO3} + C_{CaHCO3} + C_{CaCO3}$$

$$\Delta C_{Tot.Ca} = \Delta C_{Tot.CO3}$$

We equate the changes in total calcium and total carbonate. The calcium-carbonate complexes cancel from the RHS and LHS. We are left with a simpler relation:

$$C_{Ca} + C_{CaOH} - C_{Tot.Ca.i} = C_{H2CO3} + C_{HCO3} + C_{CO3} - C_{Tot.CO3.i}$$

When we substitute equivalent expressions written in our major dependent variables for the various terms and employ the definition of free calcium in terms of carbonate, we have a very usable relation:

$$\frac{\left(\frac{1}{\gamma_2} + \frac{\beta_{CaOH} \cdot K_W}{H \cdot \gamma_1}\right)}{(\beta_{S.CaCO3} \cdot CO3)} - C_{Tot.Ca.i} = CO3 \cdot \left(\beta_{H2CO3} \cdot H^2 + \frac{\beta_{HCO3} \cdot H}{\gamma_1} + \frac{1}{\gamma_2}\right) - C_{Tot.CO3.i}$$

We now collect the initial calcium and carbonate abundances on the RHS, and multiply through to dispense with the denominator of the LHS. We also divide through by $\beta_{S.CaCO_3}$ to further simplify the relation. When we collect terms and algebraically zero the function, our result is a second-order polynomial in carbonate activity. We collect the terms multiplying $\{CO_3^{-2}\}^2$, $\{CO_3^{-2}\}^1$, and $\{CO_3^{-2}\}^0$ and set the resultant function equal to 0:

$$f(CO3) = \beta_{S.CaCO3} \cdot \left(\beta_{H2CO3} \cdot H^2 + \frac{\beta_{HCO3} \cdot H}{\gamma_1} + \frac{1}{\gamma_2}\right) \cdot CO3^2 \ldots = 0$$
$$+ \beta_{S.CaCO3} \cdot (C_{Tot.Ca.i} - C_{Tot.CO3.i}) \cdot CO3 - \left(\frac{1}{\gamma_2} + \frac{\beta_{CaOH} \cdot K_W}{H \cdot \gamma_1}\right)$$

We then arrange the *a*, *b*, and *c* terms to yield the activity of carbonate as a function of pH and the initial conditions. To employ this resultant function in a MathCAD worksheet, we arrange it as a series of functions:

$$\begin{pmatrix} C_{Tot.Ca.i} \\ C_{Tot.CO3.i} \end{pmatrix} := \begin{pmatrix} 0 \\ 1.526 \times 10^{-5} \end{pmatrix} \quad \begin{pmatrix} \gamma_1 \\ \gamma_2 \end{pmatrix} := \begin{pmatrix} 1 \\ 1 \end{pmatrix} \quad H(pH) := 10^{-pH}$$

$$a(pH) := \left(\beta_{H2CO3} \cdot H(pH)^2 + \frac{\beta_{HCO3} \cdot H(pH)}{\gamma_1} + \frac{1}{\gamma_2} \right)$$

$$b(pH) := (C_{Tot.Ca.i} - C_{Tot.CO3.i}) \qquad c(pH) := \frac{-\left(\frac{1}{\gamma_2} + \frac{\beta_{CaOH} \cdot K_W}{H(pH) \cdot \gamma_1} \right)}{\beta_{S.CaCO3}}$$

$$CO3(pH) := \frac{-b(pH) + \sqrt{b(pH)^2 - 4 \cdot a(pH) \cdot c(pH)}}{2 \cdot a(pH)}$$

The initial abundances of total calcium and total carbonate as well as the activity coefficients would be defined based on the application of the function in the context of the system. If we choose to write it all as a single function, we would simply insert the RHS expressions from a(pH), b(pH), and c(pH) into the form of the quadratic equation.

The portable function written in conjunction with Example 11.10 has potential significant value for use in systems dominated by carbonate-bearing strata that can be modeled as calcite. We certainly could have written the function to yield the activity of the free calcium ion as a function of pH. We may of course, employ the same process for development of functions that yield the activities of important alternative minerals, both primary and secondary, whose chemistry would dominate either a natural or engineered system. If we examine dolomite (either ordered or disordered), we realize the process will be somewhat more involved. We have six unknowns: total calcium, magnesium, and carbonate and the activities of free calcium, free magnesium, and carbonate. We therefore need six equations:

1. The formation equilibrium (relates the activities of calcium, magnesium, and carbonate.

2–4. Mole balances for total calcium, total magnesium, and total carbonate.

5. The condition that changes in total abundances of calcium and magnesium must be equal.
6. The condition that the sum of changes in the total abundances of calcium and magnesium must equal the change in total abundance of carbonate.

We also would need to define the initial total abundances of calcium, magnesium, and carbonate. We would assume initially that the ionic strength and activity coefficients would be constant and defined externally. If the dissolution of dolomite would significantly affect the ionic strength, we have seen in examples above that we can address the additional three unknowns with three additional equations. The detailed development of this mathematical/numerical model is left as an exercise for the student.

11.7.2 Solubility of Metal Carbonates—the Controlling Solid Phase

When we examined the solubility of metal hydroxides, we were able to consider the metal and its interaction with the hydroxide ligand in the absence of other potential solids. Conversely, with metal carbonates, we will always have the potential for the preferential formation of metal hydroxides relative to metal carbonates. For any system in question, it is vital to know which solid phase will form under any given set of conditions. Thus, we will examine the formation of metal-carbonate solids in the context of a "controlling solid phase". The controlling solid phase is the one that is most likely to form under a given set of conditions, and once formed, will exert control on the solubility of the metal. In practice, we find that the controlling solid phase is the one whose equilibrium constraints result in the lowest value of the activity of the free metal ion.

Snoeyink and Jenkins (1980) use Gibbs energy en route to the development of a reaction quotient, Q. The reaction quotient is identical in formulation to the RHS of the equilibrium relation. The significant difference is that Q does not represent an equilibrium condition, but rather, the state of a system in question at a particular instant in time. The relative magnitudes of Q and K_{eq} determine the direction in which a reaction would proceed, based on the instantaneous thermodynamic driving force. If $Q=K_{eq}$, the abundances of products and reactants are consistent with the equilibrium condition. If $Q>K_{eq}$, the reaction will proceed in reverse of the written direction and if $Q<K_{eq}$ the reaction will proceed as written. We may employ this concept in analyses of the idea of a controlling solid phase.

For definition of the activity of the free metal ion, considering hydroxide and carbonate control of its abundance we have, respectively:

$$\{M_{OH}{}^{+2}\} = \frac{1}{\beta_{S.MOH2}\left(\frac{K_W}{\{H^+\}}\right)^2} \quad \text{and} \quad \{M_{CO3}{}^{+2}\} = \frac{1}{\beta_{S.MCO3}\{CO_3^{-2}\}}$$

$\{M_{OH}{}^{+2}\}$ is the activity of the free metal ion if solubility control would be exerted by the hydroxide solid and $\{M_{CO3}{}^{+2}\}$ is the activity of the free metal ion if solubility control would be exerted by the carbonate solid. We must emphasize here that the activity of the free metal ion can have one and only one value and either $\{M_{OH}{}^{+2}\}$ or $\{M_{CO3}{}^{+2}\}$ as must be $\{M^{+2}\}$. We define the respective reaction quotients:

$$Q_{MOH2} = \frac{1}{\{M^{+2}\}\left(\frac{K_W}{\{H^+\}}\right)^2} \quad \text{and} \quad Q_{MCO3} = \frac{1}{\{M^{+2}\}\{CO_3^{-2}\}}$$

Let us consider the case that the predicted $\{M_{OH}{}^{+2}\} < \{M_{CO3}{}^{+2}\}$. Let us further consider that the predicted value of $\{M_{OH}{}^{+2}\}$ leads to the condition $Q_{MOH2} = \beta_{S.MOH2}$ $\left(Q_{MOH2} / K_{eq} = 1\right)$ consistent with equilibrium of the metal-hydroxide formation reaction. If we then use $\{M_{OH}^{+2}\}$ in place of $\{M_{CO3}^{+2}\}$ to compute Q_{MCO3}, we would find that $Q_{MCO3} > \beta_{S.MCO3}\left(Q_{MCO3} / K_{eq} > 1\right)$. Le Chatelier's principle would then dictate that the system is not in equilibrium and to attain the equilibrium state, one of three conditions necessarily must change:

1. The activity of carbonate ion must be increased, corresponding with increased pH.
2. The activity of the metal-carbonate solid phase must be decreased.
3. The activity of the free metal ion must be increased.

Analyses of these three potential alternatives lead to the following reasoning:

1. Since we are examining pH as a master independent variable and adjustment of the pH would also change the predicted $\{M_{OH}^{+2}\}$ and $\{M_{CO3}^{+2}\}$, we would not choose this option.
2. A decrease in the activity of the metal-carbonate solid phase would render its activity to be less than unity. We know that the activity of the solid phase must be unity, if present. Then, if the activity must be less than unity to satisfy the equilibrium condition, the metal-carbonate solid phase cannot be present.
3. Were we to adjust the activity of the free metal ion, we would find that to re-establish equilibrium its value would necessarily be the predicted $\{M_{CO3}^{+2}\}$.
4. Were we then to use $\{M_{CO3}^{+2}\}$ to compute Q_{MOH2}, we would find that $Q_{MOH2} / \beta_{S.MOH2} < 1$ and to satisfy Le Chatelier's principle, the activity of the metal hydroxide solid phase would need to increase from its value of unity, which is of course impossible.

Thus, we would assert that when we predict the condition $\{M_{OH}^{+2}\} < \{M_{CO3}^{+2}\}$ the metal hydroxide solid phase exerts control of the activity of the free metal ion and hence of the solubility of the metal.

Let us consider the alternative case that the predicted $\{M_{CO3}^{+2}\} < \{M_{OH}^{+2}\}$. Let us further consider that the predicted value of $\{M_{CO3}^{+2}\}$ leads to the condition $Q_{MCO3} = \beta_{S.CO3}\left(Q_{MCO3} / K_{eq} = 1\right)$, consistent with equilibrium of the metal-carbonate formation reaction. If we then use $\{M_{CO3}^{+2}\}$ in place of $\{M_{OH}^{+2}\}$ to compute Q_{MOH2}, we would find that $Q_{MOH2} / \beta_{S.MOH2} > 1$. Again, Le Chatelier's principle would dictate that the system is not in equilibrium and to attain the equilibrium state, one of three conditions necessarily must change:

1. The activity of hydronium would need to be increased.
2. The activity of the metal hydroxide solid phase would need to be decreased.
3. The activity of the free metal ion would need to be increased.

Analyses of these three potential alternatives lead to the following reasoning:

1. We have our same argument that pH is the master independent variable, and a change in pH would also lead to a change in predicted $\{M_{OH}^{+2}\}$ and $\{M_{CO3}^{+2}\}$, and again, we will not choose that action.
2. A decrease in the activity of the metal-hydroxide solid phase would render its activity to be less than unity. We know that the activity of the solid phase must be unity, if present. Then, if the activity must be less than unity to satisfy the equilibrium condition, the metal-hydroxide solid phase cannot be present.
3. Were we to adjust the activity of the free metal ion, we would find that to re-establish equilibrium its value would necessarily be that predicted as $\{M_{OH}^{+2}\}$.
4. Were we then to use $\{M_{OH}^{+2}\}$ to compute Q_{MCO3}, we would find that $Q_{MCO3} / \beta_{S.MCO3} < 1$ and to satisfy Le Chatelier's principle, the activity of the metal carbonate solid phase would need to increase from its value of unity, which is of course impossible.

Thus, we would assert that when we predict the condition $\{M_{CO3}^{+2}\} < \{M_{OH}^{+2}\}$ the metal carbonate solid phase exerts control of the activity of the free metal ion and hence of the solubility of the metal.

The results of the foregoing allow us to make a simple decision. At any given pH, when we predict the activities of the free metal ion based on the presence of the metal hydroxide and the metal carbonate solid phases, we would choose the lower of the two values as the controlling value. The solid phase yielding the lower value is the controlling solid phase. In theory, we could attain the result that $\{M_{CO3}^{+2}\} = \{M_{OH}^{+2}\}$. Under such a set of conditions, the two solid phases could coexist. Let us apply this concept of the controlling solid phase in the precipitation of zinc from aqueous solution.

Example 11.11 Consider that an aqueous solution contains 0.005 M total carbonate and 0.0001 M total zinc. Background major ions are present such that the ionic strength is 0.01 M. Consider that the ionic strength would remain constant throughout the adjustment of pH to precipitate zinc either as the hydroxide or the carbonate solid. Investigate the behavior of the zinc-hydroxide and zinc-carbonate systems as the pH would be adjusted from the initial value, say 6 or so, to a final value that might approach 10.

Once we collect our information including equilibrium constants (again using the larger value of $\beta_{S.ZnOH2}$), initial conditions and known parameters, we write a concise set of functions that allow the definition of the total zinc concentration as a function of the solution pH:

$$H(pH) := 10^{-pH} \qquad CO3(pH) := \frac{C_{Tot.CO3.i}}{\beta_{H2CO3} \cdot H(pH)^2 + \frac{\beta_{HCO3} \cdot H(pH)}{\gamma_1} + \frac{1}{\gamma_2}}$$

$$Zn_{OH}(pH) := \left[\beta_{S.ZnOH2} \cdot \left(\frac{K_W}{H(pH)}\right)^2\right]^{-1} \qquad Zn_{CO3}(pH) := \left(\beta_{S.ZnCO3} \cdot CO3(pH)\right)^{-1}$$

$$Zn(pH) := \min(Zn_{OH}(pH), Zn_{CO3}(pH))$$

$$C_{Tot.Zn}(pH) := Zn(pH)\cdot\left(\frac{1}{\gamma_2} + \frac{\beta_{ZnOH}\cdot K_W}{H(pH)\cdot\gamma_1} + \frac{\beta_{ZnOH2}\cdot K_W^{\,2}}{H(pH)^2} \cdots + \frac{\beta_{ZnOH3}\cdot K_W^{\,3}}{H(pH)^3\cdot\gamma_1} + \frac{\beta_{ZnOH4}\cdot K_W^{\,4}}{H(pH)^4\cdot\gamma_2}\right)$$

We could have written:

$$OH(pH) := \frac{K_W}{H(pH)}$$

and used this in the zinc-hydroxide solid formation and zinc mole balance relations. Further, we could have written:

$$\alpha_2(pH) := \left(\beta_{H_2CO_3}\cdot H(pH)^2 + \frac{1}{\gamma_1}\beta_{HCO_3}\cdot H(pH) + \frac{1}{\gamma_2}\right)^{-1},$$

$$CO_3(pH) := \alpha_2(pH)\cdot C_{Tot.CO_3.i}$$

and used these in the carbonate mole balance, resulting in perhaps a more representative set of nine rather than six functional relations.

We have employed the min() function to select for the value of the zinc ion activity consistent with the controlling solid phase and to ensure that predicted values from solid formation equilibria are consistent with the initial total zinc abundance. A plot of the two functions yielding predictions of zinc ion activity, the controlling zinc ion activity and the predicted total aqueous zinc concentration is presented in Figure E11.11.1. Although we likely could use algebra to find the pH at which zinc would first precipitate, we choose to employ a short given-find block. In order for MathCAD's given-find solver to find the pH of first zinc solid formation, we need to remove the total zinc constraints from the Zn(pH) and $C_{Tot.Zn}(pH)$ relations:

$pH := 6.5$ Given $C_{Tot.Zn}(pH) = C_{Tot.Zn.i}$ $pH_p := Find(pH) = 6.954$

We also may compute the pH at which the zinc hydroxide and zinc carbonate would exert simultaneous control of the zinc ion abundance:

$pH := 8$ Given $Zn_{OH}(pH) = Zn_{CO3}(pH)$ $pH_{equiv} := Find(pH) = 8.511$

To this point, the analyses and the associated plots are based on the presumption that the total carbonate abundance in the aqueous solution is constant. Unfortunately,

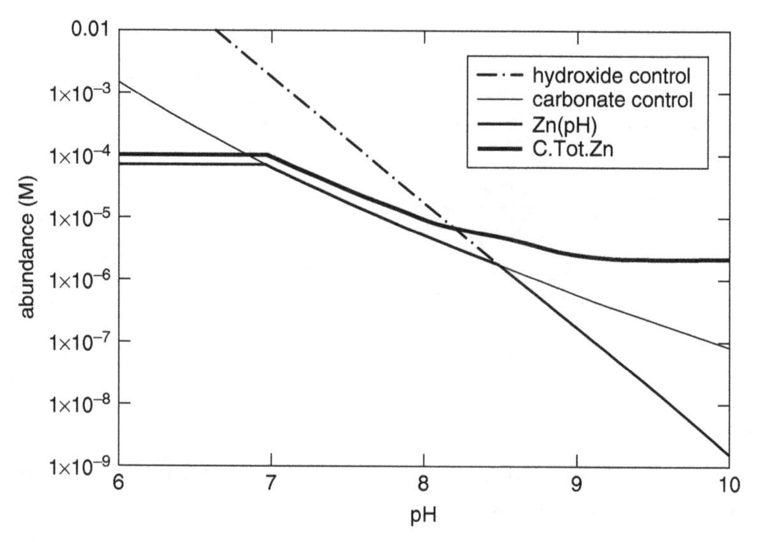

FIGURE E11.11.1 *A plot free zinc ion and total soluble zinc as predicted by equilibrium with zinc hydroxide and calcite.*

this is not the case. Zinc would begin to precipitate as zinc carbonate at pH_p. As zinc carbonate precipitates at pH above pH_p, the abundance of total carbonate, and hence the activity of the carbonate ion are affected by the quantity of zinc carbonate that has precipitated. Then, correspondingly, the predicted zinc ion activity based on the initial abundance of carbonate would be in error. Analysis of this condition cannot be accomplished using a single, explicit function. We must adjust the carbonate ion activity by the quantity of carbonate removed from the solution with zinc. The stoichiometry is 1:1 so we employ the change in the total zinc abundance expressed as the current total zinc abundance less the initial. We then rewrite the total zinc abundance using the revised relation for carbonate ion activity. We most conveniently employ a given find solve block. We can solve the system at any pH between pH_p and pH_{equiv}. The general solve block is written here and a corresponding function is defined for use later in this effort:

$$\text{Given} \quad CO32 = \frac{C_{Tot.CO3.i} + (CTotZn - C_{Tot.Zn.i})}{\beta_{H2CO3} \cdot H_a^2 + \frac{\beta_{HCO3} \cdot H_a}{\gamma_1} + \frac{1}{\gamma_2}}$$

$$CTotZn = \frac{1}{\beta_{S.ZnCO3} \cdot CO32} \cdot \left(\frac{1}{\gamma_2} + \frac{\beta_{ZnOH} \cdot K_W}{H_a \cdot \gamma_1} + \frac{\beta_{ZnOH2} \cdot K_W^2}{H_a^2} \cdots + \frac{\beta_{ZnOH3} \cdot K_W^3}{H_a^3 \cdot \gamma_1} + \frac{\beta_{ZnOH4} \cdot K_W^4}{H_a^4 \cdot \gamma_2} \right)$$

$$Sol_{Zn.CO3}(H_a, CO32, CTotZn) := \text{Find}(CO32, CTotZn)$$

We can compute the total zinc abundance as a function of the pH of the solution when pH is above pH_p. However, we do not precisely know the upper limit of the pH range of applicability.

We realize that since the total carbonate abundance is reduced by precipitation of zinc carbonate, the pH of equivalence may be different from the prediction based on the explicit relation. We add the hydronium activity as an unknown and, within the solve block equate the RHS expressions from the two solubility equilibria. The solution then includes the hydronium ion activity at which the formation of a zinc hydroxide solid phase will begin:

$$\begin{pmatrix} CO32 \\ CTotZn \\ H_a \end{pmatrix} := \begin{pmatrix} 1 \times 10^{-5} \\ 1 \times 10^{-6} \\ 1 \times 10^{-9} \end{pmatrix} \qquad \text{Given} \qquad \frac{1}{\beta_{S.ZnCO3} \cdot CO32} = \frac{H_a^{\ 2}}{\beta_{S.ZnOH2} \cdot K_W^{\ 2}}$$

$$CO32 = \frac{C_{Tot.CO3.i} + (CTotZn - C_{Tot.Zn.i})}{\beta_{H2CO3} \cdot H_a^{\ 2} + \frac{\beta_{HCO3} \cdot H_a}{\gamma_1} + \frac{1}{\gamma_2}}$$

$$CTotZn = \frac{1}{\beta_{S.ZnCO3} \cdot CO32} \cdot \left(\frac{1}{\gamma_2} + \frac{\beta_{ZnOH} \cdot K_W}{H_a \cdot \gamma_1} + \frac{\beta_{ZnOH2} \cdot K_W^{\ 2}}{H_a^{\ 2}} \cdots + \frac{\beta_{ZnOH3} \cdot K_W^{\ 3}}{H_a^{\ 3} \cdot \gamma_1} + \frac{\beta_{ZnOH4} \cdot K_W^{\ 4}}{H_a^{\ 4} \cdot \gamma_2} \right)$$

$$\begin{pmatrix} CO32 \\ CTotZn \\ H_a \end{pmatrix} := \text{Find}(CO32, CTotZn, H_a) = \begin{pmatrix} 6.387 \times 10^{-5} \\ 4.955 \times 10^{-6} \\ 3.143 \times 10^{-9} \end{pmatrix} \qquad pH_a := -\log(H_a) = 8.503$$

Our revised estimate of the pH at which the precipitation of zinc hydroxide will begin is then 8.503. Then, for pH beginning at 8.503, we have the potential for the existence of both zinc hydroxide and zinc carbonate. If we visualize the system at pH infinitesimally above 8.503, we would conclude that some infinitesimal quantity of zinc hydroxide would have formed, but also that a finite quantity of zinc carbonate solid would remain.

In the case that both solid phases would be present, each exerting its equilibrium constraints upon the system, something must give. Since we are using pH (and hence $\{OH^-\}$) as a master independent variable and since $\{Zn^{+2}\}$ can have but a sole value, the carbonate ion activity and hence the total carbonate abundance must be variable. We invoke the solve block from before with one alteration—that the total carbonate abundance returns to the initial value and compute the pH at which zinc hydroxide solid will gain full control of the zinc solubility:

$$CO32 = \frac{C_{Tot.CO3.i}}{\beta_{H2CO3} \cdot H_a^{\ 2} + \frac{\beta_{HCO3} \cdot H_a}{\gamma_1} + \frac{1}{\gamma_2}}$$

$$\begin{pmatrix} CO32 \\ CTotZn \\ H_a \end{pmatrix} := Find(CO32, CTotZn, H_a) = \begin{pmatrix} 6.633 \times 10^{-5} \\ 4.855 \times 10^{-6} \\ 3.084 \times 10^{-9} \end{pmatrix} \qquad pH_{equiv} := -\log(H_a) = 8.511$$

We are rewarded with a result matching that obtained by considering the total carbonate abundance to be constant. The previously shown zinc solubility versus pH plot is then valid at pH above pH_{equiv} as a consequence of zinc hydroxide solid phase control of zinc ion activity. We can employ the first of the two solve blocks employed above between pH_p and pH_a. Between pH_a and pH_{equiv}, the entire quantity of zinc carbonate previously precipitated is redissolved and reprecipitated as zinc hydroxide solid. Then at pH above pH_{equiv}, zinc hydroxide exerts control of zinc solubility.

To summarize, let us visualize this process as a titration of the solution with sodium or potassium hydroxide. At pH_p zinc carbonate solid would form as a consequence of the increased pH (translated directly into increased carbonate ion

$$\begin{pmatrix} CO32_0 \\ CTotZn_0 \end{pmatrix} := Sol_{Zn.CO3}(H_a, CO32, CTotZn) = \begin{pmatrix} 1.546 \times 10^{-6} \\ 1 \times 10^{-4} \end{pmatrix}$$

```
(n, pH, CO32, CTotZn) := | i ← 0
                         | ΔpH ← 0.005
                         | pH_0 ← pH_p
                         | while pH_i < pH_a - ΔpH
                         |    | i ← i + 1
                         |    | pH_i ← pH_(i-1) + ΔpH
                         |    | H ← 10^(-pH_i)
                         |    | (CO32_i, CTotZn_i) ← (CO32_(i-1), CTotZn_(i-1))
                         |    | (CO32_i, CTotZn_i) ← Sol_Zn.CO3(H, CO32_i, CTotZn_i)
                         |    | n ← i
                         | return (n, pH, CO32, CTotZn)
```

FIGURE E11.11.2 Screen capture of looping code for generating zinc and carbonate speciation versus pH for precipitation of zinc using hydroxide reagent.

abundance). As more hydroxide is added, raising the pH, additional zinc carbonate is precipitated. Then, at pH_a zinc hydroxide solid would begin to form, providing dual control of zinc solubility. At any given pH in this region ($pH_a \le pH \le pH_{equiv}$) both the hydronium and free zinc ion activities are fixed, thus the carbonate ion abundance must provide the necessary degree of freedom. Thus, as pH is increased from pH_a to pH_{equiv}, the zinc carbonate solid phase would redissolve with corresponding reprecipitation as zinc hydroxide. Then above pH_{equiv} zinc hydroxide would be the controlling solid phase. We employed the first of the two implicit solutions as a function and solved for the solubility of zinc between pH_p and pH_a. We wrote a looping program, shown in Figure E11.11.2, to step through pH values from pH_p to pH_a and stored resultant pH values, carbonate ion activities, and total zinc concentrations in three vectors (pH, CO_3^{2}, and $C_{tot.Zn}$). The maximum relative error of the predicted explicit relative to the predicted implicit total zinc solubility was computed:

$$\%diff_i := \overrightarrow{\left(\frac{C_{Tot.Zn}(pH_i) - CTotZn_i}{CTotZn_i}\cdot 100\right)}$$

The relative error and the two traces of total dissolved zinc are plotted for visual inspection in Figure E11.11.3. The two total dissolved zinc traces are almost indistinguishable, with the implicit solution predicting a slightly higher total dissolved zinc concentration. The relative difference approaches 2%. If the initial total carbonate is 2.5×10^{-3} M the %difference increases to a maximum value of 3.5%.

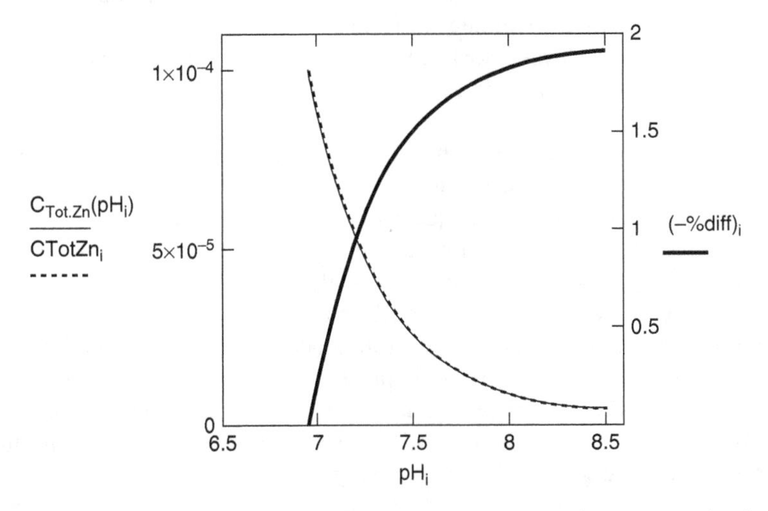

FIGURE E11.11.3 *A plot of predicted total dissolved zinc comparing an approximation that neglects the depletion of aqueous carbonate abundance with precipitation of zinc carbonate ($C_{Tot.Zn}$(pH)) with a prediction accounting for the reduction in aqueous carbonate abundance. The relative error increases with decreasing initial carbonate abundance.*

As the total carbonate abundance is decreased below 0.8×10^{-3} M, precipitation of a zinc carbonate solid becomes impossible.

We certainly could employ the BNC principles addressed previously to compute the necessary additions of base to attain each point-wise pH value, but such an exercise is left to the student.

We move away from Example 11.11 with greater understanding of the concept of solid phase control of metal ion activity. We understand that the precipitation of a metal carbonate can alter the abundance of carbonate in the aqueous solution and hence the degree of control exerted by the carbonate solid. Many industrial treatment processes rely upon precipitation of metals from solution prior to discharge. Some workers have published suggestions that precipitation as the carbonate could be more economical than precipitation as the hydroxide. Such claims must be validated by analyses such as that conducted via Example 11.11. Zinc can be quite conveniently removed in the mildly alkaline pH range as zinc carbonate. Once the system enters the pH range of carbonate solid control, the precipitation of zinc as the carbonate would be stoichiometric relative to the solid-forming reaction, considering that a reagent such as sodium carbonate would act as the precipitant. We can conduct analyses such as that accomplished for zinc in Example 11.11 for any metal that produces a carbonate solid.

We should consider one additional application of the controlling solid phase—that of a metal existing in a natural system within which a calcium-carbonate-bearing solid phase would dominate the solubility of calcium and carbonate. Such a system might be a heap leach pad employed to process ore that is a mixture of acid-forming (sulfide) and calcareous minerals. The sulfide-bearing rock would tend, through the oxidation of sulfide to sulfate, to furnish hydronium to the system and thus lower the pH from the level associated with equilibration of aqueous solution with calcite. The carbonate-bearing minerals would tend to buffer the resultant solution.

Example 11.12 Consider an abandoned heap leach pad that was used in a zinc recovery operation. The processed ore was a mixture of sulfide and carbonate-bearing rock, thus the pH of the aqueous solution resident in the pad might range anywhere from the mildly acid to the strongly alkaline range. Consider that the carbonate-bearing rock may be modeled as calcite. Concern would exist regarding the zinc content of water that might leach from the pad as a consequence of incidence of rainwater on the top of the pad and subsequent percolation through the porous media of the pad. Just to keep things simple: let us consider that the mineral composition of solid phase contained within the pad is spatially invariant. We consider that pore water is equilibrated with the solid phase within the pad and that infiltrating water displaces pore water to become leachate.

Let us first compute the upper bound of the pH of the system, as if there were no sulfide-bearing rock. We need to check whether zinc carbonate or zinc hydroxide will be the controlling solid phase. We first postulate that a solid phase zinc carbonate exists along with the calcite. For this system, we would have seven unknowns: free zinc, free calcium ion, and carbonate ion activities; total zinc, total calcium, and total carbonate abundances and the activity of hydronium (pH). We thus would write seven independent equations: (1–3) mole balances for total zinc, total calcium, and total carbonate, (4–5) equilibria for calcium and zinc ion activities, (6) a mole balance equating the changes in total zinc and total calcium to the change in total carbonate, and (7) a proton balance. We have written the first five in completing the immediately preceding examples and for clarity we have repeated the base-level relations here:

$$C_{Tot.Zn} = C_{Zn} + C_{ZnOH} + C_{ZnOH2} + C_{ZnOH3} + C_{ZnOH4}$$

$$C_{Tot.Ca} = C_{Ca} + C_{CaOH} + C_{CaCO3} + C_{Ca.HCO3}$$

$$C_{Tot.CO3} = C_{H2CO3} + C_{HCO3} + C_{CO3} + C_{CaHCO3} + C_{CaCO3}$$

$$Ca = \left(\beta_{S.CaCO3} \cdot CO3\right)^{-1} \qquad Zn = \left(\beta_{S.ZnCO3} \cdot CO3\right)^{-1}$$

We used formation equilibria to write equations (1–3) in the variables $\{Zn^{+2}\}$, $\{Ca^{+2}\}$, and $\{CO_3^{=}\}$ and then used equations (4 and 5) to write $\{Zn^{+2}\}$ and $\{Ca^{+2}\}$ using $\{CO_3^{=}\}$ as the master dependent variable for equations (1–3). For the sixth equation, if zinc carbonate controls the solubility of zinc we need to consider that for every mole of zinc and for every mole of calcium that dissolve into the solution, a mole of carbonate must also dissolve:

$$\Delta C_{tot.Zn} + \Delta C_{Tot.Ca} = \Delta C_{tot.CO3}$$

Then, since both the zinc and calcium systems join the carbonate system as participants in proton transfer reactions we need to include them in the overall proton balance. Most conveniently, we use $Ca(OH)_2^0$, $Zn(OH)_2^0$ and CO_3^{-2} (hence also $CaCO_3^0$ and $ZnCO_3^0$) as our reference species alongside water:

$$\Delta C_{OH} = 2 \cdot \Delta C_{H2CO3} + \Delta C_{HCO3} + 2 \cdot \Delta C_{Ca} + \Delta C_{CaOH} + \Delta C_{CaHCO3} \cdots$$
$$+ 2 \cdot \Delta C_{Zn} + \Delta C_{ZnOH} - \Delta C_{ZnOH3} - 2 \cdot \Delta C_{ZnOH4}$$

Since rainwater is our initial condition, we know from a previous example that the initial total carbonate, carbonic acid, and bicarbonate abundances are about 1.5×10^{-5}, 1.3×10^{-5} and 2×10^{-6} M, respectively, and that initial abundances of hydronium and hydroxide are about $10^{-5.6}$ and $10^{-8.4}$ M, respectively. All other initial concentrations save those of $CaCO_3^0$ and $ZnCO_3^0$ are zero. $ZnCO_3^0$ would undoubtedly exist, but apparently at such low abundance that we do not find its formation constant among those in Table A.2.

We use the three mole balances, all written with the carbonate ion as the master dependent variable in the equation relating the changes in total zinc, total calcium, and total carbonate abundances. The proton and carbonate activities are the master dependent variables and we write the abundances for all other species using these unknowns, using the corresponding equilibria. The seven equations are then reduced to two. We solve these two equations using the given-find of Figure E11.12.1 and use the result to predict the activity of the free zinc based on carbonate control. We compute the corresponding free zinc ion activity based on hydroxide control, at the pH of the system, and compare it with Zn_{CO3}:

$$Zn_{OH} := \left[\beta_{S.ZnOH2} \cdot \left(\frac{K_W}{H}\right)^2\right]^{-1} = 1.335 \times 10^{-10}$$

Zinc hydroxide would be the solid phase exerting control of zinc solubility at alkaline pH in this heap leach pad system.

$$\begin{pmatrix} H \\ CO3 \end{pmatrix} := \begin{pmatrix} 10^{-10} \\ 10^{-3} \end{pmatrix} \quad \text{given}$$

$$\left[\frac{\left(\frac{1}{\gamma_2} + \frac{\beta_{ZnOH} \cdot K_W}{H \cdot \gamma_1} + \frac{\beta_{ZnOH2} \cdot K_W^2}{H^2} \cdots + \frac{\beta_{ZnOH3} \cdot K_W^3}{H^3 \cdot \gamma_1} + \frac{\beta_{ZnOH4} \cdot K_W^4}{H^4 \cdot \gamma_2}\right)}{(\beta_{S.ZnCO3} \cdot CO3)} \cdots + \frac{\left(\frac{1}{\gamma_2} + \frac{\beta_{CaOH} \cdot K_W}{H \cdot \gamma_1}\right)}{(\beta_{S.CaCO3} \cdot CO3)}\right] = \left[CO3 \cdot \left(\beta_{H2CO3} \cdot H^2 \cdots + \frac{\beta_{HCO3} \cdot H}{\gamma_1} + \frac{1}{\gamma_2}\right) \cdots + -C_{Tot.CO3.i}\right]$$

$$\left[\begin{array}{l} CO3 \cdot \left[2 \cdot \left(\beta_{H2CO3} \cdot H^2\right) + \frac{\beta_{HCO3} \cdot H}{\gamma_1}\right] - 2 \cdot C_{H2CO3.i} - C_{HCO3.i} \cdots \\ + (\beta_{S.ZnCO3} \cdot CO3)^{-1} \cdot \left(\frac{2}{\gamma_2} + \frac{\beta_{ZnOH} \cdot K_W}{H \cdot \gamma_1} \cdots + \frac{\beta_{ZnOH3} \cdot K_W^3}{H^3 \cdot \gamma_1} - 2 \cdot \frac{\beta_{ZnOH4} \cdot K_W^4}{H^4 \cdot \gamma_2}\right) \cdots \\ + (\beta_{S.CaCO3} \cdot CO3)^{-1} \cdot \left(\frac{2}{\gamma_2} + \frac{\beta_{CaOH} \cdot K_W}{H \cdot \gamma_1}\right) + \frac{H}{\gamma_1} - C_{H.i} \end{array}\right] = \frac{K_W}{H \cdot \gamma_1} - C_{OH.i}$$

$$\begin{pmatrix} H \\ CO3 \end{pmatrix} := \text{find}(H, CO3) = \begin{pmatrix} 2.903 \times 10^{-11} \\ 8.742 \times 10^{-4} \end{pmatrix} \qquad pH := -\log(H) = 10.537$$

$$Zn_{CO3} := (\beta_{S.ZnCO3} \cdot CO3)^{-1} = 1.144 \times 10^{-7}$$

FIGURE E11.12.1 *A given-find solve block for determining the pH and carbonate speciation of rain water equilibrated simultaneously with zinc hydroxide and calcite.*

In the absence of zinc carbonate solid, the pH of the system would be controlled entirely by the calcium-carbonate system. We can revise our prediction of the upper bound of the pH by removing the mole balance on zinc and the zinc carbonate formation reaction from our set of seven equations. The resulting pair, with $\{CO_3^{=}\}$ and $\{H^+\}$ again as the master dependent variables, is somewhat simplified and we employ the solve block, shown in Figure E11.12.2, to obtain the predicted bounding pH value. A check corroborates that at this slightly lower pH value, zinc hydroxide remains the controlling solid phase:

$$Zn_{CO3} := (\beta_{S.ZnCO3} \cdot CO3)^{-1} = 1.805 \times 10^{-6}$$

$$Zn_{OH} := \left[\beta_{S.ZnOH2} \cdot \left(\frac{K_W}{H}\right)^2\right]^{-1} = 2.666 \times 10^{-10}$$

The lower bound of the system pH would be dictated by the abundance of sulfide-bearing minerals among the ground-up ore in the leach pad and the extent to which the sulfide would be converted to sulfate. Given long periods between infiltration events, we would expect these sulfide to sulfate reactions to release significant quantities of protons, perhaps rendering the pH well into the acid range. We might decide to employ pH 5.0 or so as the lower pH bound.

In order to analyze the system within our pH bounds, we employ pH as a master independent variable, alleviating the necessity for the proton balance, since for each pH, the hydronium ion activity is known. The system is reduced to the mole balance equating changes in total zinc and calcium abundances to the change in total carbonate abundance. We rearrange the relation into the second-order

$$\begin{pmatrix} H \\ CO3 \end{pmatrix} := \begin{pmatrix} 10^{-12} \\ 10^{-5} \end{pmatrix} \qquad \text{Given}$$

$$(\beta_{S.CaCO3} \cdot CO3)^{-1} \cdot \left(\frac{1}{\gamma_2} + \frac{\beta_{CaOH} \cdot K_W}{H \cdot \gamma_1}\right) = \left[CO3 \cdot \left(\beta_{H2CO3} \cdot H^2 + \frac{\beta_{HCO3} \cdot H}{\gamma_1} + \frac{1}{\gamma_2}\right) \cdots \atop + -C_{Tot.CO3.i}\right]$$

$$\left[CO3 \cdot \left[2 \cdot (\beta_{H2CO3} \cdot H^2) + \frac{\beta_{HCO3} \cdot H}{\gamma_1}\right] - 2 \cdot C_{H2CO3.i} - C_{HCO3.i} \cdots \atop + (\beta_{S.CaCO3} \cdot CO3)^{-1} \cdot \left(\frac{2}{\gamma_2} + \frac{\beta_{CaOH} \cdot K_W}{H \cdot \gamma_1}\right) + \frac{H}{\gamma_1} - C_{H.i}\right] = \frac{K_W}{H \cdot \gamma_1} - C_{OH.i}$$

$$\begin{pmatrix} H \\ CO3 \end{pmatrix} := \text{Find}(H, CO3) = \begin{pmatrix} 4.101 \times 10^{-11} \\ 5.539 \times 10^{-5} \end{pmatrix} \qquad pH := -\log(H) = 10.387$$

FIGURE E11.12.2 *Given-find solve block for finding the pH and speciation of rain water equilibrated with calcite.*

polynomial in carbonate ion activity and employ the quadratic formula to obtain the function describing the activity of carbonate as a function of pH:

$$H(pH) := 10^{-pH}$$

$$a(pH) := \beta_{H2CO3} \cdot H(pH)^2 + \frac{\beta_{HCO3} \cdot H(pH)}{\gamma_1} + \frac{1}{\gamma_2} \qquad b(pH) := -C_{Tot.CO3.i}$$

$$c(pH) := -\left[\frac{\left(\frac{1}{\gamma_2} + \frac{\beta_{ZnOH} \cdot K_W}{H(pH) \cdot \gamma_1} + \frac{\beta_{ZnOH2} \cdot K_W^2}{H(pH)^2} + \frac{\beta_{ZnOH3} \cdot K_W^3}{H(pH)^3 \cdot \gamma_1} + \frac{\beta_{ZnOH4} \cdot K_W^4}{H(pH)^4 \cdot \gamma_2}\right)}{\beta_{S.ZnCO3}} \cdots + \frac{\left(\frac{1}{\gamma_2} + \frac{\beta_{CaOH} \cdot K_W}{H(pH) \cdot \gamma_1}\right)}{\beta_{S.CaCO3}}\right]$$

$$CO3(pH) := \frac{-b(pH) + \sqrt{b(pH)^2 - 4 \cdot a(pH) \cdot c(pH)}}{2 \cdot a(pH)}$$

We write functions for $\{Zn^{+2}\}$ based on zinc hydroxide and zinc carbonate control of the free zinc activity:

$$Zn_{CO3}(pH) := \left(\beta_{S.ZnCO3} \cdot CO3(pH)\right)^{-1} \qquad Zn_{OH}(pH) := \left[\beta_{S.ZnOH2} \cdot \left(\frac{K_W}{H(pH)}\right)^2\right]^{-1}$$

Then the plot of these two relations against pH in Figure E11.12.3 allows us to discern the pH regions in which zinc solubility will be controlled by the respective

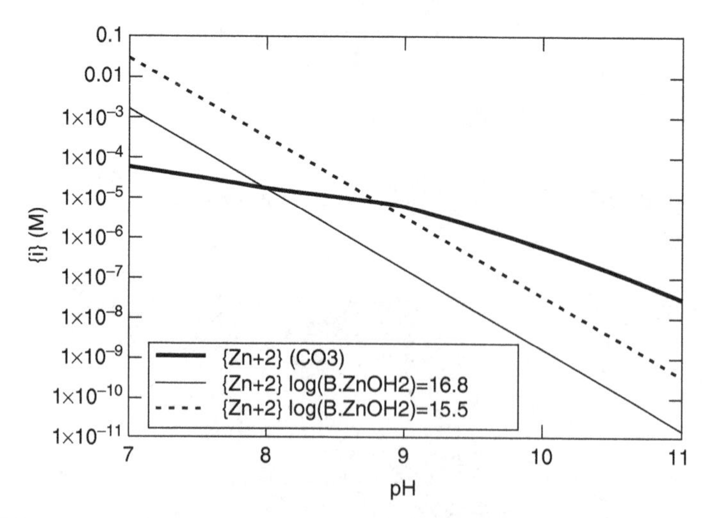

FIGURE E11.12.3 *A plot of predicted zinc ion activity in rain water infiltrating through a heap leach pad containing calcite, considering control of solubility by zinc hydroxide and zinc carbonate.*

solid phases. We have included both values of $\log \beta_{S.ZnOH2}$ from Table A.2. If we believe the larger value we would assert that $ZnCO_{3(s)}$ controls zinc solubility below pH ~7.9. If we believe the smaller value, we would assert that $ZnCO_{3(s)}$ controls zinc solubility below pH ~8.9. We observe that the predictions vary by an entire log unit (an order of magnitude in the hydronium abundance), which is not surprising since the larger value of the formation constant is about twenty times the smaller value.

We may complete an analysis such as that performed in Example 11.12 for any metal that forms a solid phase with the carbonate ion. We can also, with significant additional complexity of the solution, include multiple carbonate-forming metals in the analysis. Mole balances for each metal would be written, appropriate complex formation equilibria would be used to write each in terms of its free metal ion, and the metal-carbonate formation equilibrium would be employed to write the mole balance for the metal using $\{CO_3^=\}$ as the master dependent variable. The sum of the changes in metal abundances would be set to the change in the carbonate abundance. The pH would become a master independent variable and for each pH value the set of $\{M^{+2}\}_{CO_3}$ values would be computed. Another set of $\{M^{+2}\}_{OH}$ values would be computed based on the metal hydroxide formation equilibrium. At the stated pH value, comparison would allow removal of any of the hydroxide-controlled metals from the mole balance equating changes in metal abundance to carbonate abundance, with computation of another set of $\{M^{+2}\}_{CO_3}$ values for comparison. Once this iterative process was completed at the state pH value, the process, with the remaining metal-carbonate forming metals would be repeated at the next targeted pH value. Certainly, a piece of programming could be created to step through the pH values from low to high or from high to low, which with implementation of the min() function would yield a set of controlling $\{M^{+2}\}$ values ordered with pH.

An additional system of interest is a class of engineered systems for the control of total metals using carbonate as a precipitative reagent. The chemistry of each metal, of course, determines the capability for control of solubility by the metal carbonate. Transition and heavy metals that form divalent cations are the best targets for control by carbonate precipitation. Lead is one of the most toxic of metals, so much so that the EPA maximum contaminant level for lead in drinking water is set at zero. Obviously, for a contaminated water, zero concentration is impossible to attain, so the better definition would be a level that is below the limit of detection. Let us examine the control of the lead abundance in aqueous solution using precipitation as lead carbonate as the means for control.

Example 11.13 Consider an aqueous solution of initial pH 5.0 that contains 30 ppm_m total lead. The solution also contains 0.001 M total carbonate. In order that we can focus upon the carbonate precipitation process we will consider that significant complexing ligands are absent such that hydrolysis is the sole process enhancing the solubility of lead beyond the abundance of the free metal ion. Since zero concentration is impossible to attain, consider a target total lead level of 1 ppb_m. Determine the behavior of the system, most particularly the carbonate dose and pH value at which the target lead abundance can be attained, if possible.

Before we can write the functions for the activity of the free lead ion, as controlled by lead carbonate solid, we must define the pH dependence of the carbonate ion. Total carbonate is constant until carbonate is added or a lead carbonate solid phase would form as a consequence of dosing with carbonate salt:

$$C_{Tot.M.i} := 30 \quad C_{Tot.CO3.i} := 0.001 \quad pH_i := 5.0 \quad C_{Tot.Pb.i} := \frac{C_{Tot.M.i}}{MW_{Pb}} = 1.448 \times 10^{-4}$$

$$H(pH) := 10^{-pH} \qquad CO3(pH) := \frac{C_{Tot.CO3.i}}{\beta_{H2CO3} \cdot H(pH)^2 + \frac{\beta_{HCO3} \cdot H(pH)}{\gamma_1} + \frac{1}{\gamma_2}}$$

We write the functions for the lead ion activity based on carbonate or hydroxide control:

$$Pb_{CO3}(pH) := \left(\beta_{S.PbCO3} \cdot CO3(pH)\right)^{-1} \qquad Pb_{OH}(pH) := \left[\beta_{S.PbOH2} \cdot \left(\frac{K_W}{H(pH)}\right)^2\right]^{-1}$$

We then write functions for total soluble lead based on carbonate or hydroxide control of solubility:

$$C_{Tot.Pb.CO3}(pH) := Pb_{CO3}(pH) \cdot \left(\frac{1}{\gamma_2} + \frac{\beta_{PbOH} \cdot K_W}{H(pH) \cdot \gamma_1} + \frac{\beta_{PbOH2} \cdot K_W^2}{H(pH)^2} \dots + \frac{\beta_{PbOH3} \cdot K_W^3}{H(pH)^3 \cdot \gamma_1}\right)$$

$$C_{Tot.Pb.OH}(pH) := Pb_{OH}(pH) \cdot \left(\frac{1}{\gamma_2} + \frac{\beta_{PbOH} \cdot K_W}{H(pH) \cdot \gamma_1} + \frac{\beta_{PbOH2} \cdot K_W^2}{H(pH)^2} \dots + \frac{\beta_{PbOH3} \cdot K_W^3}{H(pH)^3 \cdot \gamma_1}\right)$$

Now we can determine which solid phase would precipitate first, should we adjust the pH upward with a reagent such as sodium or potassium hydroxide. Rather than plotting each the total soluble lead functions versus pH, here we use two very simple solve blocks:

$$pH := 5 \quad \text{Given} \quad C_{Tot.Pb.i} = C_{Tot.Pb.CO3}(pH) \qquad pH_{p.CO3} := \text{Find}(pH) = 5.325$$

$$\text{Given} \quad C_{Tot.Pb.i} = C_{Tot.Pb.OH}(pH) \qquad pH_{p.OH} := \text{Find}(pH) = 9.053$$

We find that indeed, lead carbonate would be the first lead solid to precipitate.

Since we intend to use carbonate as a reagent (as the sodium or potassium salt) our function written for carbonate ion activity based on the initial carbonate abundance is invalid at the first addition of reagent. We see that the pH must be raised to 5.325 in order to form a lead carbonate precipitate, were the carbonate abundance to remain constant. Complicating matters is the fact that the carbonate abundance is affected positively by the addition of carbonate and thereafter negatively by the formation of lead carbonate precipitate. We realize that, by dosing with a carbonate salt, the pH at which the lead carbonate solid would begin to form will actually be lower than predicted earlier, since we are increasing the carbonate abundance by the addition of carbonate reagent.

Let us then determine the pH and carbonate dose at which the lead carbonate will first form. We have four master dependent variables: $\{H^+\}$, $\{CO_3^=\}$, $\{Pb^{+2}\}$, and ΔCO_3, where ΔCO_3 is the dose ($mol_{CO_3} / L_{solution}$) of the carbonate reagent necessary to raise the pH and carbonate abundance to levels commensurate with the formation of a precipitate. We then require four equations:

1. Mole balance on carbonate, solved to yield the carbonate ion on the LHS, with total carbonate abundance as the sum of the initial total carbonate and the carbonate dose.
2. Formation equilibrium for lead carbonate solid.
3. Mole balance on lead, with total lead equal to the initial abundance.
4. Proton balance, to yield the pH at which an infinitesimal quantity of lead carbonate would form.

For the proton balance, we most conveniently use $CO_3^=$, $Pb(OH)_2^0$, and of course water as the reference species, ensuring that $PbCO_{3(s)}$ is at the same proton status as our reference species.

We must define the initial speciation in order that the changes in specie abundances may be used in the proton balance, using the stated ionic strength to compute activity coefficients:

$$H_i := 10^{-pH_i} \qquad Pb_i := \frac{C_{Tot.Pb.i}}{\left(\dfrac{1}{\gamma_2} + \dfrac{\beta_{PbOH} \cdot K_W}{H_i \cdot \gamma_1} + \dfrac{\beta_{PbOH2} \cdot K_W^2}{H_i^2} + \dfrac{\beta_{PbOH3} \cdot K_W^3}{H_i^3 \cdot \gamma_1}\right)}$$

$$\begin{pmatrix} C_{Pb.i} \\ C_{PbOH.i} \\ C_{PbOH2.i} \\ C_{PbOH3.i} \end{pmatrix} := \begin{bmatrix} \dfrac{Pb_i}{\gamma_2} \\ \dfrac{\beta_{PbOH}}{\gamma_1} \cdot Pb_i \cdot \dfrac{K_W}{H_i} \\ \beta_{PbOH2} \cdot Pb_i \cdot \left(\dfrac{K_W}{H_i}\right)^2 \\ \dfrac{\beta_{PbOH3}}{\gamma_1} \cdot Pb_i \cdot \left(\dfrac{K_W}{H_i}\right)^3 \end{bmatrix} \qquad \begin{pmatrix} C_{H.i} \\ C_{OH.i} \\ C_{H2CO3.i} \\ C_{HCO3.i} \end{pmatrix} := \begin{pmatrix} \dfrac{H_i}{\gamma_1} \\ \dfrac{K_W}{H_i \cdot \gamma_1} \\ \beta_{H2CO3} \cdot H_i^2 \cdot CO3_i \\ \dfrac{\beta_{HCO3} \cdot H_i \cdot CO3_i}{\gamma_1} \end{pmatrix}$$

$$\Sigma C_i := -\left[2 \cdot C_{H2CO3.i} + C_{HCO3.i} + 2 \cdot C_{Pb.i} + C_{PbOH.i} \cdots + (C_{H.i} - C_{OH.i} - C_{PbOH3.i})\right] = -2.253 \times 10^{-3}$$

Then we formulate and invoke the solve block:

$$\begin{pmatrix} H \\ CO3 \\ Pb \\ \Delta CO3_p \end{pmatrix} := \begin{pmatrix} 4 \cdot 10^{-6} \\ 8 \cdot 10^{-10} \\ 9 \cdot 10^{-5} \\ 2 \cdot 10^{-5} \end{pmatrix} \qquad \text{Given} \qquad CO3 = \frac{C_{Tot.CO3.i} + \Delta CO3_p}{\beta_{H2CO3} \cdot H^2 + \dfrac{\beta_{HCO3} \cdot H}{\gamma_1} + \dfrac{1}{\gamma_2}}$$

$$Pb = (\beta_{S.PbCO3} \cdot CO3)^{-1}$$

$$C_{Tot.Pb.i} = Pb \cdot \left(\frac{1}{\gamma_2} + \frac{\beta_{PbOH} \cdot K_W}{H \cdot \gamma_1} + \frac{\beta_{PbOH2} \cdot K_W^2}{H^2} + \frac{\beta_{PbOH3} \cdot K_W^3}{H^3 \cdot \gamma_1}\right)$$

$$2 \cdot \Delta CO3_p = -\left(2 \cdot \beta_{H2CO3} \cdot H^2 \cdot CO3 + \frac{\beta_{HCO3} \cdot H \cdot CO3}{\gamma_1} + 2 \cdot \frac{Pb}{\gamma_2} + \frac{\beta_{PbOH} \cdot Pb \cdot K_W}{H \cdot \gamma_1} \cdots + \frac{H}{\gamma_1} - \frac{K_W}{H \cdot \gamma_1} - \frac{\beta_{PbOH3} \cdot Pb \cdot K_W^3}{H^3 \cdot \gamma_1} + \Sigma C_i\right)$$

$$\begin{pmatrix} H_p \\ CO3_p \\ Pb_p \\ \Delta CO3_p \end{pmatrix} := \text{Find}(H, CO3, Pb, \Delta CO3_p) = \begin{pmatrix} 4.765 \times 10^{-6} \\ 8.445 \times 10^{-10} \\ 9.406 \times 10^{-5} \\ 1.361 \times 10^{-5} \end{pmatrix} \qquad pH_p := -\log(H_p) = 5.322$$

We observe that with but a small dose of carbonate reagent, the pH at which the lead carbonate precipitate would form is reduced (but only slightly) to 5.322. The result of this computation serves as the beginning point for the ensuing computations for lead solubility. We must determine the remainder of the speciation at the point of first lead carbonate precipitation:

$$Pb_p := \frac{C_{Tot.Pb.i}}{\left(\dfrac{1}{\gamma_2} + \dfrac{\beta_{PbOH} \cdot K_W}{H_p \cdot \gamma_1} + \dfrac{\beta_{PbOH2} \cdot K_W^2}{H_p^2} + \dfrac{\beta_{PbOH3} \cdot K_W^3}{H_p^3 \cdot \gamma_1}\right)} = 9.406 \times 10^{-5}$$

$$\begin{pmatrix} C_{Pb.p} \\ C_{PbOH.p} \\ C_{PbOH2.p} \\ C_{PbOH3.p} \end{pmatrix} := \begin{bmatrix} \dfrac{Pb_p}{\gamma_2} \\ \dfrac{\beta_{PbOH}}{\gamma_1} \cdot Pb_p \cdot \dfrac{K_W}{H_p} \\ \beta_{PbOH2} \cdot Pb_p \cdot \left(\dfrac{K_W}{H_p}\right)^2 \\ \dfrac{\beta_{PbOH3}}{\gamma_1} \cdot Pb_p \cdot \left(\dfrac{K_W}{H_p}\right)^3 \end{bmatrix} = \begin{pmatrix} 1.443 \times 10^{-4} \\ 4.383 \times 10^{-7} \\ 3.29 \times 10^{-11} \\ 0 \end{pmatrix}$$

$$\begin{pmatrix} C_{H2CO3.p} \\ C_{HCO3.p} \\ C_{H.p} \\ C_{OH.p} \end{pmatrix} := \begin{pmatrix} \beta_{H2CO3} \cdot H_p^{\,2} \cdot CO3_p \\ \dfrac{\beta_{HCO3} \cdot H_p \cdot CO3_p}{\gamma_1} \\ \dfrac{H_p}{\gamma_1} \\ \dfrac{K_W}{H_p \cdot \gamma_1} \end{pmatrix} = \begin{pmatrix} 9.179 \times 10^{-4} \\ 9.576 \times 10^{-5} \\ 5.304 \times 10^{-6} \\ 2.336 \times 10^{-9} \end{pmatrix}$$

$$C_{Tot.CO3.p} := C_{Tot.CO3.i} + \Delta CO3_p = 1.014 \times 10^{-3}$$

$$\Sigma C_p := -\left(2 \cdot C_{H2CO3.p} + C_{HCO3.p} + 2 \cdot C_{Pb.p} + C_{PbOH.p} \cdots + C_{H.p} - C_{OH.p} - C_{PbOH3.p}\right) = -2.226 \times 10^{-3}$$

We desire now to extend this solution to higher pH values. Our solve block serves well for that purpose but we need to adjust the numerator of the RHS of the carbonate mole balance to include loss of carbonate via the formation of lead carbonate solid by subtracting the change in the total lead abundance. All other equations remain the same. We write the solve block into a function for use with a looping program to generate the results for incremental pH values:

$$\Delta pH := 0.06 \qquad \Sigma C := \Sigma C_p$$

$$pH_1 := pH_p + \Delta pH \qquad H := 10^{-pH_1} \qquad \begin{pmatrix} \Delta CO3 \\ CO3 \\ Pb \\ C_{Tot.Pb} \end{pmatrix} := \begin{pmatrix} pH_p \\ CO3_p \\ Pb_p \\ \Delta CO3_p \end{pmatrix} \qquad \text{Given}$$

$$CO3 = \frac{C_{Tot.CO3.p} + \Delta CO3 + C_{Tot.Pb} - C_{Tot.Pb.i}}{\beta_{H2CO3} \cdot H^2 + \dfrac{\beta_{HCO3} \cdot H}{\gamma_1} + \dfrac{1}{\gamma_2}} \qquad Pb = \left(\beta_{S.PbCO3} \cdot CO3\right)^{-1}$$

$$C_{Tot.Pb} = Pb \cdot \left(\frac{1}{\gamma_2} + \frac{\beta_{PbOH} \cdot K_W}{H \cdot \gamma_1} + \frac{\beta_{PbOH2} \cdot K_W^2}{H^2} + \frac{\beta_{PbOH3} \cdot K_W^3}{H^3 \cdot \gamma_1} \right)$$

$$2 \cdot \Delta CO3 = -\left(2 \cdot \beta_{H2CO3} \cdot H^2 \cdot CO3 + \frac{\beta_{HCO3} \cdot H \cdot CO3}{\gamma_1} + 2 \cdot \frac{Pb}{\gamma_2} + \frac{\beta_{PbOH} \cdot Pb \cdot K_W}{H \cdot \gamma_1} \dots + \frac{H}{\gamma_1} - \frac{K_W}{H \cdot \gamma_1} - \frac{\beta_{PbOH3} \cdot Pb \cdot K_W^3}{H^3 \cdot \gamma_1} + \Sigma C \right)$$

$$PbCO3(H, \Sigma C, C_{Tot.Pb.i}, \Delta CO3, CO3, Pb, C_{Tot.Pb}) := Find(\Delta CO3, CO3, Pb, C_{Tot.Pb})$$

We have the zeroth solution from the point of first precipitation and the first solution from the initial invocation of the solve block. We must solve the speciation for the first solution and write that set of values into appropriate vectors using a structure exactly the same as that for computing the speciation at the pH of first precipitation. Also, in order to have the fully populated vectors for each of our species, we must write the solution from the pH of first precipitation into the zeroth element of each of the vectors. We are now ready to implement the remainder of the solution.

Initially, we are tempted to simply modify the LHS of the lead mole balance to be the target value and then solve the system. The caveat therein lies with the possibility that the system may not reach the target value. Following that route, we could chase the solution around interminably, observing results that, when pasted back into the initial guesses would beget yet different results. Two alternative approaches are available.

The carbonate dose may be used as a master independent variable. If small doses are used, the set of equations may be solved for each corresponding set of hydronium, carbonate, and free lead activity and total lead abundance. An attempt to employ this approach resulted in wandering values of pH (first up then down) and a plethora of negative abundance values for each of the four unknowns. Significant effort was expended without success in attaining a converged solution.

Alternatively, as we might have asserted initially, the pH can be used as a master independent variable (our mainstay). Beginning at the first incremented pH value, the pH is incremented in small steps with the solve block implemented at each step to determine the speciation and the carbonate dose. Each implementation of the solve block uses the result of the previous iteration as the starting point. This approach was successful. A total of n iterations beyond the zeroth and first solutions was performed to yield a range of pH values well into the alkaline range to illustrate full system behavior. The MathCAD code used to assemble the results is shown in Figure E11.13.1. A plot of selected results is shown in Figure 11.13.2 and as we can observe, the target total residual lead level (4.826×10^{-9} M) cannot be attained via carbonate precipitation.

Once the four master dependent variable vectors were populated, we created additional vectors for parameters of interest including the stepwise lead removal (ΔPb), cumulative carbonate dose (DCO_3) and total carbonate abundance. We have reversed the normal definition of ΔPb so it may be a positive quantity for plotting. We have also written a short program to search the results and find the pH value at

$n := 88$

$$
\begin{pmatrix} \Delta CO3 \\ CO3 \\ Pb \\ C_{Tot.Pb} \\ pH \end{pmatrix} :=
\left|
\begin{array}{l}
\text{for } i \in 2..n \\
\quad pH_i \leftarrow pH_{i-1} + \Delta pH \\
\quad H_i \leftarrow 10^{-pH_i} \\
\quad \begin{pmatrix} \Delta CO3_i \\ CO3_i \\ Pb_i \\ C_{Tot.Pb_i} \end{pmatrix} \leftarrow \begin{pmatrix} \Delta CO3_{i-1} \\ CO3_{i-1} \\ Pb_{i-1} \\ C_{Tot.Pb_{i-1}} \end{pmatrix} \\
\quad C_{Tot.Pb.i} \leftarrow C_{Tot.Pb_{i-1}} \\
\quad \begin{pmatrix} \Delta CO3_i \\ CO3_i \\ Pb_i \\ C_{Tot.Pb_i} \end{pmatrix} \leftarrow PbCO3\left(H_i, \Sigma C_{i-1}, C_{Tot.Pb.i}, \Delta CO3_i, CO3_i, Pb_i, C_{Tot.Pb_i}\right) \\
\quad Pb_i \leftarrow \dfrac{C_{Tot.Pb_i}}{\left[\dfrac{1}{\gamma_2} + \dfrac{\beta_{PbOH} \cdot K_W}{H_i \cdot \gamma_1} + \dfrac{\beta_{PbOH2} \cdot K_W^2}{(H_i)^2} + \dfrac{\beta_{PbOH3} \cdot K_W^3}{(H_i)^3 \cdot \gamma_1}\right]} \\
\quad \begin{pmatrix} C_{Pb_i} \\ C_{PbOH_i} \\ C_{PbOH2_i} \\ C_{PbOH3_i} \end{pmatrix} \leftarrow \begin{bmatrix} \dfrac{Pb_i}{\gamma_2} \\ \dfrac{\beta_{PbOH}}{\gamma_1} \cdot Pb_i \cdot \dfrac{K_W}{H_i} \\ \beta_{PbOH2} \cdot Pb_i \cdot \left(\dfrac{K_W}{H_i}\right)^2 \\ \dfrac{\beta_{PbOH3}}{\gamma_1} \cdot Pb_i \cdot \left(\dfrac{K_W}{H_i}\right)^3 \end{bmatrix} \\
\quad \begin{pmatrix} C_{H2CO3_i} \\ C_{HCO3_i} \\ C_{H_i} \\ C_{OH_i} \end{pmatrix} \leftarrow \begin{bmatrix} \beta_{H2CO3} \cdot (H_i)^2 \cdot CO3_i \\ \dfrac{\beta_{HCO3} \cdot H_i \cdot CO3_i}{\gamma_1} \\ \dfrac{H_i}{\gamma_1} \\ \dfrac{K_W}{H_i \cdot \gamma_1} \end{bmatrix} \\
\quad \Sigma C_i \leftarrow -\left(2 \cdot C_{H2CO3_i} + C_{HCO3_i} + 2 \cdot C_{Pb_i} + C_{PbOH_i} + C_{H_i} - C_{OH_i} - C_{PbOH3_i}\right) \\
\text{return } \begin{pmatrix} \Delta CO3 \\ CO3 \\ Pb \\ C_{Tot.Pb} \\ pH \end{pmatrix}
\end{array}
\right.
$$

FIGURE E11.13.1 *MathCAD program used to compute speciation of free carbonate, free zinc, total carbonate and total lead versus pH for precipitation of zinc from solution using carbonate reagent.*

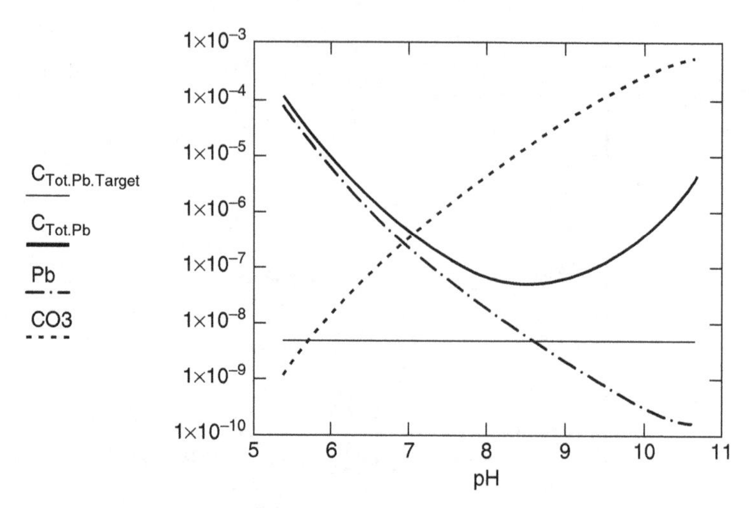

***FIGURE E11.13.2** A plot of predicted abundances for carbonate ion, lead ion, and total dissolved lead for precipitation of lead from aqueous solution using carbonate reagent.*

which total residual lead is a minimum, and associated residual total lead and carbonate dose:

```
DCO3 := | sum ← ΔCO3_p
        | for i ∈ 0.. n
        |   | sum ← sum + ΔCO3_i
        |   | DCO3_i ← sum
        | DCO3
```

$$\Delta Pb_i := \overrightarrow{\left(C_{Tot.Pb.init} - C_{Tot.Pb_i}\right)}$$

```
i_min := | for i ∈ 1.. n
         |   break if C_Tot.Pb_i ≥ C_Tot.Pb_i-1
         | return i
```

$$pH_{i_{min}} = 8.682 \qquad \Delta Pb_{i_{min}} = 1.447 \times 10^{-4}$$

$$C_{Tot.Pb_{i_{min}}} = 5.631 \times 10^{-8} \qquad DCO3_{i_{min}} = 9.314 \times 10^{-4}$$

In Figure 11.13.3, the change in total lead abundance and the dose of carbonate are plotted versus pH with the pH of minimum total residual lead identified.

We gain little removal of lead from the solution beyond pH about 6.2. We search for the pH at which removal would be 99% of the maximum removal and find the value to be ~6.5:

```
i99 := | for i ∈ 0.. n
       |   break if ΔPb_i ≥ 0.99·ΔPb_i_min
       | i
```

$$pH_{i_{99}} = 6.522$$

$$\Delta Pb_{i_{99}} = 1.434 \times 10^{-4}$$

$$DCO3_{i_{99}} = 7.125 \times 10^{-4}$$

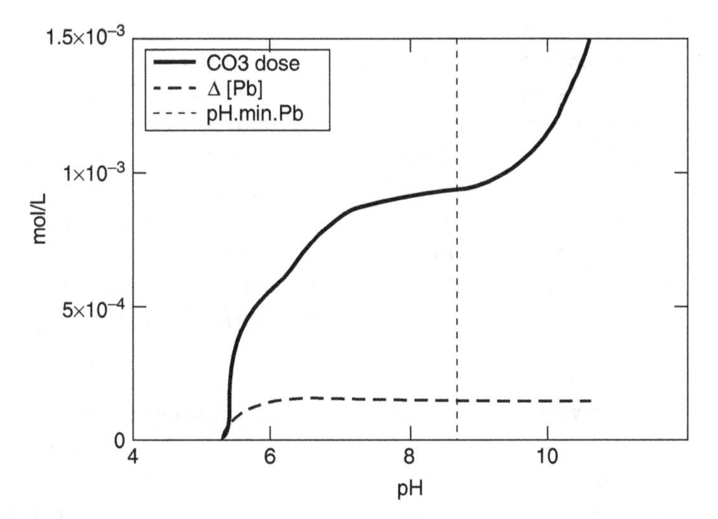

FIGURE E11.13.3 *A plot of carbonate dose and change in total lead versus pH for precipitation of lead using carbonate reagent.*

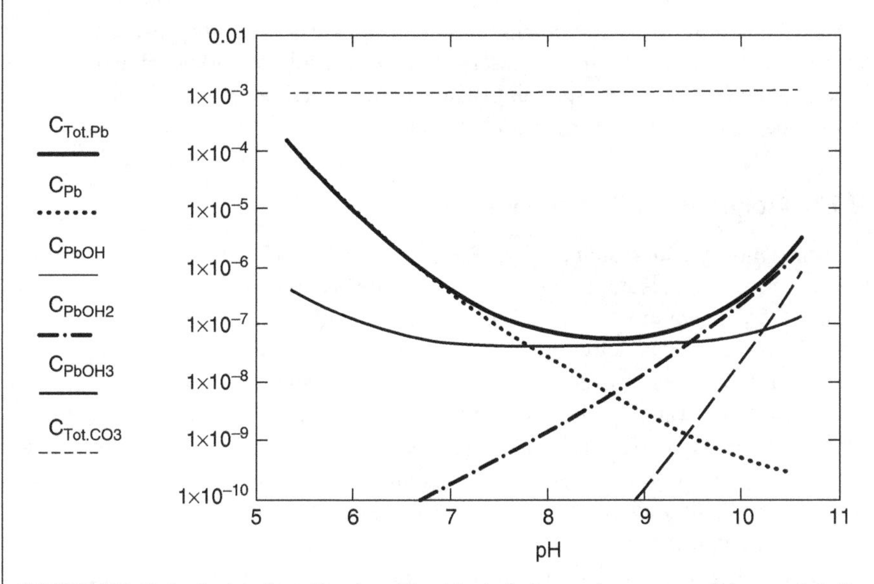

FIGURE E11.13.4 *A plot of total lead and lead hydrolysis species versus pH for precipitation of lead using carbonate reagent.*

The law of diminishing returns is quite well illustrated here. For a dose of ~7.1×10^{-4} M carbonate, we obtain 99% removal of lead. For an additional dose of ~2.2×10^{-4} M, we obtain the next 1% of the possible removal.

We finish this analysis by plotting lead specie abundances and total carbonate as a function of pH in Figure 11.13.4. We observe that carbonate abundance is relatively constant during the process and that each of the three lead hydrolysis species takes its turn along with the free metal ion as the predominant dissolved lead specie. Perhaps we could have made an assumption that the carbonate abundance would be constant at ~10^{-3} M, but we would have needed to complete the detailed solution anyway to verify the assumption. We might have reasoned that the total carbonate abundance could never be lower than the initial value, but could not have a priori predicted that it would not increase significantly beyond the initial value. These results are specific to the lead-carbonate system, however, the process can be applied to any metal carbonate system. Were several metals present in the system, our solution would necessarily have embraced their precipitative behavior simultaneously, employing the formation equilibria, mole balance, and proton balance tools we have at our disposal.

Many other analyses and comparisons can quite conveniently be made from the plethora of results obtainable from the several vectors of abundance predictions available from the MathCAD worksheet. The worksheet is large and we have attempted to include its important features, but not its entirety. Of great importance, we observe that carbonate precipitation cannot lower the abundance of lead to the targeted level. We are about an order of magnitude too high. The presence of complexing ligands beyond hydroxide, of course, would exacerbate the shortcoming. As was the case with hydroxide precipitation of zinc, we observe that as the carbonate dose is increased beyond that necessary to attain the minimum lead abundance, the lead abundance rises, corresponding with the redissolution of lead carbonate solid and increases in the total carbonate abundance.

11.7.3 Solubility of Phosphates

The phosphoric acid system, often referred to as "ortho" phosphate or dissolved reactive phosphorus is important in environmental systems as a major contributor to cultural eutrophication of water bodies. As a limiting nutrient in most systems, when introduced via activities of human society, phosphorus spurs biological activity, hastening the conversion of water bodies into shallow swamps and marshes. Along the way, water quality is often quite impaired. Once introduced into the water body, phosphorus is difficult to remove. As it is a scarce nutrient, biotic systems have evolved in many ways to hang onto every last nanogram of phosphorus possible. In the dimictic lakes of the northern and southern temperate zones, phosphorus is incorporated into green plants, is carried with detritus that falls to the lake bottoms, and is incorporated into the sediments. This cyclical process ensures that, short of removing phosphorus-laden sediments from the water body, whatever phosphorus is input into the lakes will remain. Anthropogenic releases of phosphorus from point source and nonpoint source discharges continue to endanger otherwise pristine water bodies. Recently, in the United States, focus on improving wastewater treatment for increased phosphorus removal has resulted in huge reductions in point-source phosphorus discharges. However, unfortunately, with the ever increasing demand for higher production of crops from arable lands, the applications of phosphorus fertilizers have likely increased in magnitude. Cropping practices that minimize runoff and hence mitigate release of phosphorus to surface and ground waters are not universally practiced. Thus, control of phosphorus in environmental systems, and hence, its removal from water will continue to be important in the foreseeable future.

When we peruse Table A.2 for the phosphate ligand, we find that the alkaline earth metals all form solid phases with hydrogen phosphate, that aluminum and ferric iron form 1:1 solid phases with phosphate, and most particularly ferrous iron and lead form 3:2 solid phases with phosphate. Certainly, a wider search of the broader chemical literature would turn up additional metal-phosphate solid phases. Several important such phosphate solids are identified and associated values of solid formation constants are given in Table 11.1.

If we can employ the information we have assembled in Table 11.1, we would then have high probability of successful use of information from the broader literature. Based on the general form of the solid formation reaction and associated law of mass action statement (Equation 4.8) we can write three general statements (all forms of Equation 4.8) for metal-phosphate solid formation.

For divalent metal ions forming 1:1:1 metal hydrogen phosphate solids, we have:

$$\beta_{\mathrm{S.MHPO4}} = \frac{1}{\{\mathrm{M}^{+2}\}\{\mathrm{H}^{+}\}\{\mathrm{PO}_4^{-3}\}}$$

For trivalent metal ions forming 1:1 metal phosphate solids, we have:

$$\beta_{\mathrm{S.MPO4}} = \frac{1}{\{\mathrm{M}^{+3}\}\{\mathrm{PO}_4^{-3}\}}$$

For divalent metal ions forming 3:2 metal phosphate solids, we have:

$$\beta_{\mathrm{S.M3PO42}} = \frac{1}{\{\mathrm{M}^{+3}\}^3\{\mathrm{PO}_4^{-3}\}^2}$$

TABLE 11.1 Important Phosphate Solids and Associated Formation Constants

Solid	$\log_{10}\beta_S$			
$CaHPO_{4(S)}$	6.9[a]	19[b]	6.66[c]	7.0[d]
$Ca_3(PO_4)_{2(S)}$	28.92[a]		24[c]	28.7[d]
$Ca_5(PO_4)_3OH_{(S)}$	58.333[a] (44.333)[e]		55.9[c]	
$AlPO_{4(s)}$	20.07[e]		21[c]	18.24[b]
$FePO_{4(s)}$	26.4[e]	26.4[b]	17.92[c]	21.89

[a] From Brezonik and Arnold (2011) Table 16.4 *(MINEQL v 4.6 database)*.
[b] From Table A.2.
[c] From Snoeyink and Jenkins (1980).
[d] From Dean (1992).
[e] From Brezonik and Arnold (2011) Table 10.1 *(MINEQL v 4.6 database)*.

For calcium hydroxyapatite, we have:

$$\beta_{S.Ca5PO4OH} = \frac{1}{\{Ca^{+2}\}^5 \{PO_4^{-3}\}^3 \{OH^-\}}$$

Significant interest has been expended regarding the removal of phosphorus from wastewater prior to discharge. Depending on the background conditions of the aqueous solution, precipitation as a calcium, iron or aluminum solid phase may prove the most effective. Let us examine the behavior of phosphate in aqueous solution that would be in equilibrium with the calcium phosphate solid phases.

Example 11.14 Predict the solubility of phosphate considering control by each of the respective calcium-phosphate solid phases. Use the free calcium ion abundance as the master independent variable and compare the results obtained using the equilibrium constants obtained from the various sources.

We have the requisite formation constants available from Table 11.1, and we do observe significant variation in the data obtained from the various sources. The values used herein are summarized in the following table into the four sets employed.

Set	1	2	3	4
$\log_{10} \beta_{S.CaHPO4}$	19	19	19	19
$\log_{10} \beta_{S.Ca3PO42}$	24	28.92	24	28.92
$\log_{10} \beta_{S.Ca5PO43OH}$	44.333	44.333	55.9	58.333

About the first thing we notice is the wide variation in certain of the formation constants. The two values obtained for calcium hydroxyapatite presented by Brezonik and Arnold (2011) vary by 14 orders of magnitude. Significant variation is evident in the formation constant values for tri-calcium di-phosphate. Morel and Herring's (1993) formation constant for calcium hydrogen phosphate is well removed from those of the other three sources. The variations in values notwithstanding, let us assemble the model that we would use to investigate the behavior of phosphate in systems at equilibrium with these solids.

We write functions using pH and $\{Ca^{+2}\}$ as the master independent variables:

$$H(pH) := 10^{-pH} \qquad PO4_{CaHPO4}(Ca, pH) := \left(\beta_{S.CaHPO4} \cdot H(pH) \cdot Ca\right)^{-1}$$

$$PO4_{Ca3PO42}(Ca) := \left(\beta_{S.Ca3PO42} \cdot Ca^3\right)^{-\frac{1}{2}}$$

$$PO4_{Ca5PO43OH}(Ca, pH) := \left(\beta_{S.Ca5PO43OH} \cdot Ca^5 \cdot \frac{K_W}{H(pH)}\right)^{-\frac{1}{3}}$$

We define some sort of target value against which to compare the results of our modeling effort. Typically, point source discharges, when regulated for discharge of phosphorus, are beset with a limit in the range of 0.5 mg/L total P. Let us use that as the target after converting to molar units ($\sim 1.6 \times 10^{-5}$ M). We write a second-level function that will use the minimum of the three values at any pH and calcium ion activity:

$$PO4(Ca, pH) :=$$

$$\min\left(PO4_{CaHPO4}(Ca, pH), PO4_{Ca3PO42}(Ca), PO4_{Ca5PO43OH}(Ca, pH)\right)$$

We then write a further function that will yield a predicted total phosphorus concentration:

$$\begin{pmatrix} \beta_{HPO4} \\ \beta_{H2PO4} \\ \beta_{H3PO4} \\ K_W \end{pmatrix} := \begin{pmatrix} 10^{12.35} \\ 10^{19.55} \\ 10^{21.7} \\ 10^{-14} \end{pmatrix}$$

$$C_{Tot.PO4}(Ca, pH) :=$$

$$PO4(Ca, pH) \cdot \left(\beta_{H3PO4} \cdot H(pH)^3 + \beta_{H2PO4} \cdot H(pH)^2 + \beta_{HPO4} \cdot H(pH) + 1\right)$$

We could include activity coefficients here but have not, just for simplicity. Application to a specific system would necessitate their inclusion.

We have arranged four different combinations of formation constants for analyzing the behavior of the system. For $\log_{10} \beta_{S.CaHPO4}$ values of ~7 we found no possibility for existence of $CaHPO_{4(s)}$ at any calcium ion abundance. Since our goal here is to model the systems for comparison, we opted to use Morel and Hering's formation constant for $CaHPO_{4(s)}$ in all four sets.

We employ the first set of formation constants with our set of functions at pH 7.5:

$$\begin{pmatrix} \beta_{S.CaHPO4} \\ \beta_{S.Ca3PO42} \\ \beta_{S.Ca5PO43OH} \end{pmatrix} := \begin{pmatrix} 10^{19} \\ 10^{24} \\ 10^{44.333} \end{pmatrix}$$

We then produce a plot in Figure 11.14.1 in which the predicted phosphate ion activities and total phosphate abundance are plotted:

Then, for each of the other three sets of formation constants, we have accomplished the same process:

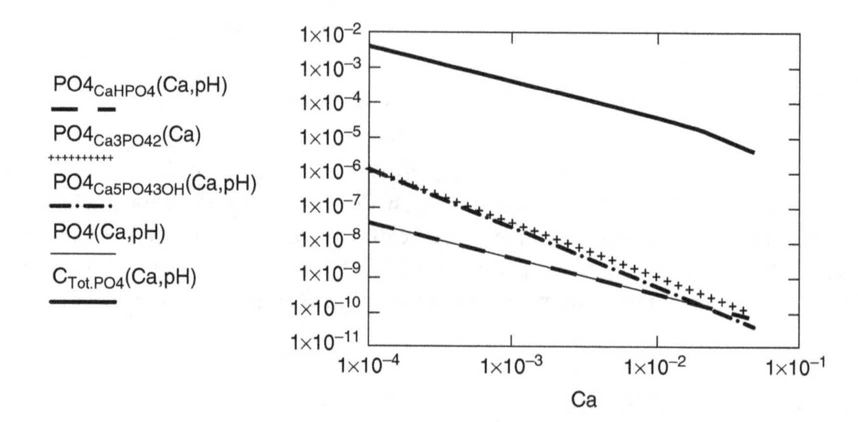

FIGURE E11.14.1 *A plot of predicted free phosphate ion and total phosphate abundance assuming control of phosphate solubility by calcium hydrogen phosphate, tri-calcium di-phosphate, and calcium hydroxyapatite.*

Set two:

$$\begin{pmatrix} \beta_{S.CaHPO4} \\ \beta_{S.Ca3PO42} \\ \beta_{S.Ca5PO43OH} \end{pmatrix} := \begin{pmatrix} 10^{19} \\ 10^{28.92} \\ 10^{44.333} \end{pmatrix}$$

Set three:

$$\begin{pmatrix} \beta_{S.CaHPO4} \\ \beta_{S.Ca3PO42} \\ \beta_{S.Ca5PO43OH} \end{pmatrix} := \begin{pmatrix} 10^{19} \\ 10^{24} \\ 10^{55.9} \end{pmatrix}$$

Set four:

$$\begin{pmatrix} \beta_{S.CaHPO4} \\ \beta_{S.Ca3PO42} \\ \beta_{S.Ca5PO43OH} \end{pmatrix} := \begin{pmatrix} 10^{19} \\ 10^{28.92} \\ 10^{58.333} \end{pmatrix}$$

We have then collected the total phosphorus abundance traces from each of the sets and plotted them together in Figure 11.14.2 to illustrate the overall result. We have included the target total P abundance for reference. For formation constant set 1, $CaHPO_{4(S)}$ is the predicted controlling solid phase through calcium ion activity ~0.025 M. For formation constant set 2, $Ca_3(PO_4)_{2(S)}$ is the predicted controlling solid phase. For formation constant sets 3 and 4 $Ca_5(PO_4)_3OH_{(S)}$ is

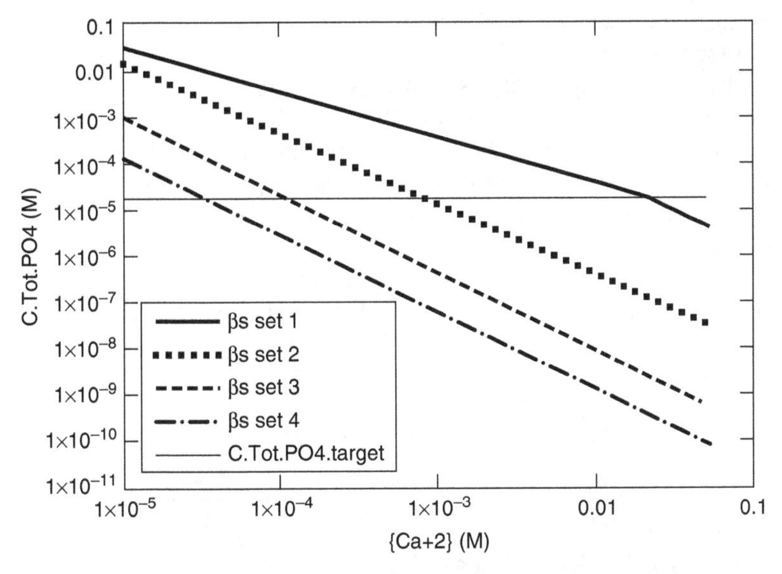

FIGURE E11.14.2 A plot of total predicted phosphate-phosphorus abundance versus aqueous free calcium for selected sets of formation constants for calcium phosphate solids.

the predicted controlling solid phase. The corresponding values of calcium ion activity necessary to control total dissolved reactive phosphorus at the target level are given as $C_{Ca.1}$ through $C_{Ca.4}$. These are of course based on the computed calcium ion activities:

$$\begin{pmatrix} C_{Ca.1} \\ C_{Ca.2} \\ C_{Ca.3} \\ C_{Ca.4} \end{pmatrix} = \begin{pmatrix} 8.3 \times 10^{2} \\ 3.2 \times 10^{1} \\ 4.1 \times 10^{0} \\ 1.3 \times 10^{0} \end{pmatrix} \quad \frac{mg_{Ca}}{L}$$

Formation of calcium-ligand complexes and nonzero ionic strength would render the total calcium concentrations to be higher than these computed values.

We are certainly less than fully satisfied with this result. Depending upon whose values of formation constants we believe, we obtain vastly different results. If we were to believe the results of sets 3 and 4, calcium hydroxyapatite control with $\log_{10} \beta_{S.Ca5PO43OH}$ values of 55.9 and 58.333, we would counsel wastewater plant operators simply to add a little calcium to their treatment basins to control phosphorus in their effluents. A well-run activated sludge plant that reduces degradable organics to levels commensurate with NPDES permits for systems for which effluent phosphorus control would be warranted will nearly fully convert organic phosphorus either to dissolved reactive phosphorus or to biomass. If we assume that biomass is nearly completely removed from the effluent, the phosphorus is nearly all in the dissolved reactive form. Experience has shown, unfortunately, that for wastewater plants receiving wastewater generated from relatively hard water (e.g., the City of Rapid City, SD, WW Renovation facility, $C_{Tot.Ca} = \sim 50\,mg/L$) effluent phosphate is not automatically controlled to the levels predicted by formation constant set 3 or 4. Even set 2 would predict that calcium in Rapid City's wastewater should control dissolved reactive phosphorus to the target level. From this analysis we could judge that the truth might lie somewhere in between sets 1 and 2.

We may easily, once the model has been assembled, investigate alternative pH values. We simply changed the master pH value and assembled the results:

pH = 7 pH = 8

$$\begin{pmatrix} C_{Ca.1} \\ C_{Ca.2} \\ C_{Ca.3} \\ C_{Ca.4} \end{pmatrix} = \begin{pmatrix} 1.44 \times 10^3 \\ 9.97 \times 10^1 \\ 1.43 \times 10^1 \\ 4.66 \times 10^0 \end{pmatrix} \quad \begin{pmatrix} C_{Ca.1} \\ C_{Ca.2} \\ C_{Ca.3} \\ C_{Ca.4} \end{pmatrix} = \begin{pmatrix} 2.88 \times 10^2 \\ 1.26 \times 10^1 \\ 1.4 \times 10^0 \\ 4.56 \times 10^{-1} \end{pmatrix} \quad \frac{mg_{Ca}}{L}$$

Examination of pH 7 and pH 8 yields the expected result that the residual calcium activity necessary to control total phosphorus at the target level is inversely related the pH of the aqueous solution. We move away from Example 11.14 with some mixed feelings.

The database (MINEQL v4.6) from which Brezonik and Arnold (2011) obtained their values of the formation constants has, over nearly three decades, received scrutiny from numerous top scholars both of scientific and engineering persuasion, as well as from investigators associated with the U.S Geological Survey. These values of equilibrium constants were derived, in some cases decades ago, via experimentation. The researchers wrote manuscripts that were reviewed by peers and when deemed sufficiently defensible, were published. In theory, these top scholars who have reviewed the database would have checked out the publications to ensure that methods used were defensible and that analyses leading to evaluation of the constants were sound. Even with this high level of scrutiny, apparently the database contains erroneous values. Given the interest in removing phosphorus from wastewaters, certainly, over the years, bench-scale and pilot-scale tests as well as full-scale operations have been conducted by industry and engineering firms employing calcium as a reagent for control of dissolved reactive phosphorus. Unfortunately, the results of many of these efforts have never been correlated into the master database. Perhaps the experimental teams were purely interested in the empirical result to tune an in-place treatment process and never invested time (resources) into experimental designs that would permit examination of the results to discern the controlling solid phase. Perhaps in some cases knowledge of the application of chemical principles was lacking and otherwise valuable data still may lie in many files of results, needing only knowledgeable and experienced investigators to sift through to correlate the results into useable form to enhance and update the master database.

Iron(III) and aluminum are also used in efforts to control the solubility of phosphates. Both have the behavior that as hydroxides, their solubilities are so low that even with modern laboratory instrumentation, detection, and quantitation is often problematic. Let us examine the solubility of phosphorus in systems wherein the aqueous solution would be in equilibrium with iron(III) and aluminum phosphate solid phases.

Example 11.15 Consider an aqueous solution of ionic strength 0.01 M, pH of 7.5, measured alkalinity of 200 mg/L as $CaCO_3$, and containing total phosphate-phosphorus of 31 mg/L. Compare the behaviors of iron(III) (as $FeCl_3$) and aluminum(III) (as $Al_2(SO_4)_3$, often called alum) in the reduction of total dissolved reactive phosphate abundance to the target level of Example 11.13.

We sift through sources to find appropriate values of the formation constants and from several sources, we assemble what appears to be a reasonable set of solid and complex formation constants for aluminum and iron(III). Our data set is taken from Table A.2 with additions from Dean (1992): β_{AlOH}, β_{AlOH_4} and β_{FeOH_3}; and from Snoeyink and Jenkins (1980): $\beta_{S.AlPO_4}$.

$$\begin{pmatrix} \beta_{S.FePO4} \\ \beta_{S.AlPO4} \\ \beta_{S.FeOH3} \\ \beta_{S.AlOH3} \end{pmatrix} := \begin{pmatrix} 10^{26.4} \\ 10^{21} \\ 10^{39.5} \\ 10^{33.5} \end{pmatrix} \quad \begin{pmatrix} \beta_{FeOH} \\ \beta_{FeOH2} \\ \beta_{FeOH3} \\ \beta_{FeOH4} \\ \beta_{Fe2OH2} \end{pmatrix} := \begin{pmatrix} 10^{11.8} \\ 10^{22.3} \\ 10^{29.67} \\ 10^{34.4} \\ 10^{25} \end{pmatrix} \quad \begin{pmatrix} \beta_{AlOH} \\ \beta_{AlOH4} \end{pmatrix} := \begin{pmatrix} 10^{9.27} \\ 10^{33.03} \end{pmatrix} \quad \begin{pmatrix} \beta_{FeHPO4} \\ \beta_{FeH2PO4} \end{pmatrix} := \begin{pmatrix} 10^{22.5} \\ 10^{23.2} \end{pmatrix}$$

We know from introductory environmental engineering texts, on the topic of jar tests, that when we add ferric chloride or alum to aqueous solution containing any buffering capacity at all, that we will form a ferric hydroxide or aluminum hydroxide solid phase. Such is the basis of coagulation for the removal of suspended matter from surface waters. If we read deeper into physical/chemical processes, we would find that ferric chloride and alum are also used to remove phosphate from aqueous solution. Herein, we will see that the two processes are quite highly related. By formation of the iron or aluminum hydroxide solid phase, we provide control on the solubility of iron or aluminum. In our introductory text, we likely learned that neither iron(III) nor aluminum is soluble in aqueous solutions of neutral to alkaline pH. As we did for the divalent metals, we may predict the activity of the free trivalent iron and free aluminum ions as functions of pH:

$$Al(pH) := \frac{H(pH)^3}{\beta_{S.AlOH3} \cdot K_W^3} \qquad Fe(pH) := \frac{H(pH)^3}{\beta_{S.FeOH3} \cdot K_W^3}$$

Let us also write a function for the activity of the phosphate ion considering that none would precipitate. Of course, once conditions arose such that either iron or aluminum phosphate would precipitate, this function would be invalid:

$$PO4(pH) := \frac{C_{Tot.PO4.i}}{\left(\beta_{H3PO4} \cdot H(pH)^3 + \frac{\beta_{H2PO4} \cdot H(pH)^2}{\gamma_1} + \frac{\beta_{HPO4} \cdot H(pH)}{\gamma_2} + \frac{1}{\gamma_3}\right)}$$

We plot the functions in Figure E11.15.1. Based on control by the metal hydroxide, aluminum appears to be a bit more soluble than iron(III).

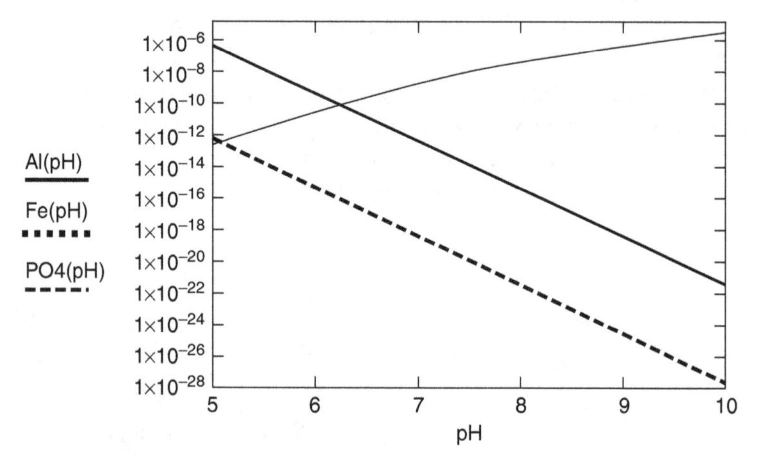

FIGURE E11.15.1 *A plot of free aluminum ion and free Fe(III) ion versus pH for aqueous solutions in equilibrium with aluminum and iron(III) hydroxide, respectively. Phosphate ion activity for total phosphate-phosphorus of 0.001 M shown for reference.*

Now, we use these three functions in combination with a re-arranged version of the metal-phosphate solid formation equilibrium ($\beta_{S.MPO4}\{M^{+3}\}\{PO_4^{-3}\} = 1$) to determine the pH value at which each precipitate would begin to form. Consider the products $\beta_{S.AlPO4} \cdot \{Al^{+3}) \cdot \{PO_4^{-3}\}$ and $\beta_{S.FePO4} \cdot \{Fe^{+3}) \cdot \{PO_4^{-3}\}$. At the point each phosphate solid would begin to form respective product would equal unity, the activity of the solid phase:

$pH := 6$ given $\beta_{S.AlPO4} \cdot Al(pH) \cdot PO4(pH) = 1$ $find(pH) \rightarrow 6.735$

$pH := 6$ Given $\beta_{S.FePO4} \cdot Fe(pH) \cdot PO4(pH) = 1$ $Find(pH) \rightarrow 6.242$

Aluminum phosphate would form at pH 6.735 and ferric phosphate would form at pH 6.242. In between the initial pH and each of these pH values, considerable metal hydroxide solid would form. The reactions leading to the formation of aluminum hydroxide and ferric hydroxide from alum or ferric chloride are of course alkalinity-consuming. For every mole of aluminum or iron that finds its way to the hydroxide form, three moles of protons are released:

$$Al_2(SO_4)_3 + 6H_2O \Leftrightarrow 2Al(OH)_3 + 6H^+ SO_4^{-2}$$

$$FeCl_3 + 3H_2O \Leftrightarrow Fe(OH)_3 + 3H^+ + 3Cl^-$$

We therefore would expect the addition of alum or ferric chloride (the free aluminum and iron(III) ions are acids) to depress the pH. In order to fully understand this process we need to examine the solubility of iron and aluminum with increased hydronium abundance. We write the mole balances accounting for aluminum and iron species:

$$C_{Tot.Al} = C_{Al} + C_{AlOH} + C_{AlOH4}$$

$$C_{Tot.Fe} = C_{Fe} + C_{FeOH} + C_{FeOH2} + C_{FeOH3} + C_{FeOH4} + 2 \cdot C_{Fe2OH2}$$

The di-hydroxo and tri-hydroxo aluminum complexes certainly would exist in the aluminum-hydroxide system, but apparently have abundances so small relative to the mono- and tetrahydroxo complexes that the formation constants perhaps have not been measured. We flesh these out these mole balances using the formation equilibria:

$$C_{Tot.Al}(pH) := Al(pH) \cdot \left[\frac{1}{\gamma_3} + \frac{\beta_{AlOH}}{\gamma_1} \cdot \frac{K_W}{H(pH)} + \frac{\beta_{AlOH4}}{\gamma_1} \cdot \left(\frac{K_W}{H(pH)} \right)^4 \right]$$

$$C_{Tot.Fe}(pH) := Fe(pH) \cdot \left[\frac{1}{\gamma_3} + \beta_{FeOH} \cdot \frac{K_W}{H(pH)} + \beta_{FeOH2} \cdot \left(\frac{K_W}{H(pH)} \right)^2 \ldots \\ + \beta_{FeOH3} \cdot \left(\frac{K_W}{H(pH)} \right)^3 + \beta_{FeOH4} \cdot \left(\frac{K_W}{H(pH)} \right)^4 \ldots \\ + 2 \cdot \beta_{Fe2OH2} \cdot Fe(pH) \cdot \left(\frac{K_W}{H(pH)} \right)^2 \right]$$

Then in Figure 11.15.2 we plot total aluminum and total iron against pH. We observe that both aluminum and iron hydroxide will remain in the system as solid phases as the pH is adjusted to below 4. We can then examine the solubility of phosphate, as would be controlled by the solubility of aluminum or iron at pH values below those at which each respective precipitate is predicted to form.

Let us address aluminum control of phosphate solubility first. We exploit the aluminum hydroxide formation equilibrium as the control on the activity of free aluminum. We arrange the aluminum hydroxide and aluminum phosphate

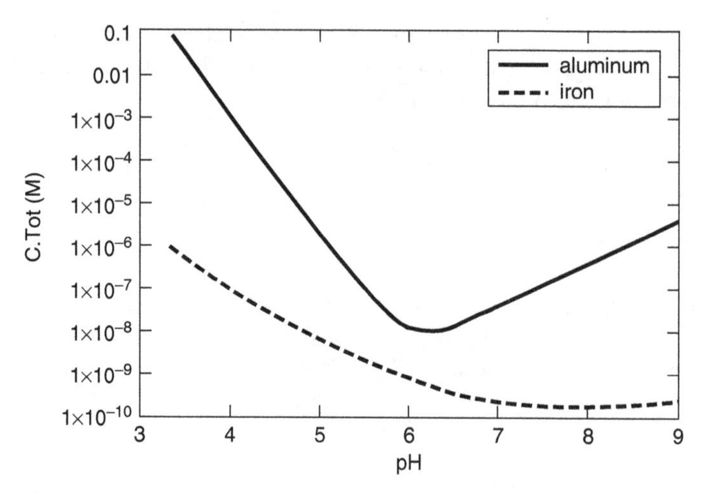

FIGURE E11.15.2 *Predicted total abundance of aluminum and iron(III) versus pH for aqueous solutions in equilibrium with aluminum hydroxide and iron(III) hydroxide, respectively. Abundances of hydrolysis species are not shown.*

formation equilibria to isolate $\{Al^{+3}\}$ on the LHS. Then since $\{Al^{+3}\}$ can have but a single value in the aqueous solution we may equate the RHS of the $Al(OH)_{3(S)}$ relation with the RHS of the $AlPO_{4(S)}$ relation to obtain the relation for the activity of phosphate with pH. Minor algebraic manipulations are required to yield the desired function:

$$PO4_{Al}(pH) := \frac{\beta_{S.AlOH3}}{\beta_{S.AlPO4}} \cdot \left(\frac{K_W}{H(pH)}\right)^3$$

A relation for the total abundance of dissolved reactive phosphorus can then be written:

$$C_{Tot.PO4.Al}(pH) := \min\left[C_{Tot.PO4.i}, PO4_{Al}(pH) \cdot \left(\beta_{H3PO4} \cdot H(pH)^3 \ldots + \frac{\beta_{H2PO4} \cdot H(pH)^2}{\gamma_1} \ldots + \frac{\beta_{HPO4} \cdot H(pH)}{\gamma_2} + \frac{1}{\gamma_3}\right)\right]$$

The function previously written for the activity of phosphate versus pH is valid only at pH higher than that at which aluminum phosphate would begin to form. The function directly written for total phosphate is valid throughout the entire pH range of the system, owing to the use of the min() function. A plot of these two functions versus pH in Figure 11.15.3 yields a pictorial solution to the question at hand.

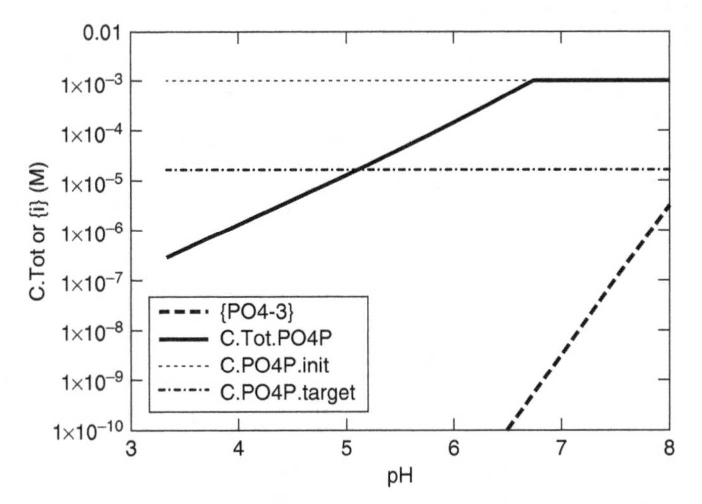

FIGURE E11.15.3 *A plot of total predicted phosphate-phosphorus versus pH for an aqueous solution in simultaneous equilibrium with aluminum hydroxide and aluminum phosphate. Phosphate provides the degree of freedom and its abundance is reduced by the precipitation of aluminum phosphate solid. The assumed initial abundance of phosphate-phosphorus is 0.001 M (31 ppm$_m$) and the target level is 0.5 ppm$_m$.*

We use a given-find ($C_{Tot.PO_4.Al} = PO_{4,target}$) to determine that at pH 5.106 the solubility of phosphate is predicted to equal the targeted value (0.5 mg/L total PO_4–P).

$pH := 5 \quad \text{Given} \quad PO4_{limit} = C_{Tot.PO4.Al}(pH) \quad \text{Find}(pH) = 5.106$

For iron, we write a function similar to that written for aluminum, with the ferric hydroxide formation equilibrium firmly in control of the free iron(III) activity and write a mole balance for total phosphate:

$$PO4_{Fe}(pH) := \frac{\beta_{S.FeOH3}}{\beta_{S.FePO4}} \cdot \left(\frac{K_W}{H(pH)}\right)^3$$

$$C_{Tot.PO4.Fe}(pH) := \min\left[C_{Tot.PO4.i}, PO4_{Fe}(pH) \cdot \left[\beta_{H3PO4} \cdot H(pH)^3 + \frac{\beta_{H2PO4} \cdot H(pH)^2}{\gamma_1} + \frac{\beta_{HPO4} \cdot H(pH)}{\gamma_2} + \frac{1}{\gamma_3} + \beta_{FeHPO4} \cdot Fe(pH) \cdot H(pH) + \beta_{FeH2PO4} \cdot (Fe(pH) \cdot H(pH))^2\right]\right]$$

In Table A.2, we find formation constants for two ferric-phosphate complexes and include these in the mole balance on iron(III). A plot of the iron-phosphate system shown in Figure E11.15.4 is very similar to that for the aluminum-phosphate system.

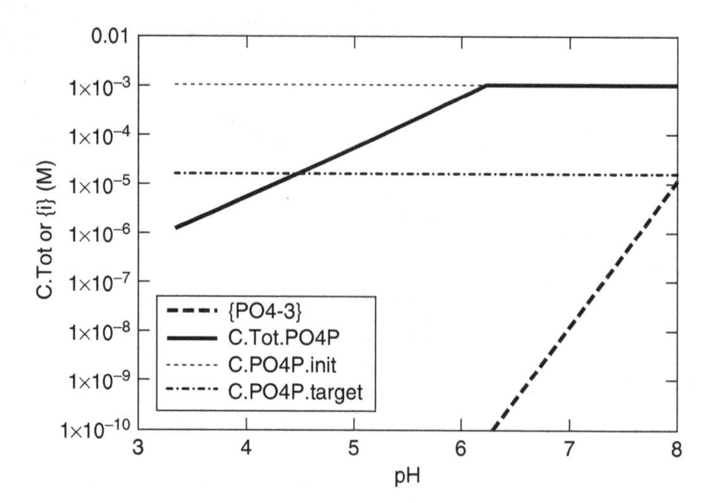

FIGURE E11.15.4 *A plot of total predicted phosphate-phosphorus versus pH for an aqueous solution in simultaneous equilibrium with ferric hydroxide and ferric phosphate. Phosphate provides the degree of freedom and its abundance is reduced by the precipitation of ferric phosphate solid. The assumed initial abundance of phosphate-phosphorus is 0.001 M (31 ppm_m) and the target level is 0.5 ppm_m.*

From a given-find we observe that the pH at which the abundance of total phosphate phosphorus is predicted to be equal to the target level is 4.508, notably lower than that predicted based on aluminum phosphate control:

$pH := 5 \quad \text{Given} \quad PO4_{limit} = C_{Tot.PO4.Fe}(pH) \quad \text{Find}(pH) = 4.508$

We could continue this exercise by computing the doses of either alum or ferric chloride necessary to provide the targeted control of the phosphate phosphorus solubility. These would be acid neutralization computations and are left as end-of-chapter exercises for the student. We also might wish to compare the efficacy of calcium versus aluminum or iron (III) as a precipitative reagent for phosphorus control. One major difference is that as pH increases calcium-phosphate solids render phosphorus to be less soluble. Conversely as pH decreases aluminum and iron phosphate solids render phosphorus to be less soluble.

Given the higher pH at which alum would cause formation of an aluminum phosphate solid and the higher pH at which resultant phosphate abundance would reach targeted levels, alum has historically found more favor for use in efforts to control phosphorus than have iron salts. Ferrous sulfate has also been used as a coagulant/precipitant. The iron(II) must first be converted to iron(III) requiring an oxidant, usually molecular oxygen supplied via aeration. Otherwise the chemistry of ferrous sulfate is identical to that for ferric chloride.

Other metals that routinely exist as the free M^{+3} ion (e.g., bismuth, cerium, chromium, and lanthanum) would act in much the same manner as aluminum and iron with regard to formation of hydroxide and phosphate solids. Dean (1992) gives formation constants for the tri-hydroxide and 1:1 phosphate solids for these metals. Perhaps, if we dug more deeply, we could locate published values of formation constants for both hydroxide and phosphate solids of additional metals that form free M^{+3} ions.

11.8 SOLUBILITY OF OTHER METAL–LIGAND SOLIDS

When we peruse Table A.2, we find formation constants for metal–ligand solids that include sulfate (SO_4^{-2}), halogens (Cl^-, Br^-, F^-), sulfide (S^{-2}), and cyanide (CN^-). When we peruse the solubility products presented by Dean (1992), we find many more. Among these, both the oxidized and reduced forms of arsenic and selenium are of keen environmental significance. Arsenate (AsO_4^{-3}) and selenate (SeO_4^{-2}) solids are typically much more soluble than are arsenite (AsO_3^{-3}) and selenite (SeO_3^{-2}) solids. In either engineered or natural environmental systems, if the oxidized form can be reduced, the arsenic or selenium can be immobilized more easily as the solid. In the next chapter, we will model oxidation and reduction as an equilibrium process, expanding our ability to model systems. In any regard, if we can identify, from the chemical literature, the formation equilibria, and the magnitudes of the formation constants for any solid formation reactions, we can model the interactions of those solids with corresponding aqueous species using the methodologies described in detail in the preceding sections of this chapter.

PROBLEMS

For the end of chapter problems that follow, the a convenient platform from which to assemble the mathematical models for each of the problems is a MathCAD worksheet. MS Excel or other software may certainly be employed but additional algebraic manipulations or structured programming may be necessary. Certainly, also, simplifying assumptions may be invoked to render pencil/paper/calculator approximations. Certainly, even graphical approximations of the solution may be assembled:

1. For the following metals, consider the hydrolysis reactions and determine the abundance of the respective hydrolysis species for total metal abundance of ~10^{-5} M. Assemble the computational/mathematical model from which the respective abundances may be computed. Consider that the aqueous solution is infinitely dilute.
   a. Show your results graphically.
   b. Determine as many stepwise acid dissociation constants as is possible from the formation constants of Table A.2. Comment as to whether the set of stepwise dissociation constants is consistent with our understanding of successive deprotonations of fully protonated conjugate acids.
      i. Calcium
      ii. Magnesium
      iii. Cadmium
      iv. Copper (Dean (1992) gives 7.0, 13.68, 17.0, and 18.5 as $\log_{10}\beta_{CuOH} - \log_{10}\beta_{CuOH_4}$)
      v. Lead
      vi. Mercury
      vii. Cobalt
      viii. Iron(II)

ix. Manganese
x. Nickel
xi. Zinc
xii. Silver
xiii. Iron(III) (Dean, 1992, gives $\log_{10} \beta_{FeOH_3}$ as 29.67)
xiv. Aluminum(III)
xv. Chromium(III)

2. For the following metals, consider that the hydroxide solid phase is present in equilibrium with an aqueous solution and that ionic strength of the solution may be approximated as 0.01 M throughout the pH range of interest. Assemble the computational/mathematical model from which the abundance of each of the hydrolysis species as well as the abundance of the total metal in aqueous solution can be computed. Consider the system in the absence of complexing ligands other than hydroxide (i.e., consider only metal hydrolysis) as a function of pH. Produce plots indicating the abundance of the predominant species as well as the total metal abundance. If possible, based on the formation constants available from Table A.2 and those given in problem 1, suggest the upper and lower pH bounds above or below which the metal hydroxide solid likely could not exist.

   a. Calcium
   b. Magnesium
   c. Cadmium
   d. Copper
   e. Lead
   f. Mercury
   g. Cobalt
   h. Iron(II)
   i. Manganese
   j. Nickel
   k. Zinc
   l. Silver
   m. Iron(III)
   n. Aluminum(III)
   o. Chromium(III)

3. For the following metal–ligand systems, consider that the metal is present in abundance sufficiently low that precipitates will not occur (e.g., $C_{Tot.M} = 10^{-6} - 10^{-5}$ M) when pH is adjusted upward or downward into the strongly alkaline or strongly acidic ranges and that ionic strength of the solution may be approximated as 0.01 M throughout the pH range of interest. Consider that the total ligand abundances would be in the range of 10^{-4} M. Assemble the computational/mathematical model that can be used to determine the aqueous speciation of the system. Use cumulative formation constants from Table A.2 as well as those given in problem 1, to determine the speciation of the metal and its metal–ligand complexes over the pH range from 3 to 12. Present your results graphically.

The effects of the various ligands on the speciation of the metal can certainly be investigated by adjusting the abundances of the ligands.

a. Silver, weak acid dissociable cyanide, ethylene di-amine tetra acetic acid
b. Mercury, ethylene di-amine tetra acetic acid, chloride
c. Calcium ($C_{Tot.Ca} = 10^{-4}$ M), phosphate-phosphorus (10^{-3} M), inorganic carbon (10^{-3} M)
d. Magnesium ($C_{Tot.Mg} = 10^{-4}$ M), phosphate-phosphorus (10^{-3} M), inorganic carbon (10^{-3} M)
e. Iron(II), phosphate phosphorus, citrate
f. Nickel, weak acid dissociable cyanide, citrate
g. Iron(III), phosphate phosphorus, acetate
h. Cadmium, weak acid dissociable cyanide, ammonia
i. Lead, inorganic carbon, acetate
j. Lead, sulfate-sulfur, citrate
k. Manganese, inorganic carbon, citrate
l. Zinc, sulfate-sulfur, glutamate
m. Copper, ammonia, inorganic carbon
n. Copper, citrate, sulfate

4. For the following metal–ligand systems, consider that the ligands are present at total abundances of 10^{-3} M and that the metal is present at an abundance dictated by the presence of the metal hydroxide solid phase. Assume that the ionic strength of the aqueous solution can be approximated as ~0.01 throughout the pH range of interest. Assemble the computational/mathematical model from which the speciation and total aqueous metal abundance may be computed. Present your results graphically over the range of pH values within which the metal hydroxide solid phase could likely exist.

   The effects of the various ligands on the speciation of the metal can certainly be investigated by adjusting the abundances of the ligands.

   a. Silver, weak acid dissociable cyanide, ethylene di-amine tetra acetic acid
   b. Mercury, ethylene di-amine tetra acetic acid, chloride
   c. Calcium, phosphate-phosphorus (0.01 M), inorganic carbon (0.01 M)
   d. Magnesium, phosphate-phosphorus, inorganic carbon
   e. Iron(II), phosphate phosphorus, citrate
   f. Nickel, weak acid dissociable cyanide, citrate
   g. Iron(III), phosphate phosphorus, acetate
   h. Cadmium, weak acid dissociable cyanide, ammonia
   i. Lead, inorganic carbon, acetate
   j. Lead, sulfate-sulfur, citrate
   k. Manganese, inorganic carbon, citrate
   l. Zinc, sulfate-sulfur, glutamate
   m. Copper, ammonia, inorganic carbon
   n. Copper, citrate, sulfate

5. For each of the metal systems of problem 2 (except for 2.(a) and 2.(b)), consider that the initial total concentration of the targeted metal is 300 ppm_m as each

respective metal and the initial pH of the solution is 5.0 (2.0 for Hg, 1.5 for Fe(III), 3.0 for Al, 4.0 for Cr(III)).

a. Consider precipitation of the metal using hydroxide (e.g., dosed as either sodium or potassium hydroxide) and determine the pH at which the solid phase would first form in the aqueous system.
b. Determine the pH value at which the total metal abundance would be at a minimum value.
c. Research the literature to determine the EPA maximum contaminant level (MCL, either primary or secondary) for drinking water. Consider precipitation of the metal using hydroxide (e.g., dosed as either sodium or potassium hydroxide) and determine the lowest pH at which the total metal abundance would reach the MCL (if possible).
d. Compute the necessary reagent addition to bring the pH value to the level at which the respective metal hydroxide solid phase would form.
e. Compute the necessary reagent addition to bring the abundance of total metal in the aqueous solution to the pH at which the abundance is predicted to be minimum.

6. A heap leach pad situated at an abandoned mine site generates a leach solution of alkaline pH (typically in the range of 7.5–10.5) and ionic strength of 0.01 M containing cadmium, copper and zinc at unknown concentrations and total cyanide (sum of all CN-containing species) at a concentration of 0.00001 – 0.005 M. Consider that solid phase calcite is present in the leach pad. Recall that this mineral may be employed as a surrogate for whatever carbonate-bearing mineral might be present.

   a. The water that infiltrates the leach pad falls as precipitation and is in contact with the normal atmosphere at a total pressure of 0.88 atm. You will need to characterize the initial condition of the rain water in order to specify the exact relation between $\Delta[CO_{3T}]$ and $\Delta[Ca_T]$ written to determine $\{CO_3^=\}$ from the $CaCO_{3(s)}$ equilibrium.
   b. Develop functions of pH to define $\{Cu^{+2}\}$, $\{Cd^{+2}\}$, and $\{Zn^{+2}\}$ over the expected pH range of the system. *Table A.2 gives β_s values for two copper hydroxo carbonate solids – $Cu(OH)_2CO_{3(s)}$ and $Cu_2(OH)_2(CO_3)_{2(s)}$ – knowing $\{OH^-\}$ and $\{CO_3^=\}$ as functions of pH, these can be fairly easily included, but ignore them until you have a solution for hydroxide and calcite. Also, Table A.2 gives a formation constant for solid, $Zn(CN)_{2(s)}$, – also ignore this until you have a solution based on hydroxide and calcite as the potential solid phases.*
   c. Develop *an* additional set of functions that will allow the specification of the respective free metal ion activities as predicted from the controlling solid phase. The controlling solid phase produces the lowest value of the metal activity. MathCAD's min() function will prove quite useful in specifying the activities of the metal ions based on the controlling solid phase.
   d. Produce plots of the controlling values of $\{Cu^{+2}\}$, $\{Cd^{+2}\}$, and $\{Zn^{+2}\}$ over the pH range of the system and identify which solid phase (the hydroxide or the carbonate) is responsible for the controlling value.

e. Write the mole balance equations for total zinc, copper, and cadmium as coupled with the mole balance equation for total cyanide. These may of course be solved simultaneously using a MathCAD given-find block. Generalize your solution so that pH and $[CN_T]$ may be used as master variables to determine resultant values of $[Cd_T]$, $[Cu_T]$, $[Zn_T]$, and $\{CN^-\}$ for a given set of values of pH and $[CN_T]$.
f. For $[CN_T] = 0.005\,M$, solve the system at pH values ranging from 7.5 to 10.5 using an increment of 0.25 pH unit. Collect the results (matrix output will prove useful) and produce a plot of Cu_T, Cd_T, and Zn_T versus pH.

7. One method employed in attempts to restore the trophic state of lakes to mesotrophic from eutrophic involves a treatment of the lake with aluminum sulfate and sodium aluminate at doses sufficient to form a cap of aluminum hydroxide overlying phosphorus-laden sediments.

   In these sediments, organically bound phosphorus is mineralized to become "ortho" (or dissolved reactive) phosphorus (PO_4–P) by the breakdown of organic sediments by the anaerobic bacteria present. This ortho phosphorus tends to migrate as a consequence of molecular diffusion from the PO_4–P rich pores of the sediments into the (relatively) PO_4–P poor water above the sediment-water interface. The system and associated phosphorus migration are depicted in Figure P11.7. When the upper layer of the sediment is anaerobic, most of the phosphorus is taken up by aerobic bacteria growing in this layer. However, if the water body is sufficiently productive (eutrophic), under stratified conditions dissolved oxygen in the deeper regions of the lake may be completely depleted. The aerobic process gives way to anoxic and then to anaerobic processes. Anaerobic bacteria have yields that are far lower than those of aerobic bacteria and they are thus unable to incorporate the migrating phosphorus into their cell matter and the phosphorus is released to the water. This process is called internal cycling and is responsible for the continued eutrophic state of many lakes well after anthropogenic inputs of phosphorus have been curtailed.

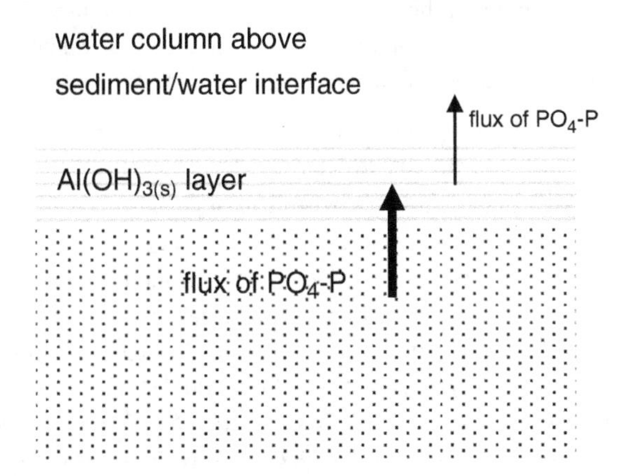

FIGURE P11.7 *Sketch of phosphorus release from sediments and capture by aluminum hydroxide layer above sediment-water interface.*

The presence of aluminum hydroxide solids in the capping layer provides control on the solubility of aluminum (i.e., $\{Al^{+3}\}$). Migration of PO_4–P into this layer creates the opportunity for formation of a second solid phase, aluminum phosphate ($AlPO_{4(s)}$). We know that if present, the equilibrium for this second solid must obeyed along with that of the first solid. Then, if pH is the master independent variable, the activity of the phosphate ion ($\{PO_4^{-3}\}$) may be determined as a function of the system pH. Through mass balance, then, the total solubility of phosphate phosphorus may also be determined as a function of pH.

Develop the relation that relates phosphate solubility and pH, assuming the presence of both aluminum hydroxide and aluminum phosphate, and produce a plot of $[PO_4–P_T]$ as a function of the system pH. The pH of these sediments is expected to vary between 6.5 and 8.5. Consider that the pore water has ionic strength of 0.005 M and that although not correct the temperature is 25 °C (*let us not deal with T corrections for equilibrium constants for this system as I am not sure the necessary database of enthalpy of formation values is at all readily accessible*). Note that *Lange's Handbook*, 14th ed., lists the pK_{sp} value for aluminum phosphate to be 18.24. Recall that the K_{sp} concept visualizes the equilibrium as the dissolution of the solid rather than the formation of the solid. *Conversely, Stumm and Morgan, in their Table 7.5 give a value of* $\log(K_{sp})$ *for the dissolution of variscite* ($AlPO_4{\cdot}2H_2O_{(s)} = Al^{+3} + PO_4^{-3} + 2H_2O$) *as −21. I think this value might yield results that will fall much closer to the observations made of behaviors of aluminum hydroxide in both lake restoration systems and wastewater treatment systems focused upon removal of phosphorus.*

8. An industrial plating process operated at 25 °C creates a waste stream containing 300 ppm_m hexavalent chromium (the chromic acid system). Due to mixing of other waste streams into the chromium laden flow stream the solution also will contain total acetate and total sulfate at concentrations of 0.01 and 0.02 M, respectively and will have an ionic strength of 0.05 M.

   One option for treatment of this waste stream would involve use of a sacrificial iron anode for creation of ferrous iron as an agent to reduce hexavalent chromium to trivalent chromium, with concomitant conversion of ferrous iron to ferric iron.

   The second phase of the process would use pH adjustment to precipitate the trivalent chromium as chromium (tri)hydroxide. The target level for total chromium in the effluent from this process is 100 $\mu g_{CrT}/L$.

   a. Assume that the iron/chromium oxidation/reduction process can virtually entirely convert Cr(VI) to Cr(III) on a stoichiometric basis and compute the quantity of iron necessary to accomplish the process in units of grams of iron per liter of waste water treated. *(We will leave this computation until we have examined the equilibria of electron transfer reactions. The solution to 4.(a) is not necessary to the solution of 4.(b)). For 4.(b), let us simply assume that all hexavalent chromium has been converted to trivalent chromium.)*
   b. Determine if this treatment objective may be met by the process and, if so, the lowest pH level (to the nearest 0.01 pH unit) at which the objective may be met. Also determine the lowest level to which total chromium can

be rendered and the pH (again, to the nearest 0.01 pH unit) at which this minimum would occur.

9. The EPA maximum contaminant levels set for mercury, cadmium, lead, and copper in drinking water are 2, 5, and 0 ppb_m, and 1.5 ppm_m, respectively, as the total metal. The hydroxide precipitation systems investigated in problems 2 and again, with complexing ligands other than hydroxide, in problem 4 likely cannot lower the total metal levels for mercury, cadmium, and lead to the respective MCLs. Certainly the presence of complexing ligands, particularly cyanide, lowers the probability that the MCL can be attained for copper. From the results of problem 5, verify these assertions.

   It is often claimed that precipitation of metals can generally be accomplished "stoichiometrically" (*i.e., addition of the sulfide reagent would result in nearly equal removal of the metal)* using sulfide to create the sulfide precipitate of the divalent metal ion. Consider the metal sulfide and develop a model that can be used to predict the total solubility of the metal and the sulfide as a function of pH, based on the addition of the metal sulfide to water of ionic strength equal to 0.01 M. Then, use the MCL as the total metal abundance to solve for the combination of metal ion and sulfide ion activities that satisfy the solid formation equilibrium. Use a mole balance on changes is metal and sulfide to determine the necessary quantity of added sulfide (e.g., NaHS) to attain the total metal at the MCL.

   a. $HgS_{(S)}$
   b. $CdS_{(S)}$
   c. $PbS_{(S)}$ *(Since zero is an unattainable value for abundance, in order to perform computations, the effective value of the MCL might be set at 1 or even 0.1 ppb_m. One might even search for the method detection limit for total lead using inductively coupled plasma spectroscopy and use that value.)*
   d. $CuS_{(S)}$

   Hint: If a MathCAD worksheet is to be employed, a solve block will be required. Since the values sought herein for the metal and sulfide activities are very small, if the formation equilibrium is written with the formation constant on the LHS, the solution might be rather unstable. However, if the activity of the sulfide ion is isolated on the LHS, the solve block should be quite stable.

10. Consider selected systems of problem 2: the initial total metal and total carbonate abundances are as stated in the table below and that the initial pH is 5. The target total metal abundances are based on the industrial pretreatment program of the City of Orlando, FL (City of Orlando, 2012).
    a. Cadmium: $C_{Tot.Cd.i}$ = 20 ppm_m, $C_{Tot.CO_3.i} = 0.0005$ M, $C_{Tot.Cd.Target}$ = 0.25 ppm_m.
    b. Copper: $C_{Tot.Cu.i}$ = 30 ppm_m, $C_{Tot.CO_3.i} = 0.001$ M, $C_{Tot.Cu.Target}$ = 0.75 ppm_m.
    c. Cobalt: $C_{Tot.Co.i}$ = 30 ppm_m, $C_{Tot.CO_3.i} = 0.001$ M, $C_{Tot.Co.Target}$ = 0.65 ppm_m.
    d. Nickel: $C_{Tot.Ni.i}$ = 30 ppm_m, $C_{Tot.CO_3.i} = 0.005$ M, $C_{Tot.Ni.Target}$ = 1.1 ppm_m.
    e. Zinc: $C_{Tot.Zn.i}$ = 30 ppm_m, $C_{Tot.CO_3.i} = 0.001$ M, $C_{Tot.Zn.Target}$ = 1.4 ppm_m.

f. Silver: $C_{Tot.Ag.i} = 30\ ppm_m$, $C_{Tot.CO_3.i} = 0.005\ M$, $C_{Tot.Ag.Target} = 5\ ppm_m$.

Note that the City of Orlando specifies 0.12 ppm_m as the limit for silver. The alternative level is attainable. An analysis such as that completed for Example 10.13 would be necessary to determine the behavior of the silver system, were we to examine the potential to reach the City of Orlando's proposed limit via carbonate precipitation.

Consider that the ionic strength of the aqueous solution is 0.01 M and can be considered constant. This is not true. As we have seen in Chapter 10, we can define the major background ions and include ionic strength and activity coefficients as unknowns. Here, let us focus upon the precipitation and complexation issues. Each of the aforementioned target total metal abundance levels should be attainable and therefore, two solve blocks should be written: one to obtain the pH and conditions at the point of metal carbonate solid formation and a second to determine the pH and conditions at the targeted total metal abundance:

a. Develop a mathematical model from which the pH at which a metal carbonate solid phase may be predicted based on the initial abundances of metal and carbonate. Ensure that initial total carbonate abundance and initial pH are left as master independent variables along with final system pH. Compare this with the pH at which a metal hydroxide solid phase would first form and determine whether initial control of the metal abundance would be through a carbonate or a hydroxide solid phase. If the metal hydroxide formation equilibrium would control the pH through the alkaline range, there is really no advantage to metal carbonate precipitation, and you need not consider parts (b) and (c) for the metal in question.
b. If the carbonate control would yield a lower total metal abundance, extend your model to predict the pH at which the carbonate solid phase would precipitate if sodium carbonate were used as the precipitative reagent. *Hints: (1) the total carbonate abundance must be the sum of the initial carbonate and the dose of sodium carbonate (and for part c the change in total metal abundance), and the carbonate ion activity may be defined using the total carbonate abundance, (2) the activities of the free metal and carbonate ions are related through the solid formation equilibrium, (3) the metal ion can be defined using the initial (or for part c the target) total metal abundance, and (4) the base neutralization capacity of the system between the initial state and the state at which the precipitate would form (or between the pH of precipitate formation and the target pH) is related to the quantity of carbonate added (i.e., the carbonate dose). Water, $M(OH)_2^0$ ($AgOH^0$ for the silver system), and carbonate make excellent choices as reference species for the proton balance defining the BNC.*
c. Extend the model of part b to predict the pH at which the target metal abundance will be obtained. Along with pH, the model of course will predict the activities of carbonate and of the free metal ion as well as the dose of carbonate necessary to attain the target total metal abundance.

Chapter 12

Oxidation and Reduction

12.1 PERSPECTIVE

In Chapters 6 and 10, we examined acids and bases. Then, in Chapter 11, we incorporated our understandings of acids and bases with the named processes of complexation and solubility/dissolution. Our next logical step is to address electron transfers, often called oxidation and reduction or simply redox processes. Elements exist in oxidation states other than their elemental form (in which valence is 0). In the elemental state, of course, the number of electrons in orbitals about the nucleus equals the number of protons held in the nucleus. Elements other than the noble gases tend to gain or lose electrons to render their electron shells more like those of the noble gases. When electrons are gained (the element becomes reduced), the charge becomes more negative, as the number of electrons in orbitals exceeds the number of protons in the nucleus. Conversely, when electrons are donated (the elements become oxidized), the charge becomes positive, as the number of protons in the nucleus now exceeds the number of electrons in orbitals. Then, in the vernacular of the processes, elements that have become reduced become potential reducing agents, as they have the capacity to donate electrons. Conversely, elements that have become oxidized become oxidizing agents as they have the capacity to accept donated electrons.

Complicating the whole set of understandings is the fact that once reduced or oxidized, the altered element is most often combined with other elements to form cations, anions, ion pairs, molecules, and crystalline or amorphous solids. One

Environmental Process Analysis: Principles and Modeling, First Edition. Henry V. Mott.

notable exception is sulfur, which when reduced can exist solely as the sulfide ion. When combined with other elements, a further complication exists in that electron pairs are most often shared between dissimilar elements. The electronegativity of each element can help us understand the probability that shared electrons would reside with that element when electrons are shared. The higher the electronegativity, the greater is the probability that shared electrons will reside in orbitals associated with the target element. The chemists have devised descriptions of electron "shells" for various electron-sharing scenarios. We could dig deeply into this probability aspect of the broad and deep topic of electron sharing. Were we to do so, this chapter would be much longer, and perhaps more interesting to selected readers. However, we will not go there. Many chemistry books already contain that information and, if needed, we know how to find it. Our intent here is to become users of the system chemistry has provided for us. The chemists have assembled their collective knowledge about electron transfers into a system, similar in many ways to that of acids and bases. It is this system that we strive to master in order that we may employ it for modeling of environmental processes.

For example, sulfur commonly exists in four distinct oxidation states: −2, +2, +4, and +6. From somewhere in the body of chemical knowledge, we can gain understandings as to why this is so. The chemists know why these are the preferred oxidation states. We will accept this behavior as fact, not necessarily needing to know why. However, we would like to know what causes sulfur to take on two or donate two, four, or six electrons and to use that understanding to model the existence of sulfur in its four oxidation states in natural or engineered systems. That "cause" is, of course, the electron availability or potential. The chemists tell us that electrons, much like protons, cannot exist as electrons in systems at equilibrium. The proton (or hydrogen ion), as we discussed in Chapter 10, can associate with a water molecule and exist, even under equilibrium conditions, as this association. The electron has no similar property. The chemists have developed a concept, called electron availability, that may be employed much as if it were the chemical activity of the electron as a distinct chemical specie. In acquiring our working knowledge of this concept, we should ensure we have a solid understanding of oxidation-reduction reactions.

12.2 REDOX HALF REACTIONS

A redox reaction involves the oxidation of one specie with simultaneous reduction of another. Electrons are transferred from the oxidized (originally the reduced) specie to the reduced (originally the oxidized) specie. Many of the reactions through which redox processes are accomplished are biological in nature. Biological entities are equipped with systems that efficiently carry out these redox processes via multiple steps. Fermentation to produce consumable alcohol is an important example. The major end products of such fermentation are ethanol and carbon dioxide. The original pool of electrons is available from sugar (let us consider glucose). We can follow the electrons from the various donors to the various acceptors throughout the process.

Then some of the carbon is reduced to be incorporated into ethanol, and some is oxidized to be incorporated into carbon dioxide. Then, we have two apparently simultaneous redox reactions occurring: reduction of glucose to become ethanol and oxidation of glucose to become carbon dioxide. More specifically, we might suggest that some of the carbon of glucose is reduced to become ethanol carbon and some of the carbon of glucose is oxidized to become carbon dioxide carbon (we have also called this inorganic carbon). We would break this overall process down into two (overall) half reactions—reduction of glucose to form ethanol and oxidation of glucose to form carbon dioxide. If we study further, we would find that all redox processes are the combination of a reduction and an oxidation. If we can identify the operative half reactions, we can assemble them into the overall reaction.

12.2.1 Assigning Oxidation States

Before we can write a half reaction involving an oxidized and a reduced specie, we must understand the oxidation state of the element, common to both the oxidized and reduced species, that experiences a loss or gain of electrons. The chemists have developed a set of rules, based mostly upon the magnitude of the electronegativity of the element, for assigning electrons to elements within species comprised of multiple elements.

1. Due to its low electronegativity, when hydrogen is combined with other elements, its electrons are always assigned to the element with which those electrons are shared. Then, with the exception of its combination with metals in metal hydrides, the oxidation state of hydrogen is always taken as +1. Note that hydrogen is covalently bonded in organic compounds. In this case, the carbon has slightly higher electronegativity than hydrogen and we most often assign the shared electrons to the carbon.
2. Due to its large electronegativity, when oxygen is combined with other elements, the shared electrons are assigned to oxygen. With the exception of peroxides, the oxidation state of oxygen when combined with other elements is always taken as −2. In peroxides, the oxidation state of oxygen is taken as −1. We can, armed with these two tools, determine the oxidation state of most elements in most combinations with oxygen and hydrogen.
3. A third rule, corollary to the first two, is that when combined in natural organic matter, the oxidation state of nitrogen is most often −3, as a consequence of its incorporation into the organic matter as an amine.
4. A given element combined with other elements to form a target specie can have an oxidation state that is not an integer. In such cases, the sharing of electrons cannot be attributed to individual covalent bonds.
5. When we algebraically add the products of the stoichiometric coefficients and oxidation states of the elements forming a defined chemical specie with a defined empirical chemical formula, the sum must equal the residual charge of the target chemical specie.

Let us put these rules into play.

Example 12.1 Consider the compounds ethanol, glucose, carbon dioxide, and phenol; the ions tetrathionate and citrate; and the empirical cell formula for the volatile portion of aerobic biomass, which we will call volatile biomass solids (VBS). Use the aforementioned five rules to assign oxidation states (average if necessary) to the elements other than oxygen and hydrogen comprising the target species.

First, we will obtain the chemical formulae for the targeted species: ethanol, CH_3CH_2OH; glucose, $C_6H_{12}O_6$; carbon dioxide, CO_2; phenol, C_6H_5OH; tetrathionate, $S_4O_6^=$; citrate, $(CO_2CH_2)_2CO_2COH^{-3}$ (or $C_6H_5O_7^{-3}$); and VBS, $C_5H_7O_2N$ (Tchobanoglous et al., 2003).

Noting that the oxidation states of oxygen, hydrogen, and nitrogen are −2, +1, and −3, respectively, we can assign the remainder. For the oxidation state (or valence, V) of carbon in ethanol for the empirical formula C_2H_6O, we have:

$$2 \cdot V_C + 6 \cdot (+1) + (-2) = 0; \quad V_C = -\frac{6(+1)+(-2)}{2} = -2$$

For glucose, we have:

$$6 \cdot V_C + 12 \cdot (+1) + 6 \cdot (-2) = 0; \quad V_C = -\frac{12 \cdot (+1) + 6 \cdot (-2)}{6} = 0$$

For carbon dioxide, we have:

$$V_C + 2 \cdot (-2) = 0; \quad V_C = -\frac{2 \cdot (-2)}{1} = +4$$

For phenol, we have:

$$6 \cdot V_C + 6 \cdot (+1) + (-2) = 0; \quad V_C = -\frac{6(+1)+(-2)}{6} = -\frac{2}{3}$$

For tetrathionate, we have:

$$4 \cdot V_S + 6 \cdot (-2) = -2; \quad V_S = \frac{-2-6(-2)}{4} = +\frac{5}{2}$$

For citrate (minus three protons from citric acid), we have:

$$6 \cdot V_C + 5 \cdot (+1) + 7 \cdot (-2) = -3; \quad V_C = \frac{-3-5 \cdot (+1) - 7 \cdot (-2)}{6} = +1$$

For the empirical cell formula for VBS, we have:

$$5 \cdot V_C + 7 \cdot (+1) + 2 \cdot (-2) + (-3) = 0; \quad V_C = -\frac{7 \cdot (+1) + 2 \cdot (-2) + (-3)}{5} = 0$$

12.2.2 Writing Half Reactions

Now that we can assign oxidation states to elements in target species, we are now able to better understand the writing of half reactions. Many of these have been written by the chemists, and we might be tempted to simply take what they have to offer and proceed. However, it is of great value to understand how half reactions arise since, in many environmental systems, we might be interested in particular species for which the half reactions are not necessarily easily found.

We begin by identifying the element that gains or loses electrons as a consequence of the targeted electron transfer reaction we wish to write. Most often we see that half reactions are written as reductions, with the oxidized specie on the LHS and the reduced specie on the RHS. Since this is the chemists' way, we will adopt their convention. Then, the specie containing the reduced element is situated on the RHS of the reaction and the specie containing the oxidized element is situated on the LHS.

1. We determine the oxidation states of the target element as present with the reduced and oxidized species.
2. We balance the target element between the LHS and RHS to ensure that we have an atomic balance.
3. Once the target element is balanced, we add the requisite number of electrons on the LHS to balance the change in the overall charge of the target element from the LHS to the RHS.
4. We balance elements other than oxygen and hydrogen, taking care to ensure that the most probable species are shown on the LHS and RHS of the half reaction.
5. We then balance the excess oxygen from the LHS with water on the RHS.
6. Lastly, we balance excess hydrogen from the RHS with protons on the LHS.

When we are finished, we will have both atomic and electron balances between the LHS and RHS. Most often the resultant reaction is normalized to a single electron transferred by dividing each stoichiometric coefficient by the number of electrons transferred. Sometimes this is most convenient, and other times using the half reaction as written may be the most convenient. User preference governs as long as the reaction is correctly employed. Let us get some practice with this algorithm.

Example 12.2 Develop the half reactions for the following redox couples and, where possible, compare your results with published half reactions: bicarbonate–methane; nitrate–nitrogen gas; sulfate–bisulfide; glucose–carbon dioxide; glucose–methane; VBS–methane (applicable in anaerobic digestion of wastewater biosolids); and VBS/carbon dioxide.

For the bicarbonate–methane couple, the oxidation state of carbon is +4 in bicarbonate and −4 in methane. We have one carbon in each specie, thus we begin with bicarbonate on the LHS and methane on the RHS and require eight electrons to balance the oxidation states of carbon in the two species:

$$HCO_3^- + 8e^- + \Leftrightarrow CH_{4(g)} +$$

We have no elements to balance other than carbon, oxygen, and hydrogen, so we balance the oxygen from the LHS with water on the RHS:

$$HCO_3^- + 8e^- + \Leftrightarrow CH_{4(g)} + 3H_2O$$

Lastly, we balance the excess hydrogen from the RHS with hydrogen ions on the LHS:

$$HCO_3^- + 8e^- + 9H^+ \Leftrightarrow CH_{4(g)} + 3H_2O$$

If we wish to make this look like the most often published half reaction, we simply add the reaction for the deprotonation of carbonic acid to yield bicarbonate and the reaction for the dissolution of carbon dioxide to form carbonic acid to the initially written half reaction. In doing so, we need to ensure that we multiply the equilibrium constants (or add the $\log_{10}K$ values to obtain the overall $\log_{10}K$ or add the pK values to obtain the overall pK):

$$
\begin{array}{llll}
HCO_3^- + 8e^- + 9H^+ & \Leftrightarrow CH_{4(g)} + 3H_2O & K_{HCO3.CH4} \\
H_2CO_3^* & \Leftrightarrow HCO_3^- + H^+ & K_{A1.CO3} \\
CO_{2(g)} + H_2O & \Leftrightarrow H_2CO_3^* & K_{H.CO2} \\
\hline
CO_{2(g)} + 8e^- + 8H^+ & \Leftrightarrow CH_{4(g)} + 2H_2O & K_{CO2.CH4} = K_{HCO3.CH4} \cdot K_{A1.CO3} \cdot K_{H.CO2}
\end{array}
$$

For the nitrate–nitrogen gas couple, we have two nitrate ions on the LHS to balance the single diatomic nitrogen atom on the RHS. The oxidation state of nitrogen in nitrate is +5 and that of nitrogen in elemental nitrogen gas is 0, so ten electrons are necessary to balance the oxidation states:

$$2NO_3^- + 10e^- + \Leftrightarrow N_{2(g)} +$$

Nitrogen is the only element other than oxygen and hydrogen, so we balance excess LHS oxygen with RHS water:

$$2NO_3^- + 10e^- + \Leftrightarrow N_{2(g)} + 6H_2O$$

Lastly, we balance excess RHS hydrogen with LHS hydrogen ions:

$$2NO_3^- + 10e^- + 12H^+ \Leftrightarrow N_{2(g)} + 6H_2O \quad K_{NO3.N2}$$

For the sulfate–bisulfide couple, we have a single sulfur on each side of the reaction with oxidation states of +6 in sulfate and −2 in bisulfide, requiring eight electrons on the LHS:

$$SO_4^{=} + 8e^- + \quad \Leftrightarrow HS^- +$$

Sulfur is the only element other that oxygen and hydrogen, so we balance excess oxygen from the LHS with water on the RHS:

$$SO_4^{=} + 8e^- + \quad \Leftrightarrow HS^- + 4H_2O$$

Lastly, we balance excess hydrogen from the RHS with hydrogen ions on the LHS:

$$SO_4^{=} + 8e^- + 9H^+ \Leftrightarrow HS^- + 4H_2O \quad K_{SO4.HS}$$

For the glucose–carbon dioxide couple, we have six carbons on the RHS in glucose, with average oxidation state of 0, thus we need six carbon dioxide molecules on the LHS, in which carbon has a +4 oxidation state. We therefore need 24 electrons on the LHS:

$$6CO_{2(g)} + 24e^- + \Leftrightarrow C_6H_{12}O_6 +$$

Carbon is our only element other than oxygen and hydrogen, so we may get right to the balancing of the excess LHS oxygen with RHS water:

$$6CO_{2(g)} + 24e^- + \Leftrightarrow C_6H_{12}O_6 + 6H_2O$$

We finish this off by adding two dozen hydrogen ions to the LHS to balance the excess RHS hydrogen:

$$6CO_{2(g)} + 24e^- + 24H^+ \Leftrightarrow C_6H_{12}O_6 + 6H_2O \quad K_{CO2.glu}$$

For the glucose–methane couple, we have six carbons at 0 average oxidation state in glucose on the LHS and thus need six methane molecules on the RHS in which carbon has a –4 oxidation state. Twenty-four electrons are necessary to balance the oxidation states:

$$C_6H_{12}O_6 + 24e^- + \quad \Leftrightarrow 6CH_{4(g)} +$$

Carbon is our target element beyond oxygen and hydrogen, and we may balance the excess LHS oxygen with RHS water and the resultant excess RHS hydrogen with LHS hydrogen ions:

$$C_6H_{12}O_6 + 24e^- + 24H^+ \Leftrightarrow 6CH_{4(g)} + 6H_2O \quad K_{glu.CH4}$$

For the VBS–methane couple, we have five carbon atoms with average oxidation state equal to 0 on the LHS requiring five methane molecules in which carbon has a –4 oxidation state on the RHS. Twenty electrons are needed on the LHS:

$$C_5H_7O_2N + 20e^- + \quad \Leftrightarrow 5CH_{4(g)} +$$

We have two elements (C and N) other than oxygen and hydrogen. We expect the organic nitrogen, present in the biomass as amines, will be released as ammonium given neutral pH of the process:

$$C_5H_7O_2N + 20e^- + \quad \Leftrightarrow 5CH_{4(g)} + NH_4^+ +$$

We can now balance the excess LHS oxygen with RHS water and the excess RHS hydrogen with LHS hydrogen ions:

$$C_5H_7O_2N + 20e^- + 21H^+ \quad \Leftrightarrow 5CH_{4(g)} + NH_4^+ + 2H_2O \quad K_{VBS.CH4}$$

For the VBS–carbon dioxide couple, we have five carbons with average oxidation state equal to 0 on the RHS, requiring five carbon dioxide molecules, in which carbon has a +4 oxidation state, on the LHS. We need twenty electrons to reduce the five carbons of carbon dioxide to the five carbons of VBS:

$$5CO_{2(g)} + 20e^- ++ \quad \Leftrightarrow C_5H_7O_2N +$$

We know that nitrogen will have an oxidation state of −3 in VBS, and since the process would occur at neutral pH, we need an ammonium on the LHS:

$$5CO_{2(g)} + 20e^- + NH_4^+ + \Leftrightarrow C_5H_7O_2N +$$

We can now balance the excess LHS oxygen with RHS water:

$$5CO_{2(g)} + 20e^- + NH_4^+ + \Leftrightarrow C_5H_7O_2N + 8H_2O$$

Then we balance excess RHS hydrogen with LHS hydrogen ions:

$$5CO_{2(g)} + 20e^- + NH_4^+ + 21H^+ \quad \Leftrightarrow C_5H_7O_2N + 8H_2O \quad K_{CO2.VBS}$$

We observe that we may quite easily write half reactions as long as we know the residual charge of both the oxidized and reduced species and can identify the oxidation state of elements other than oxygen and hydrogen that are associated with the reactants or products.

12.2.3 Adding Half Reactions

Overall redox reactions involve the transfer of electrons. Thus, we need an electron donor and an electron acceptor to write an overall redox reaction. Once we know the donor and the acceptor, we can select (or write if necessary) the corresponding half reactions. Since half reactions are generally written as reductions, we must reverse the direction of the half reaction when we are considering the donor. We then algebraically add the two reactions ensuring that the electrons necessary for the reduction, present on the LHS, will balance the electrons donated, present on the RHS. This process is best explained and illustrated by example.

Example 12.3 Write the overall reaction in which VBS as a carbon source would be oxidized to carbon dioxide in a biological process. Then VBS would be the electron donor. Consider an aerobic biological process in which oxygen would be the electron acceptor alongside an anaerobic process, in which carbon dioxide would be the electron acceptor. In the aerobic process, oxygen is a reactant, consumed during the biological process, producing carbon dioxide. In the anaerobic process, carbon dioxide is a reactant, consumed by the reaction, and methane is the product produced by the biological process.

We must write (or certainly we could find it in Table A.3) one additional half reaction—that for the reduction of molecular oxygen to water. The process of Example 12.2 is applied. Oxygen has an oxidation state of 0 in molecular oxygen:

$$O_{2(g)} + 4e^- + 4H^+ \Leftrightarrow 2H_2O \qquad K_{O2.H2O}$$

For the aerobic redox reaction, we must reverse the reduction of carbon dioxide:

$$C_5H_7O_2N + 8H_2O \Leftrightarrow 5CO_{2(g)} + 20e^- + NH_4^+ + 21H^+ \quad 1/K_{CO2.VBS}$$

Then, to balance the electrons donated with the electrons accepted, we need five moles of oxygen:

$$5O_{2(g)} + 20e^- + 20H^+ \Leftrightarrow 10H_2O \qquad (K_{O2.H2O})^5$$

We may now add the two reactions to obtain the overall reaction, cancelling as necessary such that species are not indicated on both sides of the reaction:

$$C_5H_7O_2N + 5O_{2(g)} \rightarrow 5CO_{2(g)} + 2H_2O + NH_4^+ + H^+$$

We observe that five moles of oxygen are required to oxidize one unit empirical formula of VBS. When we convert this stoichiometry to mass units, we obtain the result that 1.415 g of oxygen are required to oxidize each gram of VBS fully to carbon dioxide and water, releasing a mole of ammonia and a mole of protons. This overall redox reaction has importance in determining oxygen requirements for aerobic biological processes. Of note, we have not included the equilibrium constant for the derived redox reaction. There would be no point in doing so as this is not really an equilibrium relationship but merely a stoichiometric representation of the overall reaction.

For the anaerobic reaction, considering carbon dioxide as an electron acceptor, we may use the half reaction for reduction of carbon dioxide as written:

$$CO_{2(g)} + 8e^- + 8H^+ \Leftrightarrow CH_{4(g)} + 2H_2O$$

$$K_{CO2.CH4} = K_{HCO3.CH4} \cdot K_{A1.CO3} \cdot K_{H.CO2}$$

In order to balance electrons from the RHS and LHS, we need 2½ moles of carbon dioxide to accept the twenty electrons donated by VBS:

$$C_5H_7O_2N + 8H_2O \Leftrightarrow 5CO_{2(g)} + 20e^- + NH_4^+ + 21H^+ \qquad 1/K_{CO2.VBS}$$

$$2\tfrac{1}{2}CO_{2(g)} + 20e^- + 20H^+ \Leftrightarrow 2\tfrac{1}{2}CH_{4(g)} + 5H_2O$$

$$K_{CO2.CH4} = K_{HCO3.CH4} \cdot K_{A1.CO3} \cdot K_{H.CO2}$$

We add the two half reactions in the same manner as for the aerobic redox reaction, neglecting the overall reaction equilibrium constant, as again, this result is useful strictly for stoichiometry:

$$C_5H_7O_2N + 3H_2O \rightarrow 2\tfrac{1}{2}CO_{2(g)} + 2\tfrac{1}{2}CH_{4(g)} + NH_4^+ + H^+$$

We observe that carbon dioxide is both a reactant and a product, but in the end, overall, a product. We observe also that each unit cell formula of VBS can yield 2½ moles of methane gas. This stoichiometry is the basis for the anaerobic digestion of wastewater biosolids at wastewater treatment facilities. When we convert the stoichiometry to the mass units often preferred in engineering computations, we find that full conversion of each kilogram of volatile biosolids via anaerobic digestion would yield ~0.532 standard cubic meters (scm) of methane ($0.385\ kg_{CH_4}/kg_{VBS}$;~ 8.54 standard cubic feet, scf, per pound). Note that the engineers use 20 °C as their standard temperature. A standard liter is the quantity of gas occupying a volume of one liter at standard conditions, and hence actually a molar (or mass) quantity expressed as a standardized volume. When we consider that the potential heating value of methane is ~3.73×10^4 kJ/scm (~1000 BTU/scf), we immediately understand the interest in energy recovery from anaerobic digestion of wastewater biosolids.

We observe that half reactions are used algebraically in much the same manner as we have used all of our other varieties of reactions. The exception here is that by combining two half reactions to form an overall redox reaction, we do not obtain a result that is useful for equilibrium computations. Nonetheless, the stoichiometric relations that can be developed using redox half reactions prove quite valuable in many process or system modeling efforts. The process employed in Example 12.3 can be used to obtain the stoichiometry for conversion of any biodegradable organic substance to methane and carbon dioxide, as long as the elemental composition and oxidation state of the substance are known. In the absence of the chemical formula, we would resort to an alternative characterization of the organic substance, most often the chemical oxygen demand (COD).

The test for COD is used throughout the wastewater treatment industry as a means to characterize the "strength" of oxygen-demanding wastes. Similarly, the test for biochemical oxygen demand (BOD) is also employed. We recall that the test for 5-day BOD is conducted over a 5-day period and relies upon either seeded or indigenous bacteria to oxidize organic material contained in a sample over the 5-day period.

We measure the initial and final dissolved oxygen levels and normalize the difference to the volume of sample incorporated into the BOD bottle. Conversely, for the COD test, a sample is combined with a strong oxidizing agent (dichromate) in a highly acidic medium and digested at 150 °C for 2 h. The COD is then based upon the consumption of the dichromate during the test. Dichromate (Cr(VI)) is reduced to Cr(III) by electrons released from the organic matter during the test. The organic matter, in the highly oxidizing and acidic environment at high temperature, is completely oxidized to carbon dioxide and products, releasing its electrons to the provided acceptor.

Example 12.4 Use glucose as the model organic matter and develop the stoichiometric relation between dichromate and oxygen as electron acceptors for measurement of the "strength" of aqueous samples containing biodegradable organic matter.

We must write (or obtain from the literature) the half reaction for the reduction of chromium (VI) of dichromate ($Cr_2O_7^=$) to chromium (III). We will write it, applying the algorithm to obtain the desired half reaction:

$$Cr_2O_7^= + 6e^- + 14H^+ \Leftrightarrow 2Cr^{+3} + 7H_2O$$

We combine this half reaction with that written for the oxidation of glucose to obtain the overall stoichiometric relation. Since glucose is to be oxidized, we reverse the direction. Since a mole of glucose donates 24 electrons and a mole of dichromate accepts but 6, we need 4 mol of dichromate per mole of glucose:

$$C_6H_{12}O_6 + 6H_2O \Leftrightarrow 6CO_{2(g)} + 24e^- + 24H^+$$
$$4Cr_2O_7^= + 24e^- + 56H^+ \Leftrightarrow 8Cr^{+3} + 28H_2O$$
$$C_6H_{12}O_6 + 4Cr_2O_7^= + 32H^+ \rightarrow 6CO_{2(g)} + 22H_2O$$

So now we have what we might term as the chemical dichromate demand—four moles of dichromate per mole of glucose.

Then for the conversion to COD, we must employ oxygen as the electron acceptor and combine the oxidation of glucose with the reduction of oxygen to water. The aforementioned process, repeated for the glucose–oxygen combination, yields a second stoichiometric relation:

$$C_6H_{12}O_6 + 6H_2O \Leftrightarrow 6CO_{2(g)} + 24e^- + 24H^+$$
$$6O_{2(g)} + 24e^- + 24H^+ \Leftrightarrow 12H_2O$$
$$C_6H_{12}O_6 + 6O_{2(g)} \rightarrow 6CO_{2(g)} + 6H_2O$$

We can find this particular result in a plethora of textbooks. The resulting stoichiometry yields the theoretical oxygen demand associated with the complete biological oxidation of glucose.

In understanding the basis for the COD test, we need not bother with the results other than the ratio of oxygen consumption to dichromate consumption. Six moles of oxygen (192 g) are needed to do the same job as four moles (864 g) of dichromate—

completely oxidize one mole (180 g) of glucose to carbon dioxide. A nuance here, often missed in explanations, is that in using dichromate to oxidize the glucose, the entire theoretical oxygen demand is satisfied. Thus, in most cases, the theoretical oxygen demand and COD are identical. One is a theoretical calculation and the other is a chemical measurement. In the absence of the theoretical calculation, which cannot be performed unless we know the exact chemical formula of the oxidized organic, we may consider that the COD is the theoretical oxygen demand.

Conversion of chemical dichromate demand to COD is straightforward. A quantity of four moles (864 g) of dichromate is equivalent in electron accepting ability to 6 mol (192 g) of oxygen. Thus, the conversion factor is 1.5 mol theoretical oxygen demand per mole of dichromate. In mass units, the conversion factor would then be 0.222 g theoretical oxygen demand per gram dichromate consumed.

In Example 12.4, we have illustrated the calibration of oxygen demand equivalent of dichromate as an oxidant using glucose as a target compound. We could have used any known biodegradable organic substance. It is all about the electrons transferred and accepted. The developers of the specific COD test suggest the use of biphthalate (hydrogen phthalate) in connection with the COD test. By oxidizing known quantities of biphthalate alongside unknown quantities of organic matter of undefinable chemical formula employing samples of equal size, the test and reagents may be calibrated using the principles and process employed in this example.

12.2.4 Equilibrium Constants for Redox Half Reactions

In Chapter 10, we introduced the relationship between the standard Gibbs energy of reaction $\left(\Delta G^{\circ}_{\text{rxn}}\right)$ and the equilibrium constant. One example was presented, and since most equilibrium constants for proton transfer reactions are known and published, we decided to forgo further implementation, opting to wait and apply Gibbs energy principles in examination of redox half reactions. Given that we might wish to develop a half reaction based on the chemical species we can identify in a particular natural or engineered system, we desire the flexibility to obtain equilibrium constants for those specific reactions. Significant Gibbs energy data are available from Table A.1. Additionally, Gibbs energy data for selected geochemical systems have been assembled in Table A.4. There may be some overlap and some disagreement for specific species between the two tables as data are derived from varying sources. We have included Tables A.1 and A.4 with this text for the convenience of the learner. Certainly, Gibbs energy data are available for many additional systems beyond those given in Tables A.1 and A.4. In order to obtain values for equilibrium constants for any half reactions we might write, we need only use data such as those in Tables A.1 and A.4 with Equation 10.7b

$$\Delta G^{\circ}_{\text{rxn}} = \sum_i \nu_i \Delta G^{\circ}_{f,i_{\text{products}}} - \sum_i \nu_i \Delta G^{\circ}_{f,i_{\text{reactants}}} \qquad (10.7\text{b})$$

and employ the computed $\Delta G^{\circ}_{\text{rxn}}$ in Equation 10.10b:

$$K_{\text{eq}} = e^{-\frac{\Delta G^{\circ}_{\text{rxn}}}{RT}} \qquad (10.10\text{b})$$

Certainly, also, we may find equilibrium constants for specific redox half reactions in the scientific literature. We have populated Table A.5 with some key half reactions, including those for reduction of metals to their elemental or other solid states and others that address systems common in environmental systems.

Example 12.5 Determine, using Gibbs energy data, the equilibrium constants for the first four half reactions written with Example 12.2, restated here for convenience. Compare the results with published values that might be available from Tables A.5:

$$HCO_3^- + 8e^- + 9H^+ \Leftrightarrow CH_{4(g)} + 3H_2O \quad K_{HCO3.CH4}$$
$$CO_{2(g)} + 8e^- + 8H^+ \Leftrightarrow CH_{4(g)} + 2H_2O \quad K_{CO2.CH4} = K_{HCO3.CH4} \cdot K_{A1.CO3} \cdot K_{H.CO2}$$
$$2NO_3^- + 10e^- + 12H^+ \Leftrightarrow N_{2(g)} + 6H_2O \quad K_{NO3.N2}$$
$$SO_4^{=} + 8e^- + 9H^+ \Leftrightarrow HS^- + 4H_2O \quad K_{SO4.HS}$$

We assemble the Gibbs energy data (all are available from Table A.1) and assign the values to scalar variables ΔG_i° (not shown herein). Note that the unit of energy in Table A.1 is kJ while that in Table A.4 it is kcal. The value we use for the gas constant should match the energy unit for G_f°. We recall that by definition the Gibbs energy of formation for elements, a proton, and an electron are 0. We apply Equation 10.7b:

$$\begin{pmatrix} \Delta G^\circ_{HCO3.CH4} \\ \Delta G^\circ_{CO2.CH4} \\ \Delta G^\circ_{NO3.N2} \\ \Delta G^\circ_{SO4.HS} \end{pmatrix} := \begin{pmatrix} \Delta G^\circ_{CH4} + 3\cdot\Delta G^\circ_{H2O} - \Delta G^\circ_{HCO3} \\ \Delta G^\circ_{CH4} + 2\cdot\Delta G^\circ_{H2O} - \Delta G^\circ_{CO2} \\ 6\cdot\Delta G^\circ_{H2O} - 2\cdot\Delta G^\circ_{NO3} \\ \Delta G^\circ_{HS} + 4\cdot\Delta G^\circ_{H2O} - \Delta G^\circ_{SO4} \end{pmatrix} = \begin{pmatrix} -175.5 \\ -130.8 \\ -1200.5 \\ -192.1 \end{pmatrix} \frac{kJ}{mol}$$

Once we have the standard Gibbs energy of reaction, we may compute the equilibrium constants using Equation 10.10b and convert them to log10 values for comparison with Table A.5:

$$\begin{pmatrix} K_{HCO3.CH4} \\ K_{CO2.CH4} \\ K_{NO3.N2} \\ K_{SO4.HS} \end{pmatrix} := \begin{pmatrix} e^{-\frac{\Delta G^\circ_{HCO3.CH4}}{R_{kJ}\cdot T}} \\ e^{-\frac{\Delta G^\circ_{CO2.CH4}}{R_{kJ}\cdot T}} \\ e^{-\frac{\Delta G^\circ_{NO3.N2}}{R_{kJ}\cdot T}} \\ e^{-\frac{\Delta G^\circ_{SO4.HS}}{R_{kJ}\cdot T}} \end{pmatrix} \quad \begin{pmatrix} \log(K_{HCO3.CH4}) \\ \log(K_{CO2.CH4}) \\ \log(K_{NO3.N2}) \\ \log(K_{SO4.HS}) \end{pmatrix} = \begin{pmatrix} 30.75 \\ 22.91 \\ 210.33 \\ 33.65 \end{pmatrix}$$

We observe fairly close agreement 22.91 versus 22.96 for the second reaction, 210.33 versus 210.5 for the third reaction, and 33.65 versus 34.0 for the fourth reaction. In order to compare our result for the first reaction with data from Table A.5, we need to subtract two reactions to convert the second half reaction to the first. When we add reactions, we multiply the corresponding equilibrium constants. Conversely, when we subtract one reaction from another, we must divide the equilibrium constant of the first by the equilibrium constant of the second. We have set this up somewhat in Example 12.2: $K_{CO2.CH4} = K_{HCO3.CH4} \cdot K_{A1.CO3} \cdot K_{H.CO2}$. Thus, for application here we rearrange to solve for $K_{HCO3.CH4}$:

$$\begin{pmatrix} K_{A1.CO3} \\ K_{H.CO2} \end{pmatrix} := \begin{pmatrix} 10^{-6.35} \\ 0.0339 \end{pmatrix} \qquad K_{HCO3.CH4} := \frac{K_{CO2.CH4}}{K_{A1.CO3} \cdot K_{H.CO2}}$$

$$\log(K_{HCO3.CH4}) = 30.73$$

We observe that our $\log_{10} K_{HCO3.CH4}$ value of 30.75 computed from Gibbs energy matches closely with our derived value of 30.73 based on the tabulated values of $\log_{10} K_{CO2.CH4}$, $pK_{1.CO3}$ and $K_{H.CO2}$.

The chemists have chosen to organize their redox equilibrium data for the most part by writing their half reactions based on the transfer of one electron. The symbol most often used to represent the equilibrium constant is $pE°$ or some variation thereof. The precise definition is as follows:

$$pE° = \frac{1}{n} \log_{10}\left(K_{ox.red}\right) \tag{12.1a}$$

The integer n is the number of electrons transferred when the reaction is written as we have done in Example 12.2. $K_{ox.red}$ is the equilibrium constant that would be obtained from Gibbs energy for the reaction as written in Example 12.2: the reduction of the oxidized specie to the reduced specie. Prior to the age of the computers, when computations beyond simple arithmetic to more than one or two significant figures were cumbersome at best, logarithms were used extensively in computations. The use of logarithms to represent equilibrium constants was more than simple presentation convenience. Much of the computation was actually completed using logarithms. In employing redox equilibria for environmental process modeling, we find it very useful (and often necessary) to use the $K_{ox.red}$ values directly in computations. In converting equilibrium constants from existing data sources, the inverse of Equation 12.1a is conveniently employed:

$$K_{ox.red} = 10^{n \cdot pE°} \tag{12.1b}$$

We have the choice of employing the integer n in the antilogarithms to obtain equilibrium constants for the full half reactions as written in Table A.5, or for direct use of $pE°$, we may employ fractional stoichiometric coefficients in the half reactions as used by Stumm and Morgan (1996) in their Table 8.6a. Each user of the system may certainly make the individual choice.

12.3 THE NERNST EQUATION

Natural and engineered environmental systems are often viewed as being similar to the direct current electrical circuits of standard electrochemical cells. Here, we will briefly review the concept of the electrochemical cell. We encourage the reader to peruse excellent discussions of standard electrochemical cells presented by Snoeyink and Jenkins (1980), Morel and Hering (1993), and Stumm and Morgan (1996) to gain deeper understandings. In electrochemical cells, the oxidation occurs at the anode. Electrons are donated corresponding with the oxidation and dissolution of the anode, typically of elemental metal. The reduction occurs at the cathode. Electrons are usually accepted by a metal ion in solution with associated deposition upon the cathode, also usually of solid elemental metal. The cell potential arises as a consequence of the combination of the two electrode half reactions. The availability of electrons is related to both the cell potential and the abundances of the oxidized and reduced species in the aqueous solutions of the cells. The well-known Nernst equation is most often cited as the basis for determining electron availability of electrochemical cells:

$$E_{\mathrm{H}} = E_{\mathrm{H}}^{\circ} + \frac{R_{\mathrm{CV}} \cdot T \cdot \ln(10)}{n \cdot F} \log\left(\frac{\prod_{\mathrm{ox}} \{i_{\mathrm{ox}}\}^{i}}{\prod_{\mathrm{red}} \{i_{\mathrm{red}}\}^{i}}\right) \tag{12.2}$$

E_{H} is the overall cell potential (V). E_{H}° is the standard potential (all reactants and products would be at 1 M activity). R_{CV} is the gas constant; a joule is a coulomb·volt, thus $R = 8.3144 \dfrac{\mathrm{C \cdot V}}{\mathrm{mol \cdot {}^{\circ}K}}$. The integer n is the number of electrons transferred via the overall redox reaction. F is the Faraday constant (96,485 coulombs per electron mole). The chemical species, other than electrons, involved in the oxidation and reduction sides of the electrochemical cell are i_{ox} and i_{red}, respectively. The ν_{i} are the stoichiometric coefficients of the respective species. The ratio $\dfrac{R_{\mathrm{CV}} \cdot T \cdot \ln(10)}{F}$ then yields the volt as the derived unit. The symbols Π_{ox} and Π_{red} indicate the products of the LHS oxidized species and RHS reduced species, respectively.

For electrochemistry, E_{H} is most conveniently measured in volts. For environmental systems, we would prefer a measurement that is analogous to chemical activity. The LHS and RHS of the Nernst equation are multiplied by the factor $\dfrac{F}{R_{\mathrm{CV}} \cdot T \cdot \ln(10)} (= 16.91)$ to seemingly render each of the three terms unitless:

$$\mathrm{p}E^{\circ} = \frac{F}{R_{\mathrm{CV}} \cdot T \cdot \ln(10)} E_{\mathrm{H}}^{\circ} \quad \text{and} \quad \mathrm{p}E = \frac{F}{R_{\mathrm{CV}} \cdot T \cdot \ln(10)} E_{\mathrm{H}} \tag{12.3}$$

Then the Nernst equation can be restated:

$$\mathrm{p}E = \mathrm{p}E^{\circ} + \frac{1}{n} \log\left(\frac{\prod_{\mathrm{ox}} \{i_{\mathrm{ox}}\}^{i}}{\prod_{\mathrm{red}} \{i_{\mathrm{red}}\}^{i}}\right) \tag{12.4}$$

In the Nernst equation, pE appears to be a dimensionless quantity, much like the equilibrium constant. However, we remember from Chapter 4 that with a law of mass action statement for a homogeneous equilibrium, the equilibrium constant indeed would have units, but the chemists have chosen to ignore the fact as we always express all abundances as chemical activities. Then, in the Nernst equation, pE represents $-\log\{e^-\}$ and therefore $\{e^-\}$ must have units of moles per liter of chemical activity.

Example 12.6 Reconcile the Nernst equation with the law of mass action for a half reaction, such as that for the reduction of sulfate to form hydrogen sulfide gas.

This half reaction is included in Table A.5. We write the law of mass action statement for the reduction of sulfate to hydrogen sulfide gas:

$$K_{SO_4.H_2S.g} = \frac{\prod_{red}\{i_{red}\}^i}{\prod_{ox}\{i_{ox}\}^i} = \frac{P_{H_2S}\cdot\{H_2O\}^4}{\{SO_4^=\}\cdot\{H^+\}^{10}\cdot\{e^-\}^8} = \frac{P_{H_2S}}{\{SO_4^=\}\cdot\{H^+\}^{10}\cdot\{e^-\}^8}$$

We will consider that we have a fairly dilute aqueous solution, so the activity of water is unity and can be dropped from the relation. We drop the subscript on K for simplicity, separate the RHS into the product of two fractions, and take the logarithms of both sides of the relation:

$$\log(K) = \log\left(\frac{P_{H_2S}}{\{SO_4^=\}\cdot\{H^+\}^{10}}\right) - 8\ \log\{e^-\}$$

Dividing through by 8 ($n = 8$) allows us to isolate $\log\{e^-\}$:

$$\frac{1}{8}\log(K) = \frac{1}{8}\log\left(\frac{P_{H_2S}}{\{SO_4^=\}\cdot\{H^+\}^{10}}\right) - \log\{e^-\}$$

Generalization of the number of electrons, substitution of p$E°$ for $(1/n)\log(K)$ and pE for $-\log\{e^-\}$, rearranging the relation, and inverting the argument of the log to eliminate the negative sign yield the revised form of the Nernst equation, written for the sulfate–hydrogen sulfide half reaction:

$$pE = pE°_{SO4.H2S} + \frac{1}{8}\log\left(\frac{\{SO_4^=\}\cdot\{H^+\}^{10}}{P_{H_2S}}\right)$$

Before we had convenient access to computers, the logarithmic form of the Nernst equation was thought to be the most convenient. Nowadays, with powerful computational capacity available at our fingertips, implementable via the click of a mouse or tap of a finger on a touch pad, direct use of the law of mass action relation seems most straightforward.

12.4 ELECTRON AVAILABILITY IN ENVIRONMENTAL SYSTEMS

In order that we may gain understandings regarding electron availability, we should paint a usable and understandable picture of the flow of electrons in systems. If we can categorize the condition of the system as steady-state or even quasi-steady-state, whether we are addressing a natural system or an engineered system matters little. We know that conditions within natural systems always change, while in many engineered systems we can impose the steady-state condition. The Earth rotates on its axis leading to day and night; it revolves around the sun, leading to the four seasons, specific to each given point on the Earth's surface. Thus, natural systems are inherently transient in their behaviors. However, the presence of a water column or set of soil strata overlying a system of interest dampens the natural temporal effect. As a consequence of the temporal nature of our Earth, there are many systems that we could never categorize as even quasi-steady-state. Most of these exist directly at the Earth's surface. Fortunately, there are also many that change sufficiently slowly that we can call them quasi- or near-steady-state.

The standard electrochemical cell is the basis of the battery and is inherently a batch process. The oxidation products of the anodic reaction build up in the anode half of the cell, and the reactants necessary to the reduction in the cathodic reaction of cathode half of the cell are depleted in the aqueous solution of the cathode. However, if we could arrange for the oxidation products to be flushed from the system and for the reduction reactants to be continuously supplied, the process could be brought to the steady- or quasi-steady-state condition. Metallurgical engineers use this concept to great benefit in metal recovery and purification. As long as the anode remains present (typically a solid metal rod that exerts unit activity across the liquid–solid interface), the process could proceed at a steady rate. In a targeted environmental system (e.g., an engineered anaerobic biological reactor, natural sediments situated beneath the sediment–water interface of a eutrophic lake, or porous media down-gradient of a chemical release through which ground water affected by an organic input would flow), we can imagine the system to be similar to a replenished electrochemical cell. Electron donors (reduced carbon of biodegradable organic matter) are the anodes, and electron acceptors are the cathodes.

In extracting energy from electron-rich donors, biological life forms utilize the most energy-efficient electron acceptors. We would gauge the thermodynamic ease with which electron acceptors accept electrons by computing the Gibbs energy of the acceptance half reaction. The larger the release of energy per electron accepted, the more preferred is the electron acceptor. We may use the results of Example 12.5 and, using the same process, determine the standard Gibbs energy for the reductions of molecular oxygen to water, of nitrate to nitrogen gas, of sulfate to hydrogen sulfide gas, and of protons to molecular hydrogen. In this comparison, it is best to compare all five electron acceptors across the vapor–liquid interface. Then when we normalize the standard Gibbs energies of reaction to the number of electrons transferred, we obtain the following:

$$\begin{pmatrix} \Delta G^\circ_{e.O2.H2O} \\ \Delta G^\circ_{e.NO3.N2} \\ \Delta G^\circ_{e.SO4.H2S.g} \\ \Delta G^\circ_{e.CO2.CH4} \\ \Delta G^\circ_{e.H.H2} \end{pmatrix} := \begin{pmatrix} \frac{\Delta G^\circ_{O2.H2O}}{4} \\ \frac{\Delta G^\circ_{NO3.N2}}{10} \\ \frac{\Delta G^\circ_{SO4.H2S.g}}{8} \\ \frac{\Delta G^\circ_{CO2.CH4}}{8} \\ \frac{\Delta G^\circ_{H.H2}}{2} \end{pmatrix} = \begin{pmatrix} -237.2 \\ -120 \\ -29.7 \\ -16.3 \\ 0 \end{pmatrix} \frac{\text{kJ}}{\text{mol}\cdot\text{electron}}$$

We quickly observe that, of the five, molecular oxygen would be the most preferred electron acceptor and that the proton would be the least preferred. Elemental halogens as well as oxyanions such as permanganate, persulfate, perchlorate, and dichromate are more preferred than is oxygen as electron acceptors, but we seldom find these existing under even quasi-steady-state conditions in natural systems. Within the preference range between oxygen and carbon dioxide, we can find myriad electron acceptors, which, if present, will contribute to the overall flow of electrons.

Organic matter is most often the source of electrons. In most cases, we do not know the Gibbs energy of formation for natural organic matter as we cannot generally define a molecular unit. Glucose is an easily degraded organic compound containing carbon with an oxidation state near that of most natural organic matter—zero. As a surrogate for natural organic matter, we can compute the Gibbs energy change necessary to wrest each mole of electrons from glucose in its conversion to carbon dioxide. The Gibbs energy of formation for glucose (–910.4 kJ/mol) is available from Dean (1992):

$$\Delta G^\circ_{e.glu.CO2} := \frac{6\cdot\Delta G^\circ_{CO2} - \Delta G^\circ_{glu} - 6\cdot\Delta G^\circ_{H2O}}{24} = -1.364 \ \frac{\text{kJ}}{\text{mol}\cdot\text{electron}}$$

When we add each of the electron acceptance reactions from the aforementioned equation with the oxidation reaction for glucose to obtain the standard Gibbs energy of reaction, we can simply add the Gibbs energies of the two reactions. We quickly observe that the overall energy available biologically from oxidation of organics using oxygen as the electron acceptor is significantly greater than that available from the other four. Use of carbon dioxide in anaerobic processes is widely practiced. Much interest has been directed toward employment of protons as electron acceptors to yield hydrogen gas as the product of biological processes. While intriguing in the quest for carbon-neutral fuels, this process is difficult at best.

Under steady-state conditions, processes are ongoing; reactants are reacting and creating products. The vast majority of these reactions are biologically driven. Biological reactions are all about the transfer of electrons. Biological life forms (ranging from single-celled bacteria to humans) derive their energy by extracting electrons from electron donors and transferring them to electron acceptors. The process occurs so efficiently that the energy unlocked by transferring the electrons from higher to lower energy state is virtually entirely captured by biological entities. We cannot herein, in detail, examine the specific stepwise biochemical reactions involved in these processes. We will leave those topics to the biochemists. In systems with large supplies of electron donors and short supplies of electron acceptors, electrons are readily available. Conversely, if electron donors are in short supply and electron acceptors are plentiful, electrons become hard to come by. If we can gauge or measure the availability of electrons in systems, we can gain understandings regarding the speciation of targeted electron donors and acceptors. Conversely, if we can determine the speciation of targeted electron donors and acceptors, we can gain quantitative understandings of the availability of electrons.

12.4.1 p*E*–pH (E_H–pH) Predominance Diagrams

We desire to gain quantitative insight regarding the chemical speciation in redox systems. To that end, chemists and geochemists have developed diagrams from which we can visually discern predominant species if pH and p*E* are known. These are called p*E*–pH (or E_H–pH) predominance diagrams. In order that we can effectively utilize these diagrams, we should first understand their development. To locate a predominance boundary line for such a diagram, we return to the law of mass action statement for the equilibrium. We separate the RHS of the relation into a series of products, ensuring that one of these terms contains the activities of the species containing the element that forms the reduced product from the oxidized reactant. We then take the logarithms of both sides of the relation. For example, for the hydrogen sulfide–sulfate system of Example 12.6, we would write the following:

$$\log\left(K_{\text{SO4.H2S}}\right) = \log\left(\frac{P_{\text{H}_2\text{S}}}{\left\{\text{SO}_4^{=}\right\}}\right) - 10\cdot\log\left\{\text{H}^+\right\} - 8\log\left\{\text{e}^-\right\}$$

We may generalize the relation by substituting ox for $SO_4^{=}$ and red for H_2S and n_H and n_E for the number of involved protons and electrons, respectively:

$$\log\left(K_{\text{ox.red}}\right) = \log\left(\frac{\left\{\text{i}_{\text{red}}\right\}^{i}}{\left\{\text{i}_{\text{ox}}\right\}^{i}}\right) - n_{\text{H}}\cdot\log\left\{\text{H}^+\right\} - n_{\text{E}}\log\left\{\text{e}^-\right\}$$

We then substitute pH for $-\log\{H^+\}$ and p*E* for $-\log\{e^-\}$, divide the relation by n_E, and substitute $pE°$ for $(1/n_E)\log(K)$:

$$\text{p}E° = \frac{1}{n_{\text{E}}}\log\left(\frac{\left\{\text{i}_{\text{red}}\right\}^{\nu_{\text{red}}}}{\left\{\text{i}_{ox}\right\}^{\nu_{\text{ox}}}}\right) + \frac{n_{\text{H}}}{n_{\text{E}}}\cdot\text{pH} + \text{p}E \qquad (12.5)$$

We seek to define a line on a set of pE versus pH axes that separates the regions in which the oxidized and reduced species would be predominant with regard to the specie abundance. Most conveniently, we set the law of mass action contributions from the reduced and oxidized species to be equal. For cases in which the stoichiometric coefficients are equal, the activities would be equal. For cases in which the stoichiometric coefficients are not equal, the ratios of the activities raised to the powers of the respective stoichiometric coefficients would be equal. For cases involving a half reaction visualized to occur across the liquid–solid interface, both activities would necessarily be unity. Rather than call this a condition of equal activity, we prefer the idea that the oxidized and reduced species provide equal contributions to the RHS of the law of mass action. Under this condition, the argument of the logarithm of the first RHS term of Equation 12.5 is unity and the term goes to 0. Rearrangement of Equation 12.5 yields the form of the relation we seek:

$$\mathrm{p}E_{\left[\{\mathrm{ox}\}^{\nu_{\mathrm{ox}}}=\{\mathrm{red}\}^{\nu_{\mathrm{red}}}\right]} = \mathrm{p}E^{\circ} - \frac{n_{\mathrm{H}}}{n_{\mathrm{E}}}\cdot \mathrm{pH} \tag{12.6a}$$

Thus, to position the ox–red predominance line on the pE versus pH diagram, we need know only the magnitude of pE° and the number of electrons and protons involved in the reduction of the oxidized specie to the reduce specie. This information is available from a published half reaction or from a written half reaction and its Gibbs energy of reaction. We also must bear in mind that these diagrams are visualizations. To position the predominance line, often the activity of either the oxidized or the reduced specie or the total abundance of the targeted element is assumed. For example, we might wish to pinpoint the line based on known simultaneous presence of a solid and the metal from which the solid is formed. Unit activity of the metal ion is unrealistic, so we might set the activity of the free metal ion at a more realistic level, such as 10^{-5} M or lower. In such cases, we would retain the logarithmic term, but would know its value as both activity values would be known and constant:

$$\mathrm{p}E = \mathrm{p}E^{\circ} + \frac{1}{n_{\mathrm{E}}}\log\left(\frac{\{i_{\mathrm{ox}}\}^{\nu_{\mathrm{ox}}}}{\{i_{\mathrm{red}}\}^{\nu_{\mathrm{red}}}}\right) - \frac{n_{\mathrm{H}}}{n_{\mathrm{E}}}\cdot \mathrm{pH} \tag{12.6b}$$

Should we wish to adjust the activity of the free metal ion, the predominance line would be shifted upward or downward by a constant corresponding factor.

To determine the precise speciation among oxidized and reduced species of a system, particularly under pE and pH conditions located in the vicinity of a line, we should do our work with the law of mass action.

Lines drawn on pE versus pH diagrams representing half reactions have intercepts (at pH = 0) equal to the pE° and negative slopes equal to $n_{\mathrm{H}}/n_{\mathrm{E}}$. We could draw them all in this manner. Unfortunately, our diagrams would become very confusing. Then, in the context of visual identification of predominant species, most diagrams have vertical lines separating the conjugate acids and bases of acid–base systems, including hydrolysis species. Many also will have vertical lines separating predominant dissolved metal–ligand complexes from the companion solid. We use the intersections of these vertical lines with the sloping pE versus pH lines to end and initiate predominance lines applicable in certain pH ranges. A number of important pE versus pH (Brookins, 1988, has assembled them as E_H versus pH) diagrams have been included in the Appendix as Figures A.1 through A.9. Let us examine the construction of one of these diagrams.

Example 12.7 Reproduce Brookins' E_H versus pH predominance diagram for the iron (II)–iron (III) system.

Let us begin with the oxygen–water and proton–hydrogen couples. From the half reactions given in Table A.5, normalized to n_E and assuming that P_{O_2} is unity and P_{H_2} is numerically equal to {H⁺} of the system in question, we may write the equations separating the predominance of oxygen from water and of protons from hydrogen gas:

$$pE_{O2.H2O}(pH) := pE^\circ_{O2.H2O} - pH$$

$$pE_{H.H2}(pH) := pE^\circ_{H.H2} - pH$$

We need to determine the pE° for the $Fe(OH)_{3(s)}$–Fe^{+2} and $Fe(OH)_{3(s)}$–$Fe(OH)_{2(s)}$ couples. We write the half reactions, compute the standard Gibbs energies of the reactions, use the standard Gibbs energies to determine equilibrium constants, and then convert the equilibrium constants to pE° values. In both cases $n_E = 1$. Of significance here, in order that the Fe^{+2}–$Fe(OH)_{2(s)}$, Fe^{+2}–$Fe(OH)_{3(s)}$, and $Fe(OH)_{2(s)}$–$Fe(OH)_{3(s)}$ predominance lines would intersect cleanly, the value of $\Delta G^\circ_{f,FeOH2s}$ was adjusted from 486.6 to 480 kJ/mol:

$$\Delta G^\circ_{FeOH3s.Fe2} := \Delta G^\circ_{Fe.2} + 3\cdot\Delta G^\circ_{H2O} - \Delta G^\circ_{FeOH3s}$$

$$pE^\circ_{FeOH3s.Fe2} := \log\left(e^{\frac{-\Delta G^\circ_{FeOH3s.Fe2}}{R_{kJ}\cdot T}}\right) = 16.014$$

$$\Delta G^\circ_{FeOH3s.FeOH2s} := \Delta G^\circ_{FeOH2s} + \Delta G^\circ_{H2O} - \Delta G^\circ_{FeOH3s}$$

$$pE^\circ_{FeOH3s.FeOH2s} := \log\left(e^{\frac{-\Delta G^\circ_{FeOH3s.FeOH2s}}{R_{kJ}\cdot T}}\right) = 3.185$$

We write the predominance boundary relations:

$$pE_{FeOH3s.Fe2}(pH) := pE^\circ_{FeOH3s.Fe2} - 3\cdot pH$$

$$pE_{FeOH3s.FeOH2s}(pH) := pE^\circ_{FeOH3s.FeOH2s} - pH$$

Plotting several relations with differing ranges of the independent variable in a MathCAD graph requires that each series have its own defined dependent variable range. We may plot the oxygen–water and proton–hydrogen boundaries from pH 0.

14. We must plot the Fe(II)–Fe(III) boundary from pH = 0 to the equivalence pH of the Fe^{+3}–$Fe(OH)_{3(s)}$ couple (pH = 1.07). We determine this equivalence point by setting $\{Fe^{+3}\}$ equal to $\{Fe(OH)_{3(s)}\}$ (i.e., unity) and solve for the pH value that satisfies this constraint:

$$pH_{Fe3.FeOH3} := -\log\left(K_W \cdot \beta_{S.FeOH3}^{\frac{1}{3}}\right) = 1.067$$

We can then draw the vertical line separating Fe^{+3} from $Fe(OH)_{3(s)}$ by defining the line at pH 1.067 ranging from the predominance line for Fe^{+3}–$Fe(OH)_{3(s)}$ to that for O_2–H_2O:

$$Fe3_{.FeOH3s} := pE^{\circ}_{.Fe3.Fe2}, pE^{\circ}_{.Fe3.Fe2} + .01 .. pE_{.O2.H2O}\left(pH_{.Fe3.FeOH3}\right)$$

We can now plot the Fe^{+2}–$Fe(OH)_{3(s)}$ predominance line to the pH of equivalence between Fe^{+2} and $Fe(OH)_{2(s)}$. We find this pH in the same manner as we found that for Fe^{+3}–$Fe(OH)_{3(s)}$ equivalence:

$$pH_{Fe2.FeOH2s} := -\log\left(K_W \cdot \beta_{S.FeOH2}^{\frac{1}{2}}\right) = 6.45$$

Lastly, we may plot the $Fe(OH)_{2(s)}$–$Fe(OH)_{3(s)}$ predominance line from pH 6.45 to 14. Our first attempt did not result in a clean intersection between the Fe^{+2}–$Fe(OH)_{3(s)}$ and $Fe(OH)_{2(s)}$–$Fe(OH)_{3(s)}$ predominance lines at pH 6.45. Given that there is some uncertainty in the actual values of any of the thermodynamic parameters we have employed, we can adjust a little. If we adjust $\beta_{s.FeOH_2}$ to be $10^{16.3}$, the pH of Fe^{+2}–$Fe(OH)_{2(s)}$ equivalence shifts to 5.85 and the Fe^{+2}–$Fe(OH)_{3(s)}$ and $Fe(OH)_{2(s)}$–$Fe(OH)_{3(s)}$ predominance lines intersect the Fe^{+2}–$Fe(OH)_{2(s)}$ predominance line at that pH. If we adjust $\Delta G^{\circ}_{f,FeOH2s}$ to −480 kJ/mol, the Fe(II)–$Fe(OH)_{2(s)}$ equivalence pH remains at pH 6.45 and the Fe^{+2}–$Fe(OH)_{3(s)}$ and $Fe(OH)_{2(s)}$–$Fe(OH)_{3(s)}$ predominance lines intersect the Fe^{+2}–$Fe(OH)_{2(s)}$ predominance line at pH 6.45. In Figure E12.7.1, we have opted for the second option. Certainly, Brookins (1988) must have made some similar adjustments to produce his plot (which was likely hand-drawn) and his Gibbs energy data likely would be consistent.

When we compare Figure E12.7.1 with Figure A.5, we observe that our predominance boundary lines are generally below and to the left of those depicted therein. When we examine Brookins' figure caption, the difference is immediately evident. Rather than use the equivalence of $\{i_{red}\}^{\nu_{red}} / \{i_{ox}\}^{\nu_{ox}}$ as we have done, requiring that $\{Fe^{+2}\}$ = unity, he used an abundance of 10^{-6} M for dissolved iron.

Since we have constructed the mathematical model for the pE versus pH diagram in our MathCAD worksheet, we need only update a few functions and definitions to produce the plot similar to that of Figure A.5. The logarithmic term of the RHS of Equation 12.6a becomes $\log(1/10^{-6})$:

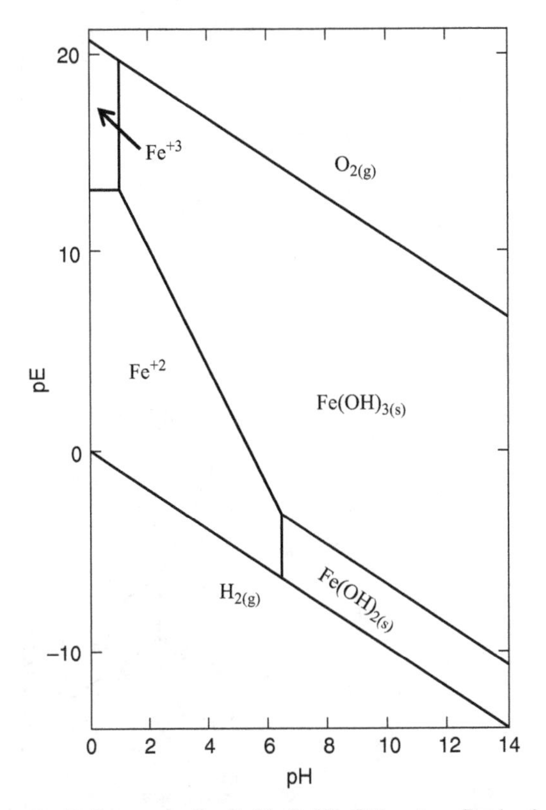

FIGURE E12.7.1 pE–pH diagram for the Fe(II)–Fe(III)–OH system. Predominance lines represent equal contributions of oxidized and reduced species of redox couples to the RHS of the law of mass action.

$$pE_{FeOH3s.Fe2}(pH) := 6 + pE°_{FeOH3s.Fe2} - 3 \cdot pH$$

$$pH_{Fe3.FeOH3} := -\log\left(10^{-2} \cdot K_W \cdot \beta_{S.FeOH3}^{\frac{1}{3}}\right) = 3.067$$

$$pH_{Fe2.FeOH2s} := -\log\left(10^{-3} \cdot K_W \cdot \beta_{S.FeOH2}^{\frac{1}{2}}\right) = 9.45$$

Our revised plot in Figure E12.7.2, employing assumed activities of dissolved iron species, more closely resembles Figure A.5. Brookins' choice of 10^{-6} M as the abundance of the dissolved free metal ion was undoubtedly based on the knowledge that Fe^{+3} and $Fe(OH)_{3(s)}$, and Fe^{+2} and $Fe(OH)_{3(s)}$, and Fe^{+2} and $Fe(OH)_{2(s)}$ would simultaneously exist at the pH and pE conditions dictated by the predominance lines. These predominance lines in truth are a little fuzzy, or perhaps wide. That separating the predominance of $Fe(OH)_{2(s)}$ and $Fe(OH)_{3(s)}$ would be somewhat

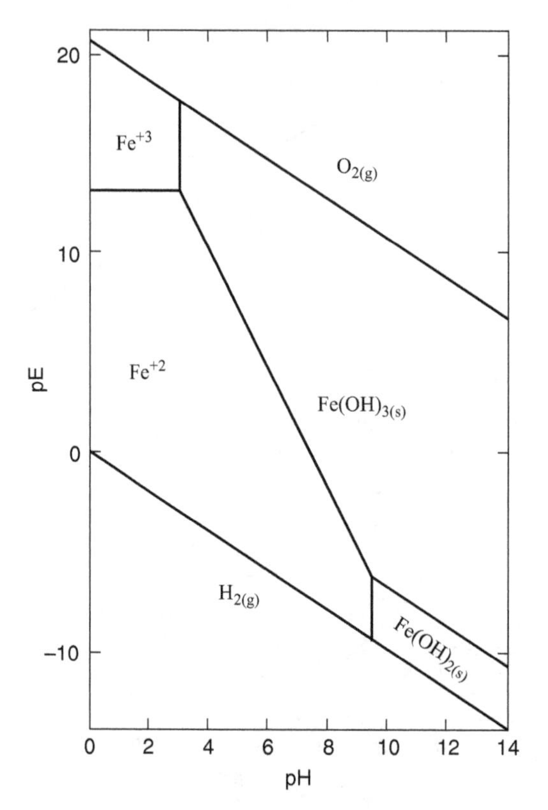

FIGURE E12.7.2 *pE–pH diagram for the Fe(II)–Fe(III)–OH system. Predominance lines represent 10^{-6} M activities of dissolved iron species.*

sharper. If the two solid phases coexist, each must have an activity of unity. Once conditions are varied even slightly, one of the solid phases must disappear in favor of the predominance of the other.

Once we have the pE versus pH diagram for a system, if we know the pH and pE (or pH and E_H) of the system, we may at a glance know the predominant specie for those conditions, except in the vicinity of a predominance line. When the combination of pH and pE situate the system in close proximity to a predominance line, implementation of the applicable half reaction equilibrium is necessary to confidently determine the speciation.

In Example 12.7, we have examined the ferrous iron as the predominant reduced specie. In many systems, carbonate is present and the predominant Fe(II) specie might be $FeCO_{3(s)}$. Let us examine the effect of the presence of dissolved carbonate in the system.

Example 12.8 Consider that dissolved carbonate is present at an abundance of 10^{-3} M and expand the pH – p*E* predominance diagram of Example 12.7. Continue with the assumption that the activity of dissolved ferrous iron would be 10^{-6} M. In order that we may focus upon the redox aspects of the system, let us assume we have infinitely dilute conditions.

We must first write the half reaction involving $Fe(OH)_{3(s)}$ and $FeCO_{3(s)}$. We find values for and employ Gibbs energy data to determine the standard Gibbs energy change for the reaction and determine the equilibrium constant (p*E*°) and write the law of mass action statement. We could substitute as necessary into the Nernst equation, but this approach is more fundamental and serves as a better illustration of the modeling process:

$$FeOH3 + e + 3 \cdot H + CO3 = FeCO3 + 3 \cdot H2O$$

$$pE^{\circ}_{FeOH3.FeCO3} := \log\left[e^{\frac{-\left(\Delta G^{\circ}_{FeCO3} + 3 \cdot \Delta G^{\circ}_{H2O} - \Delta G^{\circ}_{FeOH3s} - \Delta G^{\circ}_{CO3}\right)}{R_{kJ} \cdot T}}\right] - 26.514$$

We write the law of mass action and use unit activities for $Fe(OH)_{3(s)}$ and $FeCO_{3(s)}$ to yield the relation from which we can determine the $Fe(OH)_{3(s)}$–$FeCO_{3(s)}$ predominance line:

$$10^{pE^{\circ}_{FeOH3.FeCO3}} = \frac{1}{E \cdot H^3 \cdot CO3}$$

A nuance here, but important, in our MathCAD worksheet we represent the activity $\{e^-\}$ of electrons with an upper-case *E* to avoid redefining MathCAD's e, which is the base for the exponential function.

We take the logarithms of the LHS and RHS, making substitutions to employ p*E* and pH:

$$pE^{\circ}_{FeOH3.FeCO3} = \log\left(\frac{1}{CO3}\right) + pE + 3 \cdot pH$$

Since we have specified total dissolved carbonate, we can use $\alpha_{2.CO_3}$ to express the carbonate ion activity in terms of total carbonate abundance, acid dissociation constants, and pH. We make this substitution and rearrange the result to yield the relation describing the predominance line. Along the way we need to define H(pH) as 10^{-pH}:

$$pE_{FeOH3.FeCO3}(pH) := pE^{\circ}_{FeOH3.FeCO3} - \log\left[\frac{\left(\frac{H(pH)^2}{K_1 \cdot K_2} + \frac{H(pH)}{K_2} + 1\right)}{C_{Tot.CO3}}\right] - 3 \cdot pH$$

We then define the pH at which Fe^{+2} in equilibrium with ferrous carbonate would have activity equal to 10^{-6} M:

$$pH := 8 \qquad \text{Given} \qquad \frac{\left(\frac{H(pH)^2}{K_1 \cdot K_2} + \frac{H(pH)}{K_2} + 1\right)}{\beta_{S.FeCO3} \cdot C_{Tot.CO3}} = 10^{-6} \qquad pH_{Fe2.FeCO3} := \text{Find}(pH) \rightarrow 8.641$$

We must also find the pH at which ferrous carbonate and ferrous hydroxide solids would simultaneously be in equilibrium with the ferrous iron. Since both solids must be in equilibrium with a single unique abundance of Fe^{+2}, we simply rearrange the solid formation equilibrium expressions to isolate $\{Fe^{+2}\}$ on the LHS and equate the resultant RHS expressions:

$$pH := 6$$

$$\text{Given} \qquad \frac{H(pH)^2}{\beta_{S.FeOH2} \cdot K_W^2} = \frac{\left(\frac{H(pH)^2}{K_1 \cdot K_2} + \frac{H(pH)}{K_2} + 1\right)}{\beta_{S.FeCO3} \cdot C_{Tot.CO3}}$$

$$pH_{FeCO3.FeOH2} := \text{Find}(pH) \rightarrow 10.077$$

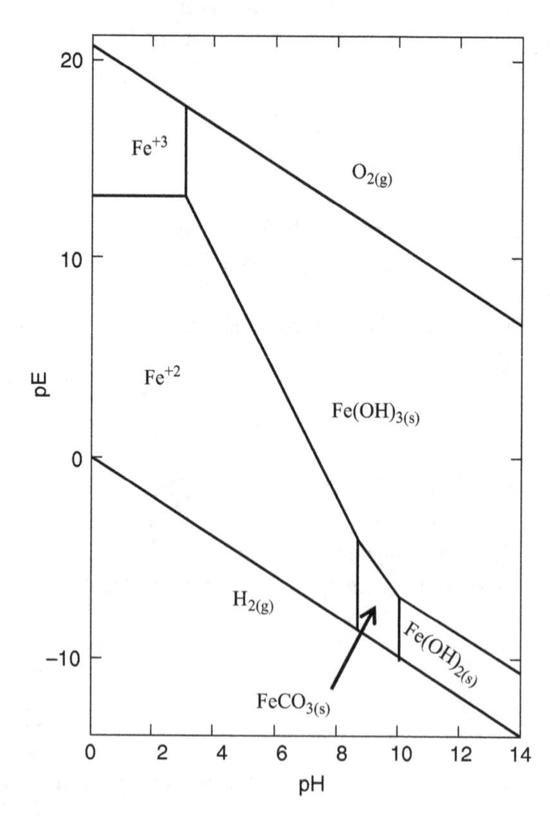

FIGURE E12.8.1 *pE–pH diagram for the Fe(II)–Fe(III)–OH^-–$CO_3^=$ system. Dissolved iron specie activities are 10^{-6} M.*

In Figure E12.8.1, we plot the $Fe(OH)_{3(s)}$–Fe^{+2} and $Fe(OH)_{3(s)}$–$Fe(OH)_{2(s)}$ predominance lines with the revised ending ($pH_{Fe_2.FeCO_3}$) and beginning ($pH_{FeCO_3.FeOH_2}$) points, respectively, and plot the $Fe(OH)_{3(s)}$–$FeCO_{3(s)}$ predominance line from $pH_{Fe_2.FeCO_3}$ to $pH_{FeCO_3.FeOH_2}$. We have merely added a region in which solid ferrous carbonate would be the predominant Fe(II) specie.

We could work further with the MathCAD worksheet adjusting equilibrium constants or Gibbs energy data to provide sharper intersections, but we would only be guessing regarding which parameters to adjust. For example, were we to use $\beta_{S.FeCO_3} = 10^{10.3}$ rather than the value $10^{10.7}$ from Table A.2, the region of solid ferrous carbonate would narrow slightly and the predominance line would rise, allowing for sharper intersection points. Certainly, Brookins performed many of the computations by hand to identify the intersection points and indeed must have adjusted his Gibbs energy data accordingly. Most certainly, his plotted lines were hand-drawn, with much more flexibility for adjustment than is available for plotting functions using either MathCAD's *x*–*y* graphs or Excel's scatter plots.

We can visualize the utility of p*E* versus pH (or E_H versus pH) diagrams in determining predominant speciation of environmental systems. They are quite useful in understanding speciation when the p*E* and pH of the system fall solidly within a predominance region. Given readily available, powerful means to perform computations, hereinafter we dig a little deeper into the utility of redox couples.

12.4.2 Effect of p*E* on Redox Couple Speciation

We can draw vertical lines on p*E* versus pH predominance diagrams, cutting the predominance lines at known pH values. We can then investigate the speciation of the redox couple along each of these lines. We can write a mole balance on the element through which the electrons are transferred and couple the mole balance with the law of mass action for the half reaction. We can write two typical variations of the law of mass action: one isolating the reduced product on the LHS and one isolating the oxidized reactant on the LHS:

$$\{i_{\text{red}}\} = \left(K_{\text{ox.red}}\{i_{\text{ox}}\}^{\nu_{i.\text{ox}}}\{\text{H}^+\}^{n_\text{H}}\{\text{e}^-\}^{n_\text{E}}\right)^{\frac{1}{\nu_{i.\text{red}}}} \tag{12.7}$$

$$\{i_{ox}\} = \left(\frac{\{i_{red}\}^{\nu_{i.red}}}{K_{ox.red}\{H^+\}^{n_H}\{e^-\}^{n_E}}\right)^{\frac{1}{\nu_{i.ox}}} \tag{12.8}$$

When species other than the oxidized and reduced species containing the element through which the electrons are transferred, protons, and electrons are present, they simply need to be included as appropriate in the law of mass action relation for the half reaction. Let us apply Equations 12.7 and 12.8 along some vertical constant pH lines cutting some predominance lines.

Example 12.9 Draw two vertical lines (lines of constant pH) on the plot developed in Example 12.8. With the first line cut the H^+–$H_{2(g)}$, $Fe(OH)_{3(s)}$–Fe(II), and $O_{2(g)}$–H_2O predominance lines at pH 6.0. With the second, cut the Fe(III)–Fe(II) predominance line at pH_2. Draw a third line cutting the sulfate–bisulfide predominance line of Figure A.9 at pH 10.0. Determine the abundances of the oxidized and reduced species as functions of p*E* along each of these vertical lines.

For the proton—hydrogen gas system, the proton activity is fixed at 10^{-6} M and the mole balance is trivial. We write a function with pH and p*E* as arguments. We can then use it at any value of pH. The value of $p\overset{\circ}{E}_{H.H2}$ from Table A.5 is 0:

$$pH := 6 \quad pE := -6.5, -6.4 .. -4.5 \quad P_{H2}(pH, pE) := \left(10^{2 \cdot pE^\circ_{H.H2}}\right) \cdot \left(10^{-pH}\right)^2 \cdot \left(10^{-pE}\right)^2$$

Then at pH 6.0 in Figure E12.9.1, we have plotted the predicted abundance of hydrogen gas, expressed as its equivalent partial pressure in atmospheres. We observe that the predicted partial pressure of hydrogen gas increases well above one atm as p*E* is lowered below the predominance line. Conversely, for each unit increase of p*E* above the line, the predicted partial pressure of hydrogen gas falls two orders of magnitude, in accord with the law of mass action statement. Certainly, the partial pressure of hydrogen gas in environmental systems would be limited to the range well below one atmosphere, suggesting that p*E* values below about 5.5 at pH 6.0 would be imaginary.

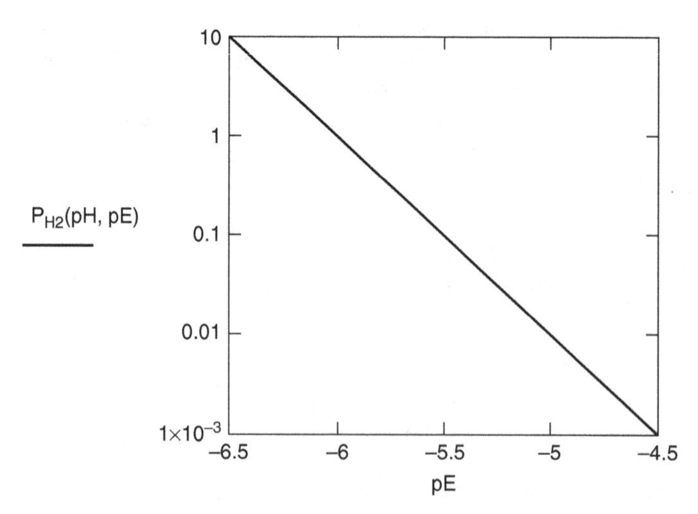

FIGURE E12.9.1 *A plot of predicted partial pressure of hydrogen gas (in atm) versus pE for redox equilibria in an aqueous solution at pH = 6.0.*

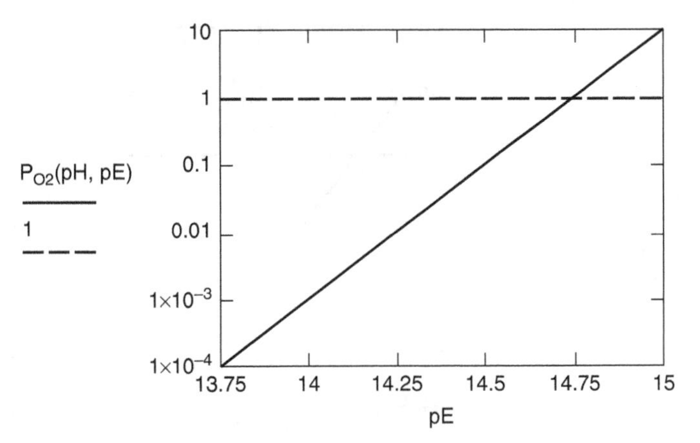

FIGURE E12.9.2 *A plot of predicted partial pressure of oxygen (in atm) versus pE for redox equilibria in an aqueous solution of pH = 6.0.*

For the oxygen–water system, the activity of water is taken as unity, again rendering the mole balance on oxygen to be trivial. From the law of mass action relation, we write a function of pH and p*E* for the partial pressure of oxygen. The value of $pE^{\circ}_{O2.H2O}$ from Table A.5 is 14.75:

$$pH := 6$$

$$pE := 13.75, 13.76 .. 15 \quad P_{O2}(pH, pE) := \left[\left(10^{4 \cdot pE^{\circ}_{O2.H2O}}\right) \cdot \left(10^{-pH}\right)^{4} \cdot \left(10^{-pE}\right)^{4}\right]^{-1}$$

In Figure E12.9.2, we plot the function versus p*E* values in the vicinity of the predominance line. The predominance line passes through the point pH = 6 and p*E* = 14.75, so as is indicated in the plot, the partial pressure of 1 atm (equivalent to the unit activity of water) occurs at p*E* = 14.75. We note that P_{O_2} would increase drastically at p*E* values beyond the predominance line and decreases four orders of magnitude with a decrease of p*E* from 14.75 to 13.75, in accord with the prediction of the law of mass action. In environmental systems, P_{O_2} really cannot be much larger than 0.01 atm, thus at pH 6 p*E* values beyond about 14.25 would be imaginary.

For the solid ferric hydroxide–ferrous iron couple, the activity of the solid is unity and again the mole balance on dissolved iron would include only Fe(II) species. We have done those in Chapter 11, so let us focus upon the activity of free ferrous iron as a function of p*E*. We write the function for the activity of ferrous iron as a function of pH and p*E*:

$$pH := 6$$

$$pE := -3.0, -2.9 .. 4 \quad Fe2(pH, pE) := 10^{pE^{\circ}_{FeOH3s.Fe2}} \cdot 10^{-pE} \cdot \left(10^{-pH}\right)^{3}$$

In Figure E12.9.3, we plot this function along our vertical line at pH 6, varying p*E*. We have included the line for $\{Fe^{+2}\} = 1$ M as a rough estimate of the maximum abundance of iron(II) in water. When we consider the complexes along with the free metal ion and

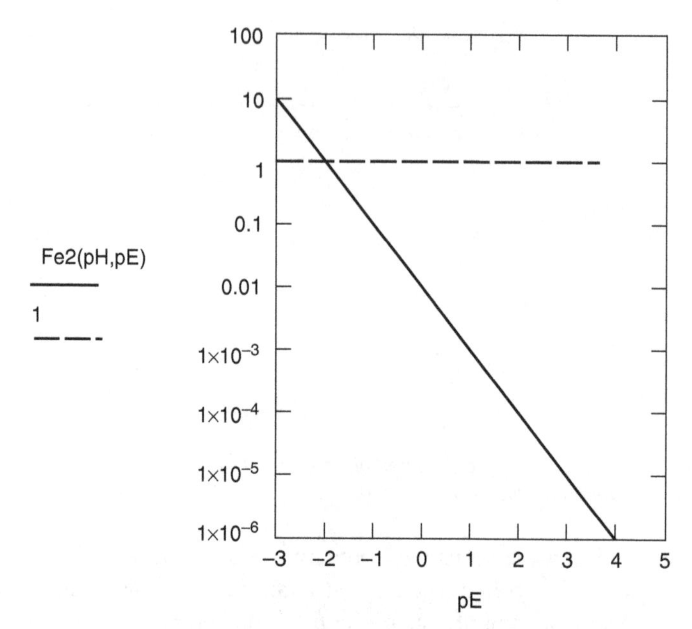

FIGURE E12.9.3 *A plot of predicted activity of* Fe^{+2} *versus* pE *for an aqueous solution at pH = 6.0 in which ferric hydroxide solid is assumed to be present.*

that the activity coefficients would be significantly lower than unity, and we ignore the potential for formation of ferrous hydroxide solid, we guess that an activity around one molar might approximate the maximum $\{Fe^{+2}\}$ relative to dissolution in water. The significance is that the ferric hydroxide solid phase can exist at pE values well into ferrous iron predominance region. The predominance line drawn in Example 12.8 considered that the ferrous iron abundance would be 10^{-6} M, commensurate with p$E \sim +4$. We observe that the ferric hydroxide solid can exist at pE values five or more orders of magnitude below the predominance line. This predominance boundary is indeed rather wide.

The dissolved Fe^{+3}–Fe^{+2} couple is rather simple. Here, we can employ the mole balance and let us suggest that the total dissolved iron (neglecting ionic strength effects) is 10^{-5} M. We write the mole balance as the sum of dissolved iron species and employ the redox equilibria much in the same manner as we would for a monoprotic acid to produce functions of pH and pE for Fe^{+3} and Fe^{+2}:

$$pH := 2 \qquad pE := 11, 11.1 .. 15$$

$$Fe2(pH, pE) := \frac{C_{Tot.Fe}}{\dfrac{1}{\left(10^{pE^\circ_{Fe3.Fe2}} \cdot 10^{-pE}\right)} + 1} \qquad Fe3(pH, pE) := \frac{C_{Tot.Fe}}{1 + 10^{pE^\circ_{Fe3.Fe2}} \cdot 10^{-pE}}$$

We plot these two functions against pE in Figure E12.9.4. We may observe that the Fe^{+3}–Fe^{+2} redox couple appears to behave in a manner relative to pE exactly as a monoprotic acid would behave relative to pH. For the Fe^{+3}–Fe^{+2} system, a single

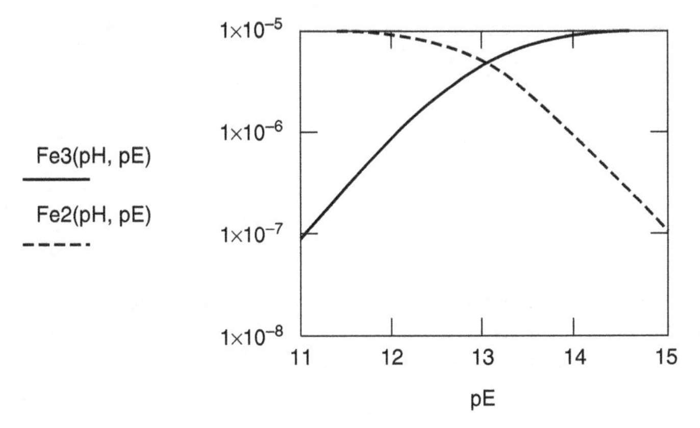

FIGURE E12.9.4 *A plot of $\{Fe^{+2}\}$ and $\{Fe^{+3}\}$ versus pE for an aqueous solution containing 10^{-5} M Fe_{Tot} at pH 6.0.*

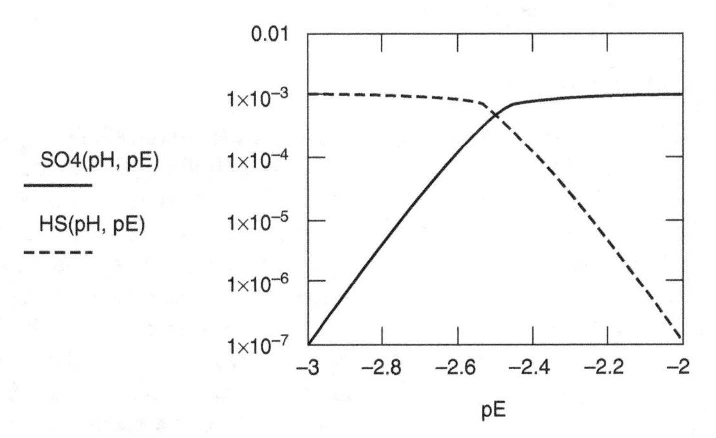

FIGURE E12.9.5 *Plot of sulfate and bisulfide speciation versus pE at pH 6 for an aqueous solution containing 10^{-3} M total sulfur.*

electron is transferred and thus the slopes of the relations are unity, in accord with the stoichiometry of the redox half reaction. We see that the predominance line separating Fe^{+3} from Fe^{+2} is also wide, with both the reduced and the oxidized species able to exist in significant abundance at least a pE unit on the opposite side of the predominance line. In general, the magnitude of the slope of the $\log\{i\}$ versus pE trace will be equal to the number of electrons transferred.

For the sulfate–bisulfide couple, we have chosen pH 10 to minimize the significance of dissolved hydrogen sulfide. Within the predominance region below sulfate, we observe a vertical line at pH ~7. This is, of course, the predominance boundary between hydrogen sulfide and bisulfide. We specify a total sulfur abundance of 10^{-3} M. We write the mole balance as the sum of sulfate and bisulfide and employ the redox equilibrium to write functions of sulfate and bisulfide activities with pH and pE as the arguments. The value of $pE^{\circ}_{SO4.HS}$ from Table A.5 is 4.25:

$$pH := 6 \qquad SO4(pH, pE) := \frac{C_{Tot.S}}{1 + \left(10^{pE^\circ_{SO4.HS}}\right)^8 \cdot \left(10^{-pH}\right)^9 \cdot \left(10^{-pE}\right)^8}$$

$$pE := -3, -2.95 .. -2$$

$$HS(pH, pE) := \frac{C_{Tot.S}}{\dfrac{1}{\left(10^{pE^\circ_{SO4.HS}}\right)^8 \cdot \left(10^{-pH}\right)^9 \cdot \left(10^{-pE}\right)^8} + 1}$$

We plot these relations versus pE in Figure E12.9.5. Abundances of sulfate and bisulfide decrease dramatically as the pE value departs from the predominance line at pH 6. As the value of pE is decreased or increased one-half unit ($\{e^-\}$ changes by a factor of 3.16 or $1/3.16$), the abundances are decreased or increased four orders of magnitude, again, consistent with the stoichiometry of the half reaction. We would describe the predominance line separating the sulfate from bisulfide predominance region as very sharp.

12.4.3 Determining System p*E*

Electrochemists and their engineering partners, the metallurgical engineers, tend to prefer using E_H (and, of course, E°_H) as their property for characterization of the electron availability in their targeted systems. They drive their processes using electrical currents and the volt is the most convenient unit for their characterization of electrical potential. Chemists tend to employ pE (and pE°) because the unit conversions from volts to the seemingly dimensionless electron-normalized equilibrium constants are cumbersome and pE is a very convenient partner to pH. A third characterization of electron availability is the oxidation–reduction potential (ORP) with units of millivolts. In fact, the ORP electrode has been developed for physically measuring the electron availability of aqueous samples. In a manner similar to that used for pH and ion-specific ion electrodes, the ORP electrode employs an electrochemical cell in order to sense the availability of electrons. The measurement is translated directly as an electrical potential. Of prime importance is the necessity to measure the ORP of samples prior to changes in sample character arising from the removal of the sample from its location within the system. When we can insert the electrode (or probe) into the aqueous solution involved with a process, we can obtain fairly accurate measurements. Also, when we can employ an ORP probe in a flow-through cell through which ground water is directed from a well, we can obtain fairly accurate measurements, as long as the path from the ground water source to the flow cell is tightly closed. We might even be able to insert an ORP probe into the sediments below the sediment–water interface of a water body and obtain fairly accurate measurements. As analytical instruments, ORP electrodes must be calibrated and the calibration status continually verified and adjusted as necessary or output readings could be positively or negatively biased. Seemingly very simple, ORP measurement using an electrode is fraught with many possibilities for error. Then, as a backup for ORP measurements, or even as the primary means of determining electron availability, whether we use E_H, pE, or ORP as our unit, physical determination of the speciation of key redox species can yield accurate determinations of electron availability.

Environmental systems generally are not at true chemical equilibrium. This is especially true in regard to overall redox reactions. Electrons are flowing, via many intermediate reactions, and the species of interest are generally the original reactants and the final products of the redox process. While not at equilibrium, processes within environmental systems are often proceeding at steady or near-steady rates, from which quasi-equilibrium conditions arise. Thus, changes in conditions happen only slowly. Conversely, electron transfers occur rapidly. Given this rate disparity, from the presence (or absence) of oxidized and reduced reactants and products of specific half reactions, we can learn much about the availability of electrons. Consider the speciation of the various redox pairs of Example 12.9 in the vicinity of the predominance line. When predominance lines are sharp, the presence of both oxidized and reduced species of a redox pair suggests that the electron availability must be at a level near the predominance line. Then, if we can measure the pH and by assay, determine the abundances of the oxidized and reduced species of a target redox pair, we can, with some degree of confidence through employment of equilibrium relations, determine the electron availability.

Example 12.10 Examine the redox conditions within an operating digester at a typical wastewater renovation facility. Such digesters are completely mixed flow reactors (CMFRs) with hydraulic residence times in the range of 15–30 days. Most do not have cell recycle, so solids residence times and hydraulic residence times are equal. Digesters are operated most efficiently at ~36 °C, and our standard database is at 25 °C. Perhaps we can find the enthalpy of formation data available to adjust the equilibrium constants for temperature, but herein let us accept the errors associated with the standard temperature in favor of a focus upon the redox aspects of the process and use values specific to 25 °C from our database. Further, we know that the aqueous solution within this digester is not infinitely dilute with regard to electrolytes, but again, to focus on the redox aspects, let us use the infinitely dilute assumption and accept the corresponding errors. Typically digesters operate efficiently in the pH range of 6.5–8.5. Other important information includes the composition of the gas phase above the digester liquid: $Y_{CO_2} =\sim 0.30$; $Y_{CH_4} =\sim 0.60$; and $Y_{H_2S} =\sim 0.01$ with $P_{Tot} \sim 1.0$ atm. The aqueous solution within the digester typically contains ~0.01 and ~0.005 M total acetate and total propionate, respectively. At the pH levels of digesters, speciation will highly favor the conjugate bases. Sulfate abundance is typically below limits of detection. Use the $CO_{2(g)}$–$CH_{4(g)}$, $CO_{2(g)}$–CH_3COO^- (Ac^-), $CO_{2(g)}$–$CH_3CH_2COO^-$ (Pr^-), and CH_3COO^-–$CH_{4(g)}$ redox couples to define the electron availability within the digester over the specified range of pH for the stated conditions. Suggest which couples might be the most reliable.

We obtain the $CO_{2(g)}$–$CH_{4(g)}$ half reaction and its p$E°$ value from Table A.5, but must write (or obtain from other sources) the half reactions and compute p$E°$ from Gibbs energy for the remaining redox couples:

$$2 \cdot CO2 + 8 \cdot E + 7 \cdot H = CH3COO + 2 \cdot H2O$$

$$3 \cdot CO2 + 14 \cdot E + 13 \cdot H = CH3CH2COO + 4 \cdot H2O$$

$$CH3COO + 8 \cdot E + 9 \cdot H = 2 \cdot CH4 + 2 \cdot H2O$$

We obtain the Gibbs energy of formation for acetate and carbon dioxide from Table A.1, and that for propanoic acid (HPr, $\Delta G_f^\circ = -383.5$ kJ / mol) from Dean (1992). We might search other databases and secure a value for the propanoate ion, but herein we will choose to add the deprotonation of propanoic acid to produce propanoate (Pr^-) to obtain the desired half reaction.

From our half reactions and Gibbs energy data, we obtain the equilibrium constants for the four selected half reactions, choosing to use the pE° format. For the carbon dioxide–acetate couple, we have the following:

$$\Delta G^\circ_{CO2.Ac} := \Delta G^\circ_{Ac} + 2 \cdot \Delta G^\circ_{H2O} - 2 \cdot \Delta G^\circ_{CO2}$$

$$pE^\circ_{CO2.Ac} := \frac{1}{8} \cdot \log\left(e^{\frac{-\Delta G^\circ_{CO2.Ac}}{R_{kJ} \cdot T}}\right) = 1.21$$

For the carbon dioxide–propanoate couple, we have the following:

$$3 \cdot CO2 + 14 \cdot E + 14 \cdot H = HPr + 4 \cdot H2O \qquad HPr = H + Pr$$

$$\Delta G^\circ_{CO2.HPr} := \Delta G^\circ_{HPr} + 4 \cdot \Delta G^\circ_{H2O} - 3 \cdot \Delta G^\circ_{CO2} = -149.11$$

$$K_{CO2.HPr} := e^{\frac{-\Delta G^\circ_{CO2.HPr}}{R_{kJ} \cdot T}} = 1.332 \times 10^{26} \qquad K_{HPr.Ac} := 10^{-4.7}$$

$$K_{CO2.Pr} := K_{CO2.HPr} \cdot K_{HPr.Ac} = 2.657 \times 10^{21} \qquad pE^\circ_{CO2.Pr} := \frac{1}{14} \cdot \log(K_{CO2.Pr}) = 1.53$$

For the acetate–methane couple, we have the following:

$$\Delta G^\circ_{Ac.CH4} := 2 \cdot \Delta G^\circ_{CH4} + 2 \cdot \Delta G^\circ_{H2O} - \Delta G^\circ_{Ac}$$

$$pE^\circ_{Ac.CH4} := \frac{1}{8} \cdot \log\left(e^{\frac{-\Delta G^\circ_{Ac.CH4}}{R_{kJ} \cdot T}}\right) = 4.518$$

We also know the abundances of species occupying both sides of the targeted redox couples:

$$\begin{pmatrix} P_{CH4} \\ P_{CO2} \\ Ac \\ Pr \end{pmatrix} = \begin{pmatrix} 0.6 \\ 0.3 \\ 0.01 \\ 0.005 \end{pmatrix} \begin{pmatrix} atm \\ atm \\ M \\ M \end{pmatrix}$$

We use these equilibrium constants, target specie abundances, and the equilibria for the corresponding half reactions to write functions for pE using pH as the master independent variable:

$$pE_{CO2.CH4}(pH) := -\log\left[\left(\frac{P_{CO2}}{P_{CH4}\cdot 10^{8\cdot pE^{\circ}_{CO2.CH4}}\cdot 10^{-8\cdot pH}}\right)^{\frac{1}{8}}\right]$$

$$pE_{CO2.Ac}(pH) := -\log\left[\left(\frac{Ac}{10^{8\cdot pE^{\circ}_{CO2.Ac}}\cdot P_{CO2}{}^{2}\cdot 10^{-7\cdot pH}}\right)^{\frac{1}{8}}\right]$$

$$pE_{CO2.Pr}(pH) := -\log\left[\left(\frac{Pr}{10^{14\cdot pE^{\circ}_{CO2.Pr}}\cdot P_{CO2}{}^{3}\cdot 10^{-13\cdot pH}}\right)^{\frac{1}{14}}\right]$$

$$pE_{Ac.CH4}(pH) := -\log\left[\left(\frac{P_{CH4}{}^{2}}{Ac\cdot 10^{8\cdot pE^{\circ}_{Ac.CH4}}\cdot 10^{-5\cdot pH}}\right)^{\frac{1}{8}}\right]$$

Each of these relations should give us an estimate of the electron availability within the digester as a function of the pH of the aqueous solution. We plot them versus pH in Figure 12.10.1 and obtain a visual result. We quickly observe close agreement between the results of the $CO_{2(g)}$–CH_3COO^- and $CO_{2(g)}$–$CH_3CH_2COO^-$ couples and reasonable agreement between results of these two couples and the $CO_{2(g)}$–$CH_{4(g)}$ couple. The results from the CH_3COO^-–$CH_{4(g)}$ couple are orders of magnitude apart from the others. We might suggest that the CH_3COO^-–$CH_{4(g)}$ is the anomaly. We can test our theory by examining the predicted abundance of sulfate based on the results from the four couples. With known vapor abundance of hydrogen sulfide and known pH we can predict aqueous hydrogen sulfide from Henry's law and bisulfide abundances from acid dissociation. From bisulfide abundance we can predict, through the half reaction, the abundance of the sulfate ion. We write four corresponding MathCAD functions:

$$SO4_{Ac.CH4}(pH) := \frac{\frac{K_{A.H2S}}{10^{-pH}}\cdot K_{H.H2S}\cdot P_{H2S}}{10^{8\cdot pE^{\circ}_{SO4.HS}}\cdot 10^{-9\cdot pH}\cdot 10^{-8\cdot pE_{Ac.CH4}(pH)}}$$

$$SO4_{CO2.CH4}(pH) := \frac{\frac{K_{A.H2S}}{10^{-pH}}\cdot K_{H.H2S}\cdot P_{H2S}}{10^{8\cdot pE^{\circ}_{SO4.HS}}\cdot 10^{-9\cdot pH}\cdot 10^{-8\cdot pE_{CO2.CH4}(pH)}}$$

$$SO4_{CO2.Ac}(pH) := \frac{\frac{K_{A.H2S}}{10^{-pH}}\cdot K_{H.H2S}\cdot P_{H2S}}{10^{8\cdot pE^{\circ}_{SO4.HS}}\cdot 10^{-9\cdot pH}\cdot 10^{-8\cdot pE_{CO2.Ac}(pH)}}$$

$$SO4_{CO2.Pr}(pH) := \frac{\frac{K_{A.H2S}}{10^{-pH}}\cdot K_{H.H2S}\cdot P_{H2S}}{10^{8\cdot pE^{\circ}_{SO4.HS}}\cdot 10^{-9\cdot pH}\cdot 10^{-8\cdot pE_{CO2.Pr}(pH)}}$$

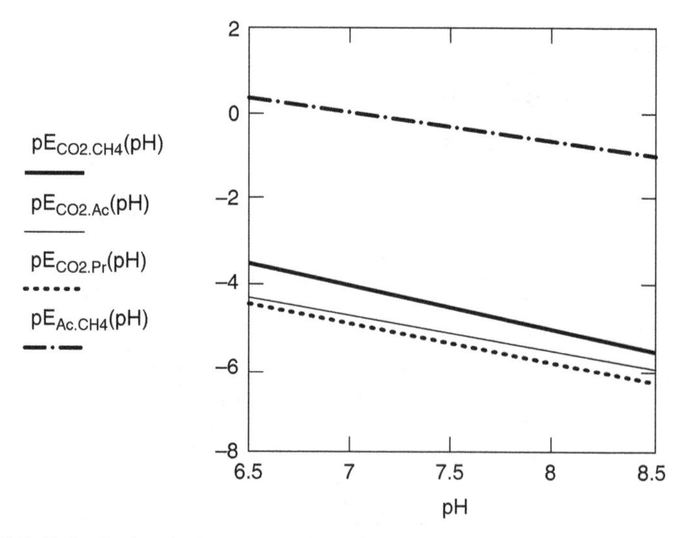

FIGURE E12.10.1 *A plot of pE predicted from CO_2–CH_4, CO_2–CH_3COO^-, CO_2–$CH_3CH_2COO^-$, and CH_3COO^-–CH_4 redox couples for typical conditions within the aqueous solution of an anaerobic digester.*

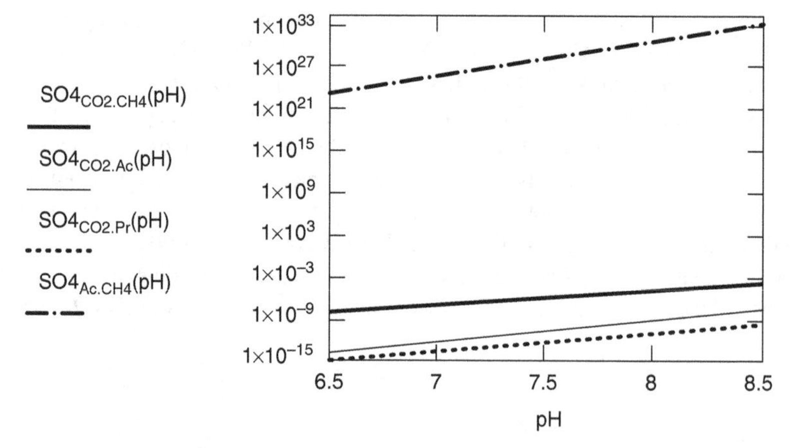

FIGURE E12.10.2 *Predicted sulfate activities from pE values derived from the CO_2–CH_4, CO_2–CH_3COO^-, CO_2–$CH_3CH_2COO^-$, and CH_3COO^-–CH_4 redox couples for typical conditions within the aqueous solution of an anaerobic digester.*

When we plot these functions in Figure E12.10.2, the resultant picture is worth the proverbial 1000 words. We immediately observe that the acetate–methane couple is the outlier as the predicted sulfate abundances are many orders of magnitude greater than is possible. From the result of the carbon dioxide–methane couple, we would predict that at pH 8.5 sulfate would be ~10^{-3} M, certainly not below limits of detection. Our conclusion is that the true electron availability is somewhere near the traces of the carbon dioxide–propanoate and carbon dioxide–acetate couples.

When we dig a little deeper into the anaerobic digestion process, we find that we may divide the process into four major subprocesses, all occurring simultaneously in the CMFR that constitutes the digester:hydrolysis (breakdown of initial substrates); acidogenesis (conversion to long-chain carboxylic acids); acetogenesis (conversion to acetic acid); and methanogenesis (conversion to methane). The microbiologists inform us that of these four subprocesses, methanogenesis is the rate-limiting step due to the slow growth of methanogenic bacteria. This, of course, means that acetic acid, which is the product of the acetogenesis subprocess, will be in abundance well beyond the levels predicted by the acetate–methane redox equilibrium. Relative to the end product methane, all intermediate byproducts would tip the reactant side of Le Chatelier's balance. Our reaction quotient, Q, would be well below unity, indicating a strong driving force for the reaction to proceed, based on electron availability. Unfortunately, the microbes are the limiting factor in the process. The acetate–methane half reaction is far-removed from its equilibrium condition. We might choose to use the results from the carbon dioxide–methane and carbon dioxide–propanoate redox couples as limiting cases for describing the pE in the targeted anaerobic digester.

In Example 12.10, we employed knowledge of the composition of a gas phase in contact with an aqueous phase together with general knowledge of the key components of the aqueous phase to estimate electron availability within the overall system. We must be sure to understand that our result is but an overall estimate.

We may employ these principles in other systems involving fully dissolved components and involving an interface between an aqueous solution and a solid phase. Estimating the pE of a ground water is one notable case we can investigate. Particularly in the Midwestern United States, ground waters may contain measureable quantities of iron. Certainly we know that at the pH values of ground waters (e.g., 6–8) the iron that is in the aqueous solution must be of the ferrous variety. Let us examine such a case.

Example 12.11 Consider that ground water may have pH between 6 and 8 and may contain total iron in the range of 1–10 ppm_m. Also consider that total inorganic carbon would be in the range of 0.001–0.005 M. Lastly, ground water almost always contains sulfur mostly in the form of sulfate. In many instances, the sulfur can be in the reduced sulfide form. Let us suggest that, upon sampling, the field personnel would detect only a very slight odor of rotten eggs (associated with hydrogen sulfide gas) from the water when sampling. However, when tested using a field test kit, the sulfide level was too low to be quantitated. Determine as closely as possible the electron availability of the ground water sampled.

Rather than a specific value or set of values correlated with pH, herein we seek to identify the region of the pE versus pH (or E_H versus pH) diagram within which the pE must lie relative to the pH of the system. Most appropriately, we will produce a plot. We must consider the potential existence of Fe(III) solids (e.g., $Fe(OH)_{3(s)}$, $FeOOH_{(s)}$, and $Fe_2O_{3(s)}$). For illustration herein, we will use $Fe(OH)_{3(s)}$.

Given the presence of aqueous carbonate, we also must consider the presence of the Fe(II) solid $FeCO_{3(s)}$. In a previous analysis, we found that $FeCO_{3(s)}$ would predominate over $Fe(OH)_{2(s)}$ in the pH region of interest herein. Again, in order that we may focus attention directly upon the redox aspects of this question, we will accept the error associated with use of constants at 25 °C and with employment of the dilute solution assumption, rendering all activity coefficients to be unity. We will omit the activity coefficients from the relations we write.

Let us first identify the pH locations of the vertical lines that would separate the predominance regions of Fe^{+2} from $FeCO_{3(s)}$, based on the maximum and minimum abundances of Fe and CO_{3Tot}. We employ the formation equilibrium for ferrous carbonate to write a function for the ferrous ion abundance:

$$Fe2_{FeCO3s}(pH, C_{Tot.CO3}) := \frac{\left(\frac{H(pH)^2}{K_1 \cdot K_2} + \frac{H(pH)}{K_2} + 1\right)}{\beta_{S.FeCO3} \cdot C_{Tot.CO3}}$$

We employ this function to compute the position of the vertical lines separating the predominance regions of Fe^{+2} and $FeCO_{3(s)}$ for the four possible combinations of iron and carbonate abundance:

$$MW_{Fe} := 55847 \frac{mg}{mol} \qquad \begin{pmatrix} CO3_{T.min} \\ CO3_{T.max} \\ Fe_{T.min} \\ Fe_{T.max} \end{pmatrix} := \begin{pmatrix} 0.001 \\ 0.005 \\ \frac{1}{MW_{Fe}} \\ \frac{10}{MW_{Fe}} \end{pmatrix} \frac{mol}{L}$$

$$pH := 7$$

Given $Fe_{T.min} = Fe2_{FeCO3s}(pH, CO3_{T.min})$ $pH_{min.min} := Find(pH) = 7.413$

Given $Fe_{T.min} = Fe2_{FeCO3s}(pH, CO3_{T.max})$ $pH_{min.max} := Find(pH) = 6.808$

Given $Fe_{T.max} = Fe2_{FeCO3s}(pH, CO3_{T.min})$ $pH_{max.min} := Find(pH) = 6.579$

Given $Fe_{T.max} = Fe2_{FeCO3s}(pH, CO3_{T.max})$ $pH_{max.max} := Find(pH) = 6.113$

Then, in the Fe(II) predominance region, we would predict that:

1. for the minimum assumed value of total iron and the minimum assumed carbonate abundance, below pH 7.413 $FeCO_{3(s)}$ would be absent.
2. for the minimum assumed value of total iron and the maximum assumed carbonate abundance, below pH 6.808 $FeCO_{3(s)}$ would be absent.
3. for the maximum assumed value of total iron and the minimum assumed carbonate abundance, below pH 6.579 $FeCO_{3(s)}$ would be absent.
4. for the maximum assumed value of total iron and the maximum assumed carbonate abundance, below pH 6.113 $FeCO_{3(s)}$ would be absent.

These pH values are used with functions written below to situate the vertical predominance boundaries on the plots developed hereinafter with this example.

We have our set of half reactions from previous examples: $Fe(OH)_{3(s)}$–Fe^{+2}; $SO_4^{=}$–$H_2S_{(aq)}$; and $SO_4^{=}$–HS^-. Let us first situate the two $Fe(OH)_{3(s)}$–Fe^{+2} predominance boundaries associated with the assumed minimum and maximum values of total dissolved iron. We write two functions, similar to those written in Example 12.8:

$$pE_{min.FeOH3s.Fe2}(pH) := -\log(Fe_{T.min}) + pE°_{FeOH3s.Fe2} - 3 \cdot pH$$

$$pE_{max.FeOH3s.Fe2}(pH) := -\log(Fe_{T.max}) + pE°_{FeOH3s.Fe2} - 3 \cdot pH$$

These will appear in the plots associated with $Fe_T = 1$ and 10 ppm_m, respectively.

In order to situate the $SO_4^{=}$–H_2S and $SO_4^{=}$–HS^- lines that would provide the lower bound on the p*E* value for the Fe(II) predominance region, we must assign some values to the sulfate and sulfide species. We research the capacity of field test kits for assay of sulfide and find a limit of detection in the range of 0.1 ppm_m (~3 × 10^{-6} M). We will use this as our total sulfide abundance. Unless overly influenced by sulfate-containing minerals, ground water might contain 10 ppm_m total sulfate–sulfur (~3 ×10^{-4} M). At the pH range of the system in question, the sulfate ion is by far the predominant specie and we will use total sulfate as the sulfate ion activity. Given the sharpness of the general predominance line separating sulfate and sulfide species found in Example 12.8, the exact abundances of sulfate and sulfide matter little as long as both are present. Then, since total sulfide was below the assumed value for the detection limit, these lines represent the absolute lower limit for the p*E* (as a function of pH) of the ground water system in question.

To define these lines, we begin with the law of mass action statements for the two half reactions, stated in a manner similar to the function we wish to write:

$$10^{8 \cdot pE°_{SO4.HS}} = \frac{HS}{SO4 \cdot 10^{-8 \cdot pE} \cdot 10^{-9 \cdot pH}}$$

$$10^{8 \cdot pE°_{SO4.H2S}} = \frac{HS}{SO4 \cdot 10^{-8 \cdot pE} \cdot 10^{-10 \cdot pH}}$$

We can treat hydrogen sulfide as a monoprotic acid in this pH region and employ the corresponding abundance fraction relation for use with total sulfide and rearrange the relations to solve for the p*E*:

$$pE_{SO4.HS}(pH) := -\log\left[\left[\frac{S_T \cdot \left(\frac{10^{-pH}}{K_{1S}} + 1\right)^{-1}}{SO4_T \cdot 10^{8 \cdot pE°_{SO4.HS}} \cdot 10^{-9 \cdot pH}}\right]^{\frac{1}{8}}\right]$$

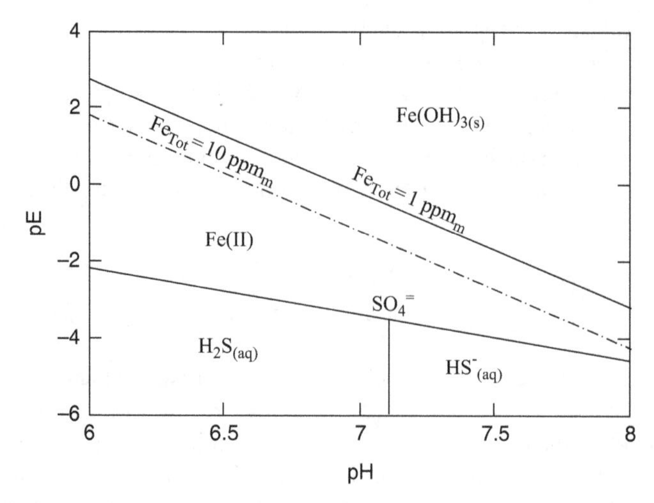

FIGURE E12.11.1 *pE–pH diagram for the Fe(III), Fe(II), $SO_4^{=}$, and $S^{=}$ system for Fe_{Tot} ranging from 1 to 10 ppm_m.*

$$pE_{SO4.H2S}(pH) := -\log\left[\left[\frac{S_T \cdot \left(1 + \frac{K_{1S}}{10^{-pH}}\right)^{-1}}{SO4_T \cdot 10^{8 \cdot pE^{\circ}_{SO4.H2S}} \cdot 10^{-10 \cdot pH}}\right]^{\frac{1}{8}}\right]$$

Then, in Figure E12.11.1, when we plot these relations along with those for the Fe(III)–Fe(II) couple on pE versus pH axes, we obtain the region in which the pE value could lie between the respective Fe(III)–Fe(II) boundaries and the defined pE versus pH line dictated by the known S(VI) and S(−II) abundances. We observe that the pE could lie anywhere in the region bounded by the sulfate–sulfide boundary and, depending upon the total iron abundance, the $Fe_{Tot} = 1$ or 10 ppm_m boundaries. Along these two boundaries, the iron abundance would be controlled by the presence of ferric hydroxide solid.

We can now employ the four pH values associated with the minimum and maximum total iron and total carbonate abundances, first the minimum total iron. In Figure E12.11.2, we superimpose conditions 1 and 2 on the pE versus pH plot of Figure E12.11.1 on which we have included only the $Fe_{Tot} = 1$ ppm_m Fe(III)–Fe(II) line to avoid overcomplicating the plot. If total dissolved inorganic carbon is 0.005 M, total iron can be 1 ppm_m or greater only in the region bounded by the sulfate–sulfide boundaries and the heavy dash-dot line. If total dissolved inorganic carbon is 0.001 M, total iron can be 1 ppm_m or greater only in the region bounded by the heavy solid line at pH ~7.4, the sulfate–sulfide boundary and the heavy solid and dash-dot lines separating iron (III) from iron (II). Then, relative to our goal of understanding the limits on the electron availability:

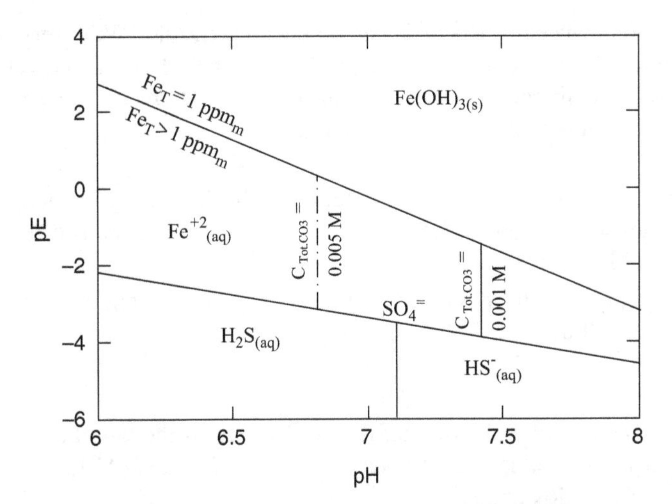

FIGURE E12.11.2 *pE–pH diagram for the Fe(III), Fe(II), $CO_3^{=}$. $SO_4^{=}$, and $S^{=}$ system for Fe_{Tot} = 1 ppm_m.*

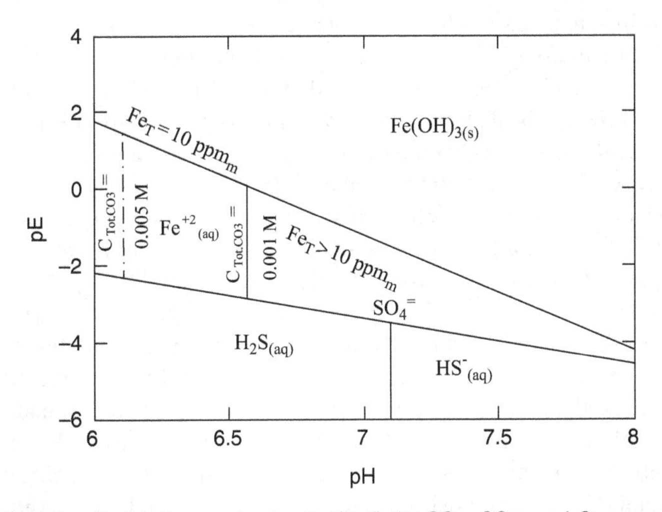

FIGURE E12.11.3 *pE–pH diagram for the Fe(III), Fe(II), $CO_3^{=}$. $SO_4^{=}$, and $S^{=}$ system for Fe_{Tot} = 10 ppm_m.*

1. If $C_{Tot.CO3} = 0.005\,M$, the pE value must lie in the region bounded by the sulfate–sulfide and Fe(II)–$Fe(OH)_{3(s)}$ pE boundaries, and to the left of the vertical line at pH ~6.8.
2. If $C_{Tot.CO3} = 0.001\,M$, the pE value must lie in the region bounded by the sulfate–sulfide and Fe(II)–$Fe(OH)_{3(s)}$ pE boundaries, and to the left of the vertical line at pH ~7.4.

In Figure E12.11.3, we superimpose conditions 3 and 4 from on the pE versus pH plot of Figure E12.11.1 on which we have included only the $Fe_{Tot} = 10\ ppm_m$ Fe(III)–

Fe(II) predominance line to avoid overcomplicating the plot. We observe the regions in which total iron could be 10 ppm_m to be similar to those for $Fe_T = 1$ ppm_m, but displaced to significantly lower pH values. Then, relative to our goal of understanding the limits of the electron availability:

3. If $C_{Tot.CO3} = 0.005\,M$, the p*E* value must lie in the region bounded by the sulfate–sulfide and Fe(II)–$Fe(OH)_{3(s)}$ p*E* boundaries, and to the left of the vertical line at pH ~6.1.
4. If $C_{Tot.CO3} = 0.001\,M$, the p*E* value must lie in the region bounded by the sulfate–sulfide and Fe(II)–$Fe(OH)_{3(s)}$ p*E* boundaries, and to the left of the vertical line at pH ~6.6.

Given that an odor of hydrogen sulfide was detected by the sampling team, we then know that the p*E* value must lie very near the p*E* versus pH line drawn for the sulfate–sulfide redox couple.

When all is said and done regarding the iron and carbonate levels in the aqueous solution of Example 12.11, we realize we should closely observe the sulfate–sulfide couple. Given the sharpness of the predominance boundary line between sulfate and sulfide, we realize that use of the sulfate–sulfide couple to define p*E* is likely the better of the two options in this situation. If both sulfate and sulfide species simultaneously exist in the aqueous solution, the p*E* must lie very close to the predominance line. Then, in any investigations of anoxic systems in which electron availability is of interest, determination of both total sulfate and total sulfide would be useful and vital. We must also bear in mind that had we employed the $FeOOH_{(s)}$–Fe^{+2} or the $Fe_2O_{3(s)}$–Fe^{+2} redox couples, our results likely would be different. We could gain insight as to the solid phase controlling iron in ground water systems from such analyses. Certainly, also, the iron-bearing mineral controlling the solubility of iron might be far more complex than the simple iron(III) solids used or mentioned. We might even use a modeling effort such as that of Example 12.11 to seek the identity of the iron-controlling solid phase among a sampling of iron-bearing minerals whose metal–ligand compositions are known for which the solid formation equilibria can be written. If formation constants are known, these iron-bearing minerals can simply replace the ferric hydroxide and/or ferrous carbonate of Example 12.11.

12.4.4 Speciation Using Electron Availability

Once we quantitatively know the electron availability of a system as p*E*, we may employ $\{e^-\}$ in computations with the law of mass action to compute speciation of other systems that might be present. We can employ mole balances along with equilibria in much the same manner as with acid–base and complex formation equilibria to determine speciation. The process is best illustrated by example.

Example 12.12 Consider an aquatic sediment system from which core samples were obtained. Some of the cores were centrifuged to extract pore water and assayed for sulfur species, pH, and major ions. Results indicated that total sulfide–sulfur and total sulfate–sulfur were 10 ppb_m and 100 ppm_m, respectively, pH was 8.5, and ionic strength was 0.01 M. Great care was exercised to completely isolate core samples and derived pore waters from atmospheric oxygen. It is known that these sediments are contaminated with arsenic and a parallel sample of the sediments was dried, pulverized, homogenized, subsampled, and tested for total arsenic with the result that abundance of total arsenic was found to be 2 grams total arsenic per kilogram dried sediment (2 g_{As}/kg_{DS}). We would like to know the speciation of arsenic in the sediments. We know that these sediment solids have a density of 2 kg/L and that the void fraction of the sediments is 0.40 L_{void}/L_{tot}.

Since we have both sides of the sulfate–bisulfide redox couple and the pH, we may compute a direct estimate of the pE of the sediment pore water. We retrieve our function written for Example 12.11 and populate it with known parameters to compute the value of pE. Then, for comparison with the E_H versus pH diagram of Figure A.2, we convert our value of pE to E_H:

$$A := 0.5115 \qquad I := 0.01 \qquad \gamma_1 := 10^{\frac{-A\cdot\sqrt{I}}{1+\sqrt{I}}} = 0.898 \qquad \gamma_2 := 10^{\frac{-A\cdot 4\cdot\sqrt{I}}{1+\sqrt{I}}} = 0.652$$

$$pK_{1S} := 7.1 \quad K_{1S} := 10^{-pK_{1S}} \quad S_T := \frac{10}{32000000} \quad SO4_T := \frac{100}{32000} \quad pH_{sys} := 8.5$$

$$pE_{SO4.HS}(pH) := -\log\left[\left[\frac{S_T\cdot\left(\frac{10^{-pH}}{K_{1S}}+\frac{1}{\gamma_1}\right)^{-1}}{\gamma_2\cdot SO4_T\cdot 10^{8\cdot pE^\circ_{SO4.HS}}\cdot 10^{-9\cdot pH}}\right]^{\frac{1}{8}}\right]$$

$$pE_{sys} := pE_{SO4.HS}(pH_{sys}) = -4.828$$

$$E_{H.SO4.HS}(pH) := \frac{pE_{SO4.HS}(pH)}{16.91} \qquad E_{H.SO4.HS}(8.5) = -0.286 \quad V$$

We pinpoint this E_H versus pH condition on Figure A.1 (showing the result in Figure E12.12.1) and determine that we are near the predominance boundary between elemental arsenic and arsenous acid. We are also not far from the predominance boundary between arsenous acid and dihydrogen arsenite. Lastly, the region of predominance of the arsenic (III) species is not wide and we may have the potential that arsenic (V) in the form of $HAsO_4^{=}$ could be of significant abundance. We are sufficiently close to the predominance boundaries that we will need to perform the speciation computations.

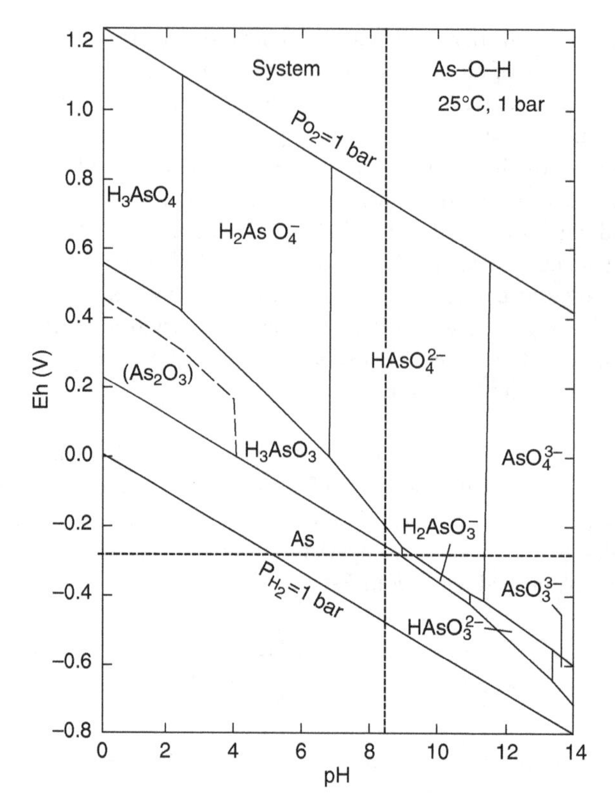

FIGURE E12.12.1 *An E_H–pH condition located on Brookins' As–O–H predominance plot for specified conditions in aquatic sediments.*

Here, the temperature is not likely 25 °C, but we have accomplished these adjustments in previous examples, and in order to concentrate here on the redox behavior, we will not adjust equilibrium constants for temperature.
We write a mole balance on arsenic species:

$$C_{Tot.As} = C_{H3AsO3} + C_{H2AsO3} + C_{HAsO4}$$

We quickly observe that all four terms are unknowns. We are sure not to include elemental arsenic in our mole balance as, if present, it would be a solid. We require three more independent equations to fully define the mathematical system. The acid–base equilibrium between arsenous acid and dihydrogen arsenite provides the second equation, the redox half reaction between elemental arsenic and either arsenous acid or dihydrogen arsenite provides the third, and the redox half reaction between either arsenous acid or dihydrogen arsenite and hydrogen arsenate provides the fourth:

HAsO4 + 2·E + 4·H = H3AsO3 + H2O

H3AsO3 + 3·E + 3·H = As + 3·H2O

H3AsO3 = H + H2AsO3

We will need the equilibrium constants so let us compute them from Gibbs energy data from Table A.4:

$$\begin{pmatrix} \Delta G^{\circ}_{H2O} \\ \Delta G^{\circ}_{HAsO4} \\ \Delta G^{\circ}_{H3AsO3} \\ \Delta G^{\circ}_{H2AsO3} \end{pmatrix} := \begin{pmatrix} -56.69 \\ -170.69 \\ -152.92 \\ -140.33 \end{pmatrix} \frac{kcal}{mol} \qquad R_{kcal} := 0.001987 \frac{kcal}{mol \cdot {}^{\circ}K} \qquad T := 298.15 \ {}^{\circ}K$$

$$K_{HAsO4.H3AsO3} := \exp\left[\frac{-(\Delta G^{\circ}_{H3AsO3} + \Delta G^{\circ}_{H2O} - \Delta G^{\circ}_{HAsO4})}{R_{kcal} \cdot T}\right] = 3.4 \times 10^{28}$$

$$K_{H3AsO3.As} := \exp\left[\frac{-(3 \cdot \Delta G^{\circ}_{H2O} - \Delta G^{\circ}_{H3AsO3})}{R_{kcal} \cdot T}\right] = 3.735 \times 10^{12}$$

$$K_{H3AsO3.H2AsO3} := \exp\left[\frac{-(\Delta G^{\circ}_{H2AsO3} - \Delta G^{\circ}_{H3AsO3})}{R_{kcal} \cdot T}\right] = 5.895 \times 10^{-10}$$

We could include all four equations in a given-find block or we could perform some algebra and obtain a single explicit relation yielding one of the dependent variables with which we can compute the values of the remaining variables. Let us do the algebra.

We have several options available for consolidating the system of equations. Of these, the most straightforward is the computation of the activity of arsenous acid, based on the presence of elemental arsenic, and subsequent computation of the remaining species and of total dissolved arsenic via summation. If this computation yields a computed total arsenic that is greater than the measured value, we can be assured that elemental arsenic does not exist. Then the measured total arsenic would be used to determine the total dissolved arsenic abundance and the computations would be repeated. If the computed total dissolved arsenic is less than the measured value, solid elemental arsenic is among the arsenic species present:

$$H3AsO3 := \frac{1}{K_{H3AsO3.As} \cdot 10^{-3 \cdot pE_{sys}} \cdot 10^{-3 \cdot pH_{sys}}} = 0.028$$

$$H2AsO3 := \frac{K_{H3AsO3.H2AsO3}}{10^{-pH_{sys}}} \cdot H3AsO3 = 5.177 \times 10^{-3}$$

$$HAsO4 := \frac{H3AsO3}{K_{HAsO4.H3AsO3} \cdot 10^{-2 \cdot pE_{sys}} \cdot 10^{-4 \cdot pH_{sys}}} = 1.803 \times 10^{-6}$$

$$C_{Tot.As} := H3AsO3 + \frac{H2AsO3}{\gamma_1} + \frac{HAsO4}{\gamma_2} = 0.034 \quad \frac{mol}{L}$$

Now we have a prediction for the abundance of arsenic in the pore water and can determine the portion of the total measured arsenic that is dissolved. We consult Section 8.2.2 for some volume–mass relations and compute the mass of arsenic dissolved in the volume of aqueous solution driven off during the drying process, leaving solid arsenic behind:

$$\rho_{sol} := 2\frac{kg}{L} \quad \varepsilon_w := 0.4 \quad \frac{L_W}{L_{Tot}} \quad V_{sol} := \frac{1}{\rho_{sol}} = 0.5 \quad V_w := \frac{\varepsilon_w}{1-\varepsilon_w} \cdot V_{sol} = 0.333 \quad \frac{L}{kg_{DS}}$$

$$Mol_{As} := V_w \cdot C_{Tot.As} = 0.011 \quad \frac{mol}{kg_{DS}} \qquad Mass_{As} := Mol_{As} \cdot 74.9 = 0.837 \quad \frac{g}{kg_{DS}}$$

We determine that, assuming presence of elemental arsenic in the sediments, just under a gram of the 2 g_{As}/kg_{DS} would be dissolved in the water. One of two conditions must then exist. Either solid elemental arsenic is present in the sediments or adsorption of the arsenic species to the sediment solids accounts for over a gram of As per kg of dry solids. Since the dissolved arsenic species are oxyanions of negative charge and since at the moderately alkaline pH of the sediments, the sediment solids would likely have a net negative charge, sorption of arsenic oxyanions would be minimal. We would conclude that elemental arsenic is likely present under the stated conditions of these sediments.

PROBLEMS

For the end-of-chapter problems that follow, the convenient platform from which to assemble the mathematical models for each of the problems is a MathCAD worksheet. MS Excel or other software may certainly be employed, but additional algebraic manipulations or structured programming may be necessary. Certainly, also, simplifying assumptions may be invoked to render pencil/paper/calculator approximations. Certainly, even graphical approximations of the solution may be assembled:

1. The gas phase within an anaerobic digester ($P_T = 1$ atm) typically consists of ~65% methane and ~35% carbon dioxide on a molar basis. Consider that the solution has pH of 7.3
   a. Consider that total soluble sulfur is 10^{-3} M and determine the speciation of sulfur (neglecting elemental S) in the aqueous solution of the digester. Figure A.8 will prove helpful.
   b. Determine the solubility of selenium in the aqueous solution. Note that elemental selenium could be the controlling solid phase. Figure A.9 will prove helpful.

2. The chromium system is very important in a number of industrial applications and Cr(VI) is a contaminant of interest in many environmental systems affected by rogue industrial discharges. This system is applicable to the removal of chromium from metal plating wastewaters that employ chromium. (Fortunately, the U.S. auto industry is getting away from metal plating as certainly there is much less "chrome" on U.S. cars than in recent past—it is all on Harley-Davidson motorcycles.)

   a. Write the half reactions for the reduction of Cr^{+6} to Cr^{+3}. Consider that the primary reactants are bichromate ($HCrO_4^-$) and/or chromate ($CrO_4^=$) and the primary product is Cr^{+3}. Use Gibbs free energy of formation from the appendix and equilibria as necessary from Chapter 6, the Appendix, or other sources to determine the $p\varepsilon^\circ$ for this half reaction.
   b. Expand the half reaction as written earlier, consider that the primary reactants are bichromate ($HCrO_4^-$) and/or chromate ($CrO_4^=$) and the primary product is $Cr_2O_{3(s)}$. Use Gibbs free energy of formation from Appendix and equilibria as necessary from Chapter 6, the Appendix, or other sources to determine the $p\varepsilon^\circ$ for this half reaction.
   c. Find the equations of the lines you would plot on a figure, such as those contained in Figure A.1, Figure A.2, Figure A.3, Figure A.4, Figure A.5, Figure A.6, Figure A.7, Figure A.8, and Figure A.9, that would yield the predominance boundaries for $\{HCrO_4^-\}$, $\{CrO_4^=\}$, and $\{Cr^{+3}\}$as a function of pH and $p\varepsilon$. Plot these lines either in MathCAD or in an Excel chart.
   d. Find the equations of the lines you would plot on a figure, such as those contained in Figure A.1, Figure A.2, Figure A.3, Figure A.4, Figure A.5, Figure A.6, Figure A.7, Figure A.8, and Figure A.9, that would yield the predominance boundaries for $\{HCrO_4^-\}$, $\{CrO_4^=\}$, and $\{Cr_2O_{3(s)}\}$as a function of pH and $p\varepsilon$. Plot these lines either in MathCAD or in the Excel chart created for the previous exercise.
   e. Indicate in the plot completed for part d the pH versus $p\varepsilon$ region where solid phase Cr_2O_3 would have a good probability of existence in environmental systems. Back this up with some reasoning and some computations.

3. A dilute aqueous system contains (among other constituents) dissolved inorganic carbon, nitrogen, and selenium (Se) species. Assays of the solution yielded the following results:

pH	6.5	$[CH_4(aq)]$	Below detection limit
$[O_2(aq)]$	Below detection limit	$[NH_3\text{–}N]$	3.5×10^{-4} M (= $[C_T, NH_3]$)
$[C_{inorganic}]$	0.003 M (= $[C_T, CO_3]$)	$[NO_3\text{–}N]$	1.0×10^{-3} M (= $[NO_3^-]$)
Se_{Tot}	10^{-5} M	S_{Tot}	10^{-4} M

   a. Determine the $p\varepsilon$ of the water.
   b. Determine the speciation of selenium. The E_H–pH predominance diagram for the Se–O–H system and Gibbs energy data in the Appendix will prove useful.

c. Determine the speciation of sulfur.
d. Suggest (include supporting computations) whether $MnO_{2(s)}$ could be present with aquifer solids.
e. Suggest (include supporting computations) whether $Fe(OH)_{3(s)}$ could be present with aquifer solids.

4. Consider the sulfate–sulfide system, most particularly under mildly acidic pH conditions wherein sulfate and hydrogen sulfide would be the predominant sulfur species depending on the pε of the system. You may consider that the aqueous system is infinitely dilute and thus all activity coefficients have the value of unity.

   a. Write the half reaction for the sulfate–aqueous hydrogen sulfide couple. This might be quite similar to the reaction of Table A.5, but note that hydrogen sulfide must be an aqueous rather than a gaseous specie.
   b. Determine the pε° value for the reaction. Perform this computation using two methods: (1) use Gibbs free energy of formation to obtain the equilibrium constant and from K_{eq} determine pε°, and (2) begin with the reaction 12 Table A.5 and appropriately add the reaction for the dissolution of hydrogen sulfide gas in water to determine the equilibrium constant and associated pε° using the addition of Gibbs free energies (addition of log(K) values).
   c. Develop the equations (p$\varepsilon = f$(pH)) of the lines that separate the regions of $SO_4^{=}$ and $H_2S_{(aq)}$ and $SO_4^{=}$ and HS^- predominance. Plot these on a single set of axes and compare with Figure A.8.
   d. Explain (and use Gibbs energy to position) the vertical line on Figure A.8 that separates H_2S from HS^-
   e. Write a set of MathCAD functions that will allow the computation of the three potentially significant sulfur species for specific values of pH and for a range of pε values corresponding with Eh values between −0.3 and +0.3. Use this set of functions to produce plots of the activities of sulfate, aqueous hydrogen sulfide, and bisulfide as functions of Eh (or pε) over the specified range of Eh values. Test your set of functions at pH values of 6.1, 7.1, and 8.1, for a total sulfur concentration of 0.001 M by producing appropriate plots.

5. *Lange's Handbook* (Dean, 1992) gives the Gibbs energy of formation for permanganate (MnO_4^-) as −447.3 kJ/mol. The permanganate–manganese dioxide redox couple is important in the removal of manganese from natural waters.

   a. Write the half reaction for the permanganate–manganese dioxide redox couple and determine the pε° for the half reaction. You will need to employ the standard enthalpy and entropy to obtain Gibbs energy of formation for MnO_4^- in order to determine the standard Gibbs energy of the reaction and hence an equilibrium constant for the reaction you have written.
   b. Determine the equation (p$\varepsilon = f$(pH)) of the line separating the predominance regions of permanganate from manganese dioxide and carefully sketch the line on the manganese diagram from the Appendix.

6. Develop the overall redox reactions for the oxidation of Mn(II) to $MnO_{2(s)}$ first using permanganate and then using molecular oxygen as the oxidant. Determine the equilibrium constants for each of the reactions. Here, you must write the half reaction for the Mn(II)–$MnO_{2(s)}$ couple and determine the standard Gibbs free energy of the reaction, write the half reaction for the oxygen–water couple and determine the standard Gibbs free energy of reaction, and write the half reaction for the permanganate–$MnO_{2(s)}$ couple and determine the standard Gibbs free energy of reaction. Then, write the Mn(II)–$MnO_{2(s)}$ reaction as the oxidation of Mn(II), appropriately add each of the reduction reactions, appropriately sum the Gibbs energies, and compute the equilibrium constants from the Gibbs energies. One can use the log values of the equilibrium constants in lieu of the Gibbs energy, remembering that the negative logarithms (or *p* values) are additive as is the Gibbs energy.

   a. Use the equilibrium constants and law of mass action to determine the theoretical value of $\{Mn^{+2}\}$ for a partial pressure of oxygen of 0.189 atm over the range of pH from 6 to 8. This is, of course, very dependent upon pH. Thus, a MathCAD function is highly appropriate.
   b. Set the target residual (equilibrium) value of $\{MnO_4^-\}$ to be 10^{-10} M, use this as a scalar value in MathCAD and compute the corresponding theoretical value of $\{Mn^{+2}\}$ over the pH range of 6–8. This is, of course, also very dependent upon pH and writing of a MathCAD function is highly appropriate.
   c. Vary the residual permanganate abundance to yield a result for residual Mn(II) similar to that for oxidation with molecular oxygen.

7. An aqueous stream emanating from an industrial process has a pH of 6.0 and contains 0.002 M sulfate, 0.001 M total inorganic carbon, 0.002 M Na^+, 0.0015 M Mg^{++}, and 10^{-6} M total arsenic (As). A sample taken from the vapor space in the industrial reactor contained ~1% oxygen ($O_{2(g)}$), 78% nitrogen ($N_{2(g)}$), ~20% carbon dioxide ($CO_{2(g)}$), 0.1 ppb_v (part per billion by volume) hydrogen sulfide gas ($H_2S_{(g)}$), and no detectable methane gas ($CH_{4(g)}$). The total pressure of the vapor space was measured to be 1 atm. Determine the $p\varepsilon$ of the aqueous solution and the predominant arsenic specie(s). You may assume that the solution is infinitely dilute. The attached E_H-pH diagram for the As–O–H system and associated Gibbs energy data information from the Appendix should prove useful for this determination.

   a. Perform a charge balance and assume that the imbalance can be ascribed to monovalent cations or anions, whichever the case may be.
   b. Use available information to determine the pE of the system and use reasoning as necessary to determine which value of pE would be correct.
   c. Investigate whether elemental arsenic could exist in this system.

8. You are examining a river system, located in the southeastern United States, which has potentially been impacted by industrial releases containing arsenic. Understanding of the magnitude of the potential problem would be aided by determination of the speciation of the arsenic that might be in the sediments. As part of another study, a set of core samples was obtained from the sediments sit-

uated near the discharge of this river to a major fresh water reservoir and characterized. The sediment samples were obtained using special coring equipment, handled with virtually no contact between the atmosphere and the sediments, and transported to an analytical laboratory. Once in the lab sediments were centrifuged to separate pore water from sediment solids. The pore water was then analyzed and found to have the following character. *Unfortunately, aqueous As was not quantitated.*

pH – 8.05	Total format – ~1 ppb_m
Alkalinity[a] – 75 mg/L as $CaCO_3$	Ammonia nitrogen (NH_3–N) – 8.2 ppm_m
Total calcium (Ca_T) – 28 ppm_m	Nitrate nitrogen (NO_3^-–N) ND[b]
Total sodium (Na_T) – 21.5 ppm_m	Methane ($CH_{4(g)}$,[c]) – 1 ppm_v (@ $P_T = 1$ atm)
Total sulfate sulfur ($SO_4^=$–S) – 1 ppm_m	Total sulfide sulfur ($S^=$–S) – 35 ppm_m

[a]The inorganic carbon system is virtually entirely responsible for the alkalinity of this sample.
[b]ND—below method detection limits.
[c]The partial pressure of methane in equilibrium with the pore water, by solid-phase micro-extraction.

Figure A.2 and associated Gibbs energy data from Tables A.1 and A.4 will be immensely useful herein.

a. Use available information to determine the pE of the aqueous solution. Note that three potential redox couples have known abundances for oxidized and reduced species. Use judgments to ascertain which of the results is the most representative. Note that either an implicit or iterative successive substitutions solution will be necessary to determine activity coefficients for aqueous speciation.
b. If arsenic is present in the sediments as a solid phase, given the presence of sulfide, determine whether the most probable specie is $As_{S(s)}$ or $As_2S_{3(s)}$. Use Gibbs energy data from Table A.4 to determine the value for the second deprotonation constant for the sulfide system, if used.
c. Based on the outcome of part b, determine the probable total abundance of arsenic in the pore water of the sediments.

9. A sediment sample was obtained from a depth of ~1 ft beneath the sediment–water interface in the inlet bay of a southwest Minnesota lake. The sample was centrifuged and the supernatant solution was assayed and found to contain the following:

pH – 6.05	
Ammonia nitrogen (NH_3–N) – 8.2 ppm_m	Nitrate nitrogen (NO_3^-–N) – ND[a]
Methane ($CH_{4(aq)}$,) – 1 ppm_m	Alkalinity[b] – 45 mg/L *as $CaCO_3$*

[a]ND – below method detection limits
[b]You may assume that the inorganic carbon system is virtually entirely responsible for the alkalinity of this sample.

Were it possible to measure the ORP of the in-place sediments, what value would you expect it to be? Fortify your assertions with appropriate computations.

10. A water sample taken from an aquifer was tested and found to contain alkalinity of 100 mg/L *as* $CaCO_3$, be of pH 7.1, and contain total iron (Fe_T) of 2 ppm_m. Of interest would be the nature of the solid phase, if any, controlling the solubility of iron in this water. The iron is known to be divalent (Fe^{+2}). The water is from an aquifer of very low total dissolved solids content, such that it may be considered infinitely dilute.

    Which solid phase, $Fe(OH)_{2(s)}$ or $FeCO_{3(s)}$ is the most probable control on the solubility of iron(II). Support your assertion with computations and associated reasoning.

11. A water sample was taken from a shallow aquifer underlying a cultivated field in California's Central Valley. Field measurements of ORP and pH yielded values of +125 mV and 7.60, respectively. Irrigation waters applied to this particular field over a century of irrigation have contained measurable levels of selenium and environmental regulators fear that the aquifer solids might contain solid phase selenium. Thus, the water sample was assayed for selenium and found to contain 1.5×10^{-5} M total selenium (Se_T).

    Based on the stated information and appropriate computations, suggest whether or not solid phase elemental selenium might exist with the aquifer solids. Your computations should include those enabling the determination of $p\varepsilon^\circ$ for appropriate half reactions employed in your computations.

12. A dilute (you may assume infinitely so) aqueous solution obtained from a ground water source contains (among other constituents) dissolved inorganic carbon, nitrogen, sulfur (S), and selenium (Se) species. A portion of the sample obtained was allowed to stand on the laboratory bench for several days to fully equilibrate with atmospheric oxygen. The solution remained clear over the entire period, indicating that iron and manganese were not present in the aqueous solution. Assays of the solution yielded the following results:

pH	6.5	ORP	+405 mV	$[CH_4(aq)]$	Trace (below quantitation limit)
$[O_{2(aq)}]$	Below detection limit			$[NH_3\text{–}N]$	3.5×10^{-4} M
$[C_{inorganic}]$	0.003 M			$[NO_3\text{–}N]$	1.0×10^{-3} M
Se_T	10^{-5} M			S_T	10^{-4} M

a. Assume that the water was at equilibrium when the sample was taken and corroborate the ORP reading taken of the water. Look carefully at the half reactions relating inorganic carbon and methane and relating nitrate–nitrogen to ammonia–nitrogen.

b. From the pε of (a) determine the speciation of selenium given the stated total concentration.
c. From the pε of (a) determine the speciation of sulfur given the stated total concentration.
d. From the pε of (a) suggest (include relevant computations) whether $MnO_{2(s)}$ and/or $Fe(OH)_{3(s)}$ could be present with aquifer solids. Compute $\{Mn^{+2}\}$ and $\{Fe^{+2}\}$ associated with the stated ORP based on the presence of the respective solids. Are these computed activities reasonable?

Appendices

TABLE A.1 Gibbs Energy of Formation $\overline{G}_f^\circ$, Enthalpy of Formation $\overline{H}_f^\circ$, and Entropy of Formation $\overline{S}_f^\circ$ Values for Common Chemical Species in Aquatic Systems[a]: Valid at 25°C, 1 atm Pressure, and Standard States[b]

	Formation from the elements		Entropy	
Species	$\overline{G}_f^\circ$ (kJ/mol)	$\overline{H}_f^\circ$ (kJ/mol)	$\overline{S}_f^\circ$ (J/mol/K)	References[c]
Ag (Silver)				
Ag (Metal)	0	0	42.6	NBS
Ag^+ (aq)	77.12	105.6	73.4	NBS
AgBr	−96.9	−100.6	107	NBS
AgCl	−109.8	−127.1	96	NBS
AgI	−66.2	−61.84	115	NBS
$Ag_2S(\alpha)$	−40.7	−29.4	14	NBS
AgOH(aq)	−92			NBS
$Ag(OH)_2^-$(aq)	−260.2			NBS
AgCl(aq)	−72.8	−72.8	154	NBS
$AgCl_2^-$(aq)	−215.5	−245.2	231	NBS
Al (Aluminum)				
Al	0	0	28.3	R

(*Continued*)

Environmental Process Analysis: Principles and Modeling, First Edition. Henry V. Mott.

TABLE A.1 (*Continued*)

Species	Formation from the elements		Entropy	
	$\bar{G}_f^\circ$ (kJ/mol)	$\bar{H}_f^\circ$ (kJ/mol)	$\bar{S}_f^\circ$ (J/mol/K)	References[c]
Al^{3+} (aq)	−489.4	−531.0	−308	R
$AlOH^{2+}$ (aq)	−698			S
$Al(OH)_2^+$(aq)	−911			S
$Al(OH)_3$(aq)	−1115			S
$Al(OH)_4^-$(aq)	−1325			S
$Al(OH)_3$ (amorph)	−1139			R
Al_2O_3 (Corundum)	−1582	−1676	50.9	R
AlOOH (Boehmite)	−922	−1000	17.8	R
$Al(OH)_3$ (Gibbsite)	−1155	−1293	68.4	R
$Al_2Si_2O_5(OH)_4$ (Kaolinite)	−3799	−4120	203	R
$KAl_3Si_3O_{10}(OH)_2$ (Muscovite)	−1341			R
$Mg_5Al_2Si_3O_{10}(OH)_8$ (Chlorite)	−1962			R
$CaAl_2Si_2O_8$ (Anorthite)	−4017.3	−4243.0	199	R
$NaAlSi_3O_8$ (Albite)	−3711.7	−3935.1		R
As (Arsenic)				
As (α−Metal)	0	0	35.1	NBS
H_3AsO_4(aq)	−766.0	−898.7	206	NBS
$H_2AsO_4^-$(aq)	−748.5	−904.5	117	NBS
$HAsO_4^{2-}$(aq)	−707.1	−898.7	3.8	NBS
AsO_4^{3-}(aq)	−636.0	−870.3	−145	NBS
$H_2AsO_3^-$(aq)	−587.4			NBS
Ba (Barium)				
Ba^{2+} (aq)	−560.7	−537.6	9.6	R
$BaSO_4$ (Barite)	−1362	−1473	132	R
$BaCO_3$ (Witherite)	−1132	−1211	112	R
Be (Beryllium)				
Be^{2+} (aq)	−380	−382	−130	NBS
$Be(OH)_2$(α)	−815.0	−902	51.9	NBS
$Be_3(OH)_3^{3+}$	−1802			NBS
B (Boron)				
H_3BO_3(aq)	−968.7	−1072	162	NBS
$B(OH)_4^-$(aq)	−1153.3	−1344	102	NBS
Br (Bromide)				
Br_2(l)	0	0	152	NBS
Br_2(aq)	3.93	−259	130.5	NBS
Br^- (aq)	−104.0	−121.5	82.4	NBS
HBrO(aq)	−82.2	−113.0	147	NBS
BrO^-(aq)	−33.5	−94.1	42	NBS
C (Carbon)				
C (Graphite)	0	0	152	NBS
C (Diamond)	3.93	−2.59	130.5	NBS
CO_2(g)	−394.37	−393.5	213.6	NBS
$H_2CO_3^*$(aq)	−623.2	−699.6	187.0	R[d]
H_2CO_3(aq) ("true")	~ −607.1			S
HCO_3^-(aq)	−586.8	−692.0	91.2	S

TABLE A.1 (*Continued*)

Species	Formation from the elements		Entropy	
	$\bar{G}_f^\circ$ (kJ/mol)	$\bar{H}_f^\circ$ (kJ/mol)	$\bar{S}_f^\circ$ (J/mol/K)	References[c]
CO_3^{2-}(aq)	−527.9	−677.1	−56.9	NBS
CH_4(g)	−50.79	−74.80	186	NBS
CH_4(aq)	−34.39	−89.04	83.7	NBS
CH_3OH(aq)	−175.4	−245.9	133	NBS
HCOOH(aq)	−372.3	−425.4	163	NBS
$HCOO^-$(aq)	−351.0	−425.6	92	NBS
CH_2O(aq)	−129.7			
CH_2O(g)	−110.0	−116.0	218.6	S
HCN(aq)	112.0	105.0	129	NBS
CN^-(aq)	166.0	151.0	118	NBS
COS(g)	−169.2	−137.2	234.5	NBS
CNS^-(aq)	88.7	72.0		S
$H_2C_2O_4$(aq)	−697.0	−818.26		S
$HC_2O_4^-$(aq)	−690.86	−818.8		S
$C_2O_4^{2-}$(aq)	−674.04	−818.8	45.6	S
Ca (Calcium)				
Ca^{2+} (aq)	−553.54	−542.83	−53	R
$CaOH^+$ (aq)	−718.4			NBS
$Ca(OH)_2$(aq)	−868.1	−1003	−74.5	NBS
$Ca(OH)_2$ (Portlandite)	−898.4	−986.0	83	R
$CaCO_3$ (Calcite)	−1128.8	−1207.4	91.7	R
$CaCO_3$ (Aragonite)	−1127.8	−1207.4	88.0	R
$CaMg(CO_3)_2$ (Dolomite)	−2161.7	−2324.5	155.2	R
$CaSiO_3$ (Wollastonite)	−1549.9	−1635.2	82.0	R
$CaSO_4$ (Anhydrite)	−1321.7	−1434.1	106.7	R
$CaSO_4 \cdot 2H_2O$ (Gypsum)	−1797.2	−2022.6	194.1	R
$Ca_5(PO_4)_3OH$ (Hydroxyapatite)	−6338.4	−6721.6	390.4	R
Cd (Cadmium)				
Cd (γ−Metal)				
Cd^{2+} (aq)	−77.58	−75.90	−73.2	R
$CdOH^+$ (aq)	−284.5			R
$Cd(OH)_3^-$(aq)	−600.8			R
$Cd(OH)_4^{2-}$ (aq)	−758.5			R
$Cd(OH)_2$(aq)	−392.2			R
CdO (s)	−228.4	−258.1	54.8	
$Cd(OH)_2$ (precip.)	−473.6	−560.6	96.2	R
$CdCl^+$ (aq)	−224.4	−240.6	43.5	R
$CdCl_2$ (aq)	−340.1	−410.2	39.8	R
$CdCl_3^-$(aq)	−487.0	−561.0	203	R
$CdCO_3$(s)	−669.4	−750.6	92.5	R
Cl (Chlorine)				
Cl^-(aq)	−131.3	−167.2	56.5	NBS
Cl_2(g)	0	0	223.0	NBS
Cl_2(aq)	6.90	−23.4	121	NBS
HClO(aq)	−79.9	−120.9	142	NBS

(*Continued*)

TABLE A.1 (*Continued*)

Species	Formation from the elements		Entropy	
	$\bar{G}_f^\circ$ (kJ/mol)	$\bar{H}_f^\circ$ (kJ/mol)	$\bar{S}_f^\circ$ (J/mol/K)	References[c]
ClO^-(aq)	−36.8	−107.1	42	NBS
ClO_2(aq)	117.6	74.9	173	NBS
ClO_2^-(aq)	17.1	−66.5	101	NBS
ClO_3^-(aq)	−3.35	−99.2	162	NBS
ClO_4^-(aq)	−8.62	−129.3	182	NBS
Co (Cobalt)				
Co (Metal)	0	0	30.04	R
Co^{2+} (aq)	−54.4	−58.2	−113	R
Co^{3+} (aq)	−134	−92	−305	R
$HCoO_2^-$(aq)	−407.5			NBS
$Co(OH)_2$(aq)	−369	−518	134	NBS
$Co(OH)_2$ (blue precip.)	−450			NBS
CoO(s)	−214.2	−237.9	53.0	R
Co_3O_4 (Cobalt Spinel)	−725.5	−891.2	102.5	R
Cr (Chromium)				
Cr (Metal)	0	0	23.8	NBS
Cr^{2+} (aq)		−143.5		NBS
Cr^{3+} (aq)	−215.5	−256.0	308	NBS
Cr_2O_3 (Eskolaite)	−1053	−1135	81	R
$HCrO_4^-$(aq)	−764.8	−878.2	184	R
CrO_4^{2-}(aq)	−727.9	−881.1	50	R
$Cr_2O_7^{2-}$(aq)	−1301	−1490	262	R
$Cr(OH)_3$ (hydrous)	−858	−984	(1051)	Bard et al.
$Cr(OH)^{2+}$	−430	−495	(−156)	Bard et al.
$Cr(OH)_2^+$	−653	−748	(−27)	Bard et al.
$Cr(OH)_4^-$	−1013	−1169	(238)	Bard et al.
Cu (Copper)				
Cu (Metal)	0	0	33.1	NBS
Cu^+ (aq)	50.0	71.7	40.6	NBS
Cu^{2+} (aq)	65.5	64.8	−99.6	NBS
$Cu(OH)_2$(aq)	−249.1	−395.2	−121	NBS
$HCuO_2^-$(aq)	−258			
CuS (Covellite)	−53.6	−53.1	66.5	NBS
Cu_2S (α)	−86.2	−79.5	121	NBS
CuO (Tenorite)	−129.7	−157.3	43	NBS
$CuCO_3 \cdot Cu(OH)_2$ (Malachite)	−893.7	−1051.4	186	NBS
$2CuCO_3 \cdot Cu(OH)_2$ (Azurite)		−1632		NBS
F (Fluorine)				
F_2(g)	0	0	202	NBS
F^-(aq)	−278.8	−332.6	−13.8	NBS
HF(aq)	−296.8	320.0	88.7	NBS
HF_2^-(aq)	−578.1	−650	92.5	NBS
Fe (Iron)				
Fe (Metal)	0	0	27.3	NBS
Fe^{2+} (aq)	−78.87	−89.10	−138	NBS
$FeOH^+$ (aq)	−277.4	324.7	29	NBS

TABLE A.1 (*Continued*)

Species	Formation from the elements		Entropy	
	$\bar{G}_f^\circ$ (kJ/mol)	$\bar{H}_f^\circ$ (kJ/mol)	$\bar{S}_f^\circ$ (J/mol/K)	References[c]
$Fe(OH)_2(aq)$	−441.0	–	–	NBS
Fe^{3+} (aq)	−4.60	−48.5	−316	NBS
$FeOH^{2+}$ (aq)	−229.4	−324.7	−29.2	NBS
$Fe(OH)_2^+(aq)$	−438	−250.8	142.0	NBS
$Fe(OH)_3(aq)$	−659.4	–	–	NBS
$Fe(OH)_4^-(aq)$	−842.2	–	34.5	NBS
$Fe_2(OH)_2^{4+}(aq)$	−467.27	612.1	356.0	NBS
FeS_2 (Pyrite)	−160.2	−171.5	52.9	R
FeS_2 (Marcasite)	−158.4	−169.4	53.9	R
FeO(s)	−251.1	−272.0	59.8	R
$Fe(OH)_2$ (precip.)	−486.6	−569	87.9	NBS
$\alpha-Fe_2O_3$ (Hematite)[e]	−742.7	−824.6	87.4	R
Fe_3O_4 (Magnetite)	−1012.6	−1115.7	146	R
$\alpha-FeOOH$ (Goethite)[e]	−488.6	−559.3	60.5	R
FeOOH (amorph)[e]	−462			S
$Fe(OH)_3$ (amorph)[e]	−699(−712)			S
$FeCO_3$(Siderite)	−666.7	−737.0	105	R
Fe_2SiO_4 (Fayalite)	−1379.4	−1479.3	148	R
H (Hydrogen)				
$H_2(g)$	0	0	130.6	NBS
$H_2(aq)$	17.57	−4.18	57.7	NBS
$H^+(aq)$	0	0	0	NBS
$H_2O(l)$	−237.18	−285.83	69.91	NBS
$H_2O(g)$	−228.57	−241.8	188.72	R
$H_2O_2(aq)$	−134.1	−191.17	143.9	NBS
$HO_2^-(aq)$	−67.4	−160.33	23.8	NBS
Hg (Mercury)				
Hg(l)	0	0	76.0	NBS
$Hg_2^{2+}(aq)$	153.6	172.4	84.5	NBS
Hg^{2+} (aq)	164.4	171.0	−32.2	NBS
Hg_2Cl_2 (Calomel)	−210.8	265.2	192.4	NBS
HgO(red)	−58.5	−90.8	70.3	NBS
HgS (Metacinnabar)	−43.3	−46.7	96.2	NBS
HgI_2 (red)	−101.7	−105.4	180	NBS
$HgCl^+(aq)$	−5.44	−18.8	75.3	NBS
$HgCl_2(aq)$	−173.2	−216.3	155	NBS
$HgCl_3^-(aq)$	−309.2	−388.7	209	NBS
$HgCl_4^{2-}(aq)$	−446.8	−554.0	293	NBS
$HgOH^+(aq)$	−52.3	−84.5	71	NBS
$Hg(OH)_2(aq)$	−274.9	−355.2	142	NBS
$HgO_2^-(aq)$	−190.3			NBS
I (Iodine)				
I_2 (Crystal)	0	0	116	NBS
$I_2(aq)$	16.4	22.6	137	NBS
$I^-(aq)$	−51.59	−55.19	111	NBS

(*Continued*)

TABLE A.1 (*Continued*)

Species	Formation from the elements		Entropy	
	$\bar{G}_f^\circ$ (kJ/mol)	$\bar{H}_f^\circ$ (kJ/mol)	$\bar{S}_f^\circ$ (J/mol/K)	References[c]
I_3^-(aq)	−51.5	−51.5	239	NBS
HIO(aq)	−99.2	−138	95.4	NBS
IO^-(aq)	−38.5	−107.5	−5.4	NBS
HIO_3(aq)	−132.6	−211.3	167	NBS
IO_3^-	−128.0	−221.3	118	NBS
Mg (Magnesium)				
Mg (Metal)	0	0	32.7	R
Mg^{2+}(aq)	−454.8	−466.8	−138	R
$MgOH^+$(aq)	−626.8			S
$Mg(OH)_2$(aq)	−769.4	−926.8	−149	NBS
$Mg(OH)_2$ (Brucite)	−833.5	−924.5	63.2	R
Mn (Manganese)				
Mn (Metal)	0	0	32.0	R
Mn^{2+} (aq)	−228.0	−220.7	−73.6	R
$Mn(OH)_2$ (precip.)	−616			S
Mn_3O_4 (Hausmannite)	−1281			S
MnOOH (α−Manganite)	−557.7			S
MnO_2 (Manganate) (IV) ($MnO_{1.7}$-MnO_2)	−453.1			S
MnO_2 (Pyrolusite)	−465.1	−520.0	53	R
$MnCO_3$ (Rhodochrosite)	−816.0	−889.3	100	R
MnS (Albandite)	−218.1	−213.8	87	R
$MnSiO_3$ (Rhodonite)	−1243	−1319	131	R
N (Nitrogen)				
N_2(g)	0	0	191.5	NBS
NO(g)	86.57	90.25	210.6	S
NO_2(g)	51.3	33.2	240.0	S
N_2O(g)	104.2	82.0	220	NBS
NH_3(g)	−16.48	−46.1	192	NBS
NH_3(aq)	−26.57	−80.29	111	NBS
NH_4^+(aq)	−79.37	−132.5	113.4	NBS
HNO_2(aq)	−42.97	−119.2	153	NBS
NO_2^-(aq)	−37.2	−104.6	140	NBS
HNO_3(aq)	−111.3	−207.3	146	NBS
NO_3^-(aq)	−111.3	−207.3	146.4	NBS
Ni (Nickel)				
Ni^{2+} (aq)	−45.6	−54.0	−129	R
NiO (Bunsenite)	−211.6	−239.7	38	R
NiS (Millerite)	−86.2	−84.9	66	R
O (Oxygen)				
O_2(g)	0	0	205	NBS
O_2(aq)	16.32	−11.71	111	NBS
O_3(g)	163.2	142.7	239	NBS
O_3(aq)		125.9		NBS
$O_2^{\bullet-}$	31.84			NBS

TABLE A.1 (*Continued*)

Species	Formation from the elements		Entropy	
	$\bar{G}_f^\circ$ (kJ/mol)	$\bar{H}_f^\circ$ (kJ/mol)	$\bar{S}_f^\circ$ (J/mol/K)	References[c]
$HO_2^\bullet$(aq)	4.44			NBS
H_2O_2(g)	−105.6	−136.31	232.6	NBS
H_2O_2(aq)	−134.1	−191.17	143.9	NBS
HO_2^-(aq)	−67.4	−160.33	23.8	NBS
$OH^\bullet$(g)	34.22	38.95	183.64	NBS
$OH^\bullet$(aq)	7.74			NBS
OH^-(aq)	−157.29	−230.0	−10.75	NBS
P (Phosphorus)				
P (α, white)	0	0	41.1	
PO_4^{3-}(aq)	−1018.8	−1277.4	−222	NBS
HPO_4^{2-}(aq)	−1089.3	−1292.1	−33.4	NBS
$H_2PO_4^-$(aq)	−1130.4	−1296.3	90.4	NBS
H_3PO_4(aq)	−1142.6	−1288.3	158	NBS
Pb (Lead)				
Pb (Metal)	0	0	64.8	NBS
Pb^{2+}(aq)	−24.39	−1.67	10.5	NBS
$PbOH^+$(aq)	−226.3			NBS
$Pb(OH)_3^-$(aq)	−575.7			NBS
$Pb(OH)_2$ (precip.)	−452.2			NBS
PbO (yellow)	−187.9	−217.3	68.7	NBS
PbO_2	−217.4	−277.4	68.6	NBS
Pb_3O_4	−601.2	−718.4	211	NBS
PbS	−98.7	−100.4	91.2	NBS
$PbSO_4$	−813.2	−920.0	149	NBS
$PbCO_3$ (Cerussite)	−625.5	−699.1	131	NBS
S (Sulfur)				
S (rhombic)	0	0	31.8	NBS
SO_2(g)	−300.2	−296.8	248	NBS
SO_3(g)	−371.1	−395.7	257	NBS
H_2S(g)	−33.56	−20.63	205.7	NBS
H_2S(aq)	−27.87	−39.75	121.3	NBS
S^{2-}(aq)	85.8[f]	33.0	−14.6	NBS
HS^-(aq)	12.05	−17.6	62.8	NBS
SO_3^{2-}(aq)	−486.6	−635.5	−29	NBS
HSO_3^-(aq)	−527.8	−626.2	140	NBS
$H_2SO_3^*$	−537.9	−608.8	232	NBS[g]
H_2SO_3(aq) ("true")	~ −534.5			S
SO_4^{2-}(aq)	−744.6	−909.2	20.1	NBS
HSO_4^-(aq)	−756.0	−887.3	132	NBS
Se (Selenium)				
Se (black)	0	0	42.4	NBS
SeO_3^{2-}(aq)	−369.9	−509.2	12.6	NBS
$HSeO_3^-$(aq)	−431.5	−514.5	135	NBS
H_2SeO_3(aq)	−426.2	−507.5	208	NBS

(*Continued*)

TABLE A.1 (*Continued*)

Species	Formation from the elements		Entropy	
	$\bar{G}_f^\circ$ (kJ/mol)	$\bar{H}_f^\circ$ (kJ/mol)	$\bar{S}_f^\circ$ (J/mol/K)	References[3]
SeO_4^{2-} (aq)	−441.4	−599.1	54.0	NBS
$HSeO_4^-$ (aq)	−452.3	−581.6	149	NBS
Si (Silicon)				
Si (Metal)	0	0	18.8	NBS
SiO_2 (α, Quartz)	−856.67	−910.94	41.8	NBS
SiO_2 (α, Cristobalite)	−855.88	−909.48	42.7	NBS
SiO_2 (α, Tridymite)	−855.29	−909.06	43.5	NBS
SiO_2 (amorph)	−850.73	−903.49	46.9	NBS
H_4SiO_4(aq)	−1308.0[h]	−1468.6	180	NBS
Sr (Strontium)				
Sr^{2+} (aq)	−559.4	−545.8	−33	R
$SrOH^+$(aq)	−721			NBS
$SrCO_3$ (Strontianite)	−1137.6	−1218.7	97	R
$SrSO_4$ (Celestite)	−1341.0	−1453.2	118	R
Zn (Zinc)				
Zn (Metal)	0	0	29.3	NBS
Zn^{2+} (aq)	−147.0	−153.9	112	NBS
$ZnOH^+$ (aq)	−330.1			NBS
$Zn(OH)_2$ (aq)	−522.3			NBS
$Zn(OH)_3^-$(aq)	−694.3			NBS
$Zn(OH)_4^{2-}$ (aq)	−858.7			NBS
ZnO (solid)	−318.32	−348.28	43.64	NBS
$Zn(OH)_2$ (solid *β*)	−553.6	−641.9	81.2	NBS
$ZnCl^+$ (aq)	−275.3			NBS
$ZnCl_2$(aq)	−403.8			NBS
$ZnCl_3^-$(aq)	−540.6			NBS
$ZnCl_4^{2-}$ (aq)	−666.1			S
$ZnCO_3$ (Smithsonite)	−731.6	−812.8	82.4	NBS

[a]The quality of the data is highly variable; the authors do not claim to have critically selected the "best" data. For information on precision of the data and for a more complete compendium, which includes less common substances, the reader is referred to the references. For research work, the original literature should be consulted.

[b]Thermodynamic properties taken from Robie, Hemingway, and Fisher are based on a reference state of the elements in their standard states at 1 bar (10^5 *P* = 0.987 atm). This change in reference pressure has a negligible effect on the tabulated values for the condensed phases. (For gas phases only data from NBS (reference state = 1 atm) are given.)

[c]NBS: D. D. Wagman et al., Selected Values of Chemical Thermodynamic Properties, U.S. National Bureau of Standards, Technical Notes 270–3 (1968), 270–4 (1969), 270–5 (1971). R: R. A. Robie, B. S. Hemingway, and J. R. Fisher, *Thermodynamic Properties of Minerals and Related Substances at 298.15 K and 1 Bar (10^5 Pascals) Pressure and at Higher Temperatures*, Geological Survey Bulletin No. 1452, Washington, DC, 1978. Bard et al.: Bard, A. J., R. Parsons and D. L. Parkhurst, *Standard Potentials in Aqueous Solution*, Marcel Dekker, New York (1985). S: Other sources (e.g., computed from data in *Stability Constants*).

[d]$[H_2CO_3^*] = [CO_2(aq)]$ + "true" $[H_2CO_3]$.

[e]The thermodynamic stability of oxides, hydroxides, or oxyhydroxides of Fe(III) depends on mode of preparation, age, and molar surface. Reported solubility products ($K_{s0} = \{Fe^{3+}\}\{OH^-\}^3$) range from $10^{-37.3}$ to $10^{-43.7}$. Correspondingly, FeOOH may have G_f° values between −452 J/mol (freshly precipitated amorphous FeOOH) and −489 J/mol (aged goethite). If the precipitate is written as $Fe(OH)_3$, its G_f° values vary from −692 to −729 J/mol.

[f]The value for this specie appears too low, on the basis of recently reported pK_2 values for H_2S(aq).

[g]$[H_2SO_3^*] = [SO_2(aq)]$ + "true" $[H_2SO_3]$.

[h]Value from reference R yields a solubility constant for quartz more in accord with observation.

TABLE A.2 Stability Constants for Formation of Complexes and Solids from Metals and Ligands[a]

	OH^-		CO_3^{2-}		SO_4^{2-}		Cl^-		Br^-		F^-		NH_3		$B(OH)_4^-$	
H^+			HL	10.33	HL	1.99					HL	3.2	HL	9.24	HL	9.24
	HL·w	14.00	H_2L	16.68									L·g	−1.8	HL_3	10.4
			H_2L·g	18.14											H_2L_3	20.4
															H_2L_4	21.0
															H_4L_5	38.8
Na^+			NaL	1.27	NaL	1.06										
			NaHL	10.08												
K^+					KL	0.96										
Ca^{2+}	CaL	1.15	CaL	3.2	CaL	2.31										
			CaHL	11.59	CaL·s	4.62					CaL	1.1				
	CaL_2·s	5.19	CaL·s	8.22							CaL_2·s	10.4				
			CaL·s	8.35												
Mg^{2+}	MgL	2.56	MgL	3.4	MgL	2.36					MgL	1.8				
	Mg_4L_4	16.28	MgHL	11.49							MgL_2·s	8.2				
	MgL_2·s	11.16	MgL·s	4.54												
			MgL·s	7.45												
Sr^{+2}			SrL·s	9.0	SrL	2.6					SrL_2·s	8.5				
					SrL·s	6.5										
Ba^{2+}			BaL	2.8	BaL	2.7					BaL_2·s	5.8				
			BaL·s	8.3	BaL·s	10.0										
Cr^{3+}	CrL	10.0			CrL	3.0	CrL	0.23			CrL	5.2				
	CrL_2	18.3									CrL_2	9.2				
	CrL_3	24.0									CrL_3	12.0				
	CrL_4	28.6														
	Cr_3L_4	47.8														
	CrL_3·s	30.0														
Al^{3+}	AlL	9.0														
	AlL_2	18.7									AlL	7.0				
	AlL_3	27.0									AlL_2	12.6				
	AlL_4	33.0									AlL_3	16.7				

(Continued)

TABLE A.2 (Continued)

	OH^-		CO_3^{2}		SO_4^{2}		Cl^-		Br^-		F^-		NH_3		$B(OH)_4^-$	
	Al_3L_4	42.1									AlL_4	19.1				
	$AlL_3 \cdot s$	33.5														
Fe^{3+}	FeL	11.8			FeL	4.0	FeL	1.5	FeL	0.6	FeL	6.0				
	FeL_2	22.3			FeL_2	5.4	FeL_2	2.1			FeL_2	10.6				
	FeL_4	34.4									FeL_3	13.7				
	Fe_2L_2	25.0														
	$FeL_3 \cdot s$	42.7														
	$FeL_3 \cdot s$	38.8														
Mn^{2+}	MnL	3.4	$MnHL$	12.1	MnL	2.3	MnL	0.6			MnL	1.3	MnL	1.0		
	MnL_2	5.8	$MnL \cdot s$	9.3									MnL_2	1.5		
	MnL_3	7.2														
	MnL_4	7.7														
	$MnL_2 \cdot s$	12.8														
Fe^{2+}	FeL	4.5	$FeL \cdot s$	10.7	FeL	2.2					FeL	1.4				
	FeL_2	7.4														
	FeL_3	11.0														
	$FeL_2 \cdot s$	15.1														
Co^{2+}	CoL	4.3	$CoL \cdot s$	10.0	CoL	2.4	CoL	0.5			CoL	1.0	CoL	2.0		
	CoL_2	9.2											CoL_2	3.5		
	CoL_3	10.5											CoL_3	4.4		
	$CoL_2 \cdot s$	15.7											CoL_4	5.0		
Ni^{2+}	NiL	4.1	$NiL \cdot s$	6.9	NiL	2.3	NiL	0.6			NiL	1.1	NiL	2.7		
	NiL_2	9.0											NiL_2	4.9		
	NiL_3	12.0											NiL_3	6.6		
	$NiL_2 \cdot s$	17.2											NiL_4	7.7		
													NiL_5	8..3		
Cu^{2+}	CuL	6.3	CuL	6.7	CuL	2.4	CuL	0.5			CuL	1.5	CuL	4.0		
	CuL_2	11.8	CuL_2	10.2	$Cu_4(OH)_6L \cdot s$	68.6							CuL_2	7.5		
	CuL_4	16.4	$CuL \cdot s$	9.6									CuL_3	10.3		
	Cu_2L_2	17.7	$Cu_2(OH)_2L \cdot s$	33.8									CuL_4	11.8		
	$CuL_2 \cdot s$	19.3	$Cu_3(OH)_2L_2 \cdot s$	46.0												
	$CuL_2 \cdot s$	20.4														

25	5.71	6.56	7.41	8.26	9.12	5.39	6.2	7	7.81	8.61	5.09	5.85	6.61	7.38	8.14
26	5.6	6.44	7.27	8.11	8.95	5.29	6.08	6.88	7.67	8.46	5	5.75	6.5	7.25	8
27	5.49	6.32	7.14	7.97	8.79	5.19	5.97	6.75	7.53	8.31	4.91	5.65	6.39	7.12	7.86
28	5.39	6.2	7.02	7.83	8.64	5.1	5.87	6.64	7.4	8.17	4.82	5.55	6.28	7	7.73
29	5.29	6.09	6.89	7.69	8.49	5.01	5.76	6.52	7.28	8.03	4.74	5.45	6.17	6.89	7.6
30	5.19	5.98	6.77	7.56	8.35	4.92	5.66	6.41	7.15	7.9	4.65	5.36	6.07	6.77	7.48
31	5.1	5.88	6.65	7.43	8.21	4.83	5.56	6.3	7.04	7.77	4.57	5.27	5.97	6.66	7.36
32	5.01	5.77	6.54	7.3	8.07	4.74	5.47	6.19	6.92	7.65	4.49	5.18	5.87	6.55	7.24
33	4.92	5.67	6.43	7.18	7.94	4.66	5.37	6.09	6.81	7.52	4.41	5.09	5.77	6.45	7.13
34	4.83	5.57	6.32	7.06	7.81	4.58	5.28	5.99	6.7	7.4	4.34	5.01	5.68	6.35	7.02
35	4.74	5.48	6.21	6.95	7.68	4.5	5.19	5.89	6.59	7.29	4.26	4.93	5.59	6.25	6.91
36	4.66	5.38	6.11	6.84	7.56	4.42	5.11	5.8	6.48	7.17	4.19	4.84	5.5	6.15	6.8
37	4.58	5.29	6.01	6.73	7.44	4.34	5.02	5.7	6.38	7.06	4.12	4.77	5.41	6.06	6.7
38	4.5	5.2	5.91	6.62	7.33	4.27	4.94	5.61	6.28	6.95	4.05	4.69	5.33	5.96	6.6
39	4.42	5.12	5.82	6.52	7.21	4.19	4.86	5.52	6.19	6.85	3.98	4.61	5.24	5.87	6.5
40	4.34	5.03	5.72	6.41	7.1	4.12	4.78	5.43	6.09	6.75	3.91	4.54	5.16	5.78	6.41

TABLE A.3 (*Continued*)

t	Salinity = 30‰ (~$ppth_m$) Absolute pressure (atm)					Salinity = 40‰ (~$ppth_m$) Absolute pressure (atm)				
°C	0.7	0.8	0.9	1	1.1	0.7	0.8	0.9	1	1.1
0	8.28	9.47	10.66	11.85	13.04	7.72	8.83	9.94	11.05	12.16
1	8.06	9.23	10.39	11.55	12.71	7.52	8.61	9.69	10.77	11.86
2	7.86	8.99	10.12	11.26	12.39	7.34	8.39	9.45	10.51	11.57
3	7.66	8.77	9.87	10.98	12.08	7.16	8.19	9.22	10.25	11.29
4	7.47	8.55	9.63	10.71	11.79	6.98	7.99	9	10.01	11.02
5	7.29	8.34	9.4	10.45	11.5	6.82	7.81	8.79	9.78	10.76
6	7.12	8.15	9.18	10.21	11.23	6.66	7.62	8.59	9.55	10.51
7	6.95	7.96	8.96	9.97	10.98	6.51	7.45	8.39	9.34	10.28
8	6.79	7.78	8.76	9.74	10.73	6.36	7.29	8.21	9.13	10.05
9	6.64	7.6	8.56	9.53	10.49	6.22	7.13	8.03	8.93	9.83
10	6.49	7.43	8.37	9.32	10.26	6.09	6.97	7.86	8.74	9.62
11	6.35	7.27	8.19	9.12	10.04	5.96	6.82	7.69	8.56	9.42
12	6.21	7.12	8.02	8.92	9.83	5.83	6.68	7.53	8.38	9.23
13	6.08	6.97	7.85	8.74	9.63	5.71	6.54	7.38	8.21	9.04
14	5.95	6.82	7.69	8.56	9.43	5.6	6.41	7.23	8.05	8.86
15	5.83	6.68	7.54	8.39	9.24	5.48	6.29	7.09	7.89	8.69
16	5.71	6.55	7.39	8.22	9.06	5.38	6.16	6.95	7.74	8.52
17	5.6	6.42	7.24	8.06	8.88	5.27	6.04	6.82	7.59	8.36
18	5.49	6.3	7.1	7.91	8.72	5.17	5.93	6.69	7.45	8.21
19	5.38	6.18	6.97	7.76	8.55	5.07	5.82	6.57	7.31	8.06
20	5.28	6.06	6.84	7.62	8.4	4.98	5.71	6.45	7.18	7.92
21	5.18	5.95	6.71	7.48	8.24	4.89	5.61	6.33	7.05	7.78
22	5.08	5.84	6.59	7.34	8.1	4.8	5.51	6.22	6.93	7.64
23	4.99	5.73	6.47	7.21	7.96	4.71	5.41	6.11	6.81	7.51
24	4.9	5.63	6.36	7.09	7.82	4.63	5.32	6.01	6.69	7.38
25	4.81	5.53	6.25	6.97	7.69	4.54	5.22	5.9	6.58	7.26
26	4.72	5.43	6.14	6.85	7.56	4.47	5.13	5.8	6.47	7.14

27	4.64	5.34	6.04	6.73	7.43	4.39	5.05	5.71	6.37	7.03
28	4.56	5.25	5.94	6.62	7.31	4.31	4.96	5.61	6.26	6.92
29	4.48	5.16	5.84	6.52	7.19	4.24	4.88	5.52	6.16	6.81
30	4.4	5.07	5.74	6.41	7.08	4.17	4.8	5.43	6.07	6.7
31	4.33	4.99	5.65	6.31	6.97	4.1	4.72	5.35	5.97	6.6
32	4.25	4.91	5.56	6.21	6.86	4.03	4.65	5.26	5.88	6.5
33	4.18	4.83	5.47	6.11	6.75	3.96	4.57	5.18	5.79	6.4
34	4.11	4.75	5.38	6.02	6.65	3.9	4.5	5.1	5.7	6.3
35	4.04	4.67	5.3	5.92	6.55	3.83	4.43	5.02	5.62	6.21
36	3.98	4.6	5.21	5.83	6.45	3.77	4.36	4.95	5.53	6.12
37	3.91	4.52	5.13	5.75	6.36	3.71	4.29	4.87	5.45	6.03
38	3.84	4.45	5.05	5.66	6.26	3.65	4.22	4.8	5.37	5.95
39	3.78	4.38	4.98	5.58	6.17	3.59	4.16	4.72	5.29	5.86
40	3.72	4.31	4.9	5.49	6.08	3.53	4.09	4.65	5.22	5.78

TABLE A.4 Gibbs Energy of Formation for Selected Geochemical Species[a]

Specie (state[b])	ΔG_f° (kcal/$_g$mol)	Specie (state[b])	ΔG_f° (kcal/$_g$mol)
Water		Iron (continued)	
H_2O (liq)	−56.69	$Fe(OH)_3$ (c)	−166.47
OH^- (aq)	−37.59	$Fe(OH)_2$ (c)	−116.30
H^+ (aq)	0	FeO·OH (c)	−116.77
H_2 (g)	0	FeS_2 (c)	−39.89
O_2 (g)	0	FeS (c)	−24.00
H_2O (g)	−54.63	$FeSiO_3$ (c)	−267.16
Arsenic		$FeCO_3$ (c)	−159.34
As (c)	0	$FeSi_3O_3(OH)_8^0$ (aq)	−898.00
AsS (c)	−16.80	Manganese	
As_2S_3 (c)	−40.30	Mn^{+2} (aq)	−54.52
As_2O_3 (c)	−137.66	MnO (c)	−86.74
H_3AsO_4 (aq)	−183.08	MnO_2 (c)	−111.17
$H_2AsO_4^-$ (aq)	−180.01	Mn_2O_3 (c)	−210.59
$HAsO_4^=$ (aq)	−170.69	Mn_3O_4 (c)	−306.69
AsO_4^{-3} (aq)	−154.97	$MnOH^+$ (aq)	−96.80
H_3AsO_3 (aq)	−152.92	$Mn(OH)_2$ (c)	−146.99
$H_2AsO_3^-$ (aq)	−140.33	$Mn(OH)_3^-$ (aq)	−177.87
$HAsO_3^=$ (aq)	−125.31	MnS (c)	−52.20
AsO_3^{-3} (aq)	−107.00	$MnCO_3$ (c)	−195.20
Carbon		Nitrogen	
CH_4 (aq)	−8.28	N_2 (g)	0
CH_2O (aq)	−31.00	NO_3^- (aq)	−25.99
H_2CO_3 (aq)	−149.00	NH_4^+ (aq)	−18.96
HCO_3^- (aq)	−140.24	NH_3 (g)	−3.98
$CO_3^=$ (aq)	−126.15	Sulfur	
Chromium		$S^=$ (aq)	+20.51
Cr^{+3} (aq)	−51.50	HS^- (aq)	+2.89
Cr_2O_3 (c)	−252.89	H_2S (aq)	−6.65
$CrOH^{+2}$ (aq)	−103.00	$SO_3^=$ (aq)	−116.28
$Cr(OH)_2^+$ (aq)	−151.20	HSO_3^- (aq)	−126.13
CrO_2^- (aq)	−128.00	$SO_4^=$ (aq)	−177.95
$CrO_4^=$ (aq)	−173.94	HSO_4^- (aq)	−180.67
$HCrO_4^-$ (aq)	−182.77	Selenium	
$Cr_2O_7^=$ (aq)	−310.97	$Se^=$	+30.90
Iron		HSe^- (aq)	+10.50
Fe^{+2} (aq)	$SeO_3^=$ (aq)	H_2Se (aq)	+3.80
Fe^{+3} (aq)	−1.12	$SeO_3^=$ (aq)	−88.38
$FeO_2^=$ (aq)	−70.58	$HSeO_3^-$ (aq)	−98.34
Fe_2O_3 (c)	−177.39	H_2SeO_3 (aq)	−101.85
Fe_3O_4 (c)	−242.69	$SeO_4^=$ (aq)	−105.47
		$HSeO_4^-$ (aq)	−108.08

[a]Data are from tables 1, 3, 4, 7, 9, 12, 30, 37, and 40 of Brookins (1988).

[b](c) refers to the solid (crystalline, microcrystalline, or amorphous) state.

TABLE A.5 Selected Redox Half Reactions

Full, balanced half reaction	$\log_{10}K$	[a]$pE°$	[b]E_H° (V)
$Ag^+ + e^- \Leftrightarrow Ag_{(s)}$	+13.5	+13.5	+0.80
$Cu^+ + e^- \Leftrightarrow Cu_{(s)}$	+8.80	+8.80	+0.52
$Cu^{+2} + 2e^- \Leftrightarrow Cu_{(s)}$	+11.4	+5.70	+0.34
$2H^+ + 2e^- \Leftrightarrow H_{2(g)}$	0	0	0
$Co^{+2} + 2e^- \Leftrightarrow Co_{(s)}$	−9.5	−4.75	−0.28
$Fe^{+2} + 2e^- \Leftrightarrow Fe_{(s)}$	−14.9	−7.45	−0.44
$Fe^{+3} + e^- \Leftrightarrow Fe^{+2}$	+13.03	+13.03	+0.769
$Zn^{+2} + 2e^- \Leftrightarrow Zn_{(s)}$	−26.0	−13.0	−0.76
$Mg^{+2} + 2e^- \Leftrightarrow Mg_{(s)}$	−79.4	−39.7	−2.35
$Na^+ + e^- \Leftrightarrow Na_{(s)}$	−46.0	−46.0	−2.71
$2NO_3^- + 12H^+ + 10e^- \Leftrightarrow N_{2(g)} + 6H_2O$	+210.5	+21.05	+1.245
$MnO_{2(s)} + 2e^- + 4H^+ \Leftrightarrow Mn^{+2} + 2H_2O$	+41.60	+20.80	+1.227
$O_{2(g)} + 4H^+ + 4e^- \Leftrightarrow 2H_2O$	+83	+20.75	+1.23
$NO_2^- + 8H^+ + 6e^- \Leftrightarrow NH_4^+ + 2H_2O$	+90.84	+15.14	+0.896
$NO_3^- + 10H^+ + 8e^- \Leftrightarrow NH_4^+ + 3H_2O$	+119.2	+14.90	+0.881
$NO_3^- + 2H^+ + 2e^- \Leftrightarrow NO_2^- + H_2O$	+28.3	+14.15	+0.837
$CH_3OH + 2H^+ + 2e^- \Leftrightarrow CH_{4(g)} + H_2O$	+19.76	+9.88	+0.584
$CH_2O + 4H^+ + 4e^- \Leftrightarrow CH_{4(g)} + H_2O$	+27.76	+6.94	+0.411
$SO_4^= + 8H^+ + 6e^- \Leftrightarrow S_{(s)} + 4H_2O$	+36.18	+6.03	+0.357
$SO_4^= + 10H^+ + 8e^- \Leftrightarrow H_2S_{(g)} + 4H_2O$	+42.0	+5.25	+0.311
$N_{2(g)} + 8H^+ + 6e^- \Leftrightarrow 2NH_4^+$	+28.08	+4.68	+0.277
$SO_4^= + 9H^+ + 8e^- \Leftrightarrow HS^- + 4H_2O$	+34.0	+4.25	+0.251
$CH_2O + 2H^+ + 2e^- \Leftrightarrow CH_3OH$	+7.98	+3.99	+0.236
$S_{(s)} + 2H^+ + 2e^- \Leftrightarrow H_2S_{(g)}$	+5.78	+2.89	+0.171
$CO_{2(g)} + 8H^+ + 8e^- \Leftrightarrow CH_{4(g)} + 2H_2O$	+22.96	+2.87	+0.170
$HCO_2^- + 3H^+ + 2e^- \Leftrightarrow CH_2O + H_2O$	+5.64	+2.82	+0.167
$6CO_{2(g)} + 24H^+ + 24e^- \Leftrightarrow C_6H_{12}O_6 + 6H_2O$	−4.80	−0.20	−0.0118
$CO_{2(g)} + 4H^+ + 4e^- \Leftrightarrow CH_2O + H_2O$	−4.80	−1.20	−0.0710
$CO_{2(g)} + H^+ + 2e^- \Leftrightarrow HCO_2^-$	−9.66	−4.83	−0.286

[a]Values from Stumm and Morgan (1996), $pE° = \frac{1}{n_E}\log_{10} K$, n_E is the number of electrons transferred.

[b]$E_H^\circ = \frac{pE^\circ RT \ln(10)}{F}$, F is the Faraday constant (96,485 C/mol), and R is the universal gas constant $\left(8.3144 \frac{CV}{mol°K}\right)$.

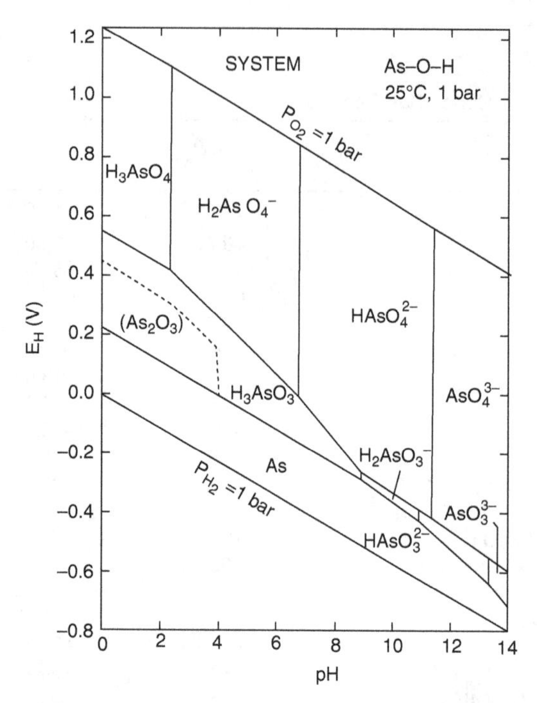

FIGURE A.1 *E_H – pH diagram for part of the system As–O–H. The assumed activity of dissolved As = 10^{-6} M. Reprinted from Brookins (1988) by permission of Springer-Verlag GmbH.*

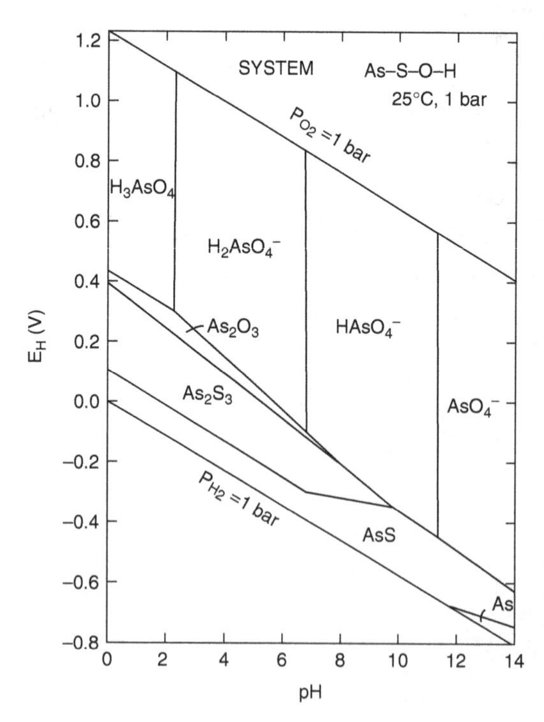

FIGURE A.2 *E_H – pH diagram for part of the system As–S–O–H. The assumed activities of dissolved species are: As = 10^{-6} M, S = 10^{-3} M. Reprinted from Brookins (1988) by permission of Springer-Verlag GmbH.*

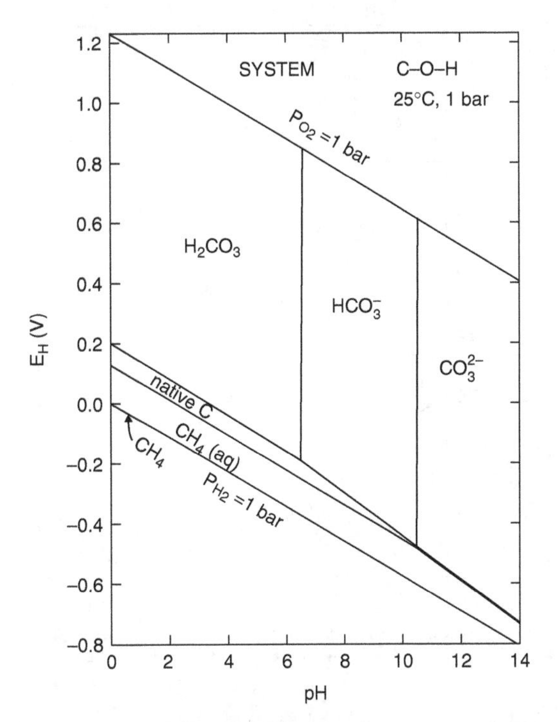

FIGURE A.3 *E_H – pH diagram for part of the system C–O–H. The assumed activity of dissolved C = 10^{-3} M. Reprinted from Brookins (1988) by permission of Springer-Verlag GmbH.*

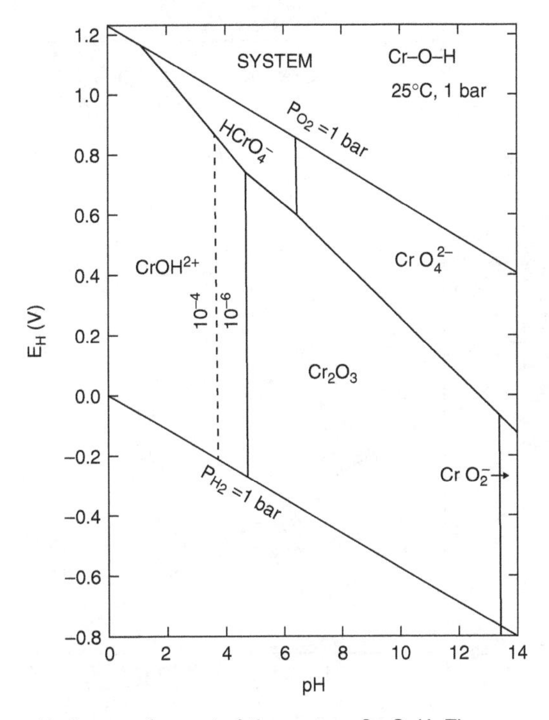

FIGURE A.4 *E_H – pH diagram for part of the system Cr–O–H. The assumed activity of dissolved Cr = 10^{-6} M. Reprinted from Brookins (1988) by permission of Springer-Verlag GmbH.*

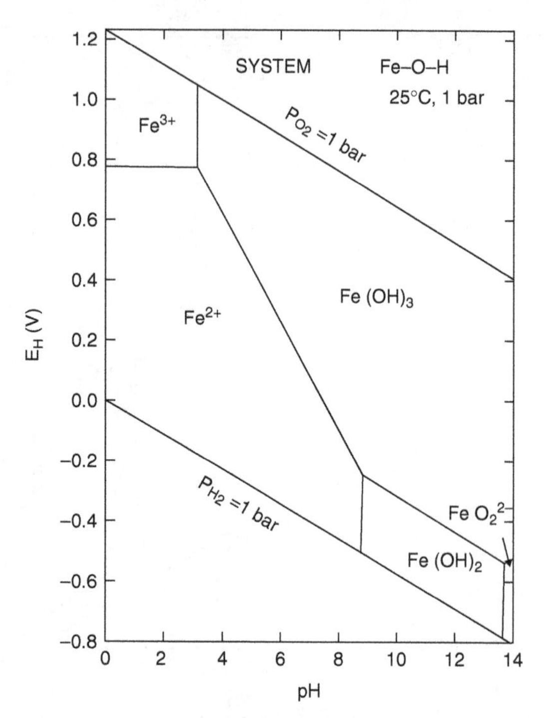

FIGURE A.5 *E_H – pH diagram for part of the system Fe–O–H assuming $Fe(OH)_{3(s)}$ as the stable Fe(III) phase. The assumed activity of dissolved Fe = 10^{-6} M. Reprinted from Brookins (1988) by permission of Springer-Verlag GmbH.*

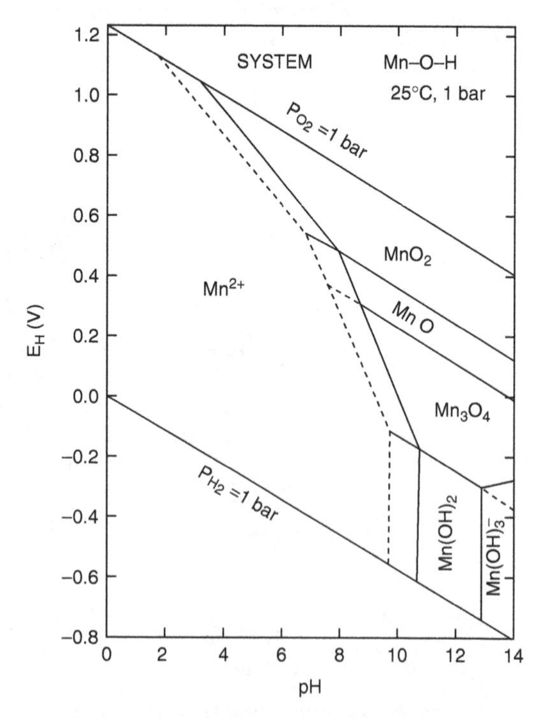

FIGURE A.6 *E_H – pH diagram for part of the system Mn–O–H. The assumed activity of dissolved Mn = 10^{-6} M. Reprinted from Brookins (1988) by permission of Springer-Verlag GmbH.*

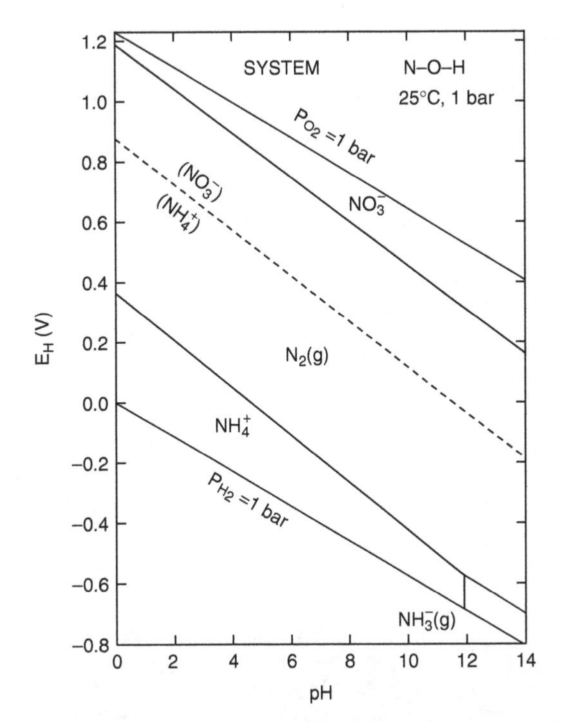

FIGURE A.7 *E_H – pH diagram for part of the system N–O–H. The assumed activity of dissolved nitrogen = $10^{-3.3}$ M* (P_{N_2} = 0.8 bar). *Reprinted from Brookins (1988) by permission of Springer-Verlag GmbH.*

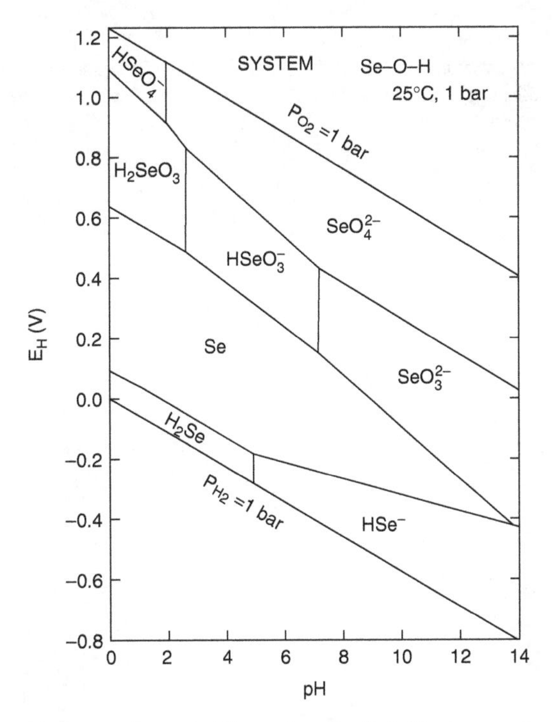

FIGURE A.8 *E_H – pH diagram for part of the system Se–O–H. The assumed activity of dissolved Se = 10^{-6} M. Reprinted from Brookins (1988) by permission of Springer-Verlag GmbH.*

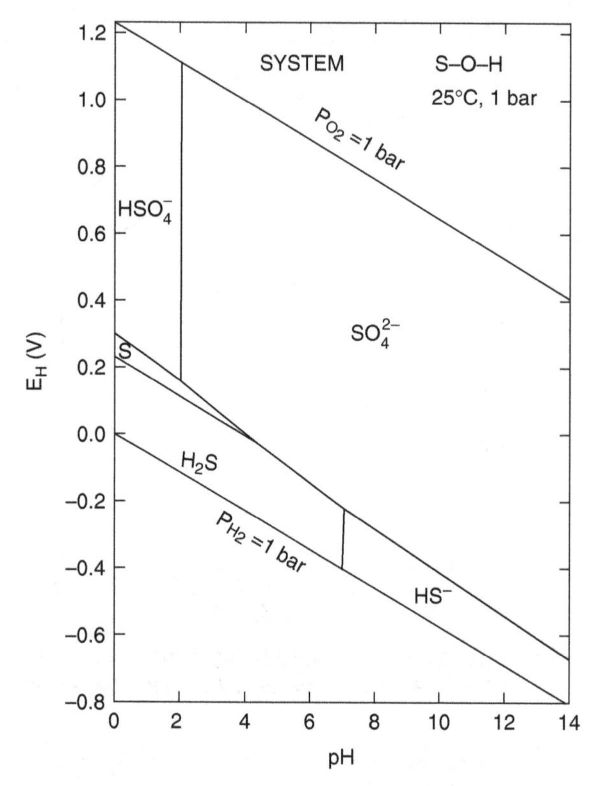

FIGURE A.9 *E_H – pH diagram for part of the system S–O–H. The assumed activity of dissolved S = 10^{-3} M (roughly 32 ppm_m) for convenience. Reprinted from Brookins (1988) by permission of Springer-Verlag GmbH.*

References

Alken Murray, Inc. 2002Toxicity of hydrogen sulfide gas. http://www.alken-murray.com/H2SREM9.HTM. Accessed on 2013 May 31.

Asadi M. *Tables, in Beet-Sugar Handbook*. Hoboken: Wiley; 2005.

Baes C, Mesmer R. *The Hydrolysis of Cations*. Malabar: Krieger; 1976.

Baes C, Mesmer R. The thermodynamics of cation hydrolysis. Am J Sci 1981;281:935–962.

Balzhiser R, Wass A, Samuels M, Eliassen J. *Chemical Engineering Thermodynamics*. Upper Saddle River: Prentice-Hall; 1972.

Bird RB, Stewart WE, Lightfoot EN. *Transport Phenomena*. New York: Wiley; 1960.

Bohn H, McNeal B, O'Connor G. *Soil Chemistry*. New York: Wiley; 1979.

Brezonik P, Arnold W. *Water Chemistry*. New York: Oxford University Press; 2011.

Brookins D. *Eh – pH Diagrams for Geochemistry*. New York: Springer-Verlag; 1988.

Carnahan B, Luther HA, Wilkes JO. *Applied Numerical Methods*. New York: Wiley; 1969.

Coello Oviedo MD, Sales Márquez D, Quiroga Alonso JM. Toxic effects of metals on microbial activity in the activated sludge process. Chem Bioch Eng 2002;16(3):139–144.

Correa A, Comesana JF, Correa JM, Sereno AM. Measurement and prediction of water activity in electrolyte solutions by a modified ASOG group contribution method. Fluid Phase Equilibria 1977;129:267–283.

Crank J. *The Mathematics of Diffusion*. 2nd ed. New York: Oxford University Press; 1979.

Crittenden JD, Trussel RR, Hand DW, Howe KJ, Tchobanoglous G. *Water Treatment Principles and Design*. 2nd ed. New York: Wiley; 2005.

Cussler EL. *Diffusion: Mass Transfer in Fluid Systems*. New York: Cambridge University Press; 1984.

Environmental Process Analysis: Principles and Modeling, First Edition. Henry V. Mott.

Danckwerts PV. Continuous flow systems: distribution of residence times. Chem Eng Sci 1953;2:1–13.

Dean J, editor. *Lange's Handbook of Chemistry*. 14th ed. New York: McGraw-Hill; 1992.

Dutkiewicz E, Jakubowska A. Water activity in aqueous solutions of homogeneous electrolytes: the effect of ions on the structure of water. Chemphyschem 2002;2:221–224.

Finnemore J, Franzini J. *Fluid Mechanics with Engineering Applications*. 10th ed. New York: McGraw-Hill; 2002.

Fogler HS. *Elements of Chemical Reaction Engineering*. 4th ed. Upper Saddle River: Prentice-Hall; 2005.

Froment GF, Bischoff KB, Juray DW. *Chemical Reactor Analysis and Design*. 3rd ed. New York: Wiley; 2011.

Haynes WM, editor. *CRC Handbook of Chemistry and Physics*. Boca Raton: CRC Press; 2012.

Hulbert HH. Chemical processes in continuous-flow systems: reaction kinetics. Ind Eng Chem 1944;36:1012–1017.

Kojima K, Tochigi K. *Prediction of Vapor–Liquid Equilibria by the ASOG Method*. New York: Elsevier; 1979.

Levenspiel O. *Chemical Reaction Engineering*. 2nd ed. New York: Wiley; 1972.

Levenspiel O. *Chemical Reaction Engineering*. 3rd ed. New York: Wiley; 1999.

Levine I. *Physical Chemistry*. 3rd ed. New York: McGraw-Hill; 1988.

Lymann WJ, Rheehl WF, Rosenblatt DH. *Handbook of Chemical Property Estimation Methods*. New York: McGraw-Hill; 1982.

Mercer JW, Skipp DC, Giffen D. Basics of pump and treat ground water remediation technology. Environmental Protection Agency EPA/600/8-90/003; 1990.

Morel F, Hering J. *Principles and Applications of Aquatic Chemistry*. New York: Wiley; 1993.

Munson BR, Young DF, Okiishi TH. *Fundamentals of Fluid Mechanics*. 3rd ed. New York: Wiley; 1998.

Perry R, Chilton C. *Chemical Engineers' Handbook*. 5th ed. New York: McGraw-Hill; 1973.

Perry R, Green DW. *Chemical Engineers' Handbook*. 8th ed. New York: McGraw-Hill; 2007.

Reid RC, Prausnitz JM, Poling BE. *The Properties of Gases and Liquids*. 4th ed. New York: McGraw-Hill; 1987.

Robinson RA, Stokes RH. *Electrolyte Solutions*. London: Butterworths; 1959.

Rogers PSZ, Pitzer KS. Volumetric properties of aqueous sodium chloride solutions. J Phys Chem 1982;11(1):15–81.

Schwarzenbach RP, Gschwend PM, Imboden DM. *Environmental Organic Chemistry*. 2nd ed. New York: Wiley; 2002.

Snoeyink V, Jenkins D. *Water Chemistry*. New York: Wiley; 1980.

Sposito G. *The Surface Chemistry of Soils*. New York: Oxford University Press; 1984.

Stumm W, Morgan J. *Aquatic Chemistry*. 3rd ed. New York: Wiley; 1996.

Tchobanoglous G, Buron FL, Stensel DH. *Wastewater Engineering: Treatment and Reuse*. 4th ed. New York: McGraw-Hill; 2003.

Treybal R. *Mass-Transfer Operations*. 3rd ed. New York: McGraw-Hill; 1980.

U.S. Geological Survey. Dissolved oxygen solubility tables. 2013. http://water.usgs.gov/software/DOTABLES/. Accessed on 2013 May 31.

Wagner RJ, Boulger Jr RW, Oblinger CJ, Smith BA. 2006, Guidelines and standard procedures for continuous water-quality monitors—station operation, record computation, and data reporting: U.S. Geological Survey Techniques and Methods 1–D3, 51, p +8 attachments; Available at http://pubs.water.usgs.gov/tm1d3. Accessed 2006 Apr 10.

Weber Jr WJ. *Physicochemical Processes for Water Quality Control*. New York: Wiley; 1972.

Weber Jr WJ, DiGiano FA. *Process Dynamics in Environmental Systems*. New York: Wiley; 1996.

Wehner JF, Wilhelm RH. Boundary conditions of flow reactor. Chem Eng Sci 1956;6:89–93.

Williams V, Mattice W, Williams H. *Physical Chemistry for the Life Sciences*. 3rd ed. San Francisco: W.H. Freeman and Company; 1978.

Wylie Jr CR. *Advanced Engineering Mathematics*. 3rd ed. New York: McGraw-Hill; 1966.

Index

Environmental Process Analysis: Principles and Modeling, First Edition. Henry V. Mott.

CPSIA information can be obtained
at www.ICGtesting.com
Printed in the USA
JSHW062303060723
43514JS00002BA/47

9 781118 115015